Partial Differential Equations in Sobolev and Analytic Spaces

Partial Differential Equations in Sobolev and Analytic Spaces

Aissa Boukarou
University of Science and Technology Houari Boumediene, Algeria

Khaled Zennir
Qassim University, Saudi Arabia

Svetlin G Georgiev
Sorbonne University, France

World Scientific

NEW JERSEY • LONDON • SINGAPORE • BEIJING • SHANGHAI • HONG KONG • TAIPEI • CHENNAI

Published by

World Scientific Publishing Co. Pte. Ltd.
5 Toh Tuck Link, Singapore 596224
USA office: 27 Warren Street, Suite 401-402, Hackensack, NJ 07601
UK office: 57 Shelton Street, Covent Garden, London WC2H 9HE

Library of Congress Control Number: 2019006783

British Library Cataloguing-in-Publication Data
A catalogue record for this book is available from the British Library.

PARTIAL DIFFERENTIAL EQUATIONS IN SOBOLEV AND ANALYTIC SPACES

ISBN 978-981-12-9850-9 (hardcover)
ISBN 978-981-12-9851-6 (ebook for institutions)
ISBN 978-981-12-9852-3 (ebook for individuals)

For any available supplementary material, please visit
https://www.worldscientific.com/worldscibooks/10.1142/13994#t=suppl

Desk Editors: Nambirajan Karuppiah/Rok Ting Tan

Typeset by Stallion Press
Email: enquiries@stallionpress.com

Preface

Partial Differential Equations are ubiquitous in mathematically oriented scientific fields such as physics and engineering. They are foundational in the modern scientific understanding of sound, heat, diffusion, electrostatics, electrodynamics, thermodynamics, fluid dynamic elasticity, general relativity and quantum mechanics. They also arise from many purely mathematical considerations such as differential geometry and the calculus of variations.

This book is devoted on some recent investigations of some classes partial differential equations in Sobolev and analytic spaces. The book contains 12 chapters. Chapter 1 is entirely devoted to the presentation of definitions and results necessary as a result of this work. We first recall a few basic results on the linear, metric, normed and Banach Spaces and its properties. These are used in particular to introduce the different concepts of weak solutions to PDEs. Chapter 2 is titled "Lebesgue Integration", it is devoted to the study of measure and integration, Lebesgue measurable functions and general measure spaces, where there are many proved results. The purpose of Chapter 3 is to present results according to the L^p spaces, which contains definitions, separability, duality and general L^p spaces with its norms. Elements of theory of distributions are introduced in Chapter 4. It is introduced the Fourier transform in the spaces $\mathscr{S}$ and $\mathscr{S}'$ and they are deducted some important properties. Sobolev spaces and Bourgain spaces are discussed in Chapter 5. Chapter 6 is devoted on the original method for KdV equations in Sobolev spaces. In Chapter 7, are investigated some fifth-order shallow water equations for local and global well posedness. Higher-order dispersive equations are studied in Chapter 8. They are given criteria for existence of local global solutions in some analytic spaces. In Chapter 9, are investigated some classes fifth-order Kadomtsev–Petviashvili I equations in Bourgain spaces for existence of solutions, continuous dependence on initial data and Gevrey regularity in t. They are studied some classes Kadomtsev–Petviashvili II equations in anisotropic Gevrey spaces. Chapter 10 is devoted on the generalized Kadomtsev–Petviashvili I equation. Coupled systems of KdV equations are introduced and studied in Chapter 11. In Chapter 12, are discussed coupled systems of generalized KdV equations.

This book is addressed to a wide audience of specialists such as mathematicians, physicists, engineers and biologists. It can be used as a textbook at the graduate level and as a reference book for several disciplines.

About the Authors

Aissa Boukarou is an associate professor at the University of Science and Technology Houari Boumediene, Algeria. His current research focuses on partial differential equations, harmonic analysis, stochastic PDEs, numerical analysis, and image processing.

Khaled Zennir earned his PhD in mathematics in 2013 from Sidi Bel Abbès University, Algeria. In 2015, he received his highest diploma of Habilitation in mathematics from Constantine University, Algeria. He is currently professor at Qassim University in the Saudi Arabia. His research interests lie in the subjects of nonlinear evolutionary partial differential equations: global existence, blowup, and longtime behavior.

Svetlin G. Georgiev is a mathematician who has worked in various areas of mathematics. His research currently focuses on harmonic analysis, functional analysis, partial differential equations, ordinary differential equations, Clifford and quaternion analysis, integral equations, and dynamic calculus on time scales. He is the author of the book series Foundations of Iso-Differential Calculus Vols. I–VI, and the author of several books, including *Real Quaternion Calculus*, *Theory of Distributions*, *Fractional Dynamic Calculus and Fractional Dynamic Equations on Time Scales*, *Fuzzy Dynamic Equations*, *Dynamic Inclusions and Optimal Control Problems on Time Scales*, *Functional Dynamic Equations on Time Scales*, *Multiple Fixed-Point Theorems and Applications in the Theory of ODEs, FDEs and PDEs*, and *Boundary Value Problems on Time Scales Volumes I and II*. He also co-authored *Conformable Dynamic Equations on Time Scales*.

Contents

Chapter 1

Preliminaries

This chapter is entirely devoted to the presentation of definitions and results necessary as a result of this work. We first recall a few basic results on the linear, metric, normed and Banach Spaces and its properties. These are used in particular to introduce the different concepts of weak solutions to PDEs. We introduce linear operators and deduct some of their properties.

1.1 Vector Spaces

With $\mathbf{F}$ we will denote the field of real numbers $\mathbb{R}$ or the field of complex numbers $\mathbf{C}$.

Definition 1.1. A vector space (also called a linear space) over the field $\mathbf{F}$ is a set $\mathbf{E}$ with two operations:

1. **Addition:** Takes any two elements x and y and assigns for them a third element of $\mathbf{E}$ which is completely written as $x + y$ and called the sum of these two elements,
2. **Scalar multiplication:** Takes any scalar a of the field $\mathbf{F}$ and an element x of $\mathbf{E}$ and gives another element $a \cdot x$ (shortly ax) of $\mathbf{E}$.

These two operations satisfy the following axioms:

(L1) (Associativity of addition)

$$x + (y + z) = (x + y) + z$$

for any elements $x, y, z \in \mathbf{E}$.

(L2) (Commutativity of addition)

$$x + y = y + x$$

for any elements $x, y \in \mathbf{E}$.

(L3) (Identity element of addition) There exists an element 0 of $\mathbf{E}$, called the zero element, such that

$$x + 0 = x$$

for any $x \in \mathbf{E}$.

(L4) (Inverse elements of addition) For any $x \in \mathbf{E}$ there exists an element $-x \in \mathbf{E}$, called the additive inverse of x, such that

$$x + (-x) = 0.$$

(L5) (Compatibility of the scalar multiplication with field multiplication)

$$(ab)x = a(bx)$$

for any $a, b \in \mathbf{F}$ and for any $x \in \mathbf{E}$.

(L6) (Distributivity of the scalar multiplication with respect to the addition of elements of $\mathbf{E}$)

$$a(x + y) = ax + ay$$

for any $a \in \mathbf{F}$ and for any $x, y \in \mathbf{E}$.

(L7) (Distributivity of scalar multiplication with respect to field addition)

$$(a + b)x = ax + bx,$$

for any $a, b \in \mathbf{F}$ and for any $x \in \mathbf{E}$.

(L8) $1 \cdot x = x$, for any $x \in \mathbf{E}$.

The elements of $\mathbf{E}$ are commonly called vectors. The elements of $\mathbf{F}$ are commonly called scalars. When the scalar field is the field of the real numbers $\mathbb{R}$, then the vector space $\mathbf{E}$ is called a real vector space. When the scalar field is the field of the complex numbers $\mathbf{C}$, we say that the vector space $\mathbf{E}$ is a complex vector space.

Below, we will write $x - y$ instead of $x + (-y)$ for any $x, y \in \mathbf{E}$.

Example 1.1 (The space m of bounded number sequences). With $\mathbf{m}$ we will denote the set of bounded number sequences $x = \{\xi_l\}_{l\in\mathbb{N}}$ implying that for every $x \in \mathbf{m}$ there exists a positive constant K_x such that$|\xi_l| \leq K_x$ for any $l \in \mathbb{N}$. For $x = \{\xi_l\}_{l\in\mathbb{N}} \in \mathbf{m}$ and $y = \{\eta_l\}_{l\in\mathbb{N}} \in \mathbf{m}$, and $a \in \mathbf{F}$, we define addition as follows

$$x + y = \{\xi_l + \eta_l\}_{l\in\mathbb{N}}, \tag{1.1}$$

and scalar multiplication by

$$ax = \{a\xi_l\}_{l\in\mathbb{N}}. \tag{1.2}$$

These operations are well-defined. Really, let $x = \{\xi_l\}_{l\in\mathbb{N}}$, $y = \{\eta_l\}_{l\in\mathbb{N}} \in \mathbf{m}$ and $a \in \mathbf{F}$ be arbitrarily chosen and fixed. Then there exist constants K_x and K_y such that

$$|\xi_l| \leq K_x \quad \text{and} \quad |\eta_l| \leq K_y \quad \text{for any } l \in \mathbb{N}.$$

Hence,

$$\begin{aligned}|\xi_l + \eta_l| &\leq |\xi_l| + |\eta_l| \\ &\leq K_x + K_y \quad \text{for any } l \in \mathbb{N},\end{aligned}$$

i.e., $x + y \in \mathbf{m}$. Also, we have

$$\begin{aligned}|a\xi_l| &= |a||\xi_l| \\ &\leq |a|K_x,\end{aligned}$$

i.e., $ax \in \mathbf{m}$. Consequently the operations (10.36) and (10.37) are well-defined. We will prove that $\mathbf{m}$ is a vector space over $\mathbf{F}$. Suppose that $a, b \in \mathbf{F}$ and $x = \{\xi_l\}_{l\in\mathbb{N}}$, $y = \{\eta_l\}_{l\in\mathbb{N}}$, $z = \{\zeta_l\}_{l\in\mathbb{N}} \in \mathbf{m}$ be arbitrarily chosen and fixed.

1. Because

   $$\xi_l + (\eta_l + \zeta_l) = (\xi_l + \eta_l) + \zeta_l \quad \text{for any } l \in \mathbb{N},$$

   we have

   $$x + (y + z) = (x + y) + z.$$

2. Since

   $$\xi_l + \eta_l = \eta_l + \xi_l \quad \text{for any } l \in \mathbb{N},$$

   we have

   $$x + y = y + x.$$

3. Let $0 = \{0\}_{l\in\mathbb{N}}$. Then $0 \in \mathbf{m}$ and
$$\begin{aligned} x + 0 &= \{\xi_l\}_{l\in\mathbb{N}} + \{0\}_{l\in\mathbb{N}} \\ &= \{\xi_l + 0\}_{l\in\mathbb{N}} \\ &= \{\xi_l\}_{l\in\mathbb{N}} \\ &= x. \end{aligned}$$
4. We define $-x = \{-\xi_l\}_{l\in\mathbb{N}}$. Then
$$\begin{aligned} x + (-x) &= \{\xi_l + (-\xi_l)\}_{l\in\mathbb{N}} \\ &= \{0\}_{l\in\mathbb{N}} \\ &= 0. \end{aligned}$$
5. We have
$$(ab)\xi_l = a(b\xi_l) \quad \text{for any } l \in \mathbb{N}.$$
Then
$$\begin{aligned} (ab)x &= \{(ab)\xi_l\}_{l\in\mathbb{N}} \\ &= \{a(b\xi_l)\}_{l\in\mathbb{N}} \\ &= a\{b\xi_l\}_{l\in\mathbb{N}} \\ &= a\left(b\{\xi_l\}_{l\in\mathbb{N}}\right) \\ &= a(bx). \end{aligned}$$
6. We have
$$\begin{aligned} a(x+y) &= a\{\xi_l + \eta_l\}_{l\in\mathbb{N}} \\ &= \{a(\xi_l + \eta_l)\}_{l\in\mathbb{N}} \\ &= \{a\xi_l + a\eta_l\}_{l\in\mathbb{N}} \\ &= \{a\xi_l\}_{l\in\mathbb{N}} + \{a\eta_l\}_{l\in\mathbb{N}} \\ &= a\{\xi_l\}_{l\in\mathbb{N}} + a\{\eta_l\}_{l\in\mathbb{N}} \\ &= ax + ay. \end{aligned}$$
7. We have
$$\begin{aligned} (a+b)x &= (a+b)\{\xi_l\}_{l\in\mathbb{N}} \\ &= \{(a+b)\xi_l\}_{l\in\mathbb{N}} \end{aligned}$$

$$\begin{aligned}
&= \{a\xi_l + b\xi_l\}_{l\in\mathbb{N}} \\
&= \{a\xi_l\}_{l\in\mathbb{N}} + \{b\xi_l\}_{l\in\mathbb{N}} \\
&= a\{\xi_l\}_{l\in\mathbb{N}} + b\{\xi_l\}_{l\in\mathbb{N}} \\
&= ax + bx.
\end{aligned}$$

8. We have

$$\begin{aligned}
1 \cdot x &= \{1 \cdot \xi_l\}_{l\in\mathbb{N}} \\
&= \{\xi_l\}_{l\in\mathbb{N}} \\
&= x.
\end{aligned}$$

Example 1.2 (The space c of convergent number sequences). With $\mathbf{c}$ we will denote the set of all convergent number sequences. For $x = \{\xi_l\}_{l\in\mathbb{N}}$, $y = \{\eta_l\}_{l\in\mathbb{N}} \in \mathbf{c}$ and $a \in \mathbf{F}$ we define the operations addition

$$x + y = \{\xi_l + \eta_l\}_{l\in\mathbb{N}} \tag{1.3}$$

and scalar multiplication

$$ax = \{a\xi_l\}_{l\in\mathbb{N}}. \tag{1.4}$$

These operations are well-defined. Really, let $x = \{\xi_l\}_{l\in\mathbb{N}} \in \mathbf{c}$ and $y = \{\eta_l\}_{l\in\mathbb{N}} \in \mathbf{c}$ be arbitrarily chosen. Then there exist $\xi, \eta \in \mathbf{F}$ such that

$$\lim_{l\to\infty} \xi_l = \xi \quad \text{and} \quad \lim_{l\to\infty} \eta_l = \eta.$$

Hence,

$$\xi + \eta = \lim_{l\to\infty} (\xi_l + \eta_l),$$

whereupon we conclude that $x + y \in \mathbf{c}$. Also,

$$a\xi = \lim_{l\to\infty} (a\xi_l), \quad a \in \mathbf{F}.$$

Therefore $ax \in \mathbf{c}$ for $a \in \mathbf{F}$. Consequently the operations (10.38) and (1.4) are well-defined.

Now, we will prove that $\mathbf{c}$ is a vector space. Take $x = \{\xi_l\}_{l\in\mathbb{N}} \in \mathbf{c}$, $y = \{\eta_l\}_{l\in\mathbb{N}} \in \mathbf{c}$ and $z = \{\zeta_l\}_{l\in\mathbb{N}} \in \mathbf{c}$ arbitrarily.

1. Since

$$\xi_l + (\eta_l + \zeta_l) = (\xi_l + \eta_l) + \zeta_l, \quad l \in \mathbb{N},$$

we conclude that

$$x + (y + z) = (x + y) + z.$$

2. Since

$$\xi_l + \eta_l = \eta_l + \xi_l, \quad l \in \mathbb{N},$$

we get

$$x + y = y + x.$$

3. Let $0 = \{0\}_{l\in\mathbb{N}}$. Then 0 is the zero element in $\mathbf{c}$. Really, we have

$$\begin{aligned} x + 0 &= \{\xi_l + 0\}_{l\in\mathbb{N}} \\ &= \{\xi_l\}_{l\in\mathbb{N}} \\ &= x. \end{aligned}$$

4. We define $-x = \{-\xi_l\}_{l\in\mathbb{N}}$. Note that $x \in \mathbf{c}$ and

$$\begin{aligned} x + (-x) &= \{\xi_l\}_{l\in\mathbb{N}} + \{-\xi_l\}_{l\in\mathbb{N}} \\ &= \{\xi_l + (-\xi_l)\}_{l\in\mathbb{N}} \\ &= \{0\}_{l\in\mathbb{N}}. \end{aligned}$$

5. Since for any $a, b \in \mathbf{F}$ we have

$$(ab)\xi_l = a(b\xi_l), \quad l \in \mathbb{N},$$

we conclude that

$$(ab)x = a(bx).$$

6. For any $a \in \mathbf{F}$ we have

$$\begin{aligned} a(x + y) &= a\{\xi_l + \eta_l\}_{l\in\mathbb{N}} \\ &= \{a\xi_l + a\eta_l\}_{l\in\mathbb{N}} \\ &= \{a\xi_l\}_{l\in\mathbb{N}} + \{a\eta_l\}_{l\in\mathbb{N}} \\ &= a\{\xi_l\}_{l\in\mathbb{N}} + a\{\eta_l\}_{l\in\mathbb{N}} \\ &= ax + ay. \end{aligned}$$

7. For any $a, b \in \mathbf{F}$ we have

$$\begin{aligned}(a+b)x &= (a+b)\{\xi_l\}_{l\in\mathbb{N}}\\ &= \{(a+b)\xi_l\}_{l\in\mathbb{N}}\\ &= \{a\xi_l + b\xi_l\}_{l\in\mathbb{N}}\\ &= \{a\xi_l\}_{l\in\mathbb{N}} + \{b\xi_l\}_{l\in\mathbb{N}}\\ &= a\{\xi_l\}_{l\in\mathbb{N}} + b\{\xi_l\}_{l\in\mathbb{N}}\\ &= ax + bx.\end{aligned}$$

8. We have

$$\begin{aligned}1\cdot x &= \{1\cdot\xi_l\}_{l\in\mathbb{N}}\\ &= \{\xi_l\}_{l\in\mathbb{N}}\\ &= x.\end{aligned}$$

Example 1.3 (The space M[0, 1] of bounded real functions). Consider the set $\mathbf{M}[0,1]$ of all bounded functions defined on $[0,1]$. Let $f, g \in \mathbf{M}([0,1])$ and $a \in \mathbf{F}$ be arbitrarily chosen. We define addition in $\mathbf{M}([0,1]$ as follows

$$(f+g)(x) = f(x) + g(x) \quad \text{for any } x \in [0,1].$$

We define scalar multiplication in $\mathbf{M}([0,1])$ as follows:

$$(af)(x) = a(f(x)) \quad \text{for any } x \in [0,1].$$

Firstly, we will prove that the operations addition and scalar multiplication are well-defined in $\mathbf{M}[0,1]$. Really, let $f, g \in \mathbf{M}[0,1]$ and $a \in \mathbf{F}$ be arbitrarily chosen. Then there exist positive constants l_1 and l_2 such that

$$|f(x)| \le l_1 \quad \text{and} \quad |g(x)| \le l_2 \quad \text{for any } x \in [0,1].$$

Hence,

$$\begin{aligned}|af(x)| &\le |a|l_1,\\ |ag(x)| &\le |a|l_2,\\ |f(x)+g(x)| &\le |f(x)| + |g(x)|\\ &\le l_1 + l_2\end{aligned}$$

for any $x \in [0,1]$.

Now, we will prove that $\mathbf{M}[0,1]$ is a vector space. Let $f, g, h \in \mathbf{M}[0,1]$ be arbitrarily chosen.

1. Let $x \in [0,1]$ be arbitrarily chosen and fixed. Then $f(x), g(x), h(x) \in \mathbf{F}$. Hence,
$$f(x) + (g(x) + h(x)) = (f(x) + g(x)) + h(x). \tag{1.5}$$
Because $x \in [0,1]$ was arbitrarily chosen, we conclude that (10.4) holds for any $x \in [0,1]$.
2. Let $x \in [0,1]$ be arbitrarily chosen and fixed. Then $f(x), g(x) \in \mathbf{F}$. Hence,
$$f(x) + g(x) = g(x) + f(x). \tag{1.6}$$
Because $x \in [0,1]$ was arbitrarily chosen, we conclude that (1.6) holds for any $x \in [0,1]$.
3. Define
$$0(x) = 0$$
for any $x \in [0,1]$. Let $x \in [0,1]$ be arbitrarily chosen and fixed. Then $f(x) \in \mathbf{F}$ and
$$0 + f(x) = f(x). \tag{1.7}$$
Because $x \in [0,1]$ was arbitrarily chosen, we get that (1.7) holds for every $x \in \mathbf{F}$.
4. Define
$$(-f)(x) = -(f(x))$$
for any $x \in [0,1]$. Let $x \in [0,1]$ be arbitrarily chosen and fixed. Then $f(x) \in \mathbf{F}$ and $-f(x) \in \mathbf{F}$. Hence,
$$f(x) + (-f(x)) = 0. \tag{1.8}$$
Because $x \in [0,1]$ was arbitrarily chosen, we obtain that (1.8) holds for any $x \in [0,1]$.
5. Let $x \in [0,1]$ and $a, b \in \mathbf{F}$ be arbitrarily chosen and fixed. Let also, $a, b \in \mathbf{F}$. Then $f(x) \in \mathbf{F}$ and
$$(ab)f(x) = a(bf(x)). \tag{1.9}$$
Because $x \in [0,1]$ was arbitrarily chosen, we conclude that (1.9) holds for every $x \in [0,1]$.

6. Let $x \in [0,1]$ and $a \in \mathbf{F}$ be arbitrarily chosen and fixed. Then $f(x), g(x) \in \mathbf{F}$ and
$$a\,(f(x)+g(x)) = af(x) + ag(x). \tag{1.10}$$
Because $x \in [0,1]$ was arbitrarily chosen, we get that (1.10) holds for any $x \in [0,1]$.
7. Let $x \in [0,1]$ be arbitrarily chosen and fixed. Let also, $a, b \in \mathbf{F}$. Then $f(x) \in \mathbf{F}$ and
$$(a+b)f(x) = af(x) + bf(x). \tag{1.11}$$
Because $x \in [0,1]$ was arbitrarily chosen, we conclude that (1.11) holds for every $x \in [0,1]$.
8. Let $x \in [0,1]$ be arbitrarily chosen and fixed. Then $f(x) \in \mathbf{F}$ and
$$1 \cdot f(x) = f(x). \tag{1.12}$$
Because $x \in [0,1]$ was arbitrarily chosen, we conclude that (1.12) holds for any $x \in [0,1]$.

Exercise 1.1. Let $\mathbf{E}_n$ be the set of all n-tuple of elements of $\mathbf{F}$. In $\mathbf{E}_n$ we introduce the operations addition
$$x + y = (\xi_1 + \eta_1, \ldots, \xi_n + \eta_n)$$
for $x = (\xi_1, \ldots, \xi_n)$, $y = (\eta_1, \ldots, \eta_n) \in \mathbf{E}_n$, and scalar multiplication
$$ax = (a\xi_1, \ldots, a\xi_n) \quad \text{for } a \in \mathbf{F} \quad \text{and} \quad x = (\xi_1, \ldots, \xi_n) \in \mathbf{E}_n.$$
Prove that $\mathbf{E}_n$ is a vector space. The space $\mathbf{E}_n$ will be called the n-dimensional Euclidean space.

Exercise 1.2. With $\mathbf{C}[0,1]$ we will denote the set of all continuous functions defined on $[0,1]$. Prove that $\mathbf{C}[0,1]$ is a vector space.

Exercise 1.3. With $\mathbf{s}$ we will denote the set of all sequences of elements of $\mathbf{F}$. Prove that $\mathbf{s}$ is a vector space.

Below with $\mathbf{E}$ we will denote a vector space.

Corollary 1.1. *The zero element is unique.*

Proof. Suppose that there are two zero elements 0_1 and 0_2 in $\mathbf{E}$. Then

$$0_1 + 0_2 = 0_2 \quad \text{and} \quad 0_2 + 0_1 = 0_1.$$

Since $\mathbf{E}$ is a vector space we have that

$$0_1 + 0_2 = 0_2 + 0_1.$$

Therefore,

$$0_1 = 0_2,$$

which completes the proof. □

Corollary 1.2. *Let $x \in \mathbf{E}$ be arbitrarily chosen. If $y \in \mathbf{E}$ be such that $x + y = 0$, then $y = -x$.*

Proof. We have

$$\begin{aligned} x + (y + (-x)) &= (x + y) + (-x) \\ &= 0 + (-x) \\ &= -x \end{aligned}$$

and

$$\begin{aligned} x + (y + (-x)) &= x + ((-x) + y) \\ &= (x + (-x)) + y \\ &= 0 + y \\ &= y. \end{aligned}$$

Therefore, $y = -x$, which completes the proof. □

Corollary 1.3. *For every $x \in \mathbf{E}$ we have*

$$0 \cdot x = 0.$$

Proof. We have

$$\begin{aligned} x &= 1 \cdot x \\ &= (1 + 0) \cdot x \\ &= 1 \cdot x + 0 \cdot x \\ &= x + 0 \cdot x. \end{aligned}$$

Hence,

$$\begin{aligned} x + (-x) &= (x + 0 \cdot x) + (-x) \\ &= x + (0 \cdot x + (-x)) \\ &= x + ((-x) + 0 \cdot x) \\ &= (x + (-x)) + 0 \cdot x \\ &= 0 + 0 \cdot x \\ &= 0 \cdot x. \end{aligned}$$

Because $x + (-x) = 0$, we conclude $0 = 0 \cdot x$. This completes the proof. □

Corollary 1.4. *For every $x \in \mathbf{E}$ we have*

$$(-1) \cdot x = -x.$$

Proof. Using Corollary 1.3, we have

$$\begin{aligned} (-1)x + x &= (-1) \cdot x + 1 \cdot x \\ &= (-1 + 1) \cdot x \\ &= 0 \cdot x \\ &= 0. \end{aligned}$$

Hence and Corollary 1.2, we obtain $(-1) \cdot x = -x$. This completes the proof. □

Corollary 1.5. *If $x \in \mathbf{E}$, $x \neq 0$, and $ax = bx$, $a, b \in \mathbf{F}$, then $a = b$.*

Proof. From $ax = bx$ and $bx - bx = 0$, we get

$$\begin{aligned} 0 &= ax - bx \\ &= (a - b)x. \end{aligned}$$

Assume that $a \neq b$. Then

$$\begin{aligned} x &= \frac{1}{a-b}\Big((a-b)\,x\Big) \\ &= \frac{1}{a-b}0 \\ &= 0, \end{aligned}$$

which is a contradiction. This completes the proof. □

Corollary 1.6. *If $a \in \mathbf{F}$, $a \neq 0$, and $ax = ay$, $x, y \in \mathbf{E}$, then $x = y$.*

Proof. We have

$$\begin{aligned} ax - ay &= ay - ay \\ &= 0, \end{aligned}$$

whereupon

$$a(x - y) = 0.$$

Hence, using that $a \neq 0$, we obtain

$$\begin{aligned} x - y &= \frac{1}{a}\left(a(x - y)\right) \\ &= \frac{1}{a}0 \\ &= 0. \end{aligned}$$

Therefore $x = y$, which completes the proof. □

Definition 1.2. The elements $x_1, \ldots, x_n$ of the vector space $\mathbf{E}$ are said to be linearly independent, if any relation

$$a_1 x_1 + \cdots + a_n x_n = 0,$$

implies that $a_1 = \cdots = a_n = 0$.

Definition 1.3. The elements $x_1, \ldots, x_n$ of the vector space $\mathbf{E}$ are said to be linearly dependent, if the relation

$$a_1 x_1 + \cdots + a_n x_n = 0,$$

is possible with at least one of the coefficients $a_1, \ldots, a_n$ is not zero.

Example 1.4. Consider $\mathbf{E}_3$ and

$$x_1 = (1, 1, 1), \quad x_2 = (2, 2, 2), \quad x_3 = (-3, -3, -3).$$

Then

$$x_1 + x_2 + x_3 = 0.$$

Here, $a_1 = a_2 = a_3 = 1$. Therefore, the elements x_1, x_2 and x_3 are linearly dependent in $\mathbf{E}_3$.

Example 1.5. Consider the set $E[a, b]$ of all real-valued functions defined on $[a, b]$ and its elements

$$1, \quad x, \ldots, \quad x^{p-1}, \quad p \in \mathbb{N}.$$

Assume that there are constants $a_0, \ldots, a_{p-1} \in \mathbf{F}$ such that

$$a_0 + a_1 x + \cdots + a_{p-1} x^{p-1} = 0 \quad \text{in} \quad [a, b], \tag{1.13}$$

and $(a_0, \ldots, a_{p-1}) \neq (0, \ldots, 0)$. Hence, the equation (1.13) could hold for at most $p - 1$ values of x, whereas it must hold for all $x \in [a, b]$. Therefore, the elements $1, \ldots, x^{p-1}$ of $\mathbf{E}[a, b]$ are linearly independent.

Example 1.6. Consider the set $E[a, b]$ and its elements

$$0 \quad \text{and} \quad \sin x.$$

Then

$$1 \cdot 0 + 0 \cdot \sin x = 0 \quad \text{for all } x \in [a, b].$$

Here, $a_0 = 1$ and $a_1 = 0$. Therefore, 0 and $\sin x$ are linearly dependent.

Exercise 1.4. Consider the set $E[a, b]$. Prove that e^x and e^{2x} are linearly independent.

Definition 1.4. The vector space $\mathbf{E}$ will be called m-dimensional if in it there are m linearly independent elements and every $m + 1$ its elements are linearly dependent. The dimension of the vector space $\mathbf{E}$ will be denoted by $\dim(\mathbf{E})$.

Example 1.7. Consider $\mathbf{E}_m$. Let

$$x_1 = (1, 0, \ldots, 0),$$
$$x_2 = (0, 1, \ldots, 0),$$
$$\vdots$$
$$x_m = (0, 0, \ldots, 1).$$

Assume that there are $a_i \in \mathbf{F}$, $i \in \{1, \ldots, m\}$, such that $\sum_{i=1}^{m} a_i x_i = 0$. Then

$$(a_1, \ldots, a_m) = (0, \ldots, 0),$$

whereupon $a_i = 0$, $i \in \{1, \ldots, m\}$. Therefore, $x_1, \ldots, x_m$ are linearly independent. Let now

$$y_1 = (y_1^1, \ldots, y_m^1),$$
$$\vdots$$
$$y_{m+1} = (y_m^{m+1}, \ldots, y_m^{m+1})$$

be arbitrarily chosen elements of $\mathbf{E}_m$. Since

$$rank \begin{pmatrix} y_1^1 & \cdots & y_1^{m+1} \\ \vdots & \vdots & \vdots \\ y_m^1 & \cdots & y_m^{m+1} \end{pmatrix} \leq m,$$

we have that $y_1, \ldots, y_{m+1}$ are linearly dependent. Therefore $\mathbf{E}_m$ is m-dimensional.

Example 1.8. Let $n \in \mathbb{N}$. With $\mathbf{P}^n$ we will denote the set of all polynomials of degree n with coefficients which are elements of the field $\mathbf{F}$. Consider 1, $t, \ldots, t^n$. Assume that there are $a_i \in \mathbf{F}$, $i \in \{1, \ldots, n+1\}$, such that

$$\sum_{i=1}^{n+1} a_i t^{i-1} = 0 \quad \text{for all } t \in \mathbb{R}. \tag{1.14}$$

Because (1.14) has at most n solutions with respect to t, we conclude that (1.14) holds if $a_i = 0$, $i \in \{1, \ldots, n+1\}$. Therefore 1, $t, \ldots, t^n$ are linearly independent. Let now $p_i \in \mathbf{P}^n$, $i \in \{1, \ldots, n+2\}$, be arbitrarily chosen and

$$p_i(t) = \sum_{j=1}^{n+1} \alpha_j^i t^{j-1}, \quad i \in \{1, \ldots, n+2\}, \quad \alpha_j^i \in \mathbf{F},$$

$i \in \{1, \ldots, n+2\}$, $j \in \{1, \ldots, n+1\}$. Assume that there are $b_i \in \mathbf{F}$, $i \in \{1, \ldots, n+2\}$, such that

$$\sum_{i=1}^{n+2} b_i p_i(t) = 0 \quad \text{for all } t \in \mathbb{R}.$$

Hence,

$$\sum_{i=1}^{n+2} b_i \alpha_j^i = 0, \quad j \in \{1, \dots, n+1\}. \tag{1.15}$$

Since

$$rank \begin{pmatrix} \alpha_1^1 & \dots & \alpha_{n+1}^1 \\ \vdots & \vdots & \vdots \\ \alpha_1^{n+2} & \dots & \alpha_{n+1}^{n+2} \end{pmatrix} \le n+1,$$

we conclude that (1.15) holds if $b_i = 0$, $i \in \{1, \dots, n+2\}$. Therefore every $n+2$ elements of $\mathbf{P}^n$ are linearly dependent. Hence, $\dim(\mathbf{P}^n) = n+1$.

Exercise 1.5. Let $m \in \mathbb{N}$. With $\mathbf{M}_m$ we will denote the set of all $m \times m$ matrices whose entries are elements of $\mathbf{F}$. Prove that $\mathbf{M}_m$ is a vector space and $\dim(\mathbf{M}_m) = m$.

Definition 1.5. Let $\mathbf{E}$ be a vector space with $\dim(\mathbf{E}) = m$. Then every system of m linearly independent elements of $\mathbf{E}$ will be called basis of $\mathbf{E}$.

Theorem 1.1. *Let* $\mathbf{E}$ *be a vector space with* $\dim(\mathbf{E}) = m$ *and* $\{e_1, \dots, e_m\}$ *be its basis. Then every* $x \in \mathbf{E}$ *can be represented in a unique way in the form*

$$x = \alpha_1 e_1 + \dots + \alpha_m e_m, \quad \alpha_i \in \mathbf{F}, \quad i \in \{1, \dots, m\}. \tag{1.16}$$

Proof. Since $e_1, \dots, e_m, x$ are linearly dependent, there are $\tilde{\alpha}_i \in \mathbf{F}$, $i \in \{1, \dots, m+1\}$, such that

$$\tilde{\alpha}_1 e_1 + \dots + \tilde{\alpha}_m e_m + \tilde{\alpha}_{m+1} x = 0, \quad (\tilde{\alpha}_1, \dots, \tilde{\alpha}_{m+1}) \neq (0, \dots, 0).$$

Assume that $\tilde{\alpha}_{m+1} = 0$. Then

$$\tilde{\alpha}_1 e_1 + \dots + \tilde{\alpha}_m e_m = 0, \quad (\tilde{\alpha}_1, \dots, \tilde{\alpha}_m) \neq (0, \dots, 0),$$

which is a contradiction because $\{e_1, \dots, e_m\}$ is a basis in $\mathbf{E}$. Therefore, $\tilde{\alpha}_{m+1} \neq 0$. Hence,

$$x = -\frac{\tilde{\alpha}_1}{\tilde{\alpha}_{m+1}} e_1 - \dots - \frac{\tilde{\alpha}_m}{\tilde{\alpha}_{m+1}} e_m.$$

We set $\alpha_i = -\frac{\tilde{\alpha}_i}{\tilde{\alpha}_{m+1}}$, $i \in \{1, \dots, m\}$, and we get the representation (1.16). Now, we assume that

$$x = \beta_1 e_1 + \cdots + \beta_m e_m, \quad \beta_i \in \mathbf{F}, \quad i \in \{1, \dots, m\}.$$

Then, using (1.16), we obtain

$$0 = (\alpha_1 - \beta_1)\, e_1 + \cdots + (\alpha_m - \beta_m)\, e_m.$$

Since $\{e_1, \dots, e_m\}$ is a basis in $\mathbf{E}$, from the last equality we conclude that $\alpha_i = \beta_i$, $i \in \{1, \dots, m\}$. This completes the proof. □

Definition 1.6. A vector space $\mathbf{E}$ is infinite-dimensional if for every natural n there is a system of n linearly independent elements in $\mathbf{E}$.

Example 1.9. Consider $\mathscr{C}([a,b])$. Because for every $n \in \mathbb{N}$ the elements

$$1,\ t, \dots, t^n$$

is linearly independent, we conclude that $\mathscr{C}([a,b])$ is infinite-dimensional vector space.

Exercise 1.6. Let $k \in \mathbb{N}$ be arbitrarily chosen. Prove that $\mathscr{C}^k([a,b])$ is infinite-dimensional vector space.

Definition 1.7. Let $\mathbf{E}$ be a vector space. Suppose that $\mathbf{V} \subseteq \mathbf{E}$. If $\mathbf{V}$ is a vector space itself with the same vector space operations as $\mathbf{E}$ has, then $\mathbf{V}$ is called a linear subspace of $\mathbf{E}$.

Theorem 1.2. *Let $\mathbf{E}$ be a m-dimensional vector space and $\mathbf{V}$ be its linear subspace. Then* $\dim(\mathbf{V}) \leq m$.

Proof. Suppose that $\dim(\mathbf{V}) > m$. Then there are $\dim(\mathbf{V})$ linearly independent elements in $\mathbf{V}$. Since $\mathbf{V}$ is a linear subspace of $\mathbf{E}$, we conclude that in $\mathbf{E}$ there are $\dim(\mathbf{V})$ linearly independent elements, which is contradiction. This completes the proof. □

Theorem 1.3. *Let $\mathbf{E}$ be a vector space and $\mathbf{V}$ be a subset of $\mathbf{E}$. Then $\mathbf{V}$ is a nonempty linear subspace if and only if $\mathbf{V}$ satisfies the following properties:*

1. $0 \in \mathbf{V}$,
2. *if $x, y \in \mathbf{V}$, then $x + y \in \mathbf{V}$,*
3. *if $x \in \mathbf{V}$ and $a \in \mathbf{F}$, then $ax \in \mathbf{V}$.*

Proof.

1. Let $\mathbf{V}$ satisfies (1), (2) and (3). By (1) it follows that $\mathbf{V}$ is nonempty. The properties (2) and (3) ensure closure of $\mathbf{V}$ under addition and scalar multiplication. Since the elements of $\mathbf{V}$ are necessarily elements of $\mathbf{E}$, the axioms (L1), (L2), (L3), (L5), (L6), (L7) and (L8) are satisfied. By the closure of $\mathbf{V}$ under scalar multiplication and Corollary 1.4, it follows that $-x = (-1) \cdot x \in \mathbf{V}$ and $x + (-x) = 0$ for every $x \in \mathbf{V}$, i.e., the axiom (L4) is satisfied.
2. Let $\mathbf{V}$ be a nonempty linear subspace of $\mathbf{E}$. Then $\mathbf{V}$ is itself a vector space under the operations induced by $\mathbf{E}$. So, the properties (2) and (3) are satisfied. By property (3) and Corollary 1.4, we have that $-x \in \mathbf{V}$ for every $x \in \mathbf{V}$. Hence, it follows that $\mathbf{V}$ is closed under subtraction as well. Because $\mathbf{V}$ is nonempty, there is an element $x \in \mathbf{V}$. For it we have that $x - x = 0 \in \mathbf{V}$, i.e., the property (1) is satisfied. This completes the proof. □

Theorem 1.4. *Let $\mathbf{E}$ be a vector space and $\mathbf{V}$ be its nonempty subset. Then $\mathbf{V}$ is a linear subspace of $\mathbf{E}$ if and only if every linear combination of finitely many elements of $\mathbf{V}$ also belongs to $\mathbf{V}$.*

Proof.

1. Let $\mathbf{V}$ be a subspace. Then the properties (1), (2) and (3) of Theorem 1.3 are satisfied. Let $x_1, \ldots, x_k \in \mathbf{V}$ and $a_1, \ldots, a_k \in \mathbf{F}$ be arbitrarily chosen. Then, by the property (3) of Theorem 1.3, it follows that

$$a_l x_l \in \mathbf{V} \quad \text{for any } l \in \{1, \ldots, k\}.$$

Hence, by the property (2) of Theorem 1.3, it follows that

$$a_1 x_1 + a_2 x_2 \in \mathbf{V}.$$

Again, by the property (2) of Theorem 1.3, it follows that

$$a_1 x_1 + a_2 x_2 + a_3 x_3 \in \mathbf{V}.$$

And so on,

$$a_1 x_1 + \cdots + a_k x_k \in \mathbf{V}.$$

2. Let $\mathbf{V}$ is closed under linear combination of finitely many its elements. Then, if $x, y \in \mathbf{V}$ and $a \in \mathbf{F}$ be arbitrarily chosen, we get

$$x + y \in \mathbf{V} \quad \text{and} \quad ax \in \mathbf{V},$$

i.e., the properties (2) and (3) of Theorem 1.3 are satisfied. Now, using the same arguments as in the second part of the proof of Theorem 1.3, we conclude that the property (1) of Theorem 1.3 is satisfied. This completes the proof. □

Definition 1.8. Let $\mathbf{E}$ be a vector space and $x_1, \ldots, x_k \in \mathbf{E}$. The set

$$Span\{x_1, \ldots, x_k\} = \{a_1x_1 + \cdots + a_kx_k : a_l \in \mathbf{F}, l \in \{1, \ldots, k\}\}$$

will be called the span of the elements $x_1, \ldots, x_k$.

Theorem 1.5. *Let* $\mathbf{E}$ *be a vector space and* $x_1, \ldots, x_k$ *be its elements. Then* $Span\{x_1, \ldots, x_k\}$ *is a linear subspace of* $\mathbf{E}$.

Proof. Note that $0 \in Span\{x_1, \ldots, x_k\}$ because

$$0 = 0x_1 + \cdots + 0x_k \in Span\{x_1, \ldots, x_k\}.$$

Let

$$x = a_1x_1 + \cdots + a_kx_k, \quad y = b_1x_1 + \cdots + b_kx_k, \quad a_l, b_l \in \mathbf{F},$$
$$l \in \{1, \ldots, k\}.$$

Then, for any $a, b \in \mathbf{F}$, we have $aa_l, bb_l \in \mathbf{F}$, $l \in \{1, \ldots, k\}$, and

$$\begin{aligned} ax + by &= aa_1x_1 + \cdots + aa_kx_k + bb_1x_1 + \cdots + bb_kx_k \\ &= (aa_1 + bb_1)\,x_1 + \cdots + (aa_k + bb_k)\,x_k \in Span\{x_1, \ldots, x_k\}. \end{aligned}$$

Also, $ax \in Span\{x_1, \ldots, x_k\}$. Hence and Theorem 1.3, it follows that $Span\{x_1, \ldots, x_k\}$ is a linear subspace. This completes the proof. □

Theorem 1.6. *The intersection of any collection of linear subspaces of a vector space* $\mathbf{E}$ *is a linear subspace of* $\mathbf{E}$.

Proof. Let $\{\mathbf{A}_n\}_{n \in \mathscr{A}}$ be a collection of linear subspaces of the vector space $\mathbf{E}$. Here $\mathscr{A}$ is an index set. We set

$$\mathbf{B} = \bigcap_{n \in \mathscr{A}} \mathbf{A}_n.$$

Since $0 \in \mathbf{A}_n$ for any $n \in \mathscr{A}$, we get that $0 \in \mathbf{B}$. Let $x, y \in \mathbf{B}$ and $a, b \in \mathbf{F}$ be arbitrarily chosen. Then $x, y \in \mathbf{A}_n$ for any $n \in \mathscr{A}$. Because $\mathbf{A}_n$, $n \in \mathscr{A}$, are linear subspaces of $\mathbf{E}$, we conclude that $ax + by, ax \in \mathbf{A}_n$ for any $n \in \mathscr{A}$. Therefore $ax + by, ax \in \mathbf{B}$. Hence and Theorem 1.3, we get that $\mathbf{B}$ is a linear subspace of $\mathbf{E}$. This completes the proof. □

Definition 1.9. Suppose that $\mathbf{E}$ is a vector space and $\mathbf{S}$ is its subset. There exist linear subspaces $\mathbf{A}_n$ containing the set $\mathbf{S}$. For example, the space $\mathbf{E}$ is a such subspace. With $\mathbf{B}$ we will denote the intersection of any linear subspaces containing the set $\mathbf{S}$. Then $\mathbf{B}$ contains $\mathbf{S}$ and it is the smallest such linear subspace. The subspace $\mathbf{B}$ is called the linear subspace spanned by $\mathbf{S}$ or the span of $\mathbf{S}$.

Theorem 1.7. *If* $\mathbf{S}$ *is any set of elements in a vector space* $\mathbf{E}$ *and if* $\mathbf{B}$ *is the linear subspace spanned by* $\mathbf{S}$, *then* $\mathbf{B}$ *is the same as the set of all linear combinations of elements of* $\mathbf{S}$.

Proof. Let $\mathbf{C}$ be the set of all linear combinations of elements of $\mathbf{S}$. Note that $\mathbf{C}$ is a linear subspace containing $\mathbf{S}$. Therefore

$$\mathbf{B} \subseteq \mathbf{C}. \tag{1.17}$$

On the other hand, since $\mathbf{B}$ is spanned by $\mathbf{S}$ it must contain all linear combinations of elements of $\mathbf{S}$. Consequently

$$\mathbf{C} \subseteq \mathbf{B}.$$

From the last inclusion and from (1.17), we conclude that $\mathbf{B} = \mathbf{C}$. This completes the proof. □

Theorem 1.8. *Let* $\mathbf{A}$ *and* $\mathbf{B}$ *be linear subspaces of a vector space* $\mathbf{E}$ *and* $\mathbf{C}$ *be the linear subspace spanned by* $\mathbf{A}$ *and* $\mathbf{B}$ *together. Then* $\mathbf{C}$ *is the same as the set of all elements of the form* $x + y$ *with* $x \in \mathbf{A}$ *and* $y \in \mathbf{B}$.

Proof. Let $\mathbf{D}$ be the set of all elements of the form $x + y$, $x \in \mathbf{A}$ and $y \in \mathbf{B}$. Then

$$\mathbf{C} \subseteq \mathbf{D}. \tag{1.18}$$

Since $\mathbf{C}$ is spanned by $\mathbf{A}$ and $\mathbf{B}$ together it must contain all elements of the form $x + y$, $x \in \mathbf{A}$ and $y \in \mathbf{B}$. Therefore,

$$D \subseteq \mathbf{C}.$$

From the last inclusion and from (1.18), we conclude that $\mathbf{C} = \mathbf{D}$. This completes the proof. □

Prompted by Theorem 1.8, we shall use the notation $\mathbf{A} + \mathbf{B}$ for the linear subspace $\mathbf{C}$ spanned by the linear subspaces $\mathbf{A}$ and $\mathbf{B}$ together.

Definition 1.10. We will say that a linear subspace $\mathbf{A}$ of a vector space $\mathbf{E}$ is a complement of a linear subspace $\mathbf{B}$ if

$$\mathbf{A} \cap \mathbf{B} = \{0\} \quad \text{and} \quad \mathbf{A} + \mathbf{B} = \mathbf{E}.$$

Example 1.10. Let $\mathbf{A}$ and $\mathbf{B}$ be finite dimensional linear subspaces of the same dimension. Suppose that $\mathbf{A} \subseteq \mathbf{B}$. We will prove that $\mathbf{A} = \mathbf{B}$. Really, let $\dim(\mathbf{A}) = \dim(\mathbf{B}) = m$. Then there are m linearly independent elements of $\mathbf{A}$, $\{x_1, \ldots, x_m\}$. Since $\mathbf{A} \subseteq \mathbf{B}$, we have that $\{x_1, \ldots, x_m\}$ are linearly independent elements of $\mathbf{B}$. Hence, using that $\dim(\mathbf{B}) = m$, we conclude that every element of $\mathbf{B}$ is a linear combination of $\{x_1, \ldots, x_m\}$. Therefore, $\mathbf{B} \subseteq \mathbf{A}$, which completes the proof.

Exercise 1.7. Let x_1, x_2 and x_3 be elements of the vector space $\mathbf{E}$ for which $x_1 + x_2 + x_3 = 0$. Prove that x_1 and x_2 span the same linear subspace as x_2 and x_3.

Exercise 1.8. A polynomial y is called even if $y(-x) = y(x)$ identically in x and odd if $y(-x) = -y(x)$ identically in x.

1. Prove that the sets $\mathbf{P}_{even}$ and $\mathbf{P}_{odd}$ of even and odd polynomials, respectively, are linear subspaces of the space $\mathbf{P}$ of all polynomials with coefficients in $\mathbf{F}$.
2. Prove that $\mathbf{P}_{even}$ and $\mathbf{P}_{odd}$ are each other's compliments.

Definition 1.11. Two vector spaces $\mathbf{U}$ and $\mathbf{V}$, over the same field $\mathbf{F}$, are called called isomorphic if there is a one-to-one correspondence between the elements x of $\mathbf{U}$ and the elements y of $\mathbf{V}$, say $y = T(x)$, such that

$$T(a_1 x_1 + a_2 x_2) = a_1 T(x_1) + a_2 T(x_2)$$

for any $a_1, a_2 \in \mathbf{F}$ and any $x_1, x_2 \in \mathbf{U}$. In other words, $\mathbf{U}$ and $\mathbf{V}$ are isomorphic if there is an isomorphism between them, where an isomorphism is a one-to-one correspondence that preserves all linear relations.

Theorem 1.9. *Let* **U** *and* **V** *be vector spaces that are isomorphic and* T *is an isomorphism between them,* $T(x) = y$, $x \in \mathbf{U}$, $y \in \mathbf{V}$. *Then* $T(0) = 0$.

Proof. For $x \in \mathbf{U}$ we have

$$\begin{aligned} T(0) &= T(x - x) \\ &= T(x) - T(x) \\ &= 0, \end{aligned}$$

which completes the proof. □

Theorem 1.10. *Let* **U** *and* **V** *be vector spaces that are isomorphic and* T *is an isomorphism between them,* $T(x) = y$, $x \in \mathbf{U}$, $y \in \mathbf{V}$.

1. *If* $\{x_1, \ldots, x_n\}$ *are linearly dependent, then* $\{T(x_1), \ldots, T(x_n)\}$ *are linearly dependent.*
2. *If* $\{x_1, \ldots, x_n\}$ *are linearly independent, then* $\{T(x_1), \ldots, T(x_n)\}$ *are linearly independent.*

Proof.

1. Since $\{x_1, \ldots, x_n\}$ are linearly dependent elements of **U**, there are constants $\alpha_1, \ldots, \alpha_n \in \mathbf{F}$ such that

   $$\alpha_1 x_1 + \cdots + \alpha_n x_n = 0, \quad (\alpha_1, \ldots, \alpha_n) \neq (0, \ldots, 0).$$

   Hence,

   $$0 = T(\alpha_1 x_1 + \cdots + \alpha_n x_n) = \alpha_1 T(x_1) + \cdots + \alpha_n T(x_n).$$

   Therefore, $\{T(x_1), \ldots, T(x_n)\}$ are linearly dependent elements of **V**.
2. Let $\{x_1, \ldots, x_n\}$ be linearly independent elements of **U**. Assume that $\{T(x_1), \ldots, T(x_n)\}$ are linearly dependent elements of **V**. Then there are $\alpha_1, \ldots, \alpha_n \in \mathbf{F}$, $(\alpha_1, \ldots, \alpha_n) \neq (0, \ldots, 0)$, such that

   $$\alpha_1 T(x_1) + \cdots + \alpha_n T(x_n) = 0.$$

   Hence,

   $$T(\alpha_1 x_1 + \cdots + \alpha_n x_n) = 0.$$

Therefore, using Theorem 1.9, we get

$$\alpha_1 x_1 + \cdots + \alpha_n x_n = 0,$$

which is a contradiction. This completes the proof. □

Corollary 1.7. *Let* $\mathbf{U}$ *and* $\mathbf{V}$ *be finite-dimensional vector spaces that are isomorphic. Then* $\dim(\mathbf{U}) = \dim(\mathbf{V})$.

Theorem 1.11. *Every* n *dimensional vector space* $\mathbf{U}$ *over the field* $\mathbf{F}$ *is isomorphic to* $\mathbf{E}_n$.

Proof. Let $\{x_1, \ldots, x_n\}$ be any basis in $\mathbf{U}$. Then every $x \in \mathbf{U}$ can be represented in the form

$$x = \xi_1 x_1 + \cdots + \xi_n x_n,$$

where $\xi_i \in \mathbf{F}$, $i \in \{1, \ldots, n\}$, are uniquely determined by x. We consider the one-to-one correspondence

$$x \to (\xi_1, \ldots, \xi_n)$$

between $\mathbf{U}$ and $\mathbf{E}_n$. If

$$y = \eta_1 x_1 + \cdots + \eta_n x_n, \quad \eta_l \in \mathbf{F}, \quad l \in \{1, \ldots, n\},$$

then for any $a, b \in \mathbf{F}$ we have

$$\begin{aligned} ax + by &= a(\xi_1 x_1 + \cdots + \xi_n x_n) + b(\eta_1 x_1 + \cdots + \eta_n x_n) \\ &= (a\xi_1 + b\eta_1) x_1 + \cdots + (a\xi_n + b\eta_n) x_n, \end{aligned}$$

this establishes the desired isomorphism. □

Theorem 1.12. *Let* $\mathbf{U}$ *and* $\mathbf{V}$ *be* n*-dimensional vector spaces. Then they are isomorphic.*

Proof. By Theorem 1.11 it follows that there exist isomorphisms T_1 and T_2 between $\mathbf{U}$ and $\mathbf{E}_n$ and $\mathbf{E}_n$ and $\mathbf{V}$, respectively. Then $T_2 \circ T_1$ is an isomorphism between $\mathbf{U}$ and $\mathbf{V}$, which completes the proof. □

Definition 1.12. A linear functional on a vector space $\mathbf{E}$ is a scalar-valued function y defined for every element $x \in \mathbf{E}$ with the property

$$y\left(a_1x_1 + a_2x_2\right) = a_1y(x_1) + a_2y(x_2)$$

for any $a_1, a_2 \in \mathbf{F}$ and any $x_1, x_2 \in \mathbf{E}$. The space of all linear functionals on $\mathbf{E}$ will be denoted by $\mathbf{E}'$. We define the zero in $\mathbf{E}'$ as follows: $y(x) = 0$ for any $x \in \mathbf{E}$. If y_1 and y_2 are two linear functionals on $\mathbf{E}$ and if a_1, $a_2 \in \mathbf{F}$, then $a_1y_1(x) + a_2y_2(x)$ is a linear functional on $\mathbf{E}$. With these concepts (zero, addition, scalar multiplication), the set $\mathbf{E}'$ forms a vector space, the dual space of $\mathbf{E}$.

Example 1.11. Consider $\mathbf{E}_n$. Let $\alpha_1, \ldots, \alpha_n, \beta \in \mathbf{F}$, $\beta \neq 0$, be fixed. For any $x \in \mathbf{E}_n$, $x = (\xi_1, \ldots, \xi_n)$, we define the scalar-valued function

$$y(x) = \alpha_1\xi_1 + \cdots + \alpha_n\xi_n + \beta.$$

If $x_1 = \left(\xi_1^1, \ldots, \xi_n^1\right)$, $x_2 = \left(\xi_1^2, \ldots, \xi_n^2\right) \in \mathbf{E}_n$ and $a_1, a_2 \in \mathbf{F}$ be arbitrarily chosen, then

$$\begin{aligned}
a_1x_1 &= \left(a_1\xi_1^1, \ldots, a_1\xi_n^1\right), \\
a_2x_2 &= \left(a_2\xi_1^2, \ldots, a_2\xi_n^2\right), \\
a_1x_1 + a_2x_2 &= \left(a_1\xi_1^1 + a_2\xi_1^2, \ldots, a_1\xi_n^1 + a_2\xi_n^2\right), \\
a_1y(x_1) &= a_1\alpha_1\xi_1^1 + \cdots + a_1\alpha_n\xi_n^1 + a_1\beta, \\
a_2y(x_2) &= a_2\alpha_1\xi_1^2 + \cdots + a_2\alpha_n\xi_n^2 + a_2\beta, \\
y\left(a_1x_1 + a_2x_2\right) &= \alpha_1\left(a_1\xi_1^1 + a_2\xi_1^2\right) + \cdots + \alpha_n\left(a_1\xi_n^1 + a_2\xi_n^2\right) + \beta.
\end{aligned}$$

Since $a_1\beta + a_2\beta \neq \beta$ unless $a_1 + a_2 = 1$, we conclude that y is not a linear functional on $\mathbf{E}_n$.

Example 1.12. Consider $\mathbf{C}$ over $\mathbb{R}$. For every $x = a + ib \in \mathbf{C}$, $a, b \in \mathbb{R}$, we define the scalar-valued function y as follows:

$$y(x) = a.$$

Let

$$x_1 = a_1 + ib_1, \quad x_2 = a_2 + ib_2, \quad a_1, a_2, b_1, b_2 \in \mathbb{R}, \quad \alpha_1, \alpha_2 \in \mathbb{R}$$

be arbitrarily chosen. Then

$$\begin{aligned}
\alpha_1 y(x_1) &= \alpha_1 a_1, \quad \alpha_2 y(x_2) \\
&= \alpha_2 a_2, \\
\alpha_1 x_1 + \alpha_2 x_2 &= (\alpha_1 a_1 + \alpha_2 a_2) + i\,(\alpha_1 b_2 + \alpha_2 b_2), \\
y\,(\alpha_1 x_1 + \alpha_2 x_2) &= \alpha_1 a_1 + \alpha_2 a_2 \\
&= \alpha_1 y(x_1) + \alpha_2 y(x_2).
\end{aligned}$$

Consequently, y is a linear functional on $\mathbf{C}$.

Exercise 1.9. Consider $\mathbf{C}$. For $x = a + ib \in \mathbf{C}$, $a, b \in \mathbb{R}$, we define the scalar-valued function y as follows

$$y(x) = \sqrt{a^2 + b^2}.$$

Check if y is a linear functional on $\mathbf{C}$.

Answer. No.

Definition 1.13. Suppose that $\mathbf{E}$ is a vector space and $\mathbf{E}'$ is its dual space. We make correspond $[x, y]$, $x \in \mathbf{E}$, $y \in \mathbf{E}'$, to be the value of y at x. In terms of the symbol $[x, y]$ the defining property of a linear functional is

$$[a_1 x_1 + a_2 x_2, y] = a_1[x_1, y] + a_2[x_2, y]$$

and the definition of the linear operations for linear functionals is

$$[x, a_1 y_1 + a_2 y_2] = a_1[x, y_1] + a_2[x, y_2].$$

The two relations together are expressed by saying that $[x, y]$ is a bilinear functional of the elements $x \in \mathbf{E}$ and $y \in \mathbf{E}'$.

Theorem 1.13. *Let $\mathbf{E}$ be a n-dimensional vector space and $\{x_1, \ldots, x_n\}$ is its basis. Let also, $\{a_1, \ldots, a_n\}$ be a set of n scalars. Then there exists unique linear functional y such that $[x_i, y] = a_i$, $i \in \{1, \ldots, n\}$.*

Proof. For any $x \in \mathbf{E}$ and any linear functional $y \in \mathbf{E}'$ we have

$$x = \xi_1 x_1 + \cdots + \xi_n x_n \quad \text{and} \quad [x, y] = \xi_1 [x_1, y] + \cdots + \xi_n [x_n, y],$$
$$\xi_l \in \mathbf{F}, \quad l \in \{1, \ldots, n\}.$$

We define a linear functional $\tilde{y}$ such that $[x_i, \tilde{y}] = \alpha_i$, $i \in \{1, \ldots, n\}$. Then the value of $[x, \tilde{y}]$ is determined by

$$[x, \tilde{y}] = \alpha_1 \xi_1 + \cdots + \alpha_n \xi_n.$$

Suppose that there are two linear functionals $\tilde{y}$ and $\tilde{\tilde{y}}$ such that

$$[x_i, \tilde{y}] = [x_i, \tilde{\tilde{y}}] = \alpha_i, \quad i \in \{1, \ldots, n\}.$$

Then $\tilde{y} - \tilde{\tilde{y}}$ is a linear functional and

$$[x, \tilde{y} - \tilde{\tilde{y}}] = 0.$$

Therefore, $\tilde{y} = \tilde{\tilde{y}}$. This completes the proof. □

Theorem 1.14. *Let* $\mathbf{E}$ *be a vector space with a basis* $\{x_1, \ldots, x_n\}$. *Then there exists a uniquely determined basis in* $\mathbf{E}'$, $\{y_1, \ldots, y_n\}$, *such that* $[x_i, y_j] = \delta_{ij}$, $i, j \in \{1, \ldots, n\}$. *Here,* $\delta_{ij} = 1$ *if* $i = j$ *and* $\delta_{ij} = 0$ *if* $i \neq j$, $i, j \in \{1, \ldots, n\}$.

Proof. By Theorem 1.13, it follows that there are uniquely determined functionals $\{y_1, \ldots, y_n\}$ such that $[x_i, y_j] = \delta_{ij}$, $i, j \in \{1, \ldots, n\}$. We will prove that $\{y_1, \ldots, y_n\}$ is a basis in $\mathbf{E}'$. Suppose that

$$a_1 y_1 + \cdots + a_n y_n = 0$$

for some $a_i \in \mathbf{F}$, $i \in \{1, \ldots, n\}$. Then for any $x \in \mathbf{E}$ we have

$$\begin{aligned}[x, a_1 y_1 + \cdots + a_n y_n] &= a_1 [x, y_1] + \cdots + a_n [x, y_n] \\ &= 0.\end{aligned}$$

In particular, if $x = x_i$, $i \in \{1, \ldots, n\}$, we get

$$\begin{aligned}[x_i, a_1 y_1 + \cdots + a_n y_n] &= \sum_{j=1}^{n} a_j [x_i, y_j] \\ &= \sum_{j=1}^{n} a_j \delta_{ij} \\ &= a_i, \quad i \in \{1, \ldots, n\}.\end{aligned}$$

Therefore, $a_i = 0$, $i \in \{1, \dots, n\}$, and hence, $\{y_1, \dots, y_n\}$ are linearly independent. Let $y \in \mathbf{E}'$ be arbitrarily chosen. We set $[x_i, y] = \alpha_i$, $i \in \{1, \dots, n\}$, and let $x = \sum_{i=1}^{n} \xi_i x_i$. Then

$$\begin{aligned}[x, y] &= \left[\sum_{i=1}^{n} \xi_i x_i, y\right] \\ &= \sum_{i=1}^{n} \xi_i [x_i, y] \\ &= \sum_{i=1}^{n} \alpha_i \xi_i.\end{aligned}$$

On the other hand,

$$\begin{aligned}[x, y_j] &= \left[\sum_{i=1}^{n} \xi_i x_i, y_j\right] \\ &= \sum_{i=1}^{n} \xi_i [x_i, y_j] \\ &= \sum_{i=1}^{n} \xi_i \delta_{ij} \\ &= \xi_j, \quad j \in \{1, \dots, n\}.\end{aligned}$$

Therefore,

$$\begin{aligned}[x, y] &= \sum_{i=1}^{n} \alpha_i [x, y_i] \\ &= \left[x, \sum_{i=1}^{n} \alpha_i y_i\right].\end{aligned}$$

Consequently, $y = \sum_{i=1}^{n} \alpha_i y_i$ and the proof of the theorem is complete. □

Theorem 1.15. *Let* $\mathbf{E}$ *be a* n*-dimensional vector space. Then for any* $x \in \mathbf{E}$, $x \neq 0$, *there corresponds a* $y \in \mathbf{E}'$ *such that* $[x, y] \neq 0$.

Proof. Let $\{x_1, \ldots, x_n\}$ be a basis in $\mathbf{E}$. Then, using Theorem 1.14, there exists uniquely determined basis in $\mathbf{E}'$, $\{y_1, \ldots, y_n\}$. If $x = \sum_{i=1}^{n} \xi_i x_i \in \mathbf{E}$, then $[x, y_j] = \xi_j$, $j \in \{1, \ldots, n\}$. Hence, if $[x, y] = 0$ for any $y \in \mathbf{E}'$, then $\xi_j = 0$, $j \in \{1, \ldots, n\}$, and hence, $x = 0$, which completes the proof. □

Definition 1.14. If $\mathbf{U}$ and $\mathbf{V}$ be vector spaces over the same field, their direct sum is the vector space $\mathbf{W}$, denoted by $\mathbf{U} \oplus \mathbf{V}$, whose elements are all ordered pairs $\langle x, z\rangle$ with $x \in \mathbf{U}$ and $z \in \mathbf{V}$, with linear operations defined by

$$a_1\langle x_1, z_1\rangle + a_2\langle x_2, z_2\rangle = \langle a_1x_1 + a_2x_2, a_1z_1 + a_2z_2\rangle.$$

Consider $\mathbf{W} = \mathbf{U} \oplus \mathbf{V}$. The set of all elements of the form $\langle u, 0\rangle$ is a linear subspace of $\mathbf{W}$. The correspondence $\langle u, 0\rangle \longmapsto u$ shows that this linear subspace is isomorphic to $\mathbf{U}$. It is convenient to identify u and $\langle u, 0\rangle$, to speak of $\mathbf{U}$ as a linear subspace of $\mathbf{W}$. Similarly, the elements $v \in \mathbf{V}$ may be identified with the elements of the form $\langle 0, v\rangle$ in $\mathbf{W}$. In the case in which $\mathbf{U}$ and $\mathbf{V}$ have no non-zero elements in common, we could have defined the direct sum of $\mathbf{U}$ and $\mathbf{V}$ as the set consisting of all $u \in \mathbf{U}$ and all $v \in \mathbf{V}$, and all those pairs $\langle u, v\rangle$ for which $u \neq 0$ and $v \neq 0$.

Theorem 1.16. *If $\mathbf{U}$ and $\mathbf{V}$ are linear subspaces of the vector space $\mathbf{E}$, then the following three conditions are equivalent:*

1. $\mathbf{E} = \mathbf{U} \oplus \mathbf{V}$,
2. $\mathbf{U} \cap \mathbf{V} = \{0\}$ *and* $\mathbf{U} + \mathbf{V} = \mathbf{E}$,
3. *Every element $z \in \mathbf{W}$ may be written in the form $z = u + v$, with $u \in \mathbf{U}$ and $v \in \mathbf{V}$, in one and only one way.*

Proof. $1 \Longrightarrow 2$. We assume that $\mathbf{E} = \mathbf{U} \oplus \mathbf{V}$. Suppose that $z = \langle u, v\rangle \in \mathbf{U}$, $z = \langle u, v\rangle \in \mathbf{V}$. Then $u = v = 0$ and $z = 0$. Therefore $\mathbf{U} \cap \mathbf{V} = \{0\}$. Since the representation

$$z = \langle u, 0\rangle + \langle 0, v\rangle$$

is valid for every z, it follows that $\mathbf{U} + \mathbf{V} = \mathbf{E}$.

$2 \Longrightarrow 3$. We assume 2. Then for every element $z \in \mathbf{E}$ we have the representation

$$z = u + v.$$

Assume that $z = u_1 + v_1$ for $u_1 \in \mathbf{U}$ and $v_1 \in \mathbf{V}$. Then

$$u + v = u_1 + v_1,$$

whereupon

$$u - u_1 = v_1 - v.$$

Since $\mathbf{U}$ and $\mathbf{V}$ are vector spaces, we have that $u - u_1 \in \mathbf{U}$ and $v_1 - v \in \mathbf{V}$. Hence,

$$\begin{aligned} u - u_1 &= 0 \\ &= v_1 - v \end{aligned}$$

or

$$\begin{aligned} u &= u_1, \\ v &= v_1. \end{aligned}$$

$3 \Longrightarrow 1$. We form the direct sum $\mathbf{U} \oplus \mathbf{V}$ and then identify $\langle u, 0\rangle$ and $\langle 0, v\rangle$ with u and v, respectively. We committed to identifying the sum

$$\langle u, v\rangle = \langle u, 0\rangle + \langle 0, v\rangle$$

with what we have assuming to be the general element $z = u + v$ of $\mathbf{E}$. From the hypothesis that the representation of z in the form $u + v$ is unique, we conclude that the correspondence between $\langle u, 0\rangle$ and u, also between $\langle 0, v\rangle$ and v, is one-to-one. This completes the proof. □

Theorem 1.17. *Let* $\mathbf{U}$ *and* $\mathbf{V}$ *be vector spaces with* $\dim(\mathbf{U}) = m$ *and* $\dim(\mathbf{V}) = n$. *Then* $\dim(\mathbf{U} \oplus \mathbf{V}) = m + n$.

Proof. Let $\{x_1, \ldots, x_m\}$ be a basis in $\mathbf{U}$ and $\{y_1, \ldots, y_n\}$ be a basis in $\mathbf{V}$. Consider the set

$$\{x_1, \ldots, x_m, y_1, \ldots, y_n\}.$$

By Theorem 1.16, assertion 3), it follows that every element $z \in \mathbf{U} \oplus \mathbf{V}$ can be represented in a unique way in the form $u + v$, $u \in \mathbf{U}$, $v \in \mathbf{V}$. Since $u \in \mathbf{U}$ and $v \in \mathbf{V}$, we have that

$$u = a_1x_1 + \cdots + a_mx_m, \quad v = b_1y_1 + \cdots + b_ny_n, \quad a_i, b_j \in \mathbf{F},$$

$i \in \{1, \ldots, m\}$, $j \in \{1, \ldots, n\}$. Hence,

$$\begin{aligned} z &= u + v \\ &= a_1x_1 + \cdots + a_mx_m + b_1y_1 + \cdots + b_ny_n. \end{aligned}$$

Consider

$$\alpha_1x_1 + \cdots + \alpha_mx_m + \beta_1y_1 + \cdots + \beta_ny_n = 0.$$

The uniqueness of the representation of 0 in the form of $u+v$ implies that

$$\begin{aligned} \alpha_1x_1 + \cdots + \alpha_mx_m &= \beta_1y_1 + \cdots + \beta_ny_n \\ &= 0, \end{aligned}$$

and hence the linear independence of $\{x_1, \ldots, x_m\}$ and $\{y_1, \ldots, y_n\}$ imply that

$$\alpha_1 = \cdots = \alpha_m = \beta_1 = \cdots = \beta_n = 0.$$

Consequently $\{x_1, \ldots, x_m, y_1, \ldots, y_n\}$ are linearly independent. This completes the proof. □

Theorem 1.18. *If* $\mathbf{E}$ *is a finite-dimensional vector space and if* $\{y_1, \ldots, y_m\}$ *is any set of linearly independent elements of* $\mathbf{E}$*, then unless* y*'s already form a basis. we can find elements* $y_{m+1}, \ldots, y_{m+p}$ *so that the totality of the* y*'s, that is*

$$\{y_1, \ldots, y_m, y_{m+1}, \ldots, y_{m+p}\},$$

is a basis.

Proof. Let $\{x_1, \ldots, x_n\}$ be a basis in $\mathbf{E}$. Consider the set

$$S = \{y_1, \ldots, y_m, x_1, \ldots, x_n\}.$$

Since y's are linear combination of x's, the set S is linearly dependent. Hence, some element of S is a linear combination of the preceding ones. Because $\{y_1, \ldots, y_m\}$ are linearly independent, this element may be different from y's. Let this element is x_i. We consider

$$S' = \{y_1, \ldots, y_m, x_1, \ldots, x_{i-1}, x_{i+1}, \ldots, x_n\}.$$

If S' is linearly independent, then we are done. If it is not, we act as above while we reach a linearly independent set containing $y_1, \ldots, y_m$, in terms of which we may express every element in $\mathbf{E}$. This completes the proof. □

Theorem 1.19. *Let* $\mathbf{E}$ *be* $n+m$*-dimensional vector space and* $\mathbf{U}$ *be its* n*-dimensional linear subspace of* $\mathbf{E}$*. Then there exists an* m*-dimensional linear subspace* $\mathbf{V}$ *of* $\mathbf{E}$ *such that* $\mathbf{E} = \mathbf{U} \oplus \mathbf{V}$*.*

Proof. Let $\{x_1, \ldots, x_n\}$ be a basis in $\mathbf{U}$. Because $\mathbf{E}$ is $m+n$-dimensional vector space, using Theorem 1.18, there is a set $\{y_1, \ldots, y_m\}$ such that $\{x_1, \ldots, x_n, y_1, \ldots, y_m\}$ is a basis of $\mathbf{E}$. Let $\mathbf{V}$ be spanned by $\{y_1, \ldots, y_m\}$. Then $\mathbf{E} = \mathbf{U} \oplus \mathbf{V}$. This completes the proof. □

Definition 1.15. Assume that $\mathbf{A}$ and $\mathbf{B}$ be two nonempty sets. The Cartesian product of $\mathbf{A}$ with $\mathbf{B}$, denoted by $\mathbf{A} \times \mathbf{B}$, is defined to be the collection of all ordered pairs (a, b), where $a \in \mathbf{A}$ and $b \in \mathbf{B}$. We consider $(a, b) = (a'.b')$ if and only if $a = a'$ and $b = b'$. For a nonempty set $\mathbf{X}$, we call a subset $\mathbb{R}$ of $\mathbf{X} \times \mathbf{X}$ a relation on $\mathbf{X}$ and write $x\mathbb{R}x'$ provided $(x, x') \in \mathbb{R}$. The relation $\mathbb{R}$ is said to be

1. reflexive provided $x\mathbb{R}x$ for all $x \in \mathbf{X}$,
2. symmetric provided $x\mathbb{R}x'$ if $x'\mathbb{R}x$,
3. transitive provided whenever $x\mathbb{R}x'$ and $x'\mathbb{R}x''$, then $x\mathbb{R}x''$.

A relation $\mathbb{R}$ on a set $\mathbf{X}$ is called an equivalence relation provided it is reflexive, symmetric, and transitive. Sometimes, an equivalence relation is denoted by $\sim$.

Definition 1.16. Let $\mathbf{E}$ be a vector space and $\mathbf{V}$ be its linear subspace. We define an equivalence relation $\sim$ on $\mathbf{E}$ by stating that $x \sim y$ if $x - y \in \mathbf{V}$. The equivalence class of x is often denoted by

$$[x] = x + \mathbf{V}.$$

We have

$$[x] = \{x + y : y \in \mathbf{V}\}.$$

The quotient space $\mathbf{E}/\mathbf{V}$ is often denoted as $\mathbf{E}/\sim$ and defined as the set of all classes over $\mathbf{E}$ by $\sim$. Scalar multiplication and addition are defined on the equivalence classes by

$$\alpha[x] = [\alpha x], \quad \alpha \in \mathbf{F}, \quad x \in \mathbf{E}, \tag{1.19}$$

$$[x] + [y] = [x + y]. \tag{1.20}$$

Exercise 1.10. Prove that the operations (1.19) and (1.20) are well-defined, i.e., do not depend on the choice of representatives.

Exercise 1.11. Prove that $\mathbf{E}/\mathbf{V}$ is a linear linear subspace of $\mathbf{E}$.

Theorem 1.20. *Let $m < n$ and $\mathbf{U}$ is an m-dimensional linear subspace of the n-dimensional vector space $\mathbf{E}$. Then $\mathbf{E}/\mathbf{U}$ has dimension $n - m$.*

Proof. By Theorem 1.19, it follows that there exists a linear subspace $\mathbf{V}$ of $\mathbf{E}$ such that $\mathbf{E} = \mathbf{U} \oplus \mathbf{V}$. Then we have that $\dim(\mathbf{V}) = n - m$ and $\mathbf{V}$ is isomorphic of $\mathbf{E}/\mathbf{U}$. Hence and Corollary 1.7, it follows that $\dim(\mathbf{E}/\mathbf{U}) = n - m$. This completes the proof. □

1.2 Metric Spaces

Definition 1.17. A metric space is an ordered pair $(\mathbf{M}, d)$, where $\mathbf{M}$ is a set and d is a metric on $\mathbf{M}$, i.e., a function $d : \mathbf{M} \times \mathbf{M} \longmapsto \mathbb{R}$ such that for any $x, y, z \in \mathbf{M}$ the following holds.

1. $d(x, y) \geq 0$ **non-negativity** or **separation axiom**,
2. $d(x, y) = 0 \iff x = y$ **identity of indiscernible**,
3. $d(x, y) = d(y, x)$ **symmetry**,
4. $d(x, z) \leq d(x, y) + d(y, z)$ **triangle inequality**.

The first condition follows from the other three. Really, for any $x, y \in \mathbf{M}$, using the triangle inequality, we have

$$d(x, x) \leq d(x, y) + d(y, x).$$

Hence, using the symmetry, we have that $d(x, y) = d(y, x)$ and

$$d(x, x) \leq 2d(x, y).$$

From here, using the identity of indiscernible, we get

$$2d(x, y) \geq 0 \quad \text{or} \quad d(x, y) \geq 0.$$

The function d is also called **distance function** or simply **distance**. When it is clear from the context what metric is used, d is omitted and one just writes $\mathbf{M}$ for a metric space.

Example 1.13. The real numbers with the distance function

$$d(x, y) = |x - y|$$

given by the absolute difference, is a metric space.

Example 1.14. The positive real numbers with the distance function

$$d(x, y) = \left|\log\left(\frac{y}{x}\right)\right|$$

is a metric space. Really,

1. $d(x, y) \geq 0$ for any $x, y \in \mathbb{R}_+$.
2.

$$\begin{aligned} d(x, y) = 0 \quad &\Longleftrightarrow \\ \left|\log\left(\frac{y}{x}\right)\right| = 0 \quad &\Longleftrightarrow \\ \log\left(\frac{y}{x}\right) = 0 \quad &\Longleftrightarrow \end{aligned}$$

$$\begin{aligned}\frac{y}{x} &= 1 \quad \text{or}\\ x &= y, \quad x, y \in \mathbb{R}_+.\end{aligned}$$

3.

$$\begin{aligned}d(x,y) &= \left|\log\left(\frac{y}{x}\right)\right|\\ &= \left|-\log\left(\frac{x}{y}\right)\right|\\ &= \left|\log\left(\frac{x}{y}\right)\right|\\ &= d(y,x), \quad x, y \in \mathbb{R}_+.\end{aligned}$$

4. For $x, y, z \in \mathbb{R}_+$ we have

$$\begin{aligned}d(x,z) &= \left|\log\left(\frac{z}{x}\right)\right|\\ &= \left|\log\left(\frac{y}{x}\cdot\frac{z}{y}\right)\right|\\ &= \left|\log\left(\frac{y}{x}\right)+\log\left(\frac{z}{y}\right)\right|\\ &\le \left|\log\left(\frac{y}{x}\right)\right|+\left|\log\left(\frac{z}{y}\right)\right|\\ &= d(x,y)+d(y,z).\end{aligned}$$

Example 1.15. Let $\mathbf{M}$ be any nonempty set. Then $\mathbf{M}$ is a metric space with the distance function

$$d(x,y) = \begin{cases} 0 & \text{if } x = y\\ 1 & \text{otherwise,}\end{cases}$$

called the **discrete metric**.

Exercise 1.12. Prove that

$$d(x,y) = \max_{i\in\{1,\dots,n\}} |x_i - y_i|, \quad x = (x_1,\dots,x_n), \quad y = (y_1,\dots,y_n) \in \mathbf{E}_n,$$

is a metric on $\mathbf{E}_n$.

If $(\mathbf{M}, d)$ is a metric space and $\mathbf{X}$ is a subset of $\mathbf{M}$, then $(\mathbf{X}, d)$ becomes a metric space by restricting the domain of d to $\mathbf{X} \times \mathbf{X}$.

Definition 1.18. An element x of a metric space $\mathbf{M}$ is said to be the limit of a sequence $\{x_n\}_{n\in\mathbb{N}}$ of elements of $\mathbf{M}$, if $d(x_n, x) \to 0$ as $n \to \infty$. In this case, we write $x_n \to x$ as $n \to \infty$ or $\lim_{n\to\infty} x_n = x$.

Theorem 1.21. *If a sequence $\{x_n\}_{n\in\mathbb{N}}$ of elements of a metric space $\mathbf{M}$ converges to an element $x \in \mathbf{M}$, then every subsequence $\{x_{n_k}\}_{k\in\mathbb{N}}$ of the sequence $\{x_n\}_{n\in\mathbb{N}}$ also converges to the same limit.*

Proof. Let $\varepsilon > 0$ be arbitrarily chosen. Then there exists $N = N(\varepsilon) > 0$ such that

$$d(x_n, x) < \varepsilon$$

for any $n > N$. In particular, when $n_k > N$, we get

$$d(x_{n_k}, x) < \varepsilon,$$

which completes the proof. □

Theorem 1.22. *A sequence $\{x_n\}_{n\in\mathbb{N}}$ of elements of a metric space $\mathbf{M}$ can converge to at most one limit.*

Proof. Assume that $x_n \to x$ and $x_n \to y$ as $n \to \infty$, $x, y \in \mathbf{M}$. Let $\varepsilon > 0$ be arbitrarily chosen. Then there exists $N = N(\varepsilon) > 0$ such that

$$d(x_n, x) < \frac{\varepsilon}{2}$$

and

$$d(x_n, y) < \frac{\varepsilon}{2}$$

for any $n > N$. Hence, for any $n > N$, we get

$$\begin{aligned} d(x, y) &\le d(x, x_n) + d(x_n, y) \\ &< \frac{\varepsilon}{2} + \frac{\varepsilon}{2} \\ &= \varepsilon. \end{aligned}$$

Because $\varepsilon > 0$ was arbitrarily chosen, we conclude that $x = y$. This completes the proof. □

Theorem 1.23. *If a sequence $\{x_n\}_{n\in\mathbb{N}}$ of elements of* $\mathbf{M}$ *converges to an element $x \in \mathbf{M}$, then the set $\{d(x_n, y) : n \in \mathbb{N}\}$ is bounded for every $y \in \mathbf{M}$.*

Proof. Let $y \in \mathbf{M}$ and $\varepsilon > 0$ be arbitrarily chosen. Then there exists $N = N(\varepsilon) > 1$, $N \in \mathbb{N}$, such that

$$d(x_n, x) < \frac{\varepsilon}{2} \quad \text{for any } n > N.$$

Hence, for any $n > N$ we have

$$\begin{aligned} d(x_n, y) &\leq d(x_n, x) + d(x, y) \\ &< \frac{\varepsilon}{2} + d(x, y). \end{aligned}$$

We set

$$K = \max\{d(x_1, y), \ldots, d(x_N, y), \frac{\varepsilon}{2} + d(x, y)\}.$$

Therefore,

$$d(x_n, y) \leq K \quad \text{for any } n \in \mathbb{N}.$$

This completes the proof. □

Definition 1.19. Let $\mathbf{M}$ be a metric space.

1. The set

   $$B_r(a) = \{x \in \mathbf{M} : d(x, a) < a\} \; (B_r[a] = \{x \in \mathbf{M} : d(x, a) \leq a\})$$

   is called an open (closed) ball with a center a and radius r.
2. Every open ball with center a point $x \in \mathbf{M}$ is called a neighbourhood of the point x.
3. A set lying in an open ball is called bounded.
4. Let $\mathbf{X} \subseteq \mathbf{M}$. Then a point $a \in \mathbf{M}$ is called an accumulation(limit) point of $\mathbf{X}$ if

   $$B_r(a) \cap (\mathbf{X} \backslash \{a\}) \neq \emptyset,$$

   for any $r > 0$. The set of all points of $\mathbf{X}$ plus the limit points of $\mathbf{X}$ is called closure of $\mathbf{X}$ and it is denoted by $\overline{\mathbf{X}}$. Evidently, $\mathbf{X} \subseteq \overline{\mathbf{X}}$. The closure of the empty set we define as the empty set.

Theorem 1.24. *Let* $\mathbf{M}$ *be a metric space and* $\mathbf{X}, \mathbf{Y} \subseteq \mathbf{M}$. *Then*

1. $\overline{X \cup Y} = \overline{X} \cup \overline{Y}$,
2. $\overline{\mathbf{X} \cap \mathbf{Y}} \subseteq \overline{\mathbf{X}} \cap \overline{\mathbf{Y}}$.

Proof.

1. Let $a \in \overline{\mathbf{X} \cup \mathbf{Y}}$ and $r > 0$ be arbitrarily chosen.
   a. Let $a \in X \cup Y$. Then
   $$a \in X \quad \text{or} \quad a \in Y.$$
   Hence, $a \in \overline{X}$ or $a \in \overline{Y}$ and then $a \in \overline{X} \cup \overline{Y}$.
   b. Let a is a limit point of $X \cup Y$. Then
   $$B_r(a) \cap ((X \cup Y)\backslash\{a\}) \neq \emptyset.$$
   Let $y \in B_r(a) \cap ((X \cup Y)\backslash\{a\})$ be arbitrarily chosen. Then
   $$y \neq a, \quad y \in X \cup Y \quad \text{and} \quad d(a, y) < r.$$
   If $y \in X$, we get
   $$y \neq a, \quad y \in \mathbf{X} \quad \text{and} \quad d(a, y) < r,$$
   i.e., $y \in B_r(a) \cap (\mathbf{X}\backslash\{a\})$. From here, we conclude that
   $$B_r(a) \cap (\mathbf{X}\backslash\{a\}) \neq \emptyset.$$
   Because $r > 0$ was arbitrarily chosen, we obtain that $a \in \overline{X}$. Hence, $a \in \overline{X} \cup \overline{Y}$. Similarly, if $y \in \mathbf{Y}$, we get that $a \in \overline{Y}$ and hence, $a \in \overline{X} \cup \overline{Y}$.

   Since $a \in \overline{\mathbf{X} \cup \mathbf{Y}}$ was arbitrarily chosen and for it we get that it is an element of $\overline{X} \cup \overline{Y}$, we obtain
   $$\overline{\mathbf{X} \cup \mathbf{Y}} \subseteq \overline{X} \cup \overline{Y}. \tag{1.21}$$
   Let now $a \in \overline{\mathbf{X}} \cup \overline{\mathbf{Y}}$ and $r > 0$ be arbitrarily chosen. Then $a \in \overline{X}$ or $a \in \overline{Y}$.
   a. Suppose that $a \in \overline{X}$. If $a \in X$, then $a \in X \cup Y$ and $a \in \overline{X \cup Y}$. Assume that a is a limit point of X. Then
   $$B_r(a) \cap (\mathbf{X}\backslash\{a\}) \neq \emptyset.$$
   There exists $y \in B - r(a) \cap (\mathbf{X}\backslash\{a\})$. Hence,
   $$y \neq a, \quad y \in \mathbf{X} \quad \text{and} \quad d(a, y) < r.$$
   From here,
   $$y \neq a, \quad y \in \mathbf{X} \cup \mathbf{Y} \quad \text{and} \quad d(a, y) < r,$$
   i.e., $B_r(a) \cap ((X \cup Y)\backslash\{a\}) \neq \emptyset$. Since $r > 0$ was arbitrarily chosen, we conclude that $y \in \overline{\mathbf{X} \cup \mathbf{Y}}$.

b. Let $a \in \overline{Y}$. As in above, we get $a \in \overline{X \cup Y}$.

Because $y \in \overline{\mathbf{X}} \cup \overline{\mathbf{Y}}$ was arbitrarily chosen and for it we obtain that it is an element of $\overline{\mathbf{X} \cup \mathbf{Y}}$, we conclude that

$$\overline{X} \cup \overline{Y} \subseteq \overline{\mathbf{X} \cup \mathbf{Y}}.$$

From the last relation and from (1.21) we obtain that $\overline{X} \cup \overline{Y} = \overline{\mathbf{X} \cup \mathbf{Y}}$.

2. Let $a \in \overline{\mathbf{X} \cap \mathbf{Y}}$ and $r > 0$ be arbitrarily chosen and fixed.
   a. Let $a \in X \cap Y$. Then $a \in X$ and $a \in Y$. Hence, $a \in \overline{X}$ and $a \in \overline{Y}$. Then $a \in \overline{X} \cap \overline{Y}$.
   b. Assume that a is a limit point of $X \cap Y$. Then

   $$B_r(a) \cap ((\mathbf{X} \cap \mathbf{Y}) \backslash \{a\}) \neq \emptyset.$$

   Let $z \in B_r(a) \cap ((\mathbf{X} \cap \mathbf{Y}) \backslash \{a\})$. Hence,

   $$z \in B_r(a), \quad z \in (\mathbf{X} \cap \mathbf{Y}) \quad \text{and} \quad z \neq a,$$

   whereupon

   $$z \in B_r(a), \quad z \in \mathbf{X}, \quad z \neq a \quad \text{and} \quad z \in B_r(a), \quad z \in Y, \quad z \neq a.$$

   Since $r > 0$ was arbitrarily chosen, we get that $a \in \overline{\mathbf{X}}$ and $a \in \overline{\mathbf{Y}}$. Therefore $a \in \overline{\mathbf{X}} \cap \overline{\mathbf{Y}}$.

   Because $a \in \overline{\mathbf{X} \cap \mathbf{Y}}$ was arbitrarily chosen and we have obtained that $a \in \overline{\mathbf{X}} \cap \overline{\mathbf{Y}}$, we conclude the desired relation. This completes the proof. □

Definition 1.20. Let $\mathbf{M}$ be a metric space and $\mathbf{X}, \mathbf{Y} \subseteq \mathbf{M}$.

1. We say that the set $\mathbf{X}$ is closed if $\mathbf{X} = \overline{\mathbf{X}}$.
2. We say that the set $\mathbf{X}$ is open if $\mathbf{M} \backslash \mathbf{X}$ is closed.
3. We say that the set $\mathbf{X}$ is dense in $\mathbf{Y}$ if $\mathbf{Y} \subseteq \overline{\mathbf{X}}$.
4. The set $\mathbf{X}$ is said to be everywhere dense in $\mathbf{M}$ if $\mathbf{M} = \overline{\mathbf{X}}$.
5. The set $\mathbf{X}$ in $\mathbf{M}$ is said to be nowhere dense in $\mathbf{M}$ if every ball of $\mathbf{M}$ contains a ball free from points of $\mathbf{X}$.

Definition 1.21. Let $(\mathbf{L}, d_{\mathbf{L}})$ and $(\mathbf{M}, d_{\mathbf{M}})$ be metric spaces. We say that the function $f : \mathbf{L} \longmapsto \mathbf{M}$ is a continuous function at the point $x_0 \in \mathbf{L}$ if for every $\varepsilon > 0$ there is a $\delta = \delta(\varepsilon) > 0$ such that

$d_{\mathbf{L}}(x, x_0) < \delta$, $x \in \mathbf{L}$, implies $d_{\mathbf{M}}(f(x), f(x_0)) < \varepsilon$. We say that the function f is continuous on $\mathbf{L}$ if it is continuous at every point of $\mathbf{L}$.

1.3 Useful Inequalities

We will give here some important inequalities. These inequalities play an important role in applied mathematics. Let $1 \leq p \leq \infty$, we denote by q the conjugate of p, i.e., $\frac{1}{p} + \frac{1}{q} = 1$.

Theorem 1.25 (Young's Inequality). *Let $a, b > 0$, $p.q \in (1, \infty)$, $\frac{1}{p} + \frac{1}{q} = 1$. Then*

$$ab \leq \frac{a^p}{p} + \frac{b^q}{q}. \tag{1.22}$$

The equality holds if and only if $a^p = b^q$.

Proof. Because the mapping $x \longmapsto e^x$ is convex, we get

$$\begin{aligned} ab &= e^{\log a + \log b} \\ &= e^{\frac{1}{p} \log a^p + \frac{1}{q} \log b^q} \\ &\leq \frac{1}{p} e^{\log a^p} + \frac{1}{q} e^{\log b^q} \\ &= \frac{a^p}{p} + \frac{b^q}{q}. \end{aligned}$$

Now, we will prove that the equality holds if and only if $a^p = b^q$. When $a = 0$ or $b = 0$ the assertion is evident. Assume that $a \neq 0$ and $b \neq 0$. We set $t = ab^{-\frac{q}{p}}$ or $a = tb^{\frac{q}{p}}$. Then

$$\begin{aligned} \frac{a^p}{p} + \frac{b^q}{q} - ab &= \frac{t^p b^q}{p} + \frac{b^q}{q} - tb^{\frac{q}{p}+1} \\ &= \frac{t^p b^q}{p} + \frac{b^q}{q} - tb^{q\left(\frac{1}{p}+\frac{1}{q}\right)} \\ &= \frac{t^p b^q}{p} + \frac{b^q}{q} - tb^q \\ &= b^q \left(\frac{t^p}{p} + \frac{1}{q} - t \right). \end{aligned}$$

Hence,

$$\begin{aligned}\frac{a^p}{p}+\frac{b^q}{q}-ab=0 &\iff\\ b^q\left(\frac{t^p}{p}+\frac{1}{q}-t\right)=0 &\iff\\ \frac{t^p}{p}+\frac{1}{q}-t=0 &\iff\\ t=1 &\iff\\ a^p=b^q.\end{aligned}$$

(because, if we take $f(t)=\frac{t^p}{p}+\frac{1}{q}-t$, we have $f(1)=0$, $f'(t)=t^{p-1}-1$, $f'(t)\geq 0$ for $t\geq 1$ and $f'(t)\leq 0$ for $t\in[0,1]$) This completes the proof. □

Theorem 1.26 (Young's Inequality with ε). *Let $a,b,\varepsilon>0$, $p,q\in(1,\infty)$, $\frac{1}{p}+\frac{1}{q}=1$. Then*

$$ab\leq\varepsilon a^p+C(\varepsilon)b^q,\quad C(\varepsilon)=\frac{1}{q\,(\varepsilon p)^{\frac{q}{p}}}.$$

The equality holds if and only if $a^p=\frac{b^q}{(\varepsilon p)^q}$.

Proof. Let

$$a_1=(\varepsilon p)^{\frac{1}{p}}a,\quad b_1=\frac{b}{(\varepsilon p)^{\frac{1}{p}}}.$$

We apply Young's inequality (1.22) for a_1 and b_1 and we get

$$\begin{aligned}ab&\leq\frac{\varepsilon p a^p}{p}+\frac{1}{(\varepsilon p)^{\frac{q}{p}}q}b^q\\&=\varepsilon a^p+C(\varepsilon)b^q.\end{aligned}$$

The equality holds if and only if $a_1^p=b_1^q$. This completes the proof. □

Theorem 1.27 (Hölder's Inequality). *Let $\xi_j,\eta_j\in\mathbf{C}$, $j\in\{1,\ldots,l\}$, $p,q>1$, $\frac{1}{p}+\frac{1}{q}=1$. Then*

$$\sum_{k=1}^{l}|\xi_k\eta_k|\leq\left(\sum_{k=1}^{l}|\xi_k|^p\right)^{\frac{1}{p}}\left(\sum_{k=1}^{l}|\eta_k|^q\right)^{\frac{1}{q}}.\tag{1.23}$$

The equality holds if and only if

$$|\xi_j|^p \left(\sum_{k=1}^{l} |\eta_k|^q \right) = |\eta_j|^q \left(\sum_{k=1}^{l} |\xi_k|^p \right)$$

for any $j \in \{1, \ldots, l\}$.

Proof. If $\sum_{k=1}^{l} |\xi_k|^p = 0$ ($\sum_{k=1}^{l} |\eta_k|^q = 0$), then $\xi_k = 0(\eta_k = 0)$ for any $k \in \{1, \ldots, l\}$ and the assertion is evident. Assume that $\sum_{k=1}^{l} |\xi_k|^p \neq 0$ and $\sum_{k=1}^{l} |\eta_k|^q \neq 0$. We set

$$a = \frac{|\xi_j|}{\left(\sum_{k=1}^{l} |\xi_k|^p\right)^{\frac{1}{p}}},$$

$$b = \frac{|\eta_j|}{\left(\sum_{k=1}^{l} |\eta_k|^q\right)^{\frac{1}{q}}}$$

for some $j \in \{1, \ldots, l\}$. We apply Young's inequality. We obtain

$$\frac{|\xi_j \eta_j|}{\left(\sum_{k=1}^{l} |\xi_k|^p\right)^{\frac{1}{p}} \left(\sum_{k=1}^{l} |\eta_k|^q\right)^{\frac{1}{q}}} \leq \frac{|\xi_j|^p}{p \sum_{k=1}^{l} |\xi_k|^p} + \frac{|\eta_j|^q}{q \sum_{k=1}^{l} |\eta_k|^q},$$

whereupon

$$\begin{aligned}
\sum_{j=1}^{l} \frac{|\xi_j \eta_j|}{\left(\sum_{k=1}^{l} |\xi_k|^p\right)^{\frac{1}{p}} \left(\sum_{k=1}^{l} |\eta_k|^q\right)^{\frac{1}{q}}} &\leq \sum_{j=1}^{l} \frac{|\xi_j|^p}{p \sum_{k=1}^{l} |\xi_k|^p} \\
&\quad + \sum_{j=1}^{l} \frac{|\eta_j|^q}{q \sum_{k=1}^{l} |\eta_k|^q} \\
&= \frac{\sum_{j=1}^{l} |\xi_j|^p}{p \sum_{k=1}^{l} |\xi_k|^p} + \frac{\sum_{j=1}^{l} |\eta_j|^q}{q \sum_{k=1}^{l} |\eta_k|^q} \\
&= \frac{1}{p} + \frac{1}{q} \\
&= 1.
\end{aligned}$$

From the last inequality we get the inequality (1.23). □

Theorem 1.28 (Minkowski's Inequality). *Let $\xi_k, \eta_k \in \mathbf{C}$, $k \in \{1, \dots, l\}$. Then for every $p \geq 1$ we have*

$$\left(\sum_{k=1}^{l} |\xi_k + \eta_k|^p\right)^{\frac{1}{p}} \leq \left(\sum_{k=1}^{l} |\xi_k|^p\right)^{\frac{1}{p}} + \left(\sum_{k=1}^{l} |\eta_k|^p\right)^{\frac{1}{p}}.$$

Proof. If $\sum_{k=1}^{l} |\xi_k + \eta_k|^p = 0$ the assertion is evident. Assume that $\sum_{k=1}^{l} |\xi_k + \eta_k|^p \neq 0$.

1. Let $p = 1$. Then

$$\sum_{k=1}^{l} |\xi_k + \eta_k| \leq \sum_{k=1}^{l} (|\xi_k| + |\eta_k|)$$
$$= \sum_{k=1}^{l} |\xi_k| + \sum_{k=1}^{l} |\eta_k|.$$

2. Let $p > 1$. We choose q be the conjugate of p. Then $q(p-1) = p$ and using Hölder's inequality, we obtain

$$\sum_{k=1}^{l} |\xi_k + \eta_k|^p = \sum_{k=1}^{l} |\xi_k + \eta_k|^{p-1} |\xi_k + \eta_k|$$
$$\leq \sum_{k=1}^{l} |\xi_k + \eta_k|^{p-1} |\xi_k| + \sum_{k=1}^{l} |\xi_k + \eta_k|^{p-1} |\eta_k|$$
$$\leq \left(\sum_{k=1}^{l} |\xi_k + \eta_k|^{(p-1)q}\right)^{\frac{1}{q}} \left(\sum_{k=1}^{l} |\xi_k|^p\right)^{\frac{1}{p}}$$
$$+ \left(\sum_{k=1}^{l} |\xi_k + \eta_k|^{(p-1)q}\right)^{\frac{1}{q}} \left(\sum_{k=1}^{l} |\eta_k|^p\right)^{\frac{1}{p}}$$
$$= \left(\sum_{k=1}^{l} |\xi_k + \eta_k|^p\right)^{\frac{1}{q}} \left(\left(\sum_{k=1}^{l} |\xi_k|^p\right)^{\frac{1}{p}} + \left(\sum_{k=1}^{l} |\eta_k|^p\right)^{\frac{1}{p}}\right),$$

whereupon

$$\left(\sum_{k=1}^{l}|\xi_k+\eta_k|^p\right)^{1-\frac{1}{q}} \le \left(\sum_{k=1}^{l}|\xi_k|^p\right)^{\frac{1}{p}} + \left(\sum_{k=1}^{l}|\eta_k|^p\right)^{\frac{1}{p}},$$

which completes the proof. □

1.4 Complete Spaces

Let $\mathbf{M}$ be a metric space.

Definition 1.22. A sequence $\{x_n\}_{n\in\mathbb{N}}$ of elements of $\mathbf{M}$ is said to be a fundamental sequence or a Cauchy sequence if for any $\varepsilon > 0$ there is an $N = N(\varepsilon) > 0$ such that $d(x_n, x_m) < \varepsilon$ whenever $m, n \ge N$.

Theorem 1.29. *Let $\{x_n\}_{n\in\mathbb{N}}$ be a sequence of elements of the metric space $\mathbf{M}$ that converges to x_0. Then it is a Cauchy sequence.*

Proof. Let $\varepsilon > 0$ be arbitrarily chosen. Since $\{x_n\}_{n\in\mathbb{N}}$ converges to x_0, there is an $N = N(\varepsilon) > 0$ such that

$$d(x_n, x_0) < \frac{\varepsilon}{2} \quad \text{whenever } n \ge N.$$

Hence,

$$\begin{aligned} d(x_n, x_m) &\le d(x_n, x_0) + d(x_0, x_m) \\ &< \frac{\varepsilon}{2} + \frac{\varepsilon}{2} \\ &= \varepsilon \quad \text{whenever } n, m \ge N. \end{aligned}$$

This completes the proof. □

Definition 1.23. The metric space $\mathbf{M}$ is said to be complete if every its Cauchy sequence is convergent to a limit point in $\mathbf{M}$.

Example 1.16. Consider the space $\mathbf{C}([a,b])$, the space of all continuous functions on $[a,b]$. We provide this space with the metric

$$d(f,g) = \max_{t\in[a,b]} |f(t) - g(t)|, \quad f, g \in \mathbf{C}([a,b]).$$

Let $\{f_n\}_{n\in\mathbb{N}}$ be a Cauchy sequence of elements of $\mathbf{C}([a,b])$, i.e., $d(f_n, f_m) \to 0$ as $n, m \to \infty$. Then the sequence $\{f_n\}_{n\in\mathbb{N}}$ satisfies the

Cauchy condition of uniform convergence on $[a,b]$. If f is the limit of this sequence, then f is continuous on $[a,b]$, i.e., $f \in \mathbf{C}([a,b])$. Therefore, $\mathbf{C}([a,b])$ is a complete metric space.

Example 1.17. Consider the space $\mathbf{m}$. In this space, we define a metric as follows

$$d(x,y) = \sup_{i\in\mathbb{N}} |\xi_i - \eta_i|, \quad x = \{\xi_l\}_{l\in\mathbb{N}}, y = \{\eta_l\}_{l\in\mathbb{N}} \in \mathbf{m}.$$

Let $\{x^n\}_{n\in\mathbb{N}}$ be a Cauchy sequence in $\mathbf{m}$, i.e., $d(x^n,x^m) \to 0$ as $n,m \to \infty$. Let also, $x^n = \{\xi_l^n\}_{l\in\mathbb{N}}$. We take $\varepsilon > 0$ arbitrarily. Then there is an $N = N(\varepsilon) > 0$ such that

$$\sup_{l\in\mathbb{N}} |\xi_l^n - \xi_l^m| < \varepsilon \quad \text{for any } m,n \geq N. \tag{1.24}$$

Since $x^n \in \mathbf{m}$ for any $n \in \mathbb{N}$, then for any $n \in \mathbb{N}$ there is a $K_n > 0$ such that

$$|\xi_l^n| \leq K_n \quad \text{for any } l \in \mathbb{N}. \tag{1.25}$$

By (1.24) we get

$$|\xi_l^n - \xi_l^m| < \varepsilon \quad \text{for any } m,n \geq N \tag{1.26}$$

and for any $l \in \mathbb{N}$. Therefore, for any $l \in \mathbb{N}$ the sequence $\{\xi_l^n\}_{n\in\mathbb{N}}$ satisfies the Cauchy condition. Hence, for any $l \in \mathbb{N}$ the sequence $\{\xi_l^n\}_{n\in\mathbb{N}}$ converges to ξ_l. Let $m \to \infty$ in (1.26). Then

$$|\xi_l^n - \xi_l| < \varepsilon \quad \text{for any } n \geq N \tag{1.27}$$

and for any $l \in \mathbb{N}$. Hence and (1.25), we get

$$\begin{aligned} |\xi_l| &= \left|\xi_l^N - \xi_l - \xi_l^N\right| \\ &\leq \left|\xi_l^N - \xi_l\right| + \left|\xi_l^N\right| \\ &< \varepsilon + K_N \end{aligned}$$

for any $l \in \mathbb{N}$. Consequently, the sequence $\{\xi_l\}_{l\in\mathbb{N}}$ is a bounded sequence. Let $x = \{\xi_l\}_{l\in\mathbb{N}}$. Then $x \in \mathbf{m}$. From (1.27) we obtain that

$$\sup_{l\in\mathbb{N}} |\xi_l^n - \xi_l| < \varepsilon \quad \text{for any } n \geq N.$$

Since $\varepsilon > 0$ was arbitrarily chosen, we conclude that $x^n \to x$ as $n \to \infty$. Therefore $\mathbf{m}$ is a complete metric space.

Example 1.18. Consider the set $\mathbf{Q}$ of rational numbers, in which a metric is defined by

$$d(r_1, r_2) = |r_1 - r_2|, \quad r_1, r_2 \in \mathbf{Q}.$$

Consider the sequence $\left\{\left(1 + \frac{1}{n}\right)^n\right\}_{n \in \mathbb{N}}$. This sequence is a Cauchy sequence. Its limit is e that is not a rational number. Therefore $\mathbf{Q}$ is not a complete metric space.

Exercise 1.13. For $p \geq 1$, $p < \infty$, with $\mathbf{l}_p$ we denote the set of all sequences $x = \{\xi_l\}_{l \in \mathbb{N}}$, $\xi_l \in \mathbf{C}$, $l \in \mathbb{N}$, such that

$$\sum_{l=1}^{\infty} |\xi_l|^p < \infty.$$

In $\mathbf{l}_p$ we define a metric by

$$d(x, y) = \left(\sum_{l=1}^{\infty} |\xi_l - \eta_l|^p\right)^{\frac{1}{p}}, \quad x = \{\xi_l\}_{l \in \mathbb{N}}, \ y = \{\eta_l\}_{l \in \mathbb{N}}. \tag{1.28}$$

1. Prove that (1.28) satisfies all axioms for a metric.
2. Prove that $\mathbf{l}_p$ is a complete metric space.

Exercise 1.14. For $k \in \mathbb{N}_0$ with $\mathbf{C}^k([a, b])$ we will denote the set of all k-times continuously-differentiable functions on $[a, b]$. In $\mathbf{C}^k([a, b])$ we define a metric by

$$d(f, g) = \sum_{l=0}^{k} \max_{x \in [a,b]} \left| f^{(l)}(x) - g^{(l)}(x) \right|, \quad f, g \in \mathbf{C}^k([a, b]). \tag{1.29}$$

1. Prove that (1.29) satisfies all axioms for a metric.
2. Prove that $\mathbf{C}^k([a, b])$ is a complete metric space.

Below we will give analogues of Cantor's lemma for contracting intervals.

Theorem 1.30. *Consider the sequence of closed balls*

$$B_{r_1}[a_1] \supset B_{r_2}[a_2] \supset \cdots \supset B_{r_n}[a_n] \supset \cdots$$

in the complete metric space $\mathbf{M}$. *If* $r_n \to 0$ *as* $n \to \infty$, *then these balls have a unique common point.*

Proof. Consider the sequence $\{a_l\}_{l\in\mathbb{N}}$ of the centers of the considered sequence of closed balls. Let $p \in \mathbb{N}$ be arbitrarily chosen. Because $B_{r_{n+p}}[a_{n+p}] \subset B_{r_n}[a_n]$, we have that $a_{n+p} \in B_{r_n}[a_n]$, whereupon

$$d(a_{n+p}, a_n) \leq r_n.$$

Using that $r_n \to 0$ as $n \to \infty$, from the last inequality we get $d(a_{n+p}, a_n) \to 0$ as $n \to \infty$. Since $p \in \mathbb{N}$ was arbitrarily chosen, we conclude that the sequence $\{a_l\}_{l\in\mathbb{N}}$ is a Cauchy sequence in the complete metric space $\mathbf{M}$. Therefore, it is convergent and let its limit is a. We have $a \in \mathbf{M}$. Let $k \in \mathbb{N}$ be arbitrarily chosen and fixed. Then

$$a_k, a_{k+1}, \ldots, \in B_{r_k}[a_k].$$

Hence, using that $B_{r_k}[a_k]$ is a closed set and $a_n \to 0$ as $n \to \infty$, we obtain that $a \in B_{r_k}[a_k]$. Because $k \in \mathbb{N}$ was arbitrarily chosen and fixed, we conclude that $a \in B_{r_l}[a_l]$ for any $l \in \mathbb{N}$. Now, we assume that $b \in \mathbf{M}$ is a common point of the considered sequence of closed balls such that $d(a,b) = \delta > 0$. Then

$$\begin{aligned}\delta &= d(a,b)\\ &\leq d(a,a_n) + d(a_n,b)\\ &\leq r_n + r_n\\ &= 2r_n \quad \text{for any } n \in \mathbb{N},\end{aligned}$$

which is a contradiction because $r_n \to 0$ as $n \to \infty$. This completes the proof. □

Definition 1.24. Let $\mathbf{X}$ be a bounded set in a metric space $\mathbf{M}$. The diameter of $\mathbf{X}$ is defined by

$$diam(X) = \sup_{x,y\in\mathbf{X}} d(x,y).$$

Using this definition, we can generalize Theorem 1.30 in the following manner.

Theorem 1.31. *Let*

$$B_{r_1}[a_1] \supset B_{r_2}[a_2] \supset \cdots \supset B_{r_n}[a_n] \supset \cdots$$

be a sequence of closed balls in the complete metric space $\mathbf{M}$ *whose diameters tend to zero. Then these balls have a unique common point.*

Proof. Since the diameters of $B_{r_n}[a_n]$ tend to zero as $n \to \infty$, we have that $r_n \to 0$ as $n \to \infty$. Hence and Theorem 1.30, we go to the desired result. This completes the proof. □

Theorem 1.32. *If in a metric space* $\mathbf{M}$ *any sequence of closed balls*

$$B_{r_1}[a_1] \supset B_{r_2}[a_2] \supset \cdots \supset B_{r_n}[a_n] \supset \cdots,$$

whose diameters tend to zero, has a nonempty intersection, then the space $\mathbf{M}$ *is complete.*

Proof. Let $\{x_n\}_{n\in\mathbb{N}}$ be arbitrarily chosen Cauchy sequence in $\mathbf{M}$. Let also, $\varepsilon \in \left(0, \frac{1}{2}\right)$ be arbitrarily chosen. Then there is an $N = N(\varepsilon) > 0$ such that

$$d(x_{n_k+p}, x_{n_k}) < \varepsilon^k$$

for any $n_k > N$ and for any $p \in \mathbb{N}$. Consider the sequence of closed balls $B_{\varepsilon^{k-1}}[x_{n_k}]$ for $n_k > N$ and $k \in \mathbb{N}$. Let $k \in \mathbb{N}$ be arbitrarily chosen and fixed, and $y \in B_{\varepsilon^k}[x_{n_{k+1}}]$ be arbitrarily chosen. Then $d\left(x_{n_{k+1}}, y\right) < \varepsilon^k$ and

$$\begin{aligned} d\left(x_{n_k}, y\right) &\le d\left(x_{n_{k+1}}, x_{n_k}\right) + d\left(x_{n_{k+1}}, y\right) \\ &< \varepsilon^k + \varepsilon^k \\ &= 2\varepsilon^k \\ &< \varepsilon^{k-1}. \end{aligned}$$

Therefore, $y \in B_{\varepsilon^{k-1}}[x_{n_k}]$. Because $y \in B_{|epsilon^k}[x_{n_{k+1}}]$ was arbitrarily chosen and we get that it is an element of $B_{\varepsilon^{k-1}}[x_{n_k}]$, we conclude that

$$B_{\varepsilon^{k-1}}[x_{n_k}] \supset B_{\varepsilon^k}[x_{n_{k+1}}].$$

Since $k \in \mathbb{N}$ was arbitrarily chosen, we obtain

$$B_1[x_{n_1}] \supset B_\varepsilon[x_{n_2}] \supset \cdots \supset B_{\varepsilon^{k-1}}[x_{n_k}] \supset \cdots.$$

By the assumptions, we have that there is an $x_0 \in B_{\varepsilon^{k-1}}[x_{n_k}]$ for any $k \in \mathbb{N}$. We have that $x_{n_k} \to x_0$ as $k \to \infty$. For $n, n_k > N$, we get

$$\begin{aligned} d(x_n, x_0) &\le d\left(x_n, x_{n_k}\right) + d(x_{n_k}, x_0) \\ &< \varepsilon^k + \varepsilon^{k-1}. \end{aligned}$$

Because $\varepsilon \in \left(0, \frac{1}{2}\right)$ was arbitrarily chosen, we conclude that $x_n \to x_0$ as $n \to \infty$. This completes the proof. □

Definition 1.25. A set $\mathbf{M}$ is said to be of the first category if it can be written as a countable union of nowhere dense sets. Otherwise, it is said to be of the second category.

Theorem 1.33. *A nonempty complete metric space* $\mathbf{M}$ *is a set of the second category.*

Proof. Assume that $\mathbf{M}$ is a set of the first category. Then

$$\mathbf{M} = \bigcup_{n\in\mathbb{N}} \mathbf{X}_n,$$

where $\mathbf{X}_n$, $n \in \mathbb{N}$, are nowhere dense sets in $\mathbf{M}$. Let $\varepsilon \in (0,1)$ be arbitrarily chosen. Then there is a ball $B_\varepsilon[a_1] \subset \mathbf{M}$ that does not contain any point of $\mathbf{X}_1$. There is a ball $B_{\varepsilon^2}[a_2] \subset B_\varepsilon[a_1]$ that does not contain any point of $\mathbf{X}_2$, and so on. In this way, we obtain the sequence of closed balls

$$B_\varepsilon[a_1] \supset B_{\varepsilon^2}[a_2] \supset \cdots \supset B_{\varepsilon^n}[a_n] \supset \cdots .$$

Because $\varepsilon^n \to 0$ as $n \to \infty$ and $\mathbf{M}$ is a complete metric space, there is $a \in B_{\varepsilon^k}[a_k]$ for any $k \in \mathbb{N}$. We have that $a \in \mathbf{M}$. On the other hand, $a \notin \mathbf{X}_k$ for any $k \in \mathbb{N}$. Hence, $a \notin \mathbf{M}$, which is a contradiction. This completes the proof. □

Definition 1.26. Let $(\mathbf{P}, d_\mathbf{P})$ and $(\mathbf{M}, d_\mathbf{M})$ be metric spaces. If there is a one-to-one correspondence f between $\mathbf{P}$ and $\mathbf{M}$, $f(\mathbf{P}) = \mathbf{M}$, and

$$d_\mathbf{P}(x,y) = d_\mathbf{M}(f(x), f(y)) \quad \text{for} \quad \text{any} \quad x, y \in \mathbf{P},$$

then the spaces $\mathbf{P}$ and $\mathbf{M}$ is said to be isometric.

Theorem 1.34 (The Completion of Metric Spaces). *Let* $\mathbf{M}$ *be a metric space that is non-complete. Then there is a complete metric space* $\mathbf{P}$, *such that it has a subset* $\mathbf{L}$ *which is everywhere dense in* $\mathbf{P}$ *and isometric to* $\mathbf{M}$.

Proof. Consider all sequences of elements of $\mathbf{M}$ which are Cauchy sequences. We associate every two Cauchy sequences $\{x_n\}_{n\in\mathbb{N}}$ and $\{x'_n\}_{n\in\mathbb{N}}$ to the same class if $d(x_n, x'_n) \to 0$ as $n \to \infty$. We consider this class $\tilde{x}$ as an element of a new set $\mathbf{P}$. Let $\tilde{x}$, $\tilde{y} \in \mathbf{P}$ and $\{x_n\}_{n\in\mathbb{N}} \in \tilde{x}$, $\{y_n\}_{n\in\mathbb{N}} \in \tilde{y}$. We have

$$\begin{aligned} d(x_n, y_n) &\le d(x_n, x_m) + d(x_m, y_n) \\ &\le d(x_n, x_m) + d(x_m, y_m) + d(y_m, y_n). \end{aligned}$$

Hence,

$$d(x_n, y_n) - d(x_m, y_m) \le d(x_n, x_m) + d(y_m, y_n). \tag{1.30}$$

If the indices m and n are interchanged, then we get

$$d(x_m, y_m) - d(x_n, y_n) \le d(x_n, x_m) + d(y_n, y_m).$$

From the last inequality and from (1.30), we obtain

$$|d(x_n, y_n) - d(x_m, y_m)| \le d(x_n, x_m) + d(y_n, y_m). \tag{1.31}$$

Since the sequences $\{x_n\}_{n\in\mathbb{N}}$ and $\{y_n\}_{n\in\mathbb{N}}$ are fundamental sequences, from the inequality (1.31), we conclude that the sequence $\{d(x_n, y_n)\}_{n\in\mathbb{N}}$ is a Cauchy sequence in $\mathbb{R}$. Therefore, there exists $\lim_{n\to\infty} d(x_n, y_n)$. Using this, we can define a metric in the space $\mathbf{P}$ as follows:

$$d(\tilde{x}, \tilde{y}) = \lim_{n\to\infty} d(x_n, y_n). \tag{1.32}$$

We will show that $d(\tilde{x}, \tilde{y})$ does not depend on the choice of the representatives $\{x_n\}_{n\in\mathbb{N}}$ and $\{y_n\}_{n\in\mathbb{N}}$ of the classes $\tilde{x}$ and $\tilde{y}$, respectively. Let $\{x'_n\}_{n\in\mathbb{N}} \in \tilde{x}$ and $\{y'_n\}_{n\in\mathbb{N}} \in \tilde{y}$. Then

$$\begin{aligned} d(x_n, y_n) &\le d(x_n, x'_n) + d(x'_n, y_n) \\ &\le d(x_n, x'_n) + d(x'_n, y'_n) + d(y'_n, y_n). \end{aligned}$$

From here,

$$\begin{aligned} \lim_{n\to\infty} d(x_n, y_n) &\le \lim_{n\to\infty} d(x_n, x'_n) + \lim_{n\to\infty} d(x'_n, y'_n) + \lim_{n\to\infty} d(y'_n, y_n) \\ &= \lim_{n\to\infty} d(x'_n, y'_n), \end{aligned}$$

i.e.,

$$\lim_{n\to\infty} d(x_n, y_n) \le \lim_{n\to\infty} d(x'_n, y'_n). \tag{1.33}$$

Also,

$$\begin{aligned} d(x'_n, y'_n) &\le d(x'_n, x_n) + d(x_n, y'_n) \\ &\le d(x'_n, x_n) + d(x_n, y_n) + d(y_n, y'_n). \end{aligned}$$

Hence,

$$\begin{aligned} \lim_{n\to\infty} d(x'_n, y'_n) &\le \lim_{n\to\infty} d(x'_n, x_n) + \lim_{n\to\infty} d(x_n, y_n) + \lim_{n\to\infty} d(y_n, y'_n) \\ &= \lim_{n\to\infty} d(x_n, y_n). \end{aligned}$$

From the last inequality and from (1.33), we obtain

$$\lim_{n\to\infty} d(x_n, y_n) = \lim_{n\to\infty} d(x'_n, y'_n).$$

Now, we will show that (1.32) satisfies all axioms for a metric. Suppose that $\tilde{x}, \tilde{y}, \tilde{z} \in \mathbf{P}$ and $\{x_n\}_{n\in\mathbb{N}} \in \tilde{x}$, $\{y_n\}_{n\in\mathbb{N}} \in \tilde{y}$ and $\{z_n\}_{n\in\mathbb{N}} \in \tilde{z}$.

1. Since $d(x_n, y_n) \ge 0$ for any $n \in \mathbb{N}$, we have
$$d(\tilde{x}, \tilde{y}) \ge 0.$$
2. We have $d(\tilde{x}, \tilde{y}) = 0$ if and only if $\lim_{n\to\infty} d(x_n, y_n) = 0$ if and only if $\{x_n\}_{n\in\mathbb{N}}$, $\{y_n\}_{n\in\mathbb{N}}$ belong to the same class, which is possible if and only if $\tilde{x} = \tilde{y}$.
3. We have
$$\begin{aligned} d(\tilde{x}, \tilde{y}) &= \lim_{n\to\infty} d(x_n, y_n) \\ &= \lim_{n\to\infty} d(y_n, x_n) \\ &= d(\tilde{y}, \tilde{x}). \end{aligned}$$
4. We have
$$\begin{aligned} d(\tilde{x}, \tilde{y}) &= \lim_{n\to\infty} d(x_n, y_n) \\ &\le \lim_{n\to\infty} d(x_n, z_n) + \lim_{n\to\infty} d(z_n, y_n) \\ &= d(\tilde{x}, \tilde{z}) + d(\tilde{z}, \tilde{y}). \end{aligned}$$

Now, we will prove that $\mathbf{P}$ is a complete metric space. Suppose that $\{\tilde{x}^n\}_{n\in\mathbb{N}}$ is a Cauchy sequence in $\mathbf{P}$. Then $d(\tilde{x}^n, \tilde{x}^m) \to 0$ as m, $n \to \infty$. Let $\{x^n_k\}_{k\in\mathbb{N}} \in \tilde{x}^n$. Since $\{x^n_k\}_{k\in\mathbb{N}}$ is a Cauchy sequence, there exists $k_n \in \mathbb{N}$ such that

$$d\left(x^n_p, x^n_{k_n}\right) \le \frac{1}{n} \quad \text{for any } p > k_n. \tag{1.34}$$

Now, we consider the sequence $\{x^l_{k_l}\}_{l\in\mathbb{N}}$. We have

$$\begin{aligned} d\left(x^n_{k_n}, x^m_{k_m}\right) &\le d\left(x^n_{k_n}, x^n_p\right) + d\left(x^n_p, x^m_{k_m}\right) \\ &\le d\left(x^n_{k_n}, x^n_p\right) + d\left(x^n_p, x^m_p\right) + d\left(x^m_p, x^m_{k_m}\right). \quad (1.35) \end{aligned}$$

Let $\varepsilon > 0$ be arbitrarily chosen and fixed. Since $\{\tilde{x}^n\}_{n\in\mathbb{N}}$ is a Cauchy sequence in $\mathbf{P}$, we have that $d(\tilde{x}^m, \tilde{x}^n) \to 0$ as $m, n \to \infty$. Therefore, there exists $n_0 \in \mathbb{N}$ such that

$$\begin{aligned} d(\tilde{x}^n, \tilde{x}^m) &= \lim_{p\to\infty} d\left(x^n_p, x^m_p\right) \\ &< \frac{\varepsilon}{2} \end{aligned}$$

whenever $m, n \ge n_0$. Hence, there is $p_1 \in \mathbb{N}$ so that

$$d\left(x^n_p, x^m_p\right) < \frac{\varepsilon}{2} \tag{1.36}$$

whenever $m, n \ge n_0$ and $p \ge p_1$. If there is a need we enlarge n_0 so that $\frac{1}{n_0} < \frac{\varepsilon}{4}$. Then for $m, n \ge n_0$, $p > \max\{k_n, k_m, p_1\}$, using (12.12), (1.35) and (1.36), we get $\frac{1}{n}, \frac{1}{m} < \frac{\varepsilon}{4}$ and

$$\begin{aligned} d\left(x^n_{k_n}, x^m_{k_m}\right) &\le \frac{1}{n} + \frac{\varepsilon}{2} + \frac{1}{m} \\ &< \frac{\varepsilon}{4} + \frac{\varepsilon}{2} + \frac{\varepsilon}{4} \qquad (1.37) \\ &= \varepsilon. \end{aligned}$$

Therefore, the sequence $\left\{x^n_{k_n}\right\}_{n\in\mathbb{N}}$ is a Cauchy sequence. We have

$$\begin{aligned} d(\tilde{x}^n, \tilde{x}) &= \lim_{p\to\infty} d\left(x^n_p, x^p_{k_p}\right) \\ &\le \lim_{p\to\infty} d\left(x^n_p, x^n_{k_n}\right) + \lim_{p\to\infty} d\left(x^n_{k_n}, x^p_{k_p}\right) \\ &\le \frac{1}{n} + \lim_{p\to\infty} d\left(x^n_{k_n}, x^p_{k_p}\right). \end{aligned}$$

Hence, and (1.37), for $n \ge n_0$, we get

$$\begin{aligned} d(\tilde{x}^n, \tilde{x}) &\le \frac{1}{n_0} + \varepsilon \\ &\le \frac{\varepsilon}{4} + \varepsilon. \end{aligned}$$

Consequently, the sequence $\{\tilde{x}^n\}_{n\in\mathbb{N}}$ converges to $\tilde{x} \in \mathbf{P}$. Therefore, $\mathbf{P}$ is a complete metric space.

Now, we consider the set $\mathbf{L}$ of all sequences of the form $\{x, \ldots, x, \ldots\}$. We have that if $\{x, \ldots, x, \ldots\} \in \tilde{x}$ and $\{y, \ldots, y, \ldots\} \in \tilde{y}$, then $d(x, y) = d(\tilde{x}, \tilde{y})$. Note that there is an one-to-one correspondence f between $\mathbf{L}$ and $\mathbf{M}$, defined by

$$f(x) = \{x, \ldots, x, \ldots\}, \quad x \in \mathbf{M}.$$

Also, $\mathbf{L}$ and $\mathbf{M}$ are isometric. Now, we will prove that $\mathbf{L}$ is everywhere dense in $\mathbf{P}$. Let $\tilde{x} \in \mathbf{P}$ be the class containing the sequence $\{x_n\}_{n\in\mathbb{N}}$. We choose $n \in \mathbb{N}$ such that $d(x_n, x_m) < \varepsilon$ for any $m > n$. We construct the sequence $\{x_n, \ldots, x_n, \ldots\}$ and denote by $\tilde{x}_\varepsilon$ the class containing this sequence. We have that $\tilde{x}_\varepsilon \in \mathbf{L}$. Also,

$$\begin{aligned} d(\tilde{x}, \tilde{x}_\varepsilon) &= \lim_{m\to\infty} d(x_m, x_n) \\ &\leq \varepsilon. \end{aligned}$$

Therefore, $\mathbf{L}$ is everywhere dense in $\mathbf{P}$.

Now, we will prove that the space $\mathbf{P}$ is uniquely defined to within isometry. Let $\mathbf{Y}$ be another complete metric space in which $\mathbf{M}$ is everywhere dense. Then every $\tilde{y} \in \mathbf{Y}$ is the limit of some sequence $\{x_n\}_{n\in\mathbb{N}} \subseteq \mathbf{M}$. Since $\{x_n\}_{n\in\mathbb{N}}$ is a Cauchy sequence, it defines some element $\tilde{x} \in \mathbf{P}$. We associate the element $\tilde{x}$ with the element $\tilde{y}$. Let now $\tilde{z} \in \mathbf{P}$ be given and $\{z_n\}_{n\in\mathbb{N}}$ be a fundamental sequence in the class $\tilde{z}$. Since $\{z_n\}_{n\in\mathbb{N}}$ belongs to $\mathbf{Y}$, it defines some element $\tilde{\tilde{z}} \in \mathbf{Y}$. Associating this element with $\tilde{z}$, we obtain an one-to-one correspondence between the elements of the spaces $\mathbf{P}$ and $\mathbf{Y}$. Also, we have

$$\begin{aligned} d(\tilde{x}, \tilde{z}) &= \lim_{n\to\infty} d(x_n, z_n) \\ &= d\left(\tilde{y}, \tilde{\tilde{z}}\right), \end{aligned}$$

i.e., the correspondence between $\mathbf{P}$ and $\mathbf{Y}$ is isometric. □

Definition 1.27. The space $\mathbf{P}$, defined by Theorem 1.34, is called the completion of the space $\mathbf{M}$.

Example 1.19. Let $\mathbf{P}([a, b])$ be the space of all polynomials defined on $[a, b]$. We provide this space with a metric

$$d(p, q) = \max_{t\in[a,b]} |p(t) - q(t)|, \quad p, q \in \mathbf{P}([a, b]).$$

We have that $\mathbf{P}([a,b])$ is not a complete metric space. Since $\mathbf{P}([a,b])$ is everywhere dense in $\mathbf{C}([a,b])$ and $\mathbf{C}([a,b])$ is complete, we have that $\mathbf{C}([a,b])$ is the completion of $\mathbf{P}([a,b])$.

1.5 Normed Spaces

It is well known that the notion of norm is of fundamental importance in discussing linear topological spaces. We shall begin with the definition of the semi-norm.

Definition 1.28. A real-valued function ν defined on a vector space $\mathbf{E}$ is called a *semi-norm* on $\mathbf{E}$, if the following conditions are satisfied:

1. $\nu(x+y) \leq \nu(x) + \nu(y)$ for any $x, y \in \mathbf{E}$, **sub-additivity**,
2. $\nu(\alpha x) = |\alpha|\nu(x)$ for any $x \in \mathbf{E}$, **homogeneity**.
3. $\nu(x) \geq 0$ for any $x \in \mathbf{E}$, **non-negativity**.

Theorem 1.35. *If $\mathbf{E}$ is a real vector space and $\nu : \mathbf{E} \longmapsto \mathbb{R}$ is a semi-norm. Then*

$$\nu(x-y) \geq |\nu(x) - \nu(y)|,$$

for any $x, y \in \mathbf{E}$.

Proof. We have

$$\nu(x) \leq \nu(x-y) + \nu(y)$$

for any $x, y \in \mathbf{E}$, so

$$\nu(x) - \nu(y) \leq \nu(x-y) \tag{1.38}$$

for any $x, y \in \mathbf{E}$. Since

$$\begin{aligned} \nu(x-y) &= |-1|\nu(y-x) \\ &\geq \nu(y) - \nu(x) \end{aligned}$$

for any $x, y \in \mathbf{E}$, we have

$$-(\nu(x) - \nu(y)) \leq \nu(x) - \nu(y) \tag{1.39}$$

for any $x, y \in \mathbf{E}$. The Inequalities (1.38) and (1.39) give the desired inequality. □

Definition 1.29. A normed space is an ordered pair $(\mathbf{E}, \|\cdot\|)$, where $\mathbf{E}$ is a vector space over $\mathbf{F}$ and $\|\cdot\|$ is a norm on $\mathbf{E}$, i.e., a function $\|\cdot\| : \mathbf{E} \longmapsto \mathbb{R}$ such that for any $x, y, z \in \mathbf{E}$ the following holds:

1. $\|x\| \geq 0$ **non-negativity**,
2. $\|x\| = 0$ iff $x = 0$ **separate points**,
3. $\|\lambda x\| = |\lambda|\|x\|$ for any $\lambda \in \mathbf{F}$ **homogeneity of the norm**,
4. $\|x + y\| \leq \|x\| + \|y\|$ **triangle inequality**.

Note that the first condition follows from the other three. To see this, we take $x \in \mathbf{E}$ arbitrarily. Then

$$\begin{aligned}\|0\| &= \|x + (-x)\| \\ &\leq \|x\| + \|-x\| \\ &= \|x\| + \|(-1)x\| \\ &= \|x\| + |-1|\|x\| \\ &= 2\|x\|.\end{aligned}$$

Hence, using the second condition, we get that $\|x\| \geq 0$.

Example 1.20. In $\mathbf{E}_n$ we define a norm as follows.

$$\|x\| = \left(\sum_{l=1}^{n} |x_l|^2\right)^{\frac{1}{2}}, \quad x_l \in \mathbf{F}, \quad l \in \{1, \ldots, n\}, \quad x = (x_1, \ldots, x_n). \tag{1.40}$$

We will check that (1.40) satisfies all axioms for a norm. Take x, $y \in \mathbf{F}$, $x = (x_1, \ldots, x_n)$, $y = (y_1, \ldots, y_n)$, arbitrarily. Then

1. $\|x\| \geq 0$.
2. $\|x\| = 0$ iff $\left(\sum_{l=1}^{n} |x_l|^2\right)^{\frac{1}{2}} = 0$ iff $x_l = 0$ for any $l \in \{1, \ldots, n\}$.
3.

$$\begin{aligned}\|\lambda x\| &= \left(\sum_{l=1}^{n} |\lambda x_l|^2\right)^{\frac{1}{2}} \\ &= \left(\sum_{l=1}^{n} |\lambda|^2 |x_l|^2\right)^{\frac{1}{2}}\end{aligned}$$

$$= |\lambda| \left(\sum_{l=1}^{n} |x_l|^2 \right)^{\frac{1}{2}}$$
$$= |\lambda| \|x\|$$

for any $\lambda \in \mathbf{F}$.

4. Applying Minkowski's inequality, we get

$$\|x + y\| = \left(\sum_{l=1}^{n} |x_l + y_l|^2 \right)^{\frac{1}{2}}$$
$$\leq \left(\sum_{l=1}^{n} |x_l|^2 \right)^{\frac{1}{2}} + \left(\sum_{l=1}^{n} |y_l|^2 \right)^{\frac{1}{2}}$$
$$= \|x\| + \|y\|.$$

Example 1.21. In the space $\mathbf{C}^k([a,b])$, we define a norm

$$\|f\| = \sum_{l=0}^{k} \max_{t\in[a,b]} \left| f^{(l)}(t) \right|, \quad f \in \mathbf{C}^k([a,b]). \tag{1.41}$$

We will check that (1.41) satisfies all axioms for a norm. Let $f, g \in \mathbf{C}^k([a,b])$ and $\lambda \in \mathbf{F}$ be chosen arbitrarily. Then

1. $\|f\| \geq 0$.
2.

$$0 = \|f\| \quad \text{iff} \quad \sum_{l=0}^{k} \max_{t\in[a,b]} \left| f^{(l)}(t) \right| = 0 \quad \text{iff}$$

$$\max_{t\in[a,b]} \left| f^{(l)}(t) \right| = 0 \quad \text{for any } l \in \{0, \ldots, k\}$$
$$\text{iff} \quad f \equiv 0 \quad \text{on } [a,b].$$

3.

$$||\lambda f|| = \sum_{l=0}^{k} \max_{t\in[a,b]} \left| (\lambda f)^{(l)}(t) \right|$$
$$= \sum_{l=0}^{k} \max_{t\in[a,b]} \left| \lambda f^{(l)}(t) \right|$$

$$= |\lambda| \sum_{l=0}^{k} \max_{t\in[a,b]} \left| f^{(l)}(t) \right|$$
$$= |\lambda| \|f\|.$$

4.

$$\begin{aligned}
\|f+g\| &= \sum_{l=0}^{k} \max_{t\in[a,b]} \left| (f+g)^{(l)}(t) \right| \\
&= \sum_{l=0}^{k} \max_{t\in[a,b]} \left| f^{(l)}(t) + g^{(l)}(t) \right| \\
&\leq \sum_{l=0}^{k} \max_{t\in[a,b]} \left| f^{(l)}(t) \right| + \sum_{l=0}^{k} \max_{t\in[a,b]} \left| g^{(l)}(t) \right| \\
&= ||f|| + ||g||.
\end{aligned}$$

Example 1.22. With $\mathbf{l}_p$, $p \geq 1$, we denote the set of all sequences $x = \{x_l\}_{l\in\mathbb{N}}$ for which $\sum_{l=1}^{\infty} |x_l|^p < \infty$. Note that $\mathbf{l}_p$ is a vector space. In $\mathbf{l}_p$, $p \geq 1$, $p < \infty$, we define

$$\|x\| = \left(\sum_{l=1}^{\infty} |x_l|^p \right)^{\frac{1}{p}}, \quad x = \{x_l\}_{l\in\mathbb{N}} \in \mathbf{l}_p. \tag{1.42}$$

We will check if (1.42) satisfies all axioms for a norm. Let $x, y \in \mathbf{l}_p$, $x = \{x_l\}_{l\in\mathbb{N}}$, $y = \{y_l\}_{l\in\mathbb{N}}$, $\lambda \in \mathbf{F}$ be arbitrarily chosen. Then

1. $\|x\| \geq 0$.
2.

$$0 = \|x\| \quad \text{iff} \quad \left(\sum_{l=1}^{\infty} |x_l|^p \right)^{\frac{1}{p}} = 0 \quad \text{iff}$$

$$\sum_{l=1}^{\infty} |x_l|^p = 0 \quad \text{iff} \quad x_l = 0 \quad \text{for any } l \in \mathbb{N} \quad \text{iff} \quad x = 0.$$

3.

$$\begin{aligned}\|\lambda x\| &= \left(\sum_{l=1}^{\infty} |\lambda x_l|^p\right)^{\frac{1}{p}} \\ &= \left(\sum_{l=1}^{\infty} |\lambda|^p\, |x_l|^p\right)^{\frac{1}{p}} \\ &= |\lambda| \left(\sum_{l=1}^{\infty} |x_l|^p\right)^{\frac{1}{p}} \\ &= |\lambda|\|x\|.\end{aligned}$$

4. Since for any $m \in \mathbb{N}$, using Minkowski's inequality, we have

$$\left(\sum_{l=1}^{m} |x_l + y_l|^p\right)^{\frac{1}{p}} \leq \left(\sum_{l=1}^{m} |x_l|^p\right)^{\frac{1}{p}} + \left(\sum_{l=1}^{m} |y_l|^p\right)^{\frac{1}{p}},$$

we conclude that

$$\left(\sum_{l=1}^{\infty} |x_l + y_l|^p\right)^{\frac{1}{p}} \leq \left(\sum_{l=1}^{\infty} |x_l|^p\right)^{\frac{1}{p}} + \left(\sum_{l=1}^{\infty} |y_l|^p\right)^{\frac{1}{p}}.$$

Therefore,

$$\begin{aligned}\|x+y\| &= \left(\sum_{l=1}^{\infty} |x_l + y_l|^p\right)^{\frac{1}{p}} \\ &\leq \left(\sum_{l=1}^{\infty} |x_l|^p\right)^{\frac{1}{p}} + \left(\sum_{l=1}^{\infty} |y_l|^p\right)^{\frac{1}{p}} \\ &= \|x\| + \|y\|.\end{aligned}$$

Exercise 1.15. Check that

$$\|f\| = |f(a)| + \left|f'(a)\right| + \max_{t\in[a,b]} \left|f''(t)\right|, \quad f \in \mathbf{C}^2([a,b]),$$

satisfies all axioms for a norm.

Example 1.23. In $\mathbf{C}^1([a,b])$ we define

$$\|f\| = \max_{t\in[a,b]} \left|f'(t)\right|. \tag{1.43}$$

Since

$$\begin{array}{ll} \|f\| \;\; = 0 \quad \text{iff} \quad \max_{t\in[a,b]} |f'(t)| = 0 \quad \text{iff} \\ f'(t) = 0 \quad \text{for any } t \in [a,b] \quad \text{iff} \quad f \equiv \text{const on } [a,b], \end{array}$$

(1.43) does not satisfy the axioms for a norm.

Exercise 1.16. Check if

$$\|f\| = \max_{t\in[a,b]} \left|f'(t)\right| + |f(b) - f(a)|$$

satisfies all axioms for a norm in $\mathbf{C}^1([a,b])$.

Answer. No.

Note that in a normed space, a metric can be defined by

$$d(x,y) = \|x - y\|.$$

It is evident that the defined metric satisfies all axioms for a metric.

Below we will suppose that $\mathbf{E}$ is a normed space with a norm $\|\cdot\|$.

Lemma 1.1. *For every $x, y \in \mathbf{E}$ the following inequality:*

$$|\|x\| - \|y\|| \le \|x - y\|$$

holds.

Proof. We have

$$\begin{aligned} \|x\| &= \|x - y + y\| \\ &\le \|x - y\| + \|y\|. \end{aligned}$$

Therefore,

$$\|x\| - \|y\| \le \|x - y\|.$$

If we interchange the places of x and y in the last inequality, we get

$$\begin{aligned} \|y\| - \|x\| &\le \|y - x\| \\ &= \|x - y\|. \end{aligned}$$

This completes the proof. □

Definition 1.30.

1. An element $x_0 \in \mathbf{E}$ will be called a limit of a sequence $\{x_n\}_{n\in\mathbb{N}} \subset \mathbf{E}$, if
$$\|x_n - x_0\| \to 0 \quad \text{as} \quad n \to \infty.$$
We will write $x_n \to x_0$ as $n \to \infty$ or $\lim_{n\to\infty} x_n = x_0$.
2. For $r > 0$ the set
$$S_r(x_0) = \{x \in \mathbf{E} : \|x - x_0\| < r\} \; (S_r[x_0] = \{x \in \mathbf{E} : \|x - x_0\| \le r\})$$
will be called an open(closed) ball with a center x_0 and radius r. Sometimes, we will say that $S_r(x_0)$ is a neighborhood of x_0.
3. A set $\mathbf{M} \subset \mathbf{E}$ is said to be bounded, if there exists a positive constant c such that $\|x\| \le c$ for any $x \in \mathbf{M}$.

Theorem 1.36. *Every convergent sequence in* $\mathbf{E}$ *is a bounded sequence.*

Proof. Let $\{x_n\}_{n\in\mathbb{N}}$ be a convergent sequence in $\mathbf{E}$ to the element $x_0 \in \mathbf{E}$. Let also, $\varepsilon > 0$ be arbitrarily chosen and fixed. Then there exists an $N = N(\varepsilon) \in \mathbb{N}$ such that
$$\|x_n - x_0\| < \varepsilon$$
for any $n > N$, $n \in \mathbb{N}$. Hence, and Lemma 1.1, we conclude that
$$\|x_n\| - \|x_0\| < \varepsilon \quad \text{or} \quad \|x_n\| < \varepsilon + \|x_0\|$$
for any $n > N$, $n \in \mathbb{N}$. Let
$$c = \max\{\|x_1\|, \dots, \|x_N\|, \varepsilon + \|x_0\|\}.$$
Then
$$\|x_n\| \le c$$
for any $n \in \mathbb{N}$. This completes the proof. □

Theorem 1.37. *Let* $\{x_n\}_{n\in\mathbb{N}}$ *be a convergent sequence in* $\mathbf{E}$ *to the element* $x_0 \in \mathbf{E}$.

1. *For any* $r > 0$ *there is an* $N = N(r) \in \mathbb{N}$ *such that* $x_n \in S_r(x_0)$ *for any* $n > N$.

2. *Every subsequence* $\{x_{n_k}\}_{k\in\mathbb{N}}$ *of the sequence* $\{x_n\}_{n\in\mathbb{N}}$ *is convergent to* x_0.
3. *If* $\{\lambda_n\}_{n\in\mathbb{N}} \subseteq \mathbf{F}$ *and* $\lambda_n \to \lambda_0$ *as* $n \to \infty$, $\lambda_0 \in \mathbf{F}$, *then* $\lambda_n x_n \to \lambda_0 x_0$ *as* $n \to \infty$.
4. *If* $\{y_n\}_{n\in\mathbb{N}} \subseteq \mathbf{E}$ *and* $y_n \to y_0$ *as* $n \to \infty$, $y_0 \in \mathbf{E}$, *then* $x_n + y_n \to x_0 + y_0$ *as* $n \to \infty$.
5. $\|x_n\| \to \|x_0\|$ *as* $n \to \infty$.
6. x_0 *is unique.*

Proof.

1. Let $r > 0$ be arbitrarily chosen and fixed. Then there is an $N = N(r) \in \mathbb{N}$ such that
$$\|x_n - x_0\| \le r$$
for any $n > N$, $n \in \mathbb{N}$, i.e., $x_n \in S_r(x_0)$ for any $n > N$, $n \in \mathbb{N}$.
2. Let $\varepsilon > 0$ be arbitrarily chosen and fixed. Then there exists an $N = N(\varepsilon) \in \mathbb{N}$ such that
$$\|x_n - x_0\| < \varepsilon \tag{1.44}$$
for any $n > N$, $n \in \mathbb{N}$. Also, there is a $K = K(\varepsilon) \in \mathbb{N}$ such that $n_k > N$ for any $k > K$, $k \in \mathbb{N}$. Hence, and (1.44), we get
$$\|x_{n_k} - x_0\| < \varepsilon$$
for any $k > K$, $k \in \mathbb{N}$.
3. Since $x_n \to x_0$ and $\lambda_n \to \lambda_0$ as $n \to \infty$, there exist positive constants c_1 and c such that
$$|\lambda_n| \le c_1 \quad \text{and} \quad \|x_n\| \le c$$
for any $n \in \mathbb{N}$.

   a. Let $\lambda_0 = 0$. Then
$$\begin{aligned}\|\lambda_n x_n\| &= |\lambda_n|\|x_n\| \\ &\le c|\lambda_n| \\ &\to 0, \quad \text{as } n \to \infty.\end{aligned}$$

   b. Let $\lambda_0 \ne 0$. Then
$$\begin{aligned}\|\lambda_n x_n - \lambda_0 x_0\| &= \|\lambda_n x_n - \lambda_n x_0 + \lambda_n x_0 - \lambda_0 x_0\| \\ &\le \|\lambda_n x_n - \lambda_n x_0\| + \|\lambda_n x_0 - \lambda_0 x_0\|\end{aligned}$$

$$\begin{aligned} &= |\lambda_n|\|x_n - x_0\| + |\lambda_n - \lambda_0|\|x_0\| \\ &\le c_1\|x_n - x_0\| + |\lambda_n - \lambda_0|\|x_0\| \\ &\to 0, \quad \text{as } n \to \infty. \end{aligned}$$

4. We have

$$\begin{aligned} \|x_n + y_n - x_0 - y_0\| &= \|(x_n - x_0) + (y_n - y_0)\| \\ &\le \|x_n - x_0\| + \|y_n - y_0\| \\ &\to 0 \quad \text{as } n \to \infty. \end{aligned}$$

5. By Lemma 1.1, we get

$$\begin{aligned} |\|x_n\| - \|x_0\|| &\le \|x_n - x_0\| \\ &\to 0, \quad \text{as } n \to \infty. \end{aligned}$$

6. Assume that there exists $y_0 \in \mathbf{E}$ such that $x_n \to y_0$ as $n \to \infty$. Then

$$\begin{aligned} \|y_0 - x_0\| &= \|y_0 - x_n + x_n - x_0\| \\ &\le \|x_n - y_0\| + \|x_n - x_0\| \\ &\to 0, \quad \text{as } n \to \infty. \end{aligned}$$

   Therefore, $x_0 = y_0$.

This completes the proof. □

Definition 1.31. A set $\mathbf{M} \subset \mathbf{E}$ will be called open, if for every $x_0 \in \mathbf{M}$ there exists $r_0 > 0$ such that $S_{r_0}(x_0) \subset \mathbf{M}$.

Theorem 1.38. *Let $A_1, \ldots, A_l \subset \mathbf{E}$ be open sets. Then $\cap_{k=1}^{l} A_k$ is an open set in $\mathbf{E}$.*

Proof. Let $x \in \cap_{k=1}^{l} A_k$ be arbitrarily chosen. Then $x \in A_k$ for any $k \in \{1, \ldots, l\}$. Since A_k, $k \in \{1, \ldots, l\}$, are open sets in $\mathbf{E}$, there are $r_k > 0$ so that $S_{r_k}(x) \subset A_k$. Let $r = \min_{1 \le k \le l} r_k$. Then $S_r(x) \subset A_k$ for any $k \in \{1, \ldots, l\}$. Therefore, $S_r(x) \subset \cap_{k=1}^{l} A_k$. This completes the proof. □

Theorem 1.39. *Let $\{A_k\}_{k \in \mathbb{N}}$ be open sets in $\mathbf{E}$. Then $\cup_{k \in \mathbb{N}} A_k$ is an open set in $\mathbf{E}$.*

Proof. Let $x \in \cup_{k\in\mathbb{N}} A_k$ be arbitrarily chosen and fixed. Then there is a $k_0 \in \mathbb{N}$ such that $x \in A_{k_0}$. Since A_{k_0} is an open set in $\mathbf{E}$, there is an $r_0 > 0$ such that $S_{r_0}(x) \subset A_{k_0}$. From here, $S_{r_0}(x) \subset \cup_{k\in\mathbb{N}} A_k$. This completes the proof. □

Definition 1.32. A point $a \in \mathbf{E}$ will be called a limit point for a set $\mathbf{M} \subset \mathbf{E}$ if for any $r > 0$ there is $x \in S_r(a) \cap M$, $x \neq a$.

Theorem 1.40. *A point $a \in \mathbf{E}$ is a limit point for the set $\mathbf{M} \subset \mathbf{E}$ if and only if there is a sequence $\{x_n\}_{n\in\mathbb{N}} \subset \mathbf{M}$ that converges to a and $x_n \neq a$ for any $n \in \mathbb{N}$.*

Proof.

1. Let $a \in \mathbf{E}$ be a limit point for the set $\mathbf{M}$. Then for any $n \in \mathbb{N}$ there are $x_n \in S_{\frac{1}{n}}(a) \cap \mathbf{M}$, $x_n \neq a$. In this way we get a sequence $\{x_n\}_{n\in\mathbb{N}}$ such that
$$\|x_n - a\| \to 0 \quad \text{as } n \to \infty, \quad x_n \neq a,$$
i.e., $x_n \to a$ as $n \to \infty$ and $x_n \neq a$.
2. Let there is a sequence $\{x_n\}_{n\in\mathbb{N}} \subset \mathbf{M}$ such that $x_n \neq a$ for any $n \in \mathbb{N}$ and $x_n \to a$ as $n \to \infty$. Hence, for any $r > 0$ there is an $N = N(r) \in \mathbb{N}$ such that $x_n \in S_r(a)$ for any $n > N$ and $x_n \neq a$.

This completes the proof. □

Definition 1.33. A set $\mathbf{M} \subset \mathbf{E}$ is said to be closed if it contains all its limit points.

Theorem 1.41. *Let $\mathbf{A}_1, \ldots, \mathbf{A}_l$ be closed sets in $\mathbf{E}$. Then $\cup_{k=1}^{l} \mathbf{A}_k$ is a closed set in $\mathbf{E}$.*

Proof. Let $a \in \mathbf{E}$ be a limit point for $\cup_{k=1}^{l} \mathbf{A}_k$. Then there exists a sequence $\{x_n\}_{n\in\mathbb{N}} \subset \cup_{k=1}^{l} \mathbf{A}_k$ such that $x_n \to a$ as $n \to \infty$. Hence, there is an $m \in \{1, \ldots, l\}$ and a subsequence $\{x_{n_s}\}_{s\in\mathbb{N}}$ of the sequence $\{x_n\}_{n\in\mathbb{N}}$ such that $\{x_{n_s}\}_{s\in\mathbb{N}} \subset \mathbf{A}_m$. We have that $x_{n_s} \to a$ as $s \to \infty$ and $x_n \neq a$. Hence, and Theorem 1.40, it follows that a is a limit point for $\mathbf{A}_m$. Because $\mathbf{A}_m$ is a closed set in $\mathbf{E}$, we conclude that $a \in \mathbf{A}_m$. Therefore, $a \in \cup_{k=1}^{l} \mathbf{A}_k$ and $\cup_{k=1}^{l} \mathbf{A}_k$ is a closed set in $\mathbf{E}$. This completes the proof. □

Theorem 1.42. *Let $\{\mathbf{A}_k\}_{k\in\mathbb{N}}$ be closed sets in $\mathbf{E}$. Then $\cap_{k\in\mathbb{N}} \mathbf{A}_k$ is a closed set in $\mathbf{E}$.*

Proof. Let $a \in \mathbf{E}$ be a limit point for $\cap_{k\in\mathbb{N}}\mathbf{A}_k$. Then there exists a sequence $\{x_n\}_{n\in\mathbb{N}} \subset \cap_{k\in\mathbb{N}}\mathbf{A}_k$ such that $x_n \to a$ as $n \to \infty$. Hence, $\{x_n\}_{n\in\mathbb{N}} \subset \mathbf{A}_k$, $x_n \to a$ as $n \to \infty$ for any $k \in \mathbb{N}$. Therefore, a is a limit point of $\mathbf{A}_k$ for any $k \in \mathbb{N}$. Because $\mathbf{A}_k$, $k \in \mathbb{N}$, are closed sets in $\mathbf{E}$, we have that $a \in \mathbf{A}_k$ for any $k \in \mathbb{N}$. Therefore, $a \in \cap_{k\in\mathbb{N}}A_k$ and $\cap_{k\in\mathbb{N}}A_k$ is a closed set in $\mathbf{E}$. This completes the proof. □

Definition 1.34. Let $\mathbf{M} \subset \mathbf{E}$.

1. The set $\mathbf{M}$ together with all of its limit points is called the closure of $\mathbf{M}$. It will be denoted by $\overline{\mathbf{M}}$.
2. The set $\mathbf{E}\backslash\mathbf{M}$ will be called the completion of the set $\mathbf{M}$ to $\mathbf{E}$.
3. A point $x_0 \in \mathbf{E}$ will be called an interior point for the set $\mathbf{M}$, if there is an $r > 0$ such that $S_r(x_0) \subset \mathbf{M}$.
4. A point $x_0 \in \mathbf{E}$ will be called an exterior point of the set $\mathbf{M}$, if there is an $r > 0$ such that $S_r(x_0) \cap \mathbf{M} = \emptyset$.
5. A point $x_0 \in \mathbf{E}$ will be called a boundary point of the set $\mathbf{M}$, if for every $r > 0$ we have
$$S_r(x_0) \cap \mathbf{M} \neq \emptyset \quad \text{and} \quad S_r(x_0) \cap (\mathbf{E}\backslash\mathbf{M}) \neq \emptyset.$$
6. The set of all boundary points of the set $\mathbf{M}$ will be called the boundary of the set $\mathbf{M}$ and it will be denoted by $\partial\mathbf{M}$.

Remark 1.1. Note that we have the following possibilities:

$$\partial\mathbf{M} \subset \mathbf{M} \quad \text{or} \quad \partial\mathbf{M} \cap \mathbf{M} = \emptyset \quad \text{or} \quad \partial\mathbf{M} \cap \mathbf{M} \neq \partial\mathbf{M}.$$

Definition 1.35. Two norms $\|\cdot\|_1$ and $\|\cdot\|_2$ in $\mathbf{E}$ will be called equivalent, if there are positive constants c_1 and c_2 such that

$$c_1\|x\|_2 \leq \|x\|_1 \leq c_2\|x\|_2$$

for any $x \in \mathbf{E}$. We will write $\|\cdot\|_1 \sim \|\cdot\|_2$.

Theorem 1.43. *In every finite-dimensional vector space every norms are equivalent.*

Proof. Let $\mathbf{U}$ be a finite-dimensional vector space. With $\{\phi_l\}_{l=1}^m$ we will denote a basis in $\mathbf{U}$. Then every $x \in \mathbf{U}$ has the following representation

$$x = \sum_{k=1}^{m} \xi_k \phi_k, \quad \xi_k \in \mathbf{F}, \quad k \in \{1, \dots, m\}.$$

In $\mathbf{U}$ we define a norm

$$||x|| = \left(\sum_{l=1}^{m} |\xi_l|^2 \right)^{\frac{1}{2}} \quad \text{for } x \in \mathbf{U}. \tag{1.45}$$

We take an arbitrary norm $\|\cdot\|_1$ in $\mathbf{U}$. Let

$$c_2 = \left(\sum_{l=1}^{m} \|\phi_l\|_1^2 \right)^{\frac{1}{2}}.$$

Then for $x = \sum_{l=1}^{m} \xi_l \phi_l$, $\xi_l \in \mathbf{F}$, $l \in \{1, \dots, m\}$, we have

$$\begin{aligned}
\|x\|_1 &= \left\| \sum_{l=1}^{m} \xi_l \phi_l \right\|_1 \\
&\leq \sum_{l=1}^{m} |\xi_l| \|\phi_l\|_1 \\
&\leq \left(\sum_{l=1}^{m} |\xi_l|^2 \right)^{\frac{1}{2}} \left(\sum_{l=1}^{m} \|\phi_l\|_1^2 \right)^{\frac{1}{2}} \\
&= c_2 \|x\|,
\end{aligned}$$

i.e.,

$$\|x\|_1 \leq c_2 \|x\|. \tag{1.46}$$

On the other hand, by Lemma 1.1 and (1.46), we get

$$\begin{aligned}
|\|x\|_1 - \|y\|_1| &\leq \|x - y\|_1 \\
&\leq c_2 \|x - y\|
\end{aligned}$$

for any $x, y \in \mathbf{U}$. Therefore, the function $\|\cdot\|_1$ is a continuous function in $\mathbf{E}_m$. Hence, there exists

$$c_1 = \inf_{\|x\|=1} \|x\|_1.$$

Consequently,

$$\left\| \frac{x}{\|x\|} \right\|_1 \geq c_1 \quad \text{or} \quad \|x\|_1 \geq c_1 \|x\|.$$

This completes the proof. □

Exercise 1.17. Prove that (1.45) satisfies all axioms for a norm.

Theorem 1.44. *Let* $\mathbf{L}$ *be a linear subspace of* $\mathbf{E}$ *which is a closed set in* $\mathbf{E}$*. Then*

$$\|l\|_{\mathbf{E}/\mathbf{L}} = \inf_{x \in l} \|x\|, \quad l \in \mathbf{E}/\mathbf{L}, \tag{1.47}$$

is a norm in $\mathbf{E}/\mathbf{L}$.

Proof. Firstly, we will prove that every $l \in \mathbf{E}/\mathbf{L}$ is a closed set. Let $\{x_n\}_{n \in \mathbb{N}}$ be a sequence of elements of l such that $x_n \to x_0$ as $n \to \infty$. We fix $m \in \mathbb{N}$ and consider $x_m - x_n$ for $n \in \mathbb{N}$. We have that $x_m - x_n \in \mathbf{L}$ for any $n \in \mathbb{N}$ and $x_m - x_n \to x_m - x_0$ as $n \to \infty$. Hence, $x_m - x_0 \in \mathbf{L}$. Because $x_m \in l$, we get that $x_0 \in l$.

Let $l, m \in \mathbf{E}/\mathbf{L}$ be arbitrarily chosen.

1. $\|l\|_{\mathbf{E}/\mathbf{L}} \geq 0$.
2. We will prove that $\|l\|_{\mathbf{E}/\mathbf{L}} = 0$ iff $l = \mathbf{L}$.
   a. Let $\|l\|_{\mathbf{E}/\mathbf{L}} = 0$. Then there exists a sequence $\{x_n\}_{n \in \mathbb{N}}$ of elements of l such that $x_n \to 0$ as $n \to \infty$. Because l is a closed set, in $\mathbf{L}$, we obtain that $0 \in l$ and hence $l = \mathbf{L}$.
   b. Let $l = \mathbf{L}$. Then $0 \in l$ and $\|l\| = 0$.
3. Let $\lambda \in \mathbf{F}$ be arbitrarily chosen. Then

$$\lambda l = \{\lambda x : x \in l\}$$

and

$$\begin{aligned} \|\lambda l\|_{\mathbf{E}/\mathbf{L}} &= \inf_{x \in l} \|\lambda x\| \\ &= |\lambda| \inf_{x \in l} \|x\| \\ &= |\lambda| \|l\|_{\mathbf{E}/\mathbf{L}}. \end{aligned}$$

4. We have

$$\begin{aligned}\|l+m\|_{\mathbf{E}/\mathbf{L}} &= \inf_{x\in l+m}\|x\|\\ &\le \inf_{\substack{x=x_1+x_2\\ x_1\in l, x_2\in m}}(\|x_1+x_2\|)\\ &\le \inf_{\substack{x=x_1+x_2\\ x_1\in l, x_2\in m}}(\|x_1\|+\|x_2\|)\\ &\le \inf_{x_1\in l}\|x_1\|+\inf_{x_2\in m}\|x_2\|\\ &= \|l\|_{\mathbf{E}/\mathbf{L}}+\|m\|_{\mathbf{E}/\mathbf{L}}.\end{aligned}$$

This completes the proof. □

Theorem 1.45. *Let* $\mathbf{L}$ *be a closed linear subspace of* $\mathbf{E}$. *Then the sequence* $\{l_n\}_{n\in\mathbb{N}}$ *of elements of* $\mathbf{E}/\mathbf{L}$ *is convergent to* l *if and only if there exists a sequence* $\{x_n\}_{n\in\mathbb{N}}$ *of elements* $x_n \in l_n$ *such that* $x_n \to x$ *as* $n \to \infty$, $x \in l$.

Proof.

1. Let $\{l_n\}_{n\in\mathbb{N}}$ be a sequence of elements of $\mathbf{E}/\mathbf{L}$ that converges to l. Then, we get

$$\|l_n - l\|_{\mathbf{E}/\mathbf{L}} \to 0, \quad \text{as } n \to \infty,$$

that is

$$\|l_n - l\|_{\mathbf{E}/\mathbf{L}} = \varepsilon_n, \quad \varepsilon_n \to 0, \quad \text{as } n \to \infty.$$

Hence, there exist $y_n \in l_n$ and $x \in l$ such that

$$\|y_n - x\| < 2\varepsilon_n.$$

Let $x_0 \in l$ be arbitrarily chosen. Then

$$\begin{aligned}\|y_n - x\| &= \|(y_n - x + x_0) - x_0\|\\ &< 2\varepsilon_n.\end{aligned}$$

Since $x_0, x \in l$, we have that $x - x_0 \in \mathbf{L}$. Therefore,

$$x_n = y_n - x + x_0 \in l_n.$$

Consequently, for every $x_0 \in l$ there exists a sequence $\{x_n\}_{n\in\mathbb{N}}$, $x_n \in l_n$, such that $x_n \to x_0$ as $n \to \infty$.

2. Let there exists a sequence $\{x_n\}_{n\in\mathbb{N}}$, $x_n \in l_n$, such that $x_n \to x_0$ as $n \to \infty$, $x_0 \in l$. Then, using (1.47),

$$\|l_n - l\| \leq \|x_n - x_0\| \\ \to 0, \quad \text{as } n \to \infty.$$

This completes the proof. □

Definition 1.36. Let $\mathbf{L}$ be a linear subspace of $\mathbf{E}$. We define a distance from $x \in \mathbf{E}$ to $\mathbf{L}$ as follows:

$$dist(x, \mathbf{L}) = \inf_{y\in\mathbf{L}} \|x - y\|.$$

By Definition 1.36, we get

1. $dist(x, \mathbf{L}) \geq 0$,
2. for any $y \in \mathbf{L}$ we have

$$dist(x, \mathbf{L}) \leq \|x - y\|,$$

3. for any $\varepsilon > 0$ there exists $y_\varepsilon \in \mathbf{L}$ such that

$$\|x - y_\varepsilon\| < \varepsilon + dist(x, \mathbf{L}).$$

Theorem 1.46. *Let $\mathbf{L}$ be a closed linear subspace of $\mathbf{E}$. If $x \notin \mathbf{L}$, then $dist(x, \mathbf{L}) > 0$.*

Proof. Assume that $dist(x, \mathbf{L}) = 0$. Then there exists a sequence $\{y_n\}_{n\in\mathbb{N}}$ of elements of $\mathbf{L}$ such that

$$\|y_n - x\| < \frac{1}{n}$$

for any $n \in \mathbb{N}$. Since $\mathbf{L}$ is closed, we get that $x \in \mathbf{L}$, which is a contradiction. This completes the proof. □

Theorem 1.47. *Let $\mathbf{L}$ be a finite-dimensional linear subspace of $\mathbf{E}$. Then for any $x \in \mathbf{E}$ there exists $x^* \in \mathbf{L}$ such that*

$$dist\,(x, \mathbf{L}) = \|x - x^*\|.$$

Proof. Suppose that $\mathbf{L}$ is m-dimensional.

1. If $x \in \mathbf{L}$, then $dist(x, \mathbf{L}) = 0$ and $x = x^*$.

2. Let $x \notin \mathbf{L}$. Then $d = dist(x, L) > 0$. We take $\{\phi_l\}_{l=1}^m$ to be a basis in $\mathbf{L}$. Then any $x \in \mathbf{L}$ can be represented in the following way

$$y = \sum_{l=1}^{m} y_l \phi_l, \quad y_l \in \mathbf{F}, \quad l \in \{1, \ldots, m\}.$$

We define a norm in $\mathbf{L}$ in the following way

$$\|y\|_c = \left(\sum_{l=1}^{m} |y_l|^2 \right)^{\frac{1}{2}} \quad \text{for } y = \sum_{l=1}^{m} y_l \phi_l \in \mathbf{L}.$$

Because $\mathbf{L}$ is finite-dimensional, all norms in $\mathbf{L}$ are equivalent. For a norm $\|\cdot\|$ in $\mathbf{L}$ there exist positive constants α and β such that

$$\alpha \|z\|_c \leq \|z\| \leq \beta \|z\|_c$$

for any $z \in \mathbf{L}$. We take

$$r = \frac{d + 1 + \|x\|}{\alpha}.$$

Let $y \in \mathbf{L}$ be arbitrarily chosen. If $\|y\|_c > r$, then

$$\begin{aligned} \|x - y\| &\geq \|y\| - \|x\| \\ &\geq \alpha \|y\|_c - \|x\| \\ &> \alpha r - \|x\| \\ &= d + 1. \end{aligned}$$

Therefore, d is achieved for $\|y_l\| \leq r$. Since $\|y\|_c \leq r$ is a closed and a bounded set in $\mathbf{L}$, and $\|x - y\|$ is a continuous function on it, there exists $x^* \in \mathbf{L}$ such that

$$\inf_{\|y\|_c \leq r} \|x - y\| = \|x - x^*\|.$$

This completes the proof. □

Definition 1.37. The normed space $\mathbf{E}$ will be called strongly normed space if the equality

$$\|x + y\| = \|x\| + \|y\|$$

holds if and only if $y = \lambda x$, $\lambda > 0$, $y, x \in \mathbf{E}$.

Theorem 1.48. *Let* $\mathbf{E}$ *be a strongly normed space and* $\mathbf{L}$ *be a finite-dimensional linear subspace of* $\mathbf{E}$*. If for* $x \in \mathbf{E}$ *there exists* $x^* \in \mathbf{L}$ *such that*

$$\|x - x^*\| = \inf_{y \in \mathbf{L}} \|x - y\|,$$

then x^* *is unique.*

Proof. If $dist(x, \mathbf{L}) = 0$, then $x = x^*$. Suppose that $d = dist(x, L) > 0$. Assume that there are $x_1^*, x_2^* \in \mathbf{L}$ such that

$$\begin{aligned} d &= \|x - x_1^*\| \\ &= \|x - x_2^*\|. \end{aligned}$$

Then

$$\begin{aligned} \left\| x - \frac{x_1^* + x_2^*}{2} \right\| &= \left\| \frac{x - x_1^*}{2} + \frac{x - x_2^*}{2} \right\| \\ &\le \frac{1}{2} \|x - x_1^*\| + \frac{1}{2} \|x - x_2^*\| \\ &= \frac{1}{2} d + \frac{1}{2} d \\ &= d. \end{aligned}$$

Hence,

$$\begin{aligned} \|2x - (x_1^* + x_2^*)\| &= 2d \\ &= \|(x - x_1^*) + (x - x_2^*)\| \\ &= \|x - x_1^*\| + \|x - x_2^*\| . \end{aligned}$$

Since $\mathbf{E}$ is a strongly normed space, we get that there exists $\lambda > 0$ such that

$$x - x_1^* = \lambda (x - x_2^*).$$

If $\lambda \neq 1$, then

$$x = \frac{1}{1 - \lambda} (x_1^* - \lambda x_2^*) \in \mathbf{L},$$

which is a contradiction. Therefore, $\lambda = 1$ and $x_1^* = x_2^*$. This completes the proof. □

Lemma 1.2 (Riesz's Lemma). *Let* $\mathbf{L}$ *be a closed linear subspace of* $\mathbf{E}$ *and* $\mathbf{L} \neq \mathbf{E}$. *Then for any* $\varepsilon \in (0,1)$ *there exists* $z_\varepsilon \notin \mathbf{L}$, $\|z_\varepsilon\| = 1$, *such that*

$$dist\,(z_\varepsilon, \mathbf{L}) > 1 - \varepsilon.$$

Proof. Since $\mathbf{L} \neq \mathbf{E}$, there exists $x \in \mathbf{E}$ and $x \notin \mathbf{L}$. Let $d = \inf_{y \in \mathbf{L}} \|x - y\|$. We have $d > 0$. Then for any $\varepsilon \in (0,1)$ there exists $y_\varepsilon \in \mathbf{L}$ such that

$$d \leq \|y_\varepsilon - x\| < \frac{d}{1-\varepsilon}.$$

Let

$$z_\varepsilon = \frac{y_\varepsilon - x}{\|y_\varepsilon - x\|}.$$

We have that $\|z_\varepsilon\| = 1$. If we suppose that $z_\varepsilon \in \mathbf{L}$, then $y_\varepsilon - x \in \mathbf{L}$. Hence, $x \in \mathbf{L}$, which is a contradiction. Therefore, $z_\varepsilon \notin \mathbf{L}$. For $y \in \mathbf{L}$ we have

$$\begin{aligned}
\|z_\varepsilon - y\| &= \left\| \frac{y_\varepsilon - x}{\|y_\varepsilon - x\|} - y \right\| \\
&= \frac{\|x - (y_\varepsilon - y\|y_\varepsilon - x\|)\|}{\|y_\varepsilon - x\|} \\
&\geq \frac{d}{\|y_\varepsilon - x\|} \\
&> 1 - \varepsilon.
\end{aligned}$$

Therefore, $dist\,(z_\varepsilon, \mathbf{L}) > 1 - \varepsilon$. This completes the proof. □

1.6 Banach Spaces

Definition 1.38. A normed vector space $\mathbf{E}$ that is complete in the sense of convergence in norm is called a Banach space.

Example 1.24. The space $\mathbf{E}_n$ is a Banach space with a norm

$$\|x\| = \left(\sum_{l=1}^{n} \xi_l^2 \right)^{\frac{1}{2}}, \quad x = (\xi_1, \ldots, \xi_n) \in \mathbf{E}_n.$$

Example 1.25. The space $\mathbf{C}([a, b])$ is a Banach space with a norm

$$\|f\| = \max_{a \le t \le b} |f(t)|.$$

Example 1.26. The vector space $\mathbf{l}_p$, $p \in (1, \infty)$, is a Banach space with a norm

$$\|x\| = \left(\sum_{l=1}^{\infty} |x_l|^p\right)^{\frac{1}{p}}.$$

Theorem 1.49. *Let* $\mathbf{E}$ *be a Banach space and* $\mathbf{L}$ *be its closed linear subspace. Then* $\mathbf{E}/\mathbf{L}$ *is a Banach space.*

Proof. Let $\{l_n\}_{n \in \mathbb{N}}$ be a Cauchy sequence in $\mathbf{E}/\mathbf{L}$. We take $x_n \in \mathbf{l}_n$ so that

$$\|x_n - x_m\| \le 2\|l_n - l_m\|_{\mathbf{E}/\mathbf{L}}.$$

In this way, we get a Cauchy sequence $\{x_n\}_{n \in \mathbb{N}}$ of elements of $\mathbf{E}$. Because $\mathbf{E}$ is a Banach space, the sequence $\{x_n\}_{n \in \mathbb{N}}$ is convergent to an element $x \in \mathbf{E}$. Let l be the class containing x. Hence and Theorem 1.45, we conclude that the sequence $\{l_n\}_{n \in \mathbb{N}}$ is convergent to l. Therefore $\mathbf{E}/\mathbf{L}$ is a Banach space. This completes the proof. □

Definition 1.39. Let $x_1, x_2, \ldots, x_n, \ldots$ be elements of a Banach space $\mathbf{E}$. An expression of the form $\sum_{l=1}^{\infty} x_l$ is called a series, made up of the elements of the space $\mathbf{E}$. Let $s_n = \sum_{l=1}^{n} x_l$. If the sequence $\{s_n\}_{n \in \mathbb{N}}$ converges, then $\sum_{l=1}^{\infty} x_l$ is said to be a convergent series.

Theorem 1.50. *Let* $a_n \in \mathbf{F}$, $n \in \mathbb{N}$, *and* $\sum_{l=1}^{\infty} a_l$ *be a convergent series. Let also,* $\mathbf{E}$ *be a Banach space and* $x_n \in \mathbf{E}$, $\|x_n\| \le |a_n|$, $n \in \mathbb{N}$. *Then* $\sum_{n=1}^{\infty} x_n$ *is a convergent series.*

Proof. For any $n, p \in \mathbb{N}$, we have

$$\begin{aligned} \|s_{n+p} - s_n\| &= \left\| \sum_{l=n+1}^{n+p} x_l \right\| \\ &\le \sum_{l=n+1}^{n+p} \|x_l\| \\ &\le \sum_{l=n+1}^{n+p} |a_l|. \end{aligned}$$

Therefore, $\{s_n\}_{n\in\mathbb{N}}$ is a Cauchy sequence in $\mathbf{E}$. Because $\mathbf{E}$ is a Banach space, we conclude that the sequence $\{s_n\}_{n\in\mathbb{N}}$ is convergent. This completes the proof. □

1.7 Inner Product Spaces

Definition 1.40. An inner product space is an ordered pair $(\mathbf{E}, (\cdot,\cdot))$, where $\mathbf{E}$ is a vector space over $\mathbf{F}$ and $(\cdot,\cdot)$ is an inner product, i.e., a map $(\cdot,\cdot) : \mathbf{E}\times\mathbf{E} \longmapsto \mathbf{F}$ that satisfies the following axioms.

1. **Positive definiteness.**
$$(x, x) \geq 0 \quad \text{for any } x \in \mathbf{E},$$
and
$$(x, x) = 0 \quad \text{iff } x = 0.$$
2. **Linearity in the first argument.**
$$(\lambda x, y) = \lambda(x, y) \quad \text{and} \quad (x + y, z) = (x, z) + (y, z)$$
for any $x, y, z \in \mathbf{E}$ and for any $\lambda \in \mathbf{F}$.
3. **Conjugate symmetry.**
$$(x, y) = \overline{(y, x)}$$
for any $x, y \in \mathbf{E}$.

When it is clear from the context what inner product is used, $(\cdot,\cdot)$ is omitted and one just writes $\mathbf{E}$ for an inner product space.

Below we suppose that $\mathbf{E}$ is an inner product space.

By Definition 1.40, we get the following:

1. $(x, x) \in \mathbb{R}$ for any $x \in \mathbf{E}$ because
$$(x, x) = \overline{(x, x)}.$$
2. For any $\lambda \in \mathbb{R}$ and for any $x, y \in \mathbf{E}$ we have
$$\begin{aligned}(x, \lambda y) &= \overline{(\lambda y, x)}\\ &= \overline{\lambda(y, x)}\\ &= \overline{\lambda}(x, y).\end{aligned}$$

3. For any $x, y, z \in \mathbf{E}$ we have

$$\begin{aligned}(x, y+z) &= \overline{(y+z, x)} \\ &= \overline{(y, x) + (z, x)} \\ &= \overline{(y, x)} + \overline{(z, x)} \\ &= (x, y) + (x, z).\end{aligned}$$

4. For any $x, y \in \mathbf{E}$ we have

$$\begin{aligned}(x+y, x+y) &= (x, x+y) + (y, x+y) \\ &= (x, x) + (x, y) + (y, x) + (y, y) \\ &= (x, x) + (x, y) + \overline{(x, y)} + (y, y) \\ &= (x, x) + 2\mathrm{Re}(x, y) + (y, y).\end{aligned}$$

Example 1.27. In $\mathbf{E}_n$ we define inner product as follows:

$$(x, y) = \sum_{l=1}^{n} x_l \overline{y_l}, \quad x = (x_1, \ldots, x_n), \quad y = (y_1, \ldots, y_n) \in \mathbf{E}_n. \tag{1.48}$$

We will check that (1.48) satisfies all axioms for inner product.

1.

$$\begin{aligned}(x, x) &= \sum_{l=1}^{n} x_l \overline{x_l} \\ &= \sum_{l=1}^{n} |x_l|^2 \geq 0\end{aligned}$$

for any $x \in \mathbf{E}_n$. Also,

$$(x, x) = 0 \quad \text{iff} \quad \sum_{l=1}^{n} |x_l|^2 = 0 \quad \text{iff } x_l = 0$$

for any $l \in \{1, \ldots, n\}$.

2. Let $\lambda \in \mathbf{F}$ and $x, y, z \in \mathbf{E}_n$,

$$x = (x_1, \ldots, x_n), \quad y = (y_1, \ldots, y_n), \quad z = (z_1, \ldots, z_n).$$

Then

$$\begin{aligned}(\lambda x, y) &= \sum_{l=1}^{n} (\lambda x_l)\overline{y_l} \\ &= \lambda \sum_{l=1}^{n} x_l \overline{y_l} \\ &= \lambda(x, y).\end{aligned}$$

Also,

$$\begin{aligned}(x + y, z) &= \sum_{l=1}^{n} (x_l + y_l)\,\overline{z_l} \\ &= \sum_{l=1}^{n} x_l \overline{z_l} + \sum_{l=1}^{n} y_l \overline{z_l} \\ &= (x, z) + (y, z).\end{aligned}$$

3. For $x, y \in \mathbf{E}_n$,

$$x = (x_1, \ldots, x_n), \quad y = (y_1, \ldots, y_n),$$

we have

$$\begin{aligned}(x, y) &= \sum_{l=1}^{n} x_l \overline{y_l} \\ &= \sum_{l=1}^{n} \overline{\overline{x_l} y_l} \\ &= \overline{\sum_{l=1}^{n} y_l \overline{x_l}} \\ &= \overline{(y, x)}.\end{aligned}$$

Exercise 1.18. In $\mathbf{l}_2$, we define an inner product as follows

$$(x, y) = \sum_{l=1}^{\infty} x_l \overline{y_l}, \quad x = \{x_l\}_{l \in \mathbb{N}}, \quad y = \{y_l\}_{l \in \mathbb{N}} \in \mathbf{l}_2. \tag{1.49}$$

Prove that (1.49) satisfies all axioms for an inner product.

In $\mathbf{E}$, we define a norm as follows:

$$\|x\| = \sqrt{(x,x)}, \quad x \in \mathbf{E}. \tag{1.50}$$

Theorem 1.51. *For any $x, y \in \mathbf{E}$ we have*

$$|(x,y)| \le \|x\|\|y\|.$$

Proof. If $x = 0$ and $y = 0$ the assertion is evident. Suppose that $x \neq 0$ or $y \neq 0$. Without loss of generality we assume that $y \neq 0$. By the axioms for an inner product we have

$$\begin{aligned} 0 &\le (x + \lambda y, x + \lambda y) \\ &= (x,x) + 2\mathrm{Re}(x, \lambda y) + (\lambda y, \lambda y) \\ &= \|x\|^2 + 2\mathrm{Re}\left(\overline{\lambda}(x,y)\right) + (\lambda y, \lambda y) \\ &= \|x\|^2 + 2\mathrm{Re}\left(\overline{\lambda}(x,y)\right) + |\lambda|^2\|y\|^2 \end{aligned} \tag{1.51}$$

for any $\lambda \in \mathbf{F}$. In particular, for $\lambda = -\frac{(x,y)}{\|y\|^2}$, we have

$$\begin{aligned} 0 &\le \|x\|^2 + 2\mathrm{Re}\left(-\frac{\overline{(x,y)}(x,y)}{\|y\|^2}\right) + \frac{\left|\overline{(x,y)}\right|^2}{\|y\|^4}\|y\|^2 \\ &= \|x\|^2 - 2\frac{\left|\overline{(x,y)}\right|^2}{\|y\|^2} + \frac{\left|\overline{(x,y)}\right|^2}{\|y\|^2} \\ &= \|x\|^2 - \frac{\left|\overline{(x,y)}\right|^2}{\|y\|^2}, \end{aligned}$$

whereupon we get the desired result. This completes the proof. □

Exercise 1.19. Prove that (1.50) satisfies all axioms for a norm.

Theorem 1.52. *Let $\mathbf{V}$ be a normed vector space over $\mathbf{C}$ with a norm $\|\cdot\|$ that satisfies*

$$\|x+y\|^2 + \|x-y\|^2 = 2\left(\|x\|^2 + \|y\|^2\right)$$

for any $x, y \in \mathbf{V}$. Then

$$(x,y) = \frac{1}{4}\left(\|x+y\|^2 - \|x-y\|^2\right) + \frac{i}{4}\left(\|x+iy\|^2 - \|x-iy\|^2\right) \tag{1.52}$$

is an inner product in $\mathbf{V}$.

Proof. We will prove that (1.52) satisfies all axioms for an inner product. Let $x, y, z \in \mathbf{V}$ and $\lambda \in \mathbf{C}$.

1.

$$\begin{aligned}(x,x) &= \frac{1}{4}\left(\|2x\|^2\right) + \frac{i}{4}\left(\|x+ix\|^2 - \|x-ix\|^2\right)\\ &= \|x\|^2\\ &\geq 0.\end{aligned}$$

$(x, x) = 0$ iff $\|x\|^2 = 0$ iff $x = 0$.

2.

$$\begin{aligned}(x+z,y) &= \frac{1}{4}(\|x+y+z\|^2 - \|x+z-y\|^2)\\ &\quad + \frac{i}{4}\left(\|x+z+iy\|^2 - \|x+z-iy\|^2\right),\\ (x,y)+(z,y) &= \frac{1}{4}(\|x+y\|^2 - \|x-y\|^2)\\ &\quad + \frac{i}{4}(\|x+iy\|^2 - \|x-iy\|^2)\\ &\quad + \frac{1}{4}(\|z+y\|^2 - \|z-y\|^2)\\ &\quad + \frac{i}{4}(\|z+iy\|^2 - \|z-iy\|^2)\\ &= \frac{1}{4}(\|x+y\|^2 + \|z+y\|^2 - \|x-y\|^2 - \|z-y\|^2)\\ &\quad + \frac{i}{4}(\|x+iy\|^2 + \|z+iy\|^2 - \|x-iy\|^2\\ &\qquad - \|z-iy\|^2)\end{aligned}$$

$$
\begin{aligned}
&= \frac{1}{4}\left(\frac{1}{2}\|x+z+2y\|^2 + \frac{1}{2}\|x-z\|^2\right.\\
&\qquad \left. - \frac{1}{2}\|x+z-2y\|^2 - \frac{1}{2}\|x-z\|^2\right)\\
&\quad + \frac{i}{4}\left(\frac{1}{2}\|x+z+2iy\|^2 + \frac{1}{2}\|x-z\|^2 - \frac{1}{2}\|x\right.\\
&\qquad \left. + z - 2iy\|^2 - \frac{1}{2}\|x-z\|^2\right)\\
&= \frac{1}{8}\left(\|x+z+2y\|^2 - \|x+z-2y\|^2\right)\\
&\quad + \frac{i}{8}\left(\|x+z+2iy\|^2 - \|x+z-2iy\|^2\right)\\
&= \frac{1}{8}\left(\|x+z+y+y\|^2 - \|x+z-y-y\|^2\right)\\
&\quad + \frac{i}{8}\left(\|x+z+iy+iy\|^2 - \|x+z-iy-iy\|^2\right)\\
&= \frac{1}{8}\Big(\|x+z+y+y\|^2 + \|x+z\|^2\\
&\qquad - \|x+z\|^2 - \|x+z-y-y\|^2\Big)\\
&\quad + \frac{i}{8}\left(\|x+z+iy+iy\|^2 + \|x+z\|^2 - \|x+z\|^2\right.\\
&\qquad \left. - \|x+z-iy-iy\|^2\right)\\
&= \frac{1}{8}\Big(2\|x+z+y\|^2 + 2\|y\|^2\\
&\quad - 2\|x+z-y\|^2 - 2\|y\|^2\Big)\\
&\quad + \frac{i}{8}\left(2\|x+z+iy\|^2 + 2\|iy\|^2 - 2\|x+z-iy\|^2\right.\\
&\qquad \left. - 2\|iy\|^2\right)\\
&= \frac{1}{4}\left(\|x+z+y\|^2 - \|x+z-y\|^2\right)\\
&\quad + \frac{i}{4}\left(\|x+z+iy\|^2 - \|x+z-iy\|^2\right).
\end{aligned}
$$

Therefore,

$$(x+z,y)=(x,y)+(z,y).$$

a. Let $\lambda \in \mathbb{N}$.

i. $\lambda = 2$. Then

$$\begin{aligned}(2x,y)&=(x+x,y)\\&=(x,y)+(x,y)\\&=2(x,y).\end{aligned}$$

ii. Assume that

$$(\lambda x,y)=\lambda(x,y)$$

for some $\lambda \in \mathbb{N}$.

iii. We will prove that

$$((\lambda+1)x,y)=(\lambda+1)(x,y).$$

Really,

$$\begin{aligned}((\lambda+1)x,y)&=(\lambda x+x,y)\\&=(\lambda x,y)+(x,y)\\&=\lambda(x,y)+(x,y)\\&=(\lambda+1)(x,y).\end{aligned}$$

Therefore, for any $\lambda \in \mathbb{N}$ we have

$$(\lambda x,y)=\lambda(x,y).$$

b. Let $\lambda \in \mathbb{N}$ and $\frac{1}{\lambda}x = z$. Hence, $x = \lambda z$ and

$$(\lambda z,y)=\lambda(z,y).$$

Therefore,

$$(x,y)=\lambda\left(\frac{1}{\lambda}x,y\right)$$

or

$$\left(\frac{1}{\lambda}x,y\right)=\frac{1}{\lambda}(x,y).$$

c. We have

$$\begin{aligned}(0,y)&=\frac{1}{4}\left(\|y\|^2-\|-y\|^2\right)+\frac{i}{4}\left(\|iy\|^2-\|-iy\|^2\right)\\&=0.\end{aligned}$$

On the other hand,

$$\begin{aligned}(0, y) &= (x + (-x), y)\\ &= (x, y) + (-x, y).\end{aligned}$$

Therefore,

$$(-x, y) = -(x, y).$$

d. Let $\lambda \in -\mathbb{N} = \{-1, -2, \dots\}$. Then $-\lambda \in \mathbb{N}$ and

$$\begin{aligned}(-\lambda x, y) &= -(\lambda x, y)\\ &= -\lambda(x, y),\end{aligned}$$

whereupon

$$(\lambda x, y) = \lambda(x, y).$$

e. Let $\lambda = \frac{p}{q}$, where $p, q \in \mathbb{N}$. We have

$$\begin{aligned}(\lambda x, y) &= \left(\frac{p}{q}x, y\right)\\ &= p\left(\frac{1}{q}x, y\right)\\ &= \frac{p}{q}(x, y)\\ &= \lambda(x, y).\end{aligned}$$

f. Let $\lambda = -\frac{p}{q}$, $p, q \in \mathbb{N}$. Then

$$\begin{aligned}(\lambda x, y) &= \left(-\frac{p}{q}x, y\right)\\ &= -\left(\frac{p}{q}x, y\right)\\ &= -\frac{p}{q}(x, y)\\ &= \lambda(x, y).\end{aligned}$$

g. Let $\lambda \in \mathbb{R}$. Then there exists a sequence $\{\lambda_n\}_{n\in\mathbb{N}}$ such that $\lambda_n \in \mathbf{Q}$ for any $n \in \mathbb{N}$ and $\lambda_n \to \lambda$ as $n \to \infty$. We have

$$(\lambda_n x, y) = \lambda_n(x, y). \tag{1.53}$$

Also,

$$\lambda_n(x, y) \to \lambda(x, y) \quad \text{as } n \to \infty. \tag{1.54}$$

On the other hand,

$$\begin{aligned}
(\lambda_n x, y) &= \frac{1}{4}\left(\|\lambda_n x + y\|^2 - \|\lambda_n x - y\|^2\right) \\
&\quad + \frac{i}{4}\left(\|\lambda_n x + iy\|^2 - \|\lambda_n x - iy\|^2\right) \\
&\to \frac{1}{4}\left(\|\lambda x + y\|^2 - \|\lambda x - y\|^2\right) \\
&\quad + \frac{i}{4}\left(\|\lambda x + iy\|^2 - \|\lambda x - iy\|^2\right) \\
&= (\lambda x, y).
\end{aligned}$$

Hence and (1.53), (1.54), we obtain

$$(\lambda x, y) = \lambda(x, y).$$

h. We have

$$\begin{aligned}
(ix, y) &= \frac{1}{4}\left(\|ix + y\|^2 - \|ix - y\|^2\right) \\
&\quad + \frac{i}{4}\left(\|ix + iy\|^2 - \|ix - iy\|^2\right) \\
&= \frac{1}{4}\left(\|x - iy\|^2 - \|x + iy\|^2\right) + \frac{i}{4}\left(\|x + y\|^2 - \|x - y\|^2\right) \\
&= i\left(\frac{1}{4}\left(\|x + y\|^2 - \|x - y\|^2\right)\right. \\
&\quad \left. + \frac{1}{4i}\left(\|x - iy\|^2 - \|x + iy\|^2\right)\right) \\
&= i\left(\frac{1}{4}\left(\|x + y\|^2 - \|x - y\|^2\right)\right. \\
&\quad \left. + \frac{i}{4}\left(\|x + iy\|^2 - \|x - iy\|^2\right)\right) = i(x, y).
\end{aligned}$$

(a) Let $\lambda = a + ib$, $a, b \in \mathbb{R}$. Then

$$\begin{aligned}(\lambda x, y) &= (ax + ibx, y)\\ &= (ax, y) + (ibx, y)\\ &= a(x, y) + i(bx, y)\\ &= a(x, y) + ib(x, y)\\ &= (a + ib)(x, y)\\ &= \lambda(x, y).\end{aligned}$$

3. We have

$$\begin{aligned}(x, y) &= \frac{1}{4}\left(\|x + y\|^2 - \|x - y\|^2\right) + \frac{i}{4}\left(\|x + iy\|^2 - \|x - iy\|^2\right)\\ &= \frac{1}{4}\left(\|y + x\|^2 - \|y - x\|^2\right) - \frac{i}{4}\left(\|y + ix\|^2 - \|y - ix\|^2\right)\\ &= \overline{(y, x)}.\end{aligned}$$

This completes the proof. □

Exercise 1.20. Let $\mathbf{E}$ be a normed vector space over $\mathbb{R}$ with a norm $\|\cdot\|$ such that

$$\|x + y\|^2 + \|x - y\|^2 = 2\left(\|x\|^2 + \|y\|^2\right).$$

Prove that

$$(x, y) = \frac{1}{2}\left(\|x + y\|^2 - \|x - y\|^2\right)$$

is an inner product in $\mathbf{E}$.

Exercise 1.21. Let $\mathbf{E}$ be a normed vector space over $\mathbb{R}$ with a norm $\|\cdot\|$. Prove that

$$\|z - x\|^2 + \|z - y\|^2 = \frac{1}{2}\|x - y\|^2 + 2\left\|z - \frac{x + y}{2}\right\|^2$$

for any $x, y, z \in \mathbf{E}$.

Theorem 1.53. *The inner product in* $\mathbf{E}$ *is a continuous function with respect to norm convergence.*

Proof. Let $\{x_n\}_{n\in\mathbb{N}}$ and $\{y_n\}_{n\in\mathbb{N}}$ be two sequences in $\mathbf{E}$ such that

$$\|x_n - x\| \to 0 \quad \text{and} \quad \|y_n - y\| \to 0 \quad \text{as } n \to \infty$$

for $x, y \in \mathbf{E}$. Note that there exists a positive constant M such that

$$\|x_n\| \leq M \quad \text{and} \quad \|y_n\| \leq M$$

for any $n \in \mathbb{N}$. Hence and Theorem 1.51, we get

$$\begin{aligned} |(x_n, y_n) - (x, y)| &= |(x_n, y_n) - (x_n, y) + (x_n, y) - (x, y)| \\ &= |(x_n, y_n - y) + (x_n - x, y)| \\ &\leq |(x_n, y_n - y)| + |(x_n - x, y)| \\ &\leq \|x_n\|\|y_n - y\| + \|x_n - x\|\|y\| \\ &\leq M\|y_n - y\| + \|y\|\|x_n - x\| \\ &\to 0 \end{aligned}$$

as $n \to \infty$. This completes the proof. □

Theorem 1.54. *For any $x, y \in \mathbf{E}$ we have*

$$\|x + y\|^2 + \|x - y\|^2 = 2\left(\|x\|^2 + \|y\|^2\right).$$

Proof. We have

$$\begin{aligned} \|x + y\|^2 &= \|x\|^2 + 2\mathrm{Re}\,((x, y)) + \|y\|^2, \\ \|x - y\|^2 &= \|x\|^2 - 2\mathrm{Re}\,((x, y)) + \|y\|^2, \end{aligned}$$

whereupon we get the desired result. This completes the proof. □

Definition 1.41.

1. Two elements $x, y \in \mathbf{E}$ will be called orthogonal if $(x, y) = 0$. We will write $x \perp y$.
2. An element $x \in \mathbf{E}$ will be called orthogonal to a linear subspace $\mathbf{L} \subset \mathbf{E}$ if it is orthogonal to every element of $\mathbf{L}$. We will write $x \perp \mathbf{L}$.
3. A system $\{e_l\}_{l\in\mathbb{N}}$ of elements of the space $\mathbf{E}$ is called orthonormal, if

$$(e_l, e_m) = \begin{cases} 1 & \text{if } l = m \\ 0 & \text{if } l \neq m. \end{cases}$$

Theorem 1.55 (Schmidt Orthogonalization System). *Any system of linearly independent elements* $\{h_l\}_{l\in\mathbb{N}}$ *can be converted into an orthonormal system.*

Proof. Because $\{h_l\}_{l\in\mathbb{N}}$ is a linearly independent system of elements, we have that $h_l \neq 0$ for any $l \in \mathbb{N}$. We set

$$g_1 = \frac{h_1}{\|h_1\|}.$$

For $k = 2, \ldots,$ we take

$$\widetilde{g}_k = h_k - c_1^k g_1 - \cdots - c_{k-1}^k g_{k-1},$$

where $g_m = \frac{\widetilde{g}_m}{\|\widetilde{g}_m\|}$ for $m \in \{1, \ldots, k-1\}$, c_l^k, $l \in \{1, \ldots, k-1\}$, are chosen so that

$$(\widetilde{g}_k, g_l) = 0 \quad \text{for } l \in \{1, \ldots, k-1\}.$$

We have

$$\begin{aligned} 0 &= (\widetilde{g}_k, g_l) \\ &= (h_k, g_l) - c_l^k, \end{aligned}$$

whereupon

$$c_l^k = (h_k, g_l), \quad l \in \{1, \ldots, k-1\}.$$

Then

$$g_k = \frac{\widetilde{g}_k}{\|\widetilde{g}_k\|}, \quad k \in \mathbb{N}.$$

This completes the proof. □

1.8 Hilbert Spaces

The most important function space in modern physics and modern analysis, are known as Hilbert spaces. We give some important results on these spaces here.

Definition 1.42 (Hilbert Space). An inner product space **H** will be called Hilbert space if it is complete in the sense of the norm

$$\|\cdot\| = \sqrt{(\cdot, \cdot)}.$$

Example 1.28. The space $\mathbf{l}_2$ is a Hilbert space.

Below we will denote by $\mathbf{H}$ a Hilbert space.

Definition 1.43. A set $\mathbf{K}$ in a space $\mathbf{V}$ will be called convex if

$$\lambda x_1 + (1-\lambda)x_2 \in \mathbf{K}$$

for any $x_1, x_2 \in \mathbf{K}$ and for any $\lambda \in [0,1]$.

Theorem 1.56. *Let* $\mathbf{M}$ *be a closed convex subset of the Hilbert space* $\mathbf{H}$ *and* $x \notin \mathbf{M}$. *Then there exists a unique element* $y \in \mathbf{M}$ *such that*

$$dist(x, \mathbf{M}) = \|x - y\|.$$

Proof. By Theorem 1.46, we have that $d = distd(x, \mathbf{M}) > 0$. For every $n \in \mathbb{N}$ there exists $y_n \in \mathbf{M}$ for which

$$d \leq \|x - y_n\| < d + \frac{1}{n}. \tag{1.55}$$

We have

$$2\|x-y_n\|^2 + 2\|x-y_m\|^2 = \|y_n - y_m\|^2 + \|2x - y_n - y_m\|^2, \quad n, m \in \mathbb{N}. \tag{1.56}$$

Since $\mathbf{M}$ is a convex subset of $\mathbf{H}$ we have that

$$\frac{y_n + y_m}{2} \in \mathbf{M}.$$

Therefore,

$$\begin{aligned} \|2x - y_n - y_m\|^2 &= 4\|x - \frac{y_n + y_m}{2}\|^2 \\ &\geq 4d^2, \quad n, m \in \mathbb{N}. \end{aligned}$$

Then, using (1.55) and (1.56), we obtain that

$$\begin{aligned} \|y_n - y_m\|^2 &= 2\|x - y_n\|^2 + 2\|x - y_m\|^2 - 4\left\|x - \frac{y_n + y_m}{2}\right\|^2 \\ &< 2\left(d + \frac{1}{n}\right)^2 + 2\left(d + \frac{1}{m}\right)^2 - 4d^2 \\ &\to 0, \quad \text{as } m, n \to \infty. \end{aligned}$$

Therefore, the sequence $\{y_n\}_{n \in \mathbb{N}}$ is a fundamental sequence in $\mathbf{M}$. Since $\mathbf{M}$ is a closed subset of the Hilbert space $\mathbf{H}$, we conclude that

$\{y_n\}_{n\in\mathbb{N}}$ is convergent to $y \in \mathbf{M}$. Hence and (1.55), we obtain

$$d = \|x - y\|.$$

Now, we suppose that there are two elements $y_1, y_2 \in \mathbf{M}$ such that

$$d = \|x - y_1\|$$

and

$$d = \|x - y_2\|.$$

Since $\mathbf{M}$ is a convex subset of the Hilbert space $\mathbf{H}$, we have that

$$\frac{y_1 + y_2}{2} \in \mathbf{M}$$

and

$$\left\| x - \frac{y_1 + y_2}{2} \right\| \geq d.$$

Then

$$\begin{aligned} 4d^2 &= 2\|x - y_1\|^2 + 2\|x - y_2\|^2 \\ &= 4\left\| x - \frac{y_1 + y_2}{2} \right\|^2 + \|y_1 - y_2\|^2 \\ &\geq 4d^2 + \|y_1 - y_2\|^2, \end{aligned}$$

whereupon

$$\|y_1 - y_2\| = 0.$$

Therefore, $y_1 = y_2$. This completes the proof. □

Corollary 1.8. *Let* $\mathbf{M}$ *be a closed linear subspace of the Hilbert space* $\mathbf{H}$ *and* $x \in \mathbf{H}\backslash\mathbf{M}$. *Then there exists an unique* $y \in \mathbf{M}$ *such that*

$$dist(x, M) = \|x - y\|.$$

Proof. Because $\mathbf{M}$ is a closed linear subspace of the Hilbert space $\mathbf{H}$, it is a closed convex subset of the Hilbert space $\mathbf{H}$. Hence and 7 Theorem 1.56, it follows the desired result. □

Theorem 1.57. *Let* $\mathbf{M}$ *be a closed linear subspace of the Hilbert space* $\mathbf{H}$, $x \in \mathbf{H}\backslash\mathbf{M}$ *and* $\|x-y\| = dist(x, \mathbf{M})$ *for some* $y \in \mathbf{M}$. *Then* $x - y \perp \mathbf{M}$.

Proof. Let $\lambda \in \mathbf{F}$ be arbitrarily chosen and fixed. Then for any $z \in \mathbf{M}$, $z \neq 0$, we have

$$\|x - y + \lambda z\| \geq \|x - y\|,$$

whereupon

$$\begin{aligned}\|x-y\|^2 &\leq \|x - y + \lambda z\|^2 \\ &= \|x-y\|^2 + 2\text{Re}\,((x-y, \lambda z)) + \|\lambda z\|^2 \\ &= \|x-y\|^2 + 2\text{Re}\,((x-y, \lambda z)) + |\lambda|^2\|z\|^2.\end{aligned}$$

In particular, when $\lambda = -\frac{(x-y,z)}{\|z\|^2}$, we get

$$\begin{aligned}0 &\leq -2\text{Re}\left(\frac{\overline{(x-y,z)}(x-y,z)}{\|z\|^2}\right) + \frac{|(x-y,z)|^2}{\|z\|^2} \\ &= -\frac{|(x-y,z)|^2}{\|z\|^2}.\end{aligned}$$

Therefore, $(x - y, z) = 0$. Because $z \in \mathbf{M}$ was arbitrarily chosen, we conclude that $x - y \perp \mathbf{M}$. This completes the proof. □

Corollary 1.9. *Let* $\mathbf{M}$ *be a closed linear subspace of the Hilbert space* $\mathbf{H}$. *Then every* $x \in \mathbf{H}$ *can be represented in an unique way in the following manner:*

$$x = y + z,$$

where $y \in \mathbf{M}$, $z \in \mathbf{H}\backslash\mathbf{M}$ *and* $z \perp \mathbf{M}$.

Proof. If $x \in \mathbf{M}$, then $y = x$ and $z = 0$. Let $x \notin \mathbf{M}$. By Corollary 1.8 it follows that there exists unique $y \in \mathbf{M}$ such that

$$\|x - y\| = dist(x, M).$$

Let $z = x - y$. By Theorem 1.57, we have that $z \perp \mathbf{M}$. This completes the proof. □

Definition 1.44. Let $\mathbf{M}$ be a closed linear subspace of the Hilbert space $\mathbf{H}$. The set of all elements of $\mathbf{H}$ which are orthogonal of $\mathbf{M}$ is called the orthogonal complement of $\mathbf{M}$. We will denote it by $\mathbf{M}^{\perp}$.

Theorem 1.58. *Let $\mathbf{M}$ be a closed linear subspace of the Hilbert space $\mathbf{H}$. Then $\mathbf{M}^{\perp}$ is a closed linear subspace of $\mathbf{H}$.*

Proof. Let $z_1, z_2 \in \mathbf{M}^{\perp}$ and $y \in \mathbf{M}$ be arbitrarily chosen. Then

$$\begin{aligned}(\lambda_1 z_1 + \lambda_2 z_2, y) &= \lambda_1 (z_1, y) + \lambda_2 (z_2, y)\\ &= 0,\end{aligned}$$

i.e., $\lambda_1 z_1 + \lambda_2 z_2 \in \mathbf{M}^{\perp}$. Note that $0 \in \mathbf{M}^{\perp}$. Hence and Theorem 1.3, we conclude that $\mathbf{M}^{\perp}$ is a linear subspace of $\mathbf{H}$. Let $\{z_n\}_{n\in\mathbb{N}}$ be a convergent sequence of elements of $\mathbf{M}$ that converges to $z \in \mathbf{H}$. Then, using Theorem 1.53,

$$\begin{aligned}0 &= (z_n, y)\\ &\to (z, y), \quad \text{as } n \to \infty.\end{aligned}$$

Therefore, $(z, y) = 0$ and $z \in \mathbf{M}^{\perp}$. Consequently $\mathbf{M}^{\perp}$ is a closed subset of $\mathbf{H}$. This completes the proof. □

Theorem 1.59. *Let $\mathbf{M}$ be a linear subspace of the Hilbert space $\mathbf{H}$. Then $\mathbf{M}$ is dense in $\mathbf{H}$ if and only if $\mathbf{M}^{\perp} = \{0\}$.*

Proof.

1. Let $\mathbf{M}$ be dense in $\mathbf{H}$. Then $\overline{\mathbf{M}} = \mathbf{H}$. Assume that there exists $z_0 \in \mathbf{H}$ such that $z_0 \perp \mathbf{M}$. Let $y \in \mathbf{H}$ be arbitrarily chosen and fixed. Let $\{y_n\}_{n\in\mathbb{N}}$ be a sequence of elements of $\mathbf{M}$ such that $y_n \to y$ as $n \to \infty$. Then, using Theorem 1.53,

$$\begin{aligned}0 &= (y_n, z_0)\\ &\to (y, z_0), \quad \text{as } n \to \infty.\end{aligned}$$

   Therefore, $(y, z_0) = 0$. Because $y \in \mathbf{H}$ was arbitrarily chosen, we can take $y = z_0$. Then $\|z_0\| = 0$. Therefore, $z_0 = 0$.
2. Let $\mathbf{M}^{\perp} = \{0\}$. Assume that $\mathbf{M}$ is not dense in $\mathbf{H}$. Then there exists $x \in \mathbf{H}\backslash\overline{\mathbf{M}}$. Hence and Corollary 1.9, we conclude that

$x = y + z$, $y \in \overline{\mathbf{M}}$ and $z = \overline{\mathbf{M}}^{\perp}$. Note that $\overline{\mathbf{M}}^{\perp} = \mathbf{M}^{\perp}$. Really, let $z_1 \in \overline{\mathbf{M}}^{\perp}$ be arbitrarily chosen and fixed. Then

$$(z_1, z_2) = 0 \tag{1.57}$$

for any $z_2 \in \overline{\mathbf{M}}$. In particular, (1.57) holds for any $z_2 \in \mathbf{M}$. Therefore, $z_1 \in \mathbf{M}^{\perp}$. Since $z_1 \in \overline{\mathbf{M}}^{\perp}$ was arbitrarily chosen and for it we get that it is an element of $\mathbf{M}^{\perp}$, we conclude that

$$\overline{\mathbf{M}}^{\perp} \subseteq \mathbf{M}^{\perp}. \tag{1.58}$$

Let $z_3 \in \mathbf{M}^{\perp}$ be arbitrarily chosen and fixed. We take $z_4 \in \overline{\mathbf{M}}$ arbitrarily. Then there exists a sequence $\{z_4^n\}_{n\in\mathbb{N}}$ of elements of $\mathbf{M}$ such that

$$z_4^n \to z_4, \quad \text{as } n \to \infty.$$

Since $z_3 \in \mathbf{M}^{\perp}$ and $z_4^n \in \mathbf{M}$, $n \in \mathbb{N}$, we have

$$(z_3, z_4^n) = 0 \quad \text{for any } n \in \mathbb{N}.$$

Hence and Theorem 1.53, we get

$$\begin{aligned} 0 &= (z_3, z_4^n) \\ &\to (z_3, z_4), \end{aligned}$$

i.e., $(z_3, z_4) = 0$. Because $z_4 \in \overline{\mathbf{M}}$ was arbitrarily chosen, we obtain $z_3 \perp \overline{\mathbf{M}}$. Therefore $z_3 \in \overline{\mathbf{M}}^{\perp}$. Since $z_3 \in \mathbf{M}^{\perp}$ was arbitrarily chosen and for it we get that it is an element of $\overline{\mathbf{M}}^{\perp}$, we obtain

$$\mathbf{M}^{\perp} \subseteq \overline{\mathbf{M}}^{\perp}.$$

From the last relation and from (1.58), we get $\mathbf{M}^{\perp} = \overline{\mathbf{M}}^{\perp}$. Then $z = 0$ and $y = x \in \overline{\mathbf{M}}$, which is a contradiction. This completes the proof. □

Let $\mathbf{M}$ be a linear subspace of the Hilbert space $\mathbf{H}$, spanned by the orthonormal system $\{\phi_k\}_{k\in\mathbb{N}}$ and $x \in \mathbf{M}$.

Definition 1.45.

1. The series

$$\sum_{k=1}^{\infty} \alpha_k \phi_k, \quad \alpha_k \in \mathbf{F},$$

will be called Fourier's series with respect to the orthonormal system $\{\phi_k\}_{k\in\mathbb{N}}$.
2. The numbers

$$\alpha_k = (x, \phi_k), \quad k \in \mathbb{N},$$

will be called Fourier's coefficients for the element x with respect to the orthonormal system $\{\phi_k\}_{k\in\mathbb{N}}$.

Let $\mathbf{M}_n = Span\{\phi_1, \ldots, \phi_n\}$ and $u_n = \sum_{k=1}^{n} c_k\phi_k \in \mathbf{M}_n$ for some $c_k \in \mathbf{F}$. Then

$$\left(x - \sum_{k=1}^{n} c_k\phi_k, x - \sum_{k=1}^{n} c_k\phi_k\right) = (x, x) - 2\sum_{k=1}^{n} \text{Re}\,(c_k(x, \phi_k)) + \sum_{k=1}^{n} |c_k|^2.$$

Let

$$\Delta_n = \|x - u_n\|.$$

Then

$$\Delta_n^2 = \|x\|^2 - 2\sum_{k=1}^{n} \text{Re}\,(\overline{c_k}(x, \phi_k)) + \sum_{k=1}^{n} |c_k|^2.$$

Hence,

$$\Delta_n^2 = \|x\|^2 - \sum_{k=1}^{n} \alpha_k\overline{c_k} - \sum_{k=1}^{n} \overline{\alpha_k}c_k + \sum_{k=1}^{n} |c_k|^2.$$

Note that

$$\begin{aligned} |\alpha_k - c_k|^2 &= (\alpha_k - c_k)\left(\overline{\alpha_k} - \overline{c_k}\right) \\ &= \alpha_k\overline{\alpha_k} - \alpha_k\overline{c_k} - c_k\overline{\alpha_k} + c_k\overline{c_k} \\ &= |\alpha_k|^2 - \alpha_k\overline{c_k} - c_k\overline{\alpha_k} + |c_k|^2, \end{aligned}$$

whereupon

$$-\alpha_k\overline{c_k} - c_k\overline{\alpha_k} + |c_k|^2 = |\alpha_k - c_k|^2 - |\alpha_k|^2.$$

Therefore,

$$\Delta_n^2 = \|x\|^2 + \sum_{k=1}^{n} |\alpha_k - c_k|^2 - \sum_{k=1}^{n} |\alpha_k|^2.$$

Let

$$d_n = dist\,(x, \mathbf{M}_n).$$

Then

$$\begin{aligned} d_n^2 &= \inf_{u_n \in \mathbf{M}_n} \|x - u_n\|^2 \\ &= \inf_{c_1,\ldots,c_n} \Delta_n^2 \\ &= \Delta_n^2\Big|_{\alpha_1=c_1,\ldots,\alpha_n=c_n} \\ &= \|x\|^2 - \sum_{k=1}^{n} |\alpha_k|^2 \end{aligned}$$

and

$$d_n = \left\| x - \sum_{k=1}^{n} \alpha_k \phi_k \right\|.$$

Theorem 1.60. *We have*

$$\sum_{k=1}^{n} |\alpha_k|^2 \leq \|x\|^2 \quad \textit{for any } n \in \mathbb{N}. \tag{1.59}$$

and $d_n < d_m$ *for any* $m < n$.

Proof. Since $d_n^2 \geq 0$ for any $n \in \mathbb{N}$, we get the inequality (1.59). For $m < n$ we have

$$\begin{aligned} d_m^2 &= \|x\|^2 - \sum_{k=1}^{m} |\alpha_k|^2 \\ &> \|x\|^2 - \sum_{k=1}^{n} |\alpha_k|^2 \\ &= d_n^2, \end{aligned}$$

whereupon $d_m > d_n$. This completes the proof. □

Theorem 1.61 (Bessel's Inequality). *We have*

$$\sum_{k=1}^{\infty} |c_k|^2 \leq \|x\|^2. \tag{1.60}$$

Proof. Since the inequality (1.59) holds for any $n \in \mathbb{N}$, we conclude that the series $\sum_{k=1}^{\infty} |\alpha_k|^2$ is convergent and (1.60) holds. This completes the proof. □

Definition 1.46 (Complete Orthonormal System). An orthonormal system $\{\phi_k\}_{k\in\mathbb{N}}$ in the Hilbert space $\mathbf{H}$ is called complete orthonormal system if for any $x \in \mathbf{H}$ the representation

$$x = \sum_{k=1}^{\infty} c_k \phi_k, \quad c_k = (x, \phi_k), \quad k \in \mathbb{N},$$

holds.

Theorem 1.62. *An orthonormal system $\{\phi_k\}_{k\in\mathbb{N}}$ in the Hilbert space $\mathbf{H}$ is complete if and only if*

$$\|x\|^2 = \sum_{k=1}^{\infty} |c_k|^2, \quad c_k = (x, \phi_k), \quad k \in \mathbb{N},$$

for any $x \in \mathbf{H}$.

Proof.

1. Let $\{\phi_k\}_{k\in\mathbb{N}}$ is a complete orthonormal system. Let also, $x \in \mathbf{H}$ be arbitrarily chosen and fixed. Then

$$x = \sum_{k=1}^{\infty} c_k \phi_k.$$

Hence,

$$\begin{aligned} \|x\|^2 &= (x, x) \\ &= \sum_{k=1}^{\infty} |c_k|^2. \end{aligned}$$

2. Let

$$\|x\|^2 = \sum_{k=1}^{\infty} |c_k|^2,$$

$$c_k = (x, \phi_k), \quad k \in \mathbb{N}.$$

Hence, using that

$$\left\| x - \sum_{k=1}^{n} c_k \phi_k \right\|^2 = \|x\|^2 - \sum_{k=1}^{n} |c_k|^2$$

for any $n \in \mathbb{N}$, we get that $x = \sum_{k=1}^{\infty} c_k \phi_k$. This completes the proof.

□

Theorem 1.63. *Let $\{\phi_k\}_{k\in\mathbb{N}}$ be an orthonormal system in the Hilbert space $\mathbf{H}$ and $\boldsymbol{M} = Span\{\{\phi_k\}_{k\in\mathbb{N}}\}$. Then $\{\phi_k\}_{k\in\mathbb{N}}$ is a complete orthonormal system if and only if $\overline{\mathbf{M}} = \mathbf{H}$.*

Proof.

1. Let $\{\phi_k\}_{k\in\mathbb{N}}$ is a complete orthonormal system in $\mathbf{H}$. Assume that $\overline{\mathbf{M}} \neq \mathbf{H}$. Then there exists $x_0 \in \mathbf{H}\backslash \left(\overline{\mathbf{M}}\right)$, $x_0 \neq 0$. Hence, $(x_0, \phi_k) = 0$ for any $k \in \mathbb{N}$. Since $\{\phi_k\}_{k\in\mathbb{N}}$ is a complete orthonormal system, we have

$$\begin{aligned} x_0 &= \sum_{k=1}^{\infty} (x_0, \phi_k)\phi_k \\ &= 0. \end{aligned}$$

   This is a contradiction. Therefore $\overline{\mathbf{M}} = \mathbf{H}$.
2. Let $\overline{\mathbf{M}} = \mathbf{H}$ and $x \in \mathbf{H}$ be arbitrarily chosen and fixed. Then for any $\varepsilon > 0$ there exists $x_\varepsilon \in \mathbf{M}$ such that

$$\|x - x_\varepsilon\| < \varepsilon.$$

   Hence, for any $\varepsilon > 0$ there exists $N = N(\varepsilon) \in \mathbb{N}$ such that

$$\begin{aligned} x_\varepsilon &= \sum_{k=1}^{N} c_k \phi_k, \\ c_k &= (x, \phi_k), \quad k \in \{1, \ldots, N\}. \end{aligned}$$

   Therefore,

$$\begin{aligned} \left\| x - \sum_{k=1}^{N} c_k \phi_k \right\| &= \|x - x_\varepsilon\| \\ &< \varepsilon. \end{aligned}$$

By Theorem 1.60, we obtain

$$\begin{aligned}\left\| x - \sum_{k=1}^{n} c_k \phi_k \right\| &\leq \left\| x - \sum_{k=1}^{N} c_k \phi_k \right\| \\ &\leq \| x - x_\varepsilon \| \\ &< \varepsilon\end{aligned}$$

for any $n > N$. Therefore,

$$x = \sum_{k=1}^{\infty} c_k \phi_k.$$

Since $x \in \mathbf{H}$ was arbitrarily chosen and fixed, we conclude that $\{\phi_k\}_{k \in \mathbb{N}}$ is a complete orthonormal system in $\mathbf{H}$. This completes the proof.

□

1.9 Separable Spaces

Definition 1.47 (Separable Space). A vector space $\mathbf{X}$ is said to be separable if there is a sequence $\{x_n\}_{n \in \mathbb{N}}$ of elements of $\mathbf{X}$ such that for any $x \in \mathbf{X}$ there exists a subsequence $\{x_{n_k}\}_{k \in \mathbb{N}}$ of the sequence $\{x_n\}_{n \in \mathbb{N}}$ that converges to x. In other words, a vector space X is said to be separable if it contains a countable everywhere dense set.

Example 1.29. Consider $\mathbf{E}_n$. Let

$$\mathbf{E}_n^0 = \{(x_1, \ldots, x_n) : x_k \in \mathbf{Q}, k \in \{1, \ldots, n\}\}.$$

Then $\mathbf{E}_n^0$ is a countable everywhere dense set in $\mathbf{E}_n$. Consequently, $\mathbf{E}_n$ is a separable space.

Example 1.30. Consider the space $\mathbf{C}([a, b])$. With $\widetilde{C}$ we will denote its linear subspace of all polynomials with rational coefficients. Then $\widetilde{C}$ is a countable everywhere dense set of the space $\mathbf{C}([a, b])$. Therefore, $\mathbf{C}([a, b])$ is a separable space.

Theorem 1.64. *If in the Hilbert space* $\mathbf{H}$ *there exists a finite or a countable orthogonal basis* $\{f_n\}$*, then* $\mathbf{H}$ *is separable.*

Proof. The set $\{\alpha_n f_n : \alpha_n \in \mathbf{Q}\}$ is a countable everywhere dense set in $\mathbf{H}$. Therefore, $\mathbf{H}$ is separable. This completes the proof. □

Theorem 1.65. *In every separable Hilbert space* $\mathbf{H}$ *there exists an orthogonal basis.*

Proof. Since $\mathbf{H}$ is a separable Hilbert space there exists a countable everywhere dense set $\mathbf{A}$. Let $e_1 \in \mathbf{A}$, $e_1 \neq 0$. Then we take $e_2 \in \mathbf{A}\backslash\{e_1\}$ so that e_1 and e_2 are linearly independent. Then we take $e_3 \in \mathbf{A}\backslash\{e_1, e_2\}$ such that e_1, e_2 and e_3 are linearly independent. And so on. In this way, we get a linearly independent system $\{e_n\}$. From here, using Theorem 1.55, we get a countable orthogonal basis in $\mathbf{H}$. This completes the proof. □

1.10 Linear Operators

An operator is generally a mapping that acts on the elements of a vector space to produce other elements of the same or other vector space. The most common operators are linear operators, which act on vector spaces.

Suppose that X and Y are vector spaces over F.

Definition 1.48. The operator $A : \mathbf{X} \longmapsto \mathbf{Y}$ will be called a linear operator, if

1. it is additive, i.e.,

$$\mathbb{A}(x_1 + x_2) = \mathbb{A}x_1 + \mathbb{A}x_2, \quad x_1, x_2 \in \mathbf{E}_x,$$

2. it is homogeneous, i.e.,

$$\mathbb{A}(\lambda x) = \lambda \mathbb{A}x, \quad \lambda \in \mathbf{F}, \quad x \in \mathbf{E}_x.$$

Example 1.31. Let $K(t,s)$ be a continuous function on the square $0 \leq t, s \leq 1$. For $x \in \mathbf{C}([0,1])$ we define the operator

$$y(t) = \int_0^1 K(t,s)x(s)ds, \quad t \in [0,1], \quad y = \mathbb{A}x.$$

Let $\mathbf{X} = \mathbf{Y} = \mathbf{C}([0,1])$. It is evident that $\mathbb{A} : \mathbf{X} \longmapsto \mathbf{Y}$. We will prove that it is a linear operator.

1. Let $x_1, x_2 \in \mathbf{X}$ be arbitrarily chosen. Then

$$\begin{aligned}
\mathbb{A}x_1(t) &= \int_0^1 K(t,s)x_1(s)ds, \\
\mathbb{A}x_2(t) &= \int_0^1 K(t,s)x_2(s)ds, \\
\mathbb{A}\left(x_1+x_2\right)(t) &= \int_0^1 K(t,s)\left(x_1(s)+x_2(s)\right)ds \\
&= \int_0^1 K(t,s)x_1(s)ds + \int_0^1 K(t,s)x_2(s)ds \\
&= \mathbb{A}x_1(t) + \mathbb{A}x_2(t), \quad t \in [0,1].
\end{aligned}$$

2. Let $\lambda \in \mathbf{F}$ and $x \in \mathbf{E}_x$ be arbitrarily chosen and fixed. Then

$$\begin{aligned}
\mathbb{A}\left(\lambda x\right)(t) &= \int_0^1 K(t,s)\lambda x(s)ds \\
&= \lambda \int_0^1 K(t,s)x(s)ds \\
&= \lambda \mathbb{A}x(t), \quad t \in [0,1].
\end{aligned}$$

Therefore, $\mathbb{A} : \mathbf{X} \longmapsto \mathbf{Y}$ is a linear operator.

Example 1.32. For $x \in \mathbf{C}^1\left([0,1]\right)$ we define the operator

$$y(t) = \frac{d}{dt}x(t), \quad t \in [0,1], \quad y = \mathbb{A}x.$$

Let $\mathbf{X} = \mathbf{C}^1\left([0,1]\right)$, $\mathbf{Y} = \mathbf{C}\left([0,1]\right)$. It is evident that $\mathbb{A} : \mathbf{X} \longmapsto \mathbf{Y}$. We will prove that it is a linear operator.

1. Let $x_1, x_2 \in \mathbf{E}_x$ be arbitrarily chosen and fixed. Then

$$\begin{aligned}
\mathbb{A}x_1(t) &= \frac{d}{dt}x_1(t), \\
\mathbb{A}x_2(t) &= \frac{d}{dt}x_2(t), \\
\mathbb{A}\left(x_1+x_2\right)(t) &= \frac{d}{dt}\left(x_1+x_2\right)(t)
\end{aligned}$$

$$= \frac{d}{dt}x_1(t) + \frac{d}{dt}x_2(t)$$
$$= \mathbb{A}x_1(t) + \mathbb{A}x_2(t), \quad t \in [0,1].$$

2. Let $\lambda \in \mathbf{F}$ and $x \in \mathbf{C}([0,1])$ be arbitrarily chosen. Then

$$\begin{aligned}\mathbb{A}(\lambda x)(t) &= \frac{d}{dt}(\lambda x)(t)\\ &= \lambda\frac{d}{dt}x(t)\\ &= \lambda\mathbb{A}x(t), \quad t \in [0,1].\end{aligned}$$

Therefore, $\mathbb{A} : \mathbf{X} \longmapsto \mathbf{Y}$ is a linear operator.

Example 1.33. For $x \in \mathbf{C}([0,1])$ we define the operator

$$y = \int_0^1 (x(t))^2 dt, \quad y = \mathbb{A}x.$$

Let $\mathbf{X} = \mathbf{C}([0,1])$, $\mathbf{Y} = \mathbf{F}$. It is evident that $\mathbb{A} : \mathbf{X} \longmapsto \mathbf{Y}$. Let

$$\begin{aligned}x_1(t) &= 1,\\ x_2(t) &= t, \quad t \in [0,1].\end{aligned}$$

Then

$$\begin{aligned}\mathbb{A}x_1 &= \int_0^1 dt\\ &= t\Big|_{t=0}^{t=1}\\ &= 1,\\ \mathbb{A}x_2 &= \int_0^1 t^2 dt\\ &= \frac{1}{3}t^3\Big|_{t=0}^{t=1}\\ &= \frac{1}{3},\end{aligned}$$

$$\begin{aligned}
\mathbb{A}x_1 + \mathbb{A}x_2 &= 1 + \frac{1}{3} \\
&= \frac{4}{3}, \\
\mathbb{A}(x_1 + x_2) &= \int_0^1 (t+1)^2\, dt \\
&= \int_0^1 \left(t^2 + 2t + 1\right) dt \\
&= \int_0^1 t^2 dt + 2\int_0^1 t dt + \int_0^1 dt \\
&= \frac{1}{3}t^3\Big|_{t=0}^{t=1} + t^2\Big|_{t=0}^{t=1} + t\Big|_{t=0}^{t=1} \\
&= \frac{1}{3} + 1 + 1 \\
&= \frac{7}{3}.
\end{aligned}$$

Therefore,

$$\mathbb{A}(x_1 + x_2) \neq \mathbb{A}x_1 + \mathbb{A}x_2.$$

Consequently, $\mathbb{A} : \mathbf{X} \longmapsto \mathbf{Y}$ is not a linear operator.

Exercise 1.22. For $x \in \mathbf{C}([0,1])$ we define the operator

$$y(t) = x\left(t^2\right), \quad t \in [0,1], \quad y = \mathbb{A}x.$$

Prove that $\mathbb{A} : \mathbf{C}([0,1]) \longmapsto \mathbf{C}([0,1])$ is a linear operator.

In this section, we suppose that $\mathbf{X}$ and $\mathbf{Y}$ are normed vector spaces. The convergence in $\mathbf{X}$ and $\mathbf{Y}$ is a norm convergence.

Definition 1.49. We say that the linear operator $\mathbb{A} : \mathbf{X} \longmapsto \mathbf{Y}$ is continuous at $x \in \mathbf{X}$, if for any $\varepsilon > 0$ there is a $\delta = \delta(\varepsilon)$ such that

$$\|\mathbb{A}x_1 - \mathbb{A}x\| < \varepsilon$$

whenever $\|x_1 - x\| < \delta$, $x_1 \in \mathbf{X}$. In other words, the linear operator $\mathbb{A} : \mathbf{X} \longmapsto \mathbf{Y}$ is said to be continuous at $x \in \mathbf{X}$ if $\mathbb{A}x_n \to \mathbb{A}x$ in $\mathbf{Y}$

as $n \to \infty$ whenever $x_n \to x$ in $\mathbf{X}$, as $n \to \infty$, where $\{x_n\}_{n\in\mathbb{N}}$ is a sequence of elements of $\mathbf{X}$. We say that the linear operator $\mathbb{A} : \mathbf{X} \longmapsto \mathbf{Y}$ is continuous in $\mathbf{X}$, if it is continuous at every point of $\mathbf{X}$.

Example 1.34. Let $\mathbf{X} = \mathbf{Y} = \mathbf{C}([0,1])$. Consider the operator

$$\mathbb{A}x(t) = t^2 \int_0^1 x(s)ds, \quad t \in [0,1], \quad x \in \mathbf{X}.$$

Let $x \in \mathbf{X}$ be arbitrarily chosen and fixed. We take $\varepsilon > 0$ arbitrarily and $x_n \in \mathbf{X}$ such that

$$\begin{aligned}\|x_n - x\| &= \max_{t\in[0,1]} |x_n(t) - x(t)| \\ &< \varepsilon.\end{aligned}$$

Hence,

$$\begin{aligned}|\mathbb{A}x_n(t) - \mathbb{A}x(t)| &= \left| t^2 \int_0^1 x_n(s)ds - t^2 \int_0^1 x(s)ds \right| \\ &= \left| t^2 \int_0^1 (x_n(s) - x(s))\, ds \right| \\ &\le t^2 \int_0^1 |x_n(s) - x(s)|\, ds \\ &\le \int_0^1 \|x_n - x\| ds \\ &< \varepsilon, \quad t \in [0,1].\end{aligned}$$

Because $\varepsilon > 0$ was arbitrarily chosen, we conclude that $\mathbb{A}$ is continuous at x. Since $x \in \mathbf{X}$ was arbitrarily chosen, we get that the operator $\mathbb{A}$ is continuous in $\mathbf{X}$.

Now, we suppose that $\mathbb{A} : \mathbf{X} \longmapsto \mathbf{Y}$ is a linear continuous operator. We take $x = y + z$, $y, z \in \mathbf{X}$. Then

$$\begin{aligned}\mathbb{A}x &= \mathbb{A}(y + z) \\ &= \mathbb{A}y + \mathbb{A}z \\ &= \mathbb{A}y + \mathbb{A}(x - y).\end{aligned}$$

Therefore,

$$\mathbb{A}(x-y) = \mathbb{A}x - \mathbb{A}y. \tag{1.61}$$

We set $x = y$ in (1.61) and we get

$$\begin{aligned}\mathbb{A}0 &= \mathbb{A}x - \mathbb{A}x \\ &= 0.\end{aligned}$$

We set $x = 0$ in (1.61) and we obtain

$$\begin{aligned}\mathbb{A}(-y) &= \mathbb{A}0 - \mathbb{A}y \\ &= -\mathbb{A}y.\end{aligned} \tag{1.62}$$

Theorem 1.66. *Let $\mathbb{A} : \mathbf{X} \longmapsto \mathbf{Y}$ be linear operator, which is continuous at a single point $x_0 \in \mathbf{X}$. Then it is continuous on the entire space $\mathbf{X}$.*

Proof. Let $\{x_n\}_{n\in\mathbb{N}}$ be a sequence of elements of $\mathbf{X}$ such that $x_n \to x$, as $n \to \infty$, in $\mathbf{X}$, $x \in \mathbf{X}$. Hence,

$$x_n - x + x_0 \to x_0, \quad \text{as } n \to \infty.$$

Therefore,

$$\mathbb{A}(x_n - x + x_0) \to \mathbb{A}x_0, \quad \text{as } n \to \infty. \tag{1.63}$$

Since $\mathbb{A} : \mathbf{X} \longmapsto \mathbf{Y}$ is a linear operator, we get

$$\mathbb{A}(x_n - x + x_0) = \mathbb{A}x_n - \mathbb{A}x + \mathbb{A}x_0, \quad n \in \mathbb{N}.$$

From here,

$$\mathbb{A}(x_n - x + x_0) \to \mathbb{A}x_n - \mathbb{A}x + \mathbb{A}x_0, \quad \text{as } n \to \infty.$$

Using (1.63), we obtain

$$\mathbb{A}x_n \to \mathbb{A}x, \quad \text{as } n \to \infty.$$

This completes the proof. □

Definition 1.50. Let $\mathbb{A}, \mathbb{B} : \mathbf{X} \longmapsto \mathbf{Y}$ be linear operators. We define the addition of the operators $\mathbb{A}$ and $\mathbb{B}$ by

$$(\mathbb{A} + \mathbb{B})x = \mathbb{A}x + \mathbb{B}x, \quad x \in \mathbf{X},$$

and the scalar multiplication by

$$(\lambda\mathbb{A})\,x = \lambda\mathbb{A}x, \quad x \in \mathbf{X}, \quad \lambda \in \mathbf{F}.$$

The zero operator $\mathbb{O}$ we define by

$$\mathbb{O}x = 0$$

for any $x \in \mathbf{X}$. The identity operator $\mathbb{I}$ we define by

$$\mathbb{I}x = x$$

for any $x \in \mathbf{X}$. We define Let $\mathbb{A}, \mathbb{B} : \mathbf{E}_x \longmapsto \mathbf{E}_x$. We define

$$\begin{aligned}
(\mathbb{A}\mathbb{B})\,x &= \mathbb{A}\,(\mathbb{B}x), \\
\mathbb{A}^2 x &= \mathbb{A}\,(\mathbb{A}x), \\
\mathbb{A}^n x &= \mathbb{A}\left(\mathbb{A}^{n-1}x\right), \quad n \geq 3, \quad x \in \mathbf{X}.
\end{aligned}$$

Remark 1.2. If $\mathbb{A}, \mathbb{B}, \mathbb{C} : \mathbf{X} \longmapsto \mathbf{Y}$, then

$$\begin{aligned}
(\mathbb{A}\mathbb{B})\,\mathbb{C} &= \mathbb{A}\,(\mathbb{B}\mathbb{C}), \\
(\mathbb{A} + \mathbb{B})\,\mathbb{C} &= \mathbb{A}\mathbb{C} + \mathbb{B}\mathbb{C}, \\
\mathbb{C}\,(\mathbb{A} + \mathbb{B}) &= \mathbb{C}\mathbb{A} + \mathbb{C}\mathbb{B}.
\end{aligned}$$

In the general case, we have

$$\mathbb{A}\mathbb{B} \neq \mathbb{B}\mathbb{A}.$$

Definition 1.51. Let $\mathbb{A} : \mathbf{X} \longmapsto \mathbf{Y}$ be a linear operator. We say that the linear operator $\mathbb{B} : \mathbf{X} \longmapsto \mathbf{Y}$ is a left inverse of the operator $\mathbb{A}$, if

$$\mathbb{B}\mathbb{A} = \mathbb{I}.$$

We say that the linear operator $\mathbb{C} : \mathbf{X} \longmapsto \mathbf{Y}$ is a right inverse of the operator $\mathbb{A}$, if

$$\mathbb{A}\mathbb{C} = \mathbb{I}.$$

Let $\mathbb{B}, \mathbb{C} : \mathbf{X} \longmapsto \mathbf{Y}$ be left and right inverse, respectively, of the linear operator $\mathbb{A} : \mathbf{X} \longmapsto \mathbf{Y}$. Then

$$\begin{aligned}
\mathbb{B} &= \mathbb{B}\mathbb{I} \\
&= \mathbb{B}\,(\mathbb{A}\mathbb{C})
\end{aligned}$$

$$\begin{aligned} &= (\mathbb{B}\mathbb{A})\,\mathbb{C} \\ &= \mathbb{I}\mathbb{C} \\ &= \mathbb{C}. \end{aligned}$$

In this case it is said that the operator $\mathbb{A}$ has an inverse denoted by $\mathbb{A}^{-1}$. Thus, if $\mathbb{A}^{-1}$ exists, we have

$$\begin{aligned} \mathbb{A}^{-1}\mathbb{A} &= \mathbb{A}\mathbb{A}^{-1} \\ &= \mathbb{I}. \end{aligned}$$

A linear operator $\mathbb{A} : \mathbf{X} \longmapsto \mathbf{Y}$ can have at most one inverse.

Theorem 1.67. *Let $\mathbb{A} : \mathbf{X} \longmapsto \mathbf{Y}$ be a linear operator that is continuous at 0. Then $\mathbb{A}$ is continuous in $\mathbf{X}$.*

Proof. Let $x \in \mathbf{X}$ be arbitrarily chosen. Since $\mathbb{A}$ is continuous at 0, we get

$$\|\mathbb{A}(x_n - x) - \mathbb{A}0\| \to 0, \quad \text{as } \|x_n - x\| \to 0,$$

or

$$\|\mathbb{A}x_n - \mathbb{A}x\| \to 0, \quad \text{as } \|x_n - x\| \to 0.$$

Therefore, $\mathbb{A}$ is continuous at x. Because $x \in \mathbf{X}$ was arbitrarily chosen, we conclude that $\mathbb{A}$ is continuous in $\mathbf{X}$. This completes the proof. □

Definition 1.52. A linear operator $\mathbb{A} : \mathbf{X} \longmapsto \mathbf{Y}$ will be called bounded if there is a constant $M \geq 0$ such that

$$\|\mathbb{A}x\| \leq M\|x\|$$

for any $x \in \mathbf{X}$.

Example 1.35. Let $\mathbf{X} = \mathbf{Y} = \mathbf{C}([0, 1])$, $a \in \mathbf{X}$. Consider the operator

$$\mathbb{A}x(t) = \int_0^t a(s)x(s)ds, \quad t \in [0, 1], \quad x \in \mathbf{C}([0, 1]).$$

We have that $\mathbb{A} : \mathbf{X} \longmapsto \mathbf{Y}$. In $\mathbf{X}$, we define a norm as follows,

$$\|x\| = \max_{t\in[0,1]} |x(t)|, \quad x \in \mathbf{X}.$$

Because $a \in \mathbf{C}([0,1])$, there is a positive constant M such that

$$|a(t)| \leq M$$

for any $t \in [0,1]$. Let $x \in \mathbf{X}$ be arbitrarily chosen. Hence, for any $t \in [0,1]$, we have

$$\begin{aligned} |\mathbb{A}x(t)| &= \left| \int_0^t a(s)x(s)ds \right| \\ &\leq \int_0^t |a(s)||x(s)|ds \\ &\leq M\|x\| \int_0^t ds \\ &\leq M\|x\|. \end{aligned}$$

From here,

$$\max_{t\in[0,1]} |\mathbb{A}x(t)| \leq M\|x\|,$$

or

$$\|\mathbb{A}x\| \leq M\|x\|.$$

Because $x \in \mathbf{X}$ was arbitrarily chosen, we conclude that $\mathbb{A} : \mathbf{X} \longmapsto \mathbf{Y}$ is a bounded operator.

Theorem 1.68. *A linear operator $\mathbb{A} : \mathbf{X} \longmapsto \mathbf{Y}$ is bounded if and only if it is continuous.*

Proof.

1. Let $\mathbb{A} : \mathbf{X} \longmapsto \mathbf{Y}$ be a continuous operator. Assume that it is not bounded. Then, there is a sequence $\{x_n\}_{n\in\mathbb{N}}$ of elements of $\mathbf{X}$ such that

   $$\|\mathbb{A}x_n\| > n\|x_n\|, \quad x_n \neq 0,$$

   for any $n \in \mathbb{N}$. We set

   $$\xi_n = \frac{x_n}{n\|x_n\|}.$$

Then

$$\|\mathbb{A}\xi_n\| > 1, \quad n \in \mathbb{N}. \tag{1.64}$$

On the other hand,

$$\begin{aligned}\|\xi_n\| &= \left\| \frac{x_n}{n\|x_n\|} \right\| \\ &= \frac{\|x_n\|}{n\|x_n\|} \\ &= \frac{1}{n} \\ &\to 0, \quad \text{as } n \to \infty.\end{aligned}$$

Because $\mathbb{A} : \mathbf{X} \longmapsto \mathbf{Y}$ is continuous, we get

$$\|\mathbb{A}\xi_n\| \to 0, \quad \text{as } n \to \infty.$$

This contradicts with (1.64). Therefore, $\mathbb{A} : \mathbf{X} \longmapsto \mathbf{Y}$ is bounded.

2. Let $\mathbb{A} : \mathbf{X} \longmapsto \mathbf{Y}$ be a bounded operator. Then there exists a positive constant M such that

$$\|\mathbb{A}x\| \leq M\|x\|$$

for any $x \in \mathbf{X}$. Let $x_n \to x$ as $n \to \infty$, i.e., $\|x_n - x\| \to 0$ as $n \to \infty$. Then

$$\begin{aligned}\|\mathbb{A}x_n - \mathbb{A}x\| &= \|\mathbb{A}(x_n - x)\| \\ &\leq M\|x_n - x\| \\ &\to 0, \quad \text{as } n \to \infty.\end{aligned}$$

Therefore, $\mathbb{A} : \mathbf{X} \longmapsto \mathbf{Y}$ is continuous. This completes the proof. □

The space of all linear bounded operators $\mathbb{A} : \mathbf{X} \longmapsto \mathbf{Y}$ will be denoted with $\mathscr{L}(\mathbf{X}, \mathbf{Y})$. Note that $\mathscr{L}(\mathbf{X}, \mathbf{Y})$. is a vector space.

Theorem 1.69. *Let $\mathbf{X}$ be a Banach space and $\mathbb{A} : \mathbf{X} \longmapsto \mathbf{Y}$ be a linear operator. By $\mathbf{X}_n$ we denote the set of those $x \in \mathbf{X}$ for which $\|\mathbb{A}x\| \leq n\|x\|$. Then $\mathbf{X} = \cup_{n=1}^{\infty}\mathbf{X}_n$ and at least one of the sets $\mathbf{X}_n$ is everywhere dense in $\mathbf{X}$.*

Proof. Note that the sets $\mathbf{X}_n$, $n \in \mathbb{N}$, are not empty because $0 \in \mathbf{X}_n$ for any $n \in \mathbb{N}$. Also, any $x \in \mathbf{X}$, $x \neq 0$, occurs in one of the sets $\mathbf{X}_n$ because it is sufficient to take n as the least integer, greater than $\frac{\|\mathbb{A}x\|}{\|x\|}$. Therefore, $\mathbf{X} \subseteq \cup_{n=1}^{\infty} \mathbf{X}_n$. By the definition of the sets $\mathbf{X}_n$ we have that $\cup_{n=1}^{\infty} \mathbf{X}_n \subseteq \mathbf{X}$. Consequently $\mathbf{X} = \cup_{n=1}^{\infty} \mathbf{X}_n$. By Theorem 1.33, it follows that a complete space $\mathbf{X}$ cannot be a countable sum of nowhere dense sets. Therefore there is an $n_0 \in \mathbb{N}$ and there is an open ball $S_r(x_0)$ containing $S_r(x_0) \cap \mathbf{X}_{n_0}$ everywhere dense. We take $x_1 \in S_r(x_0) \cap \mathbf{X}_{n_0}$ and let $S_{r_1}[x_1]$ be a closed ball such that

$$S_{r_1}[x_1] \subset S_r(x_0).$$

Let $x \in \mathbf{X}$ be an element for which $\|x\| = r_1$. Since

$$\begin{aligned} \|(x_1 + x) - x_1\| &= \|x\| \\ &= r_1, \end{aligned}$$

we conclude that $x + x_1 \in S_{r_1}[x_1]$. There is a sequence $\{z_k\}_{k\in\mathbb{N}}$ of elements of $S_{r_1}[x_1] \cap \mathbf{X}_{n_0}$, (this sequence can be stationary if $x_1 + x \in \mathbf{X}_{n_0}$) such that $z_k \to x + x_1$ as $k \to \infty$. Consequently,

$$x_k = z_k - x_1 \to x, \quad \text{as } k \to \infty.$$

Because $x_k \to x$ as $k \to \infty$ and $\|x\| = r_1$, we can assume that $\frac{r_1}{2} \leq \|x_k\| \leq r_1$ for any enough large $k \in \mathbb{N}$. Now, using that $z_k, x_1 \in \mathbf{X}_{n_0}$, we get

$$\begin{aligned} \|\mathbb{A}x_k\| &= \|\mathbb{A}z_k - \mathbb{A}x_1\| \\ &\leq \|\mathbb{A}z_k\| + \|\mathbb{A}x_1\| \\ &\leq n_0 (\|z_k\| + \|x_1\|). \end{aligned}$$

Also,

$$\begin{aligned} \|z_k\| &= \|x_k + x_1\| \\ &\leq \|x_k\| + \|x_1\| \\ &\leq r_1 + \|x_1\|. \end{aligned}$$

Therefore, for any enough large $k \in \mathbb{N}$,

$$\begin{aligned} \|\mathbb{A}x_k\| &\leq n_0 (r_1 + 2\|x_1\|) \\ &\leq \frac{2n_0 (r_1 + 2\|x_1\|)}{r_1} \|x_k\|. \end{aligned}$$

Let n be the least integer greater than $\frac{2n_0(r_1+2\|x_1\|)}{r_1}$. Then

$$\|\mathbb{A}x_k\| \leq n\|x_k\|$$

for an y enough large $k \in \mathbb{N}$. Consequently, $x_k \in \mathbf{X}_n$ for any enough large $k \in \mathbb{N}$. Thus every element $x \in \mathbf{X}$ with norm equal to r_1 can be approximated by elements of $\mathbf{X}_n$. Let now $x \in \mathbf{X}$, $x \neq 0$. be arbitrarily chosen. We set $\xi = r_1 \frac{x}{\|x\|}$. Then $\|\xi\| = r_1$. As in above, there is a sequence $\{\xi_k\}_{k\in\mathbb{N}} \subseteq \mathbf{X}_n$ convergent to ξ. Then

$$\begin{aligned} x_k &= \xi_k \frac{\|x\|}{r_1} \\ &\to x, \\ \|\mathbb{A}x_k\| &= \frac{\|x\|}{r_1}\|\mathbb{A}\xi_k\| \\ &\leq \frac{\|x\|}{r_1} n\|\xi_k\| \\ &= n\|x_k\|, \quad k \in \mathbb{N}. \end{aligned}$$

Thus, $x_k \in \mathbf{X}_n$, $k \in \mathbb{N}$. Consequently, $\mathbf{X}_n$ is everywhere dense in $\mathbf{X}$. This completes the proof. □

Definition 1.53. Let $\mathbb{A} : \mathbf{X} \longmapsto \mathbf{Y}$ be a bounded linear operator. The smallest number M for which $\|\mathbb{A}x\| \leq M\|x\|$ for any $x \in \mathbf{X}$, will be called the norm of the operator $\mathbb{A}$. It is denoted by $\|\mathbb{A}\|$.

By Definition 1.53 it follows:

1. $\|\mathbb{A}x\| \leq \|\mathbb{A}\|\|x\|$ for any $x \in \mathbf{X}$.
2. for any $\varepsilon > 0$ there is an element $x_\varepsilon \in \mathbf{X}$, $x_\varepsilon \neq 0$, such that

$$\|\mathbb{A}x_\varepsilon\| > (\|\mathbb{A}\| - \varepsilon)\|x_\varepsilon\|.$$

Theorem 1.70. *Let* $\mathbb{A} : \mathbf{X} \longmapsto \mathbf{Y}$ *be a bounded linear operator. Then*

$$\|\mathbb{A}\| = \sup_{\|x\|\leq 1} \|\mathbb{A}x\|. \tag{1.65}$$

Proof. For any $x \in \mathbf{X}$, $\|x\| \le 1$, we have

$$\begin{aligned} \|\mathbb{A}x\| &\le \|\mathbb{A}\|\|x\| \\ &\le \|\mathbb{A}\|. \end{aligned} \tag{1.66}$$

Let $\varepsilon > 0$ be arbitrarily chosen. Then there exists $x_\varepsilon \in \mathbf{X}$, $x_\varepsilon \neq 0$, such that

$$\|\mathbb{A}x_\varepsilon\| > (\|\mathbb{A}\| - \varepsilon)\,\|x_\varepsilon\|.$$

We take $\xi_\varepsilon = \frac{x_\varepsilon}{\|x_\varepsilon\|}$. Then $\|\xi_\varepsilon\| = 1$ and

$$\begin{aligned} \|\mathbb{A}\xi_\varepsilon\| &= \left\|\mathbb{A}\left(\frac{x_\varepsilon}{\|x_\varepsilon\|}\right)\right\| \\ &= \left\|\frac{1}{\|x_\varepsilon\|}\mathbb{A}x_\varepsilon\right\| \\ &= \frac{1}{\|x_\varepsilon\|}\|\mathbb{A}x_\varepsilon\| \\ &> \|\mathbb{A}\| - \varepsilon. \end{aligned}$$

Because $\|\xi_\varepsilon\| = 1$, from the last inequality, we get

$$\sup_{\|x\|\le 1} \|\mathbb{A}x\| \ge \|\mathbb{A}\| - \varepsilon.$$

Hence, using that $\varepsilon > 0$ was arbitrarily chosen, we get

$$\sup_{\|x\|\le 1} \|\mathbb{A}x\| \ge \|\mathbb{A}\|.$$

From the last inequality and from (7.71), we obtain the inequality (7.70). This completes the proof. □

Remark 1.3. By Theorem 1.70, it follows that

$$\|\mathbb{A}\| = \sup_{x\in\mathbf{X}, x\neq 0} \frac{\|\mathbb{A}x\|}{\|x\|}.$$

Example 1.36. Let $\mathbf{X} = \mathbf{Y} = \mathbf{C}([0,1])$. Consider the operator

$$\mathbb{A}x(t) = \int_0^1 x(s)ds, \quad x \in \mathbf{X}, \quad t \in [0,1].$$

In $\mathbf{X}$, we take the norm $\|x\| = \max_{t\in[0,1]} |x(t)|$. We have $\mathbb{A} : \mathbf{X} \longmapsto \mathbf{X}$. We will find $\|\mathbb{A}\|$. For any $x \in \mathbf{X}$, $\|x\| \leq 1$, we have

$$\begin{aligned}
|\mathbb{A}x(t)| &= \left| \int_0^1 x(s)ds \right| \\
&\leq \int_0^1 |x(s)|\, ds \\
&\leq \int_0^1 \max_{s\in[0,1]} |x(s)|\, ds \\
&= \|x\| \\
&\leq 1, \quad t \in [0,1].
\end{aligned}$$

Hence,

$$\max_{t\in[0,1]} |\mathbb{A}x(t)| \leq 1$$

or

$$\|\mathbb{A}x\| \leq 1.$$

Therefore,

$$\|\mathbb{A}\| \leq 1. \tag{1.67}$$

Now, we take $y(t) = 1$, $t \in [0,1]$. Then $\|y\| = 1$ and

$$\begin{aligned}
\mathbb{A}y(t) &= \int_0^1 ds \\
&= 1, \quad t \in [0,1].
\end{aligned}$$

Hence, using that $\|y\| = 1$, we get

$$\sup_{\|x\|\leq 1} \|\mathbb{A}x\| \geq 1,$$

i.e., $\|\mathbb{A}\| \geq 1$. From here and from (1.67), we obtain that $\|\mathbb{A}\| = 1$.

Example 1.37. Let $K \in \mathbf{C}([0,1] \times [0,1])$, $\mathbf{X} = \mathbf{Y} = \mathbf{C}([0,1])$. Consider the operator

$$\mathbb{A}x(t) = \int_0^1 K(t,s)x(s)ds, \quad x \in \mathbf{X}, \quad t \in [0,1].$$

In $\mathbf{X}$, we take the norm $\|x\| = \max_{t\in[0,1]} |x(t)|$. We have $\mathbb{A} : \mathbf{X} \longmapsto \mathbf{Y}$. We will find $\|\mathbb{A}\|$. Let $x \in \mathbf{X}$ be arbitrarily chosen. Then

$$\begin{aligned}
|\mathbb{A}x(t)| &= \left| \int_0^1 K(t,s)x(s)ds \right| \\
&\le \int_0^1 |K(t,s)|\,|x(s)|ds \\
&\le \|x\| \max_{t\in[0,1]} \int_0^1 |K(t,s)|\,ds, \quad t \in [0,1].
\end{aligned}$$

Hence,

$$\max_{t\in[0,1]} |\mathbb{A}x(t)| \le \|x\| \max_{t\in[0,1]} \int_0^1 |K(t,s)|ds$$

or

$$\|\mathbb{A}x\| \le \|x\| \max_{t\in[0,1]} \int_0^1 |K(t,s)|ds.$$

Then

$$\begin{aligned}
\sup_{\|x\|\le 1} \|\mathbb{A}x\| &= \sup_{\|x\|\le 1} \left(\max_{t\in[0,1]} |\mathbb{A}x(t)| \right) \\
&\le \sup_{\|x\|\le 1} \left(\|x\| \max_{t\in[0,1]} \int_0^1 |K(t,s)|\,ds \right) \\
&\le \max_{t\in[0,1]} \int_0^1 |K(t,s)|\,ds.
\end{aligned}$$

Therefore,

$$\|\mathbb{A}\| \le \max_{t\in[0,1]} \int_0^1 |K(t,s)|\,ds. \tag{1.68}$$

For each $n \in \mathbb{N}$ we take a set $\mathbf{E}_n \subseteq [0,1]$ such that

$$m(\mathbf{E}_n) \leq \frac{1}{2Mn},$$

where $M = \max_{(t,s)\in[0,1]\times[0,1]} |K(t,s)|$. Since $t \longmapsto \int_0^1 |K(t,s)|\, ds$ is a continuous function on $[0,1]$, there exists $t_0 \in [0,1]$ such that

$$\int_0^1 |K(t_0,s)|\, ds = \max_{t\in[0,1]} \int_0^1 |K(t,s)|\, ds.$$

Let

$$z_0(s) = \operatorname{sign} K(t_0,s), \quad s \in [0,1].$$

For each $n \in \mathbb{N}$ we take a continuous function x_n on $[0,1]$ such that $|x_n(s)| \leq 1$ for any $s \in [0,1]$ and $x_n(s) = z_0(s)$ for $s \in [0,1]\backslash\mathbf{E}_n$. We have that

$$\begin{aligned}
|x_n(s) - z_0(s)| &\leq |x_n(s)| + |z_0(s)| \\
&\leq 1 + 1 \\
&= 2
\end{aligned}$$

for any $s \in \mathbf{E}_n$. Then

$$\begin{aligned}
&\left| \int_0^1 K(t,s) z_0(s) ds - \int_0^1 K(t,s) x_n(s) ds \right| \\
&= \left| \int_0^1 K(t,s) \left(z_0(s) - x_n(s) \right) ds \right| \\
&\leq \int_0^1 |K(t,s)| \, |z_0(s) - x_n(s)| \, ds \\
&= \int_{\mathbf{E}_n} |K(t,s)| \, |z_0(s) - x_n(s)| \, ds \\
&\leq 2Mm(\mathbf{E}_n) \\
&\leq \frac{1}{n}
\end{aligned}$$

for any $t \in [0,1]$ and for any $n \in \mathbb{N}$. Therefore,

$$\begin{aligned}\int_0^1 K(t,s)z_0(s)ds &\le \int_0^1 K(t,s)x_n(s)ds + \frac{1}{n}\\ &= \mathbb{A}x_n(t) + \frac{1}{n}\\ &\le \|\mathbb{A}\|\|x_n\| + \frac{1}{n}\\ &\le \|\mathbb{A}\| + \frac{1}{n}\end{aligned}$$

for any $t \in [0,1]$ and for any $n \in \mathbb{N}$. In particular,

$$\int_0^1 K(t_0,s)\,z_0(s)ds \le \|\mathbb{A}\| + \frac{1}{n}$$

for any $n \in \mathbb{N}$. Therefore,

$$\int_0^1 K(t_0,s)\,z_0(s)ds \le \|\mathbb{A}\|,$$

i.e.,

$$\max_{t\in[0,1]} \int_0^1 |K(t,s)|\,ds \le \|\mathbb{A}\|.$$

From the last inequality and from (1.68), we get

$$\|\mathbb{A}\| = \max_{t\in[0,1]} \int_0^1 |K(t,s)|\,ds.$$

Example 1.38. Let $\mathbf{X} = \mathbf{Y} = \mathbf{C}([0,1])$. Consider the operator

$$\mathbb{A}x(t) = \frac{t^2}{2}\int_0^t x(s)ds, \quad t \in [0,1], \quad x \in \mathbf{X}.$$

In $\mathbf{X}$, we take the norm $\|x\| = \max_{t\in[0,1]} |x(t)|$. We have $\mathbb{A} : \mathbf{X} \longmapsto \mathbf{Y}$. We will find $\|\mathbb{A}\|$. Let $x \in \mathbf{X}$ be arbitrarily chosen. Then

$$\begin{aligned}|\mathbb{A}x(t)| &= \left|\frac{t^2}{2}\int_0^t x(s)ds\right|\\ &\le \frac{t^2}{2}\int_0^t |x(s)|\,ds\\ &\le \frac{1}{2}\|x\|\end{aligned}$$

for any $t \in [0,1]$. Hence,

$$\begin{aligned}\|\mathbb{A}x\| &= \max_{t\in[0,1]} |\mathbb{A}x(t)| \\ &\leq \frac{1}{2}\|x\|.\end{aligned}$$

Then

$$\begin{aligned}\|\mathbb{A}\| &= \sup_{|x|\leq 1} \|\mathbb{A}x\| \\ &\leq \frac{1}{2}.\end{aligned} \tag{1.69}$$

For each $n \in \mathbb{N}$ we take a set $\mathbf{E}_n \subset [0,1]$ such that $m(\mathbf{E}_n) \leq \frac{1}{n}$. Also, for each $n \in \mathbb{N}$ we take a continuous function x_n on $[0,1]$ such that $|x_n(s)| \leq 1$ for any $s \in [0,1]$ and $x_n(s) = 1$ for $s \in [0,1]\backslash\mathbf{E}_n$. Then

$$|x_n(s) - 1| \leq 2$$

for any $s \in \mathbf{E}_n$. Hence,

$$\begin{aligned}\left|\frac{t^3}{2} - \frac{t^2}{2}\int_0^t x_n(s)ds\right| &= \left|\frac{t^2}{2}\int_0^t (1 - x_n(s))\, ds\right| \\ &\leq \frac{t^2}{2}\int_0^t |1 - x_n(s)|\, ds \\ &\leq \frac{t^2}{2}\int_0^1 |1 - x_n(s)|\, ds \\ &= \frac{t^2}{2}\int_{\mathbf{E}_n} |1 - x_n(s)|\, d \\ &\leq m(\mathbf{E}_n) \\ &\leq \frac{1}{n}\end{aligned}$$

for any $t \in [0,1]$ and for any $n \in \mathbb{N}$. Hence,

$$\begin{aligned}\frac{t^3}{2} &\leq \frac{1}{n} + \frac{t^2}{2}\int_0^t x_n(s)ds \\ &= \frac{1}{n} + \mathbb{A}x_n(t)\end{aligned}$$

$$\leq \frac{1}{n} + \|\mathbb{A}\|\|x_n\|$$
$$\leq \frac{1}{n} + \|\mathbb{A}\|$$

for any $t \in [0,1]$ and for any $n \in \mathbb{N}$. In particular, for $t = 1$, we get

$$\frac{1}{2} \leq \frac{1}{n} + \|\mathbb{A}\|$$

for any $n \in \mathbb{N}$. Consequently,

$$\frac{1}{2} \leq \|\mathbb{A}\|.$$

From the last inequality and from (1.69), we get $\|\mathbb{A}\| = \frac{1}{2}$.

Exercise 1.23. Let $\mathbf{X} = \mathbf{Y} = \mathbf{C}([0,1])$, $\|x\| = \max_{t\in[0,1]} |x(t)|$, $x \in \mathbf{X}$. Find $\|\mathbb{A}\|$, where

$$\mathbb{A}x(t) = \frac{t^2}{3}\int_0^t s^2 x(s)ds, \quad x \in \mathbf{X}, \quad t \in [0,1].$$

Answer. $\frac{1}{9}$.

Theorem 1.71. *Let $\mathbb{A} : \mathbf{X} \longmapsto \mathbf{Y}$ be a linear operator. Then $\mathbb{A}$ is a bounded operator if and only if there is a constant $M \geq 0$ such that $\|\mathbb{A}\| \leq M$.*

Proof.

1. Let $\mathbb{A}$ be a bounded operator. Then there is a constant $M \geq 0$ such that

$$\|\mathbb{A}x\| \leq M\|x\|$$

for any $x \in \mathbf{X}$. Hence,

$$\|\mathbb{A}\| = \sup_{\|x\|\leq 1} \|\mathbb{A}x\|$$
$$\leq \sup_{\|x\|\leq 1} (M\|x\|)$$
$$= M.$$

2. Let there is a constant $M \geq 0$ such that $\|\mathbb{A}\| \leq M$. Then

$$\begin{aligned}\|\mathbb{A}x\| &\leq \|\mathbb{A}\|\|x\| \\ &\leq M\|x\|\end{aligned}$$

for any $x \in \mathbf{X}$. Therefore, $\mathbb{A}$ is a bounded operator. This completes the proof. □

Theorem 1.72. *Let $\mathbb{A}, \mathbb{B} : \mathbf{X} \longmapsto \mathbf{Y}$ be linear operators. Then*

$$\|\mathbb{A} + \mathbb{B}\| \leq \|\mathbb{A}\| + \|\mathbb{B}\| \quad \textit{and} \quad \|\lambda\mathbb{A}\| = |\lambda|\|\mathbb{A}\|$$

for any $\lambda \in \mathbf{F}$.

Proof. We have

$$\begin{aligned}\|\mathbb{A} + \mathbb{B}\| &= \sup_{x \in \mathbf{X}, \|x\| \leq 1} \|(\mathbb{A} + \mathbb{B})\, x\| \\ &= \sup_{x \in \mathbf{X}, \|x\| \leq 1} \|\mathbb{A}x + \mathbb{B}x\| \\ &\leq \sup_{x \in \mathbf{X}, \|x\| \leq 1} (\|\mathbb{A}x\| + \|\mathbb{B}x\|) \\ &\leq \sup_{x \in \mathbf{X}, \|x\| \leq 1} \|\mathbb{A}x\| + \sup_{x \in \mathbf{X}, \|x\| \leq 1} \|\mathbb{B}x\| \\ &= \|\mathbb{A}\| + \|\mathbb{B}\|,\end{aligned}$$

and

$$\begin{aligned}\|\lambda\mathbb{A}\| &= \sup_{x \in \mathbf{X}, \|x\| \leq 1} \|\lambda\mathbb{A}x\| \\ &= \sup_{x \in \mathbf{X}, \|x\| \leq 1} (|\lambda|\|\mathbb{A}x\|) \\ &= |\lambda| \sup_{x \in \mathbf{X}, \|x\| \leq 1} \|\mathbb{A}x\| \\ &= \lambda\|\mathbb{A}\|\end{aligned}$$

for any $\lambda \in \mathbf{F}$. This completes the proof. □

Definition 1.54. We say that a sequence $\{\mathbb{A}_n\}_{n \in \mathbb{N}}$ of elements of $\mathscr{L}(\mathbf{X}, \mathbf{Y})$ is uniformly convergent to $\mathbb{A} \in \mathscr{L}(\mathbf{X}, \mathbf{Y})$ if

$$\|\mathbb{A}_n - \mathbb{A}\| \to 0, \quad \text{as } n \to \infty.$$

We will write $\mathbb{A}_n \to \mathbb{A}$, as $n \to \infty$, or $\lim_{n \to \infty} \mathbb{A}_n = \mathbb{A}$.

Theorem 1.73. *Let* $\{\mathbb{A}_n\}_{n\in\mathbb{N}} \subseteq \mathscr{L}(\mathbf{X}, \mathbf{Y})$ *and* $\mathbb{A} \in \mathscr{L}(\mathbf{X}, \mathbf{Y})$. *Then* $\mathbb{A}_n \to \mathbb{A}$, *as* $n \to \infty$, *uniformly if and only if*

$$\|\mathbb{A}_n x - \mathbb{A}x\| \to 0, \quad \textit{as } n \to \infty,$$

for any $x \in \mathbf{X}$ *such that* $\|x\| \leq 1$.

Proof.

1. Let $\mathbb{A}_n \to \mathbb{A}$, as $n \to \infty$, uniformly. Then

$$\|\mathbb{A}_n - \mathbb{A}\| \to 0, \quad \text{as } n \to \infty.$$

Hence, for $x \in \mathbf{X}$, $\|x\| \leq 1$, we get

$$\begin{aligned} \|\mathbb{A}_n x - \mathbb{A}x\| &\leq \|\mathbb{A}_n - \mathbb{A}\|\|x\| \\ &\leq \|\mathbb{A}_n - \mathbb{A}\| \\ &\to 0, \quad \text{as } n \to \infty, \end{aligned}$$

for any $x \in \mathbf{X}$ such that $\|x\| \leq 1$.

2. Let $\|\mathbb{A}_n x - \mathbb{A}x\| \to 0$, as $n \to \infty$, for any $x \in \mathbf{X}$ such that $\|x\| \leq 1$. Hence,

$$\sup_{x\in\mathbf{X}, \|x\|\leq 1} \|\mathbb{A}_n x - \mathbb{A}x\| \leq \frac{\varepsilon}{2}$$

for any $n \geq N$. Therefore,

$$\|\mathbb{A}_n - \mathbb{A}\| \to 0, \quad \text{as } n \to \infty.$$

This completes the proof. □

Corollary 1.10. *Let* $\{\mathbb{A}_n\}_{n\in\mathbb{N}} \subseteq \mathscr{L}(\mathbf{X}, \mathbf{Y})$ *be such that* $\mathbb{A}_n \to \mathbb{A}$, *as* $n \to \infty$, *for* $\mathbb{A} \in \mathscr{L}(\mathbf{X}, \mathbf{Y})$. *Let also,* $\mathbf{U}$ *be an arbitrary bounded set in* $\mathbf{X}$. *Then*

$$\|\mathbb{A}_n x - \mathbb{A}x\| \to 0, \quad \textit{as } n \to \infty,$$

for any $x \in \mathbf{U}$.

Proof. Because $\mathbf{U}$ is a bounded set in $\mathbf{X}$, there is an $R > 0$ such that $\|x\| \leq R$ for any $x \in \mathbf{U}$. Let $\varepsilon > 0$ be arbitrarily chosen. Since $\mathbb{A}_n \to \mathbb{A}$, as $n \to \infty$, we have

$$\|\mathbb{A}_n - \mathbb{A}\| \to 0, \quad \text{as } n \to \infty.$$

Hence, for $x \in \mathbf{U}$, we have

$$\begin{aligned} \|\mathbb{A}_n x - \mathbb{A}x\| &\leq \|\mathbb{A}_n - \mathbb{A}\| \|x\| \\ &\leq R\|\mathbb{A}_n - \mathbb{A}\| \\ &\to 0, \quad \text{as } n \to \infty. \end{aligned}$$

This completes the proof. □

Definition 1.55. We say that a sequence $\{\mathbb{A}_n\}_{n\in\mathbb{N}}$ of elements of $\mathscr{L}(\mathbf{X}, \mathbf{Y})$ is a Cauchy sequence, if for any $\varepsilon > 0$ there is an $N \in \mathbb{N}$ such that

$$\|\mathbb{A}_{n+p} - \mathbb{A}_n\| < \varepsilon$$

for any $n, p > N$.

Theorem 1.74. *Let* $\mathbf{Y}$ *be a Banach space. Then* $\mathscr{L}(\mathbf{X}, \mathbf{Y})$ *is a Banach space.*

Proof. Let $\{\mathbb{A}_n\}_{n\in\mathbb{N}}$ be a Cauchy sequence of elements of $\mathscr{L}(\mathbf{X}, \mathbf{Y})$. We fix $\varepsilon > 0$. Then there is an $N \in \mathbb{N}$ such that

$$\|\mathbb{A}_{n+p} - \mathbb{A}_n\| < \varepsilon$$

for any $n, p \geq N$. Let $x \in \mathbf{X}$ be arbitrarily chosen. We have

$$\begin{aligned} \|\mathbb{A}_{n+p}x - \mathbb{A}x\| &= \|(\mathbb{A}_n - \mathbb{A})\, x\| \\ &\leq \|\mathbb{A}_n - \mathbb{A}\| \, \|x\| \end{aligned}$$

for any $n, p \geq N$. Therefore, $\{\mathbb{A}_n x\}_{n\in\mathbb{N}}$ is a Cauchy sequence in $\mathbf{Y}$. Since $\mathbf{Y}$ is a Banach space, we have that there exists $\lim_{n\to\infty} \mathbb{A}_n x$. We define the operator $\mathbb{A} : \mathbf{X} \to \mathbf{Y}$ as follows:

$$\mathbb{A}x = \lim_{n\to\infty} \mathbb{A}_n x, \quad x \text{ in } \mathbf{X}.$$

For any $\alpha, \beta \in \mathbf{F}$ and $x_1, x_2 \in \mathbf{X}$ we have

$$\begin{aligned}
\mathbb{A}(\alpha x_1 + \beta x_2) &= \lim_{n\to\infty} \mathbb{A}_n(\alpha x_1 + \beta x_2)\\
&= \lim_{n\to\infty} (\alpha \mathbb{A}_n x_1 + \beta \mathbb{A}_n x_2)\\
&= \alpha \lim_{n\to\infty} \mathbb{A}_n x_1 + \beta \lim_{n\to\infty} \mathbb{A}_n x_2\\
&= \alpha \mathbb{A} x_1 + \beta \mathbb{A} x_2.
\end{aligned}$$

Therefore, $\mathbb{A} : \mathbf{X} \longmapsto \mathbf{Y}$ is a linear operator. Note that

$$|\|\mathbb{A}_{n+p}\| - \|\mathbb{A}_n\|| \le \|\mathbb{A}_{n+p} - \mathbb{A}_n\|$$

for any $n, p \in \mathbb{N}$. Therefore, $\{\|\mathbb{A}_n\|\}_{n\in\mathbb{N}}$ is Cauchy sequence in $\mathbb{R}$. Hence, it is a bounded sequence in $\mathbb{R}$ and there exists a constant $c > 0$ such that

$$\|\mathbb{A}_n\| \le c$$

for any $n \in \mathbb{N}$. Hence,

$$\|\mathbb{A}_n x\| \le c\|x\|$$

for any $x \in \mathbf{X}$ and for any $n \in \mathbb{N}$. Consequently,

$$\|\mathbb{A} x\| \le c\|x\|$$

for any $x \in \mathbf{X}$. From here, $\mathbb{A} \in \mathscr{L}(\mathbf{X}, \mathbf{Y})$. This completes the proof. □

Definition 1.56. Let $\mathbb{A}_n \in \mathscr{L}(\mathbf{X}, \mathbf{Y})$, $n \in \mathbb{N}$.

1. The series $\sum_{n=1}^{\infty} \mathbb{A}_n$ is said to be uniformly convergent, if the sequence

$$\left\{ S_n = \sum_{k=1}^{n} \mathbb{A}_k \right\}_{n\in\mathbb{N}}$$

is uniformly convergent.
2. The series $\sum_{n=1}^{\infty} \mathbb{A}_n$ is said to be absolutely convergent if the series $\sum_{n=1}^{\infty} \|\mathbb{A}_n\|$ is convergent.

Theorem 1.75. *Let $\mathscr{L}(\mathbf{X}, \mathbf{Y})$ be a Banach space. If the series $\sum_{n=1}^{\infty} \mathbb{A}_n$ is an absolutely convergent series, then it is uniformly convergent.*

Proof. Since the series $\sum_{n=1}^{\infty} \mathbb{A}_n$ is absolutely convergent, we have that the series $\sum_{n=1}^{\infty} \|\mathbb{A}_n\|$ is convergent. We fix $\varepsilon > 0$. Then there is an $N \in \mathbb{N}$ such that

$$\sum_{n=N+1}^{N+p} \|\mathbb{A}_n\| < \varepsilon$$

for any $p \in \mathbb{N}$. Hence,

$$\begin{aligned}\|S_{N+p} - S_N\| &= \left\| \sum_{n=N+1}^{N+p} \mathbb{A}_n \right\| \\ &\leq \sum_{n=N+1}^{N+p} \|\mathbb{A}_n\| \\ &< \varepsilon\end{aligned}$$

for any $p \in \mathbb{N}$. Therefore, $\{S_n\}_{n\in\mathbb{N}}$ is a Cauchy sequence in $\mathscr{L}(\mathbf{X}, \mathbf{Y})$. Because $\mathscr{L}(\mathbf{X}, \mathbf{Y})$ is a Banach space, we have that $\{S_n\}_{n\in\mathbb{N}}$ is uniformly convergent. This completes the proof. □

Below with $\mathscr{L}(\mathbf{X})$ we will denote the space $\mathscr{L}(\mathbf{X}, \mathbf{X})$.

Exercise 1.24. Let $\mathbb{A}, \mathbb{B} \in \mathscr{L}(\mathbf{X})$. Prove that $\mathbb{A}\mathbb{B} \in \mathscr{L}(\mathbf{X})$ and $\mathbb{A}^k \in \mathscr{L}(\mathbf{X})$ for any $k \in \mathbb{N}$.

Exercise 1.25. For $\mathbb{A} \in \mathscr{L}(\mathbf{X})$ we define

$$e^{\mathbb{A}} = \sum_{k=1}^{\infty} \frac{\mathbb{A}^k}{k!}.$$

Prove that $e^{\mathbb{A}} \in \mathscr{L}(\mathbf{X})$ and $\left\|e^{\mathbb{A}}\right\| \leq e^{\|\mathbb{A}\|}$.

Theorem 1.76. *Let* $\{\mathbb{A}_n\}_{n\in\mathbb{N}}, \{\mathbb{B}_n\}_{n\in\mathbb{N}} \subset \mathscr{L}(\mathbf{X})$, $\mathbb{A}, \mathbb{B} \in \mathscr{L}(\mathbf{X})$. *If* $\mathbb{A}_n \to \mathbb{A}$, $\mathbb{B}_n \to \mathbb{B}$, *as* $n \to \infty$, *then* $\mathbb{A}_n\mathbb{B}_n \to \mathbb{A}\mathbb{B}$, *as* $n \to \infty$.

Proof. Because $\mathbb{B}_n \in \mathscr{L}(\mathbf{X})$, $n \in \mathbb{N}$, $\mathbb{B}_n \to \mathbb{B}$, as $n \to \infty$, we have that the sequence $\{\|\mathbb{B}_n\|\}_{n\in\mathbb{N}}$ is a bounded sequence. Therefore, there is a constant $M_1 > 0$ such that $\|\mathbb{B}_n\| \leq M_1$ for any $n \in \mathbb{N}$. Since $\mathbb{A} \in \mathscr{L}(\mathbf{X})$, then there is a constant $M_2 > 0$ such that $\|\mathbb{A}\| \leq M_2$.

We take $\varepsilon > 0$ arbitrarily. Since $\mathbb{A}_n \to \mathbb{A}$, $\mathbb{B}_n \to \mathbb{B}$, as $n \to \infty$, there is an $N \in \mathbb{N}$ such that

$$\|\mathbb{A}_n - \mathbb{A}\| < \frac{\varepsilon}{2M_1} \quad \text{and} \quad \|\mathbb{B}_n - \mathbb{B}\| < \frac{\varepsilon}{2M_2}$$

for any $n \geq N$. Hence,

$$\begin{aligned}\|\mathbb{A}_n\mathbb{B}_n - \mathbb{A}\mathbb{B}\| &= \|\mathbb{A}_n\mathbb{B}_n - \mathbb{A}\mathbb{B}_n + \mathbb{A}\mathbb{B}_n - \mathbb{A}\mathbb{B}\| \\ &\leq \|(\mathbb{A}_n - \mathbb{A})\,\mathbb{B}_n\| + \|\mathbb{A}\,(\mathbb{B}_n - \mathbb{B})\| \\ &\leq \|\mathbb{A}_n - \mathbb{A}\|\,\|\mathbb{B}_n\| + \|\mathbb{B}_n - \mathbb{B}\|\,\|\mathbb{A}\| \\ &< \frac{\varepsilon}{2M_1}M_1 + \frac{\varepsilon}{2M_2}M_2 \\ &= \frac{\varepsilon}{2} + \frac{\varepsilon}{2} \\ &= \varepsilon\end{aligned}$$

for any $n \geq N$. This completes the proof. □

Definition 1.57. We will say that the sequence $\{\mathbb{A}_n\}_{n\in\mathbb{N}} \subset \mathscr{L}(\mathbf{X}, \mathbf{Y})$ is strongly convergent to the operator $\mathbb{A} \in \mathscr{L}(\mathbf{X}, \mathbf{Y})$, if for any $x \in \mathbf{X}$ we have

$$\|\mathbb{A}_n x - \mathbb{A}x\| \to 0$$

as $n \to \infty$.

Theorem 1.77. *If $\{\mathbb{A}_n\}_{n\in\mathbb{N}} \subseteq \mathscr{L}(\mathbf{X}, \mathbf{Y})$ is uniformly convergent to $\mathbb{A} \in \mathscr{L}(\mathbf{X}, \mathbf{Y})$, then it is strongly convergent to $\mathbb{A}$.*

Proof. Take $\varepsilon > 0$ arbitrarily. Then there is an $N \in \mathbb{N}$ such that

$$\|\mathbb{A}_n - \mathbb{A}\| < \varepsilon$$

for any $n \geq N$. Hence,

$$\begin{aligned}\|\mathbb{A}_n x - \mathbb{A}x\| &\leq \|\mathbb{A}_n - \mathbb{A}\|\,\|x\| \\ &< \varepsilon\|x\|\end{aligned}$$

for any $n \geq N$ and for any $x \in \mathbf{X}$. This completes the proof. □

Theorem 1.78. *Let $\{\mathbb{A}_n\}_{n\in\mathbb{N}} \subset \mathscr{L}(\mathbf{X}, \mathbf{Y})$ and there exists $c > 0$ and a closed ball $S_r[x_0]$ such that $\|\mathbb{A}_n x\| \leq c$ for any $x \in S_r[x_0]$. Then the sequence $\{\|\mathbb{A}_n\|\}_{n\in\mathbb{N}}$ is bounded.*

Proof. Let $x \in \mathbf{X}$, $x \neq 0$. Then $x_0 + \frac{x}{\|x\|} r \in S_r[x_0]$. Hence,

$$\begin{aligned} c &\geq \left\| \mathbb{A}_n \left(r\frac{x}{\|x\|} + x_0 \right) \right\| \\ &= \left\| \frac{r}{\|x\|} \mathbb{A}_n x + \mathbb{A}_n x_0 \right\| \\ &\geq \left\| \frac{r}{\|x\|} \mathbb{A}_n x \right\| - \|\mathbb{A}_n x_0\| \\ &\geq \frac{r}{\|x\|} \|\mathbb{A}_n x\| - c, \end{aligned}$$

whereupon

$$\frac{r}{\|x\|} \|\mathbb{A}_n x\| \leq 2c \quad \text{or} \quad \frac{\|\mathbb{A}_n x\|}{\|x\|} \leq \frac{2c}{r}.$$

Therefore,

$$\|\mathbb{A}_n\| \leq \frac{2c}{r}.$$

This completes the proof. □

Theorem 1.79 (Uniform Boundedness Principle). *Let $\mathbf{X}$ be a Banach space and $\{\mathbb{A}_n\}_{n\in\mathbb{N}} \subset \mathscr{L}(\mathbf{X}, \mathbf{Y})$. If $\{\mathbb{A}_n x\}_{n\in\mathbb{N}}$ is bounded for any fixed $x \in \mathbf{X}$, then the sequence $\{\|\mathbb{A}_n\|\}_{n\in\mathbb{N}}$ is bounded.*

Proof. Assume the contrary. Suppose that there is a closed ball $S_r[x_0]$ such that the sequences $\{\|\mathbb{A}_n x\|\}_{n\in\mathbb{N}}$ is unbounded for some $x \in S_r[x_0]$. Then there is $x_1 \in S_r[x_0]$ and $n_1 \in \mathbb{N}$ such that $\|\mathbb{A}_{n_1} x_1\| > 1$. Since $\mathbb{A}_{n_1}$ is continuous, there is a closed ball $S_{r_1}[x_1] \subset S_r[x_0]$ such that $\|\mathbb{A}_{n_1} x\| > 1$ for any $x \in S_{r_1}[x_1]$. By Theorem 1.78, it follows that the sequences $\{\|\mathbb{A}_n x\|\}_{n\in\mathbb{N}}$ is unbounded for any $x \in S_{r_1}[x_1]$. Then there is $x_2 \in S_{r_1}[x_1]$ and $n_2 > n_1$, $n_2 \in \mathbb{N}$ such that $\|\mathbb{A}_{n_2} x_2\| > 2$. Since $\mathbb{A}_{n_2}$ is continuous, there is a closed ball

$S_{r_2}[x_2] \subset S_{r_1}[x_1]$ such that $\|\mathbb{A}_{n_2}x\| > 2$ for any $x \in S_{r_2}[x_2]$ and so on. In this way get the sequences $\{x_k\}_{k\in\mathbb{N}}$ and $\{S_{r_k}[x_k]\}_{k\in\mathbb{N}}$ such that

$$\cdots \subset S_{r_k}[x_k] \subset S_{r_{k-1}}[x_{k-1}] \subset \cdots \subset S_{r_1}[x_1]$$

and $\|\mathbb{A}_{n_k}x\| \geq k$ for any $k \in \mathbb{N}$ and for any $x \in S_{r_k}[x_k]$. Hence and Theorem 1.30, it follows that there is an $\overline{x} \in \mathbf{X}$ such that $\overline{x} \in S_{r_k}[x_k]$ for any $k \in \mathbb{N}$. Then $\|\mathbb{A}_n\overline{x}\| \geq k$ for any $k \in \mathbb{N}$, i.e., the sequence $\{\|\mathbb{A}_n\overline{x}\|\}_{k\in\mathbb{N}}$ is unbounded. This is a contradiction. Consequently, $\{\|\mathbb{A}_n\|\}_{n\in\mathbb{N}}$ is bounded. This completes the proof. □

Theorem 1.80 (Banach=Steinhaus's Theorem). *Let $\{\mathbb{A}_n\}_{n\in\mathbb{N}} \subseteq \mathscr{L}(\mathbf{X},\mathbf{Y})$ and $\mathbb{A} \in \mathscr{L}(\mathbf{X},\mathbf{Y})$. Then $\mathbb{A}_n \to \mathbb{A}$, as $n \to \infty$, strongly if and only if*

(H1) *the sequence $\{\|\mathbb{A}_n\|\}$ is bounded and*

$$\|\mathbb{A}_n x - \mathbb{A}x\| \to 0, \quad \text{as} \quad n \to \infty,$$

for any $x \in \mathbf{X}'$, where $\mathbf{X}'$ is dense in $\mathbf{X}$.

Proof.

1. Let $\mathbb{A}_n \to \mathbb{A}$, as $n \to \infty$, strongly. Then $\mathbb{A}_n x \to \mathbb{A}x$, as $n \to \infty$, for any $x \in \mathbf{X}$. Therefore $\{\|\mathbb{A}_n x\|\}_{n\in\mathbb{N}}$ is a bounded sequence for any $x \in \mathbf{X}$. Hence and Theorem 1.79, it follows that the sequence $\{\|\mathbb{A}_n\|\}_{n\in\mathbb{N}}$ is a bounded sequence. As $\mathbf{X}'$ we can take $\mathbf{X}$.
2. Suppose (**H1**) holds. We set

$$c = \sup_{n\in\mathbb{N}_0} \|\mathbb{A}_n\|,$$

where $\mathbb{A}_0 = \mathbb{A}$. We take $x \in \mathbf{X}$, $x \notin \mathbf{X}'$. Let $\varepsilon > 0$ be arbitrarily chosen. Because $\mathbf{X}'$ is dense in $\mathbf{X}$, there is an element $x_1 \in \mathbf{X}'$ such that $\|x - x_1\| < \frac{\varepsilon}{3c}$. Also, there is an $N \in \mathbb{N}$ such that

$$\|\mathbb{A}_n x_1 - \mathbb{A}x_1\| < \frac{\varepsilon}{3}$$

for any $n \geq N$. From here,

$$\begin{aligned}\|\mathbb{A}_n x - \mathbb{A}x\| &= \|\mathbb{A}_n(x - x_1) + (\mathbb{A}_n x_1 - \mathbb{A}x_1) + \mathbb{A}(x_1 - x)\| \\ &\leq \|\mathbb{A}_n(x - x_1)\| + \|\mathbb{A}_n x_1 - \mathbb{A}x_1\| + \|\mathbb{A}(x - x_1)\|\end{aligned}$$

$$
\begin{aligned}
&\leq \|\mathbb{A}_n\| \|x - x_1\| + \|\mathbb{A}_n x_1 - \mathbb{A}x_1\| + \|\mathbb{A}\| \|x - x_1\| \\
&\leq 2c \|x - x_1\| + \|\mathbb{A}_n x_1 - \mathbb{A}x_1\| \\
&< 2c\frac{\varepsilon}{3c} + \frac{\varepsilon}{3} \\
&= \varepsilon
\end{aligned}
$$

for any $n \geq N$. This completes the proof. □

Now, we suppose that $\mathbb{A} : \mathbf{D}(\mathbf{A}) \longmapsto \mathbf{Y}$ is a linear operator, where $\mathbf{D}(\mathbf{A}) \subseteq \mathbf{X}$. For $\mathbb{A}$ we can define its norm in the following way:

$$
\|\mathbb{A}\| = \sup_{\substack{x \in \mathbf{D}(\mathbf{A}) \\ \|x\| \leq 1}} \|\mathbb{A}x\|
$$

and we will say that $\mathbb{A}$ is bounded if $\|\mathbb{A}\| < \infty$.

Theorem 1.81. *Let $\mathbf{Y}$ be a Banach space and $\mathbb{A} : \mathbf{D}(\mathbf{A}) \longmapsto \mathbf{Y}$ be a linear bounded operator, where $\mathbf{D}(\mathbf{A}) \subseteq \mathbf{X}$ and $\mathbf{D}(\mathbb{A})$ is dense in $\mathbf{X}$. Then there exists a linear bounded operator $\widetilde{\mathbb{A}} : \mathbf{X} \longmapsto \mathbf{Y}$ such that $\widetilde{\mathbb{A}}x = \mathbb{A}x$ for any $x \in \mathbf{D}(\mathbf{A})$ and $\|\mathbb{A} = \|\widetilde{\mathbb{A}}\|$.*

Proof. For $x \in \mathbf{D}(\mathbf{A})$ we define $\widetilde{\mathbb{A}}x = \mathbb{A}x$. Let $x \notin \mathbf{D}(\mathbf{A})$ and $x \in \mathbf{X}$. Since $\mathbf{D}(\mathbf{A})$ is dense in $\mathbf{X}$, there exists a sequence $\{x_n\}_{n\in\mathbb{N}}$ of elements of $\mathbf{D}(\mathbf{A})$ such that $x_n \to x$, as $n \to \infty$. Then we set

$$
\widetilde{\mathbb{A}}x = \lim_{n\to\infty} \mathbb{A}x_n. \tag{1.70}
$$

We will prove that $\widetilde{\mathbb{A}}$ is well-defined, i.e., we will prove that the limit (1.70) exists and it does not depend on the choice of the sequence $\{x_n\}_{n\in\mathbb{N}}$. Note that

$$
\|\mathbb{A}x_n - \mathbb{A}x_m\| \leq \|\mathbb{A}\| \|x_n - x_m\| \to 0 \quad \text{as } m, n \to \infty.
$$

Therefore, $\{\mathbb{A}x_n\}_{n\in\mathbb{N}}$ is a Cauchy sequence in $\mathbf{Y}$. Since $\mathbf{Y}$ is a Banach space, it is convergent. Consequently (1.70) exists. Suppose that

$\{x_n\}_{n\in\mathbb{N}}$ and $\{x'_n\}_{n\in\mathbb{N}}$ are two sequences of elements of $\mathbf{X}$ such that $x_n \to x$, $x'_n \to x$, as $n \to \infty$. Let

$$\alpha = \lim_{n\to\infty} \mathbb{A}x_n,$$
$$\beta = \lim_{n\to\infty} \mathbb{A}x'_n.$$

Then

$$\begin{aligned}\|\alpha-\beta\| &= \left\|\alpha - \mathbb{A}x_n + \mathbb{A}x_n - \mathbb{A}x'_n + \mathbb{A}x'_n - \beta\right\| \\ &\le \|\alpha - \mathbb{A}x_n\| + \left\|\mathbb{A}x_n - \mathbb{A}x'_n\right\| + \left\|\beta - \mathbb{A}x'_n\right\| \\ &\le \|\mathbb{A}x_n - \alpha\| + \|\mathbb{A}\| \left\|x_n - x'_n\right\| + \left\|\mathbb{A}x'_n - \beta\right\| \\ &\to 0, \quad \text{as } n \to \infty,\end{aligned}$$

i.e., $\alpha = \beta$. Next,

$$\|\mathbb{A}x_n\| \le \|\mathbb{A}\| \|x_n\|.$$

Hence,

$$\lim_{n\to\infty} \|\mathbb{A}x_n\| \le \lim_{n\to\infty} \|\mathbb{A}\| \|x_n\|,$$

whereupon

$$\left\|\widetilde{\mathbb{A}}x\right\| \le \|\mathbb{A}\| \|x\|.$$

Consequently,

$$\left\|\widetilde{\mathbb{A}}\right\| \le \|\mathbb{A}\|. \tag{1.71}$$

On the other hand,

$$\begin{aligned}\left\|\widetilde{\mathbb{A}}\right\| &= \sup_{\substack{\|x\|\le 1,\\ x\in\mathbf{X}}} \left\|\widetilde{\mathbb{A}}x\right\| \\ &\ge \sup_{\substack{\|x\|\le 1,\\ x\in\mathbf{D}(\mathbf{A})}} \left\|\widetilde{\mathbb{A}}x\right\| \\ &= \sup_{\substack{\|x\|\le 1,\\ x\in\mathbf{D}(\mathbf{A})}} \|\mathbb{A}x\| \\ &= \|\mathbb{A}\|.\end{aligned}$$

From the last inequality and from (1.71), we get that $\|\mathbb{A}\| = \|\widetilde{\mathbb{A}}\|$. This completes the proof. □

1.11 The Banach Fixed Point Theorem

Banach's fixed-point theorem (also known as the contracting-application theorem) is a simple-to-prove theorem and applies to complete spaces and has many applications. These applications include solution existence theorems for differential equations or integral equations and the study of the convergence of certain methods, like Newton's in solving nonlinear equations.

Suppose that $\mathbf{X}$ is a metric space with a metric d. Let also, $\mathbf{A} \subseteq \mathbf{X}$.

Definition 1.58. We say that a map $\mathbb{T} : \mathbf{A} \longmapsto \mathbf{X}$ is a contraction map if there exists $k \in (0, 1)$ such that

$$d(\mathbb{T}x_1, \mathbb{T}x_2) \le kd(x_1, x_2)$$

for any $x_1, x_2 \in \mathbf{X}$.

Exercise 1.26. Let $\mathbb{T} : \mathbf{A} \longmapsto \mathbf{X}$ is a contraction map. Prove that it is continuous on $\mathbf{A}$.

Definition 1.59. We say that a map $\mathbb{T} : \mathbf{A} \longmapsto \mathbf{X}$ has a fixed point $x_0 \in \mathbf{A}$ if

$$\mathbb{T}x_0 = x_0.$$

Theorem 1.82. *Let $\mathbf{X}$ be a complete metric space and*

$$\mathbb{T} : X \longmapsto X$$

be a contraction map,

$$d(\mathbb{T}x_1, \mathbb{T}x_2) \le kd(x_1, x_2)$$

for any $x_1, x_2 \in \mathbf{X}$ and for some constant $k \in (0, 1)$. Then $\mathbb{T}$ has unique fixed point in $\mathbf{X}$.

Proof.

1. Firstly, we will prove the uniqueness of the fixed point of the map $\mathbb{T}$. Let x_1, x_2 be two fixed points of $\mathbb{T}$. Then $\mathbb{T}(x_1) = x_1$, $\mathbb{T}(x_2) = x_2$ and
$$d(\mathbb{T}x_1, \mathbb{T}x_2) \leq kd(x_1, x_2),$$
whereupon
$$d(x_1, x_2) \leq kd(x_1, x_2).$$
Hence, $d(x_1, x_2) = 0$. Therefore $x_1 = x_2$.
2. Now, we will prove existence of a fixed point of the map $\mathbb{T}$. Let $x_0 \in X$ be arbitrarily chosen. Define the sequence $\{x_n\}_{n\in\mathbb{N}}$ as follows
$$x_{n+1} = \mathbb{T}x_n, \quad n \in \mathbb{N}_0,$$
where $\mathbb{N}_0 = \mathbb{N} \cup \{0\}$. We will prove that it is a Cauchy sequence in $\mathbf{X}$. Let $p, q \in \mathbb{N}$ be such that $q > p$. Then
$$d(x_p, x_q) \leq d(x_p, x_{p+1}) + \cdots + d(x_{q-1}, x_q).$$
Thus,
$$\begin{aligned} d(x_p, x_{p+1}) &= d(f(x_{p-1}), f(x_p)) \\ &\leq kd(x_{p-1}, x_p) \\ &\leq k^2 d(x_{p-2}, x_{p-1}) \\ &\leq \cdots \\ &\leq k^p d(x_0, x_1). \end{aligned}$$
Then
$$\begin{aligned} d(x_p, x_q) &\leq (k^p + k^{p+1} + \cdots + k^{q-1}) d(x_0, x_1) \\ &= k^p d(x_0, x_1) \sum_{k=0}^{q-1} k^p \\ &< \frac{k^p}{1-k} d(x_0, x_1). \end{aligned}$$
If $d(x_0, x_1) = 0$, then $x_0 = x_1 = \mathbb{T}(x_0)$ and x_0 is a fixed point of $\mathbb{T}$. If $d(x_0, x_1) \neq 0$, then for any $\varepsilon > 0$ there exists $n_0 \in \mathbb{N}$ such

that

$$\frac{k^p}{1-k}d(x_0, x_1) < \varepsilon.$$

Therefore,

$$d(x_p, x_q) < \varepsilon d(x_0, x_1)$$

for any $p, q \in \mathbb{N}$, $q > p$. Therefore $\{x_n\}_{n\in\mathbb{N}}$ is a Chauchy sequence in $\mathbf{X}$. Because $\mathbf{X}$ is a complete metric space, we conclude that the sequence $\{x_n\}_{n\in\mathbb{N}}$ is convergent to $x \in \mathbf{X}$. Therefore

$$x_n \to x, \quad \text{as } n \to \infty,$$

and

$$\mathbb{T}x_n \to \mathbb{T}x, \quad \text{as } n \to \infty.$$

Hence,

$$\mathbb{T}(x) = x.$$

This completes the proof. □

1.12 Advanced Practical Problems

Problem 1.1. With $\mathbf{C}^k[a,b]$, $k \in \mathbf{N}_0$, $a, b \in \mathbf{R}$, $a < b$, we will denote the set of all k-times continuously-differentiable functions defined on $[a,b]$. Prove that $\mathbf{C}^k[a,b]$ is a vector space.

Problem 1.2. Let $\mathbf{D}[a,b]$ be the set of all differential operators

$$P(D) = \sum_{k=0}^{m} p_k(x)\frac{d^k}{dx^k}, \quad Q(D) = \sum_{k=0}^{m} q_k(x)\frac{d^k}{dx^k},$$

where $x \in [a,b]$, p_k, $q_k \in \mathbf{C}[a,b]$, $k \in \{1,\ldots,m\}$, $\frac{d^0}{dx^0} = 1$, with operations

$$P(D) + Q(D) = \sum_{k=0}^{m} (p_k(x) + q_k(x))\frac{d^k}{dx^k},$$

$$\alpha P(D) = \sum_{k=0}^{m} (\alpha p_k(x))\frac{d^k}{dx^k}, \quad \alpha \in \mathbf{F}.$$

Prove that $\mathbf{D}[a,b]$ is a vector space.

Problem 1.3. Consider the set $E[a,b]$. Prove that

1. xe^x, $x^2e^{x^2}$ and $x^3e^{x^3}$ are linearly independent.
2. 0, $\sin x$, $\cos x$ and $\sin x + \cos x$ are linearly dependent.

Problem 1.4. Suppose that $\mathbf{A}$, $\mathbf{B}$ and $\mathbf{C}$ are linear subspaces of a vector space $\mathbf{E}$. Prove that

$$\mathbf{A} \cap (\mathbf{B} + (\mathbf{A} \cap \mathbf{C})) = (\mathbf{A} \cap \mathbf{B}) + (\mathbf{A} \cap \mathbf{C}).$$

Problem 1.5. Prove that

$$d(x,y) = \sum_{i=1}^{n} |x_i - y_i|, \quad x = (x_1, \ldots, x_n), \quad y = (y_1, \ldots, y_n) \in \mathbf{E}_n,$$

is a metric on $\mathbf{E}_n$.

Problem 1.6. We say that a function $f : [a,b] \longmapsto \mathbf{R}$ is a function of bounded variation if there exists a positive constant c such that for every partition $P = \{a = t_1 < \cdots < t_{n_p} = b\}$ of the interval $[a,b]$ we have

$$\sum_{l=1}^{n_p} |f(t_l) - f(t_{l-1})| < c.$$

Let $\mathscr{P}$ be the set of all partitions P of the interval $[a,b]$. If $f : [a,b] \longmapsto \mathbf{R}$ is a function of bounded variation, then its total variation is defined by

$$\bigvee_a^b f = \sup_{P \in \mathscr{P}} \sum_{l=1}^{n_p} |f(t_l) - f(t_{l-1})|.$$

The set of all functions of bounded variation on $[a,b]$ is denoted by $\mathbf{V}([a,b])$. In $\mathbf{V}([a,b])$ we define a metric by

$$d(f,g) = |f(a) - g(a)| + \bigvee_a^b (f - g), \quad f, g \in \mathbf{V}([a,b]). \tag{1.72}$$

Prove that (1.72) satisfies all axioms for a metric.

Problem 1.7. Check that

$$\|f\| = |f(a)| + |f(b)| + \max_{t\in[a,b]} |f''(t)|, \quad f \in \mathbf{C}^2([a,b]),$$

satisfies all axioms for a norm.

Problem 1.8. Check if

$$\|f\| = \max_{t\in[a,b]} |f(t)| + |f(b) - f(a)|$$

satisfies all axioms for a norm in $\mathbf{C}^1([a,b])$.

Answer. Yes.

Problem 1.9. Prove that the sequence

$$\left\{ \frac{x^{n+1}}{n+1} - \frac{x^{n+2}}{n+2} \right\}$$

is convergent in $\mathbf{C}([0,1])$ and $\mathbf{C}^1([0,1])$.

Problem 1.10. Let $x_1, \ldots, x_n$ be an orthogonal system in the Hilbert space $\mathbf{H}$. Prove that

$$\left\| \sum_{k=1}^{n} x_k \right\|^2 = \sum_{k=1}^{n} \|x_k\|^2.$$

Problem 1.11. Let $\mathbf{X} = \mathbf{Y} = \mathbf{C}([0,1])$, For $x \in \mathbf{X}$ we define the operator

$$y(t) = t \int_0^1 x(s)\, ds, \quad t \in [0,1], \quad y = \mathbb{A}x.$$

Prove that $\mathbb{A} : \mathbf{X} \longmapsto \mathbf{X}$ is a linear operator.

Problem 1.12. Let $\mathbf{X} = \mathbf{Y} = \mathbf{C}([0,1])$, $\|x\| = \max_{t\in[0,1]} |x(t)|$, $x \in \mathbf{C}([0,1])$. Find $\|\mathbb{A}\|$, where

$$\mathbb{A}x(t) = t^7 \int_0^t s^4 x(s) ds, \quad x \in \mathbf{X}, \quad t \in [0,1].$$

Answer. $\frac{1}{5}$.

Problem 1.13. Let $\mathbf{X}$ be a normed vector space, $\mathbf{Y}$ and $\mathbf{Z}$ be Banach spaces. Suppose that $\mathbf{B} \in \mathscr{L}(\mathbf{X}, \mathbf{Y})$, $\mathbb{A} \in \mathscr{L}(\mathbf{Y}, \mathbf{Z})$. Prove that

$$\|\mathbb{A}\mathbb{B}\| \leq \|\mathbb{A}\| \, \|\mathbb{B}\| .$$

Problem 1.14. Let $\mathbf{X}$, $\mathbf{Y}$ and $\mathbf{Z}$ be normed vector spaces. Suppose that $\{\mathbb{B}_n\}_{n\in\mathbf{N}} \subset \mathscr{L}(\mathbf{X}, \mathbf{Y})$, $\mathbb{B} \in \mathscr{L}(\mathbf{X}, \mathbf{Y})$, $\{\mathbb{A}_n\}_{n\in\mathbf{N}} \subset \mathscr{L}(\mathbf{Y}, \mathbf{Z})$, $\mathbb{A} \in \mathscr{L}(\mathbf{Y}, \mathbf{Z})$. Suppose that $\mathbb{A}_n \to \mathbb{A}$, $\mathbb{B}_n \to \mathbb{B}$, as $n \to \infty$. Prove that $\mathbb{A}_n\mathbb{B}_n \to \mathbb{A}\mathbb{B}$, as $n \to \infty$.

Problem 1.15. Let $\mathbf{X}$ be a normed vector space and $\mathbb{A} \in \mathscr{L}(\mathbf{X})$. Define

$$\sin(\mathbb{A}) = \sum_{k=0}^{\infty} \frac{(-1)^k \mathbb{A}^{2k+1}}{(2k+1)!},$$

$$\cos(\mathbb{A}) = \sum_{k=0}^{\infty} \frac{(-1)^k \mathbb{A}^{2k}}{(2k)!},$$

$$\sinh(\mathbb{A}) = \sum_{k=0}^{\infty} \frac{\mathbb{A}^{2k+1}}{(2k+1)!},$$

$$\cosh(\mathbb{A}) = \sum_{k=0}^{\infty} \frac{\mathbb{A}^{2k}}{(2k)!}.$$

Prove that

$$e^{\mathbb{A}} = \sinh(\mathbb{A}) + \cosh(\mathbb{A}),$$

$$\|\sinh(\mathbb{A})\| \leq \sinh(\|\mathbb{A}\|),$$

$$\|\cosh(\mathbb{A})\| \leq \cosh(\|\mathbb{A}\|),$$

$$\|\sin(\mathbb{A})\| \leq \sinh(\|\mathbb{A}\|),$$

$$\|\cos(\mathbb{A})\| \leq \cosh(\|\mathbb{A}\|).$$

Chapter 2

Lebesgue Integration

In this chapter, we introduce Lebesgue measure and we deduct some of its properties.

2.1 Lebesgue Outer Measure: Measurable Sets

Definition 2.1. Let $\mathbf{I}$ be a nonempty interval of real numbers. We define its length $l(\mathbf{I})$ to be ∞ if $\mathbf{I}$ is unbounded and if $\mathbf{I}$ is bounded, we define its length to be the difference of its end points.

Below with $\{I_k\}_{k\in\mathbf{N}}$, we will denote a countable collection of nonempty open, bounded intervals.

Definition 2.2. Let $\mathbf{A}$ be a set of real numbers. Then we define the outer measure of $\mathbf{A}$ as follows:

$$m^\star(A) = \inf_{\substack{\{\mathbf{I}_k\}_{k\in\mathbf{N}} \\ \mathbf{A}\subseteq \bigcup_{k=1}^{\infty} \mathbf{I}_k}} \left(\sum_{k=1}^{\infty} l(\mathbf{I}_k)\right).$$

By the definition, it follows that $m^\star(\emptyset) = 0$.

Theorem 2.1. *If $A \subseteq B$, then $m^*(A) \leq m^*(B)$.*

Proof. We have

$$\begin{aligned} m^\star(A) &= \inf_{\substack{\{\mathbf{I}_k\}_{k\in\mathbf{N}} \\ \mathbf{A}\subseteq \bigcup_{k=1}^{\infty} \mathbf{I}_k}} \left(\sum_{k=1}^{\infty} l(\mathbf{I}_k)\right) \\ &\leq \inf_{\substack{\{\mathbf{V}_k\}_{k\in\mathbf{N}} \\ \mathbf{A}\subseteq B\subseteq \bigcup_{k=1}^{\infty} \mathbf{V}_k}} \left(\sum_{k=1}^{\infty} l(\mathbf{V}_k)\right) \\ &= m^*(B). \end{aligned}$$

This completes the proof. □

Theorem 2.2. *The outer measure of an interval is its length.*

Proof.

1. Let $\mathbf{I} = [a, b]$. Let also, $\varepsilon > 0$ be arbitrarily chosen. Since

$$[a, b] \subset (a - \varepsilon, b + \varepsilon),$$

we have

$$\begin{aligned} m^\star([a, b]) &\leq l((a - \varepsilon, b + \varepsilon)) \\ &= b - a + 2\varepsilon. \end{aligned}$$

Because $\varepsilon > 0$ was arbitrarily chosen, we conclude that

$$m^\star([a, b]) \leq b - a. \tag{2.1}$$

Let $\{\mathbf{V}_k\}_{k\in\mathbf{N}}$ be a collection of open intervals covering the interval $[a, b]$. By the Heine–Borel theorem, it follows that there exists a finite subcollection $\{\mathbf{I}_k\}_{k=1}^n$ of the collection $\{\mathbf{V}_k\}_{k\in\mathbf{N}}$ that covers $[a, b]$. There exists $l \in \{1, \ldots, n\}$ such that $a \in \mathbf{I}_l$. Let $\mathbf{I}_l = (a_1, b_1)$. We have that $a_1 < a < b_1$. If $b_1 > b$, then

$$b - a \leq \sum_{k=1}^{n} l(\mathbf{I}_k). \tag{2.2}$$

Let $b_1 \in [a, b)$. Then there exists an interval (a_2, b_2) such that $b_1 \in (a_2, b_2)$. If $b_2 > b$, then (2.1) holds. If $b_2 < b$ we continue this process while it terminates. Then we obtain a subcollection $\{(a_k, b_k)\}_{k=1}^{N}$ of the collection $\{I_k\}_{k=1}^{n}$. We have

$$a_{k+1} < b_k \quad \text{for } 1 \le k \le N-1$$

and since this process is terminated, we have $b_N > b$. Therefore, (2.1) holds. By (2.1) and (2.2), we get that

$$m^\star([a,b]) = b - a.$$

2. Let $\mathbf{I}$ is any bounded interval. Then for any $\varepsilon > 0$ there exist two closed bounded intervals $\mathbf{J}_1$ and $\mathbf{J}_2$ such that

$$\mathbf{J}_1 \le \mathbf{I} \le \mathbf{J}_2$$

and

$$\begin{aligned} l(\mathbf{I}) - \varepsilon &\le l(\mathbf{J}_1) \\ &\le l(\mathbf{J}_2) \\ &\le l(\mathbf{I}) + \varepsilon. \end{aligned}$$

Hence,

$$\begin{aligned} l(\mathbf{I}) - \varepsilon &\le m^\star(\mathbf{J}_1) \\ &\le m^*(\mathbf{I}) \\ &\le m^\star(\mathbf{J}_2) \\ &\le l(\mathbf{I}) + \varepsilon. \end{aligned}$$

Because $\varepsilon > 0$ was arbitrarily chosen, we get that

$$m^*(\mathbf{I}) = l(\mathbf{I}).$$

3. If $\mathbf{I}$ is unbounded interval, then for any $n \in \mathbf{N}$ there exists an interval $\mathbf{J} \subseteq \mathbf{I}$ such that $l(\mathbf{J}) = n$. Hence,

$$m^\star(\mathbf{I}) \ge m^\star(\mathbf{J}) = l(\mathbf{J}) = n$$

for any $n \in \mathbf{N}$. Therefore, $m^\star(\mathbf{I}) = \infty$.

This completes the proof. □

Theorem 2.3. *$m^\star(\{y\}) = 0$ for any $y \in \mathbf{R}$.*

Proof. Let $\varepsilon > 0$ be arbitrarily chosen and fixed. Then

$$\{y\} \subset (y - \varepsilon, y + \varepsilon).$$

Hence,

$$m^\star(\{y\}) \leq 2\varepsilon.$$

Because $\varepsilon > 0$ was arbitrarily chosen, we conclude that $m^\star(\{y\}) = 0$. This completes the proof. □

Theorem 2.4. *For any set $\mathbf{A}$ and for any $y \in \mathbf{R}$ we have*

$$m^\star(\mathbf{A} + y) = m^\star(\mathbf{A}).$$

Proof. Note that any countable collection $\{I_k\}_{k\in\mathbf{N}}$ of open, bounded intervals covers $\mathbf{A}$ if and only if $\{I_k + y\}_{k\in\mathbf{N}}$ covers $\mathbf{A} + y$. Also, $m^\star(I_k + y) = m^\star(I_k)$ for any $k \in \mathbf{N}$. Therefore,

$$\begin{aligned} m^\star(\mathbf{A} + y) &= \inf_{\substack{\{\mathbf{I}_k + y\}_{k\in\mathbf{N}} \\ \mathbf{A}+y \subseteq \bigcup_{k=1}^{\infty} \mathbf{I}_k + y}} \left(\sum_{k=1}^{\infty} l(\mathbf{I}_k + y) \right) \\ &= \inf_{\substack{\{\mathbf{I}_k\}_{k\in\mathbf{N}} \\ \mathbf{A} \subseteq \bigcup_{k=1}^{\infty} \mathbf{I}_k}} \left(\sum_{k=1}^{\infty} l(\mathbf{I}_k) \right) \\ &= m^\star(\mathbf{A}). \end{aligned}$$

This completes the proof. □

Theorem 2.5. *The outer measure is countable subadditive, i.e, if $\{\mathbf{E}_k\}_{k\in\mathbf{N}}$ is any countable collection of sets, disjoint or not, then*

$$m^\star\left(\bigcup_{k\in\mathbf{N}} \mathbf{E}_k \right) \leq \sum_{k=1}^{\infty} m^\star(\mathbf{E}_k).$$

Proof. If there is a $k \in \mathbf{N}$ such that $m^\star(\mathbf{E}_k) = \infty$, then the assertion is evident. Suppose that $m^\star(\mathbf{E}_k) < \infty$ for any $k \in \mathbf{N}$. We take $\varepsilon > 0$ arbitrarily. Then for any $k \in \mathbf{N}$ there exists a countable collection $\{\mathbf{I}_{k,m}\}_{m\in\mathbf{N}}$ of open, bounded intervals, such that

$$\sum_{m=1}^{\infty} l(\mathbf{I}_{k,m}) \le m^\star(\mathbf{E}_k) + \frac{\varepsilon}{2^k}.$$

Note that $\{\mathbf{I}_{k,m}\}_{k,m\in\mathbf{N}}$ is a countable collection of open, bounded intervals for which

$$\bigcup_{k\in\mathbf{N}} \mathbf{E}_k \subseteq \bigcup_{k,m\in\mathbf{N}} \mathbf{I}_{k,m}.$$

Hence and Definition 2.2, it follows that:

$$\begin{aligned} m^\star\left(\bigcup_{k\in\mathbf{N}} \mathbf{E}_k\right) &\le \sum_{k,m=1}^{\infty} l\left(\mathbf{I}_{k,m}\right) \\ &= \sum_{k=1}^{\infty}\left(\sum_{m=1}^{\infty} l\left(\mathbf{I}_{k,m}\right)\right) \\ &\le \sum_{k=1}^{\infty}\left(m^\star\left(\mathbf{E}_k\right) + \frac{\varepsilon}{2^k}\right) \\ &= \sum_{k=1}^{\infty} m^\star\left(\mathbf{E}_k\right) + \sum_{k=1}^{\infty}\frac{\varepsilon}{2^k} \\ &= \sum_{k=1}^{\infty} m^\star\left(\mathbf{E}_k\right) + \varepsilon. \end{aligned}$$

Because $\varepsilon > 0$ was arbitrarily chosen, we get to the desired result. This completes the proof. □

Corollary 2.1. *If $\{\mathbf{E}_k\}_{k=1}^m$ is any finite collection of sets, disjoint or not, then*

$$m^\star\left(\bigcup_{k=1}^{m} \mathbf{E}_k\right) \le \sum_{k=1}^{m} m^\star\left(\mathbf{E}_k\right).$$

Proof. We take $\mathbf{E}_k = \emptyset$ for $k > m$ and we apply Theorem 2.5. This completes the proof. □

Corollary 2.2. *The interval* $[0,1]$ *is not countable.*

Proof. Assume that $[0,1]$ is countable. Then there exists a sequence $\{y_k\}_{k\in\mathbf{N}}$ of elements of $[0,1]$ such that

$$[0,1] = \bigcup_{k\in\mathbf{N}} \{y_k\}.$$

Hence, using Theorems 2.3 and 2.5, it follows that:

$$\begin{aligned} 1 &= m^\star\left([0,1]\right) \\ &= m^\star\left(\bigcup_{k\in\mathbf{N}} \{y_k\}\right) \\ &\le \sum_{k=1}^{\infty} m^\star\left(\{y_k\}\right) \\ &= 0, \end{aligned}$$

which is a contradiction. □

Corollary 2.3. *The set of irrational numbers in the interval* $[0,1]$ *has outer measure* 1.

Proof. Let $\mathbf{A}$ be the set of irrational numbers in $[0,1]$ and B be the set of rational numbers in $[0,1]$. Then

$$[0,1] = \mathbf{A} \cup \mathbf{B}. \tag{2.3}$$

Since $\mathbf{A} \subset [0,1]$, using Theorem 2.1, we get

$$\begin{aligned} m^\star\left(\mathbf{A}\right) &\le m^\star\left([0,1]\right) \\ &= 1. \end{aligned} \tag{2.4}$$

On the other hand, using Theorem 2.5, we obtain

$$\begin{aligned} 1 &= m^\star\left([0,1]\right) \\ &\le m^\star\left(\mathbf{A}\right) + m^\star\left(\mathbf{B}\right). \end{aligned} \tag{2.5}$$

Because $\mathbf{B}$ is countable, there exists a sequence $\{y_k\}_{k\in\mathbf{N}}$ of elements of $\mathbf{B}$ such that

$$\mathbf{B} = \bigcup_{k=1}^{\infty}\{y_k\}.$$

Then, using Theorems 2.3 and 2.5, we obtain

$$\begin{aligned} m^\star(\mathbf{B}) &= m^\star\left(\bigcup_{k=1}^{\infty}\{y_k\}\right) \\ &\leq \sum_{k=1}^{\infty} m^\star(\{y_k\}) \\ &= 0. \end{aligned}$$

Hence and (2.5), we find

$$1 \leq m^\star(\mathbf{A}).$$

By the last inequality and (2.4), we obtain

$$m^\star(\mathbf{A}) = 1.$$

This completes the proof. □

Theorem 2.6. *Let* $m^\star(\mathbf{A}) = 0$. *Then*

$$m^\star(\mathbf{A}\cup\mathbf{B}) = m^\star(\mathbf{B})$$

for any set $\mathbf{B}$.

Proof. Since $\mathbf{B} \subseteq \mathbf{A}\cup\mathbf{B}$, we have

$$m^\star(\mathbf{B}) \leq m^\star(\mathbf{A}\cup\mathbf{B}). \tag{2.6}$$

On the other hand,

$$\begin{aligned} m^\star(\mathbf{A}\cup\mathbf{B}) &\leq m^\star(\mathbf{A}) + m^\star(\mathbf{B}) \\ &= m^\star(\mathbf{B}). \end{aligned}$$

Hence and (2.6), we conclude the desired result. This completes the proof. □

Definition 2.3. Let $\mathbf{E}$ be a subset of $\mathbf{R}$. With $\mathbf{E}^c$ we will denote the set

$$\mathbf{E}^c = \{x \in \mathbf{R} : x \notin \mathbf{E}\}.$$

The set $\mathbf{E}^c$ will be called the complement of the set $\mathbf{E}$ in $\mathbf{R}$.

Definition 2.4. A set $\mathbf{E}$ is said to be measurable provided for any set $\mathbf{A}$,

$$m^\star(\mathbf{A}) = m^\star(\mathbf{A} \cap \mathbf{E}) + m^\star(\mathbf{A} \cap \mathbf{E}^c).$$

Lemma 2.1. *For any sets* $\mathbf{A}$ *and* $\mathbf{E}$ *we have*

$$\mathbf{A} = (\mathbf{A} \cap \mathbf{E}) \cup (\mathbf{A} \cap \mathbf{E}^c).$$

Proof. Let $x \in \mathbf{A}$ be arbitrarily chosen and fixed.

1. If $x \in \mathbf{E}$, then

$$x \in \mathbf{A} \cap \mathbf{E}$$

and

$$x \in (\mathbf{A} \cap \mathbf{E}) \cup (\mathbf{A} \cap \mathbf{E}^c).$$

2. If $x \notin \mathbf{E}$, then $x \in \mathbf{E}^c$. Hence,

$$x \in \mathbf{A} \cap \mathbf{E}^c.$$

Therefore,

$$x \in (\mathbf{A} \cap \mathbf{E}) \cup (\mathbf{A} \cap \mathbf{E}^c).$$

Because $x \in \mathbf{A}$ was arbitrarily chosen and for it we get that it is an element of $(\mathbf{A} \cap \mathbf{E}) \cup (\mathbf{A} \cap \mathbf{E}^c)$, we conclude that

$$\mathbf{A} \subseteq (\mathbf{A} \cap \mathbf{E}) \cup (\mathbf{A} \cap \mathbf{E}^c). \tag{2.7}$$

Let $x \in (\mathbf{A} \cap \mathbf{E}) \cup (\mathbf{A} \cap \mathbf{E}^c)$ be arbitrarily chosen and fixed. Then

$$x \in \mathbf{A} \cap \mathbf{E} \quad \text{or} \quad x \in \mathbf{A} \cap \mathbf{E}^c.$$

1. Let $x \in \mathbf{A} \cap \mathbf{E}$. Then $x \in \mathbf{A}$.
2. Let $x \in \mathbf{A} \cap \mathbf{E}^c$. Then $x \in \mathbf{A}$.

Because $x \in (\mathbf{A} \cap \mathbf{E}) \cup (\mathbf{A} \cap \mathbf{E}^c)$ was arbitrarily chosen and for it we obtain that it is an element of $\mathbf{A}$, we conclude that

$$(\mathbf{A} \cap \mathbf{E}) \cup (\mathbf{A} \cap \mathbf{E}^c) \subseteq \mathbf{A}.$$

From the last relation and from (2.7), we get the desired result. This completes the proof. □

Remark 2.1. Using Lemma 2.1 and Corollary 2.1, we have

$$m^\star(\mathbf{A}) \leq m^\star(\mathbf{A} \cap \mathbf{E}) + m^\star(\mathbf{A} \cap \mathbf{E}^c).$$

Therefore, $\mathbf{E}$ is measurable if and only if

$$m^\star(\mathbf{A}) \geq m^\star(\mathbf{A} \cap \mathbf{E}) + m^\star(\mathbf{A} \cap \mathbf{E}^c). \tag{2.8}$$

This inequality trivially holds if $m^\star(\mathbf{A}) = \infty$. Thus, it suffices to establish (2.8) for sets $\mathbf{A}$ that have finite outer measure.

Theorem 2.7. *If $\mathbf{A}$ is a measurable set and $\mathbf{B}$ is any set disjoint from $\mathbf{A}$, then*

$$m^\star(\mathbf{A} \cup \mathbf{B}) = m^\star(\mathbf{A}) + m^\star(\mathbf{B}).$$

Proof. Since $\mathbf{A}$ is a measurable set, using Definition 2.4, we get

$$\begin{aligned} m^\star(\mathbf{A} \cup \mathbf{B}) &= m^\star((\mathbf{A} \cup \mathbf{B}) \cap \mathbf{A}) + m^\star((\mathbf{A} \cup \mathbf{B}) \cap \mathbf{A}^c) \\ &= m^\star(\mathbf{A}) + m^\star(\mathbf{B}). \end{aligned}$$

This completes the proof. □

Theorem 2.8. *Any set of outer measure zero is measurable. In particular, any countable set is measurable.*

Proof. Let $\mathbf{E}$ be a set of outer measure zero and let $\mathbf{A}$ be any set. Since

$$\mathbf{A} \cap \mathbf{E} \subseteq \mathbf{E} \quad \text{and} \quad \mathbf{A} \cap \mathbf{E}^c \subseteq \mathbf{A},$$

we have

$$\begin{aligned} m^\star(\mathbf{A} \cap \mathbf{E}) &\leq m^\star(\mathbf{E}) \\ &= 0 \end{aligned}$$

and

$$m^\star(\mathbf{A} \cap \mathbf{E}^c) \leq m^\star(\mathbf{A}).$$

Thus,

$$\begin{aligned} m^\star(\mathbf{A}) &\geq m^\star(\mathbf{A} \cap \mathbf{E}^c) \\ &= 0 + m^\star(\mathbf{A} \cap \mathbf{E}^c) \\ &= m^\star(\mathbf{A} \cap \mathbf{E}) + m^\star(\mathbf{A} \cap \mathbf{E}^c) \end{aligned}$$

and therefore $\mathbf{E}$ is measurable. Hence, using that every countable set has outer measure zero, we conclude that every countable set is measurable. This completes the proof. □

Lemma 2.2. *For any sets $\mathbf{A}$, $\mathbf{E}_1$ and $\mathbf{E}_2$ we have*

$$(\mathbf{A} \cap \mathbf{E}_1^c) \cap \mathbf{E}_2^c = \mathbf{A} \cap (\mathbf{E}_1 \cup \mathbf{E}_2)^c \tag{2.9}$$

and

$$(\mathbf{A} \cap \mathbf{E}_1) \cup (\mathbf{A} \cap \mathbf{E}_1^c \cap \mathbf{E}_2) = \mathbf{A} \cap (\mathbf{E}_1 \cup \mathbf{E}_2). \tag{2.10}$$

Proof. Firstly, we will prove the relation (2.9). Let $x \in (\mathbf{A} \cap \mathbf{E}_1^c) \cap \mathbf{E}_2^c$ be arbitrarily chosen and fixed. Then

$$x \in \mathbf{A} \cap \mathbf{E}_1^c \quad \text{and} \quad x \in \mathbf{E}_2^c.$$

Since $x \in \mathbf{A} \cap \mathbf{E}_1^c$, we get $x \in \mathbf{A}$ and $x \in \mathbf{E}_1^c$. From $x \in \mathbf{E}_1^c$ and $x \in \mathbf{E}_2^c$, we obtain $x \notin \mathbf{E}_1$ and $x \notin \mathbf{E}_2$. Therefore, $x \notin \mathbf{E}_1 \cup \mathbf{E}_2$ and

$x \in (\mathbf{E}_1 \cup \mathbf{E}_2)^c$. From $x \in \mathbf{A}$ and $x \in (\mathbf{E}_1 \cup \mathbf{E}_2)^c$, we conclude that $x \in \mathbf{A} \cap (\mathbf{E}_1 \cup \mathbf{E}_2)^c$. Because $x \in (\mathbf{A} \cap \mathbf{E}_1^c) \cap \mathbf{E}_2^c$ was arbitrarily chosen and for it we get that it is an element of the set $\mathbf{A} \cap (\mathbf{E}_1 \cup \mathbf{E}_2)^c$, we conclude that

$$(\mathbf{A} \cap \mathbf{E}_1^c) \cap \mathbf{E}_2^c \subseteq \mathbf{A} \cap (\mathbf{E}_1 \cup \mathbf{E}_2)^c. \tag{2.11}$$

Let $x \in \mathbf{A} \cap (\mathbf{E}_1 \cup \mathbf{E}_2)^c$ be arbitrarily chosen and fixed. Then $x \in \mathbf{A}$ and $x \in (\mathbf{E}_1 \cup \mathbf{E}_2)^c$. Hence, $x \notin \mathbf{E}_1 \cup \mathbf{E}_2$. Therefore, $x \notin \mathbf{E}_1$ and $x \notin \mathbf{E}_2$. From here, it follows that $x \in \mathbf{E}_1^c$ and $x \in \mathbf{E}_2^c$. Since $x \in \mathbf{A}$ and $x \in \mathbf{E}_1^c$, we obtain that $x \in \mathbf{A} \cap \mathbf{E}_1^c$. Because $x \in \mathbf{A} \cap \mathbf{E}_1^c$ and $x \in \mathbf{E}_2^c$, we get that $x \in (\mathbf{A} \cap \mathbf{E}_1^c) \cap \mathbf{E}_2^c$. Since $x \in \mathbf{A} \cap (\mathbf{E}_1 \cup \mathbf{E}_2)^c$ was arbitrarily chosen and we obtain that it is an element of $(\mathbf{A} \cap \mathbf{E}_1^c) \cap \mathbf{E}_2^c$, we conclude that

$$\mathbf{A} \cap (\mathbf{E}_1 \cup \mathbf{E}_2)^c \subseteq (\mathbf{A} \cap \mathbf{E}_1^c) \cap \mathbf{E}_2^c.$$

From the last relation and from (2.11), we obtain the relation (2.9). Now, we will prove the relation (2.10). Let $x \in (\mathbf{A} \cap \mathbf{E}_1) \cup (\mathbf{A} \cap \mathbf{E}_1^c \cap \mathbf{E}_2)$ be arbitrarily chosen and fixed. Then $x \in \mathbf{A} \cap \mathbf{E}_1$ or $x \in \mathbf{A} \cap \mathbf{E}_1^c \cap \mathbf{E}_2$.

1. Let $x \in \mathbf{A} \cap \mathbf{E}_1$. Then $x \in \mathbf{A}$ and $x \in \mathbf{E}_1$. Hence, $x \in \mathbf{E}_1 \cup \mathbf{E}_2$ and $x \in \mathbf{A} \cap (\mathbf{E}_1 \cup \mathbf{E}_2)$.
2. Let $x \in \mathbf{A} \cap \mathbf{E}_1^c \cap \mathbf{E}_2$. Then $x \in \mathbf{A}$, $x \in \mathbf{E}_1^c$ and $x \in \mathbf{E}_2$. Hence, $x \in \mathbf{A}$ and $x \in \mathbf{E}_1 \cup \mathbf{E}_2$. Therefore $x \in \mathbf{A} \cap (\mathbf{E}_1 \cup \mathbf{E}_2)$.

Because $x \in (\mathbf{A} \cap \mathbf{E}_1) \cup (\mathbf{A} \cap \mathbf{E}_1^c \cap \mathbf{E}_2)$ was arbitrarily chosen and for it we get that it is an element of $\mathbf{A} \cap (\mathbf{E}_1 \cup \mathbf{E}_2)$, we obtain

$$(\mathbf{A} \cap \mathbf{E}_1) \cup (\mathbf{A} \cap \mathbf{E}_1^c \cap \mathbf{E}_2) \subseteq \mathbf{A} \cap (\mathbf{E}_1 \cup \mathbf{E}_2). \tag{2.12}$$

Let $x \in \mathbf{A} \cap (\mathbf{E}_1 \cup \mathbf{E}_2)$ is arbitrarily chosen and fixed. Then $x \in \mathbf{A}$ and $x \in \mathbf{E}_1 \cup \mathbf{E}_2$.

1. Let $x \in \mathbf{E}_1$. Then $x \in \mathbf{A} \cap \mathbf{E}_1$ and
$$x \in (\mathbf{A} \cap \mathbf{E}_1) \cup (\mathbf{A} \cap \mathbf{E}_1^c \cap \mathbf{E}_2). \tag{2.13}$$
2. Let $x \in \mathbf{E}_2$. If $x \in \mathbf{E}_1$, then $x \in \mathbf{A} \cap \mathbf{E}_1$ and we get (2.13). If $x \notin \mathbf{E}_1$, then $x \in \mathbf{E}_1^c$ and $x \in \mathbf{A} \cap \mathbf{E}_1^c \cap \mathbf{E}_2$. Therefore, we get (2.13).

Because $x \in \mathbf{A} \cap (\mathbf{E}_1 \cup \mathbf{E}_2)$ was arbitrarily chosen and for it we obtain that it is an element of $(\mathbf{A} \cap \mathbf{E}_1) \cup (\mathbf{A} \cap \mathbf{E}_1^c \cap \mathbf{E}_2)$, we get

$$\mathbf{A} \cap (\mathbf{E}_1 \cup \mathbf{E}_2) \subseteq (\mathbf{A} \cap \mathbf{E}_1) \cup (\mathbf{A} \cap \mathbf{E}_1^c \cap \mathbf{E}_2).$$

From the last relation and from (2.13), we get the relation (2.10). This completes the proof. □

Theorem 2.9. *The union of a finite collection of measurable sets is measurable.*

Proof. Let $\mathbf{E}_1$ and $\mathbf{E}_2$ be two measurable sets. Then for any set $\mathbf{A}$ we have

$$m^\star(\mathbf{A}) = m^\star(\mathbf{A} \cap \mathbf{E}_1) + m^\star(\mathbf{A} \cap \mathbf{E}_1^c)$$

and

$$m^\star(\mathbf{A} \cap \mathbf{E}_1^c) = m^\star(\mathbf{A} \cap \mathbf{E}_1^c \cap \mathbf{E}_2) + m^\star(\mathbf{A} \cap \mathbf{E}_1^c \cap \mathbf{E}_2^c).$$

Therefore,

$$\begin{aligned} m^\star(\mathbf{A}) &= m^\star(\mathbf{A} \cap \mathbf{E}_1) + m^\star(\mathbf{A} \cap \mathbf{E}_1^c \cap \mathbf{E}_2) \\ &\quad + m^\star(\mathbf{A} \cap \mathbf{E}_1^c \cap \mathbf{E}_2^c). \end{aligned}$$

Hence and (2.9), we obtain

$$\begin{aligned} m^\star(\mathbf{A}) &= m^\star(\mathbf{A} \cap \mathbf{E}_1) + m^\star(\mathbf{A} \cap (\mathbf{E}_1 \cup \mathbf{E}_2)^c) \\ &\quad + m^\star(\mathbf{A} \cap \mathbf{E}_1^c \cap \mathbf{E}_2) \\ &\geq m^\star(\mathbf{A} \cap (\mathbf{E}_1 \cup \mathbf{E}_2)) + m^\star(\mathbf{A} \cap (\mathbf{E}_1 \cup \mathbf{E}_2)^c) \\ &\quad + m^\star(\mathbf{A} \cap \mathbf{E}_1^c \cap \mathbf{E}_2) \\ &\geq m^\star(\mathbf{A} \cap (\mathbf{E}_1 \cup \mathbf{E}_2)) + m^\star(\mathbf{A} \cap (\mathbf{E}_1 \cup \mathbf{E}_2)^c). \end{aligned}$$

Therefore, $\mathbf{E}_1 \cup \mathbf{E}_2$ is a measurable set. Assume that $\cup_{k=1}^n \mathbf{E}_k$ is a measurable set, where $\mathbf{E}_k$, $k \in \{1, \ldots, n\}$, are measurable sets. Let $\mathbf{E}_{n+1}$ is a measurable set. Then, as in above, $(\cup_{k=1}^n \mathbf{E}_k) \cup \mathbf{E}_{n+1}$ is a measurable set. Therefore $\cup_{k=1}^{n+1}\mathbf{E}_k$ is a measurable set. This completes the proof. □

Theorem 2.10. *Let* $\mathbf{A}$ *be any set and* $\{\mathbf{E}_k\}_{k=1}^n$ *be a finite disjoint collection of measurable sets. Then*

$$m^\star\left(\mathbf{A}\cap\left(\bigcup_{k=1}^n \mathbf{E}_k\right)\right)=\sum_{k=1}^n m^\star\left(\mathbf{A}\cap\mathbf{E}_k\right).$$

In particular,

$$m^\star\left(\bigcup_{k=1}^n \mathbf{E}_k\right)=\sum_{k=1}^n m^\star\left(\mathbf{E}_k\right).$$

Proof. We will use induction.

1. Let $n=1$. Then the assertion is evident.
2. Assume that the assertion is valid for some $n\in\mathbf{N}$.
3. We will prove the assertion for $n+1$. Because $\{\mathbf{E}_k\}_{k=1}^{n+1}$ is a disjoint collection, we have

$$\mathbf{A}\cap\left(\bigcup_{k=1}^{n+1}\mathbf{E}_k\right)\cap\mathbf{E}_{n+1}=\mathbf{A}\cap\mathbf{E}_{n+1}$$

   and

$$\mathbf{A}\cap\left(\bigcup_{k=1}^{n+1}\mathbf{E}_k\right)\cap\mathbf{E}_{n+1}^c=\mathbf{A}\cap\left(\bigcup_{k=1}^{n}\mathbf{E}_k\right).$$

   Hence, by the measurability of $\mathbf{E}_{n+1}$, we get

$$\begin{aligned}m^\star\left(\mathbf{A}\cap\left(\bigcup_{k=1}^{n+1}\mathbf{E}_k\right)\right)&=m^\star\left(\mathbf{A}\cap\left(\bigcup_{k=1}^{n+1}\mathbf{E}_k\right)\cap\mathbf{E}_{n+1}\right)\\&\quad+m^\star\left(\mathbf{A}\cap\left(\bigcup_{k=1}^{n+1}\mathbf{E}_k\right)\cap\mathbf{E}_{n+1}^c\right)\\&=m^\star\left(\mathbf{A}\cap\mathbf{E}_{n+1}\right)+m^\star\left(\mathbf{A}\cap\left(\bigcup_{k=1}^{n}\mathbf{E}_k\right)\right)\end{aligned}$$

$$= m^\star(\mathbf{A} \cap \mathbf{E}_{n+1}) + \sum_{k=1}^{n} m^\star(\mathbf{A} \cap \mathbf{E}_k)$$

$$= \sum_{k=1}^{n+1} m^\star(\mathbf{A} \cap \mathbf{E}_k).$$

This completes the proof. □

Lemma 2.3. *For any sets* $\mathbf{A}$, $\mathbf{B}$ *and* $\mathbf{C}$ *we have*

$$\mathbf{A}\backslash\mathbf{B} = \mathbf{A} \cap \mathbf{B}^c, \tag{2.14}$$

$$\mathbf{A} \cap (\mathbf{B}\backslash\mathbf{C}) \subseteq \mathbf{A} \cap (\mathbf{B} \cap \mathbf{C}^c), \tag{2.15}$$

$$(\mathbf{A} \cap \mathbf{B})^c = \mathbf{A}^c \cup \mathbf{B}^c. \tag{2.16}$$

Proof.

1. We will prove (2.14). Let $x \in \mathbf{A}\backslash\mathbf{B}$ be arbitrarily chosen and fixed. Then $x \in \mathbf{A}$ and $x \notin \mathbf{B}$. From $x \notin \mathbf{B}$ it follows that $x \in \mathbf{B}^c$. By $x \in \mathbf{A}$ and $x \in \mathbf{B}^c$, we obtain that $x \in \mathbf{A} \cap \mathbf{B}^c$. Because $x \in \mathbf{A}\backslash\mathbf{B}$ was arbitrarily chosen and we obtain that it is an element of $\mathbf{A} \cap \mathbf{B}^c$, we get

$$\mathbf{A}\backslash\mathbf{B} \subseteq \mathbf{A} \cap \mathbf{B}^c. \tag{2.17}$$

   Let $x \in \mathbf{A} \cap \mathbf{B}^c$ be arbitrarily chosen. Then $x \in \mathbf{A}$ and $x \in \mathbf{B}^c$. From $x \in \mathbf{B}^c$ it follows that $x \notin \mathbf{B}$. From $x \in \mathbf{A}$ and $x \notin \mathbf{B}$, we get $x \in \mathbf{A}\backslash\mathbf{B}$. Because $x \in \mathbf{A} \cap \mathbf{B}^c$ was arbitrarily chosen and fixed and we get that it is an element of the set $\mathbf{A}\backslash\mathbf{B}$, we obtain

$$\mathbf{A} \cap \mathbf{B}^c \subseteq \mathbf{A}\backslash\mathbf{B}.$$

   From the last relation and from (2.17), we obtain (2.14).
2. We will prove (2.15). Let $x \in \mathbf{A} \cap (\mathbf{B}\backslash\mathbf{C})$ is arbitrarily chosen. Then $x \in \mathbf{A}$ and $x \in \mathbf{B}\backslash\mathbf{C}$. From $x \in \mathbf{B}\backslash\mathbf{C}$, it follows that $x \in \mathbf{B}$ and $x \notin \mathbf{C}$. By $x \notin \mathbf{C}$, we get $x \in \mathbf{C}^c$. From $x \in \mathbf{A}$, $x \in \mathbf{B}$ and $x \in \mathbf{C}^c$, we get that $x \in \mathbf{A} \cap (\mathbf{B} \cap \mathbf{C}^c)$. Because $x \in \mathbf{A} \cap (\mathbf{B}\backslash\mathbf{C})$ was arbitrarily chosen, we conclude the relation (2.15).

3. We will prove (2.16). Let $x \in (\mathbf{A} \cap \mathbf{B})^c$ be arbitrarily chosen and fixed. Then $x \notin \mathbf{A} \cap \mathbf{B}$. Hence, $x \notin \mathbf{A}$ or $x \notin \mathbf{B}$.

   a. If $x \notin \mathbf{A}$, then $x \in \mathbf{A}^c$ and $x \in \mathbf{A}^c \cup \mathbf{B}^c$.
   b. If $x \notin \mathbf{B}$, then $x \in \mathbf{B}^c$ and $x \in \mathbf{A}^c \cup \mathbf{B}^c$.

   Because $x \in (\mathbf{A} \cap \mathbf{B})^c$ was arbitrarily chosen and we get that it is an element of $\mathbf{A}^c \cup \mathbf{B}^c$, we obtain

$$(\mathbf{A} \cap \mathbf{B})^c \subseteq \mathbf{A}^c \cup \mathbf{B}^c. \tag{2.18}$$

   Let $x \in \mathbf{A}^c \cup \mathbf{B}^c$ be arbitrarily chosen and fixed. Then $x \in \mathbf{A}^c$ or $x \in \mathbf{B}^c$. Hence, $x \notin \mathbf{A}$ or $x \notin \mathbf{B}$. Therefore, $x \notin \mathbf{A} \cap \mathbf{B}$ and $x \in (\mathbf{A} \cap \mathbf{B})^c$. Because $x \in \mathbf{A}^c \cup \mathbf{B}^c$ was arbitrarily chosen and for it we get that it is an element of $(\mathbf{A} \cap \mathbf{B})^c$, we obtain

$$\mathbf{A}^c \cup \mathbf{B}^c \subseteq (\mathbf{A} \cap \mathbf{B})^c.$$

   From the last relation and from (2.18), we get the relation (2.16). □

Theorem 2.11. *Let $\mathbf{E}_1$ and $\mathbf{E}_2$ be measurable sets. Then $\mathbf{E}_1 \backslash \mathbf{E}_2$ is a measurable set.*

Proof. Let $\mathbf{A}$ be any set. Since $\mathbf{E}_1$ is a measurable set, then $\mathbf{E}_1^c$ is a measurable set. Hence and Theorem 2.9, we obtain that $\mathbf{E}_1^c \cup \mathbf{E}_2$ is a measurable set. Then

$$m^\star(\mathbf{A}) \geq m^\star(\mathbf{A} \cap (\mathbf{E}_1^c \cup \mathbf{E}_2)) + m^\star(\mathbf{A} \cap (\mathbf{E}_1^c \cup \mathbf{E}_2)^c). \tag{2.19}$$

Using (2.14) and (2.16), we have

$$(\mathbf{E}_1 \backslash \mathbf{E}_2)^c = (\mathbf{E}_1 \cap \mathbf{E}_2^c)^c = \mathbf{E}_1^c \cup \mathbf{E}_2.$$

Therefore,

$$\mathbf{A} \cap (\mathbf{E}_1 \backslash \mathbf{E}_2)^c = \mathbf{A} \cap (\mathbf{E}_1^c \cup \mathbf{E}_2)$$

and

$$m^\star(\mathbf{A} \cap (\mathbf{E}_1 \backslash \mathbf{E}_2)^c) = m^\star(\mathbf{A} \cap (\mathbf{E}_1^c \cup \mathbf{E}_2)). \tag{2.20}$$

By (2.15) and (2.16), we get

$$\mathbf{A} \cap (\mathbf{E}_1 \backslash \mathbf{E}_2) \subseteq \mathbf{A} \cap (\mathbf{E}_1 \cap \mathbf{E}_2^c) = \mathbf{A} \cap (\mathbf{E}_1^c \cup \mathbf{E}_2)^c.$$

Hence and Theorem 2.1, we obtain

$$m^\star (\mathbf{A} \cap (\mathbf{E}_1 \backslash \mathbf{E}_2)) \leq m^\star (\mathbf{A} \cap (\mathbf{E}_1^c \cup \mathbf{E}_2)^c).$$

From the last inequality and from (2.19), (2.20), we go to

$$\begin{aligned} &m^\star (\mathbf{A} \cap (\mathbf{E}_1 \backslash \mathbf{E}_2)) + m^\star (\mathbf{A} \cap (\mathbf{E}_1 \backslash \mathbf{E}_2)^c) \\ &\leq m^\star (\mathbf{A} \cap (\mathbf{E}_1^c \cup \mathbf{E}_2)^c) + m^\star (\mathbf{A} \cap (\mathbf{E}_1^c \cup \mathbf{E}_2)) \\ &\leq m^\star (\mathbf{A}). \end{aligned}$$

Because $\mathbf{A}$ was arbitrarily chosen and fixed, we conclude that $\mathbf{E}_1 \backslash \mathbf{E}_2$ is a measurable set. This completes the proof. □

Theorem 2.12. *The union of a countable collection of sets is also the union of a countable disjoint collection of sets. In particular, the union of a countable collection of measurable sets is also the union of a countable disjoint collection of measurable sets.*

Proof. Let $\mathbf{E} = \cup_{k=1}^\infty \mathbf{E}'_k$. We set

$$\mathbf{E}_1 = \mathbf{E}'_1,$$

$$\mathbf{E}_k = \mathbf{E}'_k \backslash \bigcup_{l=1}^{k-1} \mathbf{E}'_l.$$

Suppose that $x \in \mathbf{E}$ is arbitrarily chosen and fixed. Assume that $x \notin \mathbf{E}_k$ for any $k \in \mathbf{N}$. Then $x \notin \mathbf{E}'_k$ for any $k \in \mathbf{N}$ and $x \notin \mathbf{E} = \cup_{k=1}^\infty \mathbf{E}'_k$, which is a contradiction. Therefore, $x \in \mathbf{E}_k$ for some $k \in \mathbf{N}$ and hence, $x \in \cup_{k=1}^\infty \mathbf{E}_k$. Because $x \in \mathbf{E}$ was arbitrarily chosen and for it we get that it is an element of $\cup_{k=1}^\infty \mathbf{E}_k$, we get

$$\mathbf{E} \subseteq \bigcup_{k=1}^\infty \mathbf{E}_k. \tag{2.21}$$

Let now $x \in \cup_{k=1}^{\infty} \mathbf{E}_k$ is arbitrarily chosen. Suppose that $x \notin \mathbf{E}'_k$ for any $k \in \mathbf{N}$. Hence, $x \notin \mathbf{E}_k$ for any $k \in \mathbf{N}$ and $x \notin \cup_{k=1}^{\infty} \mathbf{E}_k$. This is a contradiction. Therefore, $x \in \mathbf{E}'_k$ for some $k \in \mathbf{N}$ and $x \in \mathbf{E} = \cup_{k=1}^{\infty} \mathbf{E}'_k$. Because $x \in \cup_{k=1}^{\infty} \mathbf{E}_k$ was arbitrarily chosen and for it we get that it is an element of $\mathbf{E}$, we conclude that

$$\bigcup_{k=1}^{\infty} \mathbf{E}_k \subseteq \mathbf{E}.$$

From the last relation and from (2.21), we obtain $\mathbf{E} = \cup_{k=1}^{\infty} \mathbf{E}_k$. Suppose that there is $x \in \mathbf{E}$ such that $x \in \mathbf{E}_k$ and $x \in \mathbf{E}_l$ for some $k > l$. From the definition of the set $\mathbf{E}_k$ it follows that $x \notin \cup_{m=1}^{k-1} \mathbf{E}_m$. Therefore, $x \notin \mathbf{E}_m$ for any $m \in \{1, \ldots, k-1\}$. In particular, $x \notin \mathbf{E}_l$, which is a contradiction. Consequently $\{\mathbf{E}_k\}_{k \in \mathbf{N}}$ is a collection of disjoint sets.

Now, we suppose that $\mathbf{E}'_k$, $k \in \mathbf{N}$, are measurable sets. Then for $k > 2$, using Theorem 2.9, we have that $\cup_{l=1}^{k-1} \mathbf{E}'_l$ is a measurable set. Hence and Theorem 2.11, the sets $\mathbf{E}_k$, $k \in \mathbf{N}$, are measurable sets. This completes the proof. □

Theorem 2.13. *The union of countable collection of measurable sets is measurable.*

Proof. Let $\mathbf{E} = \cup_{k=1}^{\infty} \mathbf{E}'_k$ be a countable collection of measurable sets. By Theorem 2.12 it follows that there is a countable disjoint collection $\{\mathbf{E}_k\}_{k \in \mathbf{N}}$ of measurable sets such that $\mathbf{E} = \cup_{k=1}^{\infty} \mathbf{E}_k$. Let $\mathbf{A}$ be any set, $n \in \mathbf{N}$ be arbitrarily chosen and

$$\mathbf{F}_n = \bigcup_{k=1}^{n} \mathbf{E}_k.$$

Then by Theorem 2.9 we have that $\mathbf{F}_n$ is a measurable set. Also, $\mathbf{F}_n \subseteq \mathbf{E}$ and $\mathbf{F}_n^c \supseteq \mathbf{E}^c$. Hence and Theorem 2.1, we get

$$m^\star(\mathbf{A} \cap \mathbf{E}^c) \leq m^\star(\mathbf{A} \cap \mathbf{F}_n^c).$$

Then, using that $\mathbf{F}_n$ is a measurable set, we obtain

$$\begin{aligned} m^\star(\mathbf{A}) &= m^\star(\mathbf{A} \cap \mathbf{F}_n) + m^\star(\mathbf{A} \cap \mathbf{F}_n^c) \\ &\geq m^\star(\mathbf{A} \cap \mathbf{F}_n) + m^\star(\mathbf{A} \cap \mathbf{E}^c). \end{aligned} \tag{2.22}$$

By Theorem 2.10, we obtain

$$\begin{aligned} m^{\star}(\mathbf{A}\cap\mathbf{F}_n) &= m^{\star}\left(\mathbf{A}\cap\left(\bigcup_{k=1}^{n}\mathbf{E}_k\right)\right) \\ &= \sum_{k=1}^{n} m^{\star}(\mathbf{A}\cap\mathbf{E}_k). \end{aligned}$$

Hence and (2.22), we get

$$m^{\star}(\mathbf{A}) \geq \sum_{k=1}^{n} m^{\star}(\mathbf{A}\cap\mathbf{E}_k) + m^{\star}(\mathbf{A}\cap\mathbf{E}^c).$$

Because $n \in \mathbf{N}$ was arbitrarily chosen, from the last inequality we find

$$m^{\star}(\mathbf{A}) \geq \sum_{k=1}^{\infty} m^{\star}(\mathbf{A}\cap\mathbf{E}_k) + m^{\star}(\mathbf{A}\cap\mathbf{E}^c). \tag{2.23}$$

From Theorem 2.5, we have

$$\begin{aligned} m^{\star}(\mathbf{A}\cap\mathbf{E}) &= m^{\star}\left(\mathbf{A}\cap\left(\bigcup_{k=1}^{\infty}\mathbf{E}_k\right)\right) \\ &\leq m^{\star}\left(\bigcup_{k=1}^{\infty}(\mathbf{A}\cap\mathbf{E}_k)\right) \\ &\leq \sum_{k=1}^{\infty} m^{\star}(\mathbf{A}\cap\mathbf{E}_k). \end{aligned}$$

Hence and (2.23), we find

$$m^{\star}(\mathbf{A}) \geq m^{\star}(\mathbf{A}\cap\mathbf{E}) + m^{\star}(\mathbf{A}\cap\mathbf{E}^c).$$

This completes the proof. □

Definition 2.5. A collection $\mathscr{F}$ of subsets of $\mathbf{R}$ is called σ-algebra if

1. it contains $\mathbf{R}$,
2. $\mathbf{A} \in \mathscr{F}$, then $\mathbf{A}^c \in \mathscr{F}$,
3. $\mathbf{A}_k \in \mathscr{F}$, $k \in \mathbf{N}$, then $cup_{k=1}^{\infty} \mathbf{A}_k \in \mathscr{F}$.

Remark 2.2. Let $\mathscr{F}$ be a σ-algebra. Let also, $\mathbf{A}_k \in \mathscr{F}$, $k \in \mathbf{N}$. Then $\mathbf{A}_k^c \in \mathscr{F}$, $\cup_{k=1}^{\infty} \mathbf{A}_k^c \in \mathscr{F}$. Hence,

$$\left(\bigcup_{k=1}^{\infty} \mathbf{A}_k^c \right)^c = \bigcap_{k=1}^{\infty} \mathbf{A}_k \in \mathscr{F}.$$

Therefore, every σ-algebra is closed with respect to the formation of countable intersections.

Remark 2.3. By Theorems 2.11 and 2.13, it follows that the set of all measurable sets is a σ-algebra.

Remark 2.4. By Theorem 2.13 it follows that the collection $\mathscr{M}$ of measurable sets is a σ-algebra.

Theorem 2.14. *If a σ-algebra $\mathscr{F}$ of subsets of $\mathbf{R}$ contains intervals of the form (a, ∞), then it contains all intervals.*

Proof. Since $(a, \infty) \in \mathscr{F}$ and $\mathscr{F}$ is a σ-algebra, we have that

$$(-\infty, a] = (a, \infty)^c \in \mathscr{F}.$$

Because

$$(a, b]^c = (-\infty, a] \cup (b, \infty)$$

and $(-\infty, a]$, $(b, \infty) \in \mathscr{F}$ and $\mathscr{F}$ is closed with respect to the formation of finite unions, we conclude that $(a, b]^c \in \mathscr{F}$. Hence, using that $\mathscr{F}$ is closed with respect to complements, we get that $(a, b] \in \mathscr{F}$. Now, we consider $[a, \infty)$. We will prove that

$$[a, \infty) = \bigcap_{n=1}^{\infty} \left(a - \frac{1}{n}, \infty \right). \tag{2.24}$$

Let $x \in [a, \infty)$ be arbitrarily chosen and fixed. Then $x \in \left(a - \frac{1}{n}, \infty\right)$ for any $n \in \mathbf{N}$. Therefore, $x \in \cap_{n=1}^{\infty} \left(a - \frac{1}{n}, \infty\right)$. Since $x \in [a, \infty)$

was arbitrarily chosen and for it we get that it is an element of $\bigcap_{n=1}^{\infty}\left(a-\frac{1}{n},\infty\right)$, we obtain the relation

$$[a,\infty)\subseteq\bigcap_{n=1}^{\infty}\left(a-\frac{1}{n},\infty\right). \tag{2.25}$$

Let now $x\in\bigcap_{n=1}^{\infty}\left(a-\frac{1}{n},\infty\right)$ be arbitrarily chosen and fixed. Then

$$a-x<\frac{1}{n}$$

for any $n\in\mathbf{N}$. Hence, $a-x\leq 0$, i.e., $x\in[a,\infty)$. Because $x\in\bigcap_{n=1}^{\infty}\left(a-\frac{1}{n},\infty\right)$ was arbitrarily chosen and for it we get that it is an element of $[a,\infty)$, we conclude that

$$\bigcap_{n=1}^{\infty}\left(a-\frac{1}{n},\infty\right)\subseteq[a,\infty).$$

From the last relation and from (2.25), we obtain the relation (2.24). Since

$$\left(a-\frac{1}{n},\infty\right)\in\mathscr{F}$$

for any $n\in\mathbf{N}$ and $\mathscr{F}$ is closed with respect to the formation of countable intersections, we obtain that $[a,\infty)\in\mathscr{F}$. Hence, using that $\mathscr{F}$ is closed with respect to the complements, we find

$$(-\infty,a)=[a,\infty)^c\in\mathscr{F}.$$

Note that

$$[a,b)=((-\infty,a)\cup[b,\infty))^c.$$

Because $(-\infty,a)$, $[b,\infty)\in\mathscr{F}$ and $\mathscr{F}$ is closed with respect to the formation of finite unions and complements, we conclude that $[a,b)\in\mathscr{F}$. Also,

$$[a,b]=(-\infty,b]\cap[a,\infty).$$

Since $(-\infty,b]$, $[a,\infty)\in\mathscr{F}$ and $\mathscr{F}$ is closed with respect to the formation of finite intersections, we obtain $[a,b]\in\mathscr{F}$. This completes the proof. □

Theorem 2.15. *Every interval is measurable.*

Proof. We will prove that every interval of the form (a, ∞) is measurable. Let $\mathbf{A}$ be a set such that $a \notin \mathbf{A}$. Let also, $\{\mathbf{I}_k\}_{k\in\mathbf{N}}$ be any collection of open, bounded intervals that covers $\mathbf{A}$. We define

$$\mathbf{A}_1 = \mathbf{A} \cap (-\infty, a), \quad \mathbf{A}_2 = \mathbf{A} \cap (a, \infty),$$
$$\mathbf{I}'_k = \mathbf{I}_k \cap (-\infty, a), \quad \mathbf{I}''_k = \mathbf{I}_k \cap (a, \infty)$$

for each index k. Then

$$l(\mathbf{I}_k) = l(\mathbf{I}'_k) + l(\mathbf{I}''_k)$$

for each index k. Note that $\{\mathbf{I}'_k\}_{k\in\mathbf{N}}$ and $\{\mathbf{I}''_k\}_{k\in\mathbf{N}}$ are countable collections of open, bounded intervals that cover $\mathbf{A}_1$ and $\mathbf{A}_2$, respectively. Hence, using the definition for outer measure, we get

$$m^\star(\mathbf{A}_1) \le \sum_{k=1}^{\infty} l(\mathbf{I}'_k)$$

and

$$m^\star(\mathbf{A}_2) \le \sum_{k=1}^{\infty} l(\mathbf{I}''_k).$$

Therefore,

$$\begin{aligned} m^\star(\mathbf{A}_1) + m^\star(\mathbf{A}_2) &\le \sum_{k=1}^{\infty} l(\mathbf{I}'_k) + \sum_{k=1}^{\infty} l(\mathbf{I}''_k) \\ &= \sum_{k=1}^{\infty} \left(l(\mathbf{I}'_k) + l(\mathbf{I}''_k) \right) \\ &= \sum_{k=1}^{\infty} l(\mathbf{I}_k), \end{aligned}$$

i.e.,

$$m^\star(\mathbf{A}_1) + m^\star(\mathbf{A}_2) \le \sum_{k=1}^{\infty} l(\mathbf{I}_k).$$

Because $\{\mathbf{I}_k\}_{k\in\mathbf{N}}$ was arbitrarily chosen, from the last inequality we get

$$m^\star(\mathbf{A}\cap(a,\infty)) + m^\star(\mathbf{A}\cap(a,\infty)^c) \le m^\star(\mathbf{A}). \tag{2.26}$$

Let now $a \in \mathbf{A}$. Then we consider the set $\mathbf{B} = \mathbf{A}\backslash\{a\}$. As in above

$$m^\star(\mathbf{B}\cap(a,\infty)) + m^\star(\mathbf{B}\cap(a,\infty)^c) \le m^\star(\mathbf{B}). \tag{2.27}$$

By Theorem 2.4, it follows that

$$\begin{aligned} m^\star(\mathbf{B}) &= m^\star(\mathbf{A}),\\ m^\star(\mathbf{B}\cap(a,\infty)) &= m^\star(\mathbf{A}\cap(a,\infty)),\\ m^\star(\mathbf{B}\cap(a,\infty)^c) &= m^\star(\mathbf{A}\cap(a,\infty)^c). \end{aligned}$$

Hence and (2.27), we get (2.26). Therefore, (a,∞) is measurable. Because the measurable sets are σ-algebra and every interval of the form (a,∞) is measurable, using Theorem 2.14, we conclude that every interval is measurable. This completes the proof. □

Theorem 2.16. *The translate of a measurable set is a measurable set.*

Proof. Let $\mathbf{E}$ be a measurable set. Let also, $\mathbf{A}$ be any set and $y \in \mathbf{R}$. Since $\mathbf{E}$ is measurable, using Theorem 2.4, we have

$$\begin{aligned} m^\star(\mathbf{A}) &= m^\star(\mathbf{A}-y)\\ &= m^\star((\mathbf{A}-y)\cap\mathbf{E}) + m^\star((\mathbf{A}-y)\cap\mathbf{E}^c)\\ &= m^\star(\mathbf{A}\cap(\mathbf{E}+y)) + m^\star(\mathbf{A}\cap(\mathbf{E}+y)^c). \end{aligned}$$

Therefore, $\mathbf{E}+y$ is a measurable set. □

Theorem 2.17. *The intersection of an arbitrary nonempty collection of σ-algebras on $\mathbf{R}$ is a σ-algebra on $\mathbf{R}$.*

Proof. Let $\mathscr{A}_\alpha$, $\alpha \in \mathbf{I}$, are σ-algebras on $\mathbf{R}$. Here $\mathbf{I}$ is an index set. Let also,

$$\mathscr{A} = \bigcap_{\alpha\in\mathbf{I}} \mathscr{A}_\alpha.$$

1. Since $\mathscr{A}_\alpha$, $\alpha \in \mathbf{I}$, are σ-algebras on $\mathbf{R}$, we have that $\mathbf{R} \in \mathscr{A}_\alpha$ for any $\alpha \in \mathbf{I}$. Therefore, $\mathbf{R} \in \mathscr{A} = \cap_{\alpha\in\mathbf{I}}\mathscr{A}_\alpha$.
2. Let $\mathbf{A} \in \mathscr{A}$ be any set. Then $\mathbf{A} \in \mathscr{A}_\alpha$ for any $\alpha \in \mathbf{I}$. Because $\mathscr{A}_\alpha$, $\alpha \in \mathbf{I}$, are σ-algebras, we have that $\mathbf{A}^c \in \mathscr{A}_\alpha$ for any $\alpha \in \mathbf{I}$. Consequently, $\mathbf{A}^c \in \mathscr{A}$.
3. Let $\{\mathbf{A}_n\}_{n\in\mathbf{N}} \subseteq \mathscr{A}$. Then $\{\mathbf{A}_n\}_{n\in\mathbf{N}} \subseteq \mathscr{A}_\alpha$ for any $\alpha \in \mathbf{I}$. Because $\mathscr{A}_\alpha$, $\alpha \in \mathbf{N}$, are σ-algebras, we obtain that $\cup_{n=1}^{\infty}\mathbf{A}_n \in \mathscr{A}_\alpha$ for any $\alpha \in \mathbf{I}$. Therefore, $\cup_{n=1}^{\infty}\mathbf{A}_n \in \mathscr{A}$.

Consequently $\mathscr{A}$ is a σ-algebra. □

Definition 2.6. The intersection of all the σ-algebras of subsets of $\mathbf{R}$ that contain the open sets is a σ-algebra called the Borel σ-algebra. It will be denoted by $\mathscr{B}(\mathbf{R})$. The members of $\mathscr{B}(\mathbf{R})$ are called Borel sets.

The Borel σ-algebra is contained in every σ-algebra that contains all open sets. Since the measurable sets are a σ-algebra containing all open sets, every Borel set is measurable.

Definition 2.7. A set is said to be a $\mathbf{G}_\delta$ set provided it is the intersection of a countable collection of open sets.

Exercise 2.1. Prove that the translate of a $\mathbf{G}_\delta$ set is a $\mathbf{G}_\delta$ set.

Definition 2.8. A set is said to be a $\mathbf{F}_\sigma$ set provided it is the union of a countable collection of closed sets.

Exercise 2.2. Prove that the translate of an $\mathbf{F}_\sigma$ set is a $\mathbf{F}_\sigma$ set.

Theorem 2.18. *Every open set is measurable.*

Proof. Note that every open set is a disjoint union of a countable collection of open intervals. Hence, Theorems 2.15 and 2.13, it follows that every open set is measurable. This completes the proof. □

Exercise 2.3. Prove that every $\mathbf{F}_\sigma$ set and every $\mathbf{G}_\delta$ set is measurable.

Theorem 2.19. *Let* $\mathbf{E}$ *be any set of real numbers. Then* $\mathbf{E}$ *is measurable if and only if for each* $\varepsilon > 0$ *there is an open set* $\mathbf{O}$*, containing* $\mathbf{E}$*, so that* $m^\star(\mathbf{O}\backslash\mathbf{E}) < \varepsilon$.

Proof.

1. Suppose that $\mathbf{E}$ is measurable. We take $\varepsilon > 0$ arbitrarily.
   (a) Let $m^{\star}(\mathbf{E}) < \infty$. By the definition of outer measure, it follows that there is a countable collection of open intervals $\{\mathbf{I}_k\}_{k\in\mathbf{N}}$ which covers $\mathbf{E}$ and
   $$\sum_{k=1}^{\infty} l(\mathbf{I}_k) < m^{\star}(\mathbf{E}) + \varepsilon.$$
   We define
   $$\mathbf{O} = \bigcup_{k=1}^{\infty} \mathbf{I}_k.$$
   Then $\mathbf{O}$ is an open set that contains $\mathbf{E}$. Hence and Theorem 2.5, we obtain
   $$\begin{aligned} m^{\star}(\mathbf{O}) &= m^{\star}\left(\bigcup_{k=1}^{\infty} \mathbf{I}_k\right) \\ &\le \sum_{k=1}^{\infty} m^{\star}(\mathbf{I}_k) \\ &\le \sum_{k=1}^{\infty} l(\mathbf{I}_k) \\ &< m^{\star}(\mathbf{E}) + \varepsilon. \end{aligned}$$
   Therefore,
   $$m^{\star}(\mathbf{O}) - m^{\star}(\mathbf{E}) < \varepsilon.$$
   Because
   $$\mathbf{O} = (\mathbf{O}\backslash\mathbf{E}) \cup \mathbf{E} \quad \text{and} \quad (\mathbf{O}\backslash\mathbf{E}) \cap \mathbf{E} = \emptyset,$$
   using Theorem 2.7, we get
   $$m^{\star}(\mathbf{O}) = m^{\star}(\mathbf{O}\backslash\mathbf{E}) + m^{\star}(\mathbf{E}).$$
   Consequently,
   $$\begin{aligned} m^{\star}(\mathbf{O}\backslash\mathbf{E}) &= m^{\star}(\mathbf{O}) - m^{\star}(\mathbf{E}) \\ &< \varepsilon. \end{aligned}$$

(b) Let $m^\star(\mathbf{E}) = \infty$. Then $\mathbf{E}$ may be expressed as the disjoint union of a countable collection $\{\mathbf{E}_k\}_{k\in\mathbf{N}}$ of measurable sets, each of which has finite outer measure. By the first case, for each $k \in \mathbf{N}$ there is an open set $\mathbf{O}_k$ containing $\mathbf{E}_k$ such that

$$m^\star(\mathbf{O}_k \backslash \mathbf{E}_k) < \frac{\varepsilon}{2^k}.$$

Let

$$\mathbf{O} = \bigcup_{k=1}^{\infty} \mathbf{O}_k.$$

Then $\mathbf{O}$ is an open set that contains $\mathbf{E}$ and

$$\begin{aligned} m^\star(\mathbf{O}\backslash\mathbf{E}) &= m^\star\left(\left(\bigcup_{k=1}^{\infty}\mathbf{O}_k\right)\Bigg\backslash \mathbf{E}\right) \\ &\le m^\star\left(\bigcup_{k=1}^{\infty}(\mathbf{O}_k\backslash\mathbf{E}_k)\right) \\ &\le \sum_{k=1}^{\infty} m^\star(\mathbf{O}_k\backslash\mathbf{E}_k) \\ &< \sum_{k=1}^{\infty}\frac{\varepsilon}{2^k} \\ &= \varepsilon. \end{aligned}$$

2. Let for each $\varepsilon > 0$ there is an open set $\mathbf{O}$ containing $\mathbf{E}$ such that $m^\star(\mathbf{O}\backslash\mathbf{E}) < \varepsilon$. Let $\mathbf{A}$ be any set. Then, since $\mathbf{O}$ is an open set, using Theorem 2.18, it is measurable. Therefore,

$$m^\star(\mathbf{A}) \ge m^\star(\mathbf{A}\cap\mathbf{O}) + m^\star(\mathbf{A}\cap\mathbf{O}^c). \tag{2.28}$$

Using that $\mathbf{A}\cap\mathbf{E} \subseteq \mathbf{A}\cap\mathbf{O}$ and Theorem 2.1, we have

$$m^\star(\mathbf{A}\cap\mathbf{E}) \le m^\star(\mathbf{A}\cap\mathbf{O}). \tag{2.29}$$

Now, we will prove that

$$\mathbf{E} = \mathbf{O}\cap(\mathbf{O}\backslash\mathbf{E})^c. \tag{2.30}$$

Let $x \in \mathbf{E}$ be arbitrarily chosen. Then $x \in \mathbf{O}$ and $x \notin \mathbf{O}\backslash\mathbf{E}$. Therefore, $x \in (\mathbf{O}\backslash\mathbf{E})^c$. Because $x \in \mathbf{O}$ and $x \in (\mathbf{O}\backslash\mathbf{E})^c$, we get

$x \in \mathbf{O} \cap (\mathbf{O} \backslash \mathbf{E})^c$. Since $x \in \mathbf{E}$ was arbitrarily chosen and we get that it is an element of the set $\mathbf{O} \cap (\mathbf{O} \backslash \mathbf{E})^c$, we conclude that

$$\mathbf{E} \subseteq \mathbf{O} \cap (\mathbf{O} \backslash \mathbf{E})^c. \tag{2.31}$$

Let $x \in \mathbf{O} \cap (\mathbf{O} \backslash \mathbf{E})^c$ be arbitrarily chosen and fixed. Then $x \in \mathbf{O}$ and $x \in (\mathbf{O} \backslash \mathbf{E})^c$. Hence, $x \notin \mathbf{O} \backslash \mathbf{E}$ and $x \in \mathbf{E}$. Because $x \in \mathbf{O} \cap (\mathbf{O} \backslash \mathbf{E})^c$ was arbitrarily chosen and we get that it is an element of $\mathbf{E}$, we obtain

$$\mathbf{O} \cap (\mathbf{O} \backslash \mathbf{E})^c \subseteq \mathbf{E}.$$

From the last relation and from (2.31), we get the relation (2.30). By (2.30) we find

$$\begin{aligned} \mathbf{E}^c &= (\mathbf{O} \cap (\mathbf{O} \backslash \mathbf{E})^c)^c \\ &= \mathbf{O}^c \cup (\mathbf{O} \backslash \mathbf{E}). \end{aligned}$$

Now, we will prove

$$\mathbf{A} \cap \mathbf{E}^c \subseteq (\mathbf{A} \cap \mathbf{O}^c) \cup (\mathbf{O} \backslash \mathbf{E}). \tag{2.32}$$

Let $x \in \mathbf{A} \cap \mathbf{E}^c$ be arbitrarily chosen and fixed. Then $x \in \mathbf{A}$ and $x \in \mathbf{E}^c$. Hence, $x \notin \mathbf{E}$.

(a) If $x \in \mathbf{O}^c$, then $x \in \mathbf{A} \cap \mathbf{O}^c$ and $(\mathbf{A} \cap \mathbf{O}^c) \cup (\mathbf{O} \backslash \mathbf{E})$.
(b) If $x \notin \mathbf{O}^c$, then $x \in \mathbf{O}$ and hence, $x \in \mathbf{O} \backslash \mathbf{E}$ and $x \in (\mathbf{A} \cap \mathbf{O}^c) \cup (\mathbf{O} \backslash \mathbf{E})$.

Because $x \in \mathbf{A} \cap \mathbf{E}^c$ was arbitrarily chosen and for it we get that it is an element of $(\mathbf{A} \cap \mathbf{O}^c) \cup (\mathbf{O} \backslash \mathbf{E})$, we get the relation (2.32). By (2.32), we find

$$\begin{aligned} m^\star (\mathbf{A} \cap \mathbf{E}^c) &\leq m^\star ((\mathbf{A} \cap \mathbf{O}^c) \cup (\mathbf{O} \backslash \mathbf{E})) \\ &\leq m^\star (\mathbf{A} \cap \mathbf{O}^c) + m^\star (\mathbf{O} \backslash \mathbf{E}) \\ &< \varepsilon + m^\star (\mathbf{A} \cap \mathbf{O}^c). \end{aligned}$$

From the last inequality and from (2.29), (2.28), we obtain

$$\begin{aligned} m^\star(\mathbf{A}\cap\mathbf{E}) + m^\star(\mathbf{A}\cap\mathbf{E}^c) &\le m^\star(\mathbf{A}\cap\mathbf{O}) \\ &\quad + m^\star(\mathbf{A}\cap\mathbf{O}^c) + \varepsilon \\ &\le m^\star(\mathbf{A}) + \varepsilon. \end{aligned}$$

Because $\varepsilon > 0$ was arbitrarily chosen, we conclude that

$$m^\star(\mathbf{A}) \ge m^\star(\mathbf{A}\cap\mathbf{E}) + m^\star(\mathbf{A}\cap\mathbf{E}^c),$$

i.e., $\mathbf{E}$ is measurable. This completes the proof. □

Theorem 2.20. *Let $\mathbf{E}$ be any set of real numbers. Then $\mathbf{E}$ is measurable if and only if there is a $\mathbf{G}_\delta$ set $\mathbf{G}$, containing $\mathbf{E}$, such that $m^\star(\mathbf{G}\backslash\mathbf{E}) = 0$.*

Proof.

1. Let $\mathbf{E}$ be measurable. By Theorem 2.19, it follows that for any $k \in \mathbf{N}$ there is an open set $\mathbf{O}_k$, containing $\mathbf{E}$, such that

$$m^\star(\mathbf{O}_k\backslash\mathbf{E}) < \frac{1}{k}.$$

Let $\mathbf{G} = \cap_{k=1}^\infty \mathbf{O}_k$. Then $\mathbf{G}$ is a $\mathbf{G}_\delta$ set containing $\mathbf{E}$. Also,

$$\mathbf{G}\backslash\mathbf{E} \subseteq \bigcap_{k=1}^\infty (\mathbf{O}_k\backslash\mathbf{E}). \tag{2.33}$$

Really, let $x \in \mathbf{G}\backslash\mathbf{E}$ be arbitrarily chosen. Then $x \in \mathbf{G}$ and $x \notin \mathbf{E}$. By the definition of the set $\mathbf{G}$, it follows that $x \in \mathbf{O}_k$ for any $k \in \mathbf{N}$. Therefore, $x \in \mathbf{O}_k\backslash\mathbf{E}$ for any $k \in \mathbf{N}$. Hence, $x \in \cap_{k=1}^\infty(\mathbf{O}_k\backslash\mathbf{E})$. Because $x \in \mathbf{G}\backslash\mathbf{E}$ was arbitrarily chosen and for it we obtain that it is an element of the set $\cap_{k=1}^\infty(\mathbf{O}_k\backslash\mathbf{E})$, we get the relation (2.33). By (2.33) and Theorem 2.1, we find

$$\begin{aligned} m^\star(\mathbf{G}\backslash\mathbf{E}) &\le m^\star\left(\bigcap_{k=1}^{\infty}(\mathbf{O}_k\backslash\mathbf{E})\right) \\ &\le m^\star(\mathbf{O}_k\backslash\mathbf{E}) \\ &< \frac{1}{k} \end{aligned}$$

for any $k \in \mathbf{N}$. Therefore, $m^\star(\mathbf{G}\backslash\mathbf{E}) = 0$.

2. Let there is a $\mathbf{G}_\delta$ set $\mathbf{G}$ containing $\mathbf{E}$ such that $m^\star(\mathbf{G}\backslash\mathbf{E}) = 0$. Since any set of outer measure zero is measurable, we have that $\mathbf{G}\backslash\mathbf{E} \in \mathscr{M}$. Because $\mathscr{M}$ is a σ algebra, we have that $(\mathbf{G}\backslash\mathbf{E})^c \in \mathscr{M}$ and $\mathbf{G}\cap(\mathbf{G}\backslash\mathbf{E})^c \in \mathscr{M}$. From here, using that $\mathbf{E} = \mathbf{G}\cap(\mathbf{G}\backslash\mathbf{E})^c$ (one can prove as it is proved (2.30)), we conclude that $\mathbf{E}$ is measurable. This completes the proof. □

Theorem 2.21. *Let* $\mathbf{E}$ *be any set of real numbers. Then* $\mathbf{E}$ *is measurable if and only if for each* $\varepsilon > 0$ *there is a closed set* $\mathbf{B}$ *contained in* $\mathbf{E}$ *such that* $m^\star(\mathbf{E}\backslash\mathbf{B}) < \varepsilon$.

Proof.

1. Let $\mathbf{E}$ be a measurable set. Then $\mathbf{E}^c$ is a measurable set. We take $\varepsilon > 0$ arbitrarily. By Theorem 2.19, it follows that there is an open set $\mathbf{O}$ containing $\mathbf{E}^c$ such that

$$m^\star(\mathbf{O}\backslash\mathbf{E}^c) < \varepsilon. \tag{2.34}$$

Note that $\mathbf{O}^c$ is a closed set and $\mathbf{O}^c \subset \mathbf{E}$. Now, we will prove that

$$\mathbf{E}\backslash\mathbf{O}^c = \mathbf{O}\backslash\mathbf{E}^c. \tag{2.35}$$

Really, let $x \in \mathbf{E}\backslash\mathbf{O}^c$ be arbitrarily chosen. Then $x \in \mathbf{E}$ and $x \notin \mathbf{O}^c$. Hence, $x \in \mathbf{O}$ and $x \notin \mathbf{E}^c$. Therefore $x \in \mathbf{O}\backslash\mathbf{E}^c$. Since $x \in \mathbf{E}\backslash\mathbf{O}^c$ was arbitrarily chosen and for it we get that it is an element of $\mathbf{O}\backslash\mathbf{E}^c$, we obtain

$$\mathbf{E}\backslash\mathbf{O}^c \subseteq \mathbf{O}\backslash\mathbf{E}^c. \tag{2.36}$$

Let now $x \in \mathbf{O}\backslash\mathbf{E}^c$ be arbitrarily chosen. Then $x \in \mathbf{O}$ and $x \notin \mathbf{E}^c$. Hence, $x \notin \mathbf{O}^c$ and $x \in \mathbf{E}$. Therefore $x \in \mathbf{E}\backslash\mathbf{O}^c$. Because

$x \in \mathbf{O}\backslash\mathbf{E}^c$ was arbitrarily chosen and we get that it is an element of the set $\mathbf{E}\backslash\mathbf{O}^c$, we get

$$\mathbf{O}\backslash\mathbf{E}^c \subseteq \mathbf{E}\backslash\mathbf{O}^c.$$

From the last relation and from (2.36), we obtain the relation (2.35). By (2.35), Theorems 2.1 and (2.34), we find

$$\begin{aligned} m^\star(\mathbf{E}\backslash\mathbf{O}^c) &= m^\star(\mathbf{O}\backslash\mathbf{E}^c) \\ &< \varepsilon. \end{aligned}$$

We set $\mathbf{B} = \mathbf{O}^c$.

2. Let for any $\varepsilon > 0$ there is a closed set $\mathbf{B}$, contained in $\mathbf{E}$, such that $m^\star(\mathbf{E}\backslash\mathbf{B}) < \varepsilon$. By (2.35), we obtain $m^\star(\mathbf{B}^c\backslash\mathbf{E}^c) < \varepsilon$. Note that $\mathbf{B}^c$ is an open set and $\mathbf{E}^c \subseteq \mathbf{B}^c$. Hence and Theorem 2.19, we conclude that $\mathbf{E}^c$ is measurable, i.e., $\mathbf{E}^c \in \mathscr{M}$. Because $\mathscr{M}$ is a σ-algebra, we get that $\mathbf{E} \in \mathscr{M}$, i.e., $\mathbf{E}$ is a measurable set. This completes the proof. □

Theorem 2.22. *Let* $\mathbf{E}$ *be any set of real numbers. Then* $\mathbf{E}$ *is measurable if and only if there is an* F_σ *set* $\mathbf{B}$, *contained in* $\mathbf{E}$, *such that* $m^\star(\mathbf{E}\backslash\mathbf{B}) = 0$.

Proof. By Theorem 2.20, it follows that the set $\mathbf{E}^c$ is measurable if and only if there is a $\mathbf{G}_\delta$ set $\mathbf{A}$, containing $\mathbf{E}^c$, such that $m^\star(\mathbf{A}\backslash\mathbf{E}^c) = 0$. Hence, using (2.35) and using that the complement of a $\mathbf{G}_\delta$ set is an $\mathbf{F}_\sigma$ set, we get the desired result. This completes the proof. □

Theorem 2.23. *Let* $\mathbf{E}$ *be a measurable set of finite outer measure. Then for each* $\varepsilon > 0$ *there is a finite disjoint collection of open intervals* $\{\mathbf{I}_k\}_{k=1}^n$ *for which if* $\mathbf{O} = \bigcup_{k=1}^n \mathbf{I}_k$, *then*

$$m^\star(\mathbf{E}\backslash\mathbf{O}) + m^\star(\mathbf{O}\backslash\mathbf{E}) < \varepsilon.$$

Proof. By Theorem 2.19, it follows that for each $\varepsilon > 0$ there is an open set $\mathbf{B}$, containing $\mathbf{E}$, such that $m^\star(\mathbf{B}\backslash\mathbf{E}) < \frac{\varepsilon}{2}$. Since every open set is measurable, we have that $\mathbf{B}$ is measurable. Also, we have

$$\begin{aligned} m^\star(\mathbf{B}) &\le m^\star(\mathbf{E}) + m^\star(\mathbf{B}\backslash\mathbf{E}) \\ &< m^\star(\mathbf{E}) + \frac{\varepsilon}{2}. \end{aligned}$$

Therefore, $\mathbf{B}$ is a measurable set of finite outer measure. Because every open set of real numbers is a disjoint union of a countable collection of open intervals, there is a disjoint countable collection of open intervals $\{\mathbf{I}_k\}_{k\in\mathbf{N}}$ for which

$$\mathbf{B} = \bigcup_{k=1}^{\infty} \mathbf{I}_k, \quad \mathbf{I}_k \cap \mathbf{I}_l = \emptyset, \quad k \neq l, \quad k, l \in \mathbf{N}.$$

Since for any $n \in \mathbf{N}$ we have $\cup_{k=1}^{n} \mathbf{I}_k \subseteq \mathbf{B}$, using Theorem 2.1, we get

$$\begin{aligned}\sum_{k=1}^{n} l(\mathbf{I}_k) &= \sum_{k=1}^{n} m^\star(\mathbf{I}_k) \\ &= m^\star\left(\bigcup_{k=1}^{n} \mathbf{I}_k\right) \\ &\le m^\star(\mathbf{B}) \\ &< \infty.\end{aligned}$$

Therefore,

$$\sum_{k=1}^{\infty} l(\mathbf{I}_k) < \infty.$$

From here, there is an $n \in \mathbf{N}$ such that

$$\sum_{k=n+1}^{\infty} l(\mathbf{I}_k) < \frac{\varepsilon}{2}.$$

We define

$$\mathbf{O} = \bigcup_{k=1}^{n} \mathbf{I}_k.$$

Using that $\mathbf{O}\backslash\mathbf{E} \subseteq \mathbf{B}\backslash\mathbf{E}$ and Theorem 2.1, we obtain

$$\begin{aligned}m^\star(\mathbf{O}\backslash\mathbf{E}) &\le m^\star(\mathbf{B}\backslash\mathbf{E}) \\ &< \frac{\varepsilon}{2}.\end{aligned}$$

On the other hand, $\mathbf{E} \subseteq \mathbf{B}$ and

$$\mathbf{E}\backslash\mathbf{O} \subseteq \mathbf{B}\backslash\mathbf{O} = \bigcup_{k=n+1}^{\infty} \mathbf{I}_k.$$

By the definition of outer measure, we get

$$m^{\star}(\mathbf{E}\backslash\mathbf{O}) \leq \sum_{k=n+1}^{\infty} l(\mathbf{I}_k) < \frac{\varepsilon}{2}.$$

Thus,

$$\begin{aligned} m^{\star}(\mathbf{O}\backslash\mathbf{E}) + m^{\star}(\mathbf{E}\backslash\mathbf{O}) &< \frac{\varepsilon}{2} + \frac{\varepsilon}{2} \\ &= \varepsilon. \end{aligned}$$

This completes the proof. □

Exercise 2.4. Show that a set $\mathbf{E}$ is measurable if and only if for each $\varepsilon > 0$, there is a closed set $\mathbf{A}$ and an open set $\mathbf{B}$ for which

$$\mathbf{A} \subseteq \mathbf{E} \subseteq \mathbf{B} \quad \text{and} \quad m^{\star}(\mathbf{B}\backslash\mathbf{A}) < \varepsilon.$$

Solution. By Theorems 2.19 and 2.21, it follows that the set $\mathbf{E}$ is measurable if and only if for each $\varepsilon > 0$ there are a closed set $\mathbf{A}$ and an open set $\mathbf{B}$ such that

$$\mathbf{A} \subseteq \mathbf{E} \subseteq \mathbf{B}$$

and

$$m^{\star}(\mathbf{E}\backslash\mathbf{A}) < \frac{\varepsilon}{2}, \quad m^{\star}(\mathbf{B}\backslash\mathbf{E}) < \frac{\varepsilon}{2}. \tag{2.37}$$

Now, we will prove

$$\mathbf{B}\backslash\mathbf{A} = (\mathbf{B}\backslash\mathbf{E}) \cup (\mathbf{E}\backslash\mathbf{A}). \tag{2.38}$$

Let $x \in \mathbf{B}\backslash\mathbf{A}$ be arbitrarily chosen. Then $x \in \mathbf{B}$ and $x \notin \mathbf{A}$.

1. If $x \in \mathbf{E}$, then $x \in \mathbf{E}\backslash\mathbf{A}$ and $x \in (\mathbf{B}\backslash\mathbf{E}) \cup (\mathbf{E}\backslash\mathbf{A})$.
2. If $x \notin \mathbf{E}$, then $x \in \mathbf{B}\backslash\mathbf{E}$ and $x \in (\mathbf{B}\backslash\mathbf{E}) \cup (\mathbf{E}\backslash\mathbf{A})$.

Because $x \in \mathbf{B}\backslash\mathbf{A}$ was arbitrarily chosen and for it we get that it is an element of $(\mathbf{B}\backslash\mathbf{E}) \cup (\mathbf{E}\backslash\mathbf{A})$, we get

$$\mathbf{B}\backslash\mathbf{A} \subseteq (\mathbf{B}\backslash\mathbf{E}) \cup (\mathbf{E}\backslash\mathbf{A}). \tag{2.39}$$

Let now $x \in (\mathbf{B}\backslash\mathbf{E}) \cup (\mathbf{E}\backslash\mathbf{A})$ be arbitrarily chosen. Then $x \in \mathbf{B}\backslash\mathbf{E}$ or $x \in \mathbf{E}\backslash\mathbf{A}$.

1. Let $x \in \mathbf{B} \backslash \mathbf{E}$. Then $x \in \mathbf{B}$ and $x \notin \mathbf{E}$. Hence, $x \notin \mathbf{A}$ and $x \in \mathbf{B}$. Therefore $x \in \mathbf{B} \backslash \mathbf{A}$.
2. Let $x \in \mathbf{E} \backslash \mathbf{A}$. Then $x \in \mathbf{E}$ and $x \notin \mathbf{A}$. Hence, $x \in \mathbf{B}$ and $x \notin \mathbf{A}$. Therefore $x \in \mathbf{B} \backslash \mathbf{A}$.

Because $x \in (\mathbf{B} \backslash \mathbf{E}) \cup (\mathbf{E} \backslash \mathbf{A})$ was arbitrarily chosen and we get that it is an element of $\mathbf{B} \backslash \mathbf{A}$, we conclude that

$$(\mathbf{B} \backslash \mathbf{E}) \cup (\mathbf{E} \backslash \mathbf{A}) \subseteq \mathbf{B} \backslash \mathbf{A}.$$

From the last relation and from (2.39), we get the relation (2.38). By (2.38) and Theorem 2.1, we obtain

$$\begin{aligned} m^{\star}(\mathbf{B} \backslash \mathbf{A}) &= m^{\star}((\mathbf{B} \backslash \mathbf{E}) \cup (\mathbf{E} \backslash \mathbf{A})) \\ &\leq m^{\star}(\mathbf{B} \backslash \mathbf{E}) + m^{\star}(\mathbf{E} \backslash \mathbf{A}) \\ &< \frac{\varepsilon}{2} + \frac{\varepsilon}{2} \\ &= \varepsilon. \end{aligned}$$

This completes the proof. □

Exercise 2.5. Let $\mathbf{E}$ has finite outer measure. Show that there is an F_σ set $\mathbf{A}$ and a G_δ set $\mathbf{B}$ such that

$$\mathbf{A} \subseteq \mathbf{E} \subseteq \mathbf{B} \quad \text{and} \quad m^{\star}(\mathbf{A}) = m^{\star}(\mathbf{E}) = m^{\star}(\mathbf{B}).$$

2.2 The Lebesgue Measure: The Borel–Cantelli Lemma

Definition 2.9. The restriction of the set function outer measure to the class of measurable sets is called Lebesgue measure. It is denoted by m. In other words, if the set $\mathbf{E}$ is measurable, its Lebesgue measure, $m(\mathbf{E})$, is defined by

$$m(\mathbf{E}) = m^{\star}(\mathbf{E}).$$

Theorem 2.24. *The Lebesgue measure is countably additive, i.e., if $\{\mathbf{E}_k\}_{k \in \mathbf{N}}$ is a countable disjoint collection of measurable sets, then its union $\cup_{k=1}^{\infty} \mathbf{E}_k$ is also measurable and*

$$m\left(\bigcup_{k=1}^{\infty} \mathbf{E}_k\right) = \sum_{k=1}^{\infty} m(\mathbf{E}_k).$$

Proof. By Theorem 2.13, it follows that $\cup_{k=1}^{\infty}\mathbf{E}_k$ is a measurable set. By Theorem 2.5, we get

$$m\left(\bigcup_{k=1}^{\infty}\mathbf{E}_k\right) \leq \sum_{k=1}^{\infty} m(\mathbf{E}_k). \tag{2.40}$$

On the other hand, using Theorem 2.10, we obtain

$$m\left(\bigcup_{k=1}^{n}\mathbf{E}_k\right) = \sum_{k=1}^{n} m(\mathbf{E}_k). \tag{2.41}$$

Since $\cup_{k=1}^{n}\mathbf{E}_k \subseteq \cup_{k=1}^{\infty}\mathbf{E}_k$ for any $n \in \mathbf{N}$, using Theorem 2.1, we obtain

$$m\left(\bigcup_{k=1}^{n}\mathbf{E}_k\right) \leq m\left(\bigcup_{k=1}^{\infty}\mathbf{E}_k\right)$$

for any $n \in \mathbf{N}$. Hence and (2.41), we find

$$m\left(\bigcup_{k=1}^{\infty}\mathbf{E}_k\right) \geq \sum_{k=1}^{n} m(\mathbf{E}_k)$$

for any $n \in \mathbf{N}$. Therefore

$$m\left(\bigcup_{k=1}^{\infty}\mathbf{E}_k\right) \geq \sum_{k=1}^{\infty} m(\mathbf{E}_k).$$

From the last inequality and from (2.40), we get the desired result. This completes the proof. □

Remark 2.5. The set function Lebesgue measure, defined on the σ-algebra of Lebesgue measurable sets, assigns the length to any interval, is translation invariant, and is countable additive.

Definition 2.10. A countable collection of sets $\{\mathbf{E}_k\}_{k\in\mathbf{N}}$ is said to be

1. Ascending provided for each $k \in \mathbf{N}$, $\mathbf{E}_k \subseteq \mathbf{E}_{k+1}$.
2. Descending provided for each $k \in \mathbf{N}$, $\mathbf{E}_{k+1} \subseteq \mathbf{E}_k$.

Theorem 2.25 (The Continuity of Measure). *Lebesgue measure possesses the following properties.*

1. *If* $\{\mathbf{E}_k\}_{k\in\mathbf{N}}$ *is an ascendent collection of measurable sets, then*

$$m\left(\bigcup_{k=1}^{\infty}\mathbf{E}_k\right)=\lim_{k\to\infty}m(\mathbf{E}_k). \tag{2.42}$$

2. *If* $\{\mathbf{E}_k\}_{k\in\mathbf{N}}$ *is a descendent collection of measurable sets and* $m(\mathbf{E}_1)<\infty$, *then*

$$m\left(\bigcap_{k=1}^{\infty}\mathbf{E}_k\right)=\lim_{k\to\infty}m(\mathbf{E}_k). \tag{2.43}$$

Proof.

1. Suppose that there is an index l such that $m(\mathbf{E}_l)=\infty$. Then, using that

$$m\left(\bigcup_{k=1}^{\infty}\mathbf{E}_k\right)\geq m(\mathbf{E}_l)\quad\text{and}\quad m(\mathbf{E}_p)\geq m(\mathbf{E}_l)$$

for any $p\geq l$, $p\in\mathbf{N}$, we get

$$m\left(\bigcup_{k=1}^{\infty}\mathbf{E}_k\right)=\infty\quad\text{and}\quad m(\mathbf{E}_p)=\infty$$

for any $p\geq l$, $p\in\mathbf{N}$. Therefore, (2.42) holds. Now, we assume that $m(\mathbf{E}_l)<\infty$ for any $l\in\mathbf{N}$. We set $A_0=\emptyset$ and define the sets

$$\mathbf{A}_k=\mathbf{E}_k\backslash\mathbf{E}_{k-1},\quad k\in\mathbf{N}.$$

We will prove that $\{\mathbf{A}_k\}_{k\in\mathbf{N}}$ is a disjoint collection. Assume that there are $k,l\in\mathbf{N}$, $k>l$, such that there is an $x\in\mathbf{A}_k\cap\mathbf{A}_l$. By the definition of the sets $\mathbf{A}_k$, we obtain $x\notin\mathbf{E}_{k-1}$. Since $\{\mathbf{E}_k\}_{k\in\mathbf{N}}$ is an ascendent collection, we conclude that $x\notin\mathbf{E}_p$ for any $p\leq k-1$, $p\in\mathbf{N}$. Therefore, $x\notin\mathbf{A}_l$. This is a contradiction. Consequently $\{\mathbf{A}_k\}_{k\in\mathbf{N}}$ is a disjoint collection. Now, we will prove that

$$\bigcup_{k=1}^{\infty}\mathbf{A}_k=\bigcup_{k=1}^{\infty}\mathbf{E}_k. \tag{2.44}$$

Let $x\in\cup_{k=1}^{\infty}\mathbf{A}_k$ be arbitrarily chosen. Then there is an $l\in\mathbf{N}$ such that $x\in\mathbf{A}_l=\mathbf{E}_l\backslash\mathbf{E}_{l-1}$. Hence, $x\in\mathbf{E}_l$ and $x\in\cup_{k=1}^{\infty}\mathbf{E}_k$.

Because $x \in \cup_{k=1}^{\infty} \mathbf{A}_k$ was arbitrarily chosen and for it we get that it is an element of $\cup_{k=1}^{\infty} \mathbf{E}_k$, we conclude that

$$\bigcup_{k=1}^{\infty} \mathbf{A}_k \subseteq \bigcup_{k=1}^{\infty} \mathbf{E}_k. \tag{2.45}$$

Let now $x \in \cup_{k=1}^{\infty} \mathbf{E}_k$ be arbitrarily chosen. Then there is an $l \in \mathbf{N}$ such that $x \in \mathbf{E}_l$. Assume that $x \notin \mathbf{A}_k$ for any $k \in \mathbf{N}$. Then $x \notin \mathbf{E}_k$ for any $k \in \mathbf{N}$. In particular, $x \notin \mathbf{E}_l$. This is a contradiction. Therefore, there is a $p \in \mathbf{N}$ such that $x \in \mathbf{A}_p$. Hence, $x \in \cup_{k=1}^{\infty} \mathbf{A}_k$. Because $x \in \cup_{k=1}^{\infty} \mathbf{E}_k$ was arbitrarily chosen and for it we get that it is an element of $\cup_{k=1}^{\infty} \mathbf{A}_k$, we obtain that

$$\bigcup_{k=1}^{\infty} \mathbf{E}_k \subseteq \bigcup_{k=1}^{\infty} \mathbf{A}_k.$$

From the last relation and from (2.45), we get the relation (2.44). By the countable additivity of m, we have

$$\begin{aligned} m\left(\bigcup_{k=1}^{\infty} \mathbf{E}_k\right) &= m\left(\bigcup_{k=1}^{\infty} \mathbf{A}_k\right) \\ &= \sum_{k=1}^{\infty} m\left(\mathbf{A}_k\right) \\ &= \sum_{k=1}^{\infty} m\left(\mathbf{E}_k \backslash \mathbf{E}_{k-1}\right) \\ &= \sum_{k=1}^{\infty} \left(m\left(\mathbf{E}_k\right) - m\left(\mathbf{E}_{k-1}\right)\right) \\ &= \lim_{n\to\infty} \sum_{k=1}^{n} \left(m\left(\mathbf{E}_k\right) - m\left(\mathbf{E}_{k-1}\right)\right) \\ &= \lim_{n\to\infty} \left(m\left(\mathbf{E}_n\right) - m\left(\emptyset\right)\right) \\ &= \lim_{n\to\infty} m\left(\mathbf{E}_n\right). \end{aligned}$$

2. Let

$$\mathbf{B}_k = \mathbf{E}_1 \backslash \mathbf{E}_k, \quad k \in \mathbf{N}.$$

Since $\{\mathbf{E}_k\}_{k\in\mathbf{N}}$ is a descendent collection, we have that $\{\mathbf{B}_k\}_{k\in\mathbf{N}}$ is an ascendent collection. By (2.42), we get

$$m\left(\bigcup_{k=1}^{\infty}\mathbf{B}_k\right) = \lim_{n\to\infty} m(\mathbf{B}_n). \tag{2.46}$$

Now, we will prove that

$$\bigcup_{k=1}^{\infty}\mathbf{B}_k = \mathbf{E}_1 \Bigg\backslash \left(\bigcap_{k=1}^{\infty}\mathbf{E}_k\right). \tag{2.47}$$

Let $x \in \cup_{k=1}^{\infty}\mathbf{B}_k$ be arbitrarily chosen. Then there is an $l \in \mathbf{N}$ such that $x \in \mathbf{B}_l = \mathbf{E}_1\backslash\mathbf{E}_l$. Hence, $x \in \mathbf{E}_1$ and $x \notin \mathbf{E}_l$. Therefore, $x \notin \cap_{k=1}^{\infty}\mathbf{E}_k$. From here, $x \in \mathbf{E}_1\backslash(\cap_{k=1}^{\infty}\mathbf{E}_k)$. Because $x \in \cup_{k=1}^{\infty}\mathbf{B}_k$ was arbitrarily chosen and for it we get that it is an element of $\mathbf{E}_1\backslash(\cap_{k=1}^{\infty}\mathbf{E}_k)$, we conclude that

$$\bigcup_{k=1}^{\infty}\mathbf{B}_k \subseteq \mathbf{E}_1 \Bigg\backslash \left(\bigcap_{k=1}^{\infty}\mathbf{E}_k\right). \tag{2.48}$$

Let $x \in \mathbf{E}_1\backslash(\cap_{k=1}^{\infty}\mathbf{E}_k)$ be arbitrarily chosen. Then $x \in \mathbf{E}_1$ and $x \notin \cap_{k=1}^{\infty}\mathbf{E}_k$. Hence, there is an $l \in \mathbf{N}$, $l \neq 1$, so that $x \notin \mathbf{E}_l$. Therefore, $x \in \mathbf{B}_l$ and $x \in \cup_{k=1}^{\infty}\mathbf{B}_k$. Since $x \in \mathbf{E}_1\backslash(\cap_{k=1}^{\infty}\mathbf{E}_k)$ was arbitrarily chosen and for it we get that it is an element of the set $\cup_{k=1}^{\infty}\mathbf{B}_k$, we conclude that

$$\mathbf{E}_1 \Bigg\backslash \left(\bigcap_{k=1}^{\infty}\mathbf{E}_k\right) \subseteq \bigcup_{k=1}^{\infty}\mathbf{B}_k.$$

From the last relation and from (2.48), we get the relation (2.47). By the relation (2.47), we find

$$\begin{aligned} m\left(\bigcup_{k=1}^{\infty}\mathbf{B}_k\right) &= m\left(\mathbf{E}_1 \Bigg\backslash \left(\bigcap_{k=1}^{\infty}\mathbf{E}_k\right)\right) \\ &= m(\mathbf{E}_1) - m\left(\bigcap_{k=1}^{\infty}\mathbf{E}_k\right). \end{aligned}$$

Hence and (2.46), we obtain

$$\begin{aligned} m(\mathbf{E}_1) - m\left(\bigcap_{k=1}^{\infty} \mathbf{E}_k\right) &= \lim_{n\to\infty} m(\mathbf{B}_n) \\ &= \lim_{n\to\infty} m(\mathbf{E}_1 \backslash \mathbf{E}_n) \\ &= \lim_{n\to\infty} (m(\mathbf{E}_1) - m(\mathbf{E}_n)) \\ &= m(\mathbf{E}_1) - \lim_{n\to\infty} m(\mathbf{E}_n), \end{aligned}$$

whereupon we obtain (2.43). This completes the proof. □

Definition 2.11. For a measurable set $\mathbf{E}$, we say that a property holds almost everywhere on $\mathbf{E}$, or it holds for almost all $x \in \mathbf{E}$, provided there is a subset $\mathbf{E}_0$ of $\mathbf{E}$ for which $m(\mathbf{E}_0) = 0$ and the property holds for all $x \in \mathbf{E}\backslash\mathbf{E}_0$.

Lemma 2.4 (Borel–Cantelli Lemma). *Let $\{\mathbf{E}_k\}_{k\in\mathbf{N}}$ be a countable collection of measurable sets for which $\sum_{k=1}^{\infty} m(\mathbf{E}_k) < \infty$. Then almost all $x \in \mathbf{R}$ belong to at most finitely many of the $\mathbf{E}_k$'s.*

Proof. By Theorem 2.5, we have

$$m\left(\bigcup_{k=1}^{\infty} \mathbf{E}_k\right) \le \sum_{k=1}^{\infty} m(\mathbf{E}_k) < \infty.$$

Note that $\{\cup_{k=n}^{\infty}\mathbf{E}_k\}_{n\in\mathbf{N}}$ is a descendent collection. Hence and (2.43), we obtain

$$\begin{aligned} m\left(\bigcap_{n=1}^{\infty}\left(\bigcup_{k=n}^{\infty} \mathbf{E}_k\right)\right) &= \lim_{n\to\infty} m\left(\bigcup_{k=n}^{\infty} \mathbf{E}_k\right) \\ &\le \lim_{n\to\infty} \sum_{k=n}^{\infty} m(\mathbf{E}_k) \\ &= 0. \end{aligned}$$

Therefore, almost all $x \in \mathbf{R}$ fail to belong to $\cap_{n=1}^{\infty}(\cup_{k=n}^{\infty}\mathbf{E}_k)$ and therefore belong to at most finitely many $\mathbf{E}_k$'s. This completes the proof. □

Exercise 2.6. Show that if $\mathbf{E}_1$ and $\mathbf{E}_2$ are measurable, then

$$m(\mathbf{E}_1 \cup \mathbf{E}_2) + m(\mathbf{E}_1 \cap \mathbf{E}_2) = m(\mathbf{E}_1) + m(\mathbf{E}_2).$$

2.3 Nonmeasurable Sets

Theorem 2.26. *Let $\mathbf{E}$ be a bounded measurable set of real numbers. Suppose that there is a bounded countable infinite set Λ of real numbers for which the collection of translates of $\mathbf{E}$, $\{\lambda + \mathbf{E}\}_{\lambda \in \Lambda}$, is disjoint. Then $m(\mathbf{E}) = 0$.*

Proof. Assume that $m(\mathbf{E}) > 0$. Because $\mathbf{E}$ is a bounded set, we have $m(\mathbf{E}) < \infty$. Since the translate of a measurable set is a measurable set, we have that $\lambda + \mathbf{E}$ is a measurable set and $m(\mathbf{E}) = m(\lambda + \mathbf{E})$ for any $\lambda \in \Lambda$. By Theorem 2.24, it follows that

$$\begin{aligned} m\left(\cup_{\lambda \in \Lambda} (\lambda + \mathbf{E})\right) &= \sum_{\lambda \in \Lambda} m(\lambda + \mathbf{E}) \\ &= \sum_{\lambda \in \Lambda} m(\mathbf{E}). \end{aligned} \tag{2.49}$$

Because $\mathbf{E}$ and Λ are bounded sets, we have that $\cup_{\lambda \in \Lambda} (\lambda + \mathbf{E})$ is a bounded set. Hence,

$$m\left(\bigcup_{\lambda \in \Lambda} (\lambda + \mathbf{E})\right) < \infty. \tag{2.50}$$

Since $m(\mathbf{E}) < \infty$ and Λ is countable infinite, we get $\sum_{\lambda \in \Lambda} m(\mathbf{E}) = \infty$. Hence and (2.49), we obtain $m(\cup_{\lambda \in \Lambda}(\lambda + \mathbf{E})) = \infty$. This contradicts with (2.50). Therefore, $m(\mathbf{E}) = 0$. This completes the proof. □

Definition 2.12. For any nonempty set $\mathbf{E}$ of real numbers, we say that two points x and y of $\mathbf{E}$ are rationally equivalent provided their difference $x - y$ belongs to the set of rational numbers $\mathbf{Q}$. We write $x \sim_{\mathbf{Q}} y$.

Proposition 2.1. *The relation $\sim_{\mathbf{Q}}$ is an equivalence relation.*

Proof. Let $x, y, z \in \mathbf{E}$.

1. Since $0 \in \mathbf{Q}$, we have that $x - x \in \mathbf{Q}$ and $x \sim_{\mathbf{Q}} x$.
2. Let $x \sim_{\mathbf{Q}} y$. Then $x - y \in \mathbf{Q}$. Hence, $y - x \in \mathbf{Q}$ and $y \sim_{\mathbf{Q}} x$.
3. Let $x \sim_{\mathbf{Q}} y$ and $y \sim_{\mathbf{Q}} z$. Then $x - y \in \mathbf{Q}$ and $y - z \in \mathbf{Q}$. Hence,

$$x - z = (x - y) + (y - z) \in \mathbf{Q}.$$

Therefore, $x \sim_{\mathbf{Q}} z$. This completes the proof. □

For the rational equivalence relation, there is the disjoint decomposition of $\mathbf{E}$ into the collection of equivalence classes.

Definition 2.13. Let $\mathscr{F}$ be a nonempty family of nonempty sets. A choice function f on $\mathscr{F}$ is a function $f : \mathscr{F} \longmapsto \cup_{\mathbf{F} \in \mathscr{F}} \mathbf{F}$ with the property that for each set $\mathbf{F}$ in $\mathscr{F}$, $f(\mathbf{F})$ is a member of $\mathbf{F}$.

Zermelo's Axiom of Choice. Let $\mathscr{F}$ be a nonempty collection of nonempty sets. Then there is a choice function on $\mathscr{F}$.

Definition 2.14. A choice set for the rational equivalence relation on $\mathbf{E}$ we mean a set $\mathscr{C}_{\mathbf{E}}$ consisting of exactly one member of each equivalence class.

By Zermelo's axiom of choice there are such choice sets. A choice set $\mathscr{C}_{\mathbf{E}}$ has the following properties:

1. The difference of two points of $\mathscr{C}_{\mathbf{E}}$ is not rational.
2. For each point $x \in \mathbf{E}$ there is a $c \in \mathscr{C}_{\mathbf{E}}$ such that $x = c + q$ with q a rational number.

Note that for any set $\Lambda \subseteq \mathbf{Q}$ we have that $\{\lambda + \mathscr{C}_{\mathbf{E}}\}_{\lambda \in \Lambda}$ is disjoint.

Theorem 2.27 (Vitali's Theorem). *Any set* $\mathbf{E}$ *of real numbers with positive outer measure contains a subset that fails to be measurable.*

Proof. Suppose that $\mathbf{E}$ is bounded. Let $\mathscr{C}_{\mathbf{E}}$ be any choice set for the rational equivalence relation on $\mathbf{E}$. Assume that $\mathscr{C}_{\mathbf{E}}$ is measurable.

Let Λ_0 be any bounded, countable infinite set of rational numbers. By Theorem 2.26, we get $m(\mathscr{C}_{\mathbf{E}}) = 0$. Hence,

$$m\left(\bigcup_{\lambda \in \Lambda_0} (\lambda + \mathscr{C}_{\mathbf{E}})\right) = \sum_{\lambda \in \Lambda_0} m(\lambda + \mathscr{C}_{\mathbf{E}})$$
$$= 0.$$

Since $\mathbf{E}$ is bounded, there is a $b > 0$ such that $\mathbf{E} \subset [-b, b]$. We take

$$\Lambda_0 = [-2b, 2b] \cap \mathbf{Q}.$$

Then Λ_0 is bounded and countably infinite. We will prove that

$$\mathbf{E} \subseteq \bigcup_{\lambda \in [-2b,2b] \cap \mathbf{Q}} (\lambda + \mathscr{C}_{\mathbf{E}}). \tag{2.51}$$

Let $x \in \mathbf{E}$ be arbitrarily chosen. Then there is $c \in \mathscr{C}_{\mathbf{E}}$ such that $x = c + q$ with q a rational number. Since $\mathbf{E} \subset [-b, b]$, we have that $x, c \in [-b, b]$. Therefore $q \in [-2b, 2b]$ and $x \in q + \mathscr{C}_{\mathbf{E}}$. Hence, $x \in \cup_{\lambda \in [-2b,2b] \cap \mathbf{Q}} (\lambda + \mathscr{C}_{\mathbf{E}})$. Because $x \in \mathbf{E}$ was arbitrarily chosen and for it we get that it is an element of the set $\cup_{\lambda \in [-2b,2b] \cap \mathbf{Q}} (\lambda + \mathscr{C}_{\mathbf{E}})$, we obtain the relation (2.51). By (2.51), we get

$$m(\mathbf{E}) \leq \sum_{\lambda \in [-2b,2b] \cap \mathbf{Q}} m(\lambda + \mathscr{C}_{\mathbf{E}})$$
$$= 0,$$

which is a contradiction. Therefore, $\mathscr{C}_{\mathbf{E}}$ is not measurable. If $\mathbf{E}$ is unbounded, then it can be represented as an union of bounded sets $\mathbf{E}_k$. Then we take a such set $\mathbf{E}_k$ and as in above we prove that $\mathscr{C}_{\mathbf{E}_k}$ is not measurable. This completes the proof. □

Theorem 2.28. *There are disjoint sets* $\mathbf{A}$ *and* $\mathbf{B}$ *such that*

$$m^\star(\mathbf{A} \cup \mathbf{B}) < m^\star(\mathbf{A}) + m^\star(\mathbf{B}).$$

Proof. Assume that for every disjoint sets $\mathbf{A}$ and $\mathbf{B}$ we have

$$m^\star(\mathbf{A} \cup \mathbf{B}) = m^\star(\mathbf{A}) + m^\star(\mathbf{B}). \tag{2.52}$$

Let the set $\mathbf{E}$ be arbitrarily chosen set of real numbers. Then for any set $\mathbf{A}$ of real numbers we have

$$\mathbf{A} = (\mathbf{A} \cap \mathbf{E}) \cup (\mathbf{A} \cap \mathbf{E}^c) \quad \text{and} \quad (\mathbf{A} \cap \mathbf{E}) \cap (\mathbf{A} \cap \mathbf{E}^c) = \emptyset.$$

Hence and (2.52), we get

$$\begin{aligned} m^\star(\mathbf{A}) &= m^\star((\mathbf{A} \cap \mathbf{E}) \cup (\mathbf{A} \cap \mathbf{E}^c)) \\ &= m^\star(\mathbf{A} \cap \mathbf{E}) + m^\star(\mathbf{A} \cap \mathbf{E}^c). \end{aligned}$$

Therefore, $\mathbf{E}$ is a measurable set. Because $\mathbf{E}$ was arbitrarily chosen set of real numbers, we conclude that every set of real numbers is measurable. This contradicts with Theorem 2.27. This completes the proof. □

2.4 The Cantor Set: The Cantor–Lebesgue Function

Consider the interval $\mathbf{I} = [0, 1]$. We subdivide the interval $\mathbf{I}$ into three intervals, each of length $\frac{1}{3}$, and we remove the interior of the middle interval, i.e., we remove the interval $\left(\frac{1}{3}, \frac{2}{3}\right)$ from the interval $[0, 1]$. We set

$$\mathbf{C}_1 = \left[0, \frac{1}{3}\right] \cup \left[\frac{2}{3}, 1\right].$$

We now repeat this operation "open middle one-third removal" on each of the two intervals in $\mathbf{C}_1$ to obtain a closed set, which is the union of 2^2 closed intervals, each of length $\frac{1}{3^2}$,

$$\mathbf{C}_2 = \left[0, \frac{1}{9}\right] \cup \left[\frac{2}{9}, \frac{1}{3}\right] \cup \left[\frac{2}{3}, \frac{7}{9}\right] \cup \left[\frac{8}{9}, 1\right].$$

We now repeat the above operation "open middle one-third removal" on each of the four intervals of $\mathbf{C}_2$ to obtain a closed set $\mathbf{C}_3$, which is the union of 2^3 closed intervals, each of length $\frac{1}{3^3}$. We continue this operation countably many times to obtain the countable collection of sets $\{\mathbf{C}_k\}_{k \in \mathbf{N}}$.

Definition 2.15. We define the Cantor set $\mathbb{C}$ by

$$\mathbb{C} = \bigcap_{k=1}^{\infty} \mathbf{C}_k.$$

The collection $\{\mathbf{C}_k\}_{k\in\mathbf{N}}$ has the following properties:

1. It is a descending sequence of closed sets.
2. For each $k \in \mathbf{N}$, the set $\mathbf{C}_k$ is the disjoint union of 2^k closed intervals, each of length $\frac{1}{3^k}$.

Theorem 2.29. *The Cantor set $\mathbb{C}$ is a closed, uncountable set of measure zero.*

Proof. Since $\mathbf{C}_k$ are closed sets for any $k \in \mathbb{N}$ and the intersection of closed sets is a closed set, we conclude that $\mathbb{C}$ is a closed set. By Theorem 2.18 we have that every open set is measurable. Since the collection of the measurable sets is a σ-algebra, we conclude that every closed set is measurable. Therefore, $\mathbf{C}_k$ are measurable sets for any $k \in \mathbf{N}$. Because the countable intersection of measurable sets is a measurable set, we obtain that $\mathbb{C}$ is a measurable set. Also, $\mathbb{C} \subseteq \mathbf{C}_k$ for any $k \in \mathbf{N}$. Therefore,

$$m(\mathbb{C}) \leq m(\mathbf{C}_k) = \left(\frac{2}{3}\right)^k$$

for any $k \in \mathbf{N}$. Consequently, $m(\mathbb{C}) = 0$. Now, we suppose that $\mathbb{C}$ is countable. Let $\{c_k\}_{k\in\mathbf{N}}$ be an enumeration of $\mathbb{C}$. One of the two disjoint Cantor intervals whose union is $\mathbf{C}_1$ does not contain the point c_1. We denote it by $\mathbf{F}_1$. One of the two disjoint Cantor intervals in $\mathbf{C}_2$ whose union is $\mathbf{F}_1$ does not contain the point c_2. We denote it by $\mathbf{F}_2$. Continuing in this way, we get a countable collection of sets $\{\mathbf{F}_k\}_{k\in\mathbf{N}}$ such that

1. $\mathbf{F}_k$ is closed for any $k \in \mathbf{N}$,
2. $\mathbf{F}_k \subseteq \mathbf{C}_k$ for any $k \in \mathbf{N}$,
3. $\mathbf{F}_{k+1} \subseteq \mathbf{F}_k$ for any $k \in \mathbf{N}$,
4. $c_k \notin \mathbf{F}_k$ for any $k \in \mathbf{N}$.

From here and from the Nested Set Theorem, it follows that $\cap_{k=1}^{\infty}\mathbf{F}_k$ is not empty. Let $x \in \cap_{k=1}^{\infty}\mathbf{F}_k$. From the construction of the collection $\{\mathbf{F}_k\}_{k\in\mathbf{N}}$, we have

$$\bigcap_{k=1}^{\infty}\mathbf{F}_k \subseteq \bigcap_{k=1}^{\infty}\mathbf{C}_k = \mathbb{C}.$$

Therefore, $x \in \mathbb{C}$. Hence, there is an $l \in \mathbf{N}$ such that $x = c_l$. Because $x \in \mathbf{F}_k$ for any $k \in \mathbf{N}$, we get that $x \in \mathbf{F}_l$. Hence, $c_l \in \mathbf{F}_l$. This is a contradiction. Consequently, $\mathbb{C}$ is uncountable. This completes the proof. □

Now, we define the sets

$$\mathbf{O}_k = [0,1]\backslash\mathbf{C}_k, \quad k \in \mathbf{N},$$
$$\mathbf{O} = \bigcup_{k=1}^{\infty}\mathbf{O}_k.$$

We have

$$\mathbf{O}_1 = \left(\frac{1}{3}, \frac{2}{3}\right),$$

$$\mathbf{O}_2 = \left(\frac{1}{9}, \frac{2}{9}\right) \cup \left(\frac{1}{3}, \frac{2}{3}\right) \cup \left(\frac{7}{9}, \frac{8}{9}\right),$$

$$\mathbf{O}_3 = \left(\frac{1}{27}, \frac{2}{27}\right) \cup \left(\frac{1}{9}, \frac{2}{9}\right) \cup \left(\frac{7}{27}, \frac{8}{27}\right) \cup \left(\frac{1}{3}, \frac{2}{3}\right) \cup \left(\frac{19}{27}, \frac{20}{27}\right)$$
$$\cup \left(\frac{7}{9}, \frac{8}{9}\right) \cup \left(\frac{25}{27}, \frac{26}{27}\right).$$

Now, we will prove that

$$\mathbb{C} = [0,1]\backslash\mathbf{O}. \tag{2.53}$$

Let $x \in \mathbb{C}$ be arbitrarily chosen. Then $x \in \mathbf{C}_k$ for any $k \in \mathbf{C}_k$. Hence, $x \notin \mathbf{O}_k$ for any $k \in \mathbf{N}$. Therefore, $x \in [0,1]$ and $x \notin \mathbf{O}$. Consequently,

$x \in [0,1]\backslash\mathbf{O}$. Because $x \in \mathbb{C}$ was arbitrarily chosen and we get that it is an element of $[0,1]\backslash\mathbf{O}$, we obtain

$$\mathbb{C} \subseteq [0,1]\backslash\mathbf{O}. \tag{2.54}$$

Let $x \in [0,1]\backslash\mathbf{O}$ be arbitrarily chosen. Then $x \in [0,1]$ and $x \notin \mathbf{O}$. Hence, $x \notin \mathbf{O}_k$ for any $k \in \mathbf{N}$. Therefore, $x \in \mathbf{C}_k$ for any $k \in \mathbf{N}$. From here, $x \in \mathbf{C}$. Because $x \in [0,1]\backslash\mathbf{O}$ was arbitrarily chosen and for it we get that it is an element of $\mathbf{C}$, we obtain the relation

$$[0,1]\backslash\mathbf{O} \subseteq \mathbb{C}.$$

From the last relation and from the relation (2.54), we get the relation (2.53).

Definition 2.16 (The Cantor–Lebesgue Function). Fix a natural number k. Define ϕ on $\mathbf{O}_k$ to be the increasing function on $\mathbf{O}_k$ which is a constant on each of its $2^k - 1$ intervals and takes the $2^k - 1$ values

$$\frac{1}{2^k}, \quad \frac{2}{2^k}, \quad \frac{3}{2^k}, \ldots, \quad \frac{2^k-1}{2^k}.$$

We extend ϕ to all $[0,1]$ by defining it on $\mathbf{C}$ as follows:

$$\phi(0) = 0 \quad \text{and} \quad \phi(x) = \sup\{\phi(t) : t \in \mathbf{O} \cap [0,x)\} \quad \text{if } x \in \mathbf{C}\backslash\{0\}.$$

Thus, on $\mathbf{O}_1$ we have

$$\phi(x) = \frac{1}{2} \quad \text{for } x \in \left(\frac{1}{3}, \frac{2}{3}\right),$$

on $\mathbf{O}_2$ we have

$$\phi(x) = \begin{cases} \dfrac{1}{4} & \text{if } x \in \left(\dfrac{1}{9}, \dfrac{2}{9}\right) \\[2ex] \dfrac{1}{2} & \text{if } x \in \left(\dfrac{1}{3}, \dfrac{2}{3}\right), \\[2ex] \dfrac{3}{4} & \text{if } x \in \left(\dfrac{7}{9}, \dfrac{8}{9}\right) \end{cases}$$

on $\mathbf{O}_3$ we have

$$\phi(x) = \begin{cases} \frac{1}{8} & \text{if } x \in \left(\frac{1}{27}, \frac{2}{27}\right) \\ \frac{1}{4} & \text{if } x \in \left(\frac{1}{9}, \frac{2}{9}\right) \\ \frac{3}{8} & \text{if } x \in \left(\frac{7}{27}, \frac{8}{27}\right) \\ \frac{1}{2} & \text{if } x \in \left(\frac{1}{3}, \frac{2}{3}\right), \\ \frac{5}{8} & \text{if } x \in \left(\frac{19}{27}, \frac{20}{27}\right) \\ \frac{3}{4} & \text{if } x \in \left(\frac{7}{9}, \frac{8}{9}\right) \\ \frac{7}{8} & \text{if } x \in \left(\frac{25}{27}, \frac{26}{27}\right) \end{cases}$$

and so on.

Theorem 2.30. *The Cantor–Lebesgue function ϕ is an increasing continuous function that maps $[0,1]$ onto $[0,1]$. Its derivative exists on $\mathbf{O}$ and*

$$\phi' = 0 \quad \text{on } \mathbf{O} \quad \text{while } m(\mathbf{O}) = 1.$$

Proof. Since $m(\mathbb{C}) = 0$, we get

$$\begin{aligned} 1 &= m([0,1]) \\ &= m(\mathbb{C} \cup \mathbf{O}) \\ &= m(\mathbb{C}) + m(\mathbf{O}) \\ &= m(\mathbf{O}). \end{aligned}$$

By the definition, ϕ is increasing on $\mathbf{O}$ and $\phi(x) \geq 0$ for any $x \in [0,1]$. Let $x_1, x_2 \in \mathbb{C}$, $x_1 \geq x_2 > 0$. Then

$$\mathbf{O} \cap [0, x_2) \subseteq \mathbf{O} \cap [0, x_1)$$

and

$$\begin{aligned}\phi(x_1) &= \sup\{\phi(t) : t \in \mathbf{O} \cap [0, x_1)\}\\ &\geq \sup\{\phi(t) : t \in \mathbf{O} \cap [0, x_2)\}\\ &= \phi(x_2).\end{aligned}$$

Therefore, ϕ is increasing on $\mathbb{C}$ and it is increasing on $[0, 1]$. Let now $x \in \mathbf{O}$ be arbitrarily chosen. Then x belongs to an open interval on which it is a constant. Therefore, ϕ is continuous at x. Because $x \in \mathbf{O}$ was arbitrarily chosen, we conclude that ϕ is continuous on $\mathbf{O}$. Now we take $x_0 \in \mathbb{C}$, $x_0 \neq 0, 1$. Then x_0 is not a member of the $2^k - 1$ intervals removed in the first k stages of the removal process, whose union is denoted by $\mathbf{O}_k$. We take k large enough so that x_0 lies between two consecutive intervals (a_{1k}, a_k) and (b_{1k}, b_k), $a_k < b_{1k}$, in $\mathbf{O}_k$. By the definition of ϕ on $\mathbf{O}_k$, it follows that

$$\phi(b_k) - \phi(a_k) = \frac{1}{2^k}, \quad a_k < x_0 < b_{1k}.$$

Hence, if $x_1 < x_0$, $x_1 \in (a_k, b_{1k})$, using that ϕ is increasing,

$$\begin{aligned}0 &\leq \phi(x_0) - \phi(x_1)\\ &\leq \phi(b_k) - \phi(a_k)\\ &= \tfrac{1}{2^k}.\end{aligned} \tag{2.55}$$

If $x_1 > x_0$, $x_1 \in (a_k, b_{1k})$, using that ϕ is increasing, we have

$$\begin{aligned}0 &\leq \phi(x_1) - \phi(x_0)\\ &\leq \phi(b_k) - \phi(a_k)\\ &= \frac{1}{2^k}.\end{aligned}$$

By the last inequalities and the inequalities (2.55), we conclude that ϕ is continuous at x_0. Because $x_0 \in \mathbb{C}$ was arbitrarily chosen, we get that ϕ is continuous on $\mathbb{C}$. Therefore, ϕ is continuous on $[0, 1]$. Because $\phi(0) = 0$ and $\phi(1) = 1$, and ϕ is increasing and continuous on $[0, 1]$, we conclude that ϕ maps $[0, 1]$ onto $[0, 1]$. Since ϕ is a constant on each of the intervals removed at any stage of the removal process, its derivative exists and it is 0 at each point of $\mathbf{O}$. This completes the proof. □

Theorem 2.31. *Let ϕ be the Cantor–Lebesgue function and define the function ψ as follows:*

$$\psi(x) = \phi(x) + x, \quad x \in [0,1].$$

Then

1. *ψ is strictly increasing continuous function that maps $[0,1]$ onto $[0,2]$,*
2. *ψ maps the Cantor set $\mathbb{C}$ onto a measurable set of positive measure,*
3. *ψ maps a measurable set, subset of the Cantor set $\mathbb{C}$, onto a non-measurable set.*

Proof.

1. Because ψ is a sum of two continuous functions on $[0,1]$, we conclude that ψ is continuous on $[0,1]$. Let $x_1, x_2 \in [0,1]$, $x_1 > x_2$. Then $\phi(x_1) \geq \phi(x_2)$ and

$$\begin{aligned} \psi(x_1) &= \phi(x_1) + x_1 \\ &\geq \phi(x_2) + x_1 \\ &> \phi(x_2) + x_2 \\ &= \psi(x_2). \end{aligned}$$

Consequently ψ is strictly increasing on $[0,1]$. Since $\psi(0) = 0$, $\psi(1) = 2$ and ψ is strictly increasing continuous function on $[0,1]$, we conclude that ψ maps $[0,1]$ onto $[0,2]$.

2. We have that $[0,1] = \mathbb{C} \cup \mathbf{O}$. We will prove that

$$[0,2] = \psi(\mathbf{C}) \cup \psi(\mathbf{O}). \tag{2.56}$$

Let $y \in [0,2]$ be arbitrarily chosen. Since $\psi([0,1]) = [0,2]$, there is an $x \in [0,1]$ such that $y = \psi(x)$. If $x \in \mathbf{O}$, then $y \in \psi(\mathbf{O})$ and $y \in \psi(\mathbb{C}) \cup \psi(\mathbf{O})$. If $x \in \mathbb{C}$, then $y \in \psi(\mathbb{C})$ and $y \in \psi(\mathbb{C}) \cup \psi(\mathbf{O})$. Because $y \in [0,2]$ was arbitrarily chosen and for it we get that it is an element of $\psi(\mathbb{C}) \cup \psi(\mathbf{O})$, we obtain the relation

$$[0,2] \subseteq \psi(\mathbb{C}) \cup \psi(\mathbf{O}). \tag{2.57}$$

Let $y \in \psi(\mathbb{C}) \cup \psi(\mathbf{O})$ be arbitrarily chosen. If $y \in \psi(\mathbb{C})$, then there is an $x \in \mathbb{C}$ such that $y = \psi(x)$. Because $\mathbb{C} \subset [0,1]$, we get $x \in [0, 1]$ and $y \in \psi([0, 1]) = [0, 2]$. If $y \in \psi(\mathbf{O})$, then there is an $x \in \mathbf{O}$ so that $y = \psi(x)$. Since $\mathbf{O} \subset [0, 1]$, we obtain that $y \in \psi([0, 1]) = [0, 2]$. Because $y \in \psi(\mathbb{C}) \cup \psi(\mathbf{O})$ was arbitrarily chosen and for it we get that it is an element of $[0, 2]$, we get the relation

$$\psi(\mathbb{C}) \cup \psi(\mathbf{O}) \subseteq [0, 2].$$

From the last relation and from the relation (2.57), we obtain the relation (2.56). Assume that there is an $y \in \psi(\mathbf{O})$ and $y \in \psi(\mathbb{C})$. Then there exist $x_1 \in \mathbf{O}$ and $x_2 \in \mathbf{C}$ such that

$$y = \psi(x_1) = \psi(x_2).$$

This is a contradiction because $x_1 \neq x_2$ and ψ is a strictly increasing function on $[0, 1]$. Consequently $\psi(\mathbb{C}) \cap \psi(\mathbf{O}) = \emptyset$. Since ψ is strictly increasing continuous on $[0, 1]$ it has a continuous inverse. Therefore, $\psi(\mathbb{C})$ is closed and $\psi(\mathbf{O})$ is open. Hence, $\psi(\mathbb{C})$ and $\psi(\mathbf{O})$ are measurable. Let $\{\mathbf{I}_k\}_{k\in\mathbf{N}}$ be an enumeration of the collection of the intervals that are removed in the Cantor removal process. We have $\mathbf{O} = \cup_{k=1}^{\infty}\mathbf{I}_k$. Since ψ is one-to-one, the collection $\{\psi(\mathbf{I}_k)\}_{k\in\mathbf{N}}$ is disjoint. Because ψ is a constant on each $\mathbf{I}_k$, we have that

$$\begin{aligned} m(\psi(\mathbf{I}_k)) &= m(\mathbf{I}_k) \\ &= l(\mathbf{I}_k). \end{aligned}$$

By Theorem 2.24, we get

$$\begin{aligned} m(\psi(\mathbf{O})) &= \sum_{k=1}^{\infty} m(\psi(\mathbf{I}_k)) \\ &= \sum_{k=1}^{\infty} l(\mathbf{I}_k) \\ &= m(\mathbf{O}). \end{aligned}$$

Because $m(\mathbf{O}) = 1$, we obtain that $m(\psi(\mathbf{O})) = 1$. Hence and (2.56), we find

$$\begin{aligned} 2 &= m([0, 2]) \\ &= m(\psi(\mathbf{O}) \cup \psi(\mathbb{C})) \end{aligned}$$

$$= m(\psi(\mathbf{O})) + m(\psi(\mathbb{C}))$$
$$= 1 + m(\psi(\mathbb{C})).$$

From here, $m(\psi(\mathbb{C})) = 1$.

3. From Theorem 2.27, it follows that there is a nonmeasurable set $\mathbf{W} \subset \psi(\mathbb{C})$. Because $\psi^{-1}(\mathbf{W}) \subset \mathbb{C}$ and $m(\mathbf{C}) = 0$, we have that $\psi^{-1}(\mathbf{W})$ is measurable and has measure zero. This completes the proof. □

Exercise 2.7. Prove that a strictly increasing continuous function, defined on an interval, maps Borel sets onto Borel sets.

Theorem 2.32. *There is a measurable set, a subset of the Cantor set, that is not a Borel set.*

Proof. We take the function ψ described in Theorem 2.31. By Theorem 2.31, ψ maps a measurable set $\mathbf{A}$ onto a nonmeasurable set. If we assume that A is a Borel set, then, using Exercise 2.7, we have that $\psi(\mathbf{A})$ is a Borel set, which is a measurable set. Therefore $\mathbf{A}$ is not a Borel set. This completes the proof. □

2.5 Lebesgue Measurable Functions

Lemma 2.5. *Let* $\mathbf{E}$ *be a measurable set and* f *be a function defined on* $\mathbf{E}$. *Then the following statements are equivalent:*

1. *For each* $c \in \mathbf{R}$, *the set* $\{x \in \mathbf{E} : f(x) > c\}$ *is measurable.*
2. *For each* $c \in \mathbf{R}$, *the set* $\{x \in \mathbf{E} : f(x) \geq c\}$ *is measurable.*
3. *For each* $c \in \mathbf{R}$, *the set* $\{x \in \mathbf{E} : f(x) < c\}$ *is measurable.*
4. *For each* $c \in \mathbf{R}$, *the set* $\{x \in \mathbf{E} : f(x) \leq c\}$ *is measurable.*

Proof. Since the complement in $\mathbf{E}$ of a measurable subset of $\mathbf{E}$ is measurable and **1** and **4**, **2** and **3** are complementary in $\mathbf{E}$, we have that **1** and **4**, **2** and **3** are equivalent. Now, we assume **1**. Let $c \in \mathbf{R}$ be arbitrarily chosen. We will prove that

$$\{x \in \mathbf{E} : f(x) \geq c\} = \bigcap_{k=1}^{\infty} \left\{x \in \mathbf{E} : f(x) > c - \frac{1}{k}\right\}. \tag{2.58}$$

Let $y \in \{x \in \mathbf{E} : f(x) \geq c\}$ be arbitrarily chosen. Then $y \in \mathbf{E}$ and $f(y) \geq c$. Hence, $f(y) > c - \frac{1}{k}$ for any $k \in \mathbf{N}$. Therefore, $y \in \left\{x \in \mathbf{E} : f(x) > c - \frac{1}{k}\right\}$ for any $k \in \mathbf{N}$ and $y \in \cap_{k=1}^{\infty} \left\{x \in \mathbf{E} : f(x) > c - \frac{1}{k}\right\}$. Because $y \in \{x \in \mathbf{E} : f(x) \geq c\}$ was arbitrarily chosen and for it we get that it is an element of $\cap_{k=1}^{\infty} \left\{x \in \mathbf{E} : f(x) > c - \frac{1}{k}\right\}$, we conclude that

$$\{x \in \mathbf{E} : f(x) \geq c\} \subseteq \bigcap_{k=1}^{\infty} \left\{x \in \mathbf{E} : f(x) > c - \frac{1}{k}\right\}. \tag{2.59}$$

Let now $y \in \cap_{k=1}^{\infty} \left\{x \in \mathbf{E} : f(x) > c - \frac{1}{k}\right\}$ be arbitrarily chosen. Then

$$y \in \left\{x \in \mathbf{E} : f(x) > c - \frac{1}{k}\right\}$$

for any $k \in \mathbf{N}$. Hence, $y \in \mathbf{E}$ and $f(y) > c - \frac{1}{k}$ for any $k \in \mathbf{N}$. From here, $f(y) \geq c$ and $y \in \{x \in \mathbf{E} : f(x) \geq c\}$. Because $y \in \cap_{k=1}^{\infty} \left\{x \in \mathbf{E} : f(x) > c - \frac{1}{k}\right\}$ was arbitrarily chosen and for it we obtain that it is an element of $\{x \in \mathbf{E} : f(x) \geq c\}$, we find

$$\bigcap_{k=1}^{\infty} \left\{x \in \mathbf{E} : f(x) > c - \frac{1}{k}\right\} \subseteq \{x \in \mathbf{E} : f(x) \geq c\}.$$

From the last relation and from the relation (2.59), we obtain the relation (2.58). We have that $\left\{x \in \mathbf{E} : f(x) > c - \frac{1}{k}\right\}$ are measurable sets for any $k \in \mathbf{N}$. Since the intersection of countable collection of measurable sets is a measurable set, we get that $\cap_{k=1}^{\infty} \left\{x \in \mathbf{E} : f(x) > c - \frac{1}{k}\right\}$ is a measurable set. Hence and (2.58), we conclude that $\{x \in \mathbf{E} : f(x) \geq c\}$ is a measurable set. Now, we assume **2**. Let $c \in \mathbf{R}$ be arbitrarily chosen. We will prove

$$\{x \in \mathbf{E} : f(x) > c\} = \bigcup_{k=1}^{\infty} \left\{x \in \mathbf{E} : f(x) \geq c + \frac{1}{k}\right\}. \tag{2.60}$$

Let $y \in \{x \in \mathbf{E} : f(x) > c\}$ be arbitrarily chosen. Then $y \in \mathbf{E}$ and $f(y) > c$. Hence, there is an $l \in \mathbf{N}$ such that $f(y) \geq c + \frac{1}{l}$. Therefore, $y \in \left\{x \in \mathbf{E} : f(x) \geq c + \frac{1}{l}\right\}$ and from here, $y \in \cup_{k=1}^{\infty} \left\{x \in \mathbf{E} : f(x) \geq c + \frac{1}{k}\right\}$. Because $y \in \{x \in \mathbf{E} : f(x) > c\}$ was arbitrarily chosen and for it we get that it is an element of $\cup_{k=1}^{\infty}\left\{x \in \mathbf{E} : f(x) \geq c + \frac{1}{k}\right\}$, we get the relation

$$\{x \in \mathbf{E} : f(x) > c\} \subseteq \bigcup_{k=1}^{\infty}\left\{x \in \mathbf{E} : f(x) \geq c + \frac{1}{k}\right\}. \tag{2.61}$$

Let now $y \in \cup_{k=1}^{\infty}\left\{x \in \mathbf{E} : f(x) \geq c + \frac{1}{k}\right\}$ be arbitrarily chosen. Then there is an $l \in \mathbf{N}$ such that $y \in \left\{x \in \mathbf{E} : f(x) \geq c + \frac{1}{l}\right\}$. Hence, $y \in \mathbf{E}$ and $f(y) \geq c + \frac{1}{l} > c$. Therefore $y \in \{x \in \mathbf{E} : f(x) > c\}$. Because $y \in \cup_{k=1}^{\infty}\left\{x \in \mathbf{E} : f(x) > c + \frac{1}{k}\right\}$ was arbitrarily chosen and for it we get that it is an element of the set $\{x \in \mathbf{E} : f(x) > c\}$, we obtain the relation

$$\bigcup_{k=1}^{\infty}\left\{x \in \mathbf{E} : f(x) \geq c + \frac{1}{k}\right\} \subseteq \{x \in \mathbf{E} : f(x) > c\}.$$

From the last relation and from (2.61), we get the relation (2.60). We have that $\left\{x \in \mathbf{E} : f(x) \geq c + \frac{1}{k}\right\}$ are measurable sets for any $k \in \mathbf{N}$. Since the union of countable collection of measurable sets is a measurable set, we have that the set $\cup_{k=1}^{\infty}\left\{x \in \mathbf{E} : f(x) \geq c + \frac{1}{k}\right\}$ is a measurable set. Hence and (2.60), we conclude that $\{x \in \mathbf{E} : f(x) > c\}$ is a measurable set. This completes the proof. □

Exercise 2.8. Prove that

1.

$$\{x \in \mathbf{E} : f(x) = c\} = \{x \in \mathbf{E} : f(x) \geq c\} \cap \{x \in \mathbf{E} : f(x) \leq c\}, \tag{2.62}$$

2.

$$\{x \in \mathbf{E} : f(x) = \infty\} = \bigcap_{k=1}^{\infty}\{x \in \mathbf{E} : f(x) > k\}, \tag{2.63}$$

3.

$$\{x \in \mathbf{E} : f(x) = -\infty\} = \bigcap_{k=1}^{\infty}\{x \in \mathbf{E} : f(x) < -k\}, \tag{2.64}$$

for any $c \in \mathbf{R}$. Here $\mathbf{E}$ is a set and f is a function defined on $\mathbf{E}$.

Lemma 2.6. *Let* $\mathbf{E}$ *be a measurable set and* f *be a function defined on* $\mathbf{E}$. *Assume that one of the sets in Lemma* 2.5 *is measurable. Then for each extended real number* c, *the set* $\{x \in \mathbf{E} : f(x) = c\}$ *is a measurable set.*

Proof.

1. Let $-\infty < c < \infty$ be arbitrarily chosen. By Lemma 2.5, we have that the sets $\{x \in \mathbf{E} : f(x) \geq c\}$ and $\{x \in \mathbf{E} : f(x) \leq c\}$ are measurable sets. Because the intersection of a finite collection of measurable sets is a measurable set, we have that the set $\{x \in \mathbf{E} : f(x) \geq c\} \cap \{x \in \mathbf{E} : f(x) \leq c\}$ is a measurable set. Hence and (2.62), we conclude that the set $\{x \in \mathbf{E} : f(x) = c\}$ is a measurable set.
2. Let $c = \infty$. By Lemma 2.5, we have that the sets $\{x \in \mathbf{E} : f(x) > k\}$ are measurable sets for any $k \in \mathbf{N}$. Since the intersection of a countable collection of measurable sets is a measurable set, we get that the set $\cap_{k=1}^{\infty} \{x \in \mathbf{E} : f(x) > k\}$ is a measurable set. Hence and (2.63), we conclude that the set $\{x \in \mathbf{E} : f(x) = \infty\}$ is a measurable set.
3. Let $c = -\infty$. By Lemma 2.5, we have that the sets $\{x \in \mathbf{E} : f(x) < -k\}$ are measurable sets for any $k \in \mathbf{N}$. Because the intersection of a countable collection of measurable sets is a measurable set, we get that the set $\bigcap_{k=1}^{\infty} \{x \in \mathbf{E} : f(x) < -k\}$ is a measurable set. From here and from (2.64), we conclude that the set $\{x \in \mathbf{E} : f(x) = -\infty\}$ is a measurable set. This completes the proof. □

Definition 2.17. Let $\mathbf{E}$ be a measurable set. An extended real-valued function f defined on $\mathbf{E}$ is said to be Lebesgue measurable, or simply measurable, if it satisfies one of the statements of Lemma 2.5.

Theorem 2.33. *Let a function* f *be defined on a measurable set* $\mathbf{O}$. *Then* f *is measurable if and only if for any open set* $\mathbf{O}$ *the set* $f^{-1}(\mathbf{O}) = \{x \in \mathbf{E} : f(x) \in \mathbf{O}\}$ *is measurable.*

Proof.

1. Let f be measurable. Let also, $\mathbf{O}$ be any open set. Then it can be represented as an union of countable collection of open bounded

intervals $\{\mathbf{I}_k\}_{k\in\mathbf{N}}$ each of which is in the form $\mathbf{I}_k = \mathbf{A}_k \cap \mathbf{B}_k$, where $\mathbf{A}_k = (-\infty, a_k)$ and $B_k = (b_k, \infty)$. We will prove that

$$f^{-1}(\mathbf{O}) = \bigcup_{k=1}^{\infty} \left(f^{-1}(\mathbf{A}_k) \cap f^{-1}(\mathbf{B}_k)\right). \tag{2.65}$$

Let $x \in f^{-1}(\mathbf{O})$ be arbitrarily chosen. Then $x \in \mathbf{E}$ and $f(x) \in \mathbf{O}$. Hence, there is an $l \in \mathbf{N}$ such that $f(x) \in \mathbf{I}_l = \mathbf{A}_l \cap \mathbf{B}_l$. From here, it follows that $f(x) \in \mathbf{A}_l$ and $f(x) \in \mathbf{B}_l$. Therefore $x \in f^{-1}(\mathbf{A}_l)$ and $x \in f^{-1}(\mathbf{B}_l)$. Then we get the relations $x \in f^{-1}(\mathbf{A}_l) \cap f^{-1}(\mathbf{B}_l)$ and $x \in \cup_{k=1}^{\infty}(f^{-1}(\mathbf{A}_k) \cap f^{-1}(\mathbf{B}_k))$. Because $x \in f^{-1}(\mathbf{O})$ was arbitrarily chosen and for it we get that it is an element of the set $\cup_{k=1}^{\infty} \left(f^{-1}(\mathbf{A}_k) \cap f^{-1}(\mathbf{B}_k)\right)$, we conclude that

$$f^{-1}(\mathbf{O}) \subseteq \bigcup_{k=1}^{\infty} \left(f^{-1}(\mathbf{A}_k) \cap f^{-1}(\mathbf{B}_k)\right). \tag{2.66}$$

Let now $x \in \cup_{k=1}^{\infty} \left(f^{-1}(\mathbf{A}_k) \cap f^{-1}(\mathbf{B}_k)\right)$ be arbitrarily chosen. Then there is an $l \in \mathbf{N}$ such that $x \in f^{-1}(\mathbf{A}_l) \cap f^{-1}(\mathbf{B}_l)$. Hence, $x \in f^{-1}(\mathbf{A}_l)$ and $x \in f^{-1}(\mathbf{B}_l)$. Then $f(x) \in \mathbf{A}_l$ and $f(x) \in \mathbf{B}_l$. From here, $f(x) \in \mathbf{I}_l = \mathbf{A}_l \cap \mathbf{B}_l \subseteq \mathbf{O}$. Consequently, $x \in f^{-1}(\mathbf{O})$. Because $x \in \cup_{k=1}^{\infty} \left(f^{-1}(\mathbf{A}_k) \cap f^{-1}(\mathbf{B}_k)\right)$ was arbitrarily chosen and for it we get that it is an element of $f^{-1}(\mathbf{O})$, we get the relation

$$\bigcup_{k=1}^{\infty} \left(f^{-1}(\mathbf{A}_k) \cap f^{-1}(\mathbf{B}_k)\right) \subseteq f^{-1}(\mathbf{O}).$$

From the last relation and from the relation (2.66), it follows the relation (2.65). Since f is measurable, we have that $f^{-1}(\mathbf{A}_l)$ and $f^{-1}(\mathbf{B}_l)$ are measurable sets. Hence, $f^{-1}(\mathbf{A}_l) \cap f^{-1}(\mathbf{B}_l)$ is a measurable set and $\cup_{l=1}^{\infty} \left(f^{-1}(\mathbf{A}_l) \cap f^{-1}(\mathbf{B}_l)\right)$ is a measurable set. From here and from the relation (2.65), we conclude that $f^{-1}(\mathbf{O})$ is a measurable set.

2. Let for any open set $\mathbf{O}$ the set $f^{-1}(\mathbf{O})$ is a measurable set. Since (c, ∞) is an open set for any $c \in \mathbf{R}$, we have that $f^{-1}((c, \infty))$ is a measurable set. Therefore, the function f is measurable. This completes the proof. □

Theorem 2.34. *Let* $\mathbf{E}$ *be a measurable set and* $f : \mathbf{E} \longmapsto \mathbf{R}$ *be a continuous function. Then* f *is measurable.*

Proof. Let $\mathbf{O}$ be any open set. Since $f : \mathbf{E} \longmapsto \mathbf{R}$ is a continuous function, there is an open set $\mathbf{U}$ such that

$$f^{-1}(\mathbf{O}) = \mathbf{E} \cap \mathbf{U}.$$

Because $\mathbf{U}$ is an open set, it is measurable. Since the intersection of two measurable sets is a measurable set, we get that $\mathbf{E} \cap \mathbf{U}$ is a measurable set, or the set $f^{-1}(\mathbf{O})$ is a measurable set. Hence and Theorem 2.33, we conclude that f is a measurable function. This completes the proof. □

Exercise 2.9. Let $\mathbf{E}$ be any set of real numbers and f, g, f_l, $l \in \{1, \ldots, n\}$, be extended real-valued functions on $\mathbf{E}$. Prove

1.
$$\{x \in \mathbf{E} : g(x) > c\} = \{x \in \mathbf{A} : g(x) > c\} \cup (\{x \in \mathbf{E} : f(x) > c\} \cap (\mathbf{E} \backslash \mathbf{A})), \tag{2.67}$$
$\mathbf{A} = \{x \in \mathbf{E} : f(x) \neq g(x)\}$, for any $c \in \mathbf{R}$,

2.
$$\{x \in \mathbf{E} : f(x) > c\} = \{x \in \mathbf{D} : f(x) > c\} \cup \{x \in \mathbf{E} \backslash \mathbf{D} : f(x) > c\}, \tag{2.68}$$
$D \subseteq \mathbf{E}$, for any $c \in \mathbf{R}$,

3.
$$\{x \in \mathbf{D} : f(x) > c\} = \mathbf{D} \cap \{x \in \mathbf{E} : f(x) > c\}, \tag{2.69}$$
$D \subseteq \mathbf{E}$, for any $c \in \mathbf{R}$,

4.
$$\{x \in \mathbf{E} : |f(x)| > c\} = \{x \in \mathbf{E} : f(x) > c\} \cup \{x \in \mathbf{E} : f(x) < -c\} \tag{2.70}$$
for any $c > 0$,

5.
$$\{x \in \mathbf{E} : \alpha f(x) > c\} = \left\{x \in \mathbf{E} : f(x) > \frac{c}{\alpha}\right\} \tag{2.71}$$
for any $c \in \mathbf{R}$ and for any $\alpha > 0$,

6.

$$\{x \in \mathbf{E} : \alpha f(x) > c\} = \left\{x \in \mathbf{E} : f(x) < \frac{c}{\alpha}\right\} \tag{2.72}$$

for any $c \in \mathbf{R}$ and for any $\alpha < 0$,

7.

$$\begin{aligned}&\{x \in \mathbf{E} : f(x) + g(x) < c\}\\&\quad = \cup q \in \mathbf{Q}\left(\{x \in \mathbf{E} : g(x) < c - q\} \cap \{x \in \mathbf{E} : f(x) < q\}\right)\end{aligned} \tag{2.73}$$

for any $c \in \mathbf{R}$,

8.

$$\begin{aligned}\left\{x \in \mathbf{E} : (f(x))^2 > c\right\} &= \left\{x \in \mathbf{E} : f(x) > \sqrt{c}\right\}\\&\quad \cup \left\{x \in \mathbf{E} : f(x) < -\sqrt{c}\right\}\end{aligned} \tag{2.74}$$

for any $c \in \mathbf{R}$, $c > 0$,

9.

$$\{x \in \mathbf{E} : \max\{f_1(x), \ldots, f_n(x)\} > c\} = \bigcup_{k=1}^{n} \{x \in \mathbf{E} : f_k(x) > c\} \tag{2.75}$$

for any $c \in \mathbf{R}$,

10.

$$\{x \in \mathbf{E} : \min\{f_1(x), \ldots, f_n(x)\} > c\} = \bigcap_{k=1}^{n} \{x \in \mathbf{E} : f_k(x) > c\} \tag{2.76}$$

for any $c \in \mathbf{R}$.

Theorem 2.35. *Let* $\mathbf{E}$ *be a measurable set and* f, g *be extended real-valued functions on* $\mathbf{E}$. *If* f *is measurable on* $\mathbf{E}$ *and* $f = g$ *a.e. on* $\mathbf{E}$, *then* g *is measurable on* $\mathbf{E}$.

Proof. Let $c \in \mathbf{R}$ be arbitrarily chosen. Let also, $\mathbf{A} = \{x \in \mathbf{E} : f(x) \neq g(x)\}$. Then $m(\mathbf{A}) = 0$. Hence and Theorem 2.8, we get that the set $\mathbf{A}$ is a measurable set. Because every subset of a set of measure zero is measurable, we have that the set

$\{x \in \mathbf{A} : g(x) > c\}$ is a measurable set. Since $\mathbf{E}$ and $\mathbf{A}$ are measurable sets, using Theorem 2.11, we get that the set $\mathbf{E}\backslash\mathbf{A}$ is a measurable set. Because f is a measurable function on $\mathbf{E}$, we conclude that the set $\{x \in \mathbf{E} : f(x) > c\}$ is a measurable set. Now, using that the intersection of two measurable sets is a measurable set, we get that the set $\{x \in \mathbf{E} : f(x) > c\} \cap (\mathbf{E}\backslash\mathbf{A})$ is a measurable set. Since the union of two measurable sets is a measurable set, we obtain that the set

$$\{x \in \mathbf{A} : g(x) > c\} \cup (\{x \in \mathbf{E} : f(x) > c\} \cap (\mathbf{E}\backslash\mathbf{A}))$$

is a measurable set. Hence and (2.67), we conclude that the set $\{x \in \mathbf{E} : g(x) > c\}$ is a measurable set. Because $c \in \mathbf{R}$ was arbitrarily chosen, we obtain that the function g is a measurable function on $\mathbf{E}$. This completes the proof. $\square$

Theorem 2.36. *Let $\mathbf{E}$ and $\mathbf{D}$, $\mathbf{D} \subseteq \mathbf{E}$, be measurable sets and f be an extended real-valued function on $\mathbf{E}$. Then f is measurable on $\mathbf{E}$ if and only if its restrictions to $\mathbf{D}$ and $\mathbf{E}\backslash\mathbf{D}$ are measurable.*

Proof.

1. Let the restrictions of f to $\mathbf{D}$ and $\mathbf{E}\backslash\mathbf{D}$ are measurable. We take $c \in \mathbf{R}$ arbitrarily. Then the sets

$$\{x \in \mathbf{D} : f(x) > c\} \quad \text{and} \quad \{x \in \mathbf{E}\backslash\mathbf{D} : f(x) > c\}$$

are measurable sets. Because the union of two measurable sets is a measurable set, we have that the set

$$\{x \in \mathbf{D} : f(x) > c\} \cup \{x \in \mathbf{E}\backslash\mathbf{D} : f(x) > c\}$$

is a measurable set. Hence and (2.68), the set $\{x \in \mathbf{E} : f(x) > c\}$ is a measurable set. Because $c \in \mathbf{R}$ was arbitrarily chosen, we conclude that the function f is measurable on $\mathbf{E}$.

2. Now, we suppose that the function f is a measurable function on $\mathbf{E}$. We take $c > 0$ arbitrarily. Then the set $\{x \in \mathbf{E} : f(x) > c\}$ is a measurable set. Because $\mathbf{D}$ is a measurable set and the intersection of two measurable sets is a measurable set, we conclude that the set $\mathbf{D} \cap \{x \in \mathbf{E} : f(x) > c\}$ is a measurable set. From here and from (2.69), we obtain that the set $\{x \in \mathbf{D} : f(x) > c\}$ is a measurable

set. As in above, one can prove that the set $\{x \in \mathbf{E}\backslash\mathbf{D} : f(x) > c\}$ is a measurable set. Because $c \in \mathbf{R}$ was arbitrarily chosen, we get that the restrictions of f to $\mathbf{D}$ and $\mathbf{E}\backslash\mathbf{D}$ are measurable. This completes the proof. □

Theorem 2.37. *Let* $\mathbf{E}$ *be a measurable set and* f *be an extended real-valued measurable function on* $\mathbf{E}$. *Then* $|f|$ *is a measurable function on* $\mathbf{E}$.

Proof. Let $c \in \mathbf{R}$ be arbitrarily chosen.

1. Let $c < 0$. Then $\mathbf{E} = \{x \in \mathbf{E} : |f(x)| > c\}$ and $\{x \in \mathbf{E} : |f(x)| > c\}$ is a measurable set.
2. Let $c \geq 0$. Since f is measurable on $\mathbf{E}$, the sets

$$\{x \in \mathbf{E} : f(x) > c\} \quad \text{and} \quad \{x \in \mathbf{E} : f(x) < -c\}$$

are measurable sets. Hence, using that the union of two measurable sets is a measurable set, we get that the set

$$\{x \in \mathbf{E} : f(x) > c\} \cup \{x \in \mathbf{E} : f(x) < -c\}$$

is a measurable set. From here and from (2.70), we obtain that the set $\{x \in \mathbf{E} : |f(x)| > c\}$ is a measurable set. Because $c \in \mathbf{R}$ was arbitrarily chosen, we conclude that the function $|f|$ is a measurable function on $\mathbf{E}$. This completes the proof. □

Theorem 2.38. *Let* $\mathbf{E}$ *be a measurable set of measure zero. Then every extended real-valued function on* $\mathbf{E}$ *is measurable.*

Proof. Let f be an extended real-valued function on $\mathbf{E}$. We take $c \in \mathbf{R}$ arbitrarily. Then $\{x \in \mathbf{E} : f(x) > c\} \subseteq \mathbf{E}$. Hence, using that $m(\mathbf{E}) = 0$, we get that the set $\{x \in \mathbf{E} : f(x) > c\}$ is a measurable set. Because $c \in \mathbf{R}$ was arbitrarily chosen, we conclude that f is measurable on $\mathbf{E}$. This completes the proof. □

Theorem 2.39. *Let* $\mathbf{E}$ *be a measurable set and* f *be an extended real-valued measurable function on* $\mathbf{E}$. *Then* αf *is a measurable function on* $\mathbf{E}$ *for any* $\alpha \in \mathbf{R}$.

Proof. Let $c \in \mathbf{R}$ be arbitrarily chosen. We take $\alpha \in \mathbf{R}$ arbitrarily.

1. Let $\alpha = 0$. Then

$$\{x \in \mathbf{E} : 0 > c\} = \begin{cases} \mathbf{E} & \text{if } c < 0 \\ \emptyset & \text{if } c \geq 0. \end{cases}$$

   Since $\mathbf{E}$ and $\emptyset$ are measurable sets, we have that the set $\{\mathbf{E} : 0 > c\}$ is a measurable set.
2. Let $\alpha > 0$. Since f is a measurable function on $\mathbf{E}$, we have that the set $\left\{x \in \mathbf{E} : f(x) > \frac{c}{\alpha}\right\}$ is a measurable set. Hence and (2.71), we get that the set $\{x \in \mathbf{E} : \alpha f(x) > c\}$ is a measurable set.
3. Let $\alpha < 0$. Because f is a measurable set, the set $\left\{x \in \mathbf{E} : f(x) < \frac{c}{\alpha}\right\}$ is a measurable set. From here and from (2.72), we obtain that the set $\{x \in \mathbf{E} : \alpha f(x) > c\}$ is a measurable set.

Because $c \in \mathbf{R}$ was arbitrarily chosen and for it we get that the set $\{x \in \mathbf{E} : \alpha f(x) > c\}$ is a measurable set, we conclude that the function αf is a measurable function on $\mathbf{E}$. This completes the proof. □

Theorem 2.40. *Let $\mathbf{E}$ be a measurable set and f, g be extended real-valued measurable functions on $\mathbf{E}$. Then $f + g$ is a measurable function on $\mathbf{E}$.*

Proof. Let $c \in \mathbf{R}$ be arbitrarily chosen. Since f and g are measurable functions on $\mathbf{E}$, we have that the sets

$$\{x \in \mathbf{E} : g(x) < c - q\} \quad \text{and} \quad \{x \in \mathbf{E} : f(x) < q\}$$

are measurable sets for any $q \in \mathbf{Q}$. Because the intersection of two measurable sets is a measurable set, we obtain that the set

$$\{x \in \mathbf{E} : g(x) < c - q\} \cap \{x \in \mathbf{E} : f(x) < q\}$$

is a measurable set for any $q \in \mathbf{Q}$. Since the union of a countable collection of measurable sets is a measurable set, we have that the set

$$\cup q \in \mathbf{Q}\left(\{x \in \mathbf{E} : g(x) < c - q\} \cap \{x \in \mathbf{E} : f(x) < q\}\right)$$

is a measurable set. Hence and (2.73), we obtain that the set $\{x \in \mathbf{E} : f(x) + g(x) < c\}$ is a measurable set. Because $c \in \mathbf{R}$ was arbitrarily chosen, we conclude that $f+g$ is a measurable function on $\mathbf{E}$. This completes the proof. $\square$

Exercise 2.10. Let $\mathbf{E}$ be a measurable set and f, g be extended real-valued measurable functions on $\mathbf{E}$. Prove that $\alpha f+\beta g$ is a measurable function on $\mathbf{E}$ for any $\alpha, \beta \in \mathbf{R}$.

Corollary 2.4. *Let $\mathbf{E}$ be a measurable set and f_l, $l \in \{1, \ldots, n\}$, be extended real-valued measurable functions on $\mathbf{E}$. Then $f_1+\cdots+f_n$ is a measurable function on $\mathbf{E}$.*

Proof. If $n=2$, then the assertion follows immediately from Theorem 2.40. Let $n>2$. Since f_1 and f_2 are measurable functions on $\mathbf{E}$, by Theorem 2.40, it follows that f_1+f_2 is a measurable function on $\mathbf{E}$. Because f_1+f_2 and f_3 are measurable functions on $\mathbf{E}$, by Theorem 2.40, it follows that $f_1+f_2+f_3$ is a measurable function on $\mathbf{E}$. And so on, $f_1+\cdots+f_n$ is a measurable function on $\mathbf{E}$. This completes the proof. $\square$

Theorem 2.41. *Let $\mathbf{E}$ be a measurable set and f be an extended real-valued measurable function on $\mathbf{E}$. Then f^2 is a measurable function on $\mathbf{E}$.*

Proof. Let $c \in \mathbf{R}$ be arbitrarily chosen.

1. Let $c<0$. Then

$$\{x \in \mathbf{E} : (f(x))^2 > c\} = \mathbf{E}$$

and hence, $\{x \in \mathbf{E} : (f(x))^2 > c\}$ is a measurable set.

2. Let $c \geq 0$. Since f is a measurable function on $\mathbf{E}$, the sets

$$\left\{x \in \mathbf{E} : f(x) > \sqrt{c}\right\} \quad \text{and} \quad \left\{x \in \mathbf{E} : f(x) < -\sqrt{c}\right\}$$

are measurable sets. Hence,

$$\left\{x \in \mathbf{E} : f(x) > \sqrt{c}\right\} \cup \left\{x \in \mathbf{E} : f(x) < -\sqrt{c}\right\}$$

is a measurable set. From here and from (2.74), we get that the set

$$\{x \in \mathbf{E} : (f(x))^2 > c\}$$

is a measurable set.

Because $c \in \mathbf{R}$ was arbitrarily chosen, we conclude that f is a measurable function on $\mathbf{E}$. This completes the proof. □

Theorem 2.42. *Let* $\mathbf{E}$ *be a measurable set and* f, g *be extended real-valued measurable functions on* $\mathbf{E}$. *Then* fg *is a measurable function on* $\mathbf{E}$.

Proof. Note that

$$fg = \frac{1}{4}\left((f+g)^2 - (f-g)^2\right). \tag{2.77}$$

Since f and g are measurable functions on $\mathbf{E}$, using Theorem 2.40, we get that $f+g$ is a measurable function on $\mathbf{E}$. Hence and Theorem 2.41, it follows that $(f+g)^2$ is a measurable function on $\mathbf{E}$. Because g is a measurable function on $\mathbf{E}$, using Theorem 2.39, we obtain that $-g$ is a measurable function on $\mathbf{E}$. From here and from Theorem 2.40, we get that $f-g$ is a measurable function on $\mathbf{E}$. Hence and Theorem 2.41, it follows that $(f-g)^2$ is a measurable function on $\mathbf{E}$. Since $(f-g)^2$ is a measurable function on $\mathbf{E}$, using Theorem 2.39, it follows that $-(f-g)^2$ is a measurable function on $\mathbf{E}$. Because $(f+g)^2$ and $-(f-g)^2$ are measurable functions, by Theorem 2.40, we conclude that $(f+g)^2 - (f-g)^2$ is a measurable function on $\mathbf{E}$. From here and from Theorem 2.39, using (2.77), we obtain that the function fg is a measurable function on $\mathbf{E}$. This completes the proof. □

Theorem 2.43. *Let* $\mathbf{E}$ *be a measurable set and* $\{f_k\}_{k=1}^n$ *be a finite collection of measurable functions on* $\mathbf{E}$. *Then* $\max\{f_1, \ldots, f_n\}$ *and* $\min\{f_1, \ldots, f_n\}$ *are measurable functions on* $\mathbf{E}$.

Proof. Let $c \in \mathbf{R}$ be arbitrarily chosen. Because f_k, $k \in \{1, \ldots, n\}$, are measurable functions on $\mathbf{E}$, the sets

$$\{x \in \mathbf{E} : f_k(x) > c\}$$

are measurable sets for any $k \in \{1, \ldots, n\}$. Hence,

$$\bigcup_{k=1}^{n} \{x \in \mathbf{E} : f_k(x) > c\} \quad \text{and} \quad \bigcap_{k=1}^{n} \{x \in \mathbf{E} : f_k(x) > c\}$$

are measurable sets. Hence and (2.75), (2.76), we conclude that the sets

$$\{x \in \mathbf{E} : \max\{f_1(x), \dots, f_n(x)\} > c\} \quad \text{and}$$
$$\{x \in \mathbf{E} : \min\{f_1(x), \dots, f_n(x)\} > c\}$$

are measurable sets. Because $c \in \mathbf{R}$ was arbitrarily chosen, we conclude that the functions $\max\{f_1, \dots, f_n\}$ and $\min\{f_1, \dots, f_n\}$ are measurable functions on $\mathbf{E}$. This completes the proof. □

For a function f, defined on $\mathbf{E}$, we define

$$f^+(x) = \max\{f(x), 0\}, \quad f^-(x) = \min\{f(x), 0\}, \quad x \in \mathbf{E}.$$

Exercise 2.11. Let $\mathbf{E}$ be a measurable set and f be an extended real-valued measurable function on $\mathbf{E}$. Prove that f^+ and f^- are measurable functions on $\mathbf{E}$.

Exercise 2.12. Let $\mathbf{E}$ be a measurable set and f be an extended real-valued measurable function on $\mathbf{E}$. Prove that $|f|^p$ is a measurable function on $\mathbf{E}$ for each $p > 0$.

Theorem 2.44. *Let $\mathbf{E}$ be a measurable set, g be an extended real-valued measurable function on $\mathbf{E}$ and f be an extended real-valued continuous function on $\mathbf{R}$. Then the composition $f \circ g$ is a measurable function on $\mathbf{E}$.*

Proof. Let $\mathbf{O}$ be an open set. We will prove that

$$(f \circ g)^{-1}(\mathbf{O}) = g^{-1}\left(f^{-1}(\mathbf{O})\right). \tag{2.78}$$

Let $x \in (f \circ g)^{-1}(\mathbf{O})$ be arbitrarily chosen. Then $x \in \mathbf{E}$ and $f(g(x)) \in \mathbf{O}$. Since $g(x) \in \mathbf{R}$ and $f(g(x)) \in \mathbf{O}$, we get that $g(x) \in f^{-1}(\mathbf{O})$. Because $x \in \mathbf{E}$ and $g(x) \in f^{-1}(\mathbf{O})$, we obtain $x \in g^{-1}(f^{-1}(\mathbf{O}))$. Since $x \in (f \circ g)^{-1}(\mathbf{O})$ was arbitrarily chosen and for it we get that it is an element of $g^{-1}(f^{-1}(\mathbf{O}))$, we obtain the relation

$$(f \circ g)^{-1}(\mathbf{O}) \subseteq g^{-1}\left(f^{-1}(\mathbf{O})\right). \tag{2.79}$$

Let $x \in g^{-1}(f^{-1}(\mathbf{O}))$ be arbitrarily chosen. Then $x \in \mathbf{E}$ and $g(x) \in f^{-1}(\mathbf{O})$. Hence, $g(x) \in \mathbf{R}$ and $f(g(x)) \in \mathbf{O}$. Since $x \in \mathbf{E}$

and $f(g(x)) \in \mathbf{O}$, we conclude that $x \in (f \circ g)^{-1}(\mathbf{O})$. Because $x \in g^{-1}(f^{-1}(\mathbf{O}))$ was arbitrarily chosen and for it we get that it is an element of $(f \circ g)^{-1}(\mathbf{O})$, we obtain the relation

$$(f \circ g)^{-1}(\mathbf{O}) \subseteq (f \circ g)^{-1}(\mathbf{O}).$$

From the last relation and from (2.79), we obtain the relation (2.78). Since f is a continuous function on $\mathbf{R}$ and $\mathbf{O}$ is an open set, we have that the set $f^{-1}(\mathbf{O})$ is an open set. Because g is a measurable function on $\mathbf{E}$, using Theorem 2.33, we conclude that $g^{-1}(f^{-1}(\mathbf{O}))$ is a measurable set. Hence and the relation (2.78), we obtain that $(f \circ g)^{-1}(\mathbf{O})$ is a measurable set. From here and from Theorem 2.33, we obtain that $f \circ g$ is a measurable function on $\mathbf{E}$. This completes the proof. □

Exercise 2.13. Let $\mathbf{E}$ be a measurable set. If $f \equiv const$ on $\mathbf{E}$, prove that f is a measurable function on $\mathbf{E}$.

Definition 2.18. Let $\mathbf{E}$ be any set of real numbers and $\mathbf{A} \subseteq \mathbf{E}$. Let also, $\{f_n\}_{n\in\mathbf{N}}$ be a sequence of extended real-valued functions with common domain $\mathbf{E}$ and f be an extended real-valued function on $\mathbf{A}$.

1. We say that the sequence $\{f_n\}_{n\in\mathbf{N}}$ converges to f pointwise on $\mathbf{A}$ provided

$$\lim_{n\to\infty} f_n(x) = f(x) \quad \text{for all } x \in \mathbf{A}.$$

2. We say that the sequence $\{f_n\}_{n\in\mathbf{N}}$ converges to f pointwise a.e. on $\mathbf{A}$ provided it converges to f pointwise on $\mathbf{A}\backslash\mathbf{B}$, where $\mathbf{B} \subseteq \mathbf{A}$ and $m(\mathbf{B}) = 0$.
3. We say that the sequence $\{f_n\}_{n\in\mathbf{N}}$ converges uniformly to f on $\mathbf{A}$ provided for each $\varepsilon > 0$, there is an $N \in \mathbf{N}$ such that

$$|f_n - f| < \varepsilon \quad \text{on } \mathbf{A} \quad \text{for all } n \geq N.$$

Theorem 2.45. *Let* $\mathbf{E}$ *be a measurable set of real numbers,* $\{f_n\}_{n\in\mathbf{N}}$ *be a sequence of extended real-valued measurable functions on* $\mathbf{E}$ *that converges pointwise a.e. on* $\mathbf{E}$ *to the extended real-valued function* f*, defined on* $\mathbf{E}$*. Then* f *is measurable on* $\mathbf{E}$*.*

Proof. Let $\mathbf{A} \subseteq \mathbf{E}$ be such that the sequence $\{f_n\}_{n\in\mathbf{N}}$ converges pointwise to f on $\mathbf{A}$ and $m(\mathbf{E}\backslash\mathbf{A}) = 0$. Hence, f is measurable on $\mathbf{E}\backslash\mathbf{A}$. By Theorem 2.36, it follows that the functions f_j are measurable functions on $\mathbf{A}$ and $\mathbf{E}\backslash\mathbf{A}$ for any $j \in \mathbf{N}$. We will prove that f is measurable on $\mathbf{A}$. Let $c \in \mathbf{R}$ be arbitrarily chosen. We will prove

$$\{x \in \mathbf{A} : f(x) < c\} = \bigcup_{1\leq k,n<\infty} \left(\bigcap_{j=k}^{\infty} \left\{ x \in \mathbf{A} : f_j(x) < c - \frac{1}{n} \right\} \right). \tag{2.80}$$

Let $y \in \{x \in \mathbf{A} : f(x) < c\}$ be arbitrarily chosen. Because $f_j(y) \to f(y)$, as $j \to \infty$, and $f(y) < c$, there are $n, k \in \mathbf{N}$ such that for any $j \geq k$ we have $f_j(y) < c - \frac{1}{n}$, i.e., $y \in \left\{x \in \mathbf{A} : f_j(x) < c - \frac{1}{n}\right\}$ for any $j \geq k$. Therefore,

$$y \in \bigcap_{j=k}^{\infty} \left\{ x \in \mathbf{A} : f_j(x) < c - \frac{1}{n} \right\}$$

and from here,

$$y \in \bigcup_{1\leq k,n<\infty} \left(\bigcap_{j=k}^{\infty} \left\{ x \in \mathbf{A} : f_j(x) < c - \frac{1}{n} \right\} \right).$$

Because $y \in \{x \in \mathbf{A} : f(x) < c\}$ was arbitrarily chosen and we get that it is an element of $\cup_{1\leq k,n<\infty}(\cap_{j=k}^{\infty}\{x \in \mathbf{A} : f_j(x) < c - \frac{1}{n}\})$, we obtain the relation

$$\{x \in \mathbf{A} : f(x) < c\} \subseteq \bigcup_{1\leq k,n<\infty} \left(\bigcap_{j=k}^{\infty} \left\{ x \in \mathbf{A} : f_j(x) < c - \frac{1}{n} \right\} \right). \tag{2.81}$$

Let now $y \in \cup_{1\leq k,n<\infty}(\cap_{j=k}^{\infty}\left\{x \in \mathbf{A} : f_j(x) < c - \frac{1}{n}\right\})$ be arbitrarily chosen. Then there are $k, n \in \mathbf{N}$ such that $y \in \left\{x \in \mathbf{A} : f_j(x) < c - \frac{1}{n}\right\}$ for any $j \geq k$. Hence,

$$f_j(y) < c - \frac{1}{n}$$

for any $j \geq k$. Assume that $f(y) \geq c$. Then

$$f(y) - f_j(y) > \frac{1}{n}$$

for any $j \geq k$. This is a contradiction. Therefore $f(y) < c$ and $y \in \{x \in \mathbf{A} : f(x) < c\}$. Because $y \in \cup_{1\leq k,n<\infty}(\cap_{j=k}^{\infty}$

$\{x \in \mathbf{A} : f_j(x) < c - \frac{1}{n}\})$ was arbitrarily chosen and for it we get that it is an element of $\{x \in \mathbf{A} : f(x) < c\}$, we conclude that

$$\bigcup_{1 \le k,n < \infty} \left(\bigcap_{j=k}^{\infty} \left\{ x \in \mathbf{A} : f_j(x) < c - \frac{1}{n} \right\} \right) \subseteq \{x \in \mathbf{A} : f(x) < c\}.$$

From the last relation and from the relation (2.81), we get the relation (2.80). Since f_j are measurable functions on $\mathbf{A}$ for any $j \in \mathbf{N}$, we have that the sets $\{x \in \mathbf{A} : f_j(x) < c - \frac{1}{n}\}$ are measurable sets for any $j, n \in \mathbf{N}$. From here, the sets

$$\bigcap_{j=k}^{\infty} \left\{ x \in \mathbf{A} : f_j(x) < c - \frac{1}{n} \right\}$$

are measurable sets for any $k, n \in \mathbf{N}$. Hence,

$$\bigcup_{1 \le k,n < \infty} \left(\bigcap_{j=k}^{\infty} \left\{ x \in \mathbf{A} : f_j(x) < c - \frac{1}{n} \right\} \right)$$

is a measurable set. From here and from the relation (2.80), we conclude that the set $\{x \in \mathbf{A} : f(x) < c\}$ is a measurable set. Because $c \in \mathbf{R}$ was arbitrarily chosen, we conclude that the function f is a measurable function on $\mathbf{A}$. Since $m(\mathbf{E} \backslash \mathbf{A}) = 0$, we obtain that f is measurable on $\mathbf{E} \backslash \mathbf{A}$. Hence and Theorem 2.36, we obtain that the function f is a measurable function on $\mathbf{E}$. This completes the proof. □

Definition 2.19. Let $\mathbf{E}$ be any set of real numbers. The characteristic function of $\mathbf{E}$, $\kappa_{\mathbf{E}}$, is the function defined by

$$\kappa_{\mathbf{E}} = \begin{cases} 1 & \text{if } x \in \mathbf{E} \\ 0 & \text{if } x \notin \mathbf{E}. \end{cases}$$

Note that if $\mathbf{E}$ is a measurable set, then its characteristic function is a measurable function on $\mathbf{E}$.

Exercise 2.14. Let $\mathbf{A}$ and $\mathbf{B}$ be any sets of real numbers. Prove

1. $\kappa_{\mathbf{A} \cap \mathbf{B}} = \kappa_{\mathbf{A}} \kappa_{\mathbf{B}}$.
2. $\kappa_{\mathbf{A} \cup \mathbf{B}} = \kappa_{\mathbf{A}} + \kappa_{\mathbf{B}} - \kappa_{\mathbf{A}} \kappa_{\mathbf{B}}$.
3. $\kappa_{\mathbf{R} \backslash \mathbf{A}} = 1 - \kappa_{\mathbf{A}}$.

Definition 2.20. A real-valued function ϕ defined on a measurable set is called simple provided it is measurable and takes only a finite number of values.

If $\mathbf{E}$ is a measurable set and ϕ is a simple function on $\mathbf{E}$ and takes the distinct values $c_1, \ldots, c_n$, then

$$\phi = \sum_{k=1}^{n} c_k \kappa_{\mathbf{E}_k} \quad \text{on } \mathbf{E}, \tag{2.82}$$

where $\mathbf{E}_k = \{x \in \mathbf{E} : \phi(x) = c_k\}$, $k \in \{1, \ldots, n\}$.

Definition 2.21. The expression (2.82) of ϕ as a linear combination of characteristic functions is called the canonical representation of the simple function ϕ.

Theorem 2.46. *Let $\mathbf{E}$ be a measurable set and f be a measurable bounded real-valued function on $\mathbf{E}$. Then for each $\varepsilon > 0$, there are simple functions ϕ_ε and ψ_ε defined on $\mathbf{E}$ such that*

$$\phi_\varepsilon \le f \le \psi_\varepsilon \quad \text{and} \quad 0 \le \psi_\varepsilon - \phi_\varepsilon < \varepsilon \quad \text{on } \mathbf{E}.$$

Proof. We take $\varepsilon > 0$ arbitrarily. Let (c, d) be an open, bounded interval such that $f(\mathbf{E}) \subset (c, d)$. Consider a partition

$$c = y_0 < y_1 < \cdots < y_{n-1} < y_n = d$$

of the closed bounded interval $[c, d]$ such that $y_k - y_{k-1} < \varepsilon$ for any $k \in \{1, \ldots, n\}$. Let

$$\mathbf{I}_k = [y_{k-1}, y_k), \quad \mathbf{E}_k = f^{-1}(\mathbf{I}_k), \quad 1 \le k \le n.$$

Since f is a measurable function on $\mathbf{E}$, we have that the sets $\mathbf{E}_k$ are measurable sets for any $k \in \{1, \ldots, n\}$. Define the simple functions ϕ_ε and ψ_ε on $\mathbf{E}$ as follows

$$\phi_\varepsilon = \sum_{k=1}^{n} y_{k-1} \kappa_{\mathbf{E}_k} \quad \text{and} \quad \psi_\varepsilon = \sum_{k=1}^{n} y_k \kappa_{\mathbf{E}_k}.$$

Let $x \in \mathbf{E}$ be arbitrarily chosen. Since $f(\mathbf{E}) \subset (c, d)$, we have that $f(x) \in (c, d)$. Hence, there is a $k \in \{1, \ldots, n\}$ such that $f(x) \in \mathbf{I}_k$. Therefore

$$\phi_\varepsilon(x) = y_{k-1}, \quad \psi_\varepsilon(x) = y_k$$

and

$$\phi_\varepsilon(x) \leq f(x) < \psi_\varepsilon(x), \quad \psi_\varepsilon(x) - \phi_\varepsilon(x) = y_k - y_{k-1} < \varepsilon.$$

This completes the proof. □

Theorem 2.47 (The Simple Approximation Theorem). *An extended real-valued function f on a measurable set $\mathbf{E}$ is measurable if and only if there is a sequence $\{\phi_n\}_{n \in \mathbf{N}}$ of simple functions on $\mathbf{E}$ which converges pointwise on $\mathbf{E}$ to f and*

$$|\phi_n| \leq |f| \quad \textit{on } \mathbf{E} \textit{ for all } n \in \mathbf{N}.$$

If f is nonnegative, we can choose $\{\phi_n\}_{n \in \mathbf{N}}$ to be increasing.

Proof.

1. Let $\{\phi_n\}_{n \in \mathbf{N}}$ be a sequence of simple functions on $\mathbf{E}$ that converges pointwise on $\mathbf{E}$ to f and $|\phi_n| \leq |f|$ on $\mathbf{E}$ for all $n \in \mathbf{N}$. Because ϕ_n are measurable functions on $\mathbf{E}$ for each $n \in \mathbf{N}$, using Theorem 2.45, we get that f is a measurable function on $\mathbf{E}$.
2. Let f is a measurable function on $\mathbf{E}$. Assume that $f \geq 0$ on $\mathbf{E}$. Take $n \in \mathbf{N}$ and define

$$\mathbf{E}_n = \{x \in \mathbf{E} : f(x) \leq n\}.$$

Because f is measurable on $\mathbf{E}$, we have that $\mathbf{E}_n$ are measurable sets for each $n \in \mathbf{N}$. By Theorem 2.36, it follows that the restriction of f to $\mathbf{E}_n$ is a nonnegative bounded measurable function. We apply Theorem 2.46 to the restriction of f to $\mathbf{E}$ with the choice $\varepsilon = \frac{1}{n}$. In other words, we may choose simple functions ϕ_n and ψ_n defined on $\mathbf{E}_n$ such that

$$0 \leq \phi_n \leq f \leq \psi_n \quad \text{on } \mathbf{E}_n \quad \text{and} \quad 0 \leq \psi_n - \phi_n < \frac{1}{n} \quad \text{on } \mathbf{E}_n.$$

Note that

$$\begin{aligned}0 &\le f - \phi_n \\ &\le \psi_n - \phi_n \\ &< \frac{1}{n} \quad \text{on } \mathbf{E}_n.\end{aligned}$$

We extend ϕ_n to all $\mathbf{E}$ by setting $\phi_n(x) = n$ if $f(x) > n$, $n \in \mathbf{N}$. The function ϕ_n is a simple function defined on $\mathbf{E}$ and $0 \le \phi_n \le f$ on $\mathbf{E}$, $n \in \mathbf{N}$. Let $x \in \mathbf{E}$ be arbitrarily chosen.

(a) Assume that $f(x)$ is finite. Then there exists an $N \in \mathbf{N}$ such that $f(x) < N$. Then

$$0 \le f(x) - \phi_n(x) < \frac{1}{n} \quad \text{for all } n > N.$$

Therefore, $\phi_n(x) \to f(x)$, as $n \to \infty$.

(b) Assume that $f(x) = \infty$. Then $\phi_n(x) = n$ for all $n \in \mathbf{N}$. Therefore, $\phi_n(x) \to f(x)$, as $n \to \infty$. By replacing ϕ_n, $n \in \mathbf{N}$, with $\max\{\phi_1, \ldots, \phi_n\}$, we get that the sequence $\{\phi_n\}_{n\in\mathbf{N}}$ is increasing.

In the general case, if f is not non-negative, we take the representation $f = f^+ - f^-$ on $\mathbf{E}$ and apply the above for f^+ and f^-. This completes the proof. □

Exercise 2.15. Let $\mathbf{E}$ be a measurable set of real numbers and f be a bounded measurable function on $\mathbf{E}$. Prove that there are sequences $\{\phi_n\}_{n\in\mathbf{N}}$ and $\{\psi_n\}_{n\in\mathbf{N}}$ of simple functions on $\mathbf{E}$ such that $\{\phi_n\}_{n\in\mathbf{N}}$ is increasing and $\{\psi_n\}_{n\in\mathbf{N}}$ is decreasing and each of these sequences converges to f uniformly on $\mathbf{E}$.

Theorem 2.48. *Assume that $\mathbf{E}$ is a measurable set with a finite measure and $\{f_n\}_{n\in\mathbf{N}}$ be a sequence of bounded measurable functions on $\mathbf{E}$ that converges pointwise on $\mathbf{E}$ to the real-valued function f. Then for each $\varepsilon > 0$ and $\delta > 0$, there are a measurable subset $\mathbf{A}$ of $\mathbf{E}$ and an index N for which*

$$|f_n - f| < \varepsilon \quad \text{on } \mathbf{A} \quad \text{for all } n \ge N \quad \text{and} \quad m(\mathbf{E}\backslash\mathbf{A}) < \delta.$$

Proof. By Theorem 2.45, it follows that f is a measurable function on $\mathbf{E}$. Hence, using Theorem 2.39, the function $-f$ is a measurable function on $\mathbf{E}$. From Theorem 2.40, it follows that $f_n - f$ are measurable functions on $\mathbf{E}$ for each $n \in \mathbf{N}$. From here and from Theorem 2.37, we obtain that $|f_n - f|$ are measurable functions on $\mathbf{E}$ for all $n \in \mathbf{N}$. We take $\varepsilon > 0$ and $\delta > 0$ arbitrarily. Note that the sets

$$\mathbf{E}_n = \{x \in \mathbf{E} : |f(x) - f_k(x)| < \varepsilon \text{ for all } k \geq n\}$$

are measurable sets for all $n \in \mathbf{N}$. Then the sequence $\{\mathbf{E}_n\}_{n\in\mathbf{N}}$ is an ascending collection of measurable sets and since $\{f_n\}_{n\in\mathbf{N}}$ converges pointwise to f on $\mathbf{E}$, we have $\mathbf{E} = \cup_{n=1}^{\infty}\mathbf{E}_n$. Hence and Theorem 2.25, we get

$$m(\mathbf{E}) = \lim_{n\to\infty} m(\mathbf{E}_n).$$

Since $m(\mathbf{E}) < \infty$, we can choose an index N such that

$$m(\mathbf{E}_N) > m(\mathbf{E}) - \delta.$$

Let $\mathbf{A} = \mathbf{E}_N$. Then

$$\begin{aligned} m(\mathbf{E}\backslash\mathbf{A}) &= m(\mathbf{E}) - m(\mathbf{E}_N) \\ &< m(\mathbf{E}) - m(\mathbf{E}) + \delta \\ &= \delta. \end{aligned}$$

This completes the proof. □

Theorem 2.49 (Egoroff's Theorem). *Assume that $\mathbf{E}$ is a measurable set with finite measure and $\{f_n\}_{n\in\mathbf{N}}$ be a sequence of measurable functions on $\mathbf{E}$ that converges pointwise on $\mathbf{E}$ to the real-valued function f. Then for each $\varepsilon > 0$, there is a closed set $\mathbf{A}$ contained in $\mathbf{E}$ for which $f_n \to f$, as $n \to \infty$, uniformly on $\mathbf{A}$ and $m(\mathbf{E}\backslash\mathbf{A}) < \varepsilon$.*

Proof. Take $\varepsilon > 0$ arbitrarily. By Theorem 2.48, it follows that for each $n \in \mathbf{N}$ there is a measurable subset $\mathbf{A}_n$ of $\mathbf{E}$ and an index $N(n)$

such that

$$|f_l - f| < \frac{1}{n} \quad \text{on } \mathbf{A}_n \quad \text{for all } l \geq N \quad \text{and} \quad m(\mathbf{E} \backslash \mathbf{A}_n) < \frac{\varepsilon}{2^{n+1}}.$$

Define the set

$$\widetilde{\mathbf{A}} = \bigcap_{n=1}^{\infty} \mathbf{A}_n.$$

Then, using Theorem 2.5, we get

$$\begin{aligned} m\left(\mathbf{E} \backslash \widetilde{\mathbf{A}}\right) &= m\left(\mathbf{E} \Bigg\backslash \left(\bigcap_{n=1}^{\infty} \mathbf{A}_n\right)\right) \\ &= m\left(\bigcup_{n=1}^{\infty} (\mathbf{E} \backslash \mathbf{A}_n)\right) \\ &\leq \sum_{n=1}^{\infty} m(\mathbf{E} \backslash \mathbf{A}_n) \\ &< \sum_{n=1}^{\infty} \frac{\varepsilon}{2^{n+1}} \\ &= \frac{\varepsilon}{2}. \end{aligned}$$

Now, we choose an index n_0 such that $\frac{1}{n_0} < \varepsilon$. Then

$$|f_l - f| < \frac{1}{n_0} \quad \text{on } \mathbf{A}_{n_0} \quad \text{for all } l \geq N(n_0).$$

Since $\widetilde{\mathbf{A}} \subseteq \mathbf{A}_{n_0}$ and $\frac{1}{n_0} < \varepsilon$. we get

$$|f_l - f| < \varepsilon \quad \text{on } \widetilde{\mathbf{A}} \quad \text{for all } l \geq N(n_0).$$

Therefore, $\{f_n\}_{n \in \mathbf{N}}$ converges uniformly to f on $\widetilde{\mathbf{A}}$ and $m(\mathbf{E} \backslash \widetilde{\mathbf{A}}) < \frac{\varepsilon}{2}$. By Theorem 2.21, it follows that there is a closed set

$\mathbf{A}$ contained in $\widetilde{\mathbf{A}}$ such that $m(\widetilde{\mathbf{A}}\backslash\mathbf{A}) < \frac{\varepsilon}{2}$. Thus $\{f_n\}_{n\in\mathbf{N}}$ converges uniformly to f on $\mathbf{A}$ and

$$\begin{aligned} m\left(\mathbf{E}\backslash\mathbf{A}\right) &= m\left(\mathbf{E}\backslash\widetilde{\mathbf{A}}\right) + m\left(\widetilde{\mathbf{A}}\backslash\mathbf{A}\right) \\ &< \frac{\varepsilon}{2} + \frac{\varepsilon}{2} \\ &= \varepsilon. \end{aligned}$$

This completes the proof. □

Theorem 2.50. *Let* $\mathbf{E}$ *be a measurable set,* f *be a simple function defined on* $\mathbf{E}$*. Then for each* $\varepsilon > 0$*, there is a continuous function* g *on* $\mathbb{R}$ *and a closed set* $\mathbf{A}$ *contained in* $\mathbf{E}$ *for which*

$$f = g \quad \text{on } \mathbf{A} \quad \textit{and} \quad m\left(\mathbf{E}\backslash\mathbf{A}\right) < \varepsilon.$$

Proof. Let $a_1, \ldots, a_n$ be the finite numbers of distinct values taken by f, and let them be taken on the sets $\mathbf{E}_1, \ldots, \mathbf{E}_n$, respectively. Since a_k's are distinct, the collection $\{\mathbf{E}_k\}_{k=1}^n$ is disjoint. By Theorem 2.21, there are closed sets $A_1, \ldots, A_n$ such that

$$\mathbf{A}_k \subseteq \mathbf{E}_k \quad \text{and} \quad m\left(\mathbf{E}_k\backslash\mathbf{A}_k\right) < \frac{\varepsilon}{n}$$

for any $k \in \{1, \ldots, n\}$. Then the set $\mathbf{A} = \cup_{k=1}^n \mathbf{A}_k$ is a closed set and since $\{\mathbf{A}_k\}_{k=1}^n$ is disjoint, we get

$$\begin{aligned} m\left(\mathbf{E}\backslash\mathbf{A}\right) &= m\left(\bigcup_{k=1}^n \left(\mathbf{E}\backslash\mathbf{A}_k\right)\right) \\ &= \sum_{k=1}^n m\left(\mathbf{E}\backslash\mathbf{A}_k\right) \\ &< \sum_{k=1}^n \frac{\varepsilon}{n} \\ &= \varepsilon. \end{aligned}$$

Define g on $\mathbf{A}$ to take the values a_k on $\mathbf{A}_k$, $k \in \{1, \ldots, n\}$. Since the collection $\{\mathbf{A}_k\}_{k=1}^n$ is disjoint, the function g is properly defined and continuous. We extend g from a continuous function on $\mathbf{A}$ to a continuous function on all $\mathbf{R}$. This completes the proof. □

Theorem 2.51 (Luzin's Theorem). *Let* $\mathbf{E}$ *be a measurable set and* f *be a real-valued measurable function on* $\mathbf{E}$*. Then for each* $\varepsilon > 0$*, there is a continuous function* g *on* $\mathbf{R}$ *and a closed set* $\mathbf{A}$ *contained in* $\mathbf{E}$ *for which*

$$f = g \quad \text{on } \mathbf{A} \quad \text{and} \quad m\left(\mathbf{E} \backslash \mathbf{A}\right) < \varepsilon.$$

Proof.

1. Let $m\left(\mathbf{E}\right) < \infty$. By Theorem 2.47, there is a sequence $\{f_n\}_{n \in \mathbf{N}}$ of simple functions defined on $\mathbf{E}$ that converges to f pointwise on $\mathbf{E}$. Take $n \in \mathbf{N}$ arbitrarily. By Theorem 2.50, it follows that there is a continuous function g_n on $\mathbf{R}$ and a closed set $\mathbf{A}_n$ contained in $\mathbf{E}$ for which

$$g_n = f_n \quad \text{on } \mathbf{A}_n \quad \text{and} \quad m\left(\mathbf{E} \backslash \mathbf{A}_n\right) < \frac{\varepsilon}{2^{n+1}}.$$

By Egoroff's Theorem, there is a closed set $\mathbf{A}_0$ contained in $\mathbf{E}$ such that $\{f_n\}_{n \in \mathbf{N}}$ converges to f uniformly on $\mathbf{A}_0$ and $m(\mathbf{E} \backslash \mathbf{A}_0) < \frac{\varepsilon}{2}$. Define $\mathbf{A} = \cap_{n=0}^{\infty} \mathbf{A}_n$. Then

$$\begin{aligned}
m\left(\mathbf{E} \backslash \mathbf{A}\right) &= m\left(\left(\mathbf{E} \backslash \mathbf{A}_0\right) \cup \left(\bigcup_{n=1}^{\infty} \left(\mathbf{E} \backslash \mathbf{A}_n\right)\right)\right) \\
&\le m\left(\mathbf{E} \backslash \mathbf{A}_0\right) + m\left(\bigcup_{n=1}^{\infty} \left(\mathbf{E} \backslash \mathbf{A}_n\right)\right) \\
&\le m\left(\mathbf{E} \backslash \mathbf{A}_0\right) + \sum_{n=1}^{\infty} m\left(\mathbf{E} \backslash \mathbf{A}_n\right) \\
&< \frac{\varepsilon}{2} + \sum_{n=1}^{\infty} \frac{\varepsilon}{2^{n+1}} \\
&= \frac{\varepsilon}{2} + \frac{\varepsilon}{2} \\
&= \varepsilon.
\end{aligned}$$

The set $\mathbf{A}$ is closed. Each f_n is continuous on $\mathbf{A}$ because $\mathbf{A} \subseteq \mathbf{A}_n$ and $f_n = g_n$ on $\mathbf{A}_n$. Since the sequence $\{f_n\}_{n \in \mathbf{N}}$ converges uniformly to f on $\mathbf{A} \subseteq \mathbf{A}_0$, we conclude that the restriction of f to $\mathbf{A}$ is continuous on $\mathbf{A}$. Let g be the restriction of f to $\mathbf{A}$.

We extend g from a continuous function on $\mathbf{A}$ to a continuous function on all $\mathbf{R}$.

2. Let $m(\mathbf{E}) = \infty$. Then $\mathbf{E}$ can be expressed as the disjoint union of a countable collection $\{\mathbf{E}_k\}_{k\in\mathbf{N}}$ of measurable sets with finite measure. For each $\mathbf{E}_k$, as in the previous case, we construct a continuous function g^k on $\mathbf{E}_k$ and a closed set $\mathbf{A}_k \subseteq \mathbf{E}_k$ such that

$$f = g^k \quad \text{on } \mathbf{A}_k \quad \text{and} \quad m(\mathbf{E}_k \backslash \mathbf{A}_k) < \frac{\varepsilon}{2^{k+1}}.$$

The collection $\{\mathbf{A}_k\}_{k\in\mathbf{N}}$ is disjoint. Let $\mathbf{A} = \cup_{k\in\mathbf{N}}\mathbf{A}_k$. Then $\mathbf{A}$ is closed. We construct a continuous function g on $\mathbf{A}$ such that its restrictions on $\mathbf{A}_k$ coincide with g_k. We extend g from a continuous function on $\mathbf{A}$ to a continuous function on all $\mathbf{R}$. Note that

$$\mathbf{E}\backslash\mathbf{A} \subseteq \bigcup_{k=1}^{\infty} (\mathbf{E}_k \backslash \mathbf{A}_k)$$

and

$$\begin{aligned} m(\mathbf{E}\backslash\mathbf{A}) &\leq m\left(\bigcup_{k=1}^{\infty} (\mathbf{E}_k \backslash \mathbf{A}_k)\right) \\ &\leq \sum_{k=1}^{\infty} m(\mathbf{E}_k \backslash \mathbf{A}_k) \\ &< \sum_{k=1}^{\infty} \frac{\varepsilon}{2^{k+1}} \\ &= \varepsilon. \end{aligned}$$

This completes the proof. □

2.6 The Riemann Integral

In this section, we recall some definitions to the Riemann integral.

Definition 2.22. Let f be a bounded real-valued function defined pon the closed bounded interval $[a, b]$. Let also, $P = \{x_0, x_1, \ldots, x_n\}$

be a partition of the interval $[a, b]$, i.e.,

$$a = x_0 < x_1 < \cdots < x_n = b.$$

Define the lower and upper Darboux sums for f with respect to P, respectively, by

$$L(f, P) = \sum_{i=1}^{n} m_i(x_i - x_{i-1})$$

and

$$U(f, P) = \sum_{i=1}^{n} M_i(x_i - x_{i-1}),$$

where

$$m_i = \inf\{f(x) : x_{i-1} < x < x_i\}, \quad M_i = \sup\{f(x) : x_{i-1} < x < x_i\},$$
$$i \in \{1, \ldots, n\}.$$

Definition 2.23. We define the lower and upper Riemann integrals of fb over $[a, b]$, respectively, by

$$(R)\underline{\int_a^b} f = \sup\left\{L(f, P) : P \text{ is a partition of } [a, b]\right\},$$

$$(R)\overline{\int_a^b} = \inf\left\{U(f, P) : P \text{ is a partition of } [a, b]\right\}.$$

Note that the lower and upper Riemann integrals are finite. The upper Riemann integral is at least as large the lower Riemann integral, and if the two are equal we say that f is Riemann integrable over $[pa, b]$ and call the common value the Riemann integral of f over $[a, b]$. We denote

$$(R)\int_a^b f.$$

Definition 2.24. A real-valued function ψ defined on $[a, b]$ is called a step function if there is a partition $P = \{x_0, x_1, \ldots, x_n\}$ of $[a, b]$ and numbers $c_1, \ldots, c_n$, such that

$$\psi(x) = c_i \quad \text{if } x_{i-1} < x < x_i, \quad i \in \{1, \ldots, n\}.$$

Suppose that ψ is a step function on $[a, b]$. Observe that

$$L(\psi, P) = \sum_{i=1}^{n} c_i(x_i - x_{i-1}) = U(\psi, P)$$

and

$$(R)\int_a^b \psi = \sum_{i=1}^{n} c_i(x_i - x_{i-1}).$$

Therefore, we may reformulate the definition of the lower and upper Riemann integrals as follows.

Definition 2.25.

$$(R)\underline{\int_a^b} = \sup\left\{(R)\int_a^b \phi : \phi \text{ is a step function, } \phi \le f \quad \text{on } [a, b]\right\},$$

$$(R)\overline{\int_a^b} = \sup\left\{(R)\int_a^b \psi : \psi \text{ is a step function, } \psi \ge f \quad \text{on } [a, b]\right\}.$$

2.7 Lebesgue Integration

2.7.1 *The Lebesgue integral of a bounded measurable function over a set of finite measure*

Suppose that $\mathbf{E}$ is a measurable set of finite measure.

Definition 2.26. Let ϕ be a simple function on $\mathbf{E}$,

$$\phi = \sum_{i=1}^{n} a_i \kappa_{\mathbf{E}_i}, \quad \mathbf{E}_i = \phi^{-1}(a_i) = \{x \in \mathbf{E} : \phi(x) = a_i\},$$

$a_i \neq a_j$, $\mathbf{E}_i \cap \mathbf{E}_j = \emptyset$, $i, j \in \{1, \dots, n\}$. We define the integral of ϕ over $\mathbf{E}$ by

$$\int_{\mathbf{E}} \phi = \sum_{i=1}^{n} a_i m(\mathbf{E}_i).$$

Theorem 2.52. *Let $\{\mathbf{E}_i\}_{i=1}^{n}$ be a finite collection of disjoint measurable subsets of $\mathbf{E}$. Let also, $a_i \in \mathbf{R}$, $i \in \{1, \dots, n\}$, and*

$$\phi = \sum_{i=1}^{n} a_i \kappa_{\mathbf{E}_i}.$$

Then

$$\int_{\mathbf{E}} \phi = \sum_{i=1}^{n} a_i m(\mathbf{E}_i).$$

Proof.

1. If a_i, $i \in \{1, \ldots, n\}$, are distinct, we get the desired result using Definition 2.26.
2. Assume that a_i, $i \in \{1, \ldots, n\}$, are not distinct. Let $\{b_1, \ldots, b_m\}$ be the distinct values taken by ϕ. We define the sets

$$B_j = \{x \in \mathbf{E} : \phi(x) = b_j\}, \quad j \in \{1, \ldots, m\}.$$

Hence, by Definition 2.26,

$$\int_{\mathbf{E}} \phi = \sum_{i=1}^{m} b_i m(\mathbf{B}_i). \tag{2.83}$$

For $j \in \{1, \ldots, m\}$, with $\mathbf{I}_j$ we denote the set of indices $j \in \{1, \ldots, n\}$, such that $a_i = b_j$. Then $\{1, \ldots, n\} = \cup_{j=1}^{m} \mathbf{I}_j$ and $\mathbf{I}_k \cap \mathbf{I}_l = \emptyset$, $k \neq l$, $k, l \in \{1, \ldots, m\}$. From the finite additivity of the measure, we have

$$\begin{aligned} m(\mathbf{B}_j) &= m\left(\bigcup_{i \in \mathbf{I}_j} \mathbf{E}_i\right) \\ &= \sum_{i \in \mathbf{I}_j} m(\mathbf{E}_i). \end{aligned}$$

Hence, using (2.83), we find

$$\begin{aligned} \sum_{i=1}^{n} a_i m(\mathbf{E}_i) &= \sum_{j=1}^{m} \left(\sum_{i \in \mathbf{I}_j} a_i m(\mathbf{E}_i)\right) \\ &= \sum_{j=1}^{m} b_j \left(\sum_{i \in \mathbf{I}_j} m(\mathbf{E}_i)\right) \end{aligned}$$

$$= \sum_{j=1}^{m} b_j m(\mathbf{B}_j)$$

$$= \int_{\mathbf{E}} \phi.$$

This completes the proof. □

Theorem 2.53. *Let ϕ be a simple function on $\mathbf{E}$ and $\phi \geq 0$ on $\mathbf{E}$. Then $\int_{\mathbf{E}} \phi \geq 0$.*

Proof. Since ϕ is a simple function on $\mathbf{E}$, there is a finite disjoint collection $\{\mathbf{E}_i\}_{i=1}^{m}$ of subsets of $\mathbf{E}$ such that $\mathbf{E} = \bigcup_{j=1}^{m} \mathbf{E}_j$ and

$$\phi = \sum_{j=1}^{m} a_j \kappa_{\mathbf{E}_j},$$

for some $a_j \in \mathbf{R}$, $j \in \{1, \ldots, m\}$. Since $\phi \geq 0$ on $\mathbf{E}$, we have that $a_j \geq 0$ for any $j \in \{1, \ldots, m\}$. Hence,

$$\int_{\mathbf{E}} \phi = \sum_{j=1}^{m} a_j m(\mathbf{E}_j)$$

$$\geq 0.$$

This completes the proof. □

Theorem 2.54. *Let ϕ and ψ be simple functions on $\mathbf{E}$. Then for any $\alpha, \beta \in \mathbf{R}$, we have*

$$\int_{\mathbf{E}} (\alpha\phi + \beta\psi) = \alpha \int_{\mathbf{E}} \phi + \beta \int_{\mathbf{E}} \psi.$$

Moreover, if $\phi \leq \psi$, then

$$\int_{\mathbf{E}} \phi \leq \int_{\mathbf{E}} \psi.$$

Proof. Since both ϕ and ψ take finite values, there is a finite disjoint collection $\{\mathbf{E}_j\}_{j=1}^m$ of subsets of $\mathbf{E}$ such that $\mathbf{E} = \bigcup_{j=1}^m \mathbf{E}_j$ and

$$\phi = \sum_{j=1}^m a_j \kappa_{\mathbf{E}_j} \quad \text{and} \quad \psi = \sum_{j=1}^m b_j \kappa_{\mathbf{E}_j}$$

for some $a_j, b_j \in \mathbf{R}$, $j \in \{1, \ldots, m\}$. Then

$$\begin{aligned}
\alpha\phi &= \alpha \sum_{j=1}^m a_j \kappa_{\mathbf{E}_j} \\
&= \sum_{j=1}^m (\alpha a_j)\, \kappa_{\mathbf{E}_j}, \\
\beta\psi &= \beta \sum_{j=1}^m b_j \kappa_{\mathbf{E}_j} \\
&= \sum_{j=1}^m (\beta b_j)\, \kappa_{\mathbf{E}_j}, \\
\alpha\phi + \beta\psi &= \sum_{j=1}^m (\alpha a_j)\, \kappa_{\mathbf{E}_j} + \sum_{j=1}^m (\beta b_j)\, \kappa_{\mathbf{E}_j} \\
&= \sum_{j=1}^m (\alpha a_j + \beta b_j)\, \kappa_{\mathbf{E}_j}, \\
\int_{\mathbf{E}} \phi &= \sum_{j=1}^m a_j m\,(\mathbf{E}_j), \\
\int_{\mathbf{E}} \psi &= \sum_{j=1}^m b_j m\,(\mathbf{E}_j), \\
\int_{\mathbf{E}} (\alpha\phi + \beta\psi) &= \sum_{j=1}^m (\alpha a_j + \beta b_j)\, m\,(\mathbf{E}_j), \\
\alpha \int_{\mathbf{E}} \phi &= \alpha \sum_{j=1}^m a_j m\,(\mathbf{E}_j)
\end{aligned}$$

$$
\begin{aligned}
&= \sum_{j=1}^{m} (\alpha a_j)\, m\,(\mathbf{E}_j),\\
\beta \int_{\mathbf{E}} \psi &= \beta \sum_{j=1}^{m} b_j m\,(\mathbf{E}_j)\\
&= \sum_{j=1}^{m} (\beta b_j)\, m\,(\mathbf{E}_j),\\
\alpha \int_{\mathbf{E}} \phi + \beta \int_{\mathbf{E}} \psi &= \sum_{j=1}^{m} (\alpha a_j)\, m\,(\mathbf{E}_j) + \sum_{j=1}^{m} (\beta b_j)\, m\,(\mathbf{E}_j)\\
&= \sum_{j=1}^{m} (\alpha a_j + \beta b_j)\, m\,(\mathbf{E}_j).
\end{aligned}
$$

Therefore,

$$
\int_{\mathbf{E}} (\alpha\phi + \beta\psi) = \alpha \int_{\mathbf{E}} \phi + \beta \int_{\mathbf{E}} \psi.
$$

Let now $\phi \leq \psi$ on $\mathbf{E}$. Then $\psi - \phi \geq 0$ on $\mathbf{E}$. By Theorem 2.53, we get

$$
\int_{\mathbf{E}} (\psi - \phi) \geq 0.
$$

Hence,

$$
\int_{\mathbf{E}} \psi - \int_{\mathbf{E}} \phi = \int_{\mathbf{E}} (\psi - \phi) \geq 0.
$$

This completes the proof. □

Definition 2.27. Let f be a bounded real-valued function defined on $\mathbf{E}$. We define the lower and upper Lebesgue integral, respectively, to be

$$
\int_{\mathbf{E}}^{l} f = \sup\left\{ \int_{\mathbf{E}} \phi : \phi \text{ is simple and } \phi \leq f \text{ on } \mathbf{E} \right\},
$$

$$
\int_{\mathbf{E}}^{u} f = \inf\left\{ \int_{\mathbf{E}} \psi : \psi \text{ is simple and } f \leq \psi \text{ on } \mathbf{E} \right\}.
$$

Since f is a bounded function on $\mathbf{E}$, there is a constant M such that $f \leq M$ on $\mathbf{E}$. Let ϕ is a simple function on $\mathbf{E}$ such that $\phi \leq f$ on $\mathbf{E}$. Then $\phi \leq M$ on $\mathbf{E}$ and $M - \phi$ is a simple function on $\mathbf{E}$. By Theorem 2.54, we get

$$\begin{aligned} 0 &\leq \int_{\mathbf{E}} (M - \phi) \\ &= \int_{\mathbf{E}} M - \int_{\mathbf{E}} \phi \\ &= Mm(\mathbf{E}) - \int_{\mathbf{E}} \phi, \end{aligned}$$

whereupon

$$\int_{\mathbf{E}} \phi \leq Mm(\mathbf{E}).$$

Therefore,

$$\int_{\mathbf{E}}^{l} f \leq Mm(\mathbf{E}).$$

Let now ϕ and ψ be simple functions on $\mathbf{E}$ such that $\phi \leq f \leq \psi$ on $\mathbf{E}$. Hence and Theorem 2.54, we find

$$\int_{\mathbf{E}} \phi \leq \int_{\mathbf{E}} \psi.$$

Consequently,

$$\int_{\mathbf{E}}^{l} f \leq \int_{\mathbf{E}}^{u} f.$$

Definition 2.28. A bounded real-valued function f on $\mathbf{E}$ is said to be Lebesgue integrable provided its upper and lower Lebesgue integrals are equal. The common value of the upper and lower integrals is called the Lebesgue integral, or simply the integral, of f over $\mathbf{E}$. It is denoted by $(L)\int_{\mathbf{E}} f$ or simply $\int_{\mathbf{E}} f$.

Theorem 2.55. *Let f be a bounded function defined on a closed, bounded interval $\mathbf{I} = [a, b]$. If f is Riemann integrable over $\mathbf{I}$, then it is Lebesgue integrable over $\mathbf{I}$ and the two integrals are equal.*

Proof. Since f is Riemann integrable over $\mathbf{I}$, we have

$$(R)\int_{\mathbf{I}} f = \sup\left\{(R)\int_{\mathbf{I}} \phi : \phi \text{ is a step function, } \phi \le f\right\}$$
$$= \inf\left\{(R)\int_{\mathbf{I}} \psi : \psi \text{ is a step function, } f \le \psi\right\}.$$

Because every step function is a simple function and every simple function on a bounded closed interval is a step function, and $(R)\int_{\mathbf{I}} \phi = (L)\int_{\mathbf{I}} \phi$ for any simple function ϕ on $\mathbf{I}$, we get

$$\sup\left\{(R)\int_{\mathbf{I}} \phi : \phi \text{ is a step function, } \phi \le f\right\}$$
$$= \sup\left\{(L)\int_{\mathbf{I}} \phi : \phi \text{ is a simple function, } \phi \le f\right\}$$
$$= \inf\left\{(R)\int_{\mathbf{I}} \psi : \psi \text{ is a step function, } f \le \psi\right\}$$
$$= \inf\left\{(L)\int_{\mathbf{I}} \psi : \psi \text{ is a step function, } f \le \psi\right\}.$$

Hence, f is Lebesgue integrable over $\mathbf{I}$ and

$$(L)\int_{\mathbf{I}} f = (R)\int_{\mathbf{I}} f.$$

This completes the proof. □

Theorem 2.56. *Let f be a bounded measurable function on $\mathbf{E}$. Then f is integrable over $\mathbf{E}$.*

Proof. By Theorem 2.46, it follows that for any $n \in \mathbf{N}$ there are simple functions ϕ_n and ψ_n on $\mathbf{E}$ such that

$$\phi_n \le f \le \psi_n$$

and $\psi_n - \phi_n \le \frac{1}{n}$ on $\mathbf{E}$. Hence and Theorem 2.54, it follows that

$$0 \le \int_{\mathbf{E}} \psi_n - \int_{\mathbf{E}} \phi_n$$
$$= \int_{\mathbf{E}} (\psi_n - \phi_n)$$
$$\le \frac{1}{n} m(\mathbf{E})$$

for any $n \in \mathbf{N}$. From here,

$$\begin{aligned}0 &\leq \inf\left\{\int_{\mathbf{E}} \psi : \psi \quad \text{simple}, \psi \geq f\right\} - \sup\left\{\int_{\mathbf{E}} \phi : \phi \quad \text{simple}, \phi \leq f\right\} \\ &\leq \int_{\mathbf{E}} \psi_n - \int_{\mathbf{E}} \phi_n \\ &\leq \frac{1}{n} m(\mathbf{E})\end{aligned}$$

for any $n \in \mathbf{N}$. Therefore the upper and lower integrals of f over $\mathbf{E}$ are equal. Consequently the function f is integrable over $\mathbf{E}$. This completes the proof. □

Theorem 2.57. *Let f be a bounded measurable function on $\mathbf{E}$ and $m(\mathbf{E}) = 0$. Then*

$$\int_{\mathbf{E}} f = 0.$$

Proof. Let ϕ be an arbitrary simple function on $\mathbf{E}$ with canonical representation

$$\phi = \sum_{i=1}^{n} a_i \kappa_{\mathbf{E}_i},$$

where $\mathbf{E} = \cup_{i=1}^{n} \mathbf{E}_i$, $\mathbf{E}_i \cap \mathbf{E}_j = \emptyset$, $a_i \in \mathbf{R}$, $a_i \neq a_j$, $i \neq j$, $i, j \in \{1, \ldots, n\}$. Then $m(\mathbf{E}_i) = 0$, $i \in \{1, \ldots, n\}$, and

$$\begin{aligned}\int_{\mathbf{E}} \phi &= \sum_{i=1}^{n} a_i m(\mathbf{E}_i) \\ &= 0.\end{aligned}$$

Hence,

$$\begin{aligned}\int_{\mathbf{E}} f &= \sup\left\{\int_{\mathbf{E}} \phi : \phi \text{ is simple, } \phi \leq f\right\} \\ &= 0.\end{aligned}$$

This completes the proof. □

Theorem 2.58. *Let f and g be bounded measurable functions on $\mathbf{E}$. Then, for any $\alpha, \beta \in \mathbf{R}$, we have*

$$\int_{\mathbf{E}} (\alpha f + \beta g) = \alpha \int_{\mathbf{E}} f + \beta \int_{\mathbf{E}} g.$$

Moreover, if $f \geq g$ on $\mathbf{E}$, we have

$$\int_{\mathbf{E}} f \geq \int_{\mathbf{E}} g.$$

Proof. Suppose that $\alpha > 0$. Using that the Lebesgue integral is equal to the upper Lebesgue integral, we have

$$\begin{aligned}\int_{\mathbf{E}} (\alpha f) &= \inf_{\psi \geq \alpha f} \int_{\mathbf{E}} \psi \\ &= \alpha \inf_{\frac{\psi}{\alpha} \geq f} \int_{\mathbf{E}} \frac{\psi}{\alpha} \\ &= \alpha \int_{\mathbf{E}} f.\end{aligned}$$

Let $\alpha < 0$. Then, using that the Lebesgue integral is equal to the upper and lower Lebesgue integrals, we get

$$\begin{aligned}\int_{\mathbf{E}} (\alpha f) &= \inf_{\psi \geq \alpha f} \int_{\mathbf{E}} \psi \\ &= \alpha \sup_{\frac{\psi}{\alpha} \leq f} \int_{\mathbf{E}} \frac{\psi}{\alpha} \\ &= \alpha \int_{\mathbf{E}} f.\end{aligned}$$

Now, we will prove that

$$\int_{\mathbf{E}} (f + g) = \int_{\mathbf{E}} f + \int_{\mathbf{E}} g. \tag{2.84}$$

Let ϕ_1, ϕ_2, ψ_1 and ψ_2 be simple functions on $\mathbf{E}$ such that

$$\phi_1 \leq f \leq \psi_1 \quad \text{and} \quad \phi_2 \leq g \leq \psi_2 \quad \text{on } \mathbf{E}.$$

Then $\phi_1 + \phi_2$ and $\psi_1 + \psi_2$ are simple functions on $\mathbf{E}$ and

$$\phi_1 + \phi_2 \leq f + g \leq \psi_1 + \psi_2 \quad \text{on } \mathbf{E}.$$

Hence and Theorem 2.54, we get

$$\begin{aligned}\int_{\mathbf{E}} (f + g) &\leq \int_{\mathbf{E}} (\psi_1 + \psi_2) \\ &= \int_{\mathbf{E}} \psi_1 + \int_{\mathbf{E}} \psi_2.\end{aligned}$$

Then

$$\begin{aligned}\int_{\mathbf{E}} (f+g) &\le \inf_{\psi_1 \ge f} \int_{\mathbf{E}} \psi_1 + \inf_{\psi_2 \ge f} \int_{\mathbf{E}} \psi_2 \\ &= \int_{\mathbf{E}} f + \int_{\mathbf{E}} g.\end{aligned} \tag{2.85}$$

On the other hand,

$$\begin{aligned}\int_{\mathbf{E}} (f+g) &\ge \int_{\mathbf{E}} (\phi_1 + \phi_2) \\ &= \int_{\mathbf{E}} \phi_1 + \int_{\mathbf{E}} \phi_2.\end{aligned}$$

Then

$$\begin{aligned}\int_{\mathbf{E}} (f+g) &\ge \sup_{\phi_1 \le f} \int_{\mathbf{E}} \phi_1 + \sup_{\phi_2 \le f} \int_{\mathbf{E}} \phi_2 \\ &= \int_{\mathbf{E}} f + \int_{\mathbf{E}} g.\end{aligned}$$

From the last inequality and from (2.85), we get (2.84). Hence, since αf and βg are bounded measurable functions on $\mathbf{E}$ for any $\alpha, \beta \in \mathbf{R}$, we get

$$\begin{aligned}\int_{\mathbf{E}} (\alpha f + \beta g) &= \int_{\mathbf{E}} (\alpha f) + \int_{\mathbf{E}} (\beta g) \\ &= \alpha \int_{\mathbf{E}} f + \beta \int_{\mathbf{E}} g.\end{aligned}$$

Now, we assume that $f \ge g$ on $\mathbf{E}$. Then $h = f - g \ge 0$ on $\mathbf{E}$. Since 0 is a simple function on $\mathbf{E}$, we obtain

$$\int_{\mathbf{E}} h \ge 0.$$

From here,

$$\begin{aligned}0 &\le \int_{\mathbf{E}} (f-g) \\ &= \int_{\mathbf{E}} f - \int_{\mathbf{E}} g.\end{aligned}$$

This completes the proof. □

Theorem 2.59. *Let f be a bounded measurable function on* $\mathbf{E}$*. Then for any measurable subset* $\mathbf{A}$ *of* $\mathbf{E}$ *we have*

$$\int_{\mathbf{E}} f\kappa_{\mathbf{A}} = \int_{\mathbf{A}} f.$$

Proof. Note that $f\kappa_{\mathbf{A}}$ is a bounded measurable function on $\mathbf{E}$ and $\mathbf{A}$. Also, f is a bounded measurable function on $\mathbf{A}$. Then, by Theorem 2.56, we conclude that $f\kappa_{\mathbf{A}}$ is integrable on $\mathbf{E}$ and $\mathbf{A}$, and f is integrable on $\mathbf{A}$. Let ϕ be any simple function on $\mathbf{E}$ with the canonical representation

$$\phi = \sum_{i=1}^{n} a_i \kappa_{\mathbf{E}_i}, \quad a_i \in \mathbf{R}, \quad a_i \neq a_j, \quad i, j \in \{1, \ldots, n\},$$

$\mathbf{E} = \cup_{i=1}^{n} \mathbf{E}_i$, $\mathbf{E}_i \cap \mathbf{E}_j = \emptyset$, $i \neq j$, $i, j \in \{1, \ldots, n\}$. Then

$$\begin{aligned} \phi\kappa_{\mathbf{A}} &= \sum_{i=1}^{n} a_i \kappa_{\mathbf{E}_i} \kappa_{\mathbf{A}} \\ &= \sum_{i=1}^{n} a_i \kappa_{\mathbf{E}_i \cap \mathbf{A}} + 0\kappa_{\mathbf{E} \setminus (\cup_{i=1}^{n} (\mathbf{E}_i \cap \mathbf{A}))}. \end{aligned} \tag{2.86}$$

Hence, $\phi\kappa_{\mathbf{A}}$ is a simple function on $\mathbf{E}$ with the canonical representation (2.86). Then, using Definition 2.26, we obtain

$$\int_{\mathbf{E}} \phi\kappa_{\mathbf{A}} = \sum_{i=1}^{n} a_i m(\mathbf{E}_i \cap \mathbf{A}). \tag{2.87}$$

Note that $\mathbf{A} = \cup_{i=1}^{n} (\mathbf{E}_i \cap \mathbf{A})$, $(\mathbf{E}_i \cap \mathbf{A}) \cap (\mathbf{E}_j \cap \mathbf{A}) = \emptyset$. Hence, (2.86) is the canonical representation of the simple function $\phi\kappa_{\mathbf{A}}$ on $\mathbf{A}$. Then, using Definition 2.26 and using that $\phi\kappa_{\mathbf{A}}$ is the restriction of ϕ on $\mathbf{A}$, we get

$$\begin{aligned} \int_{\mathbf{A}} \phi &= \int_{\mathbf{A}} \phi\kappa_{\mathbf{A}} \\ &= \sum_{i=1}^{n} a_i m(\mathbf{E}_i \cap \mathbf{A}). \end{aligned}$$

From here and from (2.87), we obtain

$$\int_{\mathbf{E}} \phi\kappa_{\mathbf{A}} = \int_{\mathbf{A}} \phi.$$

Next,

$$\begin{aligned}\int_{\mathbf{E}} f\kappa_{\mathbf{A}} &= \sup\left\{\int_{\mathbf{E}} \phi : \phi \quad \text{simple, } \phi \le f\kappa_{\mathbf{A}}\right\}\\ &= \sup\left\{\int_{\mathbf{E}} \phi\kappa_{\mathbf{A}} : \phi \quad \text{simple, } \phi \le f\kappa_{\mathbf{A}}\right\}\\ &= \sup\left\{\int_{\mathbf{A}} \phi : \phi \quad \text{simple, } \phi \le f\kappa_{\mathbf{A}}\right\}\\ &= \int_{\mathbf{A}} f\kappa_{\mathbf{A}}\\ &= \int_{\mathbf{A}} f.\end{aligned}$$

This completes the proof. □

Theorem 2.60. *Let f be a bounded measurable function on $\mathbf{E}$. Suppose that $\mathbf{A}$ and $\mathbf{B}$ are disjoint measurable subsets of $\mathbf{E}$. Then*

$$\int_{\mathbf{A}\cup\mathbf{B}} f = \int_{\mathbf{A}} f + \int_{\mathbf{B}} f.$$

Proof. Note that $f\kappa_{\mathbf{A}\cup\mathbf{B}}$, $f\kappa_{\mathbf{A}}$ and $f\kappa_{\mathbf{B}}$ are bounded measurable functions on $\mathbf{E}$. Since $\mathbf{A}$ and $\mathbf{B}$ are disjoint, we have

$$f\kappa_{\mathbf{A}\cup\mathbf{B}} = f\kappa_{\mathbf{A}} + f\kappa_{\mathbf{B}}.$$

Hence, Theorems 2.58 and 2.59, we get

$$\begin{aligned}\int_{\mathbf{A}\cup\mathbf{B}} f &= \int_{\mathbf{A}\cup\mathbf{B}} f\kappa_{\mathbf{A}\cup\mathbf{B}}\\ &= \int_{\mathbf{A}\cup\mathbf{B}} (f\kappa_{\mathbf{A}} + f\kappa_{\mathbf{B}})\\ &= \int_{\mathbf{A}\cup\mathbf{B}} f\kappa_{\mathbf{A}} + \int_{\mathbf{A}\cup\mathbf{B}} f\kappa_{\mathbf{B}}\\ &= \int_{\mathbf{A}} f\kappa_{\mathbf{A}} + \int_{\mathbf{B}} f\kappa_{\mathbf{B}}\\ &= \int_{\mathbf{A}} f + \int_{\mathbf{B}} f.\end{aligned}$$

This completes the proof. □

Theorem 2.61. *Let f be a bounded measurable function on* $\mathbf{E}$. *Then*

$$\left|\int_{\mathbf{E}} f\right| \le \int_{\mathbf{E}} |f|.$$

Proof. By Theorem 2.37, we have that $|f|$ is a measurable function on $\mathbf{E}$. Since f is bounded on $\mathbf{E}$, the function $|f|$ is a bounded measurable function on $\mathbf{E}$. By Theorem 2.56, it follows that it is integrable over $\mathbf{E}$. Note that

$$-f \le |f| \le f \quad \text{on } \mathbf{E}.$$

Hence and Theorem 2.58, we obtain

$$-\int_{\mathbf{E}} f \le \int_{\mathbf{E}} |f| \le \int_{\mathbf{E}} f.$$

This completes the proof. □

Theorem 2.62. *Let $\{f_n\}_{n\in\mathbf{N}}$ be a sequence of bounded measurable functions on* $\mathbf{E}$. *If $f_n \to f$, as $n \to \infty$, uniformly on* $\mathbf{E}$, *then*

$$\lim_{n\to\infty} \int_{\mathbf{E}} f_n = \int_{\mathbf{E}} f.$$

Proof. Let $\varepsilon > 0$ be arbitrarily chosen. Since $\{f_n\}_{n\in\mathbf{N}}$ is uniformly convergent to f on $\mathbf{E}$, there is an $N \in \mathbf{N}$ such that

$$|f - f_n| < \frac{\varepsilon}{m(\mathbf{E})} \quad \text{on } \mathbf{E} \tag{2.88}$$

for any $n \ge N$. In particular,

$$|f - f_N| < \frac{\varepsilon}{m(\mathbf{E})} \quad \text{on } \mathbf{E}.$$

Hence,

$$-\frac{\varepsilon}{m(\mathbf{E})} < f - f_N < \frac{\varepsilon}{m(\mathbf{E})} \quad \text{on } \mathbf{E},$$

or

$$f_N - \frac{\varepsilon}{m(\mathbf{E})} < f < f_N + \frac{\varepsilon}{m(\mathbf{E})} \quad \text{on } \mathbf{E}.$$

From the last inequalities, since f_N is bounded on $\mathbf{E}$, we conclude that the function f is a bounded function on $\mathbf{E}$. Since the sequence

$\{f_n\}_{n\in\mathbf{N}}$ is uniformly convergent to f on $\mathbf{E}$, it is pointwise convergent to f on $\mathbf{E}$. Because f_n are measurable functions on $\mathbf{E}$ for any $n \in \mathbf{N}$, using Theorem 2.45, we conclude that f is measurable on $\mathbf{E}$. Therefore f is a bounded measurable function on $\mathbf{E}$. By Theorem 2.56, it follows that f is integrable on $\mathbf{E}$. Hence, Theorems 2.54, 2.61 and (2.88), we get

$$\begin{aligned}\left|\int_{\mathbf{E}} f_n - \int_{\mathbf{E}} f\right| &= \left|\int_{\mathbf{E}} (f_n - f)\right| \\ &\le \int_{\mathbf{E}} |f_n - f| \\ &< \int_{\mathbf{E}} \frac{\varepsilon}{m(\mathbf{E})} \\ &= \varepsilon\end{aligned}$$

for any $n \ge N$. This completes the proof. □

Theorem 2.63 (The Bounded Convergence Theorem). *Let $\{f_n\}_{n\in\mathbf{N}}$ be an uniformly bounded sequence of measurable functions on $\mathbf{E}$. If $f_n \to f$, as $n \to \infty$, pointwise on $\mathbf{E}$, then*

$$\lim_{n\to\infty} \int_{\mathbf{E}} f_n = \int_{\mathbf{E}} f.$$

Proof. Let M be a positive constant such that $|f_n| \le M$ on $\mathbf{E}$ for any $n \in \mathbf{N}$. Then $|f| \le M$ on $\mathbf{E}$ and

$$\begin{aligned}|f_n - f| &\le |f_n| + |f| \\ &\le M + M \\ &= 2M \quad \text{on } \mathbf{E}\end{aligned}$$

for any $n \in \mathbf{N}$. We take $\varepsilon > 0$ arbitrarily. By Egoroff's Theorem, Theorem 2.49, it follows that there exists a closed subset $\mathbf{A}$ of $\mathbf{E}$ such that $f_n \to f$, as $n \to \infty$, uniformly on $\mathbf{A}$ and $m(\mathbf{E}\backslash\mathbf{A}) < \frac{\varepsilon}{4M}$. Hence, there is an index $N \in \mathbf{N}$ such that

$$|f_n - f| < \frac{\varepsilon}{2m(\mathbf{E})} \quad \text{on } \mathbf{A}$$

for any $n \geq N$. Then, using Theorems 2.58, 2.60, and 2.61, we get

$$\begin{aligned}\left|\int_{\mathbf{E}} f_n - \int_{\mathbf{E}} f\right| &= \left|\int_{\mathbf{E}} (f_n - f)\right| \\ &\leq \int_{\mathbf{E}} |f_n - f| \\ &= \int_{\mathbf{A}} |f_n - f| + \int_{\mathbf{E}\backslash\mathbf{A}} |f_n - f| \\ &< \frac{\varepsilon}{2m(\mathbf{E})} m(\mathbf{A}) + 2Mm(\mathbf{E}\backslash\mathbf{A}) \\ &< \frac{\varepsilon}{2} + \frac{\varepsilon}{2} \\ &= \varepsilon\end{aligned}$$

for any $n \geq N$. This completes the proof. □

2.7.2 *The Lebesgue integral of a measurable nonnegative function*

Suppose that $\mathbf{E}$ is a measurable set.

Definition 2.29. A measurable function on $\mathbf{E}$ is said to vanish outside a set of finite measure if there is a subset $\mathbf{E}_0$ of $\mathbf{E}$ for which $m(\mathbf{E}_0) < \infty$ and $f \equiv 0$ on $\mathbf{E}\backslash\mathbf{E}_0$. We say that a measurable function f on $\mathbf{E}$ which vanishes outside a set of finite measure has finite support and define its support to be

$$\{x \in \mathbf{E} : f(x) \neq 0\}.$$

Definition 2.30. Let f be a bounded measurable function on $\mathbf{E}$ that has finite support $\mathbf{E}_0$, $\mathbf{E}_0 \subseteq \mathbf{E}$. We define its integral over $\mathbf{E}$ by

$$\int_{\mathbf{E}} f = \int_{\mathbf{E}_0} f.$$

The set of all bounded measurable functions on $\mathbf{E}$ which has finite support will be denoted by $\mathscr{E}$.

Definition 2.31. For a nonnegative measurable function f on $\mathbf{E}$ we define the integral of f over $\mathbf{E}$ by

$$\int_{\mathbf{E}} f = \sup\left\{\int_{\mathbf{E}} \phi : \phi \in \mathscr{E}, 0 \leq \phi \leq f \quad \text{on } \mathbf{E}\right\}.$$

Theorem 2.64 (Chebychev's Inequality). *Let f be a nonnegative measurable function on $\mathbf{E}$. Then for any $\lambda > 0$ we have*

$$m(\mathbf{E}_\lambda) \leq \frac{1}{\lambda}\int_{\mathbf{E}} f, \tag{2.89}$$

where $\mathbf{E}_\lambda = \{x \in \mathbf{E} : f(x) \geq \lambda\}$.

Proof.

1. Let $m(\mathbf{E}_\lambda) = \infty$. Define the sets $\mathbf{E}_{\lambda,n} = \mathbf{E}_\lambda \cap [-n, n]$ and the functions $\psi_n = \lambda\kappa_{\mathbf{E}_{\lambda,n}}$ for any $n \in \mathbf{N}$. Then ψ_n are bounded measurable functions on $\mathbf{E}$ with finite support $\mathbf{E}_{\lambda,n}$ for any $n \in \mathbf{N}$. Hence and Definition 2.31, we get

$$\begin{aligned}\int_{\mathbf{E}} \psi_n &= \int_{\mathbf{E}_{\lambda,n}} \lambda \\ &= \lambda m(\mathbf{E}_{\lambda,n})\end{aligned}$$

for any $n \in \mathbf{N}$. Note that $0 \leq \psi_n \leq f$ on $\mathbf{E}$ for any $n \in \mathbf{N}$. From here and from Definition 2.31, it follows

$$\int_{\mathbf{E}} \psi_n \leq \int_{\mathbf{E}} f$$

for any $n \in \mathbf{N}$. Also,

$$\mathbf{E}_\lambda = \bigcup_{n=1}^{\infty} \mathbf{E}_{\lambda,n} \quad \text{and} \quad \mathbf{E}_{\lambda,n} \subseteq \mathbf{E}_{\lambda,m}$$

for any $m, n \in \mathbf{N}$, $n \leq m$. Hence and Theorem 2.25, it follows that

$$\begin{aligned}\infty &= \lambda m(\mathbf{E}_\lambda) \\ &= \lambda \lim_{n\to\infty} m(\mathbf{E}_{\lambda,n}) \\ &= \lim_{n\to\infty} (\lambda m(\mathbf{E}_{\lambda,n})) \\ &= \lim_{n\to\infty} \int_{\mathbf{E}} \psi_n \\ &\leq \int_{\mathbf{E}} f.\end{aligned}$$

Therefore, $\int_{\mathbf{E}} f = \infty$. The equality (2.89) holds since both sides equal to ∞.

2. Let $m(\mathbf{E}_\lambda) < \infty$. Define $\phi = \lambda\kappa_{\mathbf{E}_\lambda}$. Then ϕ is a bounded measurable function on $\mathbf{E}$ of finite support $\mathbf{E}_\lambda$ and $0 \le \phi \le f$. Hence and Definition 2.31, we get

$$\begin{aligned}\lambda m(\mathbf{E}_\lambda) &= \int_{\mathbf{E}} \phi \\ &\le \int_{\mathbf{E}} f.\end{aligned}$$

   This completes the proof. □

Exercise 2.16. Let f be any function on a set $\mathbf{A}$. Prove

$$\{x \in \mathbf{A} : f(x) > 0\} = \bigcup_{n=1}^{\infty} \left\{x \in \mathbf{A} : f(x) \ge \frac{1}{n}\right\}.$$

Theorem 2.65. *Let f be a nonnegative measurable function on $\mathbf{E}$. Then $\int_{\mathbf{E}} f = 0$ if and only if $f = 0$ a.e. on $\mathbf{E}$.*

Proof.

1. Let $\int_{\mathbf{E}} f = 0$. Then for each $n \in \mathbf{N}$, using Chebychev's inequality, we get

$$m\left(\left\{x \in \mathbf{E} : f(x) \ge \frac{1}{n}\right\}\right) \le n \int_{\mathbf{E}} f = 0$$

   for any $n \in \mathbf{N}$. Using Exercise 2.16 and Theorem 2.5, we get

$$\begin{aligned}m(\{x \in \mathbf{E} : f(x) > 0\}) &\le \sum_{n=1}^{\infty} m\left(\left\{x \in \mathbf{E} : f(x) \ge \frac{1}{n}\right\}\right) \\ &= 0.\end{aligned}$$

   Because f is nonnegative on $\mathbf{E}$, we conclude that $f = 0$ a.e. on $\mathbf{E}$.
2. Let $f = 0$ a.e. on $\mathbf{E}$. Let ϕ be a simple function on $\mathbf{E}$ and ψ be a bounded measurable function on $\mathbf{E}$ of finite support such that $0 \le \phi \le \psi \le f$ on $\mathbf{E}$. Then $\phi = 0$ a.e. on $\mathbf{E}$ and $\int_{\mathbf{E}} \phi = 0$. Since this holds for all such functions ϕ, we get $\int_{\mathbf{E}} \psi = 0$. Because this holds for all such functions ψ, we obtain $\int_{\mathbf{E}} f = 0$. This completes the proof. □

Theorem 2.66. *Let f and g be any nonnegative measurable functions on $\mathbf{E}$. Then for any $\alpha > 0$ and $\beta > 0$ we have*

$$\int_{\mathbf{E}} (\alpha f + \beta g) = \alpha \int_{\mathbf{E}} f + \beta \int_{\mathbf{E}} g.$$

Moreover, if $f \leq g$ on $\mathbf{E}$, then

$$\int_{\mathbf{E}} f \leq \int_{\mathbf{E}} g.$$

Proof. Let $\alpha > 0$ and ϕ be a bounded measurable function on $\mathbf{E}$ of finite support such that $0 \leq \phi \leq f$ on $\mathbf{E}$. Then $0 \leq \alpha\phi \leq \alpha f$ on $\mathbf{E}$. Hence, using Theorem 2.58, we get

$$\alpha \int_{\mathbf{E}} \phi = \int_{\mathbf{E}} (\alpha\phi).$$

Because this holds for any such ϕ, we get

$$\alpha \int_{\mathbf{E}} f = \int_{\mathbf{E}} (\alpha f). \tag{2.90}$$

Now, we will prove that

$$\int_{\mathbf{E}} (f + g) = \int_{\mathbf{E}} f + \int_{\mathbf{E}} g. \tag{2.91}$$

Let ϕ and ψ be bounded measurable functions on $\mathbf{E}$ of finite support such that $0 \leq \phi \leq f$ and $0 \leq \psi \leq g$ on $\mathbf{E}$. Then $0 \leq \phi + \psi \leq f + g$ on $\mathbf{E}$ and using Theorem 2.58, we get

$$\begin{aligned}\int_{\mathbf{E}} f + g &\geq \int_{\mathbf{E}} (\phi + \psi) \\ &= \int_{\mathbf{E}} \phi + \int_{\mathbf{E}} \psi.\end{aligned}$$

Hence,

$$\int_{\mathbf{E}} (f + g) \geq \int_{\mathbf{E}} f + \int_{\mathbf{E}} g. \tag{2.92}$$

Let h be a bounded measurable function on $\mathbf{E}$ of finite support such that $0 \leq h \leq f + g$ on $\mathbf{E}$. We define the functions

$$l = \min\{f, h\} \quad \text{and} \quad k = h - l \quad \text{on } \mathbf{E}.$$

Let $x \in \mathbf{E}$ be arbitrarily chosen.

1. If $h(x) \leq f(x)$, then $l(x) = h(x)$ and

$$k(x) = 0 \leq g(x).$$

2. Let $h(x) > f(x)$. Then $l(x) = f(x)$ and

$$\begin{aligned} k(x) &= h(x) - f(x) \\ &\leq f(x) + g(x) - f(x) \\ &= g(x). \end{aligned}$$

Consequently, $k \leq g$ on $\mathbf{E}$. Because h is a bounded measurable function on $\mathbf{E}$ of finite support, we have that l is a bounded measurable function on $\mathbf{E}$ of finite support. From here, k is a bounded measurable function on $\mathbf{E}$ of finite support. We have

$$0 \leq l \leq f \quad \text{and} \quad 0 \leq k \leq g \quad \text{on } \mathbf{E}.$$

Hence, by Theorem 2.58, we get

$$\begin{aligned} \int_{\mathbf{E}} h &= \int_{\mathbf{E}} l + \int_{\mathbf{E}} k \\ &\leq \int_{\mathbf{E}} f + \int_{\mathbf{E}} g. \end{aligned}$$

Because the last inequality holds for all such h, we obtain

$$\int_{\mathbf{E}} (f+g) \leq \int_{\mathbf{E}} f + \int_{\mathbf{E}} g.$$

From the last inequality and from (2.92), we get (2.91). Since αf and βg are nonnegative measurable functions on $\mathbf{E}$, using (2.91) and (2.90), we get

$$\begin{aligned} \int_{\mathbf{E}} (\alpha f + \beta g) &= \int_{\mathbf{E}} (\alpha f) + \int_{\mathbf{E}} (\beta g) \\ &= \alpha \int_{\mathbf{E}} f + \beta \int_{\mathbf{E}} g. \end{aligned}$$

Let now $f \leq g$ on $\mathbf{E}$. Let ϕ be a bounded measurable function on $\mathbf{E}$ of finite support such that $0 \leq \phi \leq f$ on $\mathbf{E}$. Then $\phi \leq g$ on $\mathbf{E}$ and

$$\int_{\mathbf{E}} \phi \leq \int_{\mathbf{E}} g.$$

Since the last inequality holds for any such ϕ, we conclude that

$$\int_{\mathbf{E}} f \leq \int_{\mathbf{E}} g.$$

This completes the proof. □

Theorem 2.67. *Let f be a nonnegative measurable function on $\mathbf{E}$. If $\mathbf{A}$ and $\mathbf{B}$ are disjoint measurable subsets of $\mathbf{E}$, then*

$$\int_{\mathbf{A}\cup\mathbf{B}} f = \int_{\mathbf{A}} f + \int_{\mathbf{B}} f. \tag{2.93}$$

In particular, if $\mathbf{B}$ is a subset of $\mathbf{E}$ of measure zero, then

$$\int_{\mathbf{E}} f = \int_{\mathbf{E}\setminus\mathbf{B}} f. \tag{2.94}$$

Proof. Let $\phi \in \mathscr{E}$ be arbitrarily chosen such that $0 \leq \phi \leq f$ and its support is $\mathbf{E}_0 \subseteq \mathbf{A}\cup\mathbf{B}$. Let $\mathbf{E}_0 = \mathbf{E}_1 \cup \mathbf{E}_2$, $\mathbf{E}_1 \cap \mathbf{E}_2 = \emptyset$, $\mathbf{E}_1 \subseteq \mathbf{A}$, $\mathbf{E}_2 \subseteq \mathbf{B}$. Then, using Theorem 2.58, we get

$$\begin{aligned}\int_{\mathbf{A}\cup\mathbf{B}} \phi &= \int_{\mathbf{E}_0} \phi \\ &= \int_{\mathbf{E}_1\cup\mathbf{E}_2} \phi \\ &= \int_{\mathbf{E}_1} \phi + \int_{\mathbf{E}_2} \phi \\ &= \int_{\mathbf{A}} \phi + \int_{\mathbf{B}} \phi.\end{aligned}$$

Hence, using Definition 2.30, we obtain (2.93). Now we will prove (2.94). Since for any $\phi \in \mathscr{E}$ for which $0 \leq \phi \leq f$ we have $\int_{\mathbf{B}} \phi = 0$ and

$$\begin{aligned}\int_{\mathbf{E}} \phi &= \int_{(\mathbf{E}\setminus\mathbf{B})\cup\mathbf{B}} \phi \\ &= \int_{(\mathbf{E}\setminus\mathbf{B})} \phi + \int_{\mathbf{B}} \phi \\ &= \int_{(\mathbf{E}\setminus\mathbf{B})} \phi.\end{aligned}$$

Because $\phi \in \mathscr{E}$, $0 \leq \phi \leq f$, was arbitrarily chosen, we get (2.94). This completes the proof. □

Lemma 2.7 (Fatou's Lemma). *Let $\{f_n\}_{n\in\mathbf{N}}$ be a sequence of non-negative measurable functions on $\mathbf{E}$. If $f_n \to f$, as $n \to \infty$, pointwise a.e. on $\mathbf{E}$, then*

$$\int_{\mathbf{E}} f \le \liminf \int_{\mathbf{E}} f_n.$$

Proof. By Theorem 2.45, it follows that f is measurable on $\mathbf{E}$. Let $\mathbf{E}_0$ be such that $f_n \to f$, as $n \to \infty$, pointwise on $\mathbf{E}_0$ and $m(\mathbf{E}\backslash\mathbf{E}_0) = 0$. Let h be any bounded measurable function on $\mathbf{E}$ of finite support such that $0 \le h \le f$ on $\mathbf{E}$. Let also, M be a positive constant such that $0 \le h \le M$ on $\mathbf{E}$. Define the set

$$\mathbf{E}_1 = \{x \in \mathbf{E} : h(x) \neq 0\}.$$

Then $m(\mathbf{E}_1) < \infty$. For any $n \in \mathbf{N}$ we define the function h_n on $\mathbf{E}$ by

$$h_n = \min\{h, f_n\} \quad \text{on } \mathbf{E}.$$

Since h and f_n are measurable functions on $\mathbf{E}$, by Theorem 2.43, it follows that the functions h_n are measurable on $\mathbf{E}_0$ for any $n \in \mathbf{N}$. Also,

$$0 \le h_n \le M \quad \text{and} \quad h_n = 0 \quad \text{on } \mathbf{E}\backslash\mathbf{E}_1.$$

Furthermore, for each $x \in \mathbf{E}_0$ we have $h(x) \le f(x)$ and $h_n(x) \to h(x)$ as $n \to \infty$. Hence and Theorem 2.63, we get

$$\begin{aligned}\lim_{n\to\infty} \int_{\mathbf{E}} h_n &= \lim_{n\to\infty} \int_{\mathbf{E}_0} h_n \\ &= \int_{\mathbf{E}_0} h.\end{aligned}$$

However, for each $n \in \mathbf{N}$ we have $h_n \le f_n$ on $\mathbf{E}_0$ and using Definition 2.31, we have

$$\int_{\mathbf{E}_0} h_n \le \int_{\mathbf{E}_0} f_n.$$

Thus

$$\begin{aligned}\int_{\mathbf{E}} h &= \int_{\mathbf{E}_0} h_n \\ &\le \int_{\mathbf{E}_0} f_n\end{aligned}$$

$$= \liminf \int_{\mathbf{E}_0} f_n$$

$$= \liminf \int_{\mathbf{E}} f_n.$$

The last inequality is valid for all such h. Therefore

$$\int_{\mathbf{E}} f \leq \liminf \int_{\mathbf{E}} f_n.$$

This completes the proof. □

Theorem 2.68 (The Monotone Convergence Theorem). *Let $\{f_n\}_{n \in \mathbf{N}}$ be an increasing sequence of nonnegative measurable functions on $\mathbf{E}$. If $f_n \to f$, as $n \to \infty$, a.e. on $\mathbf{E}$, then*

$$\lim_{n \to \infty} \int_{\mathbf{E}} f_n = \int_{\mathbf{E}} f.$$

Proof. By Theorem 2.45, it follows that f is measurable on $\mathbf{E}$. By Fatou's Lemma, we have

$$\int_{\mathbf{E}} f \leq \liminf \int_{\mathbf{E}} f_n. \tag{2.95}$$

On the other hand, using that $\{f_n\}_{n \in \mathbf{N}}$ is an increasing sequence, we have that $f_n \leq f$ a.e. on $\mathbf{E}$ for each $n \in \mathbf{N}$. Hence and Theorem 2.66, we get

$$\int_{\mathbf{E}} f_n \leq \int_{\mathbf{E}} f$$

for any $n \in \mathbf{N}$ and

$$\limsup \int_{\mathbf{E}} f_n \leq \int_{\mathbf{E}} f.$$

From the last inequality and from (2.95), we get the desired result. This completes the proof. □

Corollary 2.5. *Let $\{f_n\}_{n \in \mathbf{N}}$ be a sequence of nonnegative measurable functions on $\mathbf{E}$. If*

$$f = \sum_{n=1}^{\infty} f_n \ a.e. \quad \text{on } \mathbf{E},$$

then

$$\int_{\mathbf{E}} f = \sum_{n=1}^{\infty} \int_{\mathbf{E}} f_n.$$

Proof. Let

$$S_m = \sum_{n=1}^{m} f_n, \quad m \in \mathbf{N}.$$

Then $\{S_n\}_{n\in\mathbf{N}}$ is an increasing sequence of measurable functions on $\mathbf{E}$ and $S_n \to f$, as $n \to \infty$, a.e. on $\mathbf{E}$. By Theorem 2.45, we obtain that f is measurable on $\mathbf{E}$. By the Monotone Convergence Theorem, it follows that

$$\lim_{n\to\infty} \int_{\mathbf{E}} S_n = \int_{\mathbf{E}} f,$$

whereupon we get the desired result. This completes the proof. □

Definition 2.32. A nonnegative measurable function f on $\mathbf{E}$ is said to be integrable over $\mathbf{E}$ if $\int_{\mathbf{E}} f < \infty$.

Theorem 2.69. *Let the nonnegative measurable function f be integrable over $\mathbf{E}$. Then f is finite a.e. on $\mathbf{E}$.*

Proof. Note that

$$\{x \in \mathbf{E} : f(x) = \infty\} \subseteq \{x \in \mathbf{E} : f(x) \geq n\}$$

for any $n \in \mathbf{N}$. Hence and Chebychev's inequality (Theorem 2.64), we get

$$\begin{aligned} m(\{x \in \mathbf{E} : f(x) = \infty\}) &\leq m(\{x \in \mathbf{E} : f(x) \geq n\}) \\ &\leq \frac{1}{n} \int_{\mathbf{E}} f \\ &\to 0 \quad \text{as } n \to \infty, \end{aligned}$$

because $0 \leq \int_{\mathbf{E}} f < \infty$. Consequently, $m(\{x \in \mathbf{E} : f(x) = \infty\}) = 0$. This completes the proof. □

Lemma 2.8 (Beppo Levi's Lemma). *Let $\{f_n\}_{n\in\mathbf{N}}$ be an increasing sequence of nonnegative measurable functions on $\mathbf{E}$. If the* *sequence* $\{\int_{\mathbf{E}} f_n\}_{n\in\mathbf{N}}$ *is bounded, then* $\{f_n\}_{n\in\mathbf{N}}$ *converges pointwise on* $\mathbf{E}$ *to a measurable function* f *that is finite a.e. on* $\mathbf{E}$ *and*

$$\lim_{n\to\infty}\int_{\mathbf{E}} f_n = \int_{\mathbf{E}} f < \infty.$$

Proof. There exists an extended real-valued nonnegative function f on $\mathbf{E}$ such that

$$f(x) = \lim_{n\to\infty} f_n(x), \quad x \in \mathbf{E}.$$

By Theorem 2.45, we have that f is measurable on $\mathbf{E}$. By the Monotone Convergence Theorem, it follows that

$$\lim_{n\to\infty}\int_{\mathbf{E}} f_n = \int_{\mathbf{E}} f.$$

Since the sequence $\{\int_{\mathbf{E}} f_n\}_{n\in\mathbf{N}}$ is a bounded sequence, we get that $\int_{\mathbf{E}} f < \infty$. From here and from Theorem 2.69, we have that f is finite a.e. on $\mathbf{E}$. This completes the proof. □

Theorem 2.70. *Let* f *be a nonnegative measurable function on* $\mathbf{E}$ *and* $\mathbf{A} \subseteq \mathbf{E}$ *be a measurable set. Then*

$$\int_{\mathbf{E}} f\kappa_{\mathbf{A}} = \int_{\mathbf{A}} f.$$

Proof. Let ϕ be any measurable function of finite support on $\mathbf{E}$ such that $0 \le \phi \le f$. Then $\phi\kappa_{\mathbf{A}}$ is a measurable function of finite support on $\mathbf{E}$ and $0 \le \phi\kappa_{\mathbf{A}} \le f\kappa_{\mathbf{A}}$. Then, using Theorem 2.59, we get

$$\int_{\mathbf{E}} \phi\kappa_{\mathbf{A}} = \int_{\mathbf{A}} \phi.$$

Because this holds for any such ϕ, we conclude that

$$\int_{\mathbf{E}} f\kappa_{\mathbf{A}} = \int_{\mathbf{A}} f.$$

This completes the proof. □

2.7.3 *The general Lebesgue integral*

Suppose that $\mathbf{E}$ is a measurable set.

Theorem 2.71. *Let f be a measurable function on $\mathbf{E}$. Then f^+ and f^- are integrable over $\mathbf{E}$ if and only if $|f|$ is integrable over $\mathbf{E}$.*

Proof.

1. Assume that f^+ and f^- are integrable over $\mathbf{E}$. Then

$$f^+ \geq 0, \quad f^- \geq 0 \quad \text{on } \mathbf{E},$$

and

$$\int_{\mathbf{E}} f^+ < \infty, \quad \int_{\mathbf{E}} f^- < \infty.$$

Hence, using that $|f| = f^+ + f^-$, we get

$$\begin{aligned} \int_{\mathbf{E}} |f| &= \int_{\mathbf{E}} f^+ + \int_{\mathbf{E}} f^- \\ &< \infty. \end{aligned}$$

2. Assume that $|f|$ is integrable over $\mathbf{E}$. We have

$$0 \leq f^+ \leq |f|, \quad 0 \leq f^- \leq |f| \quad \text{on } \mathbf{E}.$$

Hence and Theorem 2.66, it follows that

$$\begin{aligned} 0 &\leq \int_{\mathbf{E}} f^+ \\ &\leq \int_{\mathbf{E}} |f| \\ &< \infty, \\ 0 &\leq \int_{\mathbf{E}} f^- \\ &\leq \int_{\mathbf{E}} |f| \\ &< \infty. \end{aligned}$$

This completes the proof. □

Definition 2.33. A measurable function f on $\mathbf{E}$ is said to be integrable over $\mathbf{E}$ provided $|f|$ is integrable over $\mathbf{E}$. In this case we define

$$\int_{\mathbf{E}} f = \int_{\mathbf{E}} f^+ - \int_{\mathbf{E}} f^-.$$

Theorem 2.72. *Let f be a measurable function on $\mathbf{E}$. Suppose that g is a nonnegative function on $\mathbf{E}$ that is integrable over $\mathbf{E}$ and*

$$|f| \le g \quad \text{on } \mathbf{E}.$$

Then f is integrable over $\mathbf{E}$ and

$$\left| \int_{\mathbf{E}} f \right| \le \int_{\mathbf{E}} |f|.$$

Proof. By Theorem 2.66, we get

$$\int_{\mathbf{E}} |f| \le \int_{\mathbf{E}} g.$$

Hence, using that g is integrable over $\mathbf{E}$, we conclude that $|f|$ is integrable over $\mathbf{E}$. By Definition 2.33, it follows that f is integrable over $\mathbf{E}$. Also, using Theorem 2.66 and the triangle inequality for real numbers, we obtain

$$\begin{aligned}
\left| \int_{\mathbf{E}} f \right| &= \left| \int_{\mathbf{E}} f^+ - \int_{\mathbf{E}} f^- \right| \\
&\le \left| \int_{\mathbf{E}} f^+ \right| + \left| \int_{\mathbf{E}} f^- \right| \\
&= \int_{\mathbf{E}} f^+ + \int_{\mathbf{E}} f^- \\
&= \int_{\mathbf{E}} (f^+ + f^-) \\
&= \int_{\mathbf{E}} |f|.
\end{aligned}$$

This completes the proof. □

Theorem 2.73. *Let the functions f and g are integrable over $\mathbf{E}$. Then for any α, $\beta \in \mathbf{R}$, the function $\alpha f + \beta g$ is integrable over $\mathbf{E}$ and*

$$\int_{\mathbf{E}} (\alpha f + \beta g) = \alpha \int_{\mathbf{E}} f + \beta \int_{\mathbf{E}} g.$$

Moreover, if $f \leq g$ on $\mathbf{E}$, then

$$\int_{\mathbf{E}} f \leq \int_{\mathbf{E}} g.$$

Proof. Let $\alpha > 0$. Then

$$(\alpha f)^+ = \alpha f^+, \quad (\alpha f)^- = \alpha f^-.$$

Hence and Theorem 2.66, we obtain

$$\begin{aligned}
\int_{\mathbf{E}} (\alpha f) &= \int_{\mathbf{E}} ((\alpha f)^+ - (\alpha f)^-) \\
&= \int_{\mathbf{E}} (\alpha f)^+ - \int_{\mathbf{E}} (\alpha f)^- \\
&= \int_{\mathbf{E}} (\alpha f^+) - \int_{\mathbf{E}} (\alpha f^-) \\
&= \alpha \int_{\mathbf{E}} f^+ - \alpha \int_{\mathbf{E}} f^- \\
&= \alpha \left(\int_{\mathbf{E}} f^+ + \int_{\mathbf{E}} (-f^-) \right) \\
&= \alpha \int_{\mathbf{E}} (f^+ - f^-) \\
&= \alpha \int_{\mathbf{E}} f.
\end{aligned}$$

Let $\alpha < 0$. Then

$$(\alpha f)^+ = -\alpha f^-, \quad (\alpha f)^- = -\alpha f^+.$$

Hence and Theorem 2.66, we obtain

$$\begin{aligned}\int_{\mathbf{E}}(\alpha f) &= \int_{\mathbf{E}}\left((\alpha f)^{+}-(\alpha f)^{-}\right)\\ &= \int_{\mathbf{E}}(\alpha f)^{+}-\int_{\mathbf{E}}(\alpha f)^{-}\\ &= \int_{\mathbf{E}}(-\alpha f^{-})-\int_{\mathbf{E}}(-\alpha f^{+})\\ &= \alpha\int_{\mathbf{E}}(-f^{-})+\alpha\int_{\mathbf{E}}f^{+}\\ &= \alpha\left(\int_{\mathbf{E}}(-f^{-})+\int_{\mathbf{E}}f^{+}\right)\\ &= \alpha\int_{\mathbf{E}}(f^{+}-f^{-})\\ &= \alpha\int_{\mathbf{E}}f.\end{aligned}$$

Note that

$$\begin{aligned}(f+g)^{+}-(f+g)^{-} &= f+g\\ &= f^{+}-f^{-}+g^{+}-g^{-}.\end{aligned}$$

Hence,

$$(f+g)^{+}+f^{-}+g^{-}=(f+g)^{-}+f^{+}+g^{+},$$

and

$$\int_{\mathbf{E}}\left((f+g)^{+}+f^{-}+g^{-}\right)=\int_{\mathbf{E}}\left((f+g)^{-}+f^{+}+g^{+}\right). \tag{2.96}$$

By Theorem 2.66, we have

$$\int_{\mathbf{E}}\left((f+g)^{+}+f^{-}+g^{-}\right)=\int_{\mathbf{E}}(f+g)^{+}+\int_{\mathbf{E}}f^{-}+\int_{\mathbf{E}}g^{-} \tag{2.97}$$

and

$$\int_{\mathbf{E}}\left((f+g)^{-}+f^{+}+g^{+}\right)=\int_{\mathbf{E}}(f+g)^{-}+\int_{\mathbf{E}}f^{+}+\int_{\mathbf{E}}g^{+}. \tag{2.98}$$

From (2.96), (2.97) and (2.98), we obtain

$$\int_{\mathbf{E}} (f+g)^+ + \int_{\mathbf{E}} f^- + \int_{\mathbf{E}} g^- = \int_{\mathbf{E}} (f+g)^- + \int_{\mathbf{E}} f^+ + \int_{\mathbf{E}} g^+.$$

Hence,

$$\int_{\mathbf{E}} (f+g)^+ - \int_{\mathbf{E}} (f+g)^- = \int_{\mathbf{E}} f^+ - \int_{\mathbf{E}} f^- + \int_{\mathbf{E}} g^+ - \int_{\mathbf{E}} g^-,$$

or

$$\int_{\mathbf{E}} \left((f+g)^+ - (f+g)^-\right) = \int_{\mathbf{E}} \left(f^+ - f^-\right) + \int_{\mathbf{E}} \left(g^+ - g^-\right),$$

or

$$\int_{\mathbf{E}} (f+g) = \int_{\mathbf{E}} f + \int_{\mathbf{E}} g. \tag{2.99}$$

Because αf and βg are integrable over $\mathbf{E}$ for any $\alpha, \beta \in \mathbf{R}$, using (2.99), we obtain

$$\begin{aligned}\int_{\mathbf{E}} (\alpha f + \beta g) &= \int_{\mathbf{E}} (\alpha f) + \int_{\mathbf{E}} (\beta g)\\ &= \alpha \int_{\mathbf{E}} f + \beta \int_{\mathbf{E}} g.\end{aligned}$$

Now, we suppose that $f \leq g$ on $\mathbf{E}$. Define $h = g - f$ on $\mathbf{E}$. Then $h \geq 0$ on $\mathbf{E}$. By Theorem 2.66, we get

$$\begin{aligned}0 &\leq \int_{\mathbf{E}} h\\ &= \int_{\mathbf{E}} (g-f)\\ &= \int_{\mathbf{E}} g - \int_{\mathbf{E}} f.\end{aligned}$$

This completes the proof. □

Theorem 2.74. *Let f be integrable over $\mathbf{E}$ and $\mathbf{A} \subseteq \mathbf{E}$ be measurable. Then*

$$\int_{\mathbf{E}} f \kappa_{\mathbf{A}} = \int_{\mathbf{A}} f.$$

Proof. By Theorem 2.70, we have

$$\int_{\mathbf{E}} f^+ \kappa_{\mathbf{A}} = \int_{\mathbf{A}} f^+ \quad \text{and} \quad \int_{\mathbf{E}} f^- \kappa_{\mathbf{A}} = \int_{\mathbf{A}} f^-.$$

Hence, using Theorem 2.73, we get

$$\begin{aligned}
\int_{\mathbf{E}} f\kappa_{\mathbf{A}} &= \int_{\mathbf{E}} \left(f^+ - f^-\right) \kappa_{\mathbf{A}} \\
&= \int_{\mathbf{E}} \left(f^+ \kappa_{\mathbf{A}} - f^- \kappa_{\mathbf{A}}\right) \\
&= \int_{\mathbf{E}} f^+ \kappa_{\mathbf{A}} - \int_{\mathbf{E}} f^- \kappa_{\mathbf{A}} \\
&= \int_{\mathbf{A}} f^+ - \int_{\mathbf{A}} f^- \\
&= \int_{\mathbf{A}} \left(f^+ - f^-\right) \\
&= \int_{\mathbf{A}} f.
\end{aligned}$$

This completes the proof. □

Theorem 2.75. *Let f be integrable over $\mathbf{E}$. Assume that $\mathbf{A}$ and $\mathbf{B}$ are disjoint measurable subsets of $\mathbf{E}$. Then*

$$\int_{\mathbf{A}\cup\mathbf{B}} f = \int_{\mathbf{A}} f + \int_{\mathbf{B}} f.$$

Proof. We have that

$$|f\kappa_{\mathbf{A}}| \leq |f| \quad \text{and} \quad |f\kappa_{\mathbf{B}}| \leq |f| \quad \text{on } \mathbf{E}.$$

Then, using Theorem 2.66, we get

$$\int_{\mathbf{E}} |f\kappa_{\mathbf{A}}| \leq \int_{\mathbf{E}} |f| \quad \text{and} \quad \int_{\mathbf{E}} |f\kappa_{\mathbf{B}}| \leq \int_{\mathbf{E}} |f|.$$

Hence, $f\kappa_{\mathbf{A}}$ and $f\kappa_{\mathbf{B}}$ are integrable over $\mathbf{E}$. Because $\mathbf{A}$ and $\mathbf{B}$ are disjoint, we have

$$f\kappa_{\mathbf{A}\cup\mathbf{B}} = f\kappa_{\mathbf{A}} + f\kappa_{\mathbf{B}}.$$

From here, $f\kappa_{\mathbf{A}\cup\mathbf{B}}$ is integrable over $\mathbf{E}$ and using Theorem 2.74, we obtain

$$\begin{aligned}\int_{\mathbf{A}\cup\mathbf{B}} f &= \int_{\mathbf{A}\cup\mathbf{B}} f\kappa_{\mathbf{A}\cup\mathbf{B}} \\ &= \int_{\mathbf{A}\cup\mathbf{B}} (f\kappa_{\mathbf{A}} + f\kappa_{\mathbf{B}}) \\ &= \int_{\mathbf{A}\cup\mathbf{B}} f\kappa_{\mathbf{A}} + \int_{\mathbf{A}\cup\mathbf{B}} f\kappa_{\mathbf{B}} \\ &= \int_{\mathbf{A}} f\kappa_{\mathbf{A}} + \int_{\mathbf{B}} f\kappa_{\mathbf{B}} \\ &= \int_{\mathbf{A}} f + \int_{\mathbf{B}} f.\end{aligned}$$

This completes the proof. □

Theorem 2.76. *Let f be integrable over $\mathbf{E}$. Then f is finite a.e. on $\mathbf{E}$ and*

$$\int_{\mathbf{E}} f = \int_{\mathbf{E}_0} f \quad \textit{if} \quad \mathbf{E}_0 \subseteq \mathbf{E} \quad \textit{and} \quad m(\mathbf{E}_0) = 0.$$

Proof. By Theorem 2.69, it follows that f^+ and f^- are finite a.e. on $\mathbf{E}$. Hence, $f = f^+ - f^-$ is finite a.e. on $\mathbf{E}$. Let $\mathbf{E}_0$ be a measurable subset of $\mathbf{E}$ such that $m(\mathbf{E}\backslash\mathbf{E}_0) = 0$. Hence and Theorem 2.67, we obtain

$$\begin{aligned}\int_{\mathbf{E}} f &= \int_{\mathbf{E}} (f^+ - f^-) \\ &= \int_{\mathbf{E}} f^+ - \int_{\mathbf{E}} f^- \\ &= \int_{\mathbf{E}_0} f^+ - \int_{\mathbf{E}_0} f^- \\ &= \int_{\mathbf{E}_0} (f^+ - f^-) \\ &= \int_{\mathbf{E}_0} f.\end{aligned}$$

This completes the proof. □

Theorem 2.77 (The Lebesgue Dominated Convergence Theorem). *Let $\{f_n\}_{n\in\mathbf{N}}$ be a sequence of measurable functions on $\mathbf{E}$. Let g be a nonnegative function on $\mathbf{E}$ that is integrable over $\mathbf{E}$ and $|f_n| \le g$ on $\mathbf{E}$ for all $n \in \mathbf{N}$. If $f_n \to f$ pointwise a.e. on $\mathbf{E}$, then f is integrable over $\mathbf{E}$ and $\lim_{n\to\infty}\int_{\mathbf{E}} f_n = \int_{\mathbf{E}} f$.*

Proof. Since $|f_n| \le g$ on $\mathbf{E}$ for all $n \in \mathbf{N}$, we get that $|f| \le g$ a.e. on $\mathbf{E}$. There exists a measurable set $\mathbf{E}_0 \subseteq \mathbf{E}$ such that $|f| \le g$ on $\mathbf{E}_0$ and $m(\mathbf{E}\backslash\mathbf{E}_0) = 0$. By Theorem 2.72, it follows that f_n are integrable over $\mathbf{E}$ for all $n \in \mathbf{N}$ and f is integrable over $\mathbf{E}_0$. Because $m(\mathbf{E}\backslash\mathbf{E}_0) = 0$, the function f is integrable over $\mathbf{E}$. By Theorem 2.76, it follows that f_n, f and g are finite a.e. on $\mathbf{E}$. Therefore by possible excising from $\mathbf{E}$ a countable collection of sets of measure zero and using the countable additivity of the Lebesgue measure, we can suppose that f_n, f and g are finite on $\mathbf{E}$. The functions $g - f$, $g - f_n$ are properly defined, nonnegative and measurable on $\mathbf{E}$. Also, $\{g - f_n\}_{n\in\mathbf{N}}$ and $\{g + f_n\}_{n\in\mathbf{N}}$ converge pointwise a.e. on $\mathbf{E}$ to $g - f$ and $g + f$, respectively. By Fatou's Lemma, we get

$$\int_{\mathbf{E}} (g - f) \le \liminf \int_{\mathbf{E}} (g - f_n) \quad \text{and}$$

$$\int_{\mathbf{E}} (g + f) \le \liminf \int_{\mathbf{E}} (g + f_n).$$

On the other hand,

$$\begin{aligned}\int_{\mathbf{E}} g - \int_{\mathbf{E}} f &= \int_{\mathbf{E}} (g - f)\\ &\le \liminf \int_{\mathbf{E}} (g - f_n)\\ &= \int_{\mathbf{E}} g - \limsup \int_{\mathbf{E}} f_n.\end{aligned}$$

Therefore,

$$\limsup \int_{\mathbf{E}} f_n \le \int_{\mathbf{E}} f. \tag{2.100}$$

Also,

$$\begin{aligned}\int_{\mathbf{E}} g + \int_{\mathbf{E}} f &= \int_{\mathbf{E}} (g+f)\\ &\le \liminf \int_{\mathbf{E}} (g+f_n)\\ &= \int_{\mathbf{E}} g + \liminf \int_{\mathbf{E}} f_n.\end{aligned}$$

Hence,

$$\int_{\mathbf{E}} f \le \liminf \int_{\mathbf{E}} f_n.$$

By the last inequality and (2.100), we get

$$\int_{\mathbf{E}} f = \lim_{n\to\infty} \int_{\mathbf{E}} f_n.$$

This completes the proof. □

Theorem 2.78. *Let $\{f_n\}_{n\in\mathbf{N}}$ be a sequence of measurable functions on $\mathbf{E}$ that converges pointwise a.e. on $\mathbf{E}$ to f. Suppose that there is a sequence $\{g_n\}_{n\in\mathbf{N}}$ of nonnegative measurable functions on $\mathbf{E}$ that converges pointwise a.e. on $\mathbf{E}$ to g and $|f_n| \le g_n$ on $\mathbf{E}$ for all $n \in \mathbf{N}$. If $\lim_{n\to\infty} \int_{\mathbf{E}} g_n = \int_{\mathbf{E}} g < \infty$, then $\lim_{n\to\infty} \int_{\mathbf{E}} f_n = \int_{\mathbf{E}} f$.*

Proof. Since $\lim_{n\to\infty} \int_{\mathbf{E}} g_n = \int_{\mathbf{E}} g < \infty$, we have that g is integrable over $\mathbf{E}$ and there is an $N \in \mathbf{N}$ such that g_n are integrable over $\mathbf{E}$ for any $n > N$. By Theorem 2.72, it follows that f_n are integrable over $\mathbf{E}$ for any $n \ge N$. Because $|f_n| \le g_n$ on $\mathbf{E}$ for any $n \in \mathbf{N}$, we conclude that $|f| \le g$ a.e. on $\mathbf{E}$. Therefore, f is integrable over $\mathbf{E}$. Note that the sequences $\{f_n + g_n\}_{n\in\mathbf{N}}$ and $\{g_n - f_n\}_{n\in\mathbf{N}}$ converge pointwise a.e. on $\mathbf{E}$ to $f+g$ and $g-f$, respectively. Hence and Fatou's lemma, we get

$$\begin{aligned}\int_{\mathbf{E}} f + \int_{\mathbf{E}} g = \int_{\mathbf{E}} (f+g) &\leq \liminf \int_{\mathbf{E}} (f_n + g_n) \\ &= \liminf \int_{\mathbf{E}} f_n + \liminf \int_{\mathbf{E}} g_n \\ &= \liminf \int_{\mathbf{E}} f_n + \int_{\mathbf{E}} g,\end{aligned}$$

whereupon

$$\int_{\mathbf{E}} f \leq \liminf \int_{\mathbf{E}} f_n. \tag{2.101}$$

Also,

$$\begin{aligned}\int_{\mathbf{E}} g - \int_{\mathbf{E}} f = \int_{\mathbf{E}} (g-f) &\leq \liminf \int_{\mathbf{E}} (g_n - f_n) \\ &= \liminf \int_{\mathbf{E}} g_n - \limsup \int_{\mathbf{E}} f_n \\ &= \int_{\mathbf{E}} g - \limsup \int_{\mathbf{E}} f_n.\end{aligned}$$

Therefore,

$$\limsup \int_{\mathbf{E}} f_n \leq \int_{\mathbf{E}} f.$$

By the last inequality and from (2.101), we get

$$\int_{\mathbf{E}} f = \lim_{n\to\infty} \int_{\mathbf{E}} f_n.$$

This completes the proof. □

Theorem 2.79. *Let f be integrable over $\mathbf{E}$ and $\{\mathbf{E}_n\}_{n\in\mathbf{N}}$ be a disjoint countable collection of measurable subsets of $\mathbf{E}$ whose union is $\mathbf{E}$. Then*

$$\int_{\mathbf{E}} f = \sum_{i=1}^{\infty} \int_{\mathbf{E}_i} f.$$

Proof. Let $n \in \mathbf{N}$ be arbitrarily chosen. Let also, κ_n be the characteristic function of the measurable set $\cup_{k=1}^{n} \mathbf{E}_k$. Define $f_n = f\kappa_n$. Then f_n is measurable on $\mathbf{E}$ and

$$|f_n| \le |f| \quad \text{on } \mathbf{E}.$$

Note that $f_n \to f$, as $n \to \infty$, pointwise on $\mathbf{E}$. Hence and Theorem 2.77, we get

$$\lim_{n\to\infty} \int_{\mathbf{E}} f_n = \int_{\mathbf{E}} f. \tag{2.102}$$

By Theorem 2.75, we get

$$\begin{aligned}\int_{\mathbf{E}} f_n &= \int_{\bigcup_{k=1}^{n} \mathbf{E}_k} f \\ &= \sum_{k=1}^{n} \int_{\mathbf{E}_k} f.\end{aligned}$$

Hence and (2.102), we obtain

$$\begin{aligned}\int_{\mathbf{E}} f &= \lim_{n\to\infty} \sum_{k=1}^{n} \int_{\mathbf{E}_k} f \\ &= \sum_{k=1}^{\infty} \int_{\mathbf{E}_k} f.\end{aligned}$$

This completes the proof. □

Theorem 2.80. *Let f be integrable over $\mathbf{E}$.*

1. *If $\{\mathbf{E}_n\}_{n\in\mathbf{N}}$ is an ascending countable collection of measurable subsets of $\mathbf{E}$, then*

$$\int_{\bigcup_{k=1}^{\infty} \mathbf{E}_k} f = \lim_{n\to\infty} \int_{\mathbf{E}_n} f.$$

2. *If* $\{\mathbf{E}_n\}_{n\in\mathbf{N}}$ *is a descending countable collection of measurable subsets of* $\mathbf{E}$, *then*

$$\int_{\bigcap_{k=1}^{\infty}\mathbf{E}_k} f = \lim_{k\to\infty}\int_{\mathbf{E}_k} f.$$

Proof.

1. Let $\mathbf{A}_0 = \emptyset$, $\mathbf{A}_k = \mathbf{E}_k\backslash\mathbf{E}_{k-1}$, $k \in \mathbf{N}$. Then, using the proof of Theorem 2.25,

$$\bigcup_{k=1}^{\infty}\mathbf{E}_k = \bigcup_{k=1}^{\infty}\mathbf{A}_k, \quad \mathbf{E}_n = \bigcup_{k=1}^{n}\mathbf{A}_k, \quad \mathbf{A}_k\cap\mathbf{A}_l = \emptyset, \quad k\neq l,$$
$$k, l\in\mathbf{N}.$$

Hence and the proof of Theorem 2.79, we obtain

$$\begin{aligned}\int_{\bigcup_{k=1}^{\infty}\mathbf{E}_k} f &= \int_{\bigcup_{k=1}^{\infty}\mathbf{A}_k} f\\ &= \lim_{n\to\infty}\int_{\bigcup_{k=1}^{n}\mathbf{A}_k} f\\ &= \lim_{n\to\infty}\int_{\mathbf{E}_n} f.\end{aligned}$$

2. Let $\mathbf{B}_k = \mathbf{E}_1\backslash\mathbf{E}_k$, $k\in\mathbf{N}$. By the proof of Theorem 2.25, we have

$$\bigcup_{k=1}^{\infty}\mathbf{B}_k = \mathbf{E}_1\Bigg\backslash\left(\bigcap_{k=1}^{\infty}\mathbf{E}_k\right), \quad \mathbf{B}_k\cap\mathbf{B}_l = \emptyset, \quad k\neq l, \quad k, l\in\mathbf{N}.$$

Hence and the proof of Theorem 2.79, we get

$$\int_{\mathbf{E}_1\backslash\left(\bigcap_{k=1}^{\infty}\mathbf{E}_k\right)} f = \lim_{n\to\infty}\int_{\mathbf{E}_1\backslash\left(\bigcap_{k=1}^{n}\mathbf{E}_k\right)} f,$$

whereupon

$$\int_{\mathbf{E}_1} f - \int_{\bigcap_{k=1}^{\infty}\mathbf{E}_k} f = \int_{\mathbf{E}_1} f - \lim_{n\to\infty}\int_{\bigcap_{k=1}^{n}\mathbf{E}_k} f.$$

Therefore,

$$\int_{\bigcap_{k=1}^{\infty} \mathbf{E}_k} f = \lim_{n\to\infty} \int_{\bigcap_{k=1}^{n} \mathbf{E}_k} f.$$

This completes the proof. □

Theorem 2.81. *Let* $\mathbf{E}$ *be a set of finite measure and* $\delta > 0$. *Then* $\mathbf{E}$ *is the disjoint union of a finite collection of sets, each of which has measure less than* δ.

Proof. For any $n \in \mathbf{N}$ we define the sets

$$\mathbf{E}_n = \mathbf{E} \cap [-n, n].$$

Then $\{\mathbf{E}_n\}_{n\in\mathbf{N}}$ is a descending collection of measurable sets. Hence and Theorem 2.25, it follows that

$$\begin{aligned}\lim_{n\to\infty} m\left(\mathbf{E}\backslash[-n,n]\right) &= m\left(\bigcap_{n=1}^{\infty}\left(\mathbf{E}\backslash[-n,n]\right)\right)\\ &= m\left(\emptyset\right)\\ &= 0.\end{aligned}$$

Therefore, there exists an $n_0 \in \mathbf{N}$ such that

$$m\left(\mathbf{E}\backslash[-n_0, n_0]\right) < \delta.$$

We make a fine enough partition of $[-n_0, n_0]$ so that $\mathbf{E} \cap [-n_0, n_0]$ we represent as the disjoint union of a finite collection $\left\{\mathbf{E}^1, \ldots, \mathbf{E}^l\right\}$ of sets, each of which has measure less than δ. We have that

$$\left(\mathbf{E}\backslash\left[-n_0, n_0\right]\right) \cup \mathbf{E}^1 \cup \ldots \cup \mathbf{E}^l = \mathbf{E}.$$

This completes the proof. □

Theorem 2.82. *Let* f *be a measurable function on* $\mathbf{E}$. *If* f *is integrable over* $\mathbf{E}$, *then for each* $\varepsilon > 0$, *there is a* $\delta > 0$ *for which*

$$\text{if} \quad \mathbf{A} \subseteq \mathbf{E} \quad \text{is} \quad \text{measurable} \quad \text{and} \quad m\left(\mathbf{A}\right) < \delta, \quad \text{then}$$

$$\int_{\mathbf{A}} |f| < \varepsilon. \tag{2.103}$$

Conversely, in the case $m\left(\mathbf{E}\right) < \infty$, *if for each* $\varepsilon > 0$, *there is a* $\delta > 0$ *for which* (2.103) *holds, then* f *is integrable over* $\mathbf{E}$.

Proof.

1. Let f is integrable over $\mathbf{E}$. Let $\varepsilon > 0$ be arbitrarily chosen. Consider f^+. We have that f^+ is integrable over $\mathbf{E}$. Then there is a bounded measurable function f_ε of finite support for which
$$0 \le f_\varepsilon \le f^+ \quad \text{on } \mathbf{E} \quad \text{and} \quad 0 \le \int_{\mathbf{E}} f^+ - \int_{\mathbf{E}} f_\varepsilon < \frac{\varepsilon}{4}.$$
Since $f^+ - f_\varepsilon \ge 0$ on $\mathbf{E}$, if $\mathbf{A} \subseteq \mathbf{E}$ is measurable, then
$$\begin{aligned}\int_{\mathbf{A}} f^+ - \int_{\mathbf{A}} f_\varepsilon &= \int_{\mathbf{A}} (f^+ - f_\varepsilon)\\ &\le \int_{\mathbf{E}} (f^+ - f_\varepsilon)\\ &= \int_{\mathbf{E}} f^+ - \int_{\mathbf{E}} f_\varepsilon\\ &< \frac{\varepsilon}{4}.\end{aligned}$$
Since f_ε is bounded, there is a positive constant M for which $0 \le f_\varepsilon < M$ on $\mathbf{E}$. Therefore, if $\mathbf{A} \subseteq \mathbf{E}$ is measurable, then
$$\begin{aligned}\int_{\mathbf{A}} f^+ &< \int_{\mathbf{A}} f_\varepsilon + \frac{\varepsilon}{4}\\ &\le Mm(\mathbf{A}) + \frac{\varepsilon}{4}.\end{aligned}$$
We take $\delta = \frac{\varepsilon}{4M}$. Then
$$\begin{aligned}\int_{\mathbf{A}} f^+ &< \frac{\varepsilon}{4} + \frac{\varepsilon}{4}\\ &= \frac{\varepsilon}{2}.\end{aligned}$$
As in above,
$$\int_{\mathbf{A}} f^- < \frac{\varepsilon}{2}.$$
Therefore,
$$\begin{aligned}\int_{\mathbf{A}} |f| &= \int_{\mathbf{A}} (f^+ + f^-)\\ &= \int_{\mathbf{A}} f^+ + \int_{\mathbf{A}} f^-\end{aligned}$$

$$\begin{aligned}&< \frac{\varepsilon}{2} + \frac{\varepsilon}{2}\\&= \varepsilon.\end{aligned}$$

2. Suppose that $m(\mathbf{A}) < \infty$ and for each $\varepsilon > 0$ there is a $\delta > 0$ for which (2.103) holds. By Theorem 2.81, there is a finite disjoint collection of measurable subsets $\{\mathbf{E}_k\}_{k=1}^N$ such that $\mathbf{E} = \cup_{k=1}^N \mathbf{E}_k$, $m(\mathbf{E}_k) < \delta$, $k \in \{1, \ldots, N\}$. Then

$$\sum_{k=1}^N \int_{\mathbf{E}_k} f^+ < \varepsilon N.$$

Hence, if h is a bounded measurable function on $\mathbf{E}$ of finite support such that $0 \le h \le f^+$, we get

$$\begin{aligned}\int_{\mathbf{E}} h &= \int_{\bigcup_{k=1}^N \mathbf{E}_k} h\\&= \sum_{k=1}^N \int_{\mathbf{E}_k} h\\&\le \sum_{k=1}^N \int_{\mathbf{E}_k} f^+\\&< \varepsilon N.\end{aligned}$$

Then f^+ is integrable over $\mathbf{E}$. As in above, we conclude that f^- is integrable over $\mathbf{E}$. Therefore f is integrable over $\mathbf{E}$. This completes the proof. □

Definition 2.34. A family $\mathscr{F}$ of measurable functions on $\mathbf{E}$ is said to be uniformly integrable over $\mathbf{E}$ provided for each $\varepsilon > 0$, there is a $\delta > 0$, such that for each $f \in \mathscr{F}$,

$$\text{if} \quad \mathbf{A} \subseteq \mathbf{E} \text{ is measurable} \quad \text{and} \quad m(\mathbf{A}) < \delta, \text{ then } \int_{\mathbf{A}} |f| < \varepsilon.$$

Theorem 2.83. *Let $\{f_k\}_{k=1}^n$ be a finite collection of functions, each of which is integrable over $\mathbf{E}$. Then $\{f_k\}_{k=1}^n$ is uniformly integrable over $\mathbf{E}$.*

Proof. Let $\varepsilon > 0$ be arbitrarily chosen. Then for each $k \in \{1, \ldots, n\}$ there is a $\delta_k > 0$ such that

$$\text{if} \quad \mathbf{A} \subseteq \mathbf{E} \text{ is measurable} \quad \text{and} \quad m(\mathbf{A}) < \delta_k, \text{ then } \int_{\mathbf{A}} |f| < \varepsilon.$$

Take $\delta = \max\{\delta_1, \ldots, \delta_n\}$. Therefore $\{f_k\}_{k=1}^{n}$ is uniformly integrable over $\mathbf{E}$. This completes the proof. □

Theorem 2.84. *Assume that* $\mathbf{E}$ *has finite measure. Let the sequence* $\{f_n\}_{n\in\mathbf{N}}$ *is uniformly integrable over* $\mathbf{E}$*. If* $f_n \to f$*, as* $n \to \infty$*, pointwise a.e. on* $\mathbf{E}$*, then* f *is integrable over* $\mathbf{E}$*.*

Proof. Let $\varepsilon > 0$ be arbitrarily chosen. Let $\delta > 0$ responds to ε challenge in the uniform integrability criteria for the sequence $\{f_n\}_{n\in\mathbf{N}}$. By Theorem 2.81, there is a finite disjoint collection $\{\mathbf{E}_k\}_{k=1}^{n}$ of measurable sets such that $\mathbf{E} = \cup_{k=1}^{n}\mathbf{E}_k$ and $m(\mathbf{E}_k) < \delta$ for any $k \in \{1, \ldots, n\}$. Hence,

$$\begin{aligned}\int_{\mathbf{E}} |f_n| &= \int_{\bigcup_{k=1}^{n} \mathbf{E}_k} |f_n| \\ &= \sum_{k=1}^{n} \int_{\mathbf{E}_k} |f_n| \\ &< n\varepsilon.\end{aligned}$$

By Fatou's Lemma, it follows that

$$\begin{aligned}\int_{\mathbf{E}} |f| &\leq \liminf \int_{\mathbf{E}} |f_n| \\ &< \varepsilon N.\end{aligned}$$

Therefore, f is integrable over $\mathbf{E}$. This completes the proof. □

Theorem 2.85 (The Vitalli Convergence Theorem). *Let* $\mathbf{E}$ *be of finite measure. Suppose that the sequence* $\{f_n\}_{n\in\mathbf{N}}$ *is uniformly integrable over* $\mathbf{E}$*. If* $f_n \to f$*, as* $n \to \infty$*, pointwise a.e. on* $\mathbf{E}$*, then* f *is integrable over* $\mathbf{E}$ *and*

$$\lim_{n\to\infty} \int_{\mathbf{E}} f_n = \int_{\mathbf{E}} f.$$

Proof. By Theorem 2.84, it follows that f is integrable over $\mathbf{E}$. Hence, using Theorem 2.76, we get that f is finite a.e. on $\mathbf{E}$. Using Theorem 2.76 once more, by possible excising from $\mathbf{E}$ a set of measure zero, we can assume that $f_n \to f$, as $n \to \infty$, pointwise on all $\mathbf{E}$. Let $\mathbf{A} \subseteq \mathbf{E}$ be arbitrary measurable set and $n \in \mathbf{N}$ be arbitrarily chosen. Then

$$\begin{aligned} \left| \int_{\mathbf{E}} f_n - \int_{\mathbf{E}} f \right| &= \left| \int_{\mathbf{E}} (f_n - f) \right| \\ &\leq \int_{\mathbf{E}} |f_n - f| \\ &= \int_{\mathbf{E} \setminus \mathbf{A}} |f_n - f| + \int_{\mathbf{A}} |f_n - f| \\ &\leq \int_{\mathbf{E} \setminus \mathbf{A}} |f_n - f| + \int_{\mathbf{A}} |f_n| + \int_{\mathbf{A}} |f| . \end{aligned} \tag{2.104}$$

Let $\varepsilon > 0$ be arbitrarily chosen. Then there is a $\delta > 0$ such that if $\mathbf{A} \subseteq \mathbf{E}$ is a measurable set such that $m(\mathbf{A}) < \delta$ we have

$$\int_{\mathbf{A}} |f_n| < \frac{\varepsilon}{3}$$

for any $n \in \mathbf{N}$. Hence and Fatou's Lemma, we get

$$\int_{\mathbf{A}} |f| < \frac{\varepsilon}{3}$$

whenever $\mathbf{A} \subseteq \mathbf{E}$ and $m(\mathbf{A}) < \delta$. By Egoroff's Theorem, there is a measurable subset $\mathbf{E}_0$ of $\mathbf{E}$ for which $m(\mathbf{E}_0) < \delta$ and $f_n \to f$ uniformly on $\mathbf{E} \setminus \mathbf{E}_0$. Hence, there is an $N \in \mathbf{N}$ such that

$$|f_n - f| < \frac{\varepsilon}{3m(\mathbf{E})} \quad \text{on } \mathbf{E} \setminus \mathbf{E}_0$$

for all $n \geq N$. Take $\mathbf{A} = \mathbf{E}_0$ in (2.104) and we get

$$\begin{aligned} \left| \int_{\mathbf{E}} f_n - \int_{\mathbf{E}} f \right| &\leq \int_{\mathbf{E} \setminus \mathbf{E}_0} |f_n - f| + \int_{\mathbf{E}_0} |f_n| + \int_{\mathbf{E}_0} |f| \\ &< \frac{\varepsilon}{3m(\mathbf{E})} m(\mathbf{E} \setminus \mathbf{E}_0) + \frac{\varepsilon}{3} + \frac{\varepsilon}{3} \end{aligned}$$

$$\leq \frac{\varepsilon}{3} + \frac{2\varepsilon}{3}$$
$$= \frac{\varepsilon}{3}.$$

This completes the proof. □

Theorem 2.86. *Let* $\mathbf{E}$ *be of finite measure. Suppose that* $\{f_n\}_{n\in\mathbf{N}}$ *is a sequence of nonnegative integrable functions that converges pointwise a.e. on* $\mathbf{E}$ *to* $f = 0$. *Then*

$$\lim_{n\to\infty} \int_{\mathbf{E}} f_n = 0 \tag{2.105}$$

if and only if $\{f_n\}_{n\in\mathbf{N}}$ *is uniformly integrable over* $\mathbf{E}$.

Proof.

1. Let $\{f_n\}_{n\in\mathbf{N}}$ is uniformly integrable over $\mathbf{E}$. Hence and Vitalli's Theorem, we conclude that (2.105) holds.
2. Let (2.105) holds. Then there is an $N \in \mathbf{N}$ such that

   $$\int_{\mathbf{E}} f_n < \varepsilon$$

   for any $n \geq N$. Since $f_n \geq 0$ on $\mathbf{E}$ for any $n \in \mathbf{N}$, if $\mathbf{A} \subseteq \mathbf{E}$ is measurable and $n \geq N$, then

   $$\int_{\mathbf{A}} f_n < \varepsilon. \tag{2.106}$$

   By Theorem 2.83, it follows that $\{f_n\}_{n=1}^{N-1}$ is uniformly integrable over $\mathbf{E}$. Let $\delta > 0$ responds to the ε challenge regarding the criterion for uniform integrability of $\{f_n\}_{n=1}^{N-1}$. Then, using (2.106), we get the criterion for uniform integrability of $\{f_n\}_{n\in\mathbf{N}}$. This completes the proof. □

Theorem 2.87. *Let* f *be integrable over* $\mathbf{E}$. *Then for each* $\varepsilon > 0$ *there is a set of finite measure* $\mathbf{E}_0 \subseteq \mathbf{E}$ *such that*

$$\int_{\mathbf{E}\setminus\mathbf{E}_0} |f| < \varepsilon.$$

Proof. Let $\varepsilon > 0$ be arbitrarily chosen. Since $|f|$ is a nonnegative function on $\mathbf{E}$ that is integrable over $\mathbf{E}$, there exists a bounded nonnegative measurable function g on $\mathbf{E}$ with finite support $\mathbf{E}_0 \subseteq \mathbf{E}$ such that $0 \leq g \leq |f|$ on $\mathbf{E}$ and

$$\int_{\mathbf{E}} |f| - \int_{\mathbf{E}} g < \varepsilon.$$

Hence,

$$\begin{aligned}\int_{\mathbf{E}\backslash\mathbf{E}_0} |f| &= \int_{\mathbf{E}\backslash\mathbf{E}_0} (|f| - g)\\ &\leq \int_{\mathbf{E}} (|f| - g)\\ &< \varepsilon.\end{aligned}$$

This completes the proof. □

Definition 2.35. A family $\mathscr{F}$ of measurable functions on $\mathbf{E}$ is said to be tight over $\mathbf{E}$ provided for each $\varepsilon > 0$, there is a subset $\mathbf{E}_0$ of $\mathbf{E}$ of finite measure such that

$$\int_{\mathbf{E}\backslash\mathbf{E}_0} |f| < \varepsilon$$

for all $f \in \mathscr{F}$.

Theorem 2.88 (The Vitalli Convergence Theorem). *Let $\{f_n\}_{n\in\mathbf{N}}$ is a sequence of functions on $\mathbf{E}$ that is uniformly integrable and a tight over $\mathbf{E}$. Suppose that $f_n \to f$, as $n \to \infty$, pointwise a.e. on $\mathbf{E}$. Then f is integrable over $\mathbf{E}$ and*

$$\lim_{n\to\infty} \int_{\mathbf{E}} f_n = \int_{\mathbf{E}} f.$$

Proof. Let $\varepsilon > 0$ be arbitrarily chosen. Since $\{f_n\}_{n\in\mathbf{N}}$ is a tight, there exists a measurable set $\mathbf{E}_0$ of finite measure such that

$$\int_{\mathbf{E}\backslash\mathbf{E}_0} |f_n| < \frac{\varepsilon}{4}$$

for all $n \in \mathbf{N}$. Hence and Fatou's Lemma, we get

$$\int_{\mathbf{E}\backslash\mathbf{E}_0} |f| < \frac{\varepsilon}{4}.$$

Therefore, $|f|$ is integrable over $\mathbf{E}\backslash\mathbf{E}_0$. Hence,

$$\begin{aligned}\left|\int_{\mathbf{E}\backslash\mathbf{E}_0} (f_n - f)\right| &\le \int_{\mathbf{E}\backslash\mathbf{E}_0} |f_n - f| \\ &\le \int_{\mathbf{E}\backslash\mathbf{E}_0} (|f_n| + |f|) \\ &= \int_{\mathbf{E}\backslash\mathbf{E}_0} |f_n| + \int_{\mathbf{E}\backslash\mathbf{E}_0} |f| \\ &< \frac{\varepsilon}{4} + \frac{\varepsilon}{4} \\ &= \frac{\varepsilon}{2}\end{aligned}$$

for all $n \in \mathbf{N}$. Since $\mathbf{E}_0$ has finite measure and $\{f_n\}_{n\in\mathbf{N}}$ is uniformly integrable over $\mathbf{E}_0$, using Theorem 2.85, it follows that f is integrable over $\mathbf{E}_0$ and there is an $N \in \mathbf{N}$ such that

$$\left|\int_{\mathbf{E}_0} (f_n - f)\right| < \frac{\varepsilon}{2}$$

for all $n \ge N$. Therefore, f is integrable over $\mathbf{E}$ and

$$\begin{aligned}\left|\int_{\mathbf{E}} (f_n - f)\right| &= \left|\int_{\mathbf{E}\backslash\mathbf{E}_0} (f_n - f) + \int_{\mathbf{E}_0} (f_n - f)\right| \\ &\le \left|\int_{\mathbf{E}\backslash\mathbf{E}_0} (f_n - f)\right| + \left|\int_{\mathbf{E}_0} (f_n - f)\right| \\ &< \frac{\varepsilon}{2} + \frac{\varepsilon}{2} \\ &= \varepsilon\end{aligned}$$

for all $n \ge N$. This completes the proof. $\square$

Definition 2.36. Let $\{f_n\}_{n\in\mathbf{N}}$ be a sequence of measurable functions on $\mathbf{E}$ and f a measurable function on $\mathbf{E}$ for which f and each f_n is finite a.e. on $\mathbf{E}$. The sequence $\{f_n\}_{n\in\mathbf{N}}$ is said to converge in measure on $\mathbf{E}$ to f provided for each $\varepsilon > 0$

$$\lim_{n\to\infty} m\left(\{x \in \mathbf{E} : |f_n(x) - f(x)| > \varepsilon\}\right) = 0.$$

When we write $f_n \to f$, as $n \to \infty$, in measure, we are assuming that f and each f_n are measurable and finite a.e. on $\mathbf{E}$.

Exercise 2.17. Let $\{f_n\}_{n\in\mathbf{N}}$ be a sequence of measurable functions on $\mathbf{E}$ and f a measurable function on $\mathbf{E}$ for which f and each f_n is finite a.e. on $\mathbf{E}$. If $f_n \to f$, as $n \to \infty$, uniformly on $\mathbf{E}$, prove that $f_n \to f$, as $n \to \infty$, in measure.

Theorem 2.89. *Let $\mathbf{E}$ has finite measure. Let $\{f_n\}_{n\in\mathbf{N}}$ be a sequence of measurable functions on $\mathbf{E}$ that converges pointwise a.e. on $\mathbf{E}$ to f and f is finite a.e. on $\mathbf{E}$. Then $f_n \to f$, as $n \to \infty$, in measure on $\mathbf{E}$.*

Proof. By Theorem 2.45, we have that f is measurable on $\mathbf{E}$. Let $\varepsilon > 0$ be arbitrarily chosen. By Egoroff's Theorem, for each $\delta > 0$ there is a measurable subset $\mathbf{E}_0 \subseteq \mathbf{E}$ such that $f_n \to f$, as $n \to \infty$, uniformly on $\mathbf{E}_0$ and $m(\mathbf{E}\backslash\mathbf{E}_0) < \delta$. Then there is an index $N \in \mathbf{N}$ such that

$$|f_n - f| < \varepsilon \quad \text{on } \mathbf{E}_0$$

for each $n \geq N$. Thus, for each $n \geq N$, we have

$$\begin{aligned} m\left(\{x \in \mathbf{E} : |f_n(x) - f(x)| > \varepsilon\}\right) &\leq m(\mathbf{E}\backslash\mathbf{E}_0) \\ &< \delta. \end{aligned}$$

This completes the proof. □

Theorem 2.90 (Riesz's Theorem). *Let $f_n \to f$, as $n \to \infty$, in measure on $\mathbf{E}$. Then there is a subsequence $\{f_{n_k}\}_{k\in\mathbf{N}}$ of the sequence $\{f_n\}_{n\in\mathbf{N}}$ that converges pointwise a.e. on $\mathbf{E}$ to f.*

Proof. There is a sequence $\{n_k\}_{k\in\mathbf{N}}$ of natural numbers such that

$$m\left(\left\{x \in \mathbf{E} : |f_j(x) - f(x)| > \frac{1}{k}\right\}\right) < \frac{1}{2^k} \quad \text{for any } j \geq n_k.$$

Let

$$\mathbf{E}_k = \left\{x \in \mathbf{E} : |f_j(x) - f(x)| > \frac{1}{k}\right\}.$$

Then $m(\mathbf{E}_k) < \frac{1}{2^k}$ and $\sum_{k=1}^{\infty} m(\mathbf{E}_k) < \infty$. By Borel–Cantelli lemma, it follows that there is an index $K(x)$ such that $x \notin \mathbf{E}_k$ for any $k \geq K(x)$ and

$$|f_{n_k}(x) - f(x)| \leq \frac{1}{k} \quad \text{for all } k \geq K(x).$$

Therefore,

$$\lim_{k\to\infty} f_{n_k}(x) = f(x).$$

This completes the proof. □

Exercise 2.18. Let $\{f_n\}_{n\in\mathbf{N}}$ be a sequence of nonnegative integrable functions on $\mathbf{E}$. Then $\lim_{n\to\infty} \int_{\mathbf{E}} f_n = 0$ if and only if $f_n \to 0$, as $n \to \infty$, in measure on $\mathbf{E}$ and $\{f_n\}_{n\in\mathbf{N}}$ is uniformly integrable and a tight over $\mathbf{E}$.

2.8 Continuity and Differentiability of Monotone Functions: Lebesgue's Theorem

Definition 2.37. A real-valued function defined on a set $\mathbf{E}$ of real numbers is said to be increasing if $f(x) \leq f(x')$ whenever $x, x' \in \mathbf{E}$ and $x \leq x'$, and decreasing provided $-f$ is increasing. It is called monotone if it is either decreasing or increasing.

Theorem 2.91. *Let f be a monotone function on the open interval (a, b). Then f is continuous except possibly of a countable number of points in (a, b).*

Proof. Let f is decreasing on (a,b). Assume that (a,b) is bounded and f is decreasing on the closed interval $[a,b]$. Otherwise, we express (a,b) as the union of an ascending sequence of open bounded intervals such that their closures are contained in (a,b) and take the union of discontinuities in each of this countable collection of intervals. For $x_0 \in (a,b)$, we define

$$f\left(x_0^-\right) = \lim_{x\to x_0^-} f(x) = \sup\left\{f(x) : a < x < x_0\right\},$$

$$f\left(x_0^+\right) = \lim_{x\to x_0^+} f(x) = \inf\left\{f(x) : x_0 < x < b\right\}.$$

Because f is decreasing on $[a,b]$, we have that $f\left(x_0^+\right) \le f\left(x_0^-\right)$. Not that f is not continuous at x_0 if and only if $f\left(x_0^+\right) < f\left(x_0^-\right)$. If f is discontinuous at x_0, we define the open "jump" intervals

$$J\left(x_0\right) = \left\{y : f\left(x_0^+\right) < y < f\left(x_0^-\right)\right\}.$$

Because f is decreasing on $[a,b]$, we have the inclusion

$$J\left(x_0\right) \subset \left[f(b), f(a)\right].$$

Note that if $x_0 < x_1$, then $f\left(x_1^-\right) \le f\left(x_0^+\right)$. Therefore,

$$\left(f\left(x_1^+\right), f\left(x_1^-\right)\right) \cap \left(f\left(x_0^+\right), f\left(x_0^-\right)\right) = \emptyset.$$

Then the collection of jump intervals is disjoint. Hence, for each $n \in \mathbb{N}$, there are only a finite number of jump intervals of length greater than $\frac{1}{n}$. Consequently, the set of points of discontinuity of f is an union of countable collection of finite sets and then this set is countable. This completes the proof. □

Theorem 2.92. *Let* $\mathbf{A}$ *be a countable subset of the open interval* (a,b). *Then there is an increasing function on* (a,b) *that is continuous only at points in* $(a,b)\backslash\mathbf{A}$.

Proof. If $\mathbf{A}$ is finite, then the assertion is evident. Let $\mathbf{A}$ be countably infinite. Suppose that $\mathbf{A} = \{a_n\}_{n\in\mathbf{N}}$. Define the function

$$f(x) = \sum_{\{n\in\mathbf{N}_0 : a_n \le x\}} \frac{1}{2^n}$$

for all $a < x < b$. Since $\sum_{n\in\mathbf{N}_0} \frac{1}{2^n} < \infty$, the function f is well-defined. Also, for $x, y \in [a, b]$, $y \ge x$. we have

$$\begin{aligned} f(y) - f(x) &= \sum_{\{n\in\mathbf{N}_0 : a_n \le y\}} \frac{1}{2^n} - \sum_{\{n\in\mathbf{N}_0 : a_n \le x\}} \frac{1}{2^n} \\ &= \sum_{\{n\in\mathbf{N}_0 : x < a_n \le y\}} \frac{1}{2^n} \\ &> 0. \end{aligned} \tag{2.107}$$

Therefore, f is an increasing function on (a, b). Let $x_0 = a_k$ for some $k \in \mathbf{N}$. Then, using (2.107), we have

$$f(x_0) - f(x) \ge \frac{1}{2^k}$$

for all $x < x_0$. Hence, f is not continuous at x_0. Let $x_0 \in (a, b)\backslash\mathbf{A}$. Then there is an interval $\mathbf{I}$ such that $x_0 \in \mathbf{I}$ and $a_n \notin \mathbf{I}$ for any $n \in \mathbf{N}$. Then

$$|f(x) - f(x_0)| < \frac{1}{2^n}$$

for all $x, x_0 \in \mathbf{I}$. Therefore, f is continuous at x_0. This completes the proof. □

Definition 2.38. A closed bounded interval $[c, d]$ is said to be nondegenerate if $c < d$.

Definition 2.39. A collection $\mathscr{F}$ of closed bounded nondegenerate intervals is said to cover a set $\mathbf{E}$ in the sense of Vitalli if for each $x \in \mathbf{E}$ and for each $\varepsilon > 0$ there is an interval $\mathbf{I} \in \mathscr{F}$ such that $x \in \mathbf{I}$ and $l(\mathbf{I}) < \varepsilon$.

Lemma 2.9 (The Vitalli Covering Lemma). *Let* $\mathbf{E}$ *be a set of finite measure and* $\mathscr{F}$ *be a collection of closed bounded intervals that covers* $\mathbf{E}$ *in the sense of Vitalli. Then for each* $\varepsilon > 0$, *there is a finite disjoint subcollection* $\{\mathbf{I}_k\}_{k=1}^n$ *of* $\mathscr{F}$ *for which*

$$m^\star\left(\mathbf{E}\Big\backslash\left(\bigcup_{k=1}^n \mathbf{I}_k\right)\right) < \varepsilon.$$

Proof. Since $m^\star(\mathbf{E}) < \infty$, there is an open set $\mathbf{O}$, containing $\mathbf{E}$, such that $m(\mathbf{O}) < \infty$. Since $\mathscr{F}$ covers $\mathbf{E}$, we may assume that $\mathscr{F} \subseteq \mathbf{O}$. If $\{\mathbf{I}_k\}_{k\in\mathbf{N}} \subseteq \mathscr{F}$ is a disjoint collection of intervals in $\mathscr{F}$, then

$$\sum_{k=1}^\infty l(\mathbf{I}_k) \le m(\mathbf{O}) \tag{2.108}$$

and for $n \in \mathbf{N}$ we denote

$$\mathscr{F}_n = \left\{\mathbf{I} \in \mathscr{F} : \mathbf{I} \cap \left(\bigcup_{k=1}^n \mathbf{I}_k\right) = \emptyset\right\}.$$

We will prove that

$$\mathbf{E}\Big\backslash\left(\bigcup_{k=1}^\infty \mathbf{I}_k\right) \subseteq \bigcup_{\mathbf{I}\in\mathscr{F}_n} \mathbf{I} \tag{2.109}$$

for any $n \in \mathbf{N}$. Let $x \in \mathbf{E}\backslash(\cup_{k=1}^\infty \mathbf{I}_k)$ be arbitrarily chosen. Then $x \in \mathbf{E}$ and $x \notin \cup_{k=1}^\infty \mathbf{I}_k$. Hence, $x \notin \mathbf{I}_k$ for any $k \in \mathbf{N}$. Since $\mathscr{F}$ covers the set $\mathbf{E}$ in the sense of Vitalli and $x \in \mathbf{E}$, then there is an interval $\mathbf{I} \in \mathscr{F}$ such that $x \in \mathbf{I}$ and $\mathbf{I} \cap (\cup_{k=1}^n \mathbf{I}_k) = \emptyset$ for $n \in \mathbf{N}$. Then $x \in \cup_{\mathbf{I}\in\mathscr{F}_n}\mathbf{I}$ for $n \in \mathbf{N}$. Because $x \in \mathbf{E}\backslash(\cup_{k=1}^\infty \mathbf{I}_k)$ was arbitrarily chosen and for it we get that it is an element of $\cup_{\mathbf{I}\in\mathscr{F}_n}\mathbf{I}$ for $n \in \mathbf{N}$, we obtain the relation (2.109). If there is a finite disjoint subcollection $\{\mathbf{I}_k\}_{k=1}^n$ of $\mathscr{F}$ that covers $\mathbf{E}$, the assertion is proved. Otherwise, let $n \in \mathbf{N}$ and $\{\mathbf{I}_k\}_{k=1}^n \in \mathscr{F}$ be a disjoint collection. Since $\mathbf{E}\backslash(\cup_{k=1}^n \mathbf{I}_k) \neq \emptyset$, we have that $\mathscr{F}_n \neq \emptyset$. Let

$$s_n = \sup_{\mathbf{I}\in\mathscr{F}_n} l(\mathbf{I}).$$

We have that $l(\mathbf{I}) \le m(\mathbf{O})$ for any $\mathbf{I} \in \mathscr{F}_n$. Then $s_n < \infty$. We take $\mathbf{I}_{n+1} \in \mathscr{F}_n$ such that $l(\mathbf{I}_{n+1}) > \frac{s_n}{2}$. This inductively defines $\{\mathbf{I}_k\}_{k\in\mathbf{N}}$, a countable disjoint subcollection of $\mathscr{F}$ such that $l(\mathbf{I}_{n+1}) > \frac{l(\mathbf{I})}{2}$ if $\mathbf{I} \in \mathscr{F}$ and $\mathbf{I} \cap (\cup_{k=1}^n \mathbf{I}_k) = \emptyset$. Let $\mathbf{J}_k$ be the closed interval that has the same midpoint as $\mathbf{I}_k$ and 5 times its length. We will prove

$$\mathbf{E}\Big\backslash\left(\bigcup_{k=1}^{n} \mathbf{I}_k\right) \subseteq \bigcup_{k=n+1}^{\infty} \mathbf{J}_k \tag{2.110}$$

for any $n \in \mathbf{N}$. Let $n \in \mathbf{N}$ and $x \in \mathbf{E}\backslash (\cup_{k=1}^n \mathbf{I}_k)$ be arbitrarily chosen. Then there is $\mathbf{I} \in \mathscr{F}$ such that $x \in \mathbf{I}$ and $\mathbf{I} \cap (\cup_{k=1}^n \mathbf{I}_k) = \emptyset$. Assume that $\mathbf{I} \cap \mathbf{I}_k = \emptyset$ for any $k \in \mathbf{N}$. Then $l(\mathbf{I}_k) > \frac{l(\mathbf{I})}{2}$ for any $k \in \mathbf{N}$. This is a contradiction. Then there is $m \in \mathbf{N}$ such that $\mathbf{I}\cap\mathbf{I}_m \neq \emptyset$. We take $M > n$ to be the first natural number for which $\mathbf{I} \cap \mathbf{I}_M \neq \emptyset$. Then $\mathbf{I} \cap \left(\cup_{k=1}^{M-1} \mathbf{I}_k\right) = \emptyset$ and $l(\mathbf{I}_M) > \frac{l(\mathbf{I})}{2}$. Since $x \in \mathbf{I}$ and $\mathbf{I} \cap \mathbf{I}_M \neq \emptyset$, the distance from x to the midpoint of $\mathbf{I}_M$ is at most $l(\mathbf{I}) + \frac{1}{2} l(\mathbf{I}_M)$. Hence, using that $l(\mathbf{I}) < 2L(\mathbf{I}_M)$, the distance from x to the midpoint of $\mathbf{I}_M$ is less than $\frac{5}{2} l(\mathbf{I}_M)$. Therefore $x \in \mathbf{J}_M$ and $x \in \cup_{k=n+1}^{\infty} \mathbf{J}_k$. Because $x \in \mathbf{E}\backslash (\cup k = 1^n \mathbf{I}_k)$ was arbitrarily chosen and for it we get that it is an element of $\cup_{k=n+1}^{\infty} \mathbf{J}_k$, we obtain the relation (2.110). Let $\varepsilon > 0$ be arbitrarily chosen. Then there is an $n \in \mathbf{N}$ such that

$$\sum_{k=n+1}^{\infty} l(\mathbf{I}_k) < \frac{\varepsilon}{5}.$$

Hence,

$$\sum_{k=n+1}^{\infty} l(\mathbf{J}_k) < \varepsilon$$

and using the relation (2.110), we get

$$m^\star\left(\mathbf{E}\Big\backslash\left(\bigcup_{k=1}^{n} \mathbf{I}_k\right)\right) < \varepsilon.$$

This completes the proof. □

Definition 2.40. Let $f : \mathbf{E} \longmapsto \mathbf{R}$ and $x \in \mathbf{E}$. We define upper derivative $\overline{D}f(x)$ and lower derivative $\underline{D}f(x)$ of f at x as follows

$$\overline{D}f(x) = \lim_{h\to 0}\left(\sup_{0\le|t|\le h}\frac{f(x+t)-f(x)}{t}\right),$$

$$\underline{D}f(x) = \lim_{h\to 0}\left(\inf_{0\le|t|\le h}\frac{f(x+t)-f(x)}{t}\right).$$

We have $\underline{D}f(x) \le \overline{D}f(x)$. If $\underline{D}f(x)$ and $\overline{D}f(x)$ are finite and $\underline{D}f(x) = \overline{D}f(x)$, we say that f is differentiable at x and we define $f'(x)$ to be the common value of the upper derivative and the lower derivative.

Theorem 2.93. *Let f be an increasing function on the closed bounded interval $[a,b]$. Then for each $\alpha > 0$, we have*

$$m^\star\left(\left\{x \in (a,b) : \overline{D}f(x) \ge \alpha\right\}\right) \le \frac{1}{\alpha}\left(f(b)-f(a)\right) \tag{2.111}$$

and

$$m^\star\left(\left\{x \in (a,b) : \overline{D}f(x) = \infty\right\}\right) = 0. \tag{2.112}$$

Proof. Let $\alpha > 0$ be arbitrarily chosen and

$$\mathbf{E}_\alpha = \left\{x \in (a,b) : \overline{D}f(x) \ge \alpha\right\}.$$

We take $\alpha' \in (0,\alpha)$ and $\varepsilon > 0$ arbitrarily. With $\mathscr{F}$ we will denote the collection of closed bounded interval $[c,d] \subset (a,b)$ such that

$$f(d) - f(c) \ge \alpha'(d-c).$$

Since $\overline{D}f(x) \ge \alpha$ for $x \in \mathbf{E}_\alpha$, we have that $\mathscr{F}$ is a Vitalli's covering of $\mathbf{E}_\alpha$. Hence, using the Vitalli covering lemma, there is a finite disjoint subcollection $\{[c_k,d_k]\}_{k=1}^n$ of $\mathscr{F}$ such that

$$m^\star\left(\mathbf{E}_\alpha \Big\backslash \left(\bigcup_{k=1}^n [c_k,d_k]\right)\right) < \varepsilon.$$

Note that

$$\mathbf{E}_{\alpha} \subseteq \left(\bigcup_{k=1}^{n} [c_k, d_k] \right) \cup \left(\mathbf{E}_{\alpha} \Big\backslash \left(\bigcup_{k=1}^{n} [c_k, d_k \Big] \right) \right).$$

Therefore,

$$\begin{aligned}
m^{\star}(\mathbf{E}_{\alpha}) &\leq m^{\star}\left(\bigcup_{k=1}^{n} [c_k, d_k] \right) + m^{\star}\left(\mathbf{E}_{\alpha} \Big\backslash \left(\bigcup_{k=1}^{n} [c_k, d_k] \right) \right) \\
&< \sum_{k=1}^{n} (d_k - c_k) + \varepsilon \\
&\leq \frac{1}{\alpha'} \sum_{k=1}^{n} (f(d_k) - f(c_k)) + \varepsilon \\
&= \frac{1}{\alpha'} (f(d_1) - f(c_1) + f(d_2) - f(c_2) + \cdots + f(d_n) \\
&\quad - f(c_n)) + \varepsilon \\
&\leq \frac{1}{\alpha'} (f(c_2) - f(a) + f(c_3) - f(c_2) + \cdots + f(b) - f(c_n)) + \varepsilon \\
&= \frac{1}{\alpha'} (f(b) - f(a)) + \varepsilon.
\end{aligned}$$

Because $\varepsilon > 0$ was arbitrarily chosen, we obtain (2.111). Next, for each natural number n we have

$$\left\{ x \in (a,b) : \overline{D} f(x) = \infty \right\} \subseteq \mathbb{E}_n.$$

Consequently,

$$\begin{aligned}
m^{\star}\left(\left\{ x \in (a,b) : \overline{D} f(x) = \infty \right\} \right) &\leq m^{\star}(\mathbb{E}_n) \\
&\leq \frac{1}{n} (f(b) - f(a)) \\
&\to 0, \quad \text{as } n \to \infty,
\end{aligned}$$

i.e., we get (2.112). This completes the proof. □

Theorem 2.94 (Lebesgue's Theorem). *If the function f is monotone on (a, b), then it is differentiable almost everywhere on (a, b).*

Proof. Let f be increasing on (a, b). Assume that (a, b) is bounded. Otherwise, we express it as the union of an ascending sequence of open bounded intervals contained in (a, b). Let

$$\mathbf{E} = \left\{x \in (a, b) : \overline{D}f(x) > \underline{D}f(x)\right\},$$

$$\mathbf{E}_{\alpha,\beta} = \left\{x \in (a, b) : \overline{D}f(x) > \alpha > \beta > \underline{D}f(x)\right\}, \quad \alpha, \beta \in \mathbf{Q}, \quad \alpha > \beta.$$

Then

$$\mathbf{E} = \bigcup_{\alpha,\beta\in\mathbf{Q},\alpha>\beta} \mathbf{E}_{\alpha,\beta}.$$

Hence,

$$m^\star(\mathbf{E}) \leq \sum_{\alpha,\beta\in\mathbf{Q},\alpha>\beta} m^\star(\mathbf{E}_{\alpha,\beta}). \tag{2.113}$$

We fix $\alpha, \beta \in \mathbf{Q}$, $\alpha > \beta$. Let $\varepsilon > 0$ be arbitrarily chosen. We take an open set $\mathbf{O}$ such that

$$\mathbf{E}_{\alpha,\beta} \subseteq \mathbf{O} \subseteq (a, b) \quad \text{and} \quad m^\star(\mathbf{O}) < m^\star(\mathbf{E}_{\alpha,\beta}) + \varepsilon. \tag{2.114}$$

Let $\mathscr{F}$ be the collection of closed bounded intervals $[c, d]$ contained in $\mathbf{O}$ for which

$$f(d) - f(c) < \beta(d - c).$$

Since $\underline{D}f(x) < \beta$ for $x \in \mathbf{E}_{\alpha,\beta}$, we have that $\mathscr{F}$ is a Vitalli's covering of $\mathbf{E}_{\alpha,\beta}$. By the Vitalli covering lemma, there is a finite disjoint collection $\{[c_k, d_k]\}_{k=1}^n$ of $\mathscr{F}$ such that

$$m^\star\left(\mathbf{E}_{\alpha,\beta} \Big\backslash \left(\bigcup_{k=1}^n [c_k, d_k]\right)\right) < \varepsilon. \tag{2.115}$$

On the other hand, using (2.114), we obtain

$$\begin{aligned}\sum_{k=1}^n (f(d_k) - f(c_k)) &< \beta \sum_{k=1}^n (d_k - c_k) \\ &\leq \beta m^\star(\mathbf{O}) \\ &< \beta\left(m^\star(\mathbf{E}_{\alpha,\beta}) + \varepsilon\right).\end{aligned} \tag{2.116}$$

By Theorem 2.93, we get

$$m^{\star}\left(\mathbf{E}_{\alpha,\beta} \cap (c_k, d_k)\right) \le \frac{1}{\alpha}\left(f\left(d_k\right) - f\left(c_k\right)\right)$$

for any $k \in \{1, \ldots, n\}$. Note that

$$\mathbf{E}_{\alpha,\beta} \subseteq \left(\mathbf{E}_{\alpha,\beta} \cap \left(\bigcup_{k=1}^{n} (c_k, d_k)\right)\right) \cup \left(\mathbf{E}_{\alpha,\beta} \Bigg\backslash \left(\bigcup_{k=1}^{n} [c_k, d_k]\right)\right).$$

Therefore, using (2.115), we get

$$\begin{aligned} m^{\star}\left(\mathbf{E}_{\alpha,\beta}\right) &\le m^{\star}\left(\mathbf{E}_{\alpha,\beta} \cap \left(\bigcup_{k=1}^{n} (c_k, d_k)\right)\right) \\ &\quad + m^{\star}\left(\mathbf{E}_{\alpha,\beta} \Bigg\backslash \left(\bigcup_{k=1}^{n} [c_k, d_k]\right)\right) \\ &< m^{\star}\left(\mathbf{E}_{\alpha,\beta} \cap \left(\bigcup_{k=1}^{n} (c_k, d_k)\right)\right) + \varepsilon \\ &\le \sum_{k=1}^{n} m^{\star}\left(\mathbf{E}_{\alpha,\beta} \cap (c_k, d_k)\right) + \varepsilon \\ &\le \frac{1}{\alpha}\sum_{k=1}^{n}\left(f\left(d_k\right) - f\left(c_k\right)\right) + \varepsilon. \end{aligned}$$

Hence and (2.116), we obtain

$$m^{\star}\left(\mathbf{E}_{\alpha,\beta}\right) \le \frac{\beta}{\alpha}\left(m^{\star}\left(\mathbf{E}_{\alpha,\beta}\right) + \varepsilon\right) + \varepsilon$$

and

$$\frac{\alpha - \beta}{\alpha} m^{\star}\left(\mathbf{E}_{\alpha,\beta}\right) \le \frac{\beta}{\alpha}\varepsilon + \varepsilon.$$

Because $\varepsilon > 0$ was arbitrarily chosen, we get

$$m^{\star}\left(\mathbf{E}_{\alpha,\beta}\right) = 0.$$

Hence and (2.113), we obtain that $m^{\star}(\mathbf{E}) = 0$. This completes the proof. □

Definition 2.41. Let f be integrable over the closed bounded interval $[a,b]$ and take value $f(b)$ on $(b,b+1]$. For $0<h\le 1$, we define the divided difference function $Diff_h f$ and average value function $Av_h f$ of $[a,b]$ by

$$Diff_h f(x)=\frac{f(x+h)-f(x)}{h} \quad \text{and} \quad Av_h f(x)=\frac{1}{h}\int_x^{x+h} f,$$

respectively, for all $x\in[a,b]$.

Let $a\le y<z\le b$ and f be as in Definition (2.41). Then

$$\begin{aligned}\int_y^z Diff_h f(x) &= \int_y^z \frac{f(x+h)-f(x)}{h}\\ &= \frac{1}{h}\int_y^z f(x+h)-\frac{1}{h}\int_y^z f(x)\\ &= \frac{1}{h}\int_{y+h}^{z+h} f(x)-\frac{1}{h}\int_y^z f(x)\\ &= \frac{1}{h}\int_z^{z+h} f(x)-\frac{1}{h}\int_y^{y+h} f(x)\\ &= Av_h f(z)-Av_h f(y).\end{aligned}$$

Theorem 2.95. *Let f be an increasing function on the closed bounded interval $[a,b]$. Then f' is integrable over $[a,b]$ and*

$$\int_a^b f'\le f(b)-f(a). \tag{2.117}$$

Proof. Let f take value $f(b)$ on $(b,b+1]$. Then, for $0<h\le 1$, f is increasing on $[a,b+1]$. Therefore, f and $Diff_h f$ are measurable on $[a,b]$. By Theorem 2.94, it follows that f is almost everywhere differentiable on $[a,b]$. Therefore $\{Diff_{\frac{1}{n}} f\}_{n\in\mathbf{N}}$ is a sequence of nonnegative measurable functions that converges pointwise almost everywhere on $[a,b]$ to f'. Hence and Fatou's Lemma, we obtain

$$\int_a^b f'\le \liminf_{n\to\infty}\left(\int_a^b Diff_{\frac{1}{n}} f\right). \tag{2.118}$$

On the other hand, using that f is increasing on $[a, b+1]$, we have

$$\begin{aligned}\int_a^b \mathit{Diff}_{\frac{1}{n}} f &= Av_{\frac{1}{n}} f(b) - Av_{\frac{1}{n}} f(a) \\ &= n\int_b^{b+\frac{1}{n}} f(x) - n\int_a^{a+\frac{1}{n}} f(x) \\ &\leq f(b) - f(a).\end{aligned}$$

Thus,

$$\liminf_{n\to\infty} \int_a^b \mathit{Diff}_{\frac{1}{n}} f \leq f(b) - f(a).$$

From here and from (2.118), we obtain (2.117). This completes the proof. □

Definition 2.42. A real-valued function f on a closed bounded interval $[a, b]$ is said to be absolutely continuous on $[a, b]$ if for each $\varepsilon > 0$ there is a $\delta > 0$ such that for any finite disjoint collection $\{(a_k, b_k)\}_{k=1}^n$ of open intervals in (a, b) for which $\sum_{k=1}^n (b_k - a_k) < \delta$ we have $\sum_{k=1}^n |f(b_k) - f(a_k)| < \varepsilon$.

Theorem 2.96. *Let f be a Lipschitz function on the closed bounded interval $[a, b]$, i.e., there exists a constant $L > 0$ such that*

$$|f(x) - f(y)| \leq L|x - y|$$

for any $x, y \in [a, b]$. Then f is absolutely continuous on $[a, b]$.

Proof. Let $\varepsilon > 0$ be arbitrarily chosen and $\delta = \frac{\varepsilon}{L}$. We take a finite disjoint collection $\{(a_k, b_k)\}_{k=1}^n$ of open intervals in (a, b) such that

$$\sum_{k=1}^n (b_k - a_k) < \frac{\varepsilon}{L}.$$

Hence,

$$\begin{aligned}\sum_{k=1}^n |f(b_k) - f(a_k)| &\leq L\sum_{k=1}^n |b_k - a_k| \\ &< \varepsilon.\end{aligned}$$

This completes the proof. □

Theorem 2.97. *Let f be an absolutely continuous function on the closed bounded interval $[a,b]$. Then f is difference of two increasing absolutely continuous functions on $[a,b]$.*

Proof. Let $\varepsilon = 1$ and $\delta > 0$ responds to the criterion for absolute continuity of f. Let also, P be a partition of $[a,b]$ into N closed intervals $\{[c_k,d_k]\}_{k=1}^N$ each of length less than δ. Hence, for the total variation of f on $[c_k,d_k]$, we have

$$\bigvee_{c_k}^{d_k}(f) \le 1$$

for each $k \in \{1,\ldots,N\}$. Therefore,

$$\begin{aligned}\bigvee_a^b(f) &= \sum_{k=1}^N \bigvee_{c_k}^{d_k}(f)\\ &\le N,\end{aligned}$$

i.e., f is a function of bounded variation on $[a,b]$. Hence and Jordan's Theorem, we have

$$f(x) = \left(f(x) + \bigvee_a^x(f)\right) - \left(\bigvee_a^x(f)\right), \quad x \in [a,b], \tag{2.119}$$

where $f(x) + \bigvee_a^x(f)$ and $\bigvee_a^x(f)$ are increasing functions on $[a,b]$. Let $\varepsilon > 0$ be arbitrarily chosen and $\delta > 0$ responds to $\frac{\varepsilon}{2}$ regarding the criterion for absolute continuity of f on $[a,b]$. Let also, $\{(c_k,d_k)\}_{k=1}^n$ be a disjoint collection of open subintervals of (a,b) for which $\sum_{k=1}^n (d_k - c_k) < \delta$. For $k \in \{1,\ldots,n\}$, with P_k we will denote a partition of $[c_k,d_k]$. Then

$$\sum_{k=1}^n \bigvee_{c_k}^{d_k}(f, P_k) \le \frac{\varepsilon}{2}.$$

From here,

$$\begin{aligned}\sum_{k=1}^n \bigvee_{c_k}^{d_k}(f) &\le \frac{\varepsilon}{2}\\ &< \varepsilon.\end{aligned}$$

Consequently,

$$\sum_{k=1}^{n}\left(\bigvee_{a}^{d_k}(f)-\bigvee_{a}^{c_k}(f)\right)<\varepsilon.$$

Let

$$f_1(x)=\bigvee_{a}^{x}(f),\quad f_2(x)=f(x)+\bigvee_{a}^{x}(f),\quad x\in[a,b].$$

Therefore, f_1 is absolutely continuous on $[a,b]$. Hence, f_2 is absolutely continuous on $[a,b]$. Using (2.119), we conclude that f is difference of two increasing absolutely continuous functions on $[a,b]$. This completes the proof. □

Exercise 2.19. Let f is absolutely continuous on $[a,b]$. Prove that $\left\{\mathrm{Diff}_{\frac{1}{n}}f\right\}_{n\in\mathbf{N}}$ is uniformly integrable over $[a,b]$.

Hint: Use that

$$\begin{aligned}
\int_u^v \mathrm{Diff}_h f &= \frac{1}{h}\int_u^v f(t+h)-\frac{1}{h}\int_u^v f(t)\\
&=\frac{1}{h}\int_{u-v+h}^{h} f(t+v)-\frac{1}{h}\int_0^{v-u} f(u+t)\\
&=\frac{1}{h}\int_0^{h} f(t+v)+\frac{1}{h}\int_{u-v+h}^{0} f(t+v)\\
&\quad-\frac{1}{h}\int_0^{h} f(u+t)-\frac{1}{h}\int_h^{v-u} f(u+t)\\
&=\frac{1}{h}\int_0^{h} f(t+v)+\frac{1}{h}\int_{u-v+h}^{0} f(t+v)\\
&\quad-\frac{1}{h}\int_0^{h} f(u+t)-\frac{1}{h}\int_{u-v+h}^{0} f(t+v)\\
&=\frac{1}{h}\int_0^{h}\left(f(t+v)-f(u+t)\right)
\end{aligned}$$

for any $u,v\in[a,b]$.

Theorem 2.98. *Let f be absolutely continuous on the closed bounded interval $[a, b]$. Then f is differentiable almost everywhere on (a, b), its derivative f' is integrable over $[a, b]$ and*

$$\int_a^b f' = f(b) - f(a).$$

Proof. By Theorem 2.97, f is difference of two increasing absolutely continuous functions on $[a, b]$. Hence and Lebesgue's Theorem, it follows that f is differentiable almost everywhere on (a, b). Therefore, the sequence $\left\{Diff_{\frac{1}{n}} f\right\}_{n \in \mathbf{N}}$ converges pointwise almost everywhere on (a, b) to f'. Since $\left\{Diff_{\frac{1}{n}} f\right\}_{n \in \mathbf{N}}$ is uniformly integrable over $[a, b]$, using Theorem 2.85, we get

$$\begin{aligned} f(b) - f(a) &= \lim_{n \to \infty} \int_a^b Diff_{\frac{1}{n}} f \\ &= \int_a^b \lim_{n \to \infty} Diff_{\frac{1}{n}} f \\ &= \int_a^b f'. \end{aligned}$$

This completes the proof. □

2.9 General Measure Spaces

Definition 2.43. A collection of subsets of a set $\mathbf{X}$ that contains the empty set and it is closed with respect to the complements in $\mathbf{X}$ and with respect to the formation of countable unions, will be called a σ-algebra of subsets of $\mathbf{X}$.

Definition 2.44. A couple $(\mathbf{X}, \mathscr{M})$ consisting of a set $\mathbf{X}$ and a σ-algebra $\mathscr{M}$ of subsets of $\mathbf{X}$ will be called a measurable space. A subset $\mathbf{E}$ of $\mathbf{X}$ is called measurable(or measurable with respect to $\mathscr{M}$) if $\mathbf{E} \in \mathscr{M}$.

Definition 2.45. Let $(\mathbf{X}, \mathscr{M})$ be a measurable space. An extended real-valued function $\mu : \mathscr{M} \longmapsto [0, \infty]$ for which $\mu(\emptyset) = 0$ and which is countably additive in the sense that for any countable disjoint collection $\{\mathbf{E}_k\}_{k\in\mathbf{N}}$ of measurable sets

$$\mu\left(\bigcup_{k=1}^{\infty}\mathbf{E}_k\right)=\sum_{k=1}^{\infty}\mu\left(\mathbf{E}_k\right),$$

will be called measure.

By a measure space $(\mathbf{X},\mathscr{M},\mu)$ we mean a measurable space $(\mathbf{X},\mathscr{M})$ together with a measure μ defined on $\mathscr{M}$.

Example 2.1. $(\mathbf{R},\mathscr{L},m)$, where $\mathscr{L}$ is a collection of Lebesgue measurable sets of real numbers and m is the Lebesgue measure, is a measurable space.

Remark 2.6. Note that the idea for the proof of the assertions in this section is the same as in the case of Lebesgue measure. Therefore, we leave the proofs of the assertions in this section.

Theorem 2.99. *Let $(\mathbf{X},\mathscr{M},\mu)$ be a measure space.*

1. *For any finite disjoint collection $\{\mathbf{E}_k\}_{k=1}^{n}$ of measurable sets we have*

$$\mu\left(\bigcup_{k=1}^{n}\mathbf{E}_k\right)=\sum_{k=1}^{n}\mu\left(\mathbf{E}_k\right).$$

2. *If $\mathbf{A}$ and $\mathbf{B}$ are measurable sets and $\mathbf{A}\subseteq\mathbf{B}$, then*

$$\mu\left(\mathbf{A}\right)\le\mu\left(\mathbf{B}\right).$$

   Moreover, if $\mathbf{A}\subseteq\mathbf{B}$ and $\mu\left(\mathbf{A}\right)<\infty$, then

$$\mu\left(\mathbf{B}\backslash\mathbf{A}\right)=\mu\left(\mathbf{B}\right)-\mu\left(\mathbf{A}\right).$$

3. *For any countable collection $\{\mathbf{E}_k\}_{k\in\mathbf{N}}$ of measurable sets that covers a measurable set $\mathbf{E}$, we have*

$$\mu\left(\mathbf{E}\right)\le\sum_{k=1}^{\infty}\mu\left(\mathbf{E}_k\right).$$

Theorem 2.100. *Let* $(\mathbf{X}, \mathscr{M}, \mu)$ *be a measure space.*

1. *If* $\{\mathbf{A}_k\}_{k\in\mathbf{N}}$ *is an ascending sequence of measurable sets, then*
$$\mu\left(\bigcup_{k=1}^{\infty} \mathbf{A}_k\right) = \lim_{k\to\infty} \mu\left(\mathbf{A}_k\right).$$
2. *If* $\{\mathbf{A}_k\}_{k\in\mathbf{N}}$ *is a descending sequence of measurable sets for which* $\mu(\mathbf{A}_1) < \infty$*, then*
$$\mu\left(\bigcap_{k=1}^{\infty} \mathbf{A}_k\right) = \lim_{k\to\infty} \mu\left(\mathbf{A}_k\right).$$

Definition 2.46. For a measure space $(\mathbf{X}, \mathscr{M}, \mu)$ and a measurable subset $\mathbf{E}$ of $\mathbf{X}$, we say that a property holds almost everywhere on $\mathbf{E}$, or it holds for almost all $x \in \mathbf{E}$, if it holds on $\mathbf{E}\backslash\mathbf{E}_0$ and $\mathbf{E}_0$ is a measurable set for which $\mu\left(\mathbf{E}_0\right) = 0$.

Lemma 2.10 (The Borel–Cantelli Lemma). *Let* $(\mathbf{X}, \mathscr{M}, \mu)$ *be a measure space and* $\{\mathbf{A}_k\}_{k\in\mathbf{N}}$ *is a countable collection of measurable sets for which* $\sum_{k=1}^{\infty} \mu\left(\mathbf{A}_k\right) < \infty$*. Then almost all* $x \in \mathbf{X}$ *belong to at most a finite number of the* $\mathbf{A}_k$*'s.*

Definition 2.47. Let $(\mathbf{X}, \mathscr{M}, \mu)$ be a measure space. The measure μ is called finite if $\mu\left(\mathbf{X}\right) < \infty$, and it is called σ-finite if $\mathbf{X}$ is the union of a countable collection of measurable sets, each of which has finite measure. A measurable set $\mathbf{E}$ is said to be of finite measure if $\mu\left(\mathbf{E}\right) < \infty$, and it is said to be of σ-finite measure if $\mathbf{E}$ is the union of a countable collection of measurable sets, each of which has finite measure.

2.10 General Measurable Functions

Definition 2.48. Let $(\mathbf{X}, \mathscr{M})$ be a measurable space. An extended real-valued function f on $\mathbf{X}$ is said to be measurable(or measurable with respect to $\mathscr{M}$) if for each $c \in \mathbf{R}$, the set $\{x \in \mathbf{X} : f(x) > c\}$ is measurable.

Remark 2.7. Note that the idea for the proof of the assertions in this section is the same as the idea for the proofs of the assertions in Section 2.5. Therefore, we leave the proofs of the assertions in this section.

Theorem 2.101. *Let $(\mathbf{X}, \mathscr{M})$ be a measurable space and f is an extended real-valued function on $\mathbf{X}$. Then the following assertions are equivalent.*

1. *For each $c \in \mathbf{R}$, the set $\{x \in \mathbf{X} : f(x) > c\}$ is measurable.*
2. *For each $c \in \mathbf{R}$, the set $\{x \in \mathbf{X} : f(x) \geq c\}$ is measurable.*
3. *For each $c \in \mathbf{R}$, the set $\{x \in \mathbf{X} : f(x) < c\}$ is measurable.*
4. *For each $c \in \mathbf{R}$, the set $\{x \in \mathbf{X} : f(x) \leq c\}$ is measurable.*

Theorem 2.102. *Let $(\mathbf{X}, \mathscr{M})$ be a measurable space and f is a real-valued function on $\mathbf{X}$. Then f is measurable if and only if for each open set $\mathbf{O}$ of real numbers, the set $f^{-1}(\mathbf{O})$ is measurable.*

Theorem 2.103. *Let $(\mathbf{X}, \mathscr{M})$ be a measurable space and f and g be measurable real-valued functions. Then*

1. *$\alpha f + \beta g$ is measurable.*
2. *fg is measurable.*
3. *$\max\{f, g\}$ and $\min\{f, g\}$ are measurable.*

Theorem 2.104. *Let $(\mathbf{X}, \mathscr{M})$ be a measurable space, f is a measurable real-valued function on $\mathbf{X}$, $\phi : \mathbf{R} \longmapsto \mathbf{R}$ is continuous. Then $\phi \circ f : \mathbf{X} \longmapsto \mathbf{R}$ is measurable.*

Theorem 2.105. *Let $(\mathbf{X}, \mathscr{M}, \mu)$ be a measure space and $\{f_n\}_{n\in\mathbf{N}}$ be a sequence of measurable functions on $\mathbf{X}$ that converges pointwise to f on all of $\mathbf{X}$. Then f is measurable on $\mathbf{X}$.*

Definition 2.49. Let $(\mathbf{X}, \mathscr{M})$ be a measurable space and $\mathbf{E}$ be a measurable subset of $\mathbf{X}$. We define its characteristic function $\kappa_{\mathbf{E}}$ as follows:

$$\kappa_{\mathbf{E}}(x) = \begin{cases} 1 & \text{if } x \in \mathbf{E} \\ 0 & \text{if } x \in \mathbf{X} \backslash \mathbf{E}. \end{cases}$$

Definition 2.50. Let $(\mathbf{X}, \mathscr{M})$ be a measurable space. A real-valued function ψ on $\mathbf{X}$ is said to be simple, if there is a finite collection $\{\mathbf{E}_k\}_{k=1}^n$ of measurable sets and a corresponding set of real numbers $\{c_k\}_{k=1}^n$ for which

$$\psi = \sum_{k=1}^{n} c_k \kappa_{\mathbf{E}_k} \quad \text{on } \mathbf{X}.$$

Lemma 2.11 (The Simple Approximation Lemma). *Let $(\mathbf{X}, \mathscr{M})$ be a measurable space and f be a measurable function on $\mathbf{X}$ such that $|f| \leq M$ on $\mathbf{X}$ for some positive constant M. Then for each $\varepsilon > 0$, there are simple functions ϕ_ε and ψ_ε, such that*

$$\phi_\varepsilon \leq f \leq \psi_\varepsilon \quad \textit{and} \quad 0 \leq \psi_\varepsilon - \phi_\varepsilon < \varepsilon \quad \text{on } \mathbf{X}.$$

Theorem 2.106 (The Simple Approximation Theorem). *Let $(\mathbf{X}, \mathscr{M}, \mu)$ be a measure space and f be a measurable function on $\mathbf{X}$. Then there is a sequence $\{\phi_n\}_{n\in\mathbf{N}}$ of simple functions on $\mathbf{X}$ that converges pointwise on $\mathbf{X}$ to f and*

$$|\phi_n| \leq |f| \quad \text{on } \mathbf{X}$$

for any $n \in \mathbf{N}$. If $\mathbf{X}$ is σ-finite, then we may choose $\{\phi_n\}_{n\in\mathbf{N}}$ so that ϕ_n vanishes outside a set of finite measure. If f is nonnegative, we may choose $\{\phi_n\}_{n\in\mathbf{N}}$ to be increasing and $\psi_n \geq 0$ on $\mathbf{X}$.

Theorem 2.107 (Egoroff's Theorem). *Let $(\mathbf{X}, \mathscr{M}, \mu)$ be a finite measure space and $\{f_n\}_{n\in\mathbf{N}}$ be a sequence of measurable functions on $\mathbf{X}$ that converges pointwise a.e. on $\mathbf{X}$ to a function f that is finite a.e. on $\mathbf{X}$. Then for each $\varepsilon > 0$, there is a measurable subset $\mathbf{X}_\varepsilon$ of $\mathbf{X}$ for which $f_n \to f$, as $n \to \infty$, uniformly on $\mathbf{X}_\varepsilon$ and $\mu(\mathbf{X}\backslash\mathbf{X}_\varepsilon) < \varepsilon$.*

2.11 Integration Over General Measure Spaces

Definition 2.51. Let $(\mathbf{X}, \mathscr{M}, \mu)$ be a measure space and ψ be a nonnegative simple function on $\mathbf{X}$. If $\psi = 0$ on $\mathbf{X}$, we define

$$\int_{\mathbf{X}} \psi d\mu = 0.$$

Otherwise, let $c_1, \ldots, c_n$ be the positive values taken by ψ on $\mathbf{X}$ and

$$\mathbf{E}_k = \{x \in \mathbf{X} : \psi(x) = c_k\}, \quad k \in \{1, \ldots, n\}.$$

We define

$$\int_{\mathbf{X}} \psi d\mu = \sum_{k=1}^{n} c_k \mu(\mathbf{E}_k),$$

using the convention that the right-hand side is ∞ if $\mu(\mathbf{E}_k) = \infty$ for some $k \in \{1, \ldots, n\}$. For a measurable subset $\mathbf{E}$ of $\mathbf{X}$, we define

$$\int_{\mathbf{E}} \psi d\mu = \int_{\mathbf{X}} \psi \kappa_{\mathbf{E}} d\mu.$$

Remark 2.8. Note that the idea for the proof of the assertions in this section is the same as the idea for the proofs of the assertions in Section 2.6. Therefore we leave the proofs of the assertions in this section.

Theorem 2.108. *Let $(\mathbf{X}, \mathscr{M}, \mu)$ be a measure space and ϕ and ψ be nonnegative simple functions on $\mathbf{X}$.*

1. $\int_{\mathbf{X}} (\alpha\phi + \beta\psi)\, d\mu = \alpha \int_{\mathbf{X}} \phi d\mu + \beta \int_{\mathbf{X}} \psi d\mu$ *for any* $\alpha, \beta \in \mathbf{R}$.
2. *If $\mathbf{A}$ and $\mathbf{B}$ be disjoint measurable subsets of $\mathbf{X}$, then*

$$\int_{\mathbf{A}\cup\mathbf{B}} \phi d\mu = \int_{\mathbf{A}} \phi d\mu + \int_{\mathbf{B}} \phi d\mu.$$

3. *If $\mathbf{X}_0$ is a measurable subset of $\mathbf{X}$ and $\mu(\mathbf{X} \backslash \mathbf{X}_0) = 0$, then*

$$\int_{\mathbf{X}} \phi d\mu = \int_{\mathbf{X}_0} \phi d\mu.$$

4. *If $\phi \le \psi$ a.e. on $\mathbf{X}$, then*

$$\int_{\mathbf{X}} \phi d\mu \le \int_{\mathbf{X}} \psi d\mu.$$

Definition 2.52. Let $(\mathbf{X}, \mathscr{M}, \mu)$ be a measure space and f be a nonnegative extended real-valued measurable function on $\mathbf{X}$. We define

$$\int_{\mathbf{X}} f d\mu = \sup \left\{ \int_{\mathbf{X}} \psi d\mu : \psi \text{ is simple, } \ 0 \le \psi \le f \right\}.$$

For a measurable subset $\mathbf{E}$ of $\mathbf{X}$, we define

$$\int_{\mathbf{E}} f d\mu = \int_{\mathbf{X}} f \kappa_{\mathbf{E}} d\mu.$$

Theorem 2.109. *Let $(\mathbf{X}, \mathscr{M}, \mu)$ be a measure space, f and g be nonnegative measurable functions on $\mathbf{X}$.*

1. *If* $f \leq g$ *a.e. on* $\mathbf{X}$, *then*

$$\int_{\mathbf{X}} f d\mu \leq \int_{\mathbf{X}} g d\mu.$$

2. *If* $\mathbf{X}_0$ *is a measurable subset of* $\mathbf{X}$ *such that* $\mu(\mathbf{X} \backslash \mathbf{X}_0) = 0$, *then*

$$\int_{\mathbf{X}} f d\mu = \int_{\mathbf{X}_0} f d\mu.$$

Theorem 2.110 (Chebychev's Inequality). *Let* $(\mathbf{X}, \mathscr{M}, \mu)$ *be a measure space,* f *be nonnegative measurable function on* $\mathbf{X}$ *and* $\lambda > 0$. *Then*

$$\mu(\{x \in \mathbf{X} : f(x) \geq \lambda\}) \leq \frac{1}{\lambda} \int_{\mathbf{X}} f d\mu.$$

Theorem 2.111. *Let* $(\mathbf{X}, \mathscr{M}, \mu)$ *be a measure space and* f *be a nonnegative measurable function on* $\mathbf{X}$ *for which* $\int_{\mathbf{X}} f d\mu < \infty$. *Then* f *is finite a.e. on* $\mathbf{X}$ *and* $\{x \in \mathbf{X} : f(x) > 0\}$ *is* σ*-finite.*

Lemma 2.12 (Fatou's Lemma). *Let* $(\mathbf{X}, \mathscr{M}, \mu)$ *be a measure space and* $\{f_n\}_{n \in \mathbf{N}}$ *be a sequence of nonnegative measurable functions on* $\mathbf{X}$ *for which* $f_n \to f$, *as* $n \to \infty$, *pointwise a.e. on* $\mathbf{X}$. *Assume that* f *is measurable. Then*

$$\int_{\mathbf{X}} f d\mu \leq \liminf \int_{\mathbf{X}} f_n d\mu.$$

Theorem 2.112 (The Monotone Convergence Theorem). *Let* $(\mathbf{X}, \mathscr{M}, \mu)$ *be a measure space and* $\{f_n\}_{n \in \mathbf{N}}$ *be an increasing sequence of nonnegative measurable functions on* $\mathbf{X}$. *Define* $f(x) = \lim_{n \to \infty} f_n(x)$ *for each* $x \in \mathbf{X}$. *Then*

$$\lim_{n \to \infty} \int_{\mathbf{X}} f_n d\mu = \int_{\mathbf{X}} f d\mu.$$

Theorem 2.113 (Beppo Levi's Lemma). *Let* $(\mathbf{X}, \mathscr{M}, \mu)$ *be a measure space and* $\{f_n\}_{n \in \mathbf{N}}$ *be an increasing sequence of nonnegative measurable functions on* $\mathbf{X}$. *If the sequence*

$$\left\{ \int_{\mathbf{X}} f_n d\mu \right\}_{n \in \mathbf{N}}$$

is bounded, then $\{f_n\}_{n\in\mathbf{N}}$ *converges pointwise on* $\mathbf{X}$ *to a measurable function* f *that is finite a.e. on* $\mathbf{X}$ *and*

$$\lim_{n\to\infty}\int_{\mathbf{X}} f_n d\mu = \int_{\mathbf{X}} f d\mu < \infty.$$

Theorem 2.114. *Let* $(\mathbf{X}, \mathscr{M}, \mu)$ *be a measure space and* f *be a nonnegative measurable function on* $\mathbf{X}$*. Then there exists an increasing sequence* $\{\phi_n\}_{n\in\mathbf{N}}$ *of simple functions on* $\mathbf{X}$ *that converges pointwise on* $\mathbf{X}$ *to* f *and*

$$\lim_{n\to\infty}\int_{\mathbf{X}} \phi_n d\mu = \int_{\mathbf{X}} f d\mu.$$

Theorem 2.115. *Let* $(\mathbf{X}, \mathscr{M}, \mu)$ *be a measure space and* f *and* g *be nonnegative measurable functions on* $\mathbf{X}$*. Then*

$$\int_{\mathbf{X}} (\alpha f + \beta g)\, d\mu = \alpha\int_{\mathbf{X}} f d\mu + \beta\int_{\mathbf{X}} g d\mu$$

for any positive real numbers α *and* β*.*

Definition 2.53. Let $(\mathbf{X}, \mathscr{M}, \mu)$ be a measure space and f be a nonnegative measurable function on $\mathbf{X}$. Then f is said to be integrable over $\mathbf{X}$ with respect to μ if $\int_{\mathbf{X}} f d\mu < \infty$.

Definition 2.54. Let $(\mathbf{X}, \mathscr{M}, \mu)$ be a measure space. A measurable function f on $\mathbf{X}$ is said to be integrable over $\mathbf{X}$ with respect to μ if $|f|$ is integrable over $\mathbf{X}$ with respect to μ. For such a function, we define the integral of f over $\mathbf{X}$ with respect to μ by

$$\int_{\mathbf{X}} f d\mu = \int_{\mathbf{X}} f^+ d\mu - \int_{\mathbf{X}} f^- d\mu,$$

where $f^+ = \max\{f, 0\}$, $f^- = \max\{-f, 0\}$. For a measurable subset $\mathbf{E}$ of $\mathbf{X}$, we define

$$\int_{\mathbf{E}} f d\mu = \int_{\mathbf{X}} f\kappa_{\mathbf{E}} d\mu.$$

Theorem 2.116. *Let* $(\mathbf{X}, \mathscr{M}, \mu)$ *be a measure space and* f *be a measurable function on* $\mathbf{X}$*. If* g *is integrable over* $\mathbf{X}$ *and* $|f| \le g$ *a.e. on* $\mathbf{X}$*, then* f *is integrable over* $\mathbf{X}$ *and*

$$\left|\int_{\mathbf{X}} f d\mu\right| \le \int_{\mathbf{X}} |f| d\mu \le \int_{\mathbf{X}} g d\mu.$$

Theorem 2.117. *Let $(\mathbf{X}, \mathscr{M}, \mu)$ be a measure space and f and g be integrable over $\mathbf{X}$.*

1. $\int_{\mathbf{X}} (\alpha f + \beta g)\, d\mu = \alpha \int_{\mathbf{X}} f d\mu + \beta \int_{\mathbf{X}} g d\mu$ *for any real numbers* α *and* β.
2. *If* $f \leq g$ *a.e. on* $\mathbf{X}$, *then*

$$\int_{\mathbf{X}} f d\mu \leq \int_{\mathbf{X}} g d\mu.$$

3. *If* $\mathbf{A}$ *and* $\mathbf{B}$ *be disjoint measurable subsets of* $\mathbf{X}$, *then*

$$\int_{\mathbf{A}\cup\mathbf{B}} f d\mu = \int_{\mathbf{A}} f d\mu + \int_{\mathbf{B}} f d\mu.$$

Theorem 2.118. *Let $(\mathbf{X}, \mathscr{M}, \mu)$ be a measure space, the function f is integrable over $\mathbf{X}$, and $\{\mathbf{X}_n\}_{n\in\mathbf{N}}$ be a disjoint countable collection of measurable sets such that $\mathbf{X} = \cup_{n=1}^{\infty}\mathbf{X}_n$. Then*

$$\int_{\mathbf{X}} f d\mu = \sum_{n=1}^{\infty} \int_{\mathbf{X}_i} f d\mu.$$

Theorem 2.119. *Let $(\mathbf{X}, \mathscr{M}, \mu)$ be a measure space and the function f be integrable over $\mathbf{X}$.*

1. *If* $\{\mathbf{X}_n\}_{n\in\mathbf{N}}$ *is an ascending countable collection of measurable subsets of* $\mathbf{X}$, $\mathbf{X} = \cup_{n=1}^{\infty}\mathbf{X}_n$, *then*

$$\int_{\mathbf{X}} f d\mu = \lim_{n\to\infty} \int_{\mathbf{X}_n} f d\mu.$$

2. *If* $\{\mathbf{X}_n\}_{n\in\mathbf{N}}$ *be a descending countable collection of measurable subsets of* $\mathbf{X}$, *then*

$$\int_{\bigcap_{n=1}^{\infty}\mathbf{X}_n} f d\mu = \lim_{n\to\infty} \int_{\mathbf{X}_n} f d\mu.$$

Theorem 2.120. *Let $(\mathbf{X}, \mathscr{M}, \mu)$ be a measure space and f be a measurable function on $\mathbf{X}$. If f is bounded on $\mathbf{X}$ and vanishes outside a set of finite measure, then f is integrable over $\mathbf{X}$.*

Theorem 2.121 (The Lebesgue Dominated Convergence Theorem). *Let* $(\mathbf{X}, \mathscr{M}, \mu)$ *be a measure space and* $\{f_n\}_{n\in\mathbf{N}}$ *be a sequence of measurable functions on* $\mathbf{X}$ *for which* $f_n \to f$*, as* $n \to \infty$*, pointwise a.e. on* $\mathbf{X}$*, and the function* f *is measurable. Let also, there is a nonnegative function* g *that is integrable over* $\mathbf{X}$ *and*

$$|f_n| \leq g \quad a.e. \quad \text{on } \mathbf{X}$$

for any $n \in \mathbf{N}$*. Then* f *is integrable over* $\mathbf{X}$ *and*

$$\lim_{n\to\infty} \int_{\mathbf{X}} f_n d\mu = \int_{\mathbf{X}} f d\mu.$$

Definition 2.55. Let $(\mathbf{X}, \mathscr{M}, \mu)$ be a measure space and $\{f_n\}_{n\in\mathbf{N}}$ be a sequence of functions on $\mathbf{X}$, each of which is integrable over $\mathbf{X}$. The sequence $\{f_n\}_{n\in\mathbf{N}}$ is said to be uniformly integrable over $\mathbf{X}$ if for each $\varepsilon > 0$, there is a $\delta > 0$ such that for any $n \in \mathbf{N}$ and for any measurable subset $\mathbf{E}$ of $\mathbf{X}$ for which $\mu(\mathbf{E}) < \delta$, we have

$$\int_{\mathbf{E}} |f_n| \, d\mu < \varepsilon.$$

The sequence $\{f_n\}_{n\in\mathbf{N}}$ is said to be tight over $\mathbf{X}$ if for each $\varepsilon > 0$, there is a subset $\mathbf{X}_0$ of $\mathbf{X}$ that has finite measure and

$$\int_{\mathbf{X}\setminus\mathbf{X}_0} |f_n| \, d\mu < \varepsilon$$

for any $n \in \mathbf{N}$.

Theorem 2.122. *Let* $(\mathbf{X}, \mathscr{M}, \mu)$ *be a measure space and the function* f *is integrable over* $\mathbf{X}$*. Then for each* $\varepsilon > 0$*, there is a* $\delta > 0$ *such that for any measurable subset* $\mathbf{E}$ *of* $\mathbf{X}$ *for which* $\mu(\mathbf{E}) < \delta$*, we have*

$$\int_{\mathbf{E}} |f| \, d\mu < \varepsilon.$$

Moreover, for each $\varepsilon > 0$*, there is a subset* $\mathbf{X}_0$ *of* $\mathbf{X}$ *that has finite measure and*

$$\int_{\mathbf{X}\setminus\mathbf{X}_0} |f| \, d\mu < \varepsilon.$$

Theorem 2.123 (The Vitali Convergence Theorem). *Let* $(\mathbf{X}, \mathscr{M}, \mu)$ *be a measure space and* $\{f_n\}_{n\in\mathbf{N}}$ *be a sequence of functions on* $\mathbf{X}$ *that is uniformly integrable and tight over* $\mathbf{X}$. *Assume that* $f_n \to f$, *as* $n \to \infty$, *pointwise a.e. on* $\mathbf{X}$ *and the function* f *is integrable over* $\mathbf{X}$. *Then*

$$\lim_{n\to\infty} \int_{\mathbf{X}} f_n d\mu = \int_{\mathbf{X}} f d\mu.$$

2.12 Advanced Practical Problems

Problem 2.1. Let $\{\mathbf{E}_k\}_{k\in\mathbf{N}}$ be a countable disjoint subcollection of measurable sets and $\mathbf{A}$ be a measurable set. Prove that

$$m^\star\left(\mathbf{A} \cap \left(\bigcup k = 1^\infty \mathbf{E}_k\right)\right) = \sum_{k=1}^{\infty} m^\star\left(\mathbf{A} \cap \mathbf{E}_k\right).$$

Problem 2.2. Let f be a continuous function and $\mathbf{B}$ be a Borel set. Prove that $f^{-1}(\mathbf{B})$ is a Borel set.

Problem 2.3. Let f be a function with measurable domain $\mathbf{E}$. Prove that f is measurable if and only if the function g, defined on $\mathbf{R}$ by $g(x) = f(x)$ for $x \in \mathbf{E}$ and $g(x) = 0$ for $x \notin \mathbf{E}$, is measurable.

Problem 2.4. Let f be a measurable function on the measurable set $\mathbf{E}$ that is finite a.e. on $\mathbf{E}$ and $m(\mathbf{E}) < \infty$. Prove that for each $\varepsilon > 0$ there is a measurable set $\mathbf{E}_0$ contained in $\mathbf{E}$ such that f is bounded on $\mathbf{E}_0$ and $m(\mathbf{E}\backslash\mathbf{E}_0) < \varepsilon$.

Problem 2.5. Prove that the sum and product of two simple functions are simple.

Problem 2.6. Let f be a bounded measurable function on a set of finite measure $\mathbf{E}$. Assume that g is bounded and $f = g$ a.e. on $\mathbf{E}$. Prove that

$$\int_{\mathbf{E}} f = \int_{\mathbf{E}} g.$$

Problem 2.7. Let $\{f_n\}_{n\in\mathbf{N}}$ be a sequence of nonnegative measurable functions on $\mathbf{E}$ that converges pointwise on $\mathbf{E}$ to f. Suppose that $f_n \leq f$ on $\mathbf{E}$ for each $n \in \mathbf{N}$. Prove that

$$\lim_{n\to\infty} \int_{\mathbf{E}} f_n = \int_{\mathbf{E}} f.$$

Problem 2.8. Let f be a nonnegative measurable function on $\mathbf{R}$. Prove that

$$\lim_{n\to\infty} \int_{-n}^{n} f = \int_{\mathbf{R}} f.$$

Problem 2.9. Let $\{f_n\}_{n\in\mathbf{N}}$ be a sequence of nonnegative integrable functions on $\mathbf{E}$ and $f_n \to 0$, as $n \to \infty$, pointwise a.e. on $\mathbf{E}$. Prove that

$$\lim_{n\to\infty} \int_{\mathbf{E}} f_n = 0$$

if and only if $\{f_n\}_{n\in\mathbf{N}}$ is uniformly integrable and a tight over $\mathbf{E}$.

Chapter 3

The $\mathbf{L}^p$ Spaces

3.1 Definition

Let $\mathbf{E}$ be a measurable set and $\mathscr{F}$ be the collection of all measurable extended real-valued functions on $\mathbf{E}$ that are finite a.e. on $\mathbf{E}$.

Definition 3.1. Define two functions f and g in $\mathscr{F}$ to be equivalent, and we write $f \sim g$, provided

$$f(x) = g(x) \quad \text{for almost all } x \in \mathbf{E}.$$

The relation $\sim$ is an equivalent relation, that is, it is reflexive, symmetric and transitive. Therefore, it induces a partition of $\mathscr{F}$ into a disjoint collection of equivalence classes, which we denote by $\mathscr{F}/\sim$. For given two functions f and g in $\mathscr{F}$, their equivalence classes $[f]$ and $[g]$ and real numbers α and β, we define $\alpha[f]+\beta[g]$ to be the equivalence class of the functions in $\mathscr{F}$ that take the value $\alpha f(x)+\beta g(x)$ at points $x \in \mathbf{E}$ at which both f and g are finite. Note that these linear combinations are independent of the choice of the representatives of the equivalence classes. The zero element in $\mathscr{F}/\sim$ is the equivalence class of functions that vanish a.e. in $\mathbf{E}$. Thus $\mathscr{F}/\sim$ is a vector space.

Definition 3.2. For $1 \leq p < \infty$, we define $\mathbf{L}^p(\mathbf{E})$ to be the collection of equivalence classes $[f]$ for which

$$\int_{\mathbf{E}} |f|^p < \infty.$$

This is properly defined since if $f \sim g$, then

$$\int_{\mathbf{E}} |f|^p = \int_{\mathbf{E}} |g|^p.$$

Note that, if $[f], [g] \in \mathbf{L}^p(\mathbf{E})$, then for any real constants α and β, we have

$$\begin{aligned}\int_{\mathbf{E}} |\alpha f + \beta g|^p &\le 2^p \left(\int_{\mathbf{E}} |\alpha f|^p + \int_{\mathbf{E}} |\beta g|^p \right) \\ &= |2\alpha|^p \int_{\mathbf{E}} |f|^p + |2\beta|^p \int_{\mathbf{E}} |g|^p \\ &< \infty,\end{aligned}$$

i.e., $\alpha[f] + \beta[g] \in \mathbf{L}^p(\mathbf{E})$.

Definition 3.3. We call a function $f \in \mathscr{F}$ essentially bounded provided there is some $M \ge 0$, called an essential upper bound for f, for which

$$|f(x)| \le M \quad \text{for almost all } x \in \mathbf{E}.$$

Definition 3.4. We define $\mathbf{L}^\infty(\mathbf{E})$ to be the collection of equivalence classes $[f]$ for which f is essentially bounded.

Note that $\mathbf{L}^\infty(\mathbf{E})$ is properly defined since if $f \sim g$, then

$$|f(x)| = |g(x)| \le M \quad \text{for almost all } x \in \mathbf{E}.$$

Also, if $[f], [g] \in \mathbf{L}^\infty(\mathbf{E})$ and $\alpha, \beta \in \mathbf{R}$, there are nonnegative constants M_1 and M_2 such that

$$|f(x)| \le M_1, \quad |g(x)| \le M_2 \quad \text{for almost all } x \in \mathbf{E}.$$

Hence,

$$\begin{aligned}|\alpha f(x) + \beta g(x)| &\le |\alpha||f(x)| + |\beta||g(x)| \\ &\le |\alpha| M_1 + |\beta| M_2\end{aligned}$$

for almost all $x \in \mathbf{E}$. Therefore, $\alpha[f] + \beta[g] \in \mathbf{L}^\infty(\mathbf{E})$ and hence, $\mathbf{L}^\infty(\mathbf{E})$ is a vector space. For simplicity and convenience, we refer to the equivalence classes in $\mathscr{F}/\sim$ as functions and denote them by f rather than $[f]$. Thus $f = g$ means $f - g$ vanishes a.e. on $\mathbf{E}$.

Definition 3.5. For $1 \le p < \infty$, in $\mathbf{L}^p(\mathbf{E})$ we define

$$\|f\|_p = \left(\int_{\mathbf{E}} |f|^p\right)^{\frac{1}{p}}.$$

For $p = \infty$, we define $\|f\|_\infty$ to be the infimum of the essential upper bounds for f.

3.2 The Inequalities of Hölder and Minkowski

Theorem 3.1 (Hölder's Inequality). *Let* $\mathbf{E}$ *be a measurable set,* $1 \le p < \infty$ *and* q *be the conjugate of* p*. If* $f \in \mathbf{L}^p(\mathbf{E})$ *and* $g \in \mathbf{L}^q(\mathbf{E})$*, then the product* fg *is integrable over* $\mathbf{E}$ *and*

$$\int_{\mathbf{E}} |fg| \le \|f\|_p \|g\|_q. \tag{3.1}$$

Proof.

1. Let $p = 1$. Then $q = \infty$ and

$$\begin{aligned}\int_{\mathbf{E}} |fg| &= \int_{\mathbf{E}} |f||g| \\ &\le \left(\int_{\mathbf{E}} |f|\right) \|g\|_\infty \\ &= \|f\|_1 \|g\|_\infty.\end{aligned}$$

2. Let $p \in (1, \infty)$. If $\|f\|_p = 0$ or $\|g\|_q = 0$, then the assertion is evident. Let $\|f\|_p \neq 0$ and $\|g\|_q \neq 0$. We set

$$f_1 = \frac{f}{\|f\|_p}, \quad g_1 = \frac{g}{\|g\|_q}.$$

Then

$$\|f_1\|_p = \|g_1\|_q = 1.$$

Now, using Young's inequality, we have

$$|f_1 g_1| \le \frac{1}{p}|f_1|^p + \frac{1}{q}|g_1|^q.$$

Hence,

$$\begin{aligned}\frac{1}{\|f_1\|_p\|g_1\|_q}\int_{\mathbf{E}}|fg| &= \int_{\mathbf{E}}|f_1g_1| \\ &\le \int_{\mathbf{E}}\left(\frac{1}{p}|f_1|^p+\frac{1}{q}|g_1|^q\right) \\ &= \int_{\mathbf{E}}\frac{1}{p}|f_1|^p+\int_{\mathbf{E}}\frac{1}{q}|g_1|^q \\ &= \frac{1}{p}\int_{\mathbf{E}}|f_1|^p+\frac{1}{q}\int_{\mathbf{E}}|g_1|^q \\ &= \frac{1}{p}\|f_1\|_p^p+\frac{1}{q}\|g_1\|_q^q \\ &= \frac{1}{p}+\frac{1}{q} \\ &= 1.\end{aligned}$$

From the last inequality, we get the inequality (3.1). This completes the proof. □

Remark 3.1. When $p = q = 2$, the Hölder inequality is known as the Cauchy–Schwartz inequality.

Theorem 3.2 (Minkowski's Inequality). *Let* $\mathbf{E}$ *be a measurable set,* $1 \le p \le \infty$. *If* $f, g \in \mathbf{L}^p(\mathbf{E})$, *then*

$$\|f+g\|_p \le \|f\|_p + \|g\|_p.$$

Proof.

1. Let $p = 1$. Then

$$\begin{aligned}\|f+g\|_1 &= \int_{\mathbf{E}}|f+g| \\ &\le \int_{\mathbf{E}}(|f|+|g|) \\ &= \int_{\mathbf{E}}|f|+\int_{\mathbf{E}}|g| \\ &= \|f\|_1+\|g\|_1.\end{aligned}$$

2. Let $p = \infty$. Then

$$\|f + g\|_\infty \leq \|f\|_\infty + \|g\|_\infty.$$

3. Let $p \in (1, \infty)$ and q be its conjugate. If $\|f + g\|_p = 0$, then the assertion is evident. Assume that $\|f + g\|_p \neq 0$. Then, applying Hölder's inequality, we get

$$\begin{aligned}
\|f + g\|_p^p &= \int_{\mathbf{E}} |f + g|^p \\
&= \int_{\mathbf{E}} |f + g||f + g|^{p-1} \\
&= \int_{\mathbf{E}} (|f||f + g|^{p-1} + |g||f + g|^{p-1}) \\
&= \int_{\mathbf{E}} |f||f + g|^{p-1} + \int_{\mathbf{E}} |g||f + g|^{p-1} \\
&\leq \left(\int_{\mathbf{E}} |f|^p\right)^{\frac{1}{p}} \left(\int_{\mathbf{E}} |f + g|^{(p-1)q}\right)^{\frac{1}{q}} \\
&\quad + \left(\int_{\mathbf{E}} |g|^p\right)^{\frac{1}{p}} \left(\int_{\mathbf{E}} |f + g|^{(p-1)q}\right)^{\frac{1}{q}} \\
&= \|f + g\|_p^{\frac{p}{q}} \|f\|_p + \|f + g\|_p^{\frac{p}{q}} \|g\|_p.
\end{aligned}$$

Hence,

$$\|f + g\|_p^{p - \frac{p}{q}} \leq \|f\|_p + \|g\|_p$$

or

$$\|f + g\|_p \leq \|f\|_p + \|g\|_p.$$

This completes the proof. □

3.3 Some Properties

Theorem 3.3. *Let* $\mathbf{E}$ *be a measurable set and* $1 < p < \infty$. *Suppose that* $\mathscr{F}$ *is a family of functions in* $\mathbf{L}^p(\mathbf{E})$ *that is bounded in* $\mathbf{L}^p(\mathbf{E})$

in the sense that there is a constant $M > 0$ *such that*

$$\|f\|_p \leq M \quad \textit{for all } f \in \mathscr{F}.$$

Then the family $\mathscr{F}$ *is uniformly integrable over* $\mathbf{E}$.

Proof. Let $\varepsilon > 0$ be arbitrarily chosen. Suppose that $\mathbf{A}$ is a measurable subset of $\mathbf{E}$ of finite measure. Let also, $\frac{1}{p} + \frac{1}{q} = 1$. Define g to be identically equal to 1 on $\mathbf{A}$. Because $m(\mathbf{A}) < \infty$, we have that $g \in \mathbf{L}^q(\mathbf{A})$. By Hölder's inequality, for any $f \in \mathscr{F}$, we have

$$\begin{aligned}\int_{\mathbf{A}} |f| &= \int_{\mathbf{A}} |f| g \\ &\leq \left(\int_{\mathbf{A}} |f|^p\right)^{\frac{1}{p}} \left(\int_{\mathbf{A}} g^q\right)^{\frac{1}{q}}.\end{aligned}$$

On the other hand,

$$\left(\int_{\mathbf{A}} |f|^p\right)^{\frac{1}{p}} \leq M, \quad \left(\int_{\mathbf{A}} |g|^q\right)^{\frac{1}{q}} = (m(\mathbf{A}))^{\frac{1}{q}}$$

for any $f \in \mathscr{F}$. Therefore,

$$\int_{\mathbf{A}} |f| \leq (m(\mathbf{A}))^{\frac{1}{q}} M$$

for any $f \in \mathscr{F}$. Let $\delta = (\frac{\varepsilon}{M})^q$. Hence, if $m(\mathbf{A}) < \delta$, then

$$\begin{aligned}\int_{\mathbf{A}} |f| &\leq M(m(\mathbf{A}))^{\frac{1}{q}} \\ &< M\delta^{\frac{1}{q}} \\ &= M\frac{\varepsilon}{M} \\ &= \varepsilon\end{aligned}$$

for any $f \in \mathscr{F}$. Therefore $\mathscr{F}$ is uniformly integrable over $\mathbf{E}$. This completes the proof. □

Theorem 3.4. *Let* $\mathbf{E}$ *be a measurable set of finite measure and* $1 \leq p_1 < p_2 \leq \infty$. *Then*

$$\mathbf{L}^{p_2}(\mathbf{E}) \subseteq \mathbf{L}^{p_1}(\mathbf{E}). \tag{3.2}$$

Proof. Let $p_2 < \infty$ and $f \in \mathbf{L}^{p_2}(\mathbf{E})$ be arbitrarily chosen. Then $p = \frac{p_2}{p_1} > 1$. We take $q > 1$ such that $\frac{1}{p} + \frac{1}{q} = 1$. Then

$$\left(\int_{\mathbf{E}} |f|^{p_1 p}\right)^{\frac{1}{p}} = \left(\int_{\mathbf{E}} |f|^{p_2}\right)^{\frac{p_1}{p_2}} < \infty,$$

i.e., $|f|^{p_1} \in \mathbf{L}^p(\mathbf{E})$. Let $g = \kappa_{\mathbf{E}}$. Since $m(\mathbf{E}) < \infty$, then $g \in \mathbf{L}^q(\mathbf{E})$. Hence, using Hölder's inequality, we get

$$\begin{aligned}\int_{\mathbf{E}} |f|^{p_1} &= \int_{\mathbf{E}} |f|^{p_1} g \\ &\le \left(\int_{\mathbf{E}} |f|^{p_1 p}\right)^{\frac{1}{p}} \left(\int_{\mathbf{E}} g^q\right)^{\frac{1}{q}} \\ &= \|f\|_{p_2}^{p_1} (m(\mathbf{E}))^{\frac{1}{q}} \\ &< \infty.\end{aligned}$$

Therefore, $f \in \mathbf{L}^{p_1}(\mathbf{E})$. Because $f \in \mathbf{L}^{p_2}(\mathbf{E})$ was arbitrarily chosen and we get that it is an element of $\mathbf{L}^{p_1}(\mathbf{E})$, we obtain the relation (3.2). Let $p_2 = \infty$ and $f \in \mathbf{L}^{\infty}(\mathbf{E})$ be arbitrarily chosen. Then there is a positive constant M such that

$$|f(x)| \le M$$

for almost all $x \in \mathbf{E}$. Hence,

$$\int_{\mathbf{E}} |f|^{p_1} \le M^{p_1} m(\mathbf{E}).$$

Then $f \in \mathbf{L}^{p_1}(\mathbf{E})$. Because $f \in \mathbf{L}^{\infty}(\mathbf{E})$ was arbitrarily chosen and we get that it is an element of $\mathbf{L}^{p_1}(\mathbf{E})$, we obtain the relation (3.2). This completes the proof. □

Remark 3.2. Let $\{f_n\}_{n\in\mathbb{N}}$ be a sequence of elements of $\mathbf{L}^p(\mathbf{E})$ and $f \in \mathbf{L}^p(\mathbf{E})$. When $\|f_n - f\|_p \to 0$, as $n \to \infty$, we will say that the sequence $\{f_n\}_{n\in\mathbf{N}}$ converges to f in $\mathbf{L}^p(\mathbf{E})$.

3.4 The Riesz–Fischer Theorem

Definition 3.6. Let $\mathbf{X}$ be a normed vector space. A sequence $\{f_n\}_{n\in\mathbf{N}}$ in $\mathbf{X}$ is said to be a rapidly Cauchy sequence provided there is a convergent series of positive numbers $\sum_{k=1}^{\infty} \varepsilon_k$ for which

$$\|f_{k+1} - f_k\| \le \varepsilon_k^2$$

for all $k \in \mathbf{N}$.

Lemma 3.1. *Let $\mathbf{X}$ be a normed vector space and $\{f_n\}_{n\in\mathbf{N}}$ be a sequence in $\mathbf{X}$ such that*

$$\|f_{k+1} - f_k\| \le a_k$$

for all $k \in \mathbf{N}$, where a_k, $k \in \mathbf{N}$, are nonnegative numbers. Then

$$\|f_{n+k} - f_n\| \le \sum_{l=n}^{\infty} a_l \tag{3.3}$$

for all $k, n \in \mathbf{N}$.

Proof. For any $n, k \in \mathbf{N}$, we have

$$\begin{aligned} f_{n+k} - f_n &= f_{n+k} - f_{n+k-1} + f_{n+k-1} - \cdots + f_{n+1} - f_n \\ &= \sum_{j=n}^{n+k-1} (f_{j+1} - f_j). \end{aligned}$$

Hence,

$$\begin{aligned} \|f_{n+k} - f_n\| &= \left\| \sum_{j=n}^{n+k-1} (f_{j+1} - f_j) \right\| \\ &\le \sum_{j=n}^{n+k-1} \|f_{j+1} - f_j\| \\ &\le \sum_{j=n}^{n+k-1} a_j \\ &\le \sum_{j=n}^{\infty} a_j \end{aligned}$$

for any $k, n \in \mathbf{N}$. This completes the proof. □

Theorem 3.5. *Let* $\mathbf{X}$ *be a normed vector space. Then every rapidly Cauchy sequence in* $\mathbf{X}$ *is a Cauchy sequence in* $\mathbf{X}$. *Furthermore, every Cauchy sequence in* $\mathbf{X}$ *has a rapidly Cauchy subsequence in* $\mathbf{X}$.

Proof.

1. Let $\{f_n\}_{n\in\mathbf{N}}$ be a rapidly Cauchy sequence in $\mathbf{X}$. Then there is a convergent series $\sum_{k=1}^{\infty} \varepsilon_k$ of positive numbers such that

$$\|f_{k+1} - f_k\| \leq \varepsilon_k^2$$

for any $k \in \mathbf{N}$. Hence and (3.3), we obtain

$$\begin{aligned} \|f_{n+k} - f_k\| &\leq \sum_{l=n}^{\infty} \varepsilon_l^2 \\ &\to 0, \quad \text{as} \quad n \to \infty, \end{aligned}$$

for any $k \in \mathbf{N}$. Here, we have used that the series $\sum_{k=1}^{\infty} \varepsilon_k^2$ is a convergent series. Therefore, $\{f_n\}_{n\in\mathbf{N}}$ is a Cauchy sequence.

2. Let $\{f_n\}_{n\in\mathbf{N}}$ is a Cauchy sequence in $\mathbf{X}$. Then there are $M_1, M_2, M_3 \in \mathbf{N}$ such that

$$\begin{aligned} \left\|f_{m_1^1} - f_{m_2^1}\right\| &< \frac{1}{2} \quad \text{for any } m_1^1, m_2^1 \geq M_1, \\ \left\|f_{m_1^2} - f_{m_2^2}\right\| &< \frac{1}{2^2} \quad \text{for any } m_1^2, m_2^2 \geq M_2, \\ \left\|f_{m_1^3} - f_{m_2^3}\right\| &< \frac{1}{2^3} \quad \text{for any } m_1^3, m_2^3 \geq M_3. \end{aligned}$$

In particular, for $m_{|}^{|} \geq \max\{M_1, M_2, M_3\}$, $m_2^1 \geq \max\{M_1$ $M_2, M_3\}$, $m_2^3 \in \{M_1, M_2, M_3\}$, we have

$$\begin{aligned} \left\|f_{m_1^1} - f_{m_2^1}\right\| &< \frac{1}{2}, \\ \left\|f_{m_2^3} - f_{m_1^1}\right\| &< \frac{1}{2^2}, \\ \left\|f_{m_1^3} - f_{m_2^3}\right\| &< \frac{1}{2^3}. \end{aligned}$$

We set

$$n_1 = m_2^1, \quad n_2 = m_1^1, \quad n_3 = m_2^3$$

for $m_1^1 \geq \max\{M_1, M_2, M_3\}$, $m_2^1 \geq \max\{M_1 M_2, M_3\}$, $m_2^3 \in \{M_1, M_2, M_3\}$. Then

$$\|f_{n_2} - f_{n_1}\| < \frac{1}{2},$$

$$\|f_{n_3} - f_{n_2}\| < \frac{1}{2^2},$$

$$\left\|f_{m_1^3} - f_{n_3}\right\| < \frac{1}{2^3}$$

for $m_1^3 \geq \max\{M_1, M_2, M_3\}$. Continuing this process, we obtain a subsequence $\{f_{n_k}\}_{k\in\mathbf{N}}$ of the sequence $\{f_n\}_{n\in\mathbf{N}}$ such that

$$\|f_{n_{k+1}} - f_{n_k}\| < \frac{1}{2^k}$$

for any $k \in \mathbf{N}$. Since $\sum_{k=0}^{\infty} \frac{1}{(\sqrt{2})^k}$ is convergent, we conclude that $\{f_{n_k}\}_{k\in\mathbf{N}}$ is a rapidly Cauchy sequence in $\mathbf{X}$. This completes the proof. □

Theorem 3.6. *Let $\mathbf{E}$ be a measurable set and $1 \leq p \leq \infty$. Then every rapidly Cauchy sequence in $\mathbf{L}^p(\mathbf{E})$ converges both with respect to the $\mathbf{L}^p(\mathbf{E})$ norm and pointwise a.e. in $\mathbf{E}$ to a function f in $\mathbf{L}^p(\mathbf{E})$.*

Proof. Let $1 \leq p < \infty$. We leave the case $p = \infty$ as an exercise. Let $\{f_n\}_{n\in\mathbf{N}}$ be a rapidly convergent sequence in $\mathbf{L}^p(\mathbf{E})$. We choose $\sum_{k=1}^{\infty} \varepsilon_k$ to be a convergent series of positive numbers such that

$$\|f_{k+1} - f_k\|_p \leq \varepsilon_k^2 \tag{3.4}$$

for any $k \in \mathbf{N}$. Let

$$\mathbf{E}^k = \{x \in \mathbf{E} : |f_{k+1}(x) - f_k(x)| \geq \varepsilon_k\}, \quad k \in \mathbf{N}.$$

Then

$$\mathbf{E}^k = \{x \in \mathbf{E} : |f_{k+1}(x) - f_k(x)|^p \geq \varepsilon_k^p\}, \quad k \in \mathbf{N}.$$

Hence,

$$\int_{\mathbf{E}_k} |f_{k+1}(x) - f_k(x)|^p \geq \varepsilon_k^p m(\mathbf{E}^k).$$

Now, using (3.4), we get

$$\begin{aligned} m(\mathbf{E}^k) &\le \frac{1}{\varepsilon_k^p} \int_{\mathbf{E}} |f_{k+1} - f_k|^p \\ &= \frac{1}{\varepsilon_k^p} \|f_{k+1} - f_k\|_p^p \\ &\le \frac{1}{\varepsilon_k^p} \varepsilon_k^{2p} \\ &= \varepsilon_k^p. \end{aligned}$$

Since $p \ge 1$, the series $\sum_{k=1}^{\infty} \varepsilon_k^p$ is convergent and

$$\sum_{k=1}^{\infty} m(\mathbf{E}^k) \le \sum_{k=1}^{\infty} \varepsilon_k^p < \infty.$$

Hence and the Borel–Cantelli Lemma, it follows that there is $\mathbf{E}_0 \subset \mathbf{E}$ such that $m(\mathbf{E}_0) = 0$ and for each $x \in \mathbf{E} \backslash \mathbf{E}_0$ there is an index $K(x)$ such that

$$|f_{k+1}(x) - f_k(x)| < \varepsilon_k$$

for any $k \ge K(x)$. Let $x \in \mathbf{E} \backslash \mathbf{E}_0$. Then

$$\begin{aligned} |f_{n+k}(x) - f_n(x)| &\le \sum_{j=n}^{n+k-1} |f_{j+1}(x) - f_j(x)| \\ &\le \sum_{j=n}^{\infty} \varepsilon_j \end{aligned}$$

for all $n \ge K(x)$ and for any $k \in \mathbf{N}$. Hence, the sequence of real numbers $\{f_n(x)\}_{n \in \mathbf{N}}$ is a Cauchy sequence in $\mathbf{R}$ for any $x \in \mathbf{E} \backslash \mathbf{E}_0$. Therefore, it is convergent for any $x \in \mathbf{E} \backslash \mathbf{E}_0$ and let $\lim_{n \to \infty} f_n(x) = f(x)$ for any $x \in \mathbf{E} \backslash \mathbf{E}_0$. By (3.3) and (3.4), we obtain

$$\int_{\mathbf{E}} |f_{n+k} - f_n|^p \le \left(\sum_{j=n}^{\infty} \varepsilon_j^2 \right)^p \tag{3.5}$$

for all $n, k \in \mathbf{N}$. Since $f_n \to f$ pointwise a.e. on $\mathbf{E}$, we take the limit as $k \to \infty$ in (3.5) and using Fatou's Lemma, we get

$$\int_{\mathbf{E}} |f - f_n|^p \leq \left(\sum_{j=n}^{\infty} \varepsilon_j^2 \right)^p$$

for all $n \in \mathbf{N}$. Hence, $f \in \mathbf{L}^p(\mathbf{E})$ and $f_n \to f$, as $n \to \infty$, in $\mathbf{L}^p(\mathbf{E})$. This completes the proof. □

Theorem 3.7 (The Riesz–Fischer Theorem). *Let $\mathbf{E}$ be a measurable set and $1 \leq p \leq \infty$. Then $\mathbf{L}^p(\mathbf{E})$ is a Banach space. Moreover, if $f_n \to f$, as $n \to \infty$, in $\mathbf{L}^p(\mathbf{E})$, a subsequence of $\{f_n\}_{n\in\mathbf{N}}$ converges pointwise a.e. on $\mathbf{E}$ to f.*

Proof. Let $\{f_n\}_{n\in\mathbf{N}}$ be a Cauchy sequence in $\mathbf{L}^p(\mathbf{E})$. By Theorem 3.5, it follows that there is a subsequence $\{f_{n_k}\}_{k\in\mathbf{N}}$ of the sequence $\{f_n\}_{n\in\mathbf{N}}$ that is a rapidly Cauchy sequence in $\mathbf{L}^p(\mathbf{E})$. Hence and Theorem 3.6, it follows that $\{f_{n_k}\}_{k\in\mathbf{N}}$ converges both with respect to the $\mathbf{L}^p(\mathbf{E})$ norm and pointwise a.e. on $\mathbf{E}$. Let $f_{n_k} \to f$, as $k \to \infty$, in $\mathbf{L}^p(\mathbf{E})$. We take $\varepsilon > 0$ arbitrarily. Then there exists $K \in \mathbf{N}$ such that

$$\|f_{n_k} - f\|_p < \frac{\varepsilon}{2} \quad \text{and} \quad \|f_l - f_m\|_p < \frac{\varepsilon}{2}$$

for any $l, m, n_k \geq K$. Hence, for any $n, n_k \geq K$, we have

$$\begin{aligned}
\|f_n - f\|_p &= \|f_n - f_{n_k} + f_{n_k} - f\|_p \\
&\leq \|f_n - f_{n_k}\|_p + \|f_{n_k} - f\|_p \\
&< \frac{\varepsilon}{2} + \frac{\varepsilon}{2} \\
&= \varepsilon.
\end{aligned}$$

Therefore, $\{f_n\}_{n\in\mathbf{N}}$ is convergent in $\mathbf{L}^p(\mathbf{E})$ and $\mathbf{L}^p(\mathbf{E})$ is a Banach space. This completes the proof. □

Definition 3.7. A real-valued function f, defined on a set $\mathbf{A}$, is called convex if for each pair of points $x_1, x_2 \in \mathbf{A}$ and for each $\lambda \in [0, 1]$, we have

$$f(\lambda x_1 + (1 - \lambda)x_2) \leq \lambda f(x_1) + (1 - \lambda) f(x_2).$$

Theorem 3.8. *Let* $\mathbf{E}$ *be a measurable set and* $1 \le p < \infty$. *Suppose that* $\{f_n\}_{n\in\mathbf{N}}$ *is a sequence in* $\mathbf{L}^p(\mathbf{E})$ *that converges pointwise a.e. on* $\mathbf{E}$ *to the function* $f \in \mathbf{L}^p(\mathbf{E})$. *Then* $f_n \to f$, *as* $n \to \infty$, *in* $\mathbf{L}^p(\mathbf{E})$ *if and only if*

$$\lim_{n\to\infty} \int_{\mathbf{E}} |f_n|^p = \int_{\mathbf{E}} |f|^p. \tag{3.6}$$

Proof.

1. Let $f_n \to f$, as $n \to \infty$, in $\mathbf{L}^p(\mathbf{E})$. Then, by Minkowski's inequality, we have

$$\begin{aligned} |\|f_n\|_p - \|f\|_p| &\le \|f_n - f\|_p \\ &\to 0, \quad \text{as } n \to \infty. \end{aligned}$$

   Hence, (3.6) holds.
2. Assume (3.6). Let $\mathbf{E}_0 \subset \mathbf{E}$ be such that $m(\mathbf{E}_0) = 0$ and $f_n(x) \to f(x)$, as $n \to \infty$, for all $x \in \mathbf{E}\backslash\mathbf{E}_0$. Take $\psi(t) = |t|^p$, $t \in \mathbf{R}$. Then ψ is a convex function and

$$\psi\left(\frac{a+b}{2}\right) \le \frac{\psi(a)}{2} + \frac{\psi(b)}{2}$$

   for any $a, b \in \mathbf{R}$. Hence,

$$0 \le \frac{|a|^p + |b|^p}{2} - \left|\frac{a-b}{2}\right|^p$$

   for any $a, b \in \mathbf{R}$. For each $n \in \mathbf{N}$, we define the function

$$h_n(x) = \frac{|f_n(x)|^p + |f(x)|^p}{2} - \left|\frac{f_n(x) - f(x)}{2}\right|^p$$

   for any $x \in \mathbf{E}$. We have that $h_n(x) \to |f(x)|^p$, as $n \to \infty$, for any $x \in \mathbf{E}\backslash\mathbf{E}_0$. Hence and Fatou's Lemma, we obtain

$$\begin{aligned} \int_{\mathbf{E}} |f|^p &\le \liminf \int_{\mathbf{E}} h_n \\ &= \liminf \left(\int_{\mathbf{E}} \frac{|f_n|^p + |f|^p}{2} - \int_{\mathbf{E}} \left|\frac{f_n - f}{2}\right|^p \right) \\ &= \int_{\mathbf{E}} |f|^p - \limsup \int_{\mathbf{E}} \left|\frac{f_n - f}{2}\right|^p. \end{aligned}$$

Therefore,

$$\limsup \int_{\mathbf{E}} \left| \frac{f_n - f}{2} \right|^p \leq 0.$$

That is $f_n \to f$, as $n \to \infty$, in $\mathbf{L}^p(\mathbf{E})$. This completes the proof. □

Theorem 3.9. *Let* $\mathbf{E}$ *be a measurable set and* $1 \leq p < \infty$. *Suppose that* $\{f_n\}_{n\in\mathbf{N}}$ *is a sequence in* $\mathbf{L}^p(\mathbf{E})$ *that converges pointwise a.e. on* $\mathbf{E}$ *to the function* f *which belongs to* $\mathbf{L}^p(\mathbf{E})$. *Then* $f_n \to f$, *as* $n \to \infty$, *in* $\mathbf{L}^p(\mathbf{E})$, *if and only if* $\{|f_n|^p\}_{n\in\mathbf{N}}$ *is uniformly integrable and a tight over* $\mathbf{E}$.

Proof.

1. Let $f_n \to f$, as $n \to \infty$, in $\mathbf{L}^p(\mathbf{E})$. Hence,

$$\lim_{n\to\infty} \int_{\mathbf{E}} |f_n - f|^p = 0.$$

Therefore, $\{|f_n - f|^p\}_{n\in\mathbf{N}}$ is uniformly integrable and a tight over $\mathbf{E}$. Because

$$|f_n|^p \leq 2^p(|f_n - f|^p + |f|^p), \quad n \in \mathbf{N},$$

we conclude that $\{|f_n|^p\}_{n\in\mathbf{N}}$ is uniformly integrable and a tight over $\mathbf{E}$.

2. Let $\{|f_n|^p\}_{n\in\mathbf{N}}$ is uniformly integrable and a tight over $\mathbf{E}$. Since $f_n \to f$, as $n \to \infty$, pointwise a.e. on $\mathbf{E}$, we have that $|f_n|^p \to |f|^p$, as $n \to \infty$, pointwise a.e. on $\mathbf{E}$. Hence and the Vitalli Convergence Theorem, we conclude that $|f|^p$ is integrable over $\mathbf{E}$ and

$$\lim_{n\to\infty} \int_{\mathbf{E}} |f_n|^p = \int_{\mathbf{E}} |f|^p.$$

This completes the proof. □

Theorem 3.10 (The $\mathbf{L}^p$ Dominated Convergence Theorem). *Let* $\{f_n\}_{n\in\mathbf{N}}$ *be a sequence of measurable functions that converges pointwise a.e. on* $\mathbf{E}$ *to* f. *For* $1 \leq p < \infty$, *suppose that there is a function* $g \in \mathbf{L}^p(\mathbf{E})$ *such that* $|f_n| \leq g$ *a.e. on* $\mathbf{E}$ *for all* $n \in \mathbf{N}$. *Then* $f_n \to f$, *as* $n \to \infty$, *in* $\mathbf{L}^p(\mathbf{E})$.

Proof. Since $f_n \to f$, as $n \to \infty$, pointwise a.e. on $\mathbf{E}$, we have that $|f_n|^p \to |f|^p$, as $n \to \infty$, pointwise a.e. on $\mathbf{E}$. Also, $|f_n|^p \le g^p$ a.e. on $\mathbf{E}$ and g^p is integrable over $\mathbf{E}$. Hence and the Lebesgue Dominated Convergence Theorem, we conclude that $|f|^p$ is integrable over $\mathbf{E}$ and

$$\lim_{n\to\infty} \int_{\mathbf{E}} |f_n|^p = \int_{\mathbf{E}} |f|^p.$$

This completes the proof. □

3.5 Separability

Theorem 3.11. *Let $\mathbf{E}$ be a measurable set and $1 \le p \le \infty$. Then the subspace of simple functions in $\mathbf{L}^p(\mathbf{E})$ is dense in $\mathbf{L}^p(\mathbf{E})$.*

Proof. Let $g \in \mathbf{L}^p(\mathbf{E})$.

1. Let $p = \infty$. Then there is $\mathbf{E}_0 \subset \mathbf{E}$ such that $m(\mathbf{E}_0) = 0$ and g is bounded on $\mathbf{E}\backslash\mathbf{E}_0$. By Theorem 2.46, it follows that there is a sequence $\{f_n\}_{n\in\mathbf{N}}$ of simple functions on $\mathbf{E}\backslash\mathbf{E}_0$ that converges uniformly on $\mathbf{E}\backslash\mathbf{E}_0$ to g and therefore with respect to the $\mathbf{L}^\infty(\mathbf{E})$ norm. Consequently the subspace of simple functions in $\mathbf{L}^\infty(\mathbf{E})$ is dense in $\mathbf{L}^\infty(\mathbf{E})$.
2. Let $1 \le p < \infty$. By Theorem 2.47, it follows that there is a sequence of simple functions $\{f_n\}_{n\in\mathbf{N}}$ on $\mathbf{E}$ such that $f_n \to g$, as $n \to \infty$, pointwise on $\mathbf{E}$ and

   $$|f_n| \le |g| \quad \text{on } \mathbf{E}$$

   for any $n \in \mathbf{N}$. Since $g \in \mathbf{L}^p(\mathbf{E})$, we have that $f_n \in \mathbf{L}^p(\mathbf{E})$ for any $n \in \mathbf{N}$. Next,

   $$\begin{aligned} |f_n - g| &\le |f_n| + |g| \\ &\le 2|g|. \end{aligned}$$

   Hence and Theorem 3.10, it follows that $f_n \to g$, as $n \to \infty$, in $\mathbf{L}^p(\mathbf{E})$. This completes the proof. □

Theorem 3.12. *Let $[a, b]$ be a closed bounded interval and $1 \le p < \infty$. Then the subspace of the step functions on $[a, b]$ is dense in $\mathbf{L}^p([a, b])$.*

Proof. Let $\mathbf{A}$ be a measurable subset of $[a, b]$. Let also, $g = \kappa_{\mathbf{A}}$. Take $\varepsilon > 0$ arbitrarily. By Theorem 2.23, it follows that there is a finite disjoint collection $\{\mathbf{I}_k\}_{k=1}^n$ of open intervals such that if $\mathbf{U} = \bigcup_{k=1}^n \mathbf{I}_k$, then

$$m(\mathbf{A}\backslash\mathbf{U}) + m(\mathbf{U}\backslash\mathbf{A}) < \varepsilon^p.$$

Let $f = \kappa_{\mathbf{U}}$. Then

$$\begin{aligned}\|\kappa_{\mathbf{U}} - \kappa_{\mathbf{A}}\|_p &= \left(\int_{\mathbf{E}} |\kappa_{\mathbf{U}} - \kappa_{\mathbf{A}}|^p\right)^{\frac{1}{p}} \\ &= (m(\mathbf{A}\backslash\mathbf{U}) + m(\mathbf{U}\backslash\mathbf{A}))^{\frac{1}{p}} < \varepsilon.\end{aligned}$$

Therefore, the step functions are dense in the simple functions with respect to the $\mathbf{L}^p$ norm. Hence and Theorem 3.11, it follows that the step functions on $[a, b]$ are dense in $\mathbf{L}^p([a, b])$. This completes the proof. □

Theorem 3.13. *Let $\mathbf{E}$ be a measurable set and $1 \leq p < \infty$. Then $\mathbf{L}^p(\mathbf{E})$ is separable.*

Proof. Let $[a, b]$ be a closed bounded interval and $\mathbf{S}([a, b])$ be the collection of the step functions on $[a, b]$. Let also, $\mathbf{S}^1([a, b])$ be the subcollection of the collection $\mathbf{S}([a, b])$ consisting of the step functions ψ on $[a, b]$ that take rational values and for which there is a partition $\mathbf{P} = \{x_0, \ldots, x_n\}$ of $[a, b]$ so that ψ is a rational constant on (x_{k-1}, x_k), $1 \leq k \leq n$, and x_k, $1 \leq k \leq n-1$, are rational numbers. Using the density of the rational numbers in the real numbers, we have that $\mathbf{S}^1([a, b])$ is dense in $\mathbf{S}([a, b])$ with respect to the $\mathbf{L}^p(\mathbf{E})$ norm. Because the set of the rational numbers is countable, we have that $\mathbf{S}^1([a, b])$ is a countable set. We have that

$$\mathbf{S}^1([a, b]) \subseteq \mathbf{S}([a, b]) \subseteq \mathbf{L}^p([a, b]).$$

Since $\mathbf{S}^1([a, b])$ is dense in $\mathbf{S}([a, b])$, using Theorem 3.12, it follows that $\mathbf{S}^1([a, b])$ is dense in $\mathbf{L}^p([a, b])$. For each $n \in \mathbf{N}$, we define by $\mathscr{F}_n$ to be the collection of the functions that vanishes outside $[-n, n]$

and whose restrictions to $[-n, n]$ belong to $\mathbf{S}^1([-n, n])$. Let

$$\mathscr{F} = \bigcup_{n=1}^{\infty} \mathscr{F}_n.$$

Note that $\mathscr{F}$ is a countable collection of the functions that is dense in $\mathbf{L}^p(\mathbf{R})$. Also, using Theorem 2.80, we have

$$\lim_{n\to\infty} \int_{[-n,n]} |f|^p = \int_{\mathbf{R}} |f|^p$$

for all $f \in \mathbf{L}^p(\mathbf{R})$. Using the definition of $\mathscr{F}$, we conclude that $\mathscr{F}$ is a countable collection of functions that is dense in $\mathbf{L}^p(\mathbf{R})$. Hence, the collection of the restrictions on $\mathbf{E}$ of functions in $\mathscr{F}$ is a countable dense set of $\mathbf{L}^p(\mathbf{E})$. Consequently $\mathbf{L}^p(\mathbf{E})$ is separable. This completes the proof. □

3.6 Duality

Definition 3.8. For a normed vector space $\mathbf{X}$, a linear functional $\mathbb{T}$ on $\mathbf{X}$ is said to be bounded if there is an $M \geq 0$ such that

$$|\mathbb{T}(f)| \leq M\|f\| \tag{3.7}$$

for all $f \in \mathbf{X}$. The infimum of all such M will be called the norm of $\mathbb{T}$ and will be denoted by $\|\mathbb{T}\|_\star$.

Let $\mathbb{T}$ be a bounded linear functional on the normed vector space $\mathbf{X}$. Then, for any $f, g \in \mathbf{X}$, we have

$$|\mathbb{T}(f) - \mathbb{T}(g)| \leq \|f - g\|. \tag{3.8}$$

Hence, if $f_n \to f$, as $n \to \infty$, in $\mathbf{X}$, i.e., $f_n, f \in \mathbf{X}$, $n \in \mathbf{N}$,

$$\|f_n - f\| \to 0, \quad \text{as } n \to \infty,$$

using (3.8), we get

$$\mathbb{T}(f_n) \to \mathbb{T}(f), \quad \text{as } n \to \infty.$$

Proposition 3.1. *Let $\mathbb{T}$ be a bounded linear functional on the normed vector space $\mathbf{X}$. Then*

$$\|\mathbb{T}\|_\star = \sup_{\|f\|\leq 1} |\mathbb{T}(f)|. \tag{3.9}$$

Proof. By the inequality (3.7), we get

$$|\mathbb{T}(f)| \leq \|\mathbb{T}\|_\star \|f\|$$

for any $f \in \mathbf{X}$. Hence,

$$\sup_{\|f\|\leq 1} |\mathbb{T}(f)| \leq \|\mathbb{T}\|_\star. \tag{3.10}$$

Let $\varepsilon > 0$ be arbitrarily chosen. Then there exists $g \in \mathbf{X}$, $g \neq 0$, such that

$$|\mathbb{T}(g)| \geq (\|\mathbb{T}\|_\star - \varepsilon)\|g\|.$$

Hence,

$$\left|\mathbb{T}\left(\frac{g}{\|g\|}\right)\right| \geq \|\mathbb{T}\|_\star - \varepsilon$$

and

$$\sup_{\|f\|\leq 1} |\mathbb{T}(f)| \geq \|\mathbb{T}\|_\star - \varepsilon.$$

Because $\varepsilon > 0$ was arbitrarily chosen, from the last inequality, we obtain

$$\sup_{\|f\|\leq 1} |\mathbb{T}(f)| \geq \|\mathbb{T}\|_\star.$$

Hence and (3.10), we get (3.9). This completes the proof. □

Theorem 3.14. *Let* $\mathbf{E}$ *be a measurable set,* $1 \leq p < \infty$, $\frac{1}{p} + \frac{1}{q} = 1$, $g \in \mathbf{L}^q(\mathbf{E})$, $\|q\|_q \neq 0$. *Define the functional* $\mathbb{T}$ *on* $\mathbf{L}^p(\mathbf{E})$ *by*

$$\mathbb{T}(f) = \int_{\mathbf{E}} gf$$

for all $f \in \mathbf{L}^p(\mathbf{E})$. *Then* $\mathbb{T}$ *is a bounded linear functional on* $\mathbf{L}^p(\mathbf{E})$ *and* $\|\mathbb{T}\|_\star = \|g\|_q$.

Proof. Let $f_1, f_2 \in \mathbf{L}^p(\mathbf{E})$ and $\alpha, \beta \in \mathbf{F}$. Then

$$\begin{aligned}\mathbb{T}(\alpha f_1 + \beta f_2) &= \int_{\mathbf{E}} g(\alpha f_1 + \beta f_2)\\ &= \int_{\mathbf{E}} (\alpha g f_1 + \beta g f_2)\end{aligned}$$

$$\begin{aligned} &= \int_{\mathbf{E}} \alpha g f_1 + \int_{\mathbf{E}} \beta g f_2 \\ &= \alpha \int_{\mathbf{E}} g f_1 + \beta \int_{\mathbf{E}} g f_2 \\ &= \alpha \mathbb{T}(f_1) + \beta \mathbb{T}(f_2). \end{aligned}$$

Therefore, $\mathbb{T}$ is a linear functional on $\mathbf{L}^p(\mathbf{E})$. Also, using Hölder's inequality, we have

$$\begin{aligned} |\mathbb{T}(f)| &= \left| \int_{\mathbf{E}} g f \right| \leq \int_{\mathbf{E}} |g||f| \\ &\leq \left(\int_{\mathbf{E}} |g|^q \right)^{\frac{1}{q}} \left(\int_{\mathbf{E}} |f|^p \right)^{\frac{1}{p}} \\ &= \|g\|_q \|f\|_p \end{aligned}$$

for all $f \in \mathbf{L}^p(\mathbf{E})$. Consequently $\mathbb{T}$ is a bounded linear functional on $\mathbf{L}^p(\mathbf{E})$. By the last inequality, we get

$$\|\mathbb{T}\|_\star \leq \|g\|_q. \tag{3.11}$$

Let

$$g_1 = \|g\|_q^{1-q} sign(g) |g|^{q-1}.$$

We have

$$\begin{aligned} \int_{\mathbf{E}} |g_1|^p &= \int_{\mathbf{E}} \|g\|_q^{p(1-q)} |g|^{p(q-1)} \\ &= \|g\|_q^{-q} \int_{\mathbf{E}} |g|^q \\ &= 1, \end{aligned}$$

i.e., $g_1 \in \mathbf{L}^p(\mathbf{E})$ and $\|g_1\|_p = 1$. Next,

$$\begin{aligned} \mathbb{T}(g_1) &= \int_{\mathbf{E}} g g_1 \\ &= \int_{\mathbf{E}} g \|g\|_q^{1-q} sign(g) |g|^{q-1} \end{aligned}$$

$$= \|g\|_q^{1-q} \int_{\mathbf{E}} |g|^q$$
$$= \|g\|_q.$$

Therefore

$$\|\mathbb{T}\|_\star \geq \|g\|_q.$$

From the last inequality and from (3.11), we obtain $\|\mathbb{T}\|_\star = \|g\|_q$. This completes the proof. □

Theorem 3.15. *Let $\mathbb{T}$ and $\mathbb{S}$ be bounded linear functionals on a normed vector space $\mathbf{X}$. If $\mathbb{T} = \mathbb{S}$ on a dense subset $\mathbf{X}_0$ of $\mathbf{X}$, then $\mathbb{T} = \mathbb{S}$ on $\mathbf{X}$.*

Proof. Let $g \in \mathbf{X}$ be arbitrarily chosen. Then there exists a sequence $\{g_n\}_{n\in\mathbf{N}}$ of elements of $\mathbf{X}_0$ such that $g_n \to g$, as $n \to \infty$, in $\mathbf{X}$. Hence,

$$\mathbb{T}(g_n) \to \mathbb{T}(g), \quad \mathbb{S}(g_n) \to \mathbb{S}(g), \quad \text{as } n \to \infty,$$

and

$$\mathbb{T}(g_n) = \mathbb{S}(g_n)$$

for any $n \in \mathbf{N}$. Therefore, $\mathbb{T}(g) = \mathbb{S}(g)$. Because $g \in \mathbf{X}$ was arbitrarily chosen, we obtain $\mathbb{T} = \mathbb{S}$ on $\mathbf{X}$. This completes the proof. □

Theorem 3.16. *Let $\mathbf{E}$ be a measurable set and $1 \leq p < \infty$, $\frac{1}{p} + \frac{1}{q} = 1$, g is integrable over $\mathbf{E}$ and there is an $M \geq 0$ such that*

$$\left| \int_{\mathbf{E}} gf \right| \leq M\|f\|_p \tag{3.12}$$

for any simple function f in $\mathbf{L}^p(\mathbf{E})$. Then $g \in \mathbf{L}^q(\mathbf{E})$ and $\|g\|_q \leq M$.

Proof. Since g is integrable over $\mathbf{E}$, then it is finite a.e. on $\mathbf{E}$. By excising a set of measure zero from $\mathbf{E}$, we can assume that g is finite on all of $\mathbf{E}$.

1. Let $p > 1$. Because $|g|$ is a nonnegative measurable function on $\mathbf{E}$, by Theorem 2.47, there exists a sequence $\{\phi_n\}_{n\in\mathbf{N}}$ of measurable simple functions on $\mathbf{E}$ that converges pointwise on $\mathbf{E}$ to $|g|$ and

$0 \leq \phi_n \leq |g|$ on $\mathbf{E}$. Hence, $\{\phi_n^q\}_{n \in \mathbf{N}}$ is a sequence of nonnegative simple functions on $\mathbf{E}$ such that

$$0 \leq \phi_n^q \leq |g|^q \quad \text{on } \mathbf{E}$$

and $\phi_n^q \to |g|^q$, as $n \to \infty$, pointwise on $\mathbf{E}$. Hence and Fatou's Lemma,

$$\int_{\mathbf{E}} |g|^q \leq \int_{\mathbf{E}} \phi_n^q \tag{3.13}$$

for every $n \in \mathbf{N}$. Let $n \in \mathbf{N}$ be arbitrarily chosen. Then

$$\begin{aligned} \phi_n^q &= \phi_n \phi_n^{q-1} \\ &\leq |g| \phi_n^{q-1} \\ &= g sign(g) \phi_n^{q-1} \quad \text{on } \mathbf{E}. \end{aligned}$$

Let

$$f_n = sign(g) \phi_n^{q-1} \quad \text{on } \mathbf{E}.$$

Then f_n is a simple function. Because g is integrable over $\mathbf{E}$, we have that ϕ_n is integrable over $\mathbf{E}$. Then ϕ_n^q is integrable over $\mathbf{E}$ and

$$\int_{\mathbf{E}} |f_n|^p = \int_{\mathbf{E}} \phi_n^q.$$

Therefore, $f_n \in \mathbf{L}^p(\mathbf{E})$. Next, by (3.12), we obtain

$$\begin{aligned} \int_{\mathbf{E}} \phi_n^q &= \int_{\mathbf{E}} \phi_n \phi_n^{q-1} \\ &\leq \int_{\mathbf{E}} |g| \phi_n^{q-1} \\ &= \int_{\mathbf{E}} g sign(g) \phi_n^{q-1} \\ &= \int_{\mathbf{E}} g f_n \end{aligned}$$

$$\leq M\|f_n\|_p$$

$$= M\left(\int_{\mathbf{E}} \phi_n^q\right)^{\frac{1}{p}}.$$

From here,

$$\left(\int_{\mathbf{E}} \phi_n^q\right)^{\frac{1}{q}} \leq M,$$

i.e.,

$$\|\phi_n\|_q \leq M.$$

Hence and (3.13), we obtain

$$\|g\|_q \leq M.$$

2. Let $p = 1$. Suppose that M is not an essential upper bound for g. Then there is an $\varepsilon > 0$ such that the set

$$\mathbf{E}_\varepsilon = \{x \in \mathbf{E} : |g(x)| > M + \varepsilon\}$$

has finite positive measure. Let $f = sign(g)\kappa_{\mathbf{E}_\varepsilon}$. Then, by (3.12),

$$\begin{aligned}\left|\int_{\mathbf{E}} fg\right| &= \left|\int_{\mathbf{E}_\varepsilon} gsign(g)\right| \\ &= \int_{\mathbf{E}_\varepsilon} |g| \\ &> (M+\varepsilon)m(\mathbf{E}_\varepsilon).\end{aligned} \tag{3.14}$$

On the other hand, by (3.12),

$$\begin{aligned}\left|\int_{\mathbf{E}} fg\right| &\leq M\|f\|_1 \\ &= M\int_{\mathbf{E}} |sign(g)\kappa_{\mathbf{E}_\varepsilon}| \\ &= Mm(\mathbf{E}_\varepsilon),\end{aligned}$$

which contradicts with (3.14). Therefore M is an essential upper bound for g. This completes the proof. □

Theorem 3.17. *Let $[a, b]$ be a closed bounded interval, $1 \le p < \infty$, $\frac{1}{p} + \frac{1}{q} = 1$. Suppose that $\mathbb{T}$ is a bounded linear functional on $\mathbf{L}^p([a, b])$. Then there is a function $g \in \mathbf{L}^q([a, b])$ such that*

$$\mathbb{T}(f) = \int_a^b gf$$

for all $f \in \mathbf{L}^p([a, b])$.

Proof. Let $p > 1$. We leave the case $p = \infty$ for an exercise. For $x \in [a, b]$ we define

$$\Phi(x) = \mathbb{T}(\kappa_{[a,x)}).$$

For $[c, d] \subseteq [a, b]$, we have

$$\kappa_{[c,d)} = \kappa_{[a,d)} - \kappa_{[a,c)}.$$

Then

$$\begin{aligned}\Phi(d) - \Phi(c) &= \mathbb{T}(\kappa_{[a,d)}) - \mathbb{T}(\kappa_{[a,c)}) \\ &= \mathbb{T}(\kappa_{[a,d)} - \kappa_{[a,c)}) \\ &= \mathbb{T}(\kappa_{[c,d)}).\end{aligned}$$

If $\{(a_k, b_k)\}_{k=1}^n$ is a finite disjoint collection of intervals in (a, b), then

$$\begin{aligned}\sum_{k=1}^n |\Phi(b_k) - \Phi(a_k)| &= \sum_{k=1}^n sign(\Phi(b_k) - \Phi(a_k))(\Phi(b_k) - \Phi(a_k)) \\ &= \sum_{k=1}^n sign(\Phi(b_k) - \Phi(a_k))\mathbb{T}(\kappa_{[a_k,b_k)}) \\ &= \sum_{k=1}^n \mathbb{T}(sign(\Phi(b_k) - \Phi(a_k))\kappa_{[a_k,b_k)}) \\ &= \mathbb{T}\left(\sum_{k=1}^n sign(\Phi(b_k) - \Phi(a_k))\kappa_{[a_k,b_k)}\right).\end{aligned}$$

Consider the simple function

$$f = \sum_{k=1}^n sign(\Phi(b_k) - \Phi(a_k))\kappa_{[a_k,b_k)}.$$

Then

$$\begin{aligned}|\mathbb{T}(f)| &\leq \|\mathbb{T}\|_\star \|f\|_p \\ &= \|\mathbb{T}\|_\star \left(\sum_{k=1}^n (b_k - a_k)\right)^{\frac{1}{p}}.\end{aligned}$$

Consequently,

$$\sum_{k=1}^n |\Phi(b_k) - \Phi(a_k)| \leq \|\mathbb{T}\|_\star \left(\sum_{k=1}^n (b_k - a_k)\right)^{\frac{1}{p}}$$

and Φ is absolutely continuous on $[a, b]$. Therefore, Φ is differentiable almost everywhere on $[a, b]$, and if $g = \Phi'$, then g is integrable over $[a, b]$ and

$$\Phi(x) = \int_a^x g$$

for all $x \in [a, b]$. Therefore, for each $[c, d] \subseteq (a, b)$,

$$\begin{aligned}\mathbb{T}(\kappa_{[c,d)}) &= \Phi(d) - \Phi(c) \\ &= \int_a^b g\kappa_{[c,d)}.\end{aligned}$$

Since the functional $\mathbb{T}$ and the functional $f \longmapsto \int_a^b gf$ are linear on the vector space of step functions in $\mathbf{L}^p([a, b])$, it follows that

$$\mathbb{T}(f) = \int_a^b gf$$

for all step functions f in $\mathbf{L}^p([a, b])$. By Theorem 3.12, it follows that there is a sequence $\{\phi_n\}_{n\in\mathbf{N}}$ of step functions that converges to f in $\mathbf{L}^p([a, b])$ and also it is uniformly pointwise bounded on $[a, b]$. Hence and (3.8), we get

$$\lim_{n\to\infty} \mathbb{T}(\phi_n) = \mathbb{T}(f).$$

On the other hand, by the Lebesgue Dominated Convergence Theorem, we obtain

$$\lim_{n\to\infty} \int_a^b g\phi_n = \int_a^b gf.$$

Therefore,

$$\mathbb{T}(f) = \int_a^b gf$$

for all simple functions f in $\mathbf{L}^p([a,b])$. Since $\mathbb{T}$ is bounded,

$$\left| \int_a^b gf \right| = |\mathbb{T}(f)| \le \|\mathbb{T}\|_\star \|f\|_p$$

for all simple functions f in $\mathbf{L}^p([a,b])$. From here and from Theorem 3.16, we have that $g \in \mathbf{L}^q([a,b])$. By Theorem 3.14, the functional $f \longmapsto \int_a^b gf$ is bounded on $\mathbf{L}^p([a,b])$. This functional agree with $\mathbb{T}$ on the simple functions in $\mathbf{L}^p([a,b])$. Because the set of the simple functions in $\mathbf{L}^p([a,b])$ is dense in $\mathbf{L}^p([a,b])$(see Theorem 3.11), using Theorem 3.15, these two functionals agree on all of $\mathbf{L}^p([a,b])$. This completes the proof. □

Theorem 3.18 (The Riesz Representation Theorem). *Let* $\mathbf{E}$ *be a measurable set,* $1 \le p < \infty$, $\frac{1}{p} + \frac{1}{q} = 1$. *For each* $g \in \mathbf{L}^p(\mathbf{E})$, *define the bounded linear functional* $\mathbb{T}_g$ *on* $\mathbf{L}^p(\mathbf{E})$ *by*

$$\mathbb{T}_g(f) = \int_{\mathbf{E}} gf$$

for all $f \in \mathbf{L}^p(\mathbf{E})$. *Then for each bounded linear functional* $\mathbb{T}$ *on* $\mathbf{L}^p(\mathbf{E})$, *there is unique function* $g \in \mathbf{L}^q(\mathbf{E})$ *for which*

$$\mathbb{T}_g = \mathbb{T} \quad \textit{and} \quad \|\mathbb{T}\|_\star = \|g\|_q.$$

Proof. By Theorem 3.14, it follows that $\mathbb{T}_g$ is a bounded linear functional on $\mathbf{L}^p(\mathbf{E})$ and $\|\mathbb{T}_g\|_\star = \|g\|_q$ for each $g \in \mathbf{L}^p(\mathbf{E})$. Also, if $g_1, g_2 \in \mathbf{L}^q(\mathbf{E})$, then

$$\begin{aligned}
\mathbb{T}_{g_1-g_2}(f) &= \int_{\mathbf{E}} (g_1 - g_2)f \\
&= \int_{\mathbf{E}} (g_1 f - g_2 f) \\
&= \int_{\mathbf{E}} g_1 f - \int_{\mathbf{E}} g_2 f \\
&= \mathbb{T}_{g_1}(f) - \mathbb{T}_{g_2}(f)
\end{aligned}$$

for any $f \in \mathbf{L}^p(\mathbf{E})$. Therefore, if $\mathbb{T}_{g_1} = \mathbb{T}_{g_2}$, then $\mathbb{T}_{g_1 - g_2} = 0$, and hence, $\|g_1 - g_2\|_q = 0$, so that $g_1 = g_2$. Therefore, for a bounded linear functional $\mathbb{T}$ on $\mathbf{L}^p(\mathbf{E})$ there is at most one $g \in \mathbf{L}^p(\mathbf{E})$ for which $\mathbb{T}_g = \mathbb{T}$.

1. Suppose that $\mathbf{E} = \mathbf{R}$. Let $\mathbb{T}$ be a bounded linear functional on $\mathbf{L}^p(\mathbf{R})$. For any $n \in \mathbf{N}$, we define the linear functional $\mathbb{T}_n$ on $\mathbf{L}^p([-n, n])$ by

$$\mathbb{T}_n(f) = \mathbb{T}(\hat{f})$$

for all $f \in \mathbf{L}^p([-n, n])$, where $\hat{f}$ is the extension of f to all of $\mathbf{R}$ that vanishes outside $[-n, n]$. Then

$$\|f\|_p = \|\hat{f}\|_p$$

and

$$\begin{aligned} |\mathbb{T}_n(f)| &= |\mathbb{T}(\hat{f})| \\ &\leq \|\mathbb{T}\|_\star \|\hat{f}\|_p \\ &= \|\mathbb{T}\|_\star \|f\|_p \end{aligned}$$

for any $f \in \mathbf{L}^p([-n, n])$. Hence,

$$\|\mathbb{T}_n\|_\star \leq \|\mathbb{T}\|_\star.$$

By Theorem 3.17, it follows that there is a function $g_n \in \mathbf{L}^q([-n, n])$ for which

$$\mathbb{T}_n(f) = \int_{\mathbf{E}} g_n f$$

for all $f \in \mathbf{L}^p([-n, n])$ and

$$\|g_n\|_q = \|\mathbb{T}_n\|_\star \leq \|\mathbb{T}\|_\star. \tag{3.15}$$

Note that the restriction of g_{n+1} to $[-n, n]$ agree with g_n a.e. on $[-n, n]$. Define g to be a measurable function on $\mathbf{R}$ that agrees with g_n a.e. on $[-n, n]$ for each $n \in \mathbf{N}$. Hence, for all $f \in \mathbf{L}^p(\mathbf{R})$ that vanishes outside a bounded set,

$$\mathbb{T}(f) = \int_{\mathbf{R}} g f.$$

By (3.15),

$$\int_{-n}^{n} |g|^q \leq \|\mathbb{T}\|_\star^q.$$

Because the set of all functions of $\mathbf{L}^p(\mathbf{E})$ that vanishes outside a bounded set is dense in $\mathbf{L}^p(\mathbf{R})$, using Theorem 3.15, we conclude that $\mathbb{T}_g$ agrees with $\mathbb{T}$ on all $\mathbf{L}^p(\mathbf{R})$.

2. Let $\mathbf{E}$ be a measurable set and $\mathbb{T}$ be a bounded linear functional on $\mathbf{L}^p(\mathbf{E})$. Define the linear functional $\hat{\mathbb{T}}$ on $\mathbf{L}^p(\mathbf{E})$ by

$$\hat{\mathbb{T}}(f) = \mathbb{T}\left(f\Big|_{\mathbf{E}}\right), \quad f \in \mathbf{L}^p(\mathbf{R}).$$

Then $\hat{\mathbb{T}}$ is a bounded linear functional on $\mathbf{L}^p(\mathbf{R})$. Hence, there is a function $\hat{g} \in \mathbf{L}^q(\mathbf{R})$ for which

$$\hat{\mathbb{T}}(f) = \int_{\mathbf{R}} \hat{g} f$$

for any $f \in \mathbf{L}^p(\mathbf{R})$. Define $g = \hat{g}\Big|_{\mathbf{E}}$. Then $\mathbb{T} = \mathbb{T}_g$. This completes the proof.

□

Definition 3.9. Let $1 \leq p \leq \infty$ and q is its conjugate. The space $\mathbf{L}^q(\cdot)$ is called the dual space of the space $\mathbf{L}^p(\cdot)$.

Definition 3.10. Let $\mathbf{X}$ be a normed vector space. A sequence $\{f_n\}_{n\in\mathbf{N}}$ in $\mathbf{X}$ is said to converge weakly in $\mathbf{X}$ to $f \in \mathbf{X}$ if

$$\lim_{n\to\infty} \mathbb{T}(f_n) = \mathbb{T}(f)$$

for any linear functional $\mathbb{T}$ on $\mathbf{X}$. We will write

$$f_n \rightharpoonup f \quad \text{in } \mathbf{X}.$$

Remark 3.3. If $\mathbf{X}$ is a normed vector space, we will write $f_n \to f$ in $\mathbf{X}$ if

$$\|f_n - f\| \to 0 \quad \text{as } n \to \infty.$$

In this case, we will say that the sequence $\{f_n\}_{n\in\mathbf{N}}$ converges strongly to f in $\mathbf{X}$.

Theorem 3.19. *Let* $\mathbf{X}$ *be a normed vector space,* $\{f_n\}_{n\in\mathbf{N}}$ *be a sequence in* $\mathbf{X}$*,* $f \in \mathbf{X}$*. If* $f_n \to f$ *in* $\mathbf{X}$*, then* $f_n \rightharpoonup f$ *in* $\mathbf{X}$*.*

Proof. Since $f_n \to f$ in $\mathbf{X}$, we have

$$\|f_n - f\| \to 0 \quad \text{as } n \to \infty.$$

Let $\mathbb{T}$ be arbitrarily chosen linear functional on $\mathbf{X}$. Then

$$\begin{aligned}|\mathbb{T}(f_n) - \mathbb{T}(f)| &\le \|\mathbb{T}\|_\star \|f_n - f\| \\ &\to 0, \quad \text{as } n \to \infty.\end{aligned}$$

Because $\mathbb{T}$ was arbitrarily chosen linear functional on $\mathbf{X}$, we conclude that $f_n \rightharpoonup f$ in $\mathbf{X}$. This completes the proof. □

Theorem 3.20. *Let* $\mathbf{E}$ *be a measurable set,* $1 \le p < \infty$ *and* q *is its conjugate. Then* $f_n \rightharpoonup f$ *in* $\mathbf{L}^p(\mathbf{E})$ *if and only if*

$$\lim_{n\to\infty} \int_{\mathbf{E}} g f_n = \int_{\mathbf{E}} g f \tag{3.16}$$

for all $g \in \mathbf{L}^q(\mathbf{E})$*.*

Proof.

1. Let $f_n \rightharpoonup f$ in $\mathbf{L}^p(\mathbf{E})$. Then for every linear functional $\mathbb{T}$ we have

$$\lim_{n\to\infty} \mathbb{T}(f_n) = \mathbb{T}(f).$$

   Hence, using that $h \longmapsto \int_{\mathbf{E}} gh$, $h \in \mathbf{L}^p(\mathbf{E})$, is a linear functional for each $g \in \mathbf{L}^q(\mathbf{E})$, we get (3.16).
2. Assume that (3.16) holds. Let $\mathbb{T}$ be arbitrarily chosen linear functional on $\mathbf{L}^p(\mathbf{E})$. By the Riesz representation theorem, it follows that for any $h \in \mathbf{L}^p(\mathbf{E})$ there is a unique $g \in \mathbf{L}^q(\mathbf{E})$ such that

$$\mathbb{T}(h) = \int_{\mathbf{E}} gh$$

   Hence and (3.16), we obtain

$$\lim_{n\to\infty} \mathbb{T}(f_n) = \mathbb{T}(f).$$

   Because $\mathbb{T}$ was arbitrarily chosen linear functional on $\mathbf{L}^p(\mathbf{E})$, we conclude that $f_n \rightharpoonup f$ in $\mathbf{L}^p(\mathbf{E})$. This completes the proof. □

Theorem 3.21. *Let* $\mathbf{E}$ *be a measurable set,* $1 \leq p < \infty$*. Then a sequence in* $\mathbf{L}^p(\mathbf{E})$ *can converge weakly to at most one function in* $\mathbf{L}^p(\mathbf{E})$.

Proof. Let q be the conjugate of p. Let also, $\{f_n\}_{n\in\mathbf{N}}$ be a sequence in $\mathbf{L}^p(\mathbf{E})$ that converges weakly to $f_1, f_2 \in \mathbf{L}^p(\mathbf{E})$. Then $f_1 - f_2 \in \mathbf{L}^p(\mathbf{E})$ and

$$\begin{aligned}
&\int_{\mathbf{E}} \left| \|f_1 - f_2\|_p^{1-p} sign(f_1 - f_2)|f_1 - f_2|^{p-1} \right|^q \\
&\quad \leq \|f_1 - f_2\|_p^{1-p} \int_{\mathbf{E}} |f_1 - f_2|^{q(p-1)} \\
&\quad = \|f_1 - f_2\|_p^{1-p} \int_{\mathbf{E}} |f_1 - f_2|^p \\
&\quad = \|f_1 - f_2\|_p,
\end{aligned}$$

i.e., $\|f_1 - f_2\|_p^{1-p} sign(f_1 - f_2)|f_1 - f_2|^{p-1} \in \mathbf{L}^q(\mathbf{E})$. Hence and Theorem 3.20, we get

$$\begin{aligned}
&\int_{\mathbf{E}} \|f_1 - f_2\|_p^{1-p} sign(f_1 - f_2)|f_1 - f_2|^{p-1} f_1 \\
&\quad = \lim_{n\to\infty} \int_{\mathbf{E}} \|f_1 - f_2\|_p^{1-p} sign(f_1 - f_2)|f_1 - f_2|^{p-1} f_n \\
&\quad = \int_{\mathbf{E}} \|f_1 - f_2\|_p^{1-p} sign(f_1 - f_2)|f_1 - f_2|^{p-1} f_2.
\end{aligned}$$

Therefore,

$$\begin{aligned}
0 &= \int_{\mathbf{E}} \|f_1 - f_2\|_p^{1-p} sign(f_1 - f_2)|f_1 - f_2|^{p-1}(f_1 - f_2) \\
&= \|f_1 - f_2\|_p^{1-p} \int_{\mathbf{E}} |f_1 - f_2|^p \\
&= \|f_1 - f_2\|_p.
\end{aligned}$$

Consequently, $f_1 = f_2$. This completes the proof. □

Definition 3.11. Let $\mathbf{E}$ be a measurable set and $1 \leq p < \infty$.

1. Let $\{f_n\}_{n\in\mathbf{N}}$ be a sequence in $\mathbf{L}^p(\mathbf{E})$, $f \in \mathbf{L}^p(\mathbf{E})$ and $f_n \rightharpoonup f$ in $\mathbf{L}^p(\mathbf{E})$. The function f will be called the weak sequential limit.

2. Let $f \in \mathbf{L}^p(\mathbf{E})$. The function

$$f^\star = \|f\|_p^{1-p} sign(f)|f|^{p-1}$$

will be called the conjugate function of f. Note that $f^\star \in \mathbf{L}^q(\mathbf{E})$, where q is the conjugate of p.

Exercise 3.1. Let $\mathbf{E}$ be a measurable set, $1 \le p < \infty$, $f \in \mathbf{L}^p(\mathbf{E})$ and $f^\star$ be the conjugate function of f. Prove that

$$\|f^*\|_q = 1,$$

where q is the conjugate of p.

Remark 3.4. Let $\mathbf{X}$ be a linear normed space and $f_n \rightharpoonup f$ in $\mathbf{X}$. Suppose that the sequence $\{\|f_n\|\}_{n\in\mathbf{N}}$ is unbounded. By taking a subsequence and relabeling, we can suppose that $\|f_n\| \ge n3^n$. By taking a further subsequence and relabeling, we can suppose that

$$\frac{\|f_n\|}{n3^n} \to \alpha \in [1, \infty), \quad \text{as } n \to \infty.$$

Let $\mathbb{T}$ be arbitrarily chosen linear functional. Then there is a constant $M > 0$ such that

$$|\mathbb{T}(f_n)| \le M, \quad |\mathbb{T}(f)| \le M$$

for any $n \in \mathbf{N}$. Hence,

$$\begin{aligned}
\left|\mathbb{T}\left(\frac{n3^n}{\|f_n\|}f_n\right) - \mathbb{T}\left(\frac{1}{\alpha}f\right)\right| &= \left|\mathbb{T}\left(\frac{n3^n}{\|f_n\|}f_n\right)\right. \\
&\quad \left. -\mathbb{T}\left(\frac{1}{\alpha}f_n\right) + \mathbb{T}\left(\frac{1}{\alpha}f_n\right) - \mathbb{T}\left(\frac{1}{\alpha}f\right)\right| \\
&= \left|\mathbb{T}\left(\left(\frac{n3^n}{\|f_n\|} - \frac{1}{\alpha}\right)f_n\right) + \mathbb{T}\left(\frac{1}{\alpha}(f_n - f)\right)\right| \\
&\le \left|\frac{n3^n}{\|f_n\|} - \frac{1}{\alpha}\right| |\mathbb{T}(f_n)| + \frac{1}{\alpha}|\mathbb{T}(f_n) - \mathbb{T}(f)| \\
&\le \left|\frac{n3^n}{\|f_n\|} - \frac{1}{\alpha}\right| M + \frac{1}{\alpha}|\mathbb{T}(f_n) - \mathbb{T}(f)| \\
&\to 0, \quad \text{as } n \to \infty.
\end{aligned}$$

Therefore,

$$\frac{n3^n}{\|f_n\|} f_n \rightharpoonup \frac{1}{\alpha} f \quad \text{in } \mathbf{X}.$$

Theorem 3.22. *Let* $\mathbf{E}$ *be a measurable set and* $1 \leq p < \infty$. *Suppose that* $f_n \rightharpoonup f$ *in* $\mathbf{L}^p(\mathbf{E})$. *Then* $\{f_n\}_{n \in \mathbf{N}}$ *is bounded in* $\mathbf{L}^p(\mathbf{E})$ *and* $\|f\|_p \leq \liminf \|f_n\|_p$.

Proof. Let q be the conjugate of p and $f^\star$ be the conjugate function of f. Then $f^\star \in \mathbf{L}^q(\mathbf{E})$ and using Hölder's inequality, we get

$$\begin{aligned}\int_{\mathbf{E}} f^\star f_n &\leq \|f^\star\|_q \|f_n\|_p \\ &= \|f_n\|_p\end{aligned}$$

for any $n \in \mathbf{N}$. Since $f_n \rightharpoonup f$ in $\mathbf{L}^p(\mathbf{E})$, by Theorem 3.20, we get

$$\begin{aligned}\|f\|_p &= \int_{\mathbf{E}} f^\star f \\ &= \lim_{n \to \infty} \int_{\mathbf{E}} f^\star f_n \\ &\leq \liminf \|f_n\|_p.\end{aligned}$$

Now, we assume that the sequence $\{f_n\}_{n \in \mathbf{N}}$ is unbounded. Using Remark 3.4 and by taking scalar multiples, we can suppose that $\|f_n\|_p = n3^n$, $n \in \mathbf{N}$. Let $f_k^\star$ be the conjugate functions of f_k, $k \in \mathbf{N}$. Let also, $\varepsilon_1 = \frac{1}{3}$ and we define

$$\varepsilon_k = \frac{1}{3^k} \quad \text{if} \quad \int_{\mathbf{E}} \sum_{l=1}^{k-1} \varepsilon_l f_l^\star f_k \geq 0$$

and

$$\varepsilon_k = -\frac{1}{3^k} \quad \text{if} \quad \int_{\mathbf{E}} \sum_{l=1}^{k-1} \varepsilon_l f_l^\star f_k \leq 0$$

for $k \in \mathbf{N}$, $k \geq 2$. Then

$$\begin{aligned}
\left| \int_{\mathbf{E}} \left(\sum_{k=1}^{n} \varepsilon_k f_k^{\star} \right) f_n \right| &= \left| \int_{\mathbf{E}} \left(\sum_{k=1}^{n} \varepsilon_k \|f_k\|_p^{1-p} sign(f_k) |f_k|^{p-1} \right) f_n \right| \\
&= \left| \int_{\mathbf{E}} \sum_{k=1}^{n-1} \varepsilon_k \|f_k\|_p^{1-p} sign(f_k) |f_k|^{p-1} f_n + \int_{\mathbf{E}} \varepsilon_n \|f_n\|_p^{1-p} |f_n|^p \right| \\
&= \left| \int_{\mathbf{E}} \sum_{k=1}^{n-1} \varepsilon_k \|f_k\|_p^{1-p} sign(f_k) |f_k|^{p-1} f_n + \varepsilon_n \|f_n\|_p \right| \\
&\geq \frac{1}{3^n} \|f_n\|_p \\
&= n.
\end{aligned}$$

Also,

$$\|\varepsilon_k f_k^{\star}\|_q = \frac{1}{3^k}, \quad k \in \mathbf{N}.$$

Since the series $\sum_{k=1}^{\infty} \frac{1}{3^k}$ is a convergent series, we have that $\sum_{k=1}^{\infty} \varepsilon_k f_k^{\star}$ is a convergent series in $\mathbf{L}^q(\mathbf{E})$ and let

$$g = \sum_{k=1}^{\infty} \varepsilon_k f_k^{\star}.$$

For any $n \in \mathbf{N}$, we have

$$\begin{aligned}
\left| \int_{\mathbf{E}} g f_n \right| &= \left| \int_{\mathbf{E}} \left(\sum_{k=1}^{\infty} \varepsilon_k f_k^{\star} \right) f_n \right| \\
&= \left| \int_{\mathbf{E}} \left(\sum_{k=1}^{n} \varepsilon_k f_k^{\star} \right) f_n + \int_{\mathbf{E}} \left(\sum_{k=n+1}^{\infty} \varepsilon_k f_k^{\star} \right) f_n \right| \\
&\geq \left| \int_{\mathbf{E}} \left(\sum_{k=1}^{n} \varepsilon_k f_k^{\star} \right) f_n \right| - \left| \int_{\mathbf{E}} \left(\sum_{k=n+1}^{\infty} \varepsilon_k f_k^{\star} \right) f_n \right| \\
&\geq n - \left\| \sum_{k=n+1}^{\infty} \varepsilon_k f_k^{\star} \right\|_q \|f_n\|_p
\end{aligned}$$

$$
\begin{aligned}
&\geq n - \sum_{k=n+1}^{\infty} \|\varepsilon_k f_k^{\star}\|_q \|f_n\|_p \\
&= n - \sum_{k=n+1}^{\infty} \frac{1}{3^k} \|f_n\|_p \\
&= n - \frac{1}{2\,(3^n)} \|f_n\|_p \\
&= \frac{n}{2},
\end{aligned}
$$

which is a contradiction, because $f_n \rightharpoonup f$ in $\mathbf{L}^p(\mathbf{E})$ and $g \in \mathbf{L}^q(\mathbf{E})$. This completes the proof. □

Theorem 3.23. *Let* $\mathbf{E}$ *be a measurable set,* $1 \leq p < \infty$ *and* q *is its conjugate. Suppose that* $f_n \rightharpoonup f$ *in* $\mathbf{L}^p(\mathbf{E})$ *and* $g_n \to g$ *strongly in* $\mathbf{L}^q(\mathbf{E})$. *Then*

$$
\lim_{n\to\infty} \int_{\mathbf{E}} g_n f_n = \int_{\mathbf{E}} g f. \tag{3.17}
$$

Proof. We have

$$
\int_{\mathbf{E}} g_n f_n - \int_{\mathbf{E}} g f = \int_{\mathbf{E}} (g_n - g)\, f_n + \int_{\mathbf{E}} g(f_n - f). \tag{3.18}
$$

By Theorem 3.22, it follows that there is a constant $M > 0$ such that $\|f_n\|_p \leq M$ for any $n \in \mathbf{N}$. Then, using Hölder's inequality, we obtain

$$
\begin{aligned}
\left| \int_{\mathbf{E}} (g_n - g) f_n \right| &\leq \int_{\mathbf{E}} |g_n - g||f_n| \\
&\leq \|g_n - g\|_q \|f_n\|_p \\
&\leq M\|g_n - g\|_q \\
&\to 0, \quad \text{as} \quad n \to \infty.
\end{aligned} \tag{3.19}
$$

Since $f_n \rightharpoonup f$ in $\mathbf{L}^p(\mathbf{E})$ and $g \in \mathbf{L}^q(\mathbf{E})$, using Theorem 3.20, we obtain

$$
\int_{\mathbf{E}} g(f_n - f) \to 0 \quad \text{as } n \to \infty.
$$

Hence and (3.18), (3.19), we obtain (3.17). This completes the proof. □

Theorem 3.24. *Let* $\mathbf{E}$ *be a measurable set,* $1 \le p < \infty$ *and* q *be its conjugate. Let also,* $\mathscr{F} \subset \mathbf{L}^q(\mathbf{E})$ *and its span is dense in* $\mathbf{L}^q(\mathbf{E})$*. Suppose that* $\{f_n\}_{n\in\mathbf{N}}$ *is a bounded sequence in* $\mathbf{L}^p(\mathbf{E})$ *and* $f \in \mathbf{L}^p(\mathbf{E})$*. Then* $f_n \rightharpoonup f$ *in* $\mathbf{L}^p(\mathbf{E})$ *if and only if*

$$\lim_{n\to\infty} \int_{\mathbf{E}} f_n g = \int_{\mathbf{E}} fg \tag{3.20}$$

for all $g \in \mathscr{F}$*.*

Proof.

1. Let $f_n \rightharpoonup f$ in $\mathbf{L}^p(\mathbf{E})$. Using Theorem 3.20, we conclude that (3.20) holds.
2. Suppose that (3.20) holds. Let $g_0 \in \mathbf{L}^q(\mathbf{E})$ is arbitrarily chosen. For any $g \in \mathbf{L}^q(\mathbf{E})$ and for any $n \in \mathbf{N}$, we have

$$\int_{\mathbf{E}} (f_n - f)g_0 = \int_{\mathbf{E}} (f_n - f)(g_0 - g) + \int_{\mathbf{E}} (f_n - f)g$$

and hence, using Hölder's inequality, we obtain

$$\begin{aligned} \left|\int_{\mathbf{E}} (f_n - f)g_0\right| &= \left|\int_{\mathbf{E}} (f_n - f)(g_0 - g) + \int_{\mathbf{E}} (f_n - f)g\right| \\ &\le \left|\int_{\mathbf{E}} (f_n - f)(g_0 - g)\right| + \left|\int_{\mathbf{E}} (f_n - f)g\right| \\ &\le \|f_n - f\|_p \|g_0 - g\|_q + \left|\int_{\mathbf{E}} (f_n - f)\, g\right|. \end{aligned} \tag{3.21}$$

We take $\varepsilon > 0$ arbitrarily. Since $\{f_n\}_{n\in\mathbf{N}}$ is bounded in $\mathbf{L}^p(\mathbf{E})$ and the span of $\mathscr{F}$ is dense in $\mathbf{L}^q(\mathbf{E})$, there is $g \in \mathscr{F}$ such that

$$\|f_n - f\|_p \|g - g_0\|_q < \frac{\varepsilon}{2}$$

for any $n \in \mathbf{N}$. Hence and (3.21), we conclude that

$$\int_{\mathbf{E}} f_n g_0 \to \int_{\mathbf{E}} f g_0 \quad \text{as } n \to \infty.$$

Because $g_0 \in \mathbf{L}^q(\mathbf{E})$ was arbitrarily chosen, we obtain (3.16). Therefore, $f_n \rightharpoonup f$ in $\mathbf{L}^p(\mathbf{E})$. This completes the proof. □

3.7 General $\mathbf{L}^p$ Spaces

Let $(\mathbf{X}, \mathscr{M}, \mu)$ be a measure space. $\mathscr{F}$ be the collection of all measurable extended real-valued functions on $\mathbf{X}$ that are finite a.e. on $\mathbf{X}$.

Definition 3.12. Define two functions f and g in $\mathscr{F}$ to be equivalent, and we write $f \sim g$, provided

$$f(x) = g(x) \quad \text{for almost all } x \in \mathbf{X}.$$

The relation $\sim$ is an equivalent relation, that is, it is reflexive, symmetric and transitive. Therefore it induces a partition of $\mathscr{F}$ into a disjoint collection of equivalence classes, which we denote by $\mathscr{F}/\sim$. For given two functions f and g in $\mathscr{F}$, their equivalence classes $[f]$ and $[g]$ and real numbers α and β, we define $\alpha[f]+\beta[g]$ to be the equivalence class of the functions in $\mathscr{F}$ that take the value $\alpha f(x)+\beta g(x)$ at points $x \in \mathbf{X}$ at which both f and g are finite. Note that these linear combinations are independent of the choice of the representatives of the equivalence classes. The zero element in $\mathscr{F}/\sim$ is the equivalence class of functions that vanish a.e. in $\mathbf{X}$. Thus, $\mathscr{F}/\sim$ is a vector space.

Definition 3.13. For $1 \le p < \infty$, we define $\mathbf{L}^p(\mathbf{X}, \mu)$ to be the collection of equivalence classes $[f]$ for which

$$\int_{\mathbf{X}} |f|^p d\mu < \infty.$$

This is properly defined since if $f \sim g$, then

$$\int_{\mathbf{X}} |f|^p d\mu = \int_{\mathbf{X}} |g|^p d\mu.$$

Note that, if $[f], [g] \in \mathbf{L}^p(\mathbf{X}, \mu)$, then for any real constants α and β, we have

$$\begin{aligned}\int_{\mathbf{X}} |\alpha f + \beta g|^p d\mu &\le 2^p \left(\int_{\mathbf{X}} |\alpha f|^p d\mu + \int_{\mathbf{X}} |\beta g|^p d\mu \right) \\ &= |2\alpha|^p \int_{\mathbf{X}} |f|^p d\mu + |2\beta|^p \int_{\mathbf{X}} |g|^p d\mu < \infty,\end{aligned}$$

i.e., $\alpha[f] + \beta[g] \in \mathbf{L}^p(\mathbf{X}, \mu)$.

Definition 3.14. We call a function $f \in \mathscr{F}$ essentially bounded provided there is some $M \ge 0$, called an essential upper bound for

f, for which

$$|f(x)| \leq M \quad \text{for almost all } x \in \mathbf{X}.$$

Definition 3.15. We define $\mathbf{L}^\infty(\mathbf{X}, \mu)$ to be the collection of equivalence classes $[f]$ for which f is essentially bounded.

Note that $\mathbf{L}^\infty(\mathbf{X}, \mu)$ is properly defined since if $f \sim g$, then

$$|f(x)| = |g(x)| \leq M \quad \text{for almost all } x \in \mathbf{X}.$$

Also, if $[f], [g] \in \mathbf{L}^\infty(\mathbf{X}, \mu)$ and $\alpha, \beta \in \mathbf{R}$, there are nonnegative constants M_1 and M_2 such that

$$|f(x)| \leq M_1, \quad |g(x)| \leq M_2 \quad \text{for almost all } x \in \mathbf{X}.$$

Hence,

$$\begin{aligned} |\alpha f(x) + \beta g(x)| &\leq |\alpha||f(x)| + |\beta||g(x)| \\ &\leq |\alpha|\, M_1 + |\beta|\, M_2 \end{aligned}$$

for almost all $x \in \mathbf{X}$. Therefore $\alpha[f] + \beta[g] \in \mathbf{L}^\infty(\mathbf{X}, \mu)$ and hence, $\mathbf{L}^\infty(\mathbf{X}, \mu)$ is a vector space. For simplicity and convenience, we refer to the equivalence classes in $\mathscr{F}/\sim$ as functions and denote them by f rather than $[f]$. Thus, $f = g$ means $f - g$ vanishes a.e. on $\mathbf{X}$.

Definition 3.16. For $1 \leq p < \infty$, in $\mathbf{L}^p(\mathbf{X}, \mu)$ we define

$$\|f\|_p = \left(\int_{\mathbf{X}} |f|^p \right)^{\frac{1}{p}}.$$

For $p = \infty$, we define $\|f\|_\infty$ to be the infimum of the essential upper bounds for f.

Remark 3.5. Note that the idea for the proof of next assertions in this section is the same as the idea for the proof of the assertions in the previous sections in this chapter. Therefore, we leave the proof of the next assertions in this section.

Theorem 3.25. *Let $(\mathbf{X}, \mathscr{M}, \mu)$ be a measure space, $1 \le p < \infty$, and q be the conjugate of p. If $f \in \mathbf{L}^p(\mathbf{X}, \mu)$, $g \in \mathbf{L}^q(\mathbf{X}, \mu)$, then $fg \in \mathbf{L}^1(\mathbf{X}, \mu)$ and*

$$\int_{\mathbf{X}} |fg| d\mu \le \|f\|_p \|g\|_q.$$

Moreover, if $f \neq 0$, the function $f^\star = \|f\|_p^{1-p} sign(f)|f|^{p-1} \in \mathbf{L}^q(\mathbf{X}, \mu)$,

$$\int_{\mathbf{X}} f f^\star d\mu = \|f\|_p \quad \text{and} \quad \|f^\star\|_q = 1.$$

Theorem 3.26. *Let $(\mathbf{X}, \mathscr{M}, \mu)$ be a finite measure space and $1 \le p_1 < p_2$. Then $\mathbf{L}^{p_2}(\mathbf{X}, \mu) \subseteq \mathbf{L}^{p_1}(\mathbf{X}, \mu)$ and*

$$\|f\|_{p_1} \le c\|f\|_{p_2} \quad \text{for} \quad f \in \mathbf{L}^{p_2}(\mathbf{X}, \mu),$$

where

$$c = \begin{cases} (\mu(\mathbf{X}))^{\frac{p_2 - p_1}{p_1 p_2}} & \text{if } p_2 < \infty \\ (\mu(\mathbf{X}))^{\frac{1}{p_1}} & \text{if } p_2 = \infty. \end{cases}$$

Theorem 3.27. *Let $(\mathbf{X}, \mathscr{M}, \mu)$ be a measure space and $1 < p \le \infty$. If $\{f_n\}_{nb\in \mathbf{N}}$ is a bounded sequence of functions in $\mathbf{L}^p(\mathbf{X}, \mu)$, then $\{f_n\}_{n\in\mathbf{N}}$ is uniformly integrable over $\mathbf{X}$.*

Theorem 3.28. *Let $(\mathbf{X}, \mathscr{M}, \mu)$ be a measure space and $1 \le p \le \infty$. Then every rapidly Cauchy sequence in $\mathbf{L}^p(\mathbf{X}, \mu)$ converges to a function in $\mathbf{L}^p(\mathbf{X}, \mu)$, both with respect to the $\mathbf{L}^p(\mathbf{X}, \mu)$ norm and pointwise a.e. in $\mathbf{X}$.*

Theorem 3.29 (The Riesz–Fischer Theorem). *Let $(\mathbf{X}, \mathscr{M}, \mu)$ be a measure space and $1 \le p \le \infty$. Then $\mathbf{L}^p(\mathbf{X}, \mu)$ is a Banach space. Moreover, if a sequence in $\mathbf{L}^p(\mathbf{X}, \mu)$ converges in $\mathbf{L}^p(\mathbf{X}, \mu)$ to a function $f \in \mathbf{L}^p(\mathbf{X}, \mu)$, then a subsequence converges pointwise a.e. on $\mathbf{X}$ to f.*

Proof. Let $\{f_n\}_{n\in\mathbf{N}}$ be a Cauchy sequence of $\mathbf{L}^p(\mathbf{X}, \mu)$. Then it has rapidly Cauchy subsequence $\{f_{n_k}\}_{k\in\mathbf{N}}$. By Theorem 3.28, it follows that $\{f_{n_k}\}_{k\in\mathbf{N}}$ converges to a function in $\mathbf{L}^p(\mathbf{X}, \mu)$ both with respect to the $\mathbf{L}^p(\mathbf{X},\mu)$ norm and pointwise almost everywhere on $\mathbf{X}$. This completes the proof. □

Theorem 3.30. *Let $(\mathbf{X},\mathscr{M},\mu)$ be a measure space and $1 \le p < \infty$. Then the subspace of simple functions on $\mathbf{X}$ that vanish outside a set of finite measure is dense in $\mathbf{L}^p(\mathbf{X},\mu)$.*

Theorem 3.31 (The Vitalli $\mathbf{L}^p$ Convergence Theorem). *Let $(\mathbf{X},\mathscr{M},\mu)$ be a measure space and $1 \le p < \infty$. Suppose that $\{f_n\}_{n\in\mathbf{N}}$ is a sequence in $\mathbf{L}^p(\mathbf{X},\mu)$ that converges pointwise a.e. to $f \in \mathbf{L}^p(\mathbf{X},\mu)$. Then $f_n \to f$ in $\mathbf{L}^p(\mathbf{X},\mu)$, as $n\to\infty$, if and only if $\{|f_n|^p\}_{n\in\mathbf{N}}$ is uniformly integrable and tight.*

For $1 \le p < \infty$, let $f \in \mathbf{L}^q(\mathbf{X},\mu)$, where q is the conjugate of p. Define the linear functional $\mathbb{T}_f : \mathbf{L}^p(\mathbf{X},\mu) \longmapsto \mathbf{R}$ by

$$\mathbb{T}_f(g) = \int_{\mathbf{X}} fg\,d\mu, \quad g \in \mathbf{L}^p(\mathbf{X},\mu). \tag{3.22}$$

Theorem 3.32 (The Riesz Representation Theorem for the Dual Space of the Space $\mathbf{L}^p(\mathbf{X},\mu)$). *Let $(\mathbf{X},\mathscr{M},\mu)$ be a σ-finite measure space, $1 \le p < \infty$, and q be the conjugate to p. For $f \in \mathbf{L}^q(\mathbf{X},\mu)$, define $\mathbb{T}_f \in (\mathbf{L}^p(\mathbf{X},\mu))^\star$ by (3.22). Then $\mathbb{T}_f$ is an isometric isomorphism of $\mathbf{L}^q(\mathbf{X},\mu)$ onto the space of the linear functionals on $\mathbf{L}^p(\mathbf{X},\mu)$.*

Remark 3.6. Let $1 \le p \le \infty$. When it is clear from the context what measure is used, μ is omitted and one just writes $\mathbf{L}^p(\mathbf{X})$.

3.8 Advanced Practical Problems

Problem 3.1. Let $f \in \mathbf{L}^1([a,b])$ and define

$$\|f\| = \int_a^b x^2|f(x)|dx.$$

Prove that this is a norm in $\mathbf{L}^1([a,b])$.

Problem 3.2. For $f \in \mathbf{L}^\infty([a,b])$, prove that

$$\|f\|_\infty = \min\{M : m\{x\in[a,b] : |f(x)| > M\} = 0\}.$$

Problem 3.3. Let

$$f(x) = \frac{x^{-\frac{1}{2}}}{1 + \log x}, \quad x > 1.$$

Prove that $f \in \mathbf{L}^p(1, \infty)$ if and only if $p = 2$.

Problem 3.4. Let $f(x) = \log(\frac{1}{x})$, $x \in (0, 1]$, $1 \le p < \infty$. Prove that $f \in \mathbf{L}^p((0, 1])$ and $f \notin \mathbf{L}^\infty((0, 1])$.

Problem 3.5. Let $\mathbf{E}$ be a measurable set, $1 \le p < \infty$ and q is the conjugate of p, $f \in \mathbf{L}^p(\mathbf{E})$. Prove that $f \equiv 0$ if and only if

$$\int_{\mathbf{E}} fg = 0$$

for any $g \in \mathbf{L}^q(\mathbf{E})$.

Problem 3.6. Let $\mathbf{E}$ be a measurable set of finite measure, $1 \le p_1 < p_2 \le \infty$. Prove that if $f_n \to f$ strongly in $\mathbf{L}^{p_2}(\mathbf{E})$, then $f_n \to f$ strongly in $\mathbf{L}^{p_1}(\mathbf{E})$.

Problem 3.7. Let $\mathbf{E}$ be a measurable set, $1 \le p < \infty$, q is the conjugate of p, $\mathbf{S}$ is dense in $\mathbf{L}^q(\mathbf{E})$. Prove that if $g \in \mathbf{L}^p(\mathbf{E})$ and $\int_{\mathbf{E}} fg = 0$ for any $f \in \mathbf{S}$, then $g = 0$.

Problem 3.8. Let $\mathbf{E}$ be a measurable set, $1 \le p < \infty$. Prove that the functions in $\mathbf{L}^p(\mathbf{E})$ that vanish outside a bounded set are dense in $\mathbf{L}^p(\mathbf{E})$.

Problem 3.9. Let $[a, b]$ be a closed bounded interval and $f_n \rightharpoonup f$ in $\mathbf{C}([a, b])$. Prove that $\{f_n\}_{n \in \mathbf{N}}$ converges pointwise on $[a, b]$ to f.

Problem 3.10. Let $[a, b]$ be a closed bounded interval and $f_n \rightharpoonup f$ in $\mathbf{L}^\infty([a, b])$. Prove that

$$\lim_{n \to \infty} \int_a^x f_n = \int_a^x f$$

for any $x \in [a, b]$.

Chapter 4

Distributions: The Fourier Transform

4.1 The Spaces $\mathscr{C}_0^\infty$ and $\mathscr{S}$

With $\mathbb{N}^n \cup \{0\}$ we denote the space of multi-indices $\alpha = (\alpha_1, \alpha_2, \ldots, \alpha_n)$, $\alpha_k \in \mathbb{N} \cup \{0\}$, $k = 1, 2, \ldots, n$. For $\alpha = (\alpha_1, \alpha_2, \ldots, \alpha_n)$, $\beta = (\beta_1, \beta_2, \ldots, \beta_n) \in \mathbb{N}^n \cup \{0\}$, we will write $\alpha \le \beta$ if $\alpha_k \le \beta_k$, $k = 1, 2, \ldots, m$. Set $D = (D_1, D_2, \ldots, D_n)$, $D_k = \frac{\partial}{\partial x_k}$, $k = 1, 2, \ldots, n$, $D^\alpha = \frac{\partial^{|\alpha|}}{\partial_{x_1}^{\alpha_1} \partial_{x_2}^{\alpha_2} \ldots \partial_{x_n}^{\alpha_n}}$.

Let $X \subset \mathbb{R}^n$ be an open set. If $K \subset X$ is a compact set we shall write $K \subset\subset X$. The following conventions will also be used throughout the book: $U(x_0, R)$ is the open ball of radius R with centre at the point x_0, $S(x_0, R) = \partial U(x_0, R)$ is the sphere of radius R with centre at x_0, $\overline{U}(x_0, R) = U(x_0, R) \cup S(x_0, R)$ is the closed ball of radius R with centre at the point x_0 and $U_R = U(0, R)$, $S_R = S(0, R)$, $\overline{U}_R = \overline{U}(0, R)$.

If A and B are sets in $\mathbb{R}^n$, by $d(A, B)$ or $dist(A, B)$ we shall denote the distance between the sets A and B, that is

$$d(A, B) = \text{dist}(A, B) = inf_{x \in A, y \in B} |x - y|.$$

We shall use A^ε to denote the ε-neighbourhood of a set A, i.e., $A^\varepsilon = A + U_\varepsilon$. If A is an open set, then A_ε will designate the set of points in A that are more than ε away from the boundary ∂A, i.e., $A_\varepsilon = \{x : x \in A, dist(x, \partial A) > \varepsilon\}$.

We use *intA* to denote the set of interior points of the set A. With κ_A we will denote the characteristic function of A, i.e., $\kappa_A(x) = 1$ for $x \in A$ and $\kappa_A(x) = 0$ for $x \notin A$.

Definition 4.1. The set A is called convex if for any points x and y in A the segment

$$\lambda x + (1-\lambda)y, \quad \lambda \in [0,1],$$

lies entirely in A.

We will write chA to denote the convex hull of a set A.

Definition 4.2. We call space of basic functions the space $\mathscr{C}_0^\infty(X)$ of smooth functions with compact support defined on X.

Example 4.1. The function

$$\omega(x) = \begin{cases} Ce^{-\frac{1}{1-|x|^2}}, & |x| \leq 1, \\ 0, & |x| > 1, \end{cases}$$

where the constant C is chosen so that $\int_{\mathbb{R}^n} \omega(x)dx = 1$, belongs in $\mathscr{C}_0^\infty(\mathbb{R}^n)$.

Set $\omega_\varepsilon(x) = \frac{1}{\varepsilon^n}\omega(\frac{x}{\varepsilon})$, $x \in \mathbb{R}^n$.

Definition 4.3. The function ω_ε is called the "cap-shaped function".

We have

$$\int_{\mathbb{R}^n} \omega_\varepsilon(x)dx = \frac{1}{\varepsilon^n}\int_{\mathbb{R}^n} \omega\left(\frac{x}{\varepsilon}\right)dx = \int_{\mathbb{R}^n} \omega(x)dx = 1.$$

Definition 4.4. We say that the sequence $\{\phi_k\}_{k=1}^\infty$ of elements of $\mathscr{C}_0^\infty(X)$ converges to the function $\phi \in \mathscr{C}_0^\infty(X)$ if there exists a compact set $K \subset X$ such that $supp\phi_k \subset K$ for every $k \in \mathbb{N}$ and $\lim_{k\to\infty} D^\alpha\phi_k(x) = D^\alpha\phi(x)$ uniformly for every multi-index $\alpha \in \mathbb{N}^n \cup \{0\}$.

Example 4.2. Take $\phi \in \mathscr{C}_0^\infty(\mathbb{R})$. The sequence $\{\frac{1}{k}\phi(x)\}_{k=1}^\infty$ converges to 0 in $\mathscr{C}_0^\infty(\mathbb{R})$.

Example 4.3. Take $\phi \in \mathscr{C}_0^\infty(\mathbb{R})$. Then the sequence $\{\frac{1}{k}\phi(\frac{x}{k})\}_{k=1}^\infty$ does not converge to 0 in $\mathscr{C}_0^\infty(\mathbb{R})$.

Lemma 4.1. *For every set $X_1 \subset X$ and every $\varepsilon > 0$ there exists a function $\phi_\varepsilon \in \mathscr{C}^\infty(\mathbb{R}^n)$ such that $\phi_\varepsilon(x) = 1$ when $x \in X_1^\varepsilon$, $\phi_\varepsilon(x) = 0$ when $x \in \mathbb{R}^n \backslash X_1^{3\varepsilon}$ and $0 \le \phi_\varepsilon(x) \le 1$ when $x \in \mathbb{R}^n$.*

Proof. Let $\theta_{X_1^{2\varepsilon}}$ be the characteristic function of the set $X_1^{2\varepsilon}$, i.e., $\theta_{X_1^{2\varepsilon}}(x) = 1$ for $x \in X_1^{2\varepsilon}$ and $\theta_{X_1^{2\varepsilon}}(x) = 0$ for $x \notin X_1^{2\varepsilon}$. Define

$$\begin{aligned}\phi_\varepsilon(x) &= \int_{\mathbb{R}^n} \theta_{X_1^{2\varepsilon}}(y)\omega_\varepsilon(x-y)dy \\ &= \int_{X_1^{2\varepsilon}} \omega_\varepsilon(x-y)dy, \quad x \in \mathbb{R}^n.\end{aligned}$$

We will prove that the function ϕ_ε has the required properties. Since $\omega_\varepsilon \in \mathscr{C}^\infty(\mathbb{R}^n)$, $0 \le \omega_\varepsilon(x)$, $x \in \mathbb{R}^n$, supp$\omega_\varepsilon = \overline{U}_\varepsilon$ and

$$\int_{\mathbb{R}^n} \omega_\varepsilon(x)dx = 1,$$

we get $\phi_\varepsilon \in \mathscr{C}^\infty(\mathbb{R}^n)$ and

$$\begin{aligned}0 &\le \phi_\varepsilon(x) \\ &= \int_{X_1^{2\varepsilon}} \omega_\varepsilon(x-y)dy \\ &\le \int_{\mathbb{R}^n} \omega_\varepsilon(x-y)dy \\ &= \int_{\mathbb{R}^n} \omega_\varepsilon(y)dy \\ &= 1, \quad x \in \mathbb{R}^n,\end{aligned}$$

and

$$\begin{aligned}\phi_\varepsilon(x) &= \int_{\mathbb{R}^n} \theta_{X_1^{2\varepsilon}}(y)\omega_\varepsilon(x-y)dy \\ &= \int_{\overline{U}(x,\varepsilon)} \theta_{X_1^{2\varepsilon}}(y)\omega_\varepsilon(x-y)dy \\ &= \begin{cases} \int_{\overline{U}(x,\varepsilon)} \omega_\varepsilon(x-y)dy = \int_{\overline{U}_\varepsilon} \omega_\varepsilon(y)dy = 1, & x \in X_1^\varepsilon, \\ 0, \quad x \notin X_1^{3\varepsilon}. \end{cases}\end{aligned}$$

This completes the proof. □

Lemma 4.2 (Expansion of the Unit Function). *Let $\phi \in \mathscr{C}_0^\infty(\mathbb{R}^n)$ and suppϕ be covered by a finite number of neighborhoods $U(x_k, r_k)$ for some $x_k \in \mathbb{R}^n$, $r_k > 0$, $k = 1, 2, \ldots, m$, and for some $m \in \mathbb{N}$. Then there exist functions $\phi h_k \in \mathscr{C}_0^\infty(U(x_k, r_k))$, $k = 1, 2, \ldots, m$, such that $\sum_{k=1}^m h_k(x) = 1$ for x in a neighborhood of suppϕ and supp$(\phi h_k) \subset U(x_k, r_k)$, $k = 1, 2, \ldots, m$.*

Proof. For $k = 1, 2, \ldots, m$, take $0 < r'_k < r_k$ so that the union $\bigcup_{k=1}^m U(x_k, r'_k)$ covers suppϕ. Be Lemma 4.1, it follows that there exist functions $\eta_k \in \mathscr{C}_0^\infty(U(x_k, r_k))$, $k = 1, 2, \ldots, m$, so that $\eta_k(x) = 1$, $x \in U(x_k, r'_k)$, supp$\eta_k \subset U(x_k, r_k)$, $0 \le \eta_k(x) \le 1$, $x \in \mathbb{R}^n$, $k = 1, 2, \ldots, m$. Then, for $x \in \bigcup_{k=1}^m U(x_k, r'_k)$, we set

$$h(x) = \sum_{k=1}^m \eta_k(x), \quad h_k(x) = \frac{\eta_k(x)}{h(x)}, \quad k = 1, 2, \ldots, m,$$

and for $x \in \mathbb{R}^n \backslash (\bigcup_{k=1}^m U(x_k, r_k))$ we set $h(x) = h_k(x) = 0$, $k = 1, 2, \ldots, m$. Note that $\phi h_k \in \mathscr{C}_0^\infty(U(x_k, r_k))$, $k = 1, 2, \ldots, m$, and

$$h(x) \ge 1, \quad \sum_{k=1}^m h_k(x) = \sum_{k=1}^m \frac{\eta_k(x)}{h(x)} = 1, \quad x \in \bigcup_{k=1}^m U(x_k, r'_k).$$

This completes the proof. □

Definition 4.5. We say that the sequence $\{\eta_k\}_{k=1}^\infty$ in $\mathscr{C}_0^\infty(\mathbb{R}^n)$ converges to 1 in $\mathbb{R}^n$ if

1. for every $\alpha \in \mathbb{N}^n \cup \{0\}$ there exists a constant $c_\alpha > 0$ such that $|D^\alpha \eta_k(x)| \le c_\alpha$ for every $k \in \mathbb{N}$ and every $x \in \mathbb{R}^n$,
2. for every compact set K in $\mathbb{R}^n$ there exists $N = N(K) \in \mathbb{N}$ such that $\eta_k(x) = 1$ for every $k > N$ and $x \in K$.

Example 4.4. Choose $\eta \in \mathscr{C}_0^\infty(\mathbb{R}^n)$ so that $\eta(x) = 1$ for $|x| \le 1$. Set $\eta_k(x) = \eta(\frac{x}{k})$, $x \in \mathbb{R}^n$, $k \in \mathbb{N}$. Then the sequence $\{\eta_k\}_{k=1}^\infty$ tends to 1 in $\mathbb{R}^n$.

Definition 4.6. With $\mathscr{S}(\mathbb{R}^n)$ we denote the space of $\mathscr{C}^\infty$ functions ϕ such that

$$\sup_{x \in \mathbb{R}^n} |x|^\beta |D^\alpha \phi(x)| < \infty, \quad \forall \alpha \in \mathbb{N}^n \cup \{0\}, \quad \beta \in \mathbb{N} \cup \{0\}.$$

Here, $|x| = (x_1^2+x_2^2+\cdots+x_n^2)^{\frac{1}{2}}$ and $x = (x_1, x_2, \ldots, x_n)$. By $||\cdot||_{\mathscr{S},p}$, $p \in \mathbb{N}$, we shall indicate the norm

$$||\phi||_{\mathscr{S},p} = \sup_{x\in\mathbb{R}^n, |\alpha|\le p} (1+|x|^2)^{\frac{p}{2}}|D^\alpha\phi(x)|, \quad \phi \in \mathscr{S}(\mathbb{R}^n). \tag{4.1}$$

Example 4.5. Let $n = 1$ and $\phi(x) = e^{-x^2}$, $x \in \mathbb{R}$. Then $\phi \in \mathscr{S}(\mathbb{R})$.

Example 4.6. Let $n = 1$, ϕ be as in Example 4.5 and $p(x) = a_0x^k + a_1x^{k-1} + \cdots + a_{k-1}x + a_k$, $x \in \mathbb{R}$, where $a_j \in \mathbb{R}$, $j = 0, 1, \ldots, k$, $k \in \mathbb{N}$. Then $\phi p \in \mathscr{S}(\mathbb{R})$.

Exercise 4.1. Prove that (12.5) satisfies all axioms for a norm.

Theorem 4.1. *The space $\mathscr{C}_0^\infty(\mathbb{R}^n)$ is a proper subset of the space $\mathscr{S}(\mathbb{R}^n)$.*

Proof. Let $\phi \in \mathscr{C}_0^\infty(\mathbb{R}^n)$, $\alpha \in \mathbb{N}^n \cup \{0\}$ and $\beta \in \mathbb{N} \cup \{0\}$ be arbitrarily chosen. Set $K = \text{supp}\phi$ and $\psi(x) = |x|^\beta D^\alpha\phi(x)$, $x \in \mathbb{R}^n$. Then $\text{supp}\psi = K$ and there exists a constant $K_1 > 0$ so that

$$\sup_{x\in\mathbb{R}^n} |\psi(x)| = \max_{x\in K} |\psi(x)| \le K_1 < \infty.$$

Because $\alpha \in \mathbb{N}^n \cup \{0\}$ and $\beta \in \mathbb{N} \cup \{0\}$ were arbitrarily chosen, we conclude that $\phi \in \mathscr{S}(\mathbb{R}^n)$. Now, using that $\phi \in \mathscr{C}_0^\infty(\mathbb{R}^n)$ was arbitrarily chosen and we get that it is an element of the space $\mathscr{S}(\mathbb{R}^n)$, we obtain the inclusion $\mathscr{C}_0^\infty(\mathbb{R}^n) \subseteq \mathscr{S}(\mathbb{R}^n)$. Let $\psi_1(x) = e^{-|x|^2}$, $x \in \mathbb{R}^n$. Then $\psi_1 \in \mathscr{S}(\mathbb{R}^n)$ and $\psi_1 \notin \mathscr{C}_0^\infty(\mathbb{R}^n)$. Thus, $\mathscr{C}_0^\infty(\mathbb{R}^n) \subset \mathscr{S}(\mathbb{R}^n)$. This completes the proof. □

Theorem 4.2. *Let $\phi, \psi \in \mathscr{S}(\mathbb{R}^n)$. Then $\phi\psi \in \mathscr{S}(\mathbb{R}^n)$.*

Proof. Let $\alpha \in \mathbb{N}^n \cup \{0\}$ and $\beta \in \mathbb{N} \cup \{0\}$ be arbitrarily chosen. Since $\phi, \psi \in \mathscr{S}(\mathbb{R}^n)$, we have that

$$\sup_{x\in\mathbb{R}^n} |x|^\beta |D^\gamma\phi(x)| < \infty, \quad \sup_{x\in\mathbb{R}^n} |D^{\alpha-\gamma}\psi(x)| < \infty$$

for any $\gamma \in \mathbb{N}^n \cup \{0\}$, $\gamma \le \alpha$. Then, using Leibnitz's rule, we find

$$\sup_{x\in\mathbb{R}^n} |x|^\beta |D^\alpha(\phi\psi)(x)| = \sup_{x\in\mathbb{R}^n} |x|^\beta \left| \sum_{\gamma:\gamma\le\alpha} \binom{\alpha}{\gamma} D^\gamma\phi(x)D^{\alpha-\gamma}\psi(x)\right|$$

$$\leq \sup_{x\in\mathbb{R}^n} |x|^\beta \sum_{\gamma:\gamma\leq\alpha} \binom{\alpha}{\gamma} |D^\gamma\phi(x)||D^{\alpha-\gamma}\psi(x)|$$

$$\leq \sum_{\gamma:\gamma\leq\alpha} \binom{\alpha}{\gamma} \left(\sup_{x\in\mathbb{R}^n} |x|^\beta |D^\gamma\phi(x)|\right)$$

$$\times \left(\sup_{x\in\mathbb{R}^n} |D^{\alpha-\gamma}\psi(x)|\right)$$

$$< \infty,$$

i.e., $\phi\psi \in \mathscr{S}(\mathbb{R}^n)$. This completes the proof. □

Definition 4.7. We say that the sequence $\{\phi_k\}_{k=1}^\infty$ of elements of $\mathscr{S}(\mathbb{R}^n)$ converges to 0 in $\mathscr{S}(\mathbb{R}^n)$, if for every $p \in \mathbb{N}\cup\{0\}$ and every $\alpha \in \mathbb{N}^n \cup \{0\}$, we have

$$\lim_{k\to\infty} (1+|x|^2)^{\frac{p}{2}} |D^\alpha\phi_k(x)| = 0$$

uniformly.

Theorem 4.3. *Let $\{\phi_k\}_{k=1}^\infty$ be a sequence of elements of $\mathscr{C}_0^\infty(\mathbb{R}^n)$ such that $\phi_k \to_{k\to\infty} 0$ in $\mathscr{C}_0^\infty(\mathbb{R}^n)$. Then $\phi_k \to_{k\to\infty} 0$ in $\mathscr{S}(\mathbb{R}^n)$.*

Proof. Since $\phi_k \to_{k\to\infty} 0$ in $\mathscr{C}_0^\infty(\mathbb{R}^n)$, there exists a compact set $K \subset \mathbb{R}^n$ so that $\mathrm{supp}\phi_k \subset K$ for any $k\in\mathbb{N}$ and $D^\alpha\phi_k(x) \to_{k\to\infty} 0$ uniformly for any multi-index $\alpha \in \mathbb{N}^n\cup\{0\}$. Let $K_1 > 0$ be a constant so that $1+|x|^2 \leq K_1$ for any $x \in K$. Then, for any $p \in \mathbb{N}\cup\{0\}$, we have

$$0 \leq \lim_{k\to\infty} (1+|x|^2)^{\frac{p}{2}} |D^\alpha\phi(x)| \leq K_1^{\frac{p}{2}} \lim_{k\to\infty} |D^\alpha\phi_k(x)| = 0$$

uniformly, i.e., $\phi_k \to_{k\to\infty} 0$ in $\mathscr{S}(\mathbb{R}^n)$. This completes the proof. □

Theorem 4.4. *The space $\mathscr{C}_0^\infty(\mathbb{R}^n)$ is dense in the space $\mathscr{S}(\mathbb{R}^n)$.*

Proof. Let $\phi \in \mathscr{S}(\mathbb{R}^n)$ and $\eta \in \mathscr{C}_0^\infty(\mathbb{R}^n)$ be chosen so that $\eta(x) = 1$ for $|x| < 1$. Set $\phi_k(x) = \phi(x)\eta(\frac{x}{k})$, $x \in \mathbb{R}^n$, $k \in \mathbb{N}$. Then $\phi_k \in \mathscr{C}_0^\infty(\mathbb{R}^n)$ and $\phi_k \to_{k\to\infty} \phi$ in $\mathscr{C}_0^\infty(\mathbb{R}^n)$. Now, applying Theorem 4.3, we conclude that $\phi_k \to_{k\to\infty} \phi$ in $\mathscr{S}(\mathbb{R}^n)$ as well. This completes the proof. □

Remark 4.1. Note that if $a \in \mathscr{C}^\infty(\mathbb{R}^n)$ and $\phi \in \mathscr{S}(\mathbb{R}^n)$, it does not follow that $a\phi \in \mathscr{S}(\mathbb{R}^n)$. Take for instance $a(x) = e^{|x|^2}$, $x \in \mathbb{R}^n$, and $\phi(x) = e^{-|x|^2}$, $x \in \mathbb{R}^n$. Then $a \in \mathscr{C}^\infty(\mathbb{R}^n)$, $\phi \in \mathscr{S}(\mathbb{R}^n)$ and $a(x)\phi(x) = 1$, $x \in \mathbb{R}^n$. Consequently, $a\phi \notin \mathscr{S}(\mathbb{R}^n)$.

Definition 4.8. By Θ_M we denote the space of functions $a \in \mathscr{C}^\infty(\mathbb{R}^n)$ for which there exist constants $C_a > 0$ and $m_a \in \mathbb{N}$ such that

$$|D^\alpha a(x)| \le C_a(1 + |x|)^{m_a}, \quad x \in \mathbb{R}^n,$$

for every $\alpha \in \mathbb{N}^n \cup \{0\}$. Such functions are called multipliers of $\mathscr{S}(\mathbb{R}^n)$.

Theorem 4.5. *Let $a \in \Theta_M$ and $\phi \in \mathscr{S}(\mathbb{R}^n)$. Then $a\phi \in \mathscr{S}(\mathbb{R}^n)$.*

Proof. Let $\beta \in \mathbb{N} \cup \{0\}$ and $\alpha \in \mathbb{N}^n \cup \{0\}$ be arbitrarily chosen. Then

$$\begin{aligned}
&\sup_{x\in\mathbb{R}^n} |x|^\beta |D^\alpha(a\phi)(x)| \\
&\quad = \sup_{x\in\mathbb{R}^n} |x|^\beta \left| \sum_{\gamma\le\alpha} \binom{\alpha}{\gamma} D^\gamma a(x) D^{\alpha-\gamma}\phi(x) \right| \\
&\quad \le \sup_{x\in\mathbb{R}^n} |x|^\beta \left(\sum_{\gamma\le\alpha} \binom{\alpha}{\gamma} |D^\gamma a(x)| |D^{\alpha-\gamma}\phi(x)| \right) \\
&\quad \le \sup_{x\in\mathbb{R}^n} |x|^\beta \left(\sum_{\gamma\le\alpha} \binom{\alpha}{\gamma} C_a(1+|x|)^{m_a} |D^{\alpha-\gamma}\phi(x)| \right) \\
&\quad = \sum_{\gamma\le\alpha} \binom{\alpha}{\gamma} C_a \sup_{x\in\mathbb{R}^n} \left(|x|^\beta (1+|x|)^{m_a} |D^{\alpha-\gamma}\phi(x)| \right) \\
&\quad < \infty.
\end{aligned}$$

Thus, $a\phi \in \mathscr{S}(\mathbb{R}^n)$. This completes the proof. □

Theorem 4.6. *Let $a \in \Theta_M$. Then the map $\phi \to a\phi$ from $\mathscr{S}(\mathbb{R}^n)$ to $\mathscr{S}(\mathbb{R}^n)$ is a continuous map.*

Proof. Let $\phi \in \mathscr{S}(\mathbb{R}^n)$ and $\{\phi_k\}_{k=1}^{\infty}$ be a sequence of elements of $\mathscr{S}(\mathbb{R}^n)$ that converges to ϕ in $\mathscr{S}(\mathbb{R}^n)$. Then, for any $\beta \in \mathbb{N} \cup \{0\}$ and any $\alpha \in \mathbb{N}^n \cup \{0\}$, we have

$$\lim_{k\to\infty} (1+|x|^2)^{\frac{\beta}{2}} |D^\alpha \phi_k(x) - D^\alpha \phi(x)| = 0$$

uniformly. Hence, for any $\beta \in \mathbb{N} \cup \{0\}$ and $\alpha \in \mathbb{N}^n \cup \{0\}$, we get

$$\begin{aligned}
&\lim_{k\to\infty} (1+|x|^2)^{\frac{\beta}{2}} |D^\alpha (a\phi_k)(x) - D^\alpha (a\phi)(x)| \\
&= \lim_{k\to\infty} (1+|x|^2)^{\frac{\beta}{2}} \left| \sum_{\gamma\leq\alpha} \binom{\alpha}{\gamma} D^\gamma a(x) D^{\alpha-\gamma}\phi_k(x) \right. \\
&\qquad \left. - \sum_{\gamma\leq\alpha} \binom{\alpha}{\gamma} D^\gamma a(x) D^{\alpha-\gamma}\phi(x) \right| \\
&= \lim_{k\to\infty} (1+|x|^2)^{\frac{\beta}{2}} \left| \sum_{\gamma\leq\alpha} \binom{\alpha}{\gamma} D^\gamma a(x) (D^{\alpha-\gamma}\phi_k(x) - D^{\alpha-\gamma}\phi(x)) \right| \\
&\leq \lim_{k\to\infty} (1+|x|^2)^{\frac{\beta}{2}} \sum_{\gamma\leq\alpha} \binom{\alpha}{\gamma} |D^\gamma a(x)| |D^{\alpha-\gamma}\phi_k(x) - D^{\alpha-\gamma}\phi(x)| \\
&\leq \sum_{\gamma\leq\alpha} \binom{\alpha}{\gamma} C_a \lim_{k\to\infty} (1+|x|^2)^{\frac{\beta}{2}} (1+|x|)^{m_a} \\
&\quad \times |D^{\alpha-\gamma}\phi_k(x) - D^{\alpha-\gamma}\phi(x)| \\
&= 0
\end{aligned}$$

uniformly. Thus, $a\phi_k \to_{k\to\infty} a\phi$ in $\mathscr{S}(\mathbb{R}^n)$. This completes the proof. □

Exercise 4.2. Prove that the maps $\phi \to D^\alpha \phi$, $\alpha \in \mathbb{N}^n \cup \{0\}$, and $\phi(x) \to \phi(Ax+b)$, where A is an $n \times n$ matrix with $\det A \neq 0$, are linear and continuous maps from $\mathscr{S}(\mathbb{R}^n)$ to itself.

Definition 4.9. For $p \in \mathbb{N} \cup \{0\}$, with $\mathscr{S}_p(\mathbb{R}^n)$ we will denote the completion of the space $\mathscr{S}(\mathbb{R}^n)$ with respect to $\|\cdot\|_p$.

Definition 4.10. Let X and Y be two normed vector spaces with norms $\|\cdot\|_X$ and $\|\cdot\|_Y$, respectively. We say that X is compactly embedded in Y, and we write $X \hookrightarrow Y$ or $Y \hookleftarrow X$, if

1. X is continuously embedded in Y. i.e., there is a constant $C > 0$ such that $\|x\|_Y \leq C\|x\|_X$ for any $x \in X$,
2. any bounded set in X is totally bounded in Y, i.e., every sequence in such a bounded set has a subsequence that is a Cauchy sequence in the norm $\|\cdot\|_Y$.

Theorem 4.7. *The spaces $\mathscr{S}_p(\mathbb{R}^n), p \in \mathbb{N} \cup \{0\}$, are Banach spaces fitting in a chain of continuous and compact embedding*

$$\mathscr{S}_0(\mathbb{R}^n) \hookleftarrow \mathscr{S}_1(\mathbb{R}^n) \hookleftarrow \mathscr{S}_2(\mathbb{R}^n) \hookleftarrow \cdots. \tag{4.2}$$

Proof. Let $p \in \mathbb{N} \cup \{0\}$ be arbitrarily chosen. Take $\phi \in \mathscr{S}_{p+1}(\mathbb{R}^n)$ arbitrarily. Then

$$\begin{aligned}\|\phi\|_{\mathscr{S},p+1} &= \sup_{x\in\mathbb{R}^n, |\alpha|\leq p} (1+|x|^2)^{\frac{p+1}{2}} |D^\alpha \phi(x)| \\ &\geq \sup_{x\in\mathbb{R}^n, |\alpha|\leq p} (1+|x|^2)^{\frac{p}{2}} |D^\alpha \phi(x)| \\ &= \|\phi\|_{\mathscr{S},p}.\end{aligned}$$

Hence, $\phi \in \mathscr{S}_p(\mathbb{R}^n)$ and the embedding (4.2) are continuous. Now, let M be an infinitely bounded set in $\mathscr{S}_{p+1}(\mathbb{R}^n)$. Then there exists a constant $C > 0$ such that $||\phi||_{p+1} \leq C$ for every $\phi \in M$. Hence,

$$\left|D^\alpha \phi(x)\right| \leq C$$

for every $x \in \mathbb{R}^n$, $\alpha \in \mathbb{N}^n \cup \{0\}$, $|\alpha| \leq p$, $\phi \in M$. Therefore,

$$\begin{aligned}(1+|x|^2)^{\frac{p}{2}} |D^\alpha \phi(x)| &= \frac{(1+|x|^2)^{\frac{p+1}{2}} |D^\alpha \phi(x)|}{(1+|x|^2)^{\frac{1}{2}}} \\ &\leq \frac{C}{(1+|x|^2)^{\frac{1}{2}}} \to_{|x|\to\infty} 0.\end{aligned}$$

Let $\{R_k\}_{k=1}^\infty$ be an increasing sequence of positive numbers such that

$$(1+|x|^2)^{\frac{p}{2}} |D^\alpha \phi(x)| \leq \frac{1}{k} \quad \text{for } |x| > R_k, |\alpha| \leq p.$$

By the Arzela–Ascoli theorem, it follows that there exists a sequence $\{\phi_j^{(1)}\}_{j=1}^\infty$ of elements of M that converges in $\mathscr{S}_p(\overline{U}_{R_1})$. We may then find a sequence $\{\phi_j^{(2)}\}_{j=1}^{\infty}$ converging in $\mathscr{S}_p(\overline{U}_{R_2})$, and so on. The sequence $\{\phi_k^{(k)}\}_{k=1}^{\infty}$ converges in $\mathscr{S}_p(\mathbb{R}^n)$. This completes the proof. $\square$

Theorem 4.8. *If $\phi \in \mathscr{C}^p(\mathbb{R}^n)$ and $|x|^p D^\alpha \phi(x) \to_{|x|\to\infty} 0$ for $\alpha \in \mathbb{N}^n \cup \{0\}$, $|\alpha| \le p$, then $\phi \in \mathscr{S}_p(\mathbb{R}^n)$.*

Proof. To prove this assertion, we choose a sequence $\{\eta_k\}_{k=1}^{\infty}$ of elements in $\mathscr{C}_0^\infty(\mathbb{R}^n)$ such that $\eta_k \to_{k\to\infty} 1$ in $\mathbb{R}^n$. Fix $\varepsilon > 0$. Since $|x|^p |D^\alpha \phi(x)| \to_{|x|\to\infty} 0$ for any $\alpha \in \mathbb{N}^n \cup \{0\}$, $|\alpha| \le p$, it follows that there exists $R = R(\varepsilon) > 0$ such that the inequality

$$(1+|x|^2)^{\frac{p}{2}} |D^\alpha \phi(x)| < \varepsilon$$

holds for $|x| > R$. As $\eta_k \to_{k\to\infty} 1$ in $\mathbb{R}^n$, there exists $N \in \mathbb{N}$ such that $\eta_k(x) = 1$ for every $k > N$ and $|x| \le R+1$. Now, define

$$\phi_{\frac{1}{k}}(x) = \int_{\mathbb{R}^n} \phi(y)\omega_{\frac{1}{k}}(x-y)dy.$$

Observe that $\{\phi_{\frac{1}{k}}\eta_k\}_{k=1}^{\infty}$ is a sequence in $\mathscr{C}_0^\infty(\mathbb{R}^n)$ and there exists $N_1 \in \mathbb{N}$ so that

$$\sup_{\substack{x\in\mathbb{R}^n \\ |\alpha|\le p}} (1+|x|^2)^{\frac{p}{2}} |D^\alpha(\phi - \phi_{\frac{1}{k}}\eta_k)(x)| < \varepsilon$$

for $k > N_1$, $|x| \le R+1$. Set $N_2 = \max\{N, N_1\}$. Then, for $k > N_2$, we have

$$\begin{aligned}
||\phi - \phi_{\frac{1}{k}}\eta_k||_{\mathscr{S},p} &= \sup_{\substack{x\in\mathbb{R}^n \\ |\alpha|\le p}} (1+|x|^2)^{\frac{p}{2}} |D^\alpha(\phi - \phi_{\frac{1}{k}}\eta_k)(x)| \\
&\le \sup_{\substack{|x|\le R+1 \\ |\alpha|\le p}} (1+|x|^2)^{\frac{p}{2}} |D^\alpha(\phi - \phi_{\frac{1}{k}}\eta_k)(x)| \\
&\quad + \sup_{\substack{|x|> R+1 \\ |\alpha|\le p}} (1+|x|^2)^{\frac{p}{2}} |D^\alpha(\phi - \phi_{\frac{1}{k}}\eta_k)(x)| \\
&\le \sup_{\substack{|x|\le R+1 \\ |\alpha|\le p}} (1+|x|^2)^{\frac{p}{2}} |D^\alpha(\phi - \phi_{\frac{1}{k}}\eta_k)(x)| \\
&\quad + \sup_{\substack{|x|> R+1 \\ |\alpha|\le p}} (1+|x|^2)^{\frac{p}{2}}
\end{aligned}$$

$$\times \left(|D^\alpha \phi(x)| + \sum_{\beta \le \alpha} \binom{\alpha}{\beta} |D^\beta \phi_{\frac{1}{k}}(x) D^{\alpha-\beta} \eta_k(x)| \right)$$

$$< \varepsilon + \sup_{\substack{|x|>R+1 \\ |\alpha| \le p}} (1+|x|^2)^{\frac{p}{2}} |D^\alpha \phi(x)|$$

$$+ \sup_{\substack{|x|>R+1 \\ |\alpha| \le p}} (1+|x|^2)^{\frac{p}{2}} \sum_{\beta \le \alpha} \binom{\alpha}{\beta} |D^\beta \phi_{\frac{1}{k}}(x) D^{\alpha-\beta} \eta_k(x)|$$

$$< 2\varepsilon + \sup_{\substack{|x|>R+1 \\ |\alpha| \le p}} (1+|x|^2)^{\frac{p}{2}} |D^\alpha \phi_{\frac{1}{k}}(x)|$$

$$\le 2\varepsilon + \sup_{\substack{|x|>R+1 \\ |\alpha| \le p}} (1+|x|^2)^{\frac{p}{2}} \int_{\mathbb{R}^n} |D^\alpha \phi(x-y)| \omega_{\frac{1}{k}}(y) dy$$

$$= 2\varepsilon + \sup_{\substack{|x|>R+1 \\ |\alpha| \le p}} \int_{\mathbb{R}^n} (1+|x-y+y|^2)^{\frac{p}{2}} |D^\alpha \phi(x-y)| \omega_{\frac{1}{k}}(y) dy$$

$$\le 2\varepsilon + \sup_{\substack{|x|>R+1 \\ |\alpha| \le p}} \int_{\mathbb{R}^n} (1+(|x-y|+|y|)^2)^{\frac{p}{2}} |D^\alpha \phi(x-y)| \omega_{\frac{1}{k}}(y) dy$$

$$< 2\varepsilon + 2^{\frac{p}{2}} \sup_{\substack{|x|>R+1 \\ |\alpha| \le p}} \int_{\mathbb{R}^n} (1+|x-y|^2+|y|^2)^{\frac{p}{2}} |D^\alpha \phi(x-y)| \omega_{\frac{1}{k}}(y) dy$$

$$\le 2\varepsilon + 2^{\frac{p}{2}} \sup_{\substack{|x|>R+1 \\ |\alpha| \le p}} \int_{\mathbb{R}^n} ((1+|x-y|^2)$$

$$+ (1+|y|^2))^{\frac{p}{2}} |D^\alpha \phi(x-y)| \omega_{\frac{1}{k}}(y) dy$$

$$\le 2\varepsilon + 2^p \sup_{\substack{|x|>R+1 \\ |\alpha| \le p}} \int_{\mathbb{R}^n} ((1+|x-y|^2)^{\frac{p}{2}}$$

$$+ (1+|y|^2)^{\frac{p}{2}}) |D^\alpha \phi(x-y)| \omega_{\frac{1}{k}}(y) dy$$

$$= 2\varepsilon + 2^p \sup_{\substack{|x|>R+1 \\ |\alpha| \le p}} \int_{\mathbb{R}^n} (1+|x-y|^2)^{\frac{p}{2}} |D^\alpha \phi(x-y)| \omega_{\frac{1}{k}}(y) dy$$

$$+ 2^p \sup_{\substack{|x|>R+1 \\ |\alpha| \le p}} \int_{\mathbb{R}^n} (1+|y|^2)^{\frac{p}{2}} |D^\alpha \phi(x-y)| \omega_{\frac{1}{k}}(y) dy$$

$$< 2\varepsilon + 2^p\varepsilon \int_{\mathbb{R}^n} \omega_{\frac{1}{k}}(y)dy + 2^p\varepsilon \int_{\mathbb{R}^n} (1 + |y|^2)^{\frac{p}{2}} \omega_{\frac{1}{k}}(y)dy$$
$$\leq 2\varepsilon + 2^p\varepsilon \int_{\mathbb{R}^n} \omega_{\frac{1}{k}}(y)dy + 2^{\frac{3p}{2}}\varepsilon \int_{\mathbb{R}^n} \omega_{\frac{1}{k}}(y)dy$$
$$= \varepsilon(2 + 2^p + 2^{\frac{3p}{2}}).$$

Since $\varepsilon > 0$ was arbitrarily chosen, we conclude that $\phi_{\frac{1}{k}}\eta_k \to_{k\to\infty} \phi$ in $\mathscr{S}_p(\mathbb{R}^n)$. Now, using the fact that $\mathscr{S}_p(\mathbb{R}^n)$ is a Banach space, we conclude $\phi \in \mathscr{S}_p$. This completes the proof. □

Theorem 4.9. *We have*

$$\mathscr{S}(\mathbb{R}^n) = \bigcap_{p\in\mathbb{N}\cup\{0\}} \mathscr{S}_p(\mathbb{R}^n). \tag{4.3}$$

Proof. Let $\phi \in \mathscr{S}(\mathbb{R}^n)$ be arbitrarily chosen. By the definition of the space $\mathscr{S}(\mathbb{R}^n)$, it follows that for any $\alpha \in \mathbb{N}^n \cup \{0\}$ and any $p \in \mathbb{N} \cup \{0\}$, we have

$$\sup_{x\in\mathbb{R}^n} |x|^p |D^\alpha \phi(x)| < \infty.$$

Hence,

$$\sup_{x\in\mathbb{R}^n} (1 + |x|^2)^{\frac{p}{2}} |D^\alpha \phi(x)| < \infty \tag{4.4}$$

for any $\alpha \in \mathbb{N}^n \cup \{0\}$ and any $p \in \mathbb{N} \cup \{0\}$. In particular, by (4.4), we get

$$\sup_{x\in\mathbb{R}^n, |\alpha|\leq p} (1 + |x|^2)^{\frac{p}{2}} |D^\alpha \phi(x)| < \infty$$

for any $p \in \mathbb{N}\cup\{0\}$. From here, we obtain that $\phi \in \mathscr{S}_p(\mathbb{R}^n)$ for any $p \in \mathbb{N}\cup\{0\}$ and so, $\phi \in \bigcap_{p\in\mathbb{N}\cup\{0\}} \mathscr{S}_p(\mathbb{R}^n)$. Because $\phi \in \mathscr{S}(\mathbb{R}^n)$ was arbitrarily chosen and we get that it is an element of $\bigcap_{p\in\mathbb{N}\cup\{0\}} \mathscr{S}_p(\mathbb{R}^n)$, we find the following inclusion

$$\mathscr{S}(\mathbb{R}^n) \subseteq \bigcap_{p\in\mathbb{N}\cup\{0\}} \mathscr{S}_p(\mathbb{R}^n). \tag{4.5}$$

Let now, $\psi \in \bigcap_{p\in\mathbb{N}\cup\{0\}} \mathscr{S}_p(\mathbb{R}^n)$ be arbitrarily chosen. Then $\psi \in \mathscr{S}_p(\mathbb{R}^n)$ for any $p \in \mathbb{N}\cup\{0\}$. From here,

$$\sup_{x\in\mathbb{R}^n, |\alpha|\le p} (1+|x|^2)^{\frac{p}{2}}|D^\alpha \psi(x)| < \infty \tag{4.6}$$

for any $p \in \mathbb{N}\cup\{0\}$. Now, take $\beta \in \mathbb{N}\cup\{0\}$ and $\alpha_1 \in \mathbb{N}^n \cup \{0\}$ arbitrarily. Then there is a $\gamma \in \mathbb{N}\cup\{0\}$ so that $\gamma \ge \beta$ and $|\alpha_1| \le \gamma$. Hence,

$$\begin{aligned}\sup_{x\in\mathbb{R}^n} |x|^\beta |D^{\alpha_1}\psi(x)| &\le \sup_{x\in\mathbb{R}^n} (1+|x|^2)^{\frac{\gamma}{2}}|D^{\alpha_1}\psi(x)| \\ &\le \sup_{x\in\mathbb{R}^n, |\alpha_1|\le\gamma} (1+|x|^2)^{\frac{\gamma}{2}}|D^{\alpha_1}\psi(x)| \\ &< \infty,\end{aligned}$$

where we have used (4.6). Since $\beta \in \mathbb{N}\cup\{0\}$ and $\alpha_1 \in \mathbb{N}^n \cup \{0\}$ were arbitrarily chosen, we conclude that $\psi \in \mathscr{S}(\mathbb{R}^n)$. Because $\psi \in \bigcap_{p\in\mathbb{N}\cup\{0\}} \mathscr{S}_p(\mathbb{R}^n)$ was arbitrarily chosen and we find that it is an element of $\mathscr{S}(\mathbb{R}^n)$, we arrive at the inclusion

$$\bigcap_{p\in\mathbb{N}\cup\{0\}} \mathscr{S}_p(\mathbb{R}^n) \subseteq \mathscr{S}(\mathbb{R}^n).$$

By the last relation and by (4.5), we obtain (4.3). This completes the proof. □

4.2 Definition for a Distribution: Properties

Let X be an open set in $\mathbb{R}^n$, $n \in \mathbb{N}$ a fixed integer.

Definition 4.11. Every linear continuous map $u : \mathscr{C}_0^\infty(X) \to \mathbb{C}$ is called a distribution or generalized function. In other words, a distribution is a linear map $u : \mathscr{C}_0^\infty(X) \to \mathbb{C}$ such that $u(\phi_n) \to_{n\to\infty} u(\phi)$ for every sequence $\{\phi_n\}_{n=1}^\infty$ in $\mathscr{C}_0^\infty(X)$ converging to $\phi \in \mathscr{C}_0^\infty(X)$, as $n \to \infty$. The space of distributions on X will be denoted by $\mathscr{D}'(X)$. We will write $u(\phi)$ or (u, ϕ) for the value of the functional (generalized function, distribution) $u \in \mathscr{D}'(X)$ on the element $\phi \in \mathscr{C}_0^\infty(X)$.

Example 4.7. Suppose $0 \in X$ and take the map $u : \mathscr{C}_0^\infty(X) \to \mathbb{C}$ defined as follows

$$u(\phi) = \phi(0) \quad \text{for } \phi \in \mathscr{C}_0^\infty(X).$$

Let $\phi_1, \phi_2 \in \mathscr{C}_0^\infty(X)$ and $\alpha_1, \alpha_2 \in \mathbb{C}$. As

$$u(\phi_1) = \phi_1(0), \quad u(\phi_2) = \phi_2(0),$$

$$\begin{aligned} u(\alpha_1\phi_1 + \alpha_2\phi_2) &= (\alpha_1\phi_1 + \alpha_2\phi_2)(0) = \alpha_1\phi_1(0) + \alpha_2\phi_2(0) \\ &= \alpha_1 u(\phi_1) + \alpha_2 u(\phi_2), \end{aligned}$$

i.e., $u : \mathscr{C}_0^\infty(X) \to \mathbb{C}$ is linear. Let $\{\phi_n\}_{n=1}^\infty$ be a sequence in $\mathscr{C}_0^\infty(X)$ for which $\phi_n \to_{n\to\infty} \phi$ in $\mathscr{C}_0^\infty(X)$. Then there exists a compact set $K \subset X$ such that $supp\phi_n \subset K$ for every $n \in \mathbb{N}$ and $D^\alpha\phi_n \to D^\alpha\phi$ uniformly in X for every multi-index $\alpha \in \mathbb{N} \cup \{0\}$. In particular, $\phi_n(0) \to_{n\to\infty} \phi(0)$, and therefore $u(\phi_n) \to_{n\to\infty} u(\phi)$. Consequently the linear map $u : \mathscr{C}_0^\infty(X) \to \mathbb{C}$ is continuous. In other words, it is a distribution on X.

Exercise 4.3. Let $0 \in X$. For each multi-index α prove that the map $u : \mathscr{C}_0^\infty(X) \to \mathbb{C}$, defined by

$$u(\phi) = D^\alpha\phi(0) \quad \text{for } \phi \in \mathscr{C}_0^\infty(X),$$

is a distribution.

Exercise 4.4. Denote by δ_a or $\delta(x - a)$, $a \in \mathbb{C}^n$, Dirac's "delta" function at the point a:

$$\delta_a(\phi) = \phi(a) \quad \text{for } \phi \in \mathscr{C}_0^\infty(X).$$

Prove that δ_a is a distribution.

Exercise 4.5. Prove that the map $1 : \mathscr{C}_0^\infty(X) \to \mathbb{C}$, defined by

$$1(\phi) = \int_X \phi(x)dx \quad \text{for } \phi \in \mathscr{C}_0^\infty(X),$$

is a distribution.

Exercise 4.6. For $u \in L^p_{\text{loc}}(X)$, $p \geq 1$, we define $u : \mathscr{C}_0^\infty(X) \to \mathbb{C}$ by

$$u(\phi) = \int_X u(x)\phi(x)dx.$$

Prove that u is a distribution.

Exercise 4.7. Let $P\frac{1}{x} : \mathscr{C}_0^\infty(X) \to \mathbb{C}$ be the map defined by

$$P\frac{1}{x}(\phi) = P.V. \int_X \frac{\phi(x) - \phi(0)}{x} dx \quad \text{for } \phi \in \mathscr{C}_0^\infty(X).$$

Prove that $P\frac{1}{x} \in \mathscr{D}'(X)$.

Definition 4.12. The distributions $u, v \in \mathscr{D}'(X)$ are said to be equal if

$$u(\phi) = v(\phi)$$

for any $\phi \in \mathscr{C}_0^\infty(X)$.

Definition 4.13. The linear combination $\lambda u + \mu v$ of the distributions $u, v \in \mathscr{D}'(X)$ is the functional acting by the rule

$$(\lambda u + \mu v)(\phi) = \lambda u(\phi) + \mu v(\phi), \quad \phi \in \mathscr{C}_0^\infty(X).$$

This makes the set $\mathscr{D}'(X)$ a vector space.

Definition 4.14. Let $u \in \mathscr{D}'(X)$. We define a distribution $\overline{u} \in \mathscr{D}'(X)$, called the complex conjugate of u, by

$$\overline{u}(\phi) = \overline{u(\overline{\phi})}, \quad \phi \in \mathscr{C}_0^\infty(X).$$

The distributions

$$\mathrm{Re}(u) = \frac{u + \overline{u}}{2}, \quad \mathrm{Im}(u) = \frac{u - \overline{u}}{2i}$$

are called the real and imaginary parts of u, respectively. Equivalently,

$$u = \mathrm{Re}(u) + i\mathrm{Im}(u), \quad \overline{u} = \mathrm{Re}(u) - i\mathrm{Im}(u).$$

If $\mathrm{Im}(u) = 0$, then the distribution u is said to be a real distribution.

Exercise 4.8. Prove that the delta function is a real distribution.

Here are elementary properties of the distributions. If $u_1, u_2 \in \mathscr{D}'(X)$, then

1. $u_1 \pm u_2 \in \mathscr{D}'(X)$,
2. $\alpha u_1 \in \mathscr{D}'(X)$ for $\forall \alpha \in \mathbb{C}$.

These properties follow from the definition, so their proof is omitted.

Theorem 4.10. *$u \in \mathscr{D}'(X)$ if and only if for every compact subset K of X there exist constants C and k so that the inequality*

$$|u(\phi)| \leq C \sum_{|\alpha| \leq k} \sup_{x \in K} \Big| D^{\alpha} \phi(x) \Big| \tag{4.7}$$

holds for every $\phi \in \mathscr{C}_0^{\infty}(K)$.

Proof. Let $u \in \mathscr{D}'(X)$. We will prove that the inequality (4.7) holds. Actually, we suppose that there exists a compact set K in X so that

$$|u(\phi_n)| > n \sum_{\alpha \in \mathbb{N}^n \cup \{0\}} \sup_{x \in K} \Big| D^{\alpha} \phi_n(x) \Big| \tag{4.8}$$

holds for $\phi_n \in \mathscr{C}_0^{\infty}(K)$. We set

$$\psi_n(x) = \frac{\phi_n(x)}{n \sum_{\alpha \in \mathbb{N}^n \cup \{0\}} \sup_K \Big| D^{\alpha} \phi_n(x) \Big|}, \qquad x \in K.$$

From (4.8), we obtain

$$|u(\psi_n)| > 1. \tag{4.9}$$

By the definition of ψ_n, it follows that $\psi_n \to_{n \to \infty} 0$ in $\mathscr{C}_0^{\infty}(X)$. Since $u : \mathscr{C}_0^{\infty}(X) \to \mathbb{C}$ is continuous, we have $u(\psi_n) \to_{n \to \infty} 0$, which contradicts (4.9). Let now, $u : \mathscr{C}_0^{\infty}(X) \to \mathbb{C}$ is a linear map such that for every compact set K in X there exist constants $C > 0$ and $k \in \mathbb{N} \cup \{0\}$ for which (4.7) holds. We will prove that u is a distribution on X. To show this, we will prove that $u : \mathscr{C}_0^{\infty}(X) \to \mathbb{C}$ is continuous at 0. Let $\{\phi_n\}_{n=1}^{\infty}$ be a sequence in $\mathscr{C}_0^{\infty}(X)$ with $\phi_n \to_{n \to \infty} 0$ in $\mathscr{C}_0^{\infty}(X)$. Then

$$\sup_{x \in K} \Big| D^{\alpha} \phi_n(x) \Big| \to_{n \to \infty} 0$$

for every $|\alpha| \leq k$. Hence with (4.7), we conclude

$$u(\phi_n) \to_{n \to \infty} 0.$$

This completes the proof. □

Exercise 4.9. The function H defined by

$$H(x) = \begin{cases} 1 & \text{for } x \geq 0, \\ 0 & \text{for } x < 0, \quad x \in \mathbb{R}, \end{cases}$$

is called the Heaviside function. We define

$$H(\phi) = \int_{-\infty}^{\infty} H(x)\phi(x)dx,$$

$\phi \in \mathscr{C}_0^\infty(\mathbb{R})$. Using the inequality (4.7), prove that $H \in \mathscr{D}'(\mathbb{R})$.

Theorem 4.11. *A linear map $u : \mathscr{C}_0^\infty(X) \to \mathbb{C}$ is a distribution if and only if there exist functions $\rho_\alpha \in \mathscr{C}(X)$ such that*

$$|u(\phi)| \leq \sum_\alpha \sup_K \left|\rho_\alpha D^\alpha \phi\right|, \quad \forall \phi \in \mathscr{C}_0^\infty(K), \tag{4.10}$$

for every compact set $K \subset X$, and only a finite number of ρ_α vanish identically.

Proof.

1. Let u be a linear map from $\mathscr{C}_0^\infty(X)$ to $\mathbb{C}$ and $\rho_\alpha \in \mathscr{C}(X)$ be such that the inequality (4.10) holds for every $\phi \in \mathscr{C}_0^\infty(X)$ and every compact set K. Since $\rho_\alpha \in \mathscr{C}(X)$, there exists a constant C such that

$$\sup_K |\rho_\alpha| \leq C.$$

From this and (4.10), it follows that

$$|u(\phi)| \leq C \sum_\alpha \sup_K \left|D^\alpha \phi\right|.$$

As only finitely many ρ_α are zero, there is a constant k such that

$$|u(\phi)| \leq C \sum_{|\alpha| \leq k} \sup_K \left|D^\alpha \phi(x)\right|,$$

i.e., $u \in \mathscr{D}'(X)$.

2. Let $u \in \mathscr{D}'(X)$ and $\{K_j\}$ be compact subsets of X such that any compact subset is contained in some K_j. Take maps $\chi_j \in \mathscr{C}_0^\infty(X)$ with $\chi_j \equiv 1$ on K_j and define

$$\psi_j = \chi_j - \chi_{j-1} \quad j > 1,$$
$$\psi_1 = \chi_1.$$

Any $\phi \in \mathscr{C}_0^\infty(X)$ satisfies

$$\phi = \sum_{j=1}^{\infty} \psi_j \phi. \tag{4.11}$$

Note that only a finite number of summands in (12.4) vanish identically. Moreover,

$$\psi_j \neq 0 \quad \text{on } K_j \backslash K_{j-1} \quad \text{for } j > 1,$$
$$\psi_1 \neq 0 \quad \text{on } K_1.$$

Consequently,

$$supp(\psi_j \phi) \subset supp \psi_j.$$

As $\psi_j \phi$ has compact support, for every compact set K there are constants C and k_j such that

$$|u(\psi_j \phi)| \leq C \sum_{|\alpha| \leq k_j} \sup_K \left| D^\alpha(\psi_j \phi) \right|.$$

From this and (4.11), we obtain

$$\begin{aligned} |u(\phi)| &= \left| \sum_j u(\psi_j \phi) \right| \\ &\leq \sum_j |u(\psi_j \phi)| \\ &\leq C \sum_j \sum_{|\alpha| \leq k_j} \sup_K \left| D^\alpha(\psi_j \phi) \right| \\ &\leq C \sum_j \sum_{|\alpha| \leq k_j} \sum_{\beta \leq \alpha} \binom{\alpha}{\beta} \sup_K \left| D^\beta \psi_j \right| \sup_K \left| D^{\alpha-\beta} \phi \right|. \end{aligned}$$

If we set

$$\rho_\beta = \sum_j \sum_{|\alpha|\le k_j} \binom{\alpha}{\beta} D^\beta \psi_j,$$

we obtain

$$|u(\phi)| \le C \sum_{\beta\le\alpha} \sup_K \left|\rho_\beta D^{\alpha-\beta}\phi\right|.$$

This completes the proof. □

Theorem 4.12. *Let $u \in \mathscr{D}'(X)$ and let F be a relatively open subset of X with $supp u \subset F$. Then there exists a unique linear map $\tilde{u}$ on*

$$\{\phi : \phi \in \mathscr{C}^\infty(X), F \cap supp\phi \subset\subset X\}$$

such that

1. $\tilde{u}(\phi) = u(\phi)$ *for* $\phi \in \mathscr{C}_0^\infty(X)$,
2. $\tilde{u}(\phi) = 0$ *for* $\phi \in \mathscr{C}^\infty(X)$, $F \cap supp\phi = \emptyset$.

Proof.

1. (uniqueness) Let $\phi \in \mathscr{C}^\infty(X)$ and $F \cap supp\phi = K$. As K is compact, there exists $\psi \in \mathscr{C}_0^\infty(X)$ such that $\psi \equiv 1$ on a neighborhood of K. Let

$$\phi_0 = \psi\phi, \quad \phi_1 = (1-\psi)\phi,$$

Then

$$\phi = \phi_0 + \phi_1. \tag{4.12}$$

Therefore,

$$\tilde{u}(\phi) = \tilde{u}(\phi_0) + \tilde{u}(\phi_1).$$

Note that $\tilde{u}(\phi_1) = 0$. So,

$$\tilde{u}(\phi) = \tilde{u}(\phi_0) = u(\phi_0).$$

Now, suppose that there are two such distributions $\tilde{u}$, $\tilde{\tilde{u}}$. Then

$$\tilde{u}(\phi) = \tilde{u}(\phi_0), \quad \tilde{\tilde{u}}(\phi) = \tilde{\tilde{u}}(\phi_0).$$

Consequently $\tilde{u}(\phi) = \tilde{\tilde{u}}(\phi)$ for every $\phi \in \mathscr{C}^\infty(X)$ so that $F \cap supp\phi = K$. Therefore $\tilde{u} = \tilde{\tilde{u}}$.

2. (existence) Let

$$\phi = \phi_0' + \phi_1'$$

be another decomposition of kind (4.12). Define

$$\chi = \phi_0 - \phi_0'.$$

Then

$$\chi \in \mathscr{C}_0^\infty(X), \quad F \cap supp\chi = F \cap supp(\phi_1 - \phi_1') = \emptyset$$

and so,

$$u(\chi) = u(\phi_0) - u(\phi_0') = 0.$$

Define $\tilde{u}(\phi)$ by

$$\tilde{u}(\phi) = u(\phi_0).$$

This completes the proof. □

Theorem 4.13. *Let $\{X_i\}_{i\in I}$ be an open cover of $\mathbb{R}^n$ and suppose that $u_i \in \mathscr{D}'(X_i)$ satisfy $u_i = u_j$ on $X_i \cap X_j$. Then there exists a unique distribution $u \in \mathscr{D}'(X)$ such that $u_{|X_i} = u_i$ for every $i \in I$.*

Proof. Take $\phi \in \mathscr{C}_0^\infty(X)$ and $\phi_i \in \mathscr{C}_0^\infty(X_i)$. Define $\phi = \sum\limits_i \phi_i$ and

$$u(\phi) = \sum_i u_i(\phi_i). \tag{4.13}$$

We claim that the definition (4.13) does not depend on the choice of the sequence $\{\phi_i\}$. For this purpose, it is enough to prove that $\sum_i \phi_i = 0$ implies $u(\sum_i \phi_i) = 0$. Set $K = \bigcup_i supp\phi_i$. Then K is a compact set. There exist functions $\psi_k \in \mathscr{C}_0^\infty(X_k)$ such that $0 \le \psi_k \le 1$ and $\sum_k \psi_k = 1$ *on* K. Note that only a finite number of the above summands are different from zero. Moreover, $\psi_k\phi_i \in \mathscr{C}_0^\infty(X_k \cap X_i)$ and $u_k(\psi_k\phi_i) = u_i(\psi_k\phi_i)$. Therefore

$$\begin{aligned}\sum_i u_i(\phi_i) &= \sum_i u_i\left(\sum_k \psi_k\phi_i\right)\\ &= \sum_i\sum_k u_i(\psi_k\phi_i)\end{aligned}$$

$$= \sum_i \sum_k u_k(\psi_k \phi_i)$$

$$= \sum_k \sum_i u_k(\psi_k \phi_i)$$

$$= \sum_k u_k \left(\psi_k \sum_i \phi_i \right)$$

$$= \sum_k u_k(0) = 0.$$

Consequently, the definition (4.13) is consistent. Let $\phi \in \mathscr{C}_0^\infty(K)$. Then $\phi = \sum\limits_k \phi\psi_k$ and

$$\begin{aligned}
\Big|u(\phi)\Big| &= \left|\sum_i u_i(\psi_i\phi)\right| \leq \sum_i \Big|u_i(\phi\psi_i)\Big| \\
&\leq \sum_i C_i \sum_{|\alpha|\leq k} \sup\Big|D^\alpha(\phi\psi_i)\Big| \\
&\leq \sum_i C_i \sum_{|\alpha|\leq k} \sup\Big|D^\alpha\phi\Big| \\
&\leq C \sum_{|\alpha|\leq k} \sup\Big|D^\alpha\phi\Big|,
\end{aligned}$$

showing u is a distribution. We also have $u = u_i$ on X_i. Now, we will prove the uniqueness of u. Suppose there are two distributions u and $\tilde{u}$ with the previous properties. We conclude $u_{|X_i} = u_i, \quad \tilde{u}_{|X_i} = u_i$, so $(u - \tilde{u})_{|X_i} = 0$ for any i. Since X is open in $\mathbb{R}^n$, it follows that $u \equiv \tilde{u}$ on X, proving uniqueness. This completes the proof. □

Example 4.8. Let $f(x) = e^{\frac{1}{x}}$, $x \in \mathbb{R}\backslash\{0\}$. Suppose that $f \in \mathscr{D}'(\mathbb{R}\backslash\{0\})$. Pick $\phi_0 \in \mathscr{C}_0^\infty(\mathbb{R}\backslash\{0\})$ such that $\phi_0(x) \geq 0$ for every $x \neq 0$, $\phi_0(x) = 0$ for $x < 1$ and $x > 2$, and

$$\int_{-\infty}^{\infty} \phi_0(x)dx = 1.$$

Define the sequence $\{\phi_k\}_{k=1}^{\infty}$ by $\phi_k(x) = e^{-\frac{k}{2}}k\phi_0(kx)$, $x \in \mathbb{R}\backslash\{0\}$, $k \in \mathbb{N}$. It satisfies $\phi_k \to_{k\to\infty} 0$ in $\mathscr{C}_0^\infty(\mathbb{R}\backslash\{0\})$, so $f(\phi_k) \to_{k\to\infty} 0$. On the other hand,

$$\begin{aligned} f(\phi_k) &= \int_{-\infty}^{\infty} e^{\frac{1}{x}}\phi_k(x)dx \\ &= \int_1^2 e^{k(\frac{1}{y}-\frac{1}{2})}\phi_0(y)dy \\ &\geq \int_1^{\frac{3}{2}} e^{k(\frac{1}{y}-\frac{1}{2})}\phi_0(y)dy \\ &\geq e^{\frac{k}{6}}\int_1^{\frac{3}{2}} \phi_0(y)dy. \end{aligned}$$

By this and the definition of ϕ_0, we conclude $\lim_{k\to\infty} f(\phi_k(x)) = \infty$, which is a contradiction. Therefore, $f \notin \mathscr{D}'(\mathbb{R}\backslash\{0\})$.

Definition 4.15. If the inequality (4.7) holds for some integer k independent of the compact set $K \subset X$, the distribution u is said to be of finite order. The smallest such k is called the order of the distribution u.

The space of distributions on X of finite order is denoted by $\mathscr{D}'_F(X)$, and the space of distributions of order $\leq k$ is denoted by $\mathscr{D}'^k(X)$. Then

$$\mathscr{D}'_F(X) = \bigcup_k \mathscr{D}'^k(X).$$

Example 4.9. Dirac's "delta" function is a distribution of order 0.

Example 4.10. Let $u \in \mathscr{D}'(X)$ is defined by

$$u(\phi) = \sum_{k=1}^{\infty} \phi^{(k)}(k), \quad \phi \in \mathscr{C}_0^\infty(X).$$

Then the order of u is infinite.

Exercise 4.10. Prove that $P\frac{1}{x}$ has order 1 on $\mathbb{R}$.

Exercise 4.11. Prove that $P\frac{1}{x}$ is of order 0 on $\mathbb{R}\backslash\{0\}$.

Theorem 4.14. *Any $u \in \mathscr{D}'^k(X)$ can be extended in a unique way to a linear map on $\mathscr{C}_0^k(X)$ so that inequality (4.7) holds for every $\phi \in \mathscr{C}_0^k(X)$.*

Proof. Since the space $\mathscr{C}_0^\infty(X)$ is everywhere dense in $\mathscr{C}_0^k(X)$, for every $\phi \in \mathscr{C}_0^k(X)$ there exists a sequence $\{\phi_n\}_{n=1}^\infty$ in $\mathscr{C}_0^\infty(X)$ for which $\phi_n \to_{n\to\infty} \phi$ in $\mathscr{C}_0^k(X)$. Hence,

$$|u(\phi_n) - u(\phi_l)| \leq C \sum_{|\alpha|\leq k} \sup_K \Big| D^\alpha \phi_n - D^\alpha \phi_l \Big| \to_{n,l\to\infty} 0.$$

Therefore, $\{u(\phi_n)\}_{n=1}^\infty$ is a Cauchy sequence in $\mathbb{R}$, and as such it converges to, say,

$$u(\phi) = \lim_{n\to\infty} u(\phi_n). \tag{4.14}$$

The claim is that (4.14) is consistent. In fact, let $\{\phi_n\}_{n=1}^\infty$, $\{\psi_n\}_{n=1}^\infty$ be two sequences in $\mathscr{C}_0^\infty(X)$ for which

$$\lim_{n\to\infty} \phi_n = \lim_{n\to\infty} \psi_n = \phi$$

in $\mathscr{C}_0^k(X)$. Then $u(\phi) = \lim_{n\to\infty} u(\gamma_n) = \lim_{n\to\infty} u(\phi_n) = \lim_{n\to\infty} u(\psi_n)$, where $\{\gamma_n\}_{n=1}^\infty = \{\phi_n\}_{n=1}^\infty \bigcup \{\psi_n\}_{n=1}^\infty$. For the sequence γ_n we have

$$\Big|u(\gamma_n)\Big| \leq C \sum_{|\alpha|\leq k} \sup_K \Big| D^\alpha \gamma_n \Big|,$$

so

$$\Big|u(\phi)\Big| \leq C \sum_{|\alpha|\leq k} \sup_K \Big| D^\alpha \phi \Big|$$

when $n \to \infty$. This completes the proof. □

Let X_1 be an open set in $\mathbb{R}^n$. Take $f \in L^1_{loc}(X_1)$ and suppose that $x = Ay + b$ is a nonsingular transformation from X to X_1, i.e., A is a $n \times n$ matrix with $\det A \neq 0$ and $b \in \mathbb{R}^n$. Then, for any $\phi \in \mathscr{C}_0^\infty(X)$, we have

$$\int_X f(Ay + b)\phi(y)dy = \frac{1}{|\det A|} \int_{X_1} f(x)\phi(A^{-1}(x - b))dx.$$

This equality motivates us to give the following definition.

Definition 4.16. For any $u \in \mathscr{D}'(X)$, define

$$(u(Ay+b), \phi(y)) = \left(u(x), \frac{\phi(A^{-1}(x-b))}{|\det A|} \right), \quad x \in X, \quad y \in X_1. \tag{4.15}$$

Note that the map $\phi(x) \to \phi(A^{-1}(x-b))$, $x \in X$, is a linear and continuous map from $\mathscr{C}_0^\infty(X)$ to $\mathscr{C}_0^\infty(X_1)$. Therefore, the functional $u(Ay+b)$, $y \in X_1$, defined by (4.15), is well defined and it belongs to $\mathscr{D}'(X_1)$. In particular, for $u \in \mathscr{D}'(X)$ and $a \in \mathbb{C}^n$, $|a| \neq 0$, $b \in \mathbb{C}$, $b \neq 0$, we have following distributions

1. $u(\phi)(x+a) = u(\phi(x-a))(x)$, $\phi \in \mathscr{C}_0^\infty(X)$, $x \in X$,
2. $u(\phi)(bx) = \frac{1}{|b|^n} u(\phi(\frac{x}{b}))(x)$, $\phi \in \mathscr{C}_0^\infty(X)$, $x \in X$.

Example 4.11. For $\phi \in \mathscr{C}_0^\infty(\mathbb{R})$, we have

$$\begin{aligned}
\delta(\phi)(-x) &= \delta(\phi)(x), \\
\delta(\phi)(x+1-2i) &= \delta(\phi(x-1+2i))(x) \\
&= \phi(-1+2i), \\
\delta(\phi)(2ix) &= \frac{1}{2}\delta\left(\phi\left(\frac{x}{2i}\right)\right)(x) \\
&= \frac{1}{2}\phi(0), \quad x \in \mathbb{R}.
\end{aligned}$$

Exercise 4.12. Compute $\delta(\phi)(2x+3i)$, $x \in \mathbb{R}$, for $\phi \in \mathscr{C}_0^\infty(\mathbb{R})$.

Answer. $\frac{1}{2}\phi(-\frac{3i}{2})$.

Definition 4.17. For $a \in \mathscr{C}^1(\mathbb{R})$, define

$$\delta(a(x)) = \lim_{\varepsilon \to 0} \omega_\varepsilon(a(x)), \quad x \in \mathbb{R}.$$

Theorem 4.15. *If $a \in \mathscr{C}^1(\mathbb{R})$ has isolated simple zeros $x_1, x_2, \ldots$, then*

$$\delta(a(x)) = \sum_k \frac{\delta(x-x_k)}{|a'(x_k)|}, \quad x \in \mathbb{R}.$$

Proof. It is enough to prove the assertion on a neighbourhood of the simple zero x_k. Since x_k is an isolated simple zero of a, there exists $\varepsilon_k > 0$ such that $a(x) \neq 0$ for every $x \in (x_k - \varepsilon_k, x_k + \varepsilon_k)$, $x \neq x_k$, $a(x_k) = 0$. As

$$\begin{aligned}
(\delta(a(x)), \phi(x)) &= \int_{x_k-\varepsilon_k}^{x_k+\varepsilon_k} \delta(a(x))\phi(x)dx \\
&= \lim_{\varepsilon\to 0} \int_{x_k-\varepsilon_k}^{x_k+\varepsilon_k} \omega_\varepsilon(a(x))\phi(x)dx \\
&= \lim_{\varepsilon\to 0} \int_{a(x_k-\varepsilon_k)}^{a(x_k+\varepsilon_k)} \omega_\varepsilon(y)\frac{\phi(a^{-1}(y))}{|a'(a^{-1}(y))|}dy \\
&= \lim_{\varepsilon\to 0} \int_{a(x_k-\varepsilon_k)}^{a(x_k+\varepsilon_k)} \omega_\varepsilon(y)\frac{\phi(a^{-1}(a(x)))}{|a'(a^{-1}(a(x)))|}dy \\
&= \frac{\phi(x_k)}{|a'(x_k)|} \\
&= \left(\frac{\delta(x-x_k)}{|a'(x_k)|}, \phi(x)\right), \quad x \in (x_k - \varepsilon_k, x_k + \varepsilon_k),
\end{aligned}$$

for $\phi \in \mathscr{C}_0^\infty(x_k - \varepsilon_k, x_k + \varepsilon_k)$. Next, let $(c_1, d_1) \subset \mathbb{R}$ does not contain a single zero of x_k and $\phi \in \mathscr{C}_0^\infty(c_1, d_1)$. Then

$$\begin{aligned}
(\delta(a(x)), \phi(x)) &= \lim_{\varepsilon\to 0} \int_{c_1}^{d_1} \omega_\varepsilon(a(x))\phi(x)dx \\
&= 0, \quad x \in (c_1, d_1).
\end{aligned}$$

This completes the proof. □

Example 4.12. Let us consider $\delta(\cos x)$, $x \in \mathbb{R}$. Here $a(x) = \cos x$, $x \in \mathbb{R}$, and its isolated zeros are $x_k = \frac{(2k+1)\pi}{2}$, $k \in \mathbb{Z}$. We note that $|a'(x_k)| = 1$ for $k \in \mathbb{Z}$. So,

$$\delta(\cos x) = \sum_k \delta\left(x - \frac{(2k+1)\pi}{2}\right), \quad x \in \mathbb{R}.$$

Exercise 4.13. Compute $\delta(x^4 - 1)$, $x \in \mathbb{R}$.

Answer. $\frac{\delta(x-1)+\delta(x+1)}{4}$, $\quad x \in \mathbb{R}$.

4.3 Differentiation of Distributions

Let X be an open set in $\mathbb{R}^n$.

Definition 4.18. For $u \in \mathscr{D}'(X)$ and $\alpha \in \mathbb{N}^n \cup \{0\}$, we define $D^\alpha u$ as follows:

$$D^\alpha u(\phi) = (-1)^{|\alpha|} u(D^\alpha \phi) \tag{4.16}$$

for every $\phi \in \mathscr{C}_0^\infty(X)$.

Since the operation $\phi \to D^\alpha \phi$ is linear and continuous on $\mathscr{C}_0^\infty(X)$, the functional $D^\alpha u$, determined by (4.16), is linear and continuous, i.e., $D^\alpha u \in \mathscr{D}'(X)$. Below, we will list some of the properties of the derivatives of the distributions.

Theorem 4.16. *The operation $u \to D^\alpha u : \mathscr{D}'(X) \to \mathscr{D}'(X)$ is linear and continuous.*

Proof. We start by showing linearity. Let $\alpha_1, \alpha_2 \in \mathbb{C}$, $u_1, u_2 \in \mathscr{D}'(X)$ and $\phi \in \mathscr{C}_0^\infty(X)$. Then

$$\begin{aligned}(\alpha_1 u_1 + \alpha_2 u_2)(\phi) \to D^\alpha(\alpha_1 u_1 + \alpha_2 u_2)(\phi) \\ &= (-1)^{|\alpha|}(\alpha_1 u_1 + \alpha_2 u_2)(D^\alpha \phi) \\ &= (-1)^{|\alpha|}\alpha_1 u_1(D^\alpha \phi) + (-1)^{|\alpha|} u_2(D^\alpha \phi) \\ &= \alpha_1 D^\alpha u_1(\phi) + \alpha_2 D^\alpha u_2(\phi).\end{aligned}$$

Now, we will prove the continuity. Let $\{u_n\}_{n=1}^\infty$ be a sequence in $\mathscr{D}'(X)$ such that $u_n \to_{n\to\infty} 0$ in $\mathscr{D}'(X)$. Then

$$D^\alpha u_n(\phi) = (-1)^{|\alpha|} u_n(D^\alpha \phi) \to_{n\to\infty} 0$$

for $\phi \in \mathscr{C}_0^\infty(X)$. Consequently $D^\alpha u_n \to_{n\to\infty} 0$ in $\mathscr{D}'(X)$. This completes the proof. $\square$

Example 4.13. Let us consider the Heaviside function $H(x)$, $x \in \mathbb{R}$. For its derivatives, we have

$$\begin{aligned}D^\alpha H(\phi) &= (-)^\alpha H(D^\alpha \phi) \\ &= (-1)^\alpha \int_{-\infty}^{\infty} H(x) D^\alpha \phi(x)dx\end{aligned}$$

$$
\begin{aligned}
&= (-1)^{\alpha} \int_0^{\infty} D^{\alpha}\phi(x)dx \\
&= (-1)^{\alpha} D^{\alpha-1}\phi(x)|_{x=0}^{x=\infty} \\
&= (-1)^{\alpha+1} D^{\alpha-1}\phi(0) \qquad (4.17)
\end{aligned}
$$

for $\phi \in \mathscr{C}_0^{\infty}(\mathbb{R})$ and $\alpha \in \mathbb{N}$.

Example 4.14. Let us compute

$$
\lim_{\varepsilon\to 0} D^{\alpha}\omega_{\varepsilon}(x) \quad \text{in } \mathscr{D}'(\mathbb{R}), \quad \text{for } \alpha \in \mathbb{N}.
$$

Let $\phi \in \mathscr{C}_0^{\infty}(\mathbb{R})$. Then

$$
\begin{aligned}
\lim_{\varepsilon\to 0} D^{\alpha}\omega_{\varepsilon}(\phi) &= (-1)^{\alpha} \lim_{\varepsilon\to 0} \omega_{\varepsilon}(D^{\alpha}\phi) \\
&= (-)^{\alpha} \lim_{\varepsilon\to 0} \int_{-\infty}^{\infty} \omega_{\varepsilon}(x) D^{\alpha}\phi(x)dx \\
&= (-1)^{\alpha} \lim_{\varepsilon\to 0} \int_{|x|\le\varepsilon} e^{-\frac{\varepsilon^2}{\varepsilon^2-|x|^2}} D^{\alpha}\phi(x)dx \\
&= (-1)^{\alpha} D^{\alpha}\phi(0) \\
&= (-1)^{\alpha}\delta(D^{\alpha}\phi) \\
&= D^{\alpha}\delta(\phi).
\end{aligned}
$$

Therefore, $\lim_{\varepsilon\to 0} D^{\alpha}\omega_{\varepsilon}(x) = D^{\alpha}\delta(x)$ in $\mathscr{D}'(X)$.

Exercise 4.14. Find $\delta^{(k)}(x)$, $x \in \mathbb{R}$, $k \in \mathbb{N}$.

Answer. $(-1)^k\phi^{(k)}(0)$, $\phi \in \mathscr{C}_0^{\infty}(\mathbb{R})$.

By Theorem 4.16, it follows that if the series

$$
\sum_{k\ge 1} u_k(x) = S(x), \quad u_k \in L^1_{loc}(X),
$$

is uniformly convergent on every compact subset K of X, it may be differentiated term by term any number of times, and the resulting series will converge in $\mathscr{D}'(X)$. We have

$$
\sum_{k\ge 1} D^{\alpha}u_k(x) = D^{\alpha}S(x).
$$

Theorem 4.17. *Every distribution $u \in \mathscr{D}'(X)$ is differentiable infinitely many times.*

Proof. Since $u(D^\alpha\phi)$ exists for every $\alpha \in \mathbb{N}^n \cup \{0\}$ and every $\phi \in \mathscr{C}_0^\infty(X)$, we conclude that u is differentiable infinitely many times. This completes the proof. □

Theorem 4.18. *We have*

$$D^{\alpha+\beta}u = D^\alpha(D^\beta u)$$

for every $\alpha, \beta \in \mathbb{N}^n \cup \{0\}$ and every $u \in \mathscr{D}'(X)$.

Proof. Let in fact $\phi \in \mathscr{C}_0^\infty(X)$ be arbitrary. Then

$$\begin{aligned} D^{\alpha+\beta}u(\phi) &= (-1)^{|\alpha+\beta|}u(D^{\alpha+\beta}\phi) \\ &= (-1)^{|\alpha|}(-1)^{|\beta|}u(D^\beta(D^\alpha\phi)) \\ &= (-1)^{|\alpha|}D^\beta u(D^\alpha\phi) \\ &= D^\alpha(D^\beta u)(\phi). \end{aligned}$$

This completes the proof. □

Theorem 4.19. *We have*

$$D^\alpha(\alpha_1 u_1 + \alpha_2 u_2) = \alpha_1 D^\alpha u_1 + \alpha_2 D^\alpha u_2$$

for $u_1, u_2 \in \mathscr{D}'(X)$, $\alpha_1, \alpha_2 \in \mathbb{C}$, $\alpha \in \mathbb{N}^n \cup \{0\}$.

Proof. Let $\phi \in \mathscr{C}_0^\infty(X)$ be arbitrarily chosen. Then

$$\begin{aligned} D^\alpha(\alpha_1 u_1 + \alpha_2 u_2)(\phi) &= (-1)^{|\alpha|}(\alpha_1 u_1 + \alpha_2 u_2)(D^\alpha\phi) \\ &= (-1)^{|\alpha|}\alpha_1 u_1(D^\alpha\phi) + (-1)^{|\alpha|}\alpha_2 u_2(D^\alpha\phi) \\ &= \alpha_1 D^\alpha u_1(\phi) + \alpha_2 D^\alpha u_2(\phi) \\ &= (\alpha_1 D^\alpha u_1 + \alpha_2 D^\alpha u_2)(\phi). \end{aligned}$$

This completes the proof. □

Theorem 4.20. *If $u \in \mathscr{D}'(X)$, $a \in \mathscr{C}^\infty(X)$, we have*

$$D^\alpha(au) = \sum_{\beta\leq\alpha} \binom{\alpha}{\beta} D^\beta a D^{\alpha-\beta}u,$$

for every $\alpha \in \mathbb{N}^n \cup \{0\}$.

Proof. We will prove the simple case

$$\frac{\partial}{\partial x_i}(au) = a\frac{\partial u}{\partial x_i} + \frac{\partial a}{\partial x_i}u$$

for some $i \in \{1, 2, \ldots, n\}$. Using induction, the reader can deduce the general case. If $\phi \in \mathscr{C}_0^\infty(X)$, then

$$\begin{aligned}
\frac{\partial}{\partial x_i}(au)(\phi) &= -au\left(\frac{\partial}{\partial x_i}\phi\right)\\
&= -u\left(a\frac{\partial}{\partial x_i}\phi\right)\\
&= -u\left(\frac{\partial}{\partial x_i}(a\phi) - \frac{\partial a}{\partial x_i}\phi\right)\\
&= -u\left(\frac{\partial}{\partial x_i}(a\phi)\right) + u\left(\frac{\partial a}{\partial x_i}\phi\right)\\
&= \frac{\partial u}{\partial x_i}(a\phi) + \frac{\partial a}{\partial x_i}u(\phi)\\
&= a\frac{\partial u}{\partial x_i}(\phi) + \frac{\partial a}{\partial x_i}u(\phi)\\
&= \left(a\frac{\partial u}{\partial x_i} + \frac{\partial a}{\partial x_i}u\right)(\phi),
\end{aligned}$$

so $\frac{\partial}{\partial x_i}(au) = \frac{\partial a}{\partial x_i}u + a\frac{\partial u}{\partial x_i}$ in $\mathscr{D}'(X)$. This completes the proof. □

Theorem 4.21. *For every $u \in \mathscr{D}'(X)$ and every $\alpha \in \mathbb{N}^n \cup \{0\}$,*

$$supp D^\alpha u \subset supp u.$$

Proof. To see this, let $u \in \mathscr{D}'(X)$, $\phi \in \mathscr{C}_0^\infty(X)$ and $supp\phi \cap supp u = \emptyset$. Then $D^\alpha\phi \in \mathscr{C}_0^\infty(X)$ and $supp D^\alpha\phi \cap supp u = \emptyset$. Hence,

$$D^\alpha u(\phi) = (-1)^{|\alpha|}u(D^\alpha\phi) = 0,$$

and the assertion follows. □

Theorem 4.22. *Let f be a function defined on (a, b) that is piecewise-differentiable with continuity. Call $\{x_k\}$ the points in (a, b)*

where f or its derivative have jump discontinuities. Write

$$[f]_{x_k} = f(x_k + 0) - f(x_k - 0),$$

and denote by $\{f'\}$ the classical derivative of f at $x \in (a, b)$. Then

$$f' = \{f'\} + \sum_k [f]_{x_k} \delta(x - x_k).$$

Proof. For $\phi \in \mathscr{C}_0^\infty(a, b)$, we have

$$\begin{aligned}
f'(\phi) &= -f(\phi') \\
&= -\sum_k \int_{x_k}^{x_{k+1}} f(x)\phi'(x)dx \\
&= \sum_k \int_{x_k}^{x_{k+1}} \{f'\}(x)\phi(x)dx \\
&\quad - \sum_k [f(x_{k+1} - 0)\phi(x_{k+1}) - f(x_k + 0)\phi(x_k)] \\
&= \int_a^b \{f'\}(x)\phi(x)dx + \sum_k [f(x_k + 0) - f(x_k - 0)]\phi(x_k) \\
&= \{f'\}(\phi) + \sum_k [f]_{x_k} \delta(x - x_k)(\phi).
\end{aligned}$$

This completes the proof. □

Theorem 4.23. *The trigonometric series*

$$\sum_{k=-\infty}^{k=\infty} a_k e^{ikx}, \quad x \in \mathbb{R},$$

where $a_k \in \mathbb{C}$, $k \in \mathbb{Z}$, satisfy

$$|a_k| \le A(1 + |k|)^m, \quad k \in \mathbb{Z},$$

for some $m \in \mathbb{N} \cup \{0\}$, is convergent in $\mathscr{D}'(\mathbb{R})$.

Proof. Since the series

$$\frac{a_0 x^{2m+2}}{(m+2)!} + \sum_{k \neq 0} \frac{a_k}{(ik)^{m+2}} e^{ikx}, \quad x \in \mathbb{R},$$

is uniformly convergent in $\mathbb{R}$, the series which is a derivative of it of order $m+2$ is convergent in $\mathscr{D}'(\mathbb{R})$. This completes the proof. □

Theorem 4.24. *We have*

$$\frac{1}{2\pi}\sum_k e^{ikx} = \sum_k \delta(x-2k\pi).$$

Proof. The function

$$f_0(x) = \frac{x}{2} - \frac{x^2}{4\pi}, \quad 0 \le x < 2\pi,$$

has Fourier series

$$f_0(x) = \frac{\pi}{6} - \frac{1}{2\pi}\sum_{k\neq 0}\frac{e^{ikx}}{k^2}, \quad x \in \mathbb{R}.$$

Using Theorem 4.22, we have

$$f_0'(x) = -\frac{i}{2\pi}\sum_{k\neq 0}\frac{e^{ikx}}{k} = \frac{1}{2} - \frac{x}{2\pi}$$

and

$$f_0''(x) = \frac{1}{2\pi}\sum_{k\neq 0} e^{ikx} = -\frac{1}{2\pi} + \sum_k \delta(x-2k\pi).$$

From the last equation, we obtain

$$\frac{1}{2\pi}\sum_k e^{ikx} = \sum_k \delta(x-2k\pi).$$

This completes the proof. □

4.4 Tempered Distributions

Definition 4.19. A linear continuous functional on $\mathscr{S}(\mathbb{R}^n)$ is called a tempered distribution. The space of tempered distributions is indicated by $\mathscr{S}'(\mathbb{R}^n)$.

Definition 4.20. A sequence $\{u_n\}_{n=1}^{\infty}$ in $\mathscr{S}'(\mathbb{R}^n)$ is said to converge in $\mathscr{S}'(\mathbb{R}^n)$ to $u \in \mathscr{S}'(\mathbb{R}^n)$ if $u_n(\phi) \to_{n\to\infty} u(\phi)$ for every $\phi \in \mathscr{S}(\mathbb{R}^n)$.

Note that the convergence in $\mathscr{S}'(\mathbb{R}^n)$ implies convergence in $\mathscr{D}'(\mathbb{R}^n)$.

Definition 4.21. A set $M' \subset \mathscr{S}'(\mathbb{R}^n)$ is called weakly bounded if for every $\phi \in \mathscr{S}(\mathbb{R}^n)$ there is a constant C_ϕ such that $|u(\phi)| \leq C_\phi$ for every $u \in M'$.

Theorem 4.25. *If $M' \subset \mathscr{S}'(\mathbb{R}^n)$ is a weakly bounded set, there exist constants $K > 0$ and $m \in \mathbb{N}$ such that*

$$|u(\phi)| \leq K||\phi||_{\mathscr{S},m}, \quad u \in M', \phi \in \mathscr{S}(\mathbb{R}^n).$$

Proof. In order to show this let us suppose that the assertion is false, i.e., there exist sequences $\{u_k\}_{k=1}^\infty$ in M' and $\{\phi_k\}_{k=1}^\infty$ in $\mathscr{S}(\mathbb{R}^n)$ such that

$$|u_k(\phi_k)| > k||\phi_k||_{\mathscr{S},k}, \quad k \in \mathbb{N}. \tag{4.18}$$

We define the functions

$$\psi_k(x) = \frac{1}{\sqrt{k}} \frac{\phi_k(x)}{||\phi_k||_{\mathscr{S},k}}, \quad k \in \mathbb{N}.$$

So, $\psi_k \in \mathscr{S}(\mathbb{R}^n)$, $k \in \mathbb{N}$, and

$$||\psi_k||_{\mathscr{S},p} = \frac{1}{\sqrt{k}} \frac{||\phi_k||_{\mathscr{S},p}}{||\phi_k||_{\mathscr{S},k}}, \quad k, p \in \mathbb{N}.$$

Moreover,

$$||\phi_k||_{\mathscr{S},p} \leq ||\phi_k||_{\mathscr{S},k}$$

for every $k \geq p$. Hence,

$$||\psi_k||_{\mathscr{S},p} \leq \frac{1}{\sqrt{k}} \to_{k\to\infty} 0.$$

Since $p \in \mathbb{N}$ was arbitrary and $\mathscr{S} = \bigcap_{p \in \mathbb{N}\cup\{0\}} \mathscr{S}_p$, we have $\psi_k \to_{k\to\infty} 0$ in $\mathscr{S}(\mathbb{R}^n)$. Using techniques of the sort of [8], we conclude that

$$u_k(\psi_k) \to_{k\to\infty} 0. \tag{4.19}$$

On the other hand, using (4.18), we have

$$|u_k(\phi_k)| \geq \sqrt{k}\sqrt{k}||\phi_k||_{\mathscr{S},k}, \quad k \in \mathbb{N},$$

from which

$$|u_k(\psi_k)| \geq \sqrt{k} \to_{k\to\infty} \infty.$$

This contradicts with (4.19). This completes the proof. □

From this we also deduce that any tempered distribution u has finite order m. It can be extended to a linear continuous functional from the smallest dual space $\mathscr{S}'_m$, and

$$|u(\phi)| \leq ||u||_{\mathscr{S},-m}||\phi||_{\mathscr{S},m},$$

where $||u||_{\mathscr{S},-m}$ is a norm in $\mathscr{S}'_m$.

Example 4.15. Let u be defined on $\mathbb{R}^n$ and suppose

$$\int_{\mathbb{R}^n} \frac{|u(x)|}{(1+|x|)^m} dx < \infty$$

for some $m \geq 0$. Define the functional on $\mathscr{S}(\mathbb{R}^n)$

$$u(\phi) = \int_{\mathbb{R}^n} u(x)\phi(x)dx, \quad \phi \in \mathscr{S}(\mathbb{R}^n). \tag{4.20}$$

This is well defined. In fact, let $\phi \in \mathscr{S}(\mathbb{R}^n)$ and C be a positive constant such that

$$\sup_{x\in\mathbb{R}^n} \left((1+|x|)^m |\phi(x)|\right) \leq C.$$

Then

$$\begin{aligned} |u(\phi)| &= \left|\int_{\mathbb{R}^n} u(x)\phi(x)dx\right| \\ &\leq \int_{\mathbb{R}^n} |u(x)||\phi(x)|dx \\ &= \int_{\mathbb{R}^n} \frac{|u(x)|}{(1+|x|)^m}(1+|x|)^m|\phi(x)|dx \\ &\leq C\int_{\mathbb{R}^n} \frac{|u(x)|}{(1+|x|)^m}dx \\ &< \infty. \end{aligned}$$

It is a linear and continuous functional on $\mathscr{S}(\mathbb{R}^n)$, so $u \in \mathscr{S}'(\mathbb{R}^n)$.

Exercise 4.15. Prove that $e^x \notin \mathscr{S}'(\mathbb{R})$.

Exercise 4.16. Prove that $\cos(e^x)$ belongs to $\mathscr{S}'(\mathbb{R})$ but not to $\mathscr{S}(\mathbb{R})$.

Exercise 4.17. Show that $\mathscr{S}'(\mathbb{R}^n)$ is a $\mathbb{C}$-vector space.

4.5 The Fourier Transform in $\mathscr{S}(\mathbb{R}^n)$

Definition 4.22. The Fourier transform of $\phi \in \mathscr{S}(\mathbb{R}^n)$ is the integral

$$\mathscr{F}(\phi)(x) = \int_{\mathbb{R}^n} e^{-i(x,\xi)}\phi(\xi)d\xi, \quad x \in \mathbb{R}^n.$$

Note that $\mathscr{F}(\phi)$ is bounded and continuous on $\mathbb{R}^n$. Furthermore, $\mathscr{F}(\phi) \in \mathscr{C}^\infty(\mathbb{R}^n)$ and

$$\begin{aligned} D^\alpha \mathscr{F}(\phi)(x) &= \int_{\mathbb{R}^n} (-i\xi)^\alpha e^{-i(x,\xi)}\phi(\xi)d\xi \\ &= \mathscr{F}((-i\xi)^\alpha \phi)(x), \\ \mathscr{F}(D^\alpha\phi)(x) &= \int_{\mathbb{R}^n} D^\alpha\phi(\xi)e^{-i(x,\xi)}d\xi \\ &= (ix)^\alpha \mathscr{F}(\phi)(x), \quad x \in \mathbb{R}^n, \end{aligned}$$

for every $\alpha \in \mathbb{N}^n \cup \{0\}$, where

$$\begin{aligned} (-i\xi)^\alpha &= (-i)^{|\alpha|}\xi_1^{\alpha_1}\xi_2^{\alpha_2}\dots\xi_n^{\alpha_n}, \quad \xi = (\xi_1, \xi_2, \dots, \xi_n), \\ (ix)^\alpha &= i^{|\alpha|}x_1^{\alpha_1}x_2^{\alpha_2}\dots x_n^{\alpha_n}, \quad x = (x_1, x_2, \dots, x_n). \end{aligned}$$

In particular, $\mathscr{F}(\phi)$ is an integrable function on $\mathbb{R}^n$. Observe that every function $\phi \in \mathscr{S}(\mathbb{R}^n)$ can be represented by means of its Fourier transform $\mathscr{F}(\phi)$ and inverse Fourier transform

$$\mathscr{F}^{-1}(\phi)(\xi) = \frac{1}{(2\pi)^n}\int_{\mathbb{R}^n} e^{ix\xi}\phi(x)dx, \quad \xi \in \mathbb{R}^n,$$

as follows

$$\phi = \mathscr{F}^{-1}(\mathscr{F}(\phi)) = \mathscr{F}(\mathscr{F}^{-1}(\phi)).$$

Explicitly,

$$\phi(x) = \frac{1}{(2\pi)^n} \int_{\mathbb{R}^n} e^{i(x,\xi)} \mathscr{F}(\phi)(\xi) d\xi, \quad x \in \mathbb{R}^n.$$

As

$$\begin{aligned}
(1+|\xi|^2)^{\frac{p}{2}} |D^\alpha \mathscr{F}(\phi)(\xi)| &\le (1+|\xi|^2)^{[\frac{p+1}{2}]} |D^\alpha \mathscr{F}(\phi)(\xi)| \\
&\le \left| \int_{\mathbb{R}^n} (I-\Delta)^{[\frac{p+1}{2}]} ((-ix)^\alpha \phi(x)) e^{-i(x,\xi)} dx \right| \\
&\le C \sup_{x \in \mathbb{R}^n} (1+|x|^2)^{\frac{n+1}{2}} \left| (I-\Delta)^{[\frac{p+1}{2}]} (x^\alpha \phi(x)) \right|, \quad \xi \in \mathbb{R}^n,
\end{aligned}$$

for $p \in \mathbb{N}$, it follows that

$$||\mathscr{F}(\phi)||_{\mathscr{S},p} \le C_p ||\phi||_{\mathscr{S},p+n+1},$$

where C_p is a constant independent of ϕ. By the last estimate, we conclude that $\phi \to \mathscr{F}(\phi)$ is a linear and continuous map on $\mathscr{S}(\mathbb{R}^n)$. Every element $\phi \in \mathscr{S}(\mathbb{R}^n)$ can be represented as the Fourier transform of the function $\psi = \mathscr{F}^{-1}(\phi) \in \mathscr{S}(\mathbb{R}^n)$, where $\phi = \mathscr{F}(\psi)$. If $\mathscr{F}(\phi) = 0$, then $\phi = 0$. Therefore the map $\phi \to \mathscr{F}(\phi)$ is one-to-one on $\mathscr{S}(\mathbb{R}^n)$.

Exercise 4.18. Compute $\mathscr{F}(e^{-ax^2})$, $x \in \mathbb{R}$, $a = \text{const} > 0$.

Answer. $\sqrt{\frac{\pi}{a}} e^{-\frac{\xi^2}{4a}}$, $\xi \in \mathbb{R}$.

Exercise 4.19. Let A be a positive definite $n \times n$ matrix. Prove that

$$\mathscr{F}(e^{i(Ax,x)}) = \frac{\pi^{\frac{n}{2}}}{\sqrt{\det A}} e^{-i\frac{\pi n}{4} + \frac{i}{4}(A^{-1}\xi,\xi)}, \quad x, \xi \in \mathbb{R}^n.$$

4.6 The Fourier Transform in $\mathscr{S}'(\mathbb{R}^n)$

Definition 4.23. The Fourier transform of a distribution $u \in \mathscr{S}'(\mathbb{R}^n)$ is defined as follows

$$\mathscr{F}(u)(\phi) = u(\mathscr{F}(\phi)) \quad \text{for } \phi \in \mathscr{S}(\mathbb{R}^n).$$

Since the map $\phi \to \mathscr{F}(\phi) : \mathscr{S}(\mathbb{R}^n) \to \mathscr{S}(\mathbb{R}^n)$ is linear and continuous, the operation $u \to \mathscr{F}(u)$ is linear and continuous from $\mathscr{S}'(\mathbb{R}^n)$ to itself.

For $u \in \mathscr{S}'(\mathbb{R}^n)$, we define the operator $\mathscr{F}^{-1}$ in the following manner

$$\mathscr{F}^{-1}(u)(x) = \frac{1}{(2\pi)^n}\mathscr{F}(u)(-x).$$

As $\phi \to \mathscr{F}(\phi)$ is continuous and goes from $\mathscr{S}(\mathbb{R}^n)$ to $\mathscr{S}(\mathbb{R}^n)$, and $\mathscr{S}(\mathbb{R}^n)$ is dense in $\mathscr{S}'(\mathbb{R}^n)$, we conclude that

$$\mathscr{F}^{-1}(\mathscr{F}(u)) = \mathscr{F}(\mathscr{F}^{-1}(u)) = u, \quad u \in \mathscr{S}'(\mathbb{R}^n).$$

It follows that for every distribution $u \in \mathscr{S}'(\mathbb{R}^n)$ there exists a distribution $v \in \mathscr{S}'(\mathbb{R}^n)$ such that $v = \mathscr{F}^{-1}(u)$ and $u = \mathscr{F}(v)$. If $\mathscr{F}(u) = 0$, then $u = 0$.

Example 4.16. Let us define $\mathscr{F}(\delta)$. Take $\phi \in \mathscr{S}(\mathbb{R}^n)$, so

$$\begin{aligned}\mathscr{F}(\delta)(\phi) &= \delta(\mathscr{F}(\phi))\\ &= \mathscr{F}(\phi)(0)\\ &= \int_{\mathbb{R}^n} \phi(x)dx.\end{aligned}$$

This implies $\mathscr{F}(\delta) = 1$.

Exercise 4.20. Compute $\mathscr{F}(H(x)e^{-x})$, $x \in \mathbb{R}$.

Answer. $\frac{1}{1-i\xi}$.

Let $u(x, y) \in \mathscr{S}'(\mathbb{R}^{n+m})$, $x \in \mathbb{R}^n$, $y \in \mathbb{R}^m$. We introduce the Fourier transform $\mathscr{F}_x(u)$ with respect to the variable $x = (x_1, x_2, \ldots, x_n)$ by

$$\mathscr{F}_x(u)(\phi) = u(\mathscr{F}_\xi(\phi)), \quad \phi \in \mathscr{S}(\mathbb{R}^{n+m}),$$

where

$$\mathscr{F}_\xi(\phi)(x, y) = \int_{\mathbb{R}^n} e^{-i(\xi,x)}\phi(\xi, y)d\xi.$$

The map $\phi(\xi, y) \to \mathscr{F}_\xi(\phi)$ is an isomorphism from $\mathscr{S}(\mathbb{R}^{n+m})$ to itself, and $\mathscr{F}_x(u) \in \mathscr{S}'(\mathbb{R}^{n+m})$ for $u \in \mathscr{S}'(\mathbb{R}^{n+m})$. The inverse Fourier transform $\mathscr{F}_\xi^{-1}$ is defined by

$$\mathscr{F}_\xi^{-1}(u) = \frac{1}{(2\pi)^n}\mathscr{F}_\xi(u(-\xi, y))(x, y).$$

The map $u \to \mathscr{F}_x(u)$ is an automorphism of $\mathscr{S}'(\mathbb{R}^{n+m})$.

Example 4.17. Let $a = \text{const} \in \mathbb{R}$. Then

$$\begin{aligned}\mathscr{F}(\delta(x-a))(\phi) &= \delta(x-a)(\mathscr{F}(\phi))\\ &= \mathscr{F}(\phi)(a)\\ &= \int_{\mathbb{R}^n} e^{-i\xi a}\phi(\xi)d\xi, \quad \phi \in \mathscr{S}(\mathbb{R}).\end{aligned}$$

Since $\phi \in \mathscr{S}(\mathbb{R})$ is arbitrary, $\mathscr{F}(\delta(x-a)) = e^{-i\xi a}$.

Exercise 4.21. Prove that

$$\mathscr{F}\left(\frac{\delta(x-a)+\delta(x+a)}{2}\right) = \cos(a\xi), \quad a = \text{const},$$

in $\mathscr{S}'(\mathbb{R})$.

4.7 Properties of the Fourier Transform in $\mathscr{S}'(\mathbb{R}^n)$

1. For any $u \in \mathscr{S}'(\mathbb{R}^n), \alpha \in \mathbb{N}^n$, we have

$$D^\alpha \mathscr{F}(u) = \mathscr{F}((-ix)^\alpha u).$$

Example 4.18. Let $\phi \in \mathscr{S}(\mathbb{R}^n)$. Then

$$\mathscr{F}(x^\alpha)(\phi) = (i)^{|\alpha|}\mathscr{F}((-ix)^\alpha 1)(\phi) = (i)^{|\alpha|}D^\alpha\mathscr{F}(1)(\phi),$$

$\alpha \in \mathbb{N}^n \cup \{0\}$. On the other hand,

$$\mathscr{F}(\delta)(\phi) = 1(\phi).$$

Using the inverse Fourier transform, we get

$$\delta(\phi) = \mathscr{F}^{-1}(1)(\phi) = \frac{1}{(2\pi)^n}\mathscr{F}(1)(\phi), \text{ i.e., } \mathscr{F}(1)(\phi) = (2\pi)^n\delta(\phi).$$

Therefore,

$$\mathscr{F}(x^\alpha)(\phi) = (2\pi)^n(i)^{|\alpha|}D^\alpha\delta(\phi),$$

in other words

$$\mathscr{F}(x^\alpha) = (2\pi)^n (i)^{|\alpha|} D^\alpha \delta(\xi).$$

2. For any $u \in \mathscr{S}'(\mathbb{R}^n), \alpha \in \mathbb{N}^n \cup \{0\}$, we have

$$\mathscr{F}(D^\alpha u) = (i\xi)^\alpha \mathscr{F}(u).$$

Example 4.19. Let us find $\mathscr{F}(\delta'')$ in $\mathscr{S}'(\mathbb{R})$. Take $\phi \in \mathscr{S}(\mathbb{R})$ and compute

$$\begin{aligned}\mathscr{F}(\delta''(\phi))(\xi) &= (i\xi)^2 \mathscr{F}(\delta(\phi)) = (i\xi)^2 \delta(\mathscr{F}(\phi)) \\ &= (i\xi)^2 \delta\left(\int_{-\infty}^{\infty} e^{-ix\xi}\phi(x)dx\right) = (i\xi)^2 \int_{-\infty}^{\infty} \phi(x)dx,\end{aligned}$$

so

$$\mathscr{F}(\delta'')(x) = (ix)^2.$$

3. For any $u \in \mathscr{S}'(\mathbb{R}^n)$, we have

$$\mathscr{F}(u(x - x_0)) = e^{-i(\xi, x_0)} \mathscr{F}(u).$$

4. For any $u \in \mathscr{S}'(\mathbb{R}^n)$, we have

$$\mathscr{F}(u)(\xi + \xi_0) = \mathscr{F}(e^{-i(\xi_0, x)} u)(\xi).$$

5. For any nonsingular $n \times n$ matrix A, we have

$$\mathscr{F}(u(Ax))(\xi) = \frac{1}{|\det A|} \mathscr{F}(u)(A^{-1T}\xi), \quad u \in \mathscr{S}'(\mathbb{R}^n).$$

6. For any $u \in \mathscr{S}'(\mathbb{R}^n), v \in \mathscr{S}'(\mathbb{R}^m)$, we have

$$\begin{aligned}\mathscr{F}(u(x) \times v(y))(\xi, \eta) &= \mathscr{F}_x(u(x) \times \mathscr{F}(v)(\eta)) \\ &= \mathscr{F}_y(\mathscr{F}(u)(\xi) \times v(y)) = \mathscr{F}u(\xi) \times \mathscr{F}(v)(\eta).\end{aligned}$$

7. For any $u \in \mathscr{S}'(\mathbb{R}^{n+m}), \alpha \in \mathbb{N}^n \cup \{0\}, \beta \in \mathbb{N}^m \cup \{0\}$, we have

$$\begin{aligned}D_x^\alpha D_y^\beta \mathscr{F}_x(u) &= \mathscr{F}_x((-ix)^\alpha D_y^\beta u), \\ \mathscr{F}_x(D_x^\alpha D_y^\beta u) &= (i\xi)^\alpha D_y^\beta \mathscr{F}_x(u).\end{aligned}$$

The proofs of the above properties are left to the reader.

4.8 Advanced Practical Problems

Problem 4.1. Let $X_1 \subset \mathbb{R}^n$, $X_2 \subset \mathbb{R}^m$ be open sets. Prove that for every $\phi \in \mathscr{C}_0^\infty(X_1 \times X_2)$ there exist sequences $\{\phi_k\}_{k=1}^\infty \subset \mathscr{C}_0^\infty(X_1 \times X_2)$ and $\{N_k\}_{k=1}^\infty \subset \mathbb{N} \cup \{0\}$ such that

$$\phi_k(x, y) = \sum_{i=1}^{N_k} \phi_{ik}(x)\psi_{ik}(y), \quad (x, y) \in X_1 \times X_2,$$

and $\phi_k \to_{k\to\infty} \phi$ in $\mathscr{C}_0^\infty(X_1 \times X_2)$.

Problem 4.2. Prove that for every function $\phi_1 \in \mathscr{C}_0^\infty(\mathbb{R})$ there exists a function $\phi_2 \in \mathscr{C}_0^\infty(\mathbb{R})$ such that $\phi_1(x) = \phi_2'(x)$, for every $x \in \mathbb{R}$, if and only if $\int_{-\infty}^\infty \phi_1(x)dx = 0$.

Problem 4.3. Prove that for every $\phi \in \mathscr{C}_0^\infty(\mathbb{R})$ there exists a function $\phi_1 \in \mathscr{C}_0^\infty(\mathbb{R})$ such that

$$\phi(x) = \phi_0(x) \int_{-\infty}^\infty \phi(s)ds + \phi_1'(x), \quad x \in \mathbb{R},$$

where $\phi_0 \in \mathscr{C}_0^\infty(\mathbb{R})$, if and only if $\int_{-\infty}^\infty \phi_0(s)ds = 1$.

Problem 4.4. Prove that for every $\phi \in \mathscr{C}_0^\infty(\mathbb{R})$ there exists a function $\phi_2 \in \mathscr{C}_0^\infty(\mathbb{R})$ such that

$$\phi(x) = \phi_1'(x) \int_{-\infty}^\infty \int_{-\infty}^s \phi(\tau)d\tau ds + \phi_2''(x), \quad x \in \mathbb{R},$$

where $\phi_1 \in \mathscr{C}_0^\infty(\mathbb{R})$, if and only if $\int_{-\infty}^\infty \phi_1(x)dx = 1$.

Problem 4.5. Prove that for every $\phi \in \mathscr{C}_0^\infty(\mathbb{R})$ there exists a function $\phi_2 \in \mathscr{C}_0^\infty(\mathbb{R})$ such that

$$\phi(x) = \phi_1(x) \int_{-\infty}^\infty \int_{-\infty}^s \phi(\tau)d\tau ds + \phi_2''(x), \quad x \in \mathbb{R},$$

where $\phi_1 \in \mathscr{C}_0^\infty(\mathbb{R})$, if and only if $\int_{-\infty}^\infty \int_{-\infty}^s \phi_1(\tau)d\tau ds = 1$.

Problem 4.6. Prove that for every $\phi \in \mathscr{C}_0^\infty(\mathbb{R})$ there exists $\phi_3 \in \mathscr{C}_0^\infty(\mathbb{R})$ such that

$$\phi(x) = \phi_1(x)\int_{-\infty}^{\infty}\int_{-\infty}^{s}\phi(\tau)d\tau ds + \phi_2'(x) \times \int_{-\infty}^{\infty}\int_{-\infty}^{s}\phi(\tau)d\tau ds + \phi_3''(x), \quad x \in \mathbb{R},$$

where $\phi_1, \phi_2 \in \mathscr{C}_0^\infty(\mathbb{R})$, if and only if

$$\int_{-\infty}^{\infty}\int_{-\infty}^{s}\phi_1(\tau)d\tau ds + \int_{-\infty}^{\infty}\phi_2(x)dx = 1.$$

Problem 4.7. Let $P\frac{1}{x^2}$ be defined on $\mathscr{C}_0^\infty(\mathbb{R})$ by

$$P\frac{1}{x^2}(\phi) = P.V.\int_{-\infty}^{\infty}\frac{\phi(x)-\phi(0)}{x^2}dx.$$

Prove that $P\frac{1}{x^2} \in \mathscr{D}'(\mathbb{R})$.

Problem 4.8. Define u by

$$u(\phi) = \int_{|x|\le 1}\phi(x)dx \quad \forall \phi \in \mathscr{C}_0^\infty(\mathbb{R}^n).$$

Prove that $u \in \mathscr{D}'(\mathbb{R}^n)$.

Problem 4.9. Define

$$u(\phi) = \int_{|x|\le 1}D^\alpha\phi(x)dx \quad \forall \phi \in \mathscr{C}_0^\infty(\mathbb{R}^n),$$

where α is a multi-index. Show that $u \in \mathscr{D}'(\mathbb{R}^n)$.

Problem 4.10. Let α be a multi-index and set $u(\phi) = D^\alpha\phi(x_0)$, $\phi \in \mathscr{C}_0^\infty(X)$ for a given $x_0 \in X$. Prove that u is a distribution of order $|\alpha|$.

Problem 4.11. Compute

$$\frac{d^3}{dx^3}|x|, \quad x \in \mathbb{R}.$$

Answer. $2\delta'(x)$.

Problem 4.12. Compute

$$x^m \delta^{(k)}(x).$$

Answer.

$$x^m \delta^{(k)}(x) = \begin{cases} 0 & \text{for } k < m; \\ (-1)^m m! \delta & \text{for } k = m; \\ (-1)^m \binom{k}{m} m! \delta^{(k-m)} & \text{for } k > m. \end{cases}$$

Problem 4.13. Prove that

$$\delta(x) + \delta'(x-1) + \delta''(x-2) + \cdots$$

converges in $\mathscr{D}'(X)$ and that it has a finite order.

Problem 4.14. Compute in $\mathscr{S}(\mathbb{R})$

$$\mathscr{F}(e^{-\frac{x^2}{4}} \cos(\alpha x)), \quad \alpha = \text{const.}$$

Problem 4.15. Let $u, v \in \mathscr{S}(\mathbb{R}^n)$. Prove

$$\int_{\mathbb{R}^n} \mathscr{F}(u)(x) v(x) dx = \int_{\mathbb{R}^n} u(x) \mathscr{F}(v)(x) dx.$$

Problem 4.16. Prove that

$$\mathscr{F}\mathscr{F}^{-1} = \mathscr{F}^{-1}\mathscr{F} = I \quad in \ \mathscr{S}(\mathbb{R}^n).$$

Problem 4.17. Show

$$\mathscr{F}(f * g) = \mathscr{F}(f)\mathscr{F}(g)$$

for $f, g \in \mathscr{S}(\mathbb{R}^n)$.

Problem 4.18. Prove

$$\lim_{|\lambda| \to \infty} \mathscr{F}(f)(\lambda) = 0$$

for $f \in \mathscr{S}(\mathbb{R}^n)$.

Problem 4.19. Let $f \in \mathscr{S}(\mathbb{R})$. Prove

1. $\mathscr{F}(f(-x)) = \mathscr{F}(f)(-\xi)$,
2. $\mathscr{F}(f(ax+b)) = \frac{1}{a}e^{i\frac{b\xi}{a}}\mathscr{F}(f)(\frac{\xi}{a})$, $a, b, c = \text{const}, a > 0$,
3. $\mathscr{F}(e^{iax}f(x)) = \mathscr{F}(f)(\xi - a)$, $a = \text{const} \neq 0$,
4. $\mathscr{F}(e^{iax}f(bx+c)) = \frac{1}{b}e^{i\frac{c}{b}(\xi-a)}\mathscr{F}(f)(\frac{\xi-a}{b})$, $a, b, c = \text{const}$, $b > 0$.

Problem 4.20. In $\mathscr{S}'(\mathbb{R})$ compute $\mathscr{F}(u)$ when

1. $u = H(1-|x|)$,
2. $u = e^{-4x^2}$,
3. $u = e^{ix^2}$,
4. $u = e^{-ix^2}$,
5. $u = H(x)e^{-3x}$,
6. $u = H(-x)e^{4x}$,
7. $u = e^{-2|x|}$,
8. $u = \frac{2}{1+x^2}$,
9. $u = H(x)e^{-2x}\frac{x^{\alpha-1}}{\Gamma(\alpha)}$.

Chapter 5

Sobolev Spaces: Analytic Spaces

5.1 Definitions

Definition 5.1. Let A be an open set in $\mathbb{R}^n$, $m \in \mathbb{N}$ and $1 \leq p \leq \infty$. The Sobolev space $W^{m,p}(A)$ consists of functions in $L^p(A)$ whose partial derivatives up to order m, in the sense of distributions, can be identified with functions in $L^p(A)$.

Equivalently,

$$W^{m,p}(A) = \{u \in L^p(A) : D^\alpha u \in L^p(A)$$
$$\text{for any } \alpha \in \mathbb{N}^n \cup \{0\}, |\alpha| \leq m\}.$$

Notice that clearly $W^{0,q}(A) = L^q(A)$.

For $p = 2$, the symbol $W^{m,2}(A)$ is generally replaced by $H^m(A)$, and in the case $A = \mathbb{R}^n$, we can use the Fourier transform $\xi \to \mathscr{F}(u)(\xi)$ of $u \in L^2(A)$ to give the following characterisation

$$W^{m,2}(\mathbb{R}^n) = H^m(\mathbb{R}^n) = \{u \in L^2(\mathbb{R}^n) : \xi \to (1 + |\xi|^2)^{\frac{m}{2}}$$
$$\times \mathscr{F}(u)(\xi) \in L^2(\mathbb{R}^n)\}.$$

Example 5.1. Let $n = 2$. We seek conditions on $\beta > 0$ so that the function $u(x, y) = x(x^2 + y^2)^{-\beta}$, $x, y \in U_1$, is, away from the origin, an element of $H^1(U_1)$. In polar coordinates $x = r\cos\phi$, $y = r\sin\phi$,

$0 < r < 1$, $\phi \in [0, 2\pi]$, we have

$$\begin{aligned}
u(x,y) &= x(x^2+y^2)^{-\beta} \\
&= r^{1-2\beta}\cos\phi, \\
|u(x,y)|^2 &= r^{2-4\beta}\cos^2\phi, \\
u_x(x,y) &= (x^2+y^2)^{-\beta} - 2\beta x^2(x^2+y^2)^{-\beta-1} \\
&= r^{-2\beta}(1-2\beta\cos^2\phi), \\
|u_x(x,y)|^2 &= r^{-4\beta}(1-2\beta\cos^2\phi)^2, \\
u_y(x,y) &= -2\beta xy(x^2+y^2)^{-\beta-1} \\
&= -2\beta r^{-2\beta}\cos\phi\sin\phi, \\
|u_y(x,y)|^2 &= 4\beta^2 r^{-4\beta}\cos^2\phi\sin^2\phi, \quad x, y \in U_1,
\end{aligned}$$

substituting which produces

$$\begin{aligned}
\int_{U_1} |u(x,y)|^2 dxdy &= \int_0^1 \int_0^{2\pi} r^{3-4\beta}\cos^2\phi d\phi dr \\
&< \infty \\
&\Longleftrightarrow \beta < 1, \\
\int_{U_1} |u_x(x,y)|^2 dxdy &= \int_0^1 \int_0^{2\pi} r^{1-4\beta}(1-2\beta\cos^2\phi)^2 d\phi dr \\
&< \infty \\
&\Longleftrightarrow \beta < \frac{1}{2}, \\
\int_{U_1} |u_y(x,y)|^2 dxdy &= 4\beta^2 \int_0^1 \int_0^{2\pi} r^{1-4\beta}\cos^2\phi\sin^2\phi d\phi dr \\
&< \infty \\
&\Longleftrightarrow \beta < \frac{1}{2}.
\end{aligned}$$

Consequently, $u \in H^1(U_1)$ if $0 < \beta < \frac{1}{2}$.

Exercise 5.1. Let $n = 2$ and $u(x,y) = xy(x^2+y^2)^{-\beta}$, $(x,y) \in U_1 \backslash \{(0,0)\}$, $\beta > 0$. Find conditions on the parameter β so that $u \in H^1(U_1)$.

Exercise 5.2. Define u on U_1 by

$$u(x) = (1 - |x|)^{\beta}(-\log(1 - |x|))^{\alpha}, \quad x \in U_1,$$

where $\alpha, \beta > 0$ are real. Find conditions for α and β so that $u \in W^{1,p}(U_1)$, $1 \leq p \leq \infty$.

5.2 Elementary Properties

One can endow the Sobolev spaces with a norm

$$||u||_{W^{m,p}(A)} = \begin{cases} \left(\sum_{0\leq|\alpha|\leq p} ||D^{\alpha}u||^{p}_{L^{p}(A)}\right)^{\frac{1}{p}} & \text{if } 1 \leq p < \infty, \\ \max_{0\leq|\alpha|\leq m} ||D^{\alpha}u||_{L^{\infty}(A)} & \text{if } p = \infty. \end{cases}$$

In particular, the space $H^m(A)$ admits an inner product

$$(u, v) = \sum_{0\leq|\alpha|\leq m} (D^{\alpha}u, D^{\alpha}v).$$

Exercise 5.3. Check that $||\cdot||_{W^{m,p}(A)}$ fulfills the axioms for being a norm.

Exercise 5.4. Check that (u, v) is indeed an inner product.

Definition 5.2. A sequence $\{u_k\}_{k=1}^{\infty}$ in $W^{m,p}(A)$ converges to $u \in W^{m,p}(A)$ if

$$||u_k - u||_{W^{m,p}(A)} \to_{k\to\infty} 0.$$

Definition 5.3. A sequence $\{u_k\}_{k=1}^{\infty}$ in $W^{m,p}(A)$ converges to $u \in W^{m,p}(A)$ in $W^{m,p}_{\text{loc}}(A)$ if $u_k \to_{k\to\infty} u$ in $W^{m,p}(V)$ for every $V \subset\subset A$.

Definition 5.4. We denote by $W_0^{m,p}(A)$ the closure of $C_0^{\infty}(A)$ in $W^{m,p}(U)$.

In particular, $H_0^m(A) = W_0^{m,2}(A)$.

Exercise 5.5. Prove that $u \in W_0^{m,p}(A)$ if and only if there exist functions $u_k \in C_0^{\infty}(A)$ such that $u_k \to_{k\to\infty} u$ in $W^{m,p}(A)$.

Now, we set out to prove a number of elementary, but important properties of the Sobolev spaces.

Theorem 5.1. *Let $u \in W^{m,p}(A)$. For any $0 \le |\alpha| \le m$, we have $D^\alpha u \in W^{m-|\alpha|,p}(A)$ and*

$$D^\beta(D^\alpha u) = D^\alpha(D^\beta u) = D^{\alpha+\beta}u$$

for $0 \le |\alpha| + |\beta| \le m$.

Proof. To prove this property, we observe that $u \in W^{m,p}(A)$ implies that $D^\alpha u$ is a well-defined distribution, and $D^\alpha u \in L^p(A)$ for $0 \le |\alpha| \le m$. Moreover, $D^\alpha u \in W^{m-|\alpha|,p}(A)$ if $D^\beta(D^\alpha u)$ exists and belongs to $L^p(A)$ for $0 \le |\beta| \le m - |\alpha|$.

Pick $\phi \in \mathscr{C}_0^\infty(A)$. Then, for $0 \le |\alpha| + |\beta| \le m$, we have

$$\begin{aligned}\int_A D^\alpha u D^\beta \phi dx &= (-)^{|\alpha|} \int_A u D^\alpha(D^\beta \phi) dx \\ &= (-1)^{|\alpha|} \int_A u D^{\alpha+\beta} \phi dx \\ &= (-1)^{|\beta|} \int_A D^{\alpha+\beta} u \phi dx.\end{aligned}$$

On the other hand,

$$\int_A D^\alpha u D^\beta \phi dx = (-1)^{|\beta|} \int_A D^\beta(D^\alpha u) \phi dx,$$

so overall

$$D^\beta(D^\alpha u) = D^{\alpha+\beta}u.$$

This completes the proof. □

Theorem 5.2. *For any $u, v \in W^{m,p}(A)$ and constants λ, μ, $\lambda u + \mu v \in W^{m,p}(A)$.*

Proof. For this, let $\phi \in \mathscr{C}_0^\infty(A)$ be arbitrary. As u, $v \in W^{m,p}(A)$, for $0 \le |\alpha| \le m$ we have $D^\alpha u$, $D^\alpha v \in L^p(A)$ and

$$\begin{aligned}\int_A D^\alpha(\lambda u + \mu v)\phi dx &= (-1)^{|\alpha|} \int_A (\lambda u + \mu v) D^\alpha \phi dx \\ &= \lambda(-1)^{|\alpha|} \int_A u D^\alpha \phi dx + \mu(-1)^{|\alpha|} \int_A v D^\alpha \phi dx\end{aligned}$$

$$= \lambda \int_A D^\alpha u\phi dx + \mu \int_A D^\alpha v\phi dx$$

$$= \int_A (\lambda D^\alpha u + \mu D^\alpha v)\phi dx.$$

This completes the proof. □

Exercise 5.6. If $u \in W^{m,p}(A)$ and V is open in A, prove that $u \in W^{m,p}(V)$.

Theorem 5.3 (Leibniz Formula). *If* $\zeta \in \mathscr{C}_0^\infty(A)$ *and* $u \in W^{m,p}(A)$, *then* $\zeta u \in W^{m,p}(A)$ *and*

$$D^\alpha(\zeta u) = \sum_{\beta \le \alpha} \binom{\alpha}{\beta} D^\beta \zeta D^{\alpha-\beta} u.$$

Proof. First of all, fix $\phi \in \mathscr{C}_0^\infty(A)$. Then, for $|\alpha| = 1$, we get

$$\int_A D^\alpha(u\zeta)\phi dx = -\int_A u\zeta D^\alpha \phi dx$$

$$= -\int_A u(D^\alpha(\zeta\phi) - D^\alpha\zeta\phi)dx$$

$$= \int_A (D^\alpha u\zeta + uD^\alpha\zeta)\phi dx.$$

Let $l < m$. We suppose that the assertion holds for $|\alpha| \le l$ and prove it for $|\alpha| = l + 1$. Let $\alpha = \beta + \gamma$, where $|\beta| = l$ and $|\gamma| = 1$. Then

$$\int_A u\zeta D^\alpha\phi dx = \int_A u\zeta D^\beta(D^\gamma\phi)dx$$

$$= (-1)^{|\beta|} \int_A D^\beta(u\zeta)D^\gamma\phi dx$$

so by induction hypothesis

$$= (-1)^{|\beta|} \int_A \sum_{\sigma\le\beta} \binom{\beta}{\sigma} D^\sigma\zeta D^{\beta-\sigma} u D^\gamma \phi dx$$

$$= (-1)^{|\beta|+|\gamma|} \int_A \sum_{\sigma\le\beta} \binom{\beta}{\sigma} D^\gamma(D^\sigma\zeta D^{\beta-\sigma} u)\phi dx$$

$$= (-1)^{|\alpha|} \int_A \sum_{\sigma \le \beta} \binom{\beta}{\sigma} (D^{\gamma+\sigma} \zeta D^{\beta-\sigma} u + D^{\sigma} \zeta D^{\beta+\gamma-\sigma} u) \phi dx$$

$$= (-1)^{|\alpha|} \int_A \sum_{\sigma \le \alpha} \binom{\alpha}{\sigma} D^{\sigma} \zeta D^{\alpha-\sigma} u \phi dx,$$

proving the assertion. □

Theorem 5.4. *For every $m = 1, 2, \ldots$, $1 \le p \le \infty$, $W^{m,p}(A)$ is a Banach space.*

Proof. We shall prove the statement for $p < \infty$, and leave the reader to see to $p = \infty$. Let $\{u_l\}_{l=1}^{\infty} \subset W^{m,p}(A)$ be a Cauchy sequence. Then, for every $l \in \mathbb{N}$ and $0 \le |\alpha| \le m$, we have

$$||D^{\alpha} u_{l+q} - D^{\alpha} u_l||_{L^p(A)} \to_{q \to \infty} 0.$$

Therefore the sequence $\{D^{\alpha} u_l\}_{l=1}^{\infty}$ is fundamental in $L^p(A)$, for any α such that $0 \le |\alpha| \le m$. Since $L^p(A)$ is a Banach space, this sequence converges in $L^p(A)$ to some u_{α}, for any $0 \le |\alpha| \le m$. Let $u_0 = u$. We claim that $D^{\alpha} u = u_{\alpha}$ for any $0 \le |\alpha| \le m$. In fact,

$$\begin{aligned} \int_A u D^{\alpha} \phi &= \lim_{l \to \infty} \int_A u_l D^{\alpha} \phi dx \\ &= (-1)^{|\alpha|} \lim_{l \to \infty} \int_A D^{\alpha} u_l \phi dx \\ &= (-1)^{|\alpha|} \int_A u_{\alpha} \phi dx, \end{aligned}$$

so $D^{\alpha} u = u_{\alpha}$, $0 \le |\alpha| \le m$. As

$$||D^{\alpha} u_l - D^{\alpha} u||_{L^p(A)} \to_{l \to \infty} 0$$

for every $0 \le |\alpha| \le m$, we conclude that

$$||u_l - u||_{W^{m,p}(A)} \to_{l \to \infty} 0.$$

This completes the proof. □

Exercise 5.7. Prove that $H^m(A)$ is a Hilbert space for any $m \in \mathbb{N}$.

Exercise 5.8. Prove that if $u \in W^{1,p}(0,1)$ for some $1 < p < 1$, then

$$|u(x) - u(y)| \le |x-y|^{1-\frac{1}{p}} \left(\int_0^1 |u'(t)|^p dt \right)^{\frac{1}{p}}$$

for a.e. $x, y \in [0,1]$.

5.3 Approximation by Smooth Functions

Let $\varepsilon > 0$ and $A \subset \mathbb{R}^n$ be open. We define

$$u_\varepsilon(x) = \omega_\varepsilon \star u(x)$$

for $u \in W^{m,p}(A)$ and $x \in A_\varepsilon$.

Theorem 5.5. *If $u \in W^{m,p}(A)$, then $u_\varepsilon \in \mathscr{C}^\infty(A_\varepsilon)$.*

Proof. To see this, let α be an arbitrary multi-index. Then

$$\begin{aligned} D^\alpha u_\varepsilon(x) &= D^\alpha \int_A \omega_\varepsilon(x-y)u(y)dy \\ &= \int_A D_x^\alpha \omega_\varepsilon(x-y)u(y)dy \\ &= (-1)^{|\alpha|} \int_A D_y^\alpha \omega_\varepsilon(x-y)u(y)dy, \quad x \in A, \end{aligned}$$

from which

$$\begin{aligned} D^\alpha u_\varepsilon(x) &= \int_A \omega_\varepsilon(x-y) D^\alpha u(y)dy \\ &= \omega_\varepsilon \star D^\alpha u(x), \quad x \in A, \end{aligned}$$

and

$$\begin{aligned} |D^\alpha u_\varepsilon(x)| &= \left| \int_A D_x^\alpha \omega_\varepsilon(x-y)u(y)dy \right| \\ &\le \int_A |D_x^\alpha \omega_\varepsilon(x-y)||u(y)|dy \\ &\le \left(\int_A |D_x^\alpha \omega_\varepsilon(x-y)|^q dy \right)^{\frac{1}{q}} \left(\int_A |u(y)|^p dy \right)^{\frac{1}{p}} \\ &\le C||u||_{W^{m,p}(A)}, \quad x \in A. \end{aligned}$$

This completes the proof. $\square$

Theorem 5.6. *If $u \in W^{m,p}_{\text{loc}}(A)$, then*

$$u_\varepsilon \to_{\varepsilon \to 0} u$$

in $W^{m,p}_{\text{loc}}(A)$.

Proof. Let α be given. From the previous property $D^\alpha u_\varepsilon$ exists, and if $V \subset\subset A$, then

$$\begin{aligned} ||u_\varepsilon - u||_{W^{m,p}(V)} &= \left(\sum_{|\alpha| \le m} \int_V |D^\alpha u_\varepsilon - D^\alpha u|^p dx \right)^{\frac{1}{p}} \\ &\le \sum_{|\alpha| \le m} \left(\int_V |D^\alpha u_\varepsilon - D^\alpha u|^p dx \right)^{\frac{1}{p}} \\ &= \sum_{|\alpha| \le m} ||D^\alpha u_\varepsilon - D^\alpha u||_{L^p(V)}. \end{aligned} \tag{5.1}$$

Since $(D^\alpha u)_\varepsilon = D^\alpha u_\varepsilon$ and $D^\alpha u \in L^p(V)$, $0 \le |\alpha| \le m$, using the properties of the convolution, we see that

$$||D^\alpha u_\varepsilon - D^\alpha u||_{L^p(V)} \to_{\varepsilon \to 0} 0.$$

Hence, $u_\varepsilon \to_{\varepsilon \to 0} u$ in $W^{m,p}_{\text{loc}}(A)$. This completes the proof. □

Theorem 5.7. *If A is a bounded set in $\mathbb{R}^n$ and $u \in W^{m,p}(A)$, there exists a sequence $\{u_l\}_{l=1}^\infty$ in $\mathscr{C}^\infty(A) \cap W^{m,p}(A)$ such that*

$$u_l \to_{l \to \infty} u$$

in $W^{m,p}(A)$.

Proof. We claim $A = \bigcup_{i=1}^\infty A_i$, where $A_i = \{x \in A : \text{dist}(x, \partial A) > \frac{1}{i}\}$, $i \in \mathbb{N}$. Let $x \in A$ and $x \notin \partial A$. Therefore, there exists $i \in \mathbb{N}$ such that $\text{dist}(x, \partial A) > \frac{1}{i}$, so $x \in A_i$ and $x \in \bigcup_{i=1}^\infty A_i$. Consequently, $A \subset \bigcup_{i=1}^\infty A_i$. Conversely, if $x \in \bigcup_{i=1}^\infty A_i$ there exists $j \in \mathbb{N}$ such that $x \in A_j$. Therefore, $x \in A$ and $\text{dist}(x, \partial A) > \frac{1}{j}$, so we conclude $\bigcup_{i=1}^\infty A_i \subset A$. Let $V_i = A_{i+3} \backslash A_i$, $i = 1, 2, \ldots$, and choose $V_0 \subset\subset A$. Then, as above, $A = \bigcup_{i=0}^\infty V_i$. Let $\{\zeta_i\}_{i=1}^\infty$ be a sequence of smooth functions such that

$$\zeta_i \in \mathscr{C}_0^\infty(V_i), \quad 0 \le \zeta_i \le 1, \quad \sum_{i=0}^{\infty} \zeta_i = 1.$$

We define $\varepsilon_i > 0$ and $\delta > 0$, $u^i = (\zeta_i u) \star \omega_{\varepsilon_i}$ so that

$$\begin{aligned} &||u^i - \zeta_i u||_{W^{m,p}(A)} < \frac{\delta}{2^{i+1}}, \\ &\mathrm{supp} u^i \subset W_i = A_{i+4} \backslash \overline{A}_i \supset V_i. \end{aligned}$$

Note that $\zeta_i u \in W^{m,p}(A)$. From the properties of the convolution, we know

$$u^i \to_{\varepsilon_i \to 0} \zeta_i u$$

in $W^{m,p}(A)$. Let

$$v = \sum_{i=0}^{\infty} u^i.$$

Since $\sum_{i=0}^{\infty} \zeta_i = 1$, we have that $u = \sum_{i=0}^{\infty} \zeta_i u$. Let $V \subset\subset A$. Only a finite number of elements are different from zero in V, so $v \in \mathscr{C}^\infty(A)$. Therefore

$$\begin{aligned} ||v - u||_{W^{m,p}(A)} &= \left|\left| \sum_{i=0}^{\infty} u^i - \sum_{i=0}^{\infty} \zeta_i u \right|\right|_{W^{m,p}(V)} \\ &= \left|\left| \sum_{i=0}^{\infty} (u^i - \zeta_i u) \right|\right|_{W^{m,p}(V)} \\ &\le \sum_{i=0}^{\infty} ||u^i - \zeta_i u||_{W^{m,p}(V)} \\ &< \delta \sum_{i=0}^{\infty} \frac{1}{2^{i+1}} \\ &= \delta, \end{aligned}$$

and consequently

$$\sup_{V \subset\subset U} ||v - u||_{W^{m,p}(V)} \le \delta.$$

Let v_l be the function which corresponds to the space $W^{m,p}(V_l)$. Then

$$||v_l - u||_{W^{m,p}(A)} \to_{l\to\infty} 0.$$

This completes the proof. □

Exercise 5.9. Let A be a bounded set with $\mathscr{C}^1$ boundary ∂A. Prove that if $u \in W^{m,p}(A)$, $1 \le p < \infty$, there exists a sequence $\{u_l\}_{l=1}^{\infty}$ in $\mathscr{C}^\infty(\overline{A})$ such that

$$||u_l - u||_{W^{m,p}(A)} \to_{l\to\infty} 0.$$

Solution. Let $x^0 \in \partial A$. Since ∂A is $\mathscr{C}^1$, there exists a radius $r > 0$ and a $\mathscr{C}^1$ function $\gamma : \mathbb{R}^{n-1} \to \mathbb{R}$ such that

$$A \cap U(x^0, r) = \{x \in U(x^0, r) : x_n > \gamma(x_1, x_2, \dots, x_{n-1})\}.$$

Let $V = A \cap U(x^0, \frac{r}{2})$ and define

$$x^\varepsilon = x + \lambda\varepsilon e_n, \quad x \in V, \quad \varepsilon > 0, \quad \lambda > 0, \quad e_n = (0, 0, \dots, n).$$

There exists a large enough $\lambda > 0$ and a small enough $\varepsilon > 0$ such that $U(x^\varepsilon, \varepsilon)$ lies in $A \cap U(x^0, r)$ for every $x \in V$. Let $u^\varepsilon(x) = u(x^\varepsilon)$ and define

$$v_\varepsilon = \omega_\varepsilon \star u_\varepsilon.$$

Take $|\alpha| \le m$, so

$$\begin{aligned} ||D^\alpha v_\varepsilon - D^\alpha u||_{L^p(V)} &= ||D^\alpha v_\varepsilon - D^\alpha u_\varepsilon + D^\alpha u_\varepsilon - D^\alpha u||_{L^p(V)} \\ &\le ||D^\alpha v_\varepsilon - D^\alpha u_\varepsilon||_{L^p(V)} + ||D^\alpha u_\varepsilon - D^\alpha u||_{L^p(V)}. \end{aligned} \tag{5.2}$$

We have

$$|D^\alpha v_\varepsilon - D^\alpha u_\varepsilon||_{L^p(V)} \to_{\varepsilon\to 0} 0 \tag{5.3}$$

and

$$||D^\alpha u_\varepsilon - D^\alpha u||_{L^p(V)} \to_{\varepsilon\to 0} 0. \tag{5.4}$$

From (5.2), (5.3) and (5.4), it follows

$$||D^\alpha v_\varepsilon - D^\alpha u||_{L^p(V)} \to_{\varepsilon\to 0} 0 \quad \text{for } 0 \le |\alpha| \le m.$$

Consequently,

$$||v_\varepsilon - u||_{W^{m,p}(V)} \to_{\varepsilon\to 0} 0.$$

Since ∂A is compact, there exist finitely many points x_i^0, radii r_i and functions $v_i \in \mathscr{C}^\infty(\overline{V}_i)$ $(i = 1, 2, \ldots, N)$ for which

$$\partial A \subset \bigcup_{i=1}^{N} U\left(x_i^0, \frac{r_i}{2}\right), \quad ||v_i - u||_{W^{m,p}(V_i)} \le \delta,$$

where

$$V_i = A \cap U\left(x_i^0, \frac{r_i}{2}\right).$$

Let $V_0 \subset\subset U$. Then

$$A \subset \bigcup_{i=0}^{N} V_i.$$

There exist functions ζ_i, $i = 0, 1, \ldots, N$, such that $\zeta_i \in \mathscr{C}_0^\infty(V_i)$, $0 \le \zeta_i \le 1$ and $\sum_{i=0}^{N} \zeta_i = 1$. Now, we define the function

$$v = \sum_{i=0}^{N} \zeta_i v_i.$$

Then

$$u = \sum_{i=0}^{N} \zeta_i u$$

and

$$\begin{aligned} ||D^\alpha v - D^\alpha u||_{L^p(A)} &= \left\| \sum_{i=0}^{N} D^\alpha(\zeta_i(v_i - u)) \right\|_{L^p(A)} \\ &\le \sum_{i=0}^{N} ||D^\alpha(\zeta_i(v_i - u))||_{L^p(A)} \\ &= \sum_{i=0}^{N} ||D^\alpha v_i - D^\alpha u||_{L^p(V_i)} \end{aligned}$$

$$
\begin{aligned}
&\leq \sum_{i=0}^{N} ||v_i - u||_{W^{m,p}(V_i)} \\
&\leq \delta(N+1)
\end{aligned}
$$

for $0 \leq |\alpha| \leq m$.

5.4 Extensions

Theorem 5.8 (Extension Theorem). *Let $A \subset\subset V$ be bounded subsets of $\mathbb{R}^n$ and assume the boundary ∂A is $\mathscr{C}^1$. There exists a linear operator $E : W^{1,p}(A) \to W^{1,p}(\mathbb{R}^n)$ such that*

1. $Eu = u$ *for any* $u \in W^{1,p}(A)$,
2. $suppE \subset V$,
3. $||Eu||_{W^{1,p}(\mathbb{R}^n)} \leq C||u||_{W^{1,p}(A)}$.

Proof. Fix $x^0 \in \partial A$.

Case 1. $u \in \mathscr{C}^\infty(\overline{A})$.
First of all, we suppose that locally, around x^0, the boundary ∂A belongs to $\{x_n = 0\}$. Since ∂A is $\mathscr{C}^1$, there is a ball U such that

$$
\begin{aligned}
U^+ &= U \cap \{x_n \geq 0\} \\
&\subset \overline{A}, \\
U^- &= U \cap \{x_n \leq 0\} \\
&\subset \mathbb{R}^n \backslash A.
\end{aligned}
$$

Now define

$$
\overline{u}(x) = \begin{cases} u(x) & \text{for } x \in U^+, \\ -3u(x_1, x_2, \ldots, x_{n-1}, -x_n) \\ \quad + 4u\left(x_1, x_2, \ldots, x_{n-1}, -\dfrac{x_n}{2}\right) & \text{for } x \in U^-. \end{cases}
$$

We will show that $\overline{u} \in \mathscr{C}^1(U)$. In fact, let us define

$$
\begin{aligned}
u^+ &= \overline{u}_{|_{x_n \geq 0, x \in U^+}}, \\
u^- &= \overline{u}_{|_{x_n \leq 0, x \in U^-}}.
\end{aligned}
$$

We have

$$\begin{aligned} u^+_{|x_n=0} &= u(x_1, x_2, \ldots, x_{n-1}, 0), \\ u^-_{|x_n=0} &= -3u(x_1, x_2, \ldots, x_{n-1}, 0) + 4u(x_1, x_2, \ldots, x_{n-1}, 0) \\ &= u(x_1, x_2, \ldots, x_{n-1}, 0). \end{aligned}$$

Consequently,

$$u^+_{|x_n=0} = u^-_{|x_n=0},$$

and then

$$\frac{\partial u^+}{\partial x_i}_{|x_n=0} = \frac{\partial u^-}{\partial x_i}_{|x_n=0}, \qquad i = 1, 2, \ldots, n-1.$$

On the other hand,

$$\begin{aligned} \frac{\partial u^+}{\partial x_n}_{|x_n=0} &= \frac{\partial u}{\partial x_n}_{|x_n=0}, \\ \frac{\partial u^-}{\partial x_n}_{|x_n=0} &= 3\frac{\partial u}{\partial x_n}_{|x_n=0} - 2\frac{\partial u}{\partial x_n}_{|x_n=0} \\ &= \frac{\partial u}{\partial x_n}_{|x_n=0}, \end{aligned}$$

so overall,

$$\frac{\partial u^+}{\partial x_n}_{|x_n=0} = \frac{\partial u^-}{\partial x_n}_{|x_n=0}$$

and $D^\alpha u^-_{|x_n=0} = D^\alpha u^+_{|x_n=0}$ is well defined for $0 \leq |\alpha| \leq 1$. Hence, $\overline{u} \in \mathscr{C}^1(U)$. Additionally,

$$||\overline{u}||_{W^{1,p}(U)} \leq C||u||_{W^{1,p}(U+)} \leq C||u||_{W^{1,p}(A)}.$$

If ∂A does not belong on such hyperplane locally (in a neighborhood of x^0), since the boundary is $\mathscr{C}^1$ there exists a function Φ with inverse Ψ, say $x = \Phi(y)$, $y = \Psi(x)$, that maps a neighborhood of x^0 to a neighborhood of $\Psi(x^0)$ in such a way that, locally,

$\Psi(\partial A)$ lies on $\{y_n = 0\}$. Let $u^1(y) = u(x) = u(\Phi(y))$. As above, we construct a function $\overline{u}^1$ such that

$$\begin{aligned}||\overline{u}^1||_{W^{1,p}(U)} &\le C||u^1||_{W^{1,p}(U^+)}\\ &\le C||u^1||_{W^{1,p}(A)},\end{aligned}$$

and $\overline{u}^1 = u^1$ in U^+. If $W = \Psi(U)$, then

$$||\overline{u}||_{W^{1,p}(W)} \le C||u||_{W^{1,p}(A)}.$$

Now, we define the operator

$$Eu = \overline{u}.$$

Since $\overline{u}$ is bounded in $W^{1,p}(\overline{A})$, the map $u \to Eu$ is linear and bounded. Because ∂A is compact, there exist finitely many points x_1^0, x_2^0, ..., x_N^0, open sets W_i, extensions $\overline{u}_i$ of u on W_i such that if we take $W_0 \subset\subset A$, then we have

$$\partial A \subset \bigcup_{i=1}^{N} W_i, \quad A \subset \bigcup_{i=0}^{N} W_i.$$

Let $\{\zeta_i\}_{i=0}^N$ be the partition of unity corresponding to the system $W_0, W_1, \ldots, W_N$. Now, we put $\overline{u}_0 = u$. Then

$$\overline{u} = \sum_{i=0}^{N} \zeta_i \overline{u}_i$$

and

$$\begin{aligned}||\overline{u}||_{W^{1,p}(\mathbb{R}^n)} &= \left|\left|\sum_{i=0}^{N} \zeta_i \overline{u}_i\right|\right|_{W^{1,p}(\mathbb{R}^n)}\\ &\le \sum_{i=0}^{N} ||\zeta_i \overline{u}_i||_{W^{1,p}(\mathbb{R}^n)}\\ &= \sum_{i=0}^{N} ||\zeta_i \overline{u}_i||_{W^{1,p}(W_i)}\end{aligned}$$

$$\leq \sum_{i=0}^{N} ||\overline{u}_i||_{W^{1,p}(W_i)}$$
$$\leq \sum_{i=0}^{N} C||u||_{W^{1,p}(A)}$$
$$= CN||u||_{W^{1,p}(A)}.$$

Case 2. Let $u \in W^{1,p}(A)$. Then there exists a sequence $\{u_l\}_{l=1}^{\infty}$ in $\mathscr{C}^{\infty}(\overline{A}) \cap W^{1,p}(A)$ such that $u_l \to u$, $l \to \infty$, in $W^{1,p}(A)$. For any u_l, $l = 1, 2, \ldots$, we apply case 1. We also have

$$\begin{aligned} ||Eu_m - Eu_l||_{W^{1,p}(\mathbb{R}^n)} &= ||E(u_m - u_l)||_{W^{1,p}(\mathbb{R}^n)} \\ &\leq C||u_m - u_l||_{W^{1,p}(A)} \to_{l,m\to\infty} 0. \end{aligned}$$

Consequently, $\{Eu_m\}_{m=1}^{\infty}$ is a fundamental sequence in the Banach space $W^{1,p}(A)$, so it converges to some $\overline{u} \in W^{1,p}(A)$. But as $Eu_m = \overline{u}$ on $\overline{A}$, we have $\overline{u} = u$ on $\overline{A}$. This completes the proof.

□

Definition 5.5. We call Eu an extension of u to $\mathbb{R}^n$.

Exercise 5.10. Let $A \subset\subset V \subset \mathbb{R}^n$ be bounded sets with ∂A of class $\mathscr{C}^2$. Then there is a linear operator $E : W^{2,p}(A) \to W^{2,p}(\mathbb{R}^n)$ such that

1. $Eu = u$ for $u \in W^{2,p}(A)$,
2. $suppE \subset V$,
3. $||Eu||_{W^{2,p}(\mathbb{R}^n)} \leq C||u||_{W^{2,p}(A)}$.

Exercise 5.11. Extend $u \in W^{1,p}((0,\infty))$ onto $(-\infty, 0)$ by setting $\overline{u}(x) = u(-x)$. Prove that this extension $\overline{u}$ is an element of $W^{1,p}(\mathbb{R})$.

5.5 Traces

Theorem 5.9 (Trace Theorem). *Let A be a bounded set in $\mathbb{R}^n$ with $\mathscr{C}^1$ boundary ∂A. There exists a linear bounded operator*

$T : W^{1,p}(A) \to W^{1,p}(\partial A)$ *such that*

$$Tu = u_{|\partial A} \quad \text{if } u \in W^{1,p}(A) \cap \mathscr{C}(\overline{A})$$

and

$$||Tu||_{L^p(\partial U)} \leq C||u||_{W^{1,p}}(A).$$

Proof. We assume $u \in \mathscr{C}^1(\overline{A})$ and take $x^0 \in \partial A$. We also suppose that ∂A intersected with some neighbourhood of x^0 lies on the plane $\{x_n = 0\}$. Let $r > 0$ be such that $A \cap U(x^0, r) \subset \{x_n = 0\}$. We consider $U(x^0, \frac{r}{2})$ and call $\Gamma = \partial(A \cap U(x^0, \frac{r}{2}))$. We choose $\zeta \in \mathscr{C}_0^\infty(U(x^0, r))$ so that $0 \leq \zeta \leq 1$ on $U(x^0, r)$, $\zeta \equiv 1$ on $U(x^0, \frac{r}{2})$. By denoting

$$x' = (x_1, x_2, \ldots, x_{n-1}) \in \mathbb{R}^{n-1} = \{x_n = 0\},$$

we have

$$\begin{aligned} \int_\Gamma |u|^p dx' &\leq \int_{\{x_n=0\}} \zeta |u|^p dx \\ &= -\int_{U^+} (\zeta |u|^p)_{x_n} dx \\ &= -\int_{U^+} \zeta_{x_n} |u|^p dx - p \int_{U^+} \zeta |u|^{p-1} (\text{sign} u) u_{x_n} dx. \end{aligned} \tag{5.5}$$

Young's inequality, with $\frac{1}{p} + \frac{1}{q} = 1$, gives

$$\begin{aligned} |u|^{p-1} |u_{x_n}| &\leq \frac{(|u|^{p-1})^q}{q} + \frac{|u_{x_n}|^p}{p} \\ &\leq C(|u|^p + |Du|^p). \end{aligned}$$

From this and (5.5), we deduce

$$\int_\Gamma |u|^p dx \leq C \int_{U^+} (|u|^p + |Du|^p) dx,$$

so

$$\begin{aligned} ||u||_{L^p(\Gamma)} &\leq C||u||_{W^{1,p}(U^+)} \\ &\leq C||u||_{W^{1,p}(A)}. \end{aligned} \tag{5.6}$$

If we cannot find a neighborhood of x^0 the restriction of ∂A to which belongs in $\{x_n = 0\}$, there exist a $\mathscr{C}^1$ map Φ, with inverse Ψ, mapping a neighborhood of x^0 to a neighborhood of $y^0 = \Psi(x^0)$ so that, locally, $\Psi(\partial A)$ lies in $\{y_n = \Psi(x_n) = 0\}$. Since ∂A is compact, there exists a finite number of points $x_1^0, x_2^0, \ldots, x_N^0$ and balls $U(x_i^0, r_i) = V_i$ such that $\partial A \subset \bigcup_{i=1}^N V_i$, $V_0 \subset\subset A$ and

$$A \subset \bigcup_{i=0}^N V_i.$$

From (5.6),

$$||u||_{L^p(\Gamma_i)} \leq C||u||_{W^{1,p}(U)}.$$

Let $\{\zeta_i\}_{i=0}^N$ be the partition of unity of the system $\{V_i\}_{i=0}^N$. Then

$$\begin{aligned} ||u||_{L^p(\partial A)} &= \left|\left|\sum_{i=0}^N \zeta_i u\right|\right|_{L^p(\partial A)} \\ &\leq \sum_{i=0}^N ||\zeta_i u||_{L^p(\partial A)} \\ &= \sum_{i=0}^N ||u||_{L^p(\Gamma_i)} \\ &\leq C||u||_{W^{1,p}(A)}. \end{aligned} \tag{5.7}$$

Now, define the operator $T : W^{1,p}(A) \to W^{1,p}(\partial A)$ by

$$Tu = u_{|\partial A}.$$

Using (5.7), we see that

$$||Tu||_{L^p(\partial A)} \leq C||u||_{W^{1,p}(A)}.$$

Let $u \in W^{1,p}(A) \cap \mathscr{C}(\overline{A})$. There exists a sequence $\{u_m\}_{m=1}^\infty$ in $\mathscr{C}^\infty(\overline{A})$ such that $u_m \to u$ in $W^{1,p}(A)$, as $m \to \infty$. From (5.7), we have

$$\begin{aligned} ||Tu_m - Tu_l||_{L^p(\partial A)} &\leq C||u_m - u_l||_{W^{1,p}(A)} \\ &\to_{m,l\to\infty} 0. \end{aligned}$$

Therefore, the sequence $\{Tu_m\}_{m=1}^\infty$ is fundamental in the Banach space $L^p(\partial A)$, and as such it converges in $L^p(\partial A)$:

$$\lim_{m\to\infty} Tu_m = Tu.$$

As $Tu_m = u_{|\partial A}$, we infer that $Tu = u_{|\partial A}$ and T is a bounded operator. □

Definition 5.6. We call Tu the trace of u on ∂U.

Exercise 5.12. Let A be a bounded set in $\mathbb{R}^n$ and assume ∂A is $\mathscr{C}^1$. Prove that Tu vanishes on ∂A, provided $u \in W_0^{1,p}(A)$.

Hint. Use the fact that there exists a sequence $\{u_m\}_{m=1}^{\infty}$ in $\mathscr{C}_0^{\infty}(A)$ such that $u_m \to u$ in $W^{1,p}(A)$, as $m \to \infty$. Since Tu_m is zero on ∂A, we also have $Tu = 0$ on ∂A.

Exercise 5.13. Let A be a bounded set in $\mathbb{R}^n$ and let ∂A be $\mathscr{C}^1$. Take $u \in W^{1,p}(A)$ with $Tu = 0$ on ∂A. Prove that

$$\int_{\mathbb{R}^{n-1}} |u(x', x_n)|^p dx' \leq C x_n^{p-1} \int_0^{x_n} \int_{\mathbb{R}^{n-1}} |Du|^p dx' dt$$

for a.e. $x_n > 0$. Here $x' = (x_1, x_2, \ldots, x_{n-1})$.

Hint. Use the extension theorem, then choose $\{u_m\}_{m=1}^{\infty}$ in $\mathscr{C}^1(\overline{R_+^n})$ such that $u_m \to u$ in $W^{1,p}(\overline{R_+^n})$, as $m \to \infty$. Then use the identity

$$u_m(x', x_n) - u_m(x', 0) = \int_0^{x_n} u_{mx_n}(x', s) ds.$$

Here, $\overline{R_+^n}$ is the closure of $\{x = (x_1, x_2, \ldots, x_n) \in \mathbb{R}^n : x_n > 0\}$.

Exercise 5.14. Let A be a bounded set in $\mathbb{R}^n$ with ∂A of class $\mathscr{C}^1$. Take $u \in W^{1,p}(A)$ with $Tu = 0$ on ∂A. Prove that $u \in W_0^{1,p}(A)$.

Hint. Use the extension theorem. Consider the function $\zeta \in \mathscr{C}^{\infty}(\mathbb{R})$ such that $\zeta \equiv 1$ on $[0, 1]$, $0 \leq \zeta \leq 1$ on $\mathbb{R}$, $\zeta \equiv 0$ on $\mathbb{R} \backslash [0, 2]$, and the two sequences $\zeta_m(x) = \zeta(m x_n)$, $w_m(x) = u(x)(1 - \zeta_m(x))$, $x \in \mathbb{R}_+^n$. Prove that $w_m \to_{m \to \infty} u$ in $W^{1,p}(\mathbb{R}_+^n)$. Mollify w_m to produce functions $u_m \in \mathscr{C}_0^{\infty}(\mathbb{R}_+^n)$ such that $u_m \to_{m \to \infty} u$ in $W^{1,p}(\mathbb{R}_+^n)$.

5.6 Sobolev Inequalities

Definition 5.7. Let $1 \leq p < n$. The Sobolev conjugate p^* of p is defined by

$$\frac{1}{p^*} = \frac{1}{p} - \frac{1}{n}.$$

Theorem 5.10 (Gagliardo–Nirenberg–Sobolev Inequality). *Let* $u \in \mathscr{C}_0^1(\mathbb{R}^n)$. *Then*

$$||u||_{L^{p^*}(\mathbb{R}^n)} \leq C||Du||_{L^p(\mathbb{R}^n)}$$

for some constant $C > 0$, $1 \leq p < n$.

Proof.

Case 1. Let $p = 1$. Then $p^* = \frac{n}{n-1}$ and we have to prove that

$$||u||_{L^{\frac{n}{n-1}}(\mathbb{R}^n)} \leq C||Du||_{L^1(\mathbb{R}^n)}.$$

Our first observation is

$$\begin{aligned}
|u(x)| &= \left| \int_{-\infty}^{x_i} u_{x_i}(x_1, \ldots, x_{i-1}, y_i, x_{i+1}, \ldots, x_n) dy_i \right| \\
&\leq \int_{-\infty}^{x_i} |u_{x_i}(x_1, \ldots, x_{i-1}, y_i, x_{i+1}, \ldots, x_n)| dy_i \\
&\leq \int_{-\infty}^{\infty} |Du(y_i)| dy_i, \quad i = 1, 2, \ldots, n.
\end{aligned}$$

Then

$$|u(x)|^{\frac{1}{n-1}} \leq \left(\int_{-\infty}^{\infty} |Du(y_i)| dy_i \right)^{\frac{1}{n-1}}, \quad i = 1, 2, \ldots, n,$$

i.e.,

$$\begin{aligned}
|u(x)|^{\frac{1}{n-1}} &\leq \left(\int_{-\infty}^{\infty} |Du(y_1)| dy_1 \right)^{\frac{1}{n-1}}, \\
|u(x)|^{\frac{1}{n-1}} &\leq \left(\int_{-\infty}^{\infty} |Du(y_2)| dy_2 \right)^{\frac{1}{n-1}}, \\
&\cdots \\
|u(x)|^{\frac{1}{n-1}} &\leq \left(\int_{-\infty}^{\infty} |Du(y_n)| dy_n \right)^{\frac{1}{n-1}}.
\end{aligned}$$

We multiply the above inequalities and get

$$|u(x)|^{\frac{n}{n-1}} \leq \prod_{i=1}^{n} \left(\int_{-\infty}^{\infty} |Du(y_i)| dy_i \right)^{\frac{1}{n-1}}.$$

Now, we integrate in the variable x_1 and obtain

$$\int_{-\infty}^{\infty} |u(x)|^{\frac{n}{n-1}} dx_1 \leq \int_{-\infty}^{\infty} \prod_{i=1}^{n} \left(\int_{-\infty}^{\infty} |Du(y_i)| dy_i \right)^{\frac{1}{n-1}} dx_1$$

$$\leq \left(\int_{-\infty}^{\infty} |Du(y_1)| dy_1 \right)^{\frac{1}{n-1}} \prod_{i=2}^{n} \left(\int_{-\infty}^{\infty} \int_{-\infty}^{\infty} |Du(y_i)| dy_i dx_1 \right)^{\frac{1}{n-1}}.$$

Integrating now in x_2 gives

$$\int_{-\infty}^{\infty} \int_{-\infty}^{\infty} |u(x)|^{\frac{n}{n-1}} dx_1 dx_2$$

$$\leq \int_{-\infty}^{\infty} \left(\int_{-\infty}^{\infty} |Du(y_1)| dy_1 \right)^{\frac{1}{n-1}}$$

$$\times \prod_{i=2}^{n} \left(\int_{-\infty}^{\infty} \int_{-\infty}^{\infty} |Du(y_i)| dy_i dx_1 \right)^{\frac{1}{n-1}} dx_2$$

$$\leq \left(\int_{-\infty}^{\infty} \int_{-\infty}^{\infty} |Du(y_2)| dy_2 dx_1 \right)^{\frac{1}{n-1}}$$

$$\times \left(\int_{-\infty}^{\infty} \int_{-\infty}^{\infty} |Du(y_1)| dy_1 dx_2 \right)^{\frac{1}{n-1}}$$

$$\times \prod_{i=3}^{n} \left(\int_{-\infty}^{\infty} \int_{-\infty}^{\infty} \int_{-\infty}^{\infty} |Du(y_i)| dy_i dx_1 dx_2 \right)^{\frac{1}{n-1}}.$$

Iterating,

$$\int_{\mathbb{R}^n} |u(x)|^{\frac{n}{n-1}} dx \leq C \left(\int_{\mathbb{R}^n} |Du(x)| dx \right)^{\frac{n}{n-1}}, \tag{5.8}$$

and hence,

$$||u||_{L^{\frac{n}{n-1}}(\mathbb{R}^n)} \leq C ||Du||_{L^1(\mathbb{R}^n)}.$$

Case 2. Let $p > 1$. We put $v = |u|^\gamma$, where γ will be determined subsequently. We apply the inequality (5.8) to v and get

$$\left(\int_{\mathbb{R}^n} |u(x)|^{\frac{\gamma n}{n-1}} dx \right)^{\frac{n-1}{n}} \leq C \int_{\mathbb{R}^n} |u(x)|^{\gamma-1} |Du(x)| dx.$$

By Hölder's inequality

$$\left(\int_{\mathbb{R}^n}|u(x)|^{\frac{\gamma n}{n-1}}dx\right)^{\frac{n-1}{n}}\leq C\left(\int_{\mathbb{R}^n}|u(x)|^{\frac{(\gamma-1)p}{p-1}}dx\right)^{\frac{p-1}{p}}||Du||_{L^p(\mathbb{R}^n)}. \tag{5.9}$$

Now, we take $\gamma>0$ such that

$$\frac{\gamma n}{n-1}=\frac{(\gamma-1)p}{p-1},$$

so

$$\begin{aligned}\gamma&=\frac{p(n-1)}{n-p},\\ \frac{\gamma n}{n-1}&=\frac{pn}{n-p}\\ &=p^*.\end{aligned}$$

From (5.9), we find

$$\left(\int_{\mathbb{R}^n}|u(x)|^{p^*}dx\right)^{\frac{n-1}{n}}\leq C\left(\int_{\mathbb{R}^n}|u(x)|^{p^*}dx\right)^{\frac{p-1}{p}}||Du||_{L^p(\mathbb{R}^n)},$$

and then

$$||u||_{L^{p^*}(\mathbb{R}^n)}\leq C||Du||_{L^p(\mathbb{R}^n)}.$$

This completes the proof. □

Exercise 5.15. Let $1\leq p<n$ and consider A bounded in $\mathbb{R}^n$ with $\mathscr{C}^1$ boundary. For $u\in W^{1,p}(A)$ prove

$$||u||_{L^{p^*}(A)}\leq ||u||_{W^{1,p}(A)}.$$

Hint. Apply the Extension theorem so to ensure the existence of the extension $\overline{u}\in W^{1,p}(\mathbb{R}^n)$ of u. Then choose a sequence $\{u_m\}_{m=1}^{\infty}$ in $\mathscr{C}_0^{\infty}(\mathbb{R}^n)$ such that $u_m\to\overline{u}$ in $W^{1,p}(\mathbb{R}^n)$ as $m\to\infty$. Apply the Gagliardo–Nirenberg–Sobolev inequality to u_m-u_l and conclude that $\{u_m\}_{m=1}^{\infty}$ converges to $\overline{u}$ in $L^{p^*}(\mathbb{R}^n)$. Eventually, the Gagliardo–Nirenberg–Sobolev inequality on u_m gives the desired result.

Exercise 5.16. Let A be a bounded set in $\mathbb{R}^n$ with ∂A of class $\mathscr{C}^1$, $u\in W_0^{1,p}(A)$ for $1\leq p<n$. Prove

$$||u||_{L^q(A)}\leq C||Du||_{L^p(A)}$$

for some constant $C>0$ and for every $q\in[1,p^*]$.

Hint. Use approximation and the Gagliardo–Nirenberg–Sobolev inequality.

Theorem 5.11 (Morrey Inequality). *Let $n < p \leq \infty$. Then for every $u \in \mathscr{C}^1(\mathbb{R}^n)$ there exists a constant $C = C(n,p) > 0$ such that*

$$||u||_{\mathscr{C}^{0,\gamma}(\mathbb{R}^n)} \leq C||u||_{W^{1,p}(\mathbb{R}^n)},$$

where $\gamma = 1 - \frac{n}{p}$.

Proof. Let $U(x,r) \subset \mathbb{R}^n$ be arbitrary ball, $r > 0$, $s \in (0,r]$. Fix $w \in \partial U(0,1)$, so

$$\begin{aligned}
|u(x+sw) - u(x)| &= \left|\int_0^s \frac{d}{dt}u(x+tw)dt\right| \\
&= \left|\int_0^s Du(x+tw) \cdot wdt\right| \\
&\leq \int_0^s |Du(x+tw) \cdot w|dt \\
&\leq \int_0^s |Du(x+tw)|dt,
\end{aligned}$$

which we integrate on $\partial U(0,1)$ and get

$$\begin{aligned}
&\int_{\partial U(0,1)} |u(x+sw) - u(x)|dS \\
&\quad \leq \int_{\partial U(0,1)} \int_0^s |Du(x+tw)|dtdS \\
&\quad = \int_0^s \int_{\partial U(0,1)} |Du(x+tw)|dSdt \\
&\quad = \int_0^s \int_{\partial U(0,1)} |Du(x+tw)|\frac{t^{n-1}}{t^{n-1}}dSdt \quad (x+tw=y) \\
&\quad \leq \int_0^s \int_{\partial U(x,t)} |Du(y)|\frac{1}{|x-y|^{n-1}}dS_y dt
\end{aligned}$$

$$= \int_{U(x,s)} |Du(y)| \frac{1}{|x-y|^{n-1}} dy$$
$$\leq \int_{U(x,r)} |Du(y)| \frac{1}{|x-y|^{n-1}} dy.$$

From here,

$$s^{n-1} \int_{\partial U(0,1)} |u(x+sw)-u(x)| dS \leq s^{n-1} \int_{U(x,r)} |Du(y)| \frac{1}{|x-y|^{n-1}} dy.$$

Therefore,

$$\int_0^r s^{n-1} \int_{\partial U(0,1)} |u(x+sw) - u(x)| dS ds$$
$$\leq \int_0^r s^{n-1} \int_{U(x,r)} |Du(y)| \frac{1}{|x-y|^{n-1}} dy ds,$$

so

$$\int_0^r s^{n-1} \int_{\partial U(0,1)} |u(x+sw) - u(x)| dS ds$$
$$\leq \frac{r^n}{n} \int_{U(x,r)} |Du(y)| \frac{1}{|x-y|^{n-1}} dyd,$$

and substituting $x + sw = y$ finally

$$\int_0^r \int_{\partial U(x,s)} |u(y) - u(x)| dS_y ds \leq \frac{r^n}{n} \int_{U(x,r)} |Du(y)| \frac{1}{|x-y|^{n-1}} dy$$

and

$$\frac{1}{r^n} \int_{U(x,r)} |u(y) - u(x)| dy \leq \frac{1}{n} \int_{U(x,r)} |Du(y)| \frac{1}{|x-y|^{n-1}} dy.$$

We set

$$\frac{1}{r^n} \int_{U(x,r)} (\cdot) dy = \overline{\int}_{U(x,r)} (\cdot) dy.$$

Then there exists a constant $C > 0$ such that

$$\overline{\int}_{U(x,r)} |u(y) - u(x)| dy \leq C \int_{U(x,r)} |Du(y)| \frac{1}{|x-y|^{n-1}} dy. \tag{5.10}$$

On the other hand,

$$\begin{aligned}
|u(x)| &= \fint_{U(x,1)} |u(y)|dy = \fint_{U(x,1)} |u(x) - u(y) + u(y)|dy \\
&\leq \fint_{U(x,1)} |u(x) - u(y)|dy + \fint_{U(x,1)} |u(y)|dy \\
&\leq \fint_{U(x,1)} |u(x) - u(y)|dy + C||u||_{L^p(U(x,1))} \\
&\leq \fint_{U(x,1)} |u(x) - u(y)|dy + C||u||_{L^p(\mathbb{R}^n)} \quad \text{(by (5.10))} \\
&\leq \fint_{U(x,1)} \frac{|Du(y)|}{|x-y|^{n-1}} dy + C||u||_{L^p(\mathbb{R}^n)} \\
&\leq C \int_{U(x,1)} \frac{|Du(y)|}{|x-y|^{n-1}} dy \\
&\quad + C||u||_{L^p(\mathbb{R}^n)} \quad \text{(Hölder's inequality)} \\
&\leq C \left(\int_{U(x,1)} |Du(y)|^p dy \right)^{\frac{1}{p}} \left(\int_{U(x,1)} \frac{1}{|x-y|^{\frac{(n-1)p}{p-1}}} dy \right)^{\frac{p-1}{p}} \\
&\quad + C||u||_{L^p(\mathbb{R}^n)} \left(\frac{(n-1)p}{p-1} = \frac{p-n}{p-1} \right) \\
&\leq C(||u||_{L^p(\mathbb{R}^n)} + ||Du||_{L^p(\mathbb{R}^n)}) \leq C||u||_{W^{1,p}(\mathbb{R}^n)}.
\end{aligned}$$

Therefore,

$$\sup_{x \in \mathbb{R}^n} |u(x)| \leq C||u||_{W^{1,p}(\mathbb{R}^n)}.$$

Let x, $y \in \mathbb{R}^n$ be arbitrary points and

$$W = U(x,r) \cap U(y,r).$$

Then

$$\begin{aligned} |u(x)-u(y)| &= \fint_W |u(x)-u(y)|dz \\ &= \fint_W |u(x)-u(z)+u(z)-u(y)|dz \qquad (5.11) \\ &\le \fint_W |u(x)-u(z)|dz + \fint_W |u(z)-u(y)|dz. \end{aligned}$$

The inequality (10.10) allows us to estimate

$$\begin{aligned} \fint_W |u(x)-u(z)|dz &\le \fint_{U(x,r)} |u(x)-u(z)|dz \\ &\le C\int_{U(x,r)} \frac{|Du(z)|}{|x-z|^{n-1}}dz \quad \text{(Hölder's inequality)} \\ &\le C\left(\int_{U(x,r)} |Du(z)|^p dz\right)^{\frac{1}{p}} \\ &\quad \times \left(\int_{U(x,r)} \frac{1}{|x-z|^{\frac{(n-1)p}{p-1}}}dz\right)^{\frac{p-1}{p}} \\ &\le C\left(r^{n-\frac{(n-1)p}{p-1}}\right)^{\frac{p-1}{p}} ||Du||_{L^p(U(x,r))} \\ &\le C|x-y|^{1-\frac{n}{p}}||Du||_{L^p(\mathbb{R}^n)} \\ &= C|x-y|^{\gamma}||Du||_{L^p(\mathbb{R}^n)} \\ &\le C|x-y|^{\gamma}||Du||_{W^{1,p}(\mathbb{R}^n)}. \qquad (5.12) \end{aligned}$$

Hence, we deduce

$$\fint_W |u(y)-u(z)|dz \le C|x-y|^{\gamma}||Du||_{W^{1,p}(\mathbb{R}^n)}.$$

From the latter, (5.11) and (5.12), we get

$$|u(x)-u(y)| \le C|x-y|^{\gamma}||u||_{W^{1,p}(\mathbb{R}^n)},$$

so

$$\sup_{x \neq y, x, y \in \mathbb{R}^n} \frac{|u(x) - u(y)|}{|x - y|^\gamma} \leq C||u||_{W^{1,p}(\mathbb{R}^n)}$$

or

$$||u||_{\mathscr{C}^{0,\gamma}(\mathbb{R}^n)} \leq C||u||_{W^{1,p}(\mathbb{R}^n)}.$$

This completes the proof. □

To prove the Poincaré inequality we have a need of the following Rellich–Kondrachov compactness theorem.

Theorem 5.12 (Rellich–Kondrachov Compactness Theorem). *Let A be a bounded open set in $\mathbb{R}^n$ with $\mathscr{C}^1$ boundary and $1 \leq p < n$. Then*

$$W^{1,p}(A) \hookrightarrow L^q(A)$$

for every $1 \leq q < p^*$, where p^* is the Sobolev conjugate of p.

Exercise 5.17. Let A be a bounded open set in $\mathbb{R}^n$ with $\mathscr{C}^1$ boundary. Prove that $W^{1,p}(A) \hookrightarrow L^p(A)$.

Definition 5.8. One calls

$$(u)_A = \frac{1}{|A|} \int_A u(y)dy$$

the average of u over A.

Theorem 5.13 (Poincaré Inequality). *Let A be a bounded, connected, open set in $\mathbb{R}^n$, $1 \leq p \leq \infty$. Then for every $u \in W^{1,p}(A)$ there exists a constant $C = C(u, p, A)$ such that*

$$||u - (u)_A||_{L^p(A)} \leq C||Du||_{L^p(A)}.$$

Proof. Let us suppose that there exists a function $u_k \in W^{1,p}(A)$ such that

$$||u_k - (u_k)_A||_{L^p(A)} > k||Du_k||_{L^p(A)}. \tag{5.13}$$

Let

$$v_k = \frac{u_k - (u_k)_A}{||u_k - (u_k)_A||_{L^p(A)}}.$$

Then

$$\begin{aligned}(v_k)_A &= \frac{1}{||u_k - (u_k)_A||} \overline{\int}_A (u_k(y) - (u_k)_A) dy \\ &= \frac{1}{||u_k - (u_k)_A||} \left(\overline{\int}_A u_k(y) dy - (u_k)_A \overline{\int}_A dy \right) \\ &= \frac{1}{||u_k - (u_k)_A||} ((u_k)_A - (u_k)_A) \\ &= 0\end{aligned}$$

and

$$\begin{aligned}||v_k||_{L^p(A)} &= \left|\left| \frac{u_k - (u_k)_A}{||u_k - (u_k)_A||_{L^p(A)}} \right|\right|_{L^p(A)} \\ &= \frac{||u_k - (u_k)_A||_{L^p(A)}}{||u_k - (u_k)_A||_{L^p(A)}} \\ &= 1.\end{aligned}$$

From (5.13), we infer

$$\begin{aligned}\frac{1}{k} &> \frac{||Du_k||_{L^p(A)}}{||u_k - (u_k)_A||_{L^p(A)}} \\ &= \frac{||D(u_k - (u_k)_A)||_{L^p(A)}}{||u_k - (u_k)_A||_{L^p(A)}} \\ &= \left|\left| D \frac{u_k - (u_k)_A}{||u_k - (u_k)_A||_{L^p(A)}} \right|\right| \\ &= ||Dv_k||_{L^p(A)}.\end{aligned} \tag{5.14}$$

Consequently, $\{v_k\}_{k=1}^{\infty}$ is a bounded sequence in $L^p(A)$ and $W^{1,p}(A)$. By the Rellich–Kondrachov compactness theorem $W^{1,p}(A) \hookrightarrow L^p(A)$. Therefore, there is a subsequence $\{v_{k_j}\}_{j=1}^{\infty}$ that converges to some $v \in L^p(A)$. From (5.14), we have

$$\lim_{j\to\infty} ||Dv_{k_j}||_{L^p(A)} = 0.$$

Fix an arbitrary $\phi \in \mathscr{C}_0^\infty(A)$. Then

$$\begin{aligned}\int_A v\phi_{x_i}dx &= \lim_{j\to\infty}\int_A v_{k_j}\phi_{x_i}dx\\ &= -\lim_{j\to\infty}\int_A v_{k_jx_i}\phi dx\\ &= 0\end{aligned}$$

and $Dv = 0$. Consequently $v \in W^{1,p}(A)$. Since A is connected and $Dv = 0$, it follows that $v = const$. Let $v = l$. Because $||v_k|| = 1$, we have $||v||_{L^p(A)} = 1$. From $(v_k)_A = 0$, we conclude that $(v)_A = 0$, so $l = 0$, $||v||_{L^p(A)} = 0$, which is a contradiction. □

Exercise 5.18 (Poincaré Inequality on the Ball). Given $1 \leq p \leq \infty$, there exists a constant $C = C(u,p,A)$ such that

$$||u-(u)_{U(x,r)}||_{L^p(U(x,r))} \leq Cr||Du||_{L^p(U(x,r))}.$$

Hint. Consider the function $v(y) = u(x+ry)$, $y \in U(0,1)$. Use the Poincaré inequality for U_1 and change variables $y = \frac{z-x}{r}$.

5.7 The Space H^s

Sobolev spaces are named after the Russian mathematician Sergei Sobolev. Their importance comes from the fact that weak solutions of some important partial differential equations exist in appropriate Sobolev spaces, even when there are no strong solutions in spaces of continuous functions with the derivatives understood in the classical sense.

Natural spaces to measure the regularity of the initial data in Cauchy problems are the classical Sobolev spaces $H^s(\mathbb{R}^n), s \in \mathbb{R}$, which are defined as

Definition 5.9. Let $s \in \mathbb{R}$. We define Sobolev space of order s, denoted by $H^s(\mathbb{R}^n)$, as

$$H^s(\mathbb{R}^n) = \left\{f \in \mathscr{S}'(\mathbb{R}^n); ||f||^2_{H^s(\mathbb{R}^n)} = \int_{\mathbb{R}^n}(1+|\xi|^2)^s|\widehat{f}(\xi)|^2d\xi < \infty\right\},$$

in the line, where $\mathscr{S}'(\mathbb{R}^n)$ denotes the class of tempered distributions.

We can write

$$\|f\|_{H^s(\mathbb{R}^n)} = \left\|\langle\xi\rangle^s \widehat{f}(\xi)\right\|_{L^2_\xi(\mathbb{R}^n)},$$

where $\langle\cdot\rangle = \left(1 + |\cdot|^2\right)^{\frac{1}{2}}$.

Proposition 5.1 (Properties of $H^s(\mathbb{R}^n)$).

1. *If* $0 \leqslant s < s'$, *then*

$$H^{s'}(\mathbb{R}^n) \subset H^s(\mathbb{R}^n). \tag{5.15}$$

2. $H^s(\mathbb{R}^n)$ *is a Hilbert space with respect to the inner product* $\langle\cdot,\cdot\rangle_s$ *defined as follows:*

$$\text{If } f, g \in H^s(\mathbb{R}^n), \text{ then } \langle f, g\rangle_s = \int_{\mathbb{R}^n} (1+|\xi|^2)^{\frac{s}{2}} \widehat{f}(\xi)\overline{(1+|\xi|^2)^{\frac{s}{2}} \hat{g}(\xi)} \mathrm{d}\xi. \tag{5.16}$$

3. *For any* $s \in \mathbb{R}$, *the Schwartz space* $\mathscr{S}(\mathbb{R}^n)$ *is dense in* $H^s(\mathbb{R}^n)$.
4. *If* $s_1 \leq s \leq s_2$, *with* $s = \theta s_1 + (1-\theta)s_2, 0 \leq \theta \leq 1$, *then*

$$\|f\|_{H^s(\mathbb{R}^n)} \leq \|f\|^{\theta}_{H^{s_1}(\mathbb{R}^n)} \|f\|^{1-\theta}_{H^{s_2}(\mathbb{R}^n)}. \tag{5.17}$$

A really interesting fact is that for positive integer values of s, we can give a description of H^s without using the Fourier transform.

Theorem 5.14. *If k is a positive integer, then $H^k(\mathbb{R}^n)$ coincides with the space of functions $f \in L^2(\mathbb{R}^n)$ whose derivatives (in the distribution sense) $f^{(j)}$ belongs to $L^2(\mathbb{R}^n)$ for every $j \leq k$. In this case, the norms*

$$\|f\|_{H^k(\mathbb{R}^n)} \quad \textit{and} \quad \sum_{j=1}^{k} \left\|f^{(j)}\right\|_{L^2(\mathbb{R}^n)},$$

are equivalent.

Furthermore, the following proposition allows us to relate "weak derivatives" with derivatives in the classical sense.

Theorem 5.15 (Embedding). *If $s > k + \frac{1}{2}$, then $H^s(\mathbb{R}^n)$ is continuously embedding in $C^k_\infty(\mathbb{R}^n)$, the space of functions with k*

continuous derivatives vanishing at infinity. In other words, if $f \in H^s(\mathbb{R}^n), s > \frac{1}{2} + k$, *then* $f \in C^k_\infty(\mathbb{R}^n)$ *and*

$$\|f\|_{C^k(\mathbb{R}^n)} \lesssim_s \|f\|_{H^s(\mathbb{R}^n)}.$$

Proof.

- Case $k = 0$: we first show that if $f \in H^s(\mathbb{R}^n)$ then $\widehat{f} \in L^1(\mathbb{R}^n)$ with

$$\|\widehat{f}\|_{L^1(\mathbb{R}^n)} \lesssim_s \|f\|_{H^s(\mathbb{R}^n)} \quad \text{if } s > \frac{n}{2}.$$

  Using the Cauchy–Schwarz inequality we deduce

$$\begin{aligned}\int_{\mathbb{R}^n} |\widehat{f}(\xi)|d\xi &= \int_{\mathbb{R}^n} |\widehat{f}(\xi)|(1+|\xi|^2)^{\frac{s}{2}} \frac{d\xi}{(1+|\xi|^2)^{\frac{s}{2}}} \\ &\leq \|f\|_{H^s(\mathbb{R}^n)} \left(\int_{\mathbb{R}^n} \frac{d\xi}{(1+|\xi|^2)^s}\right)^{\frac{1}{2}} \\ &\lesssim_s \|f\|_{H^s(\mathbb{R}^n)}\end{aligned}$$

  if $s > \dfrac{n}{2}$, we conclude that

$$\begin{aligned}\|f\|_{L^\infty(\mathbb{R}^n)} &= \|\mathscr{F}^{-1}(\mathscr{F}f)\|_{L^\infty(\mathbb{R}^n)} \\ &\leq \|\widehat{f}\|_{L^1(\mathbb{R}^n)} \\ &\lesssim_s \|f\|_{H^s(\mathbb{R}^n)}.\end{aligned}$$

- Case $k \geq 1$: Using the same argument we have that if $f \in H^s(\mathbb{R}^n)$ with $s > \dfrac{n}{2} + k$, then for $\alpha \in \mathbb{Z}^n_+, |\alpha| \leq k$, it follows that $\partial^\alpha_x \widehat{f} \in L^1(\mathbb{R}^n)$ and

$$\begin{aligned}\|\partial^\alpha_x f\|_{L^\infty(\mathbb{R}^n)} &\leq \|\mathscr{F}(\partial^\alpha_x f)\|_{L^1(\mathbb{R}^n)} \\ &\leq \|(i\xi)^\alpha \widehat{f}\|_{L^1(\mathbb{R}^n)} \\ &\lesssim_s \|f\|_{H^s(\mathbb{R}^n)}.\end{aligned}$$

□

Corollary 5.1. *If* $s = \frac{n}{2} + k + \rho$, *with* $\rho \in (0,1)$, *then* $H^s(\mathbb{R}^n)$ *is continuously embedded in* $C^{k+\rho}(\mathbb{R}^n)$.

Proposition 5.2 (Sobolev Lemma). *For $s > \frac{1}{2}$, we have*

$$\|u\|_{L^\infty(\mathbb{R}^n)} \lesssim_s \|u\|_{H^s(\mathbb{R}^n)}.$$

Proof. Using the inverse Fourier transform formula and Cauchy–Schwartz inequality, we can find Sobolev Lemma. □

Theorem 5.16. *If $s \in (0, \frac{n}{2})$, then $H^s(\mathbb{R}^n)$ is continuously embedded in $L^p(\mathbb{R}^n)$ with $p = \frac{2n}{n-2s}$, i.e., $s = n\left(\frac{1}{2} - \frac{1}{p}\right)$. Moreover, for $f \in H^s(\mathbb{R}^n), s \in (0, \frac{n}{2})$,*

$$\|f\|_{L^p(\mathbb{R}^n)} \lesssim_{n,s} \|D^s f\|_{L^2(\mathbb{R}^n)} \leq c\|f\|_{H^s(\mathbb{R}^n)},$$

where

$$D^s f = \mathscr{F}^{-1}(|\xi|^s \mathscr{F} f).$$

Proof. We define

$$D^s f = g \quad \text{or} \quad f = D^{-s} g = C_{n,s}\mathscr{F}^{-1}\left(\frac{1}{|\xi|^s}\mathscr{F} g\right) = \frac{C_{n,s}}{|x|^{n-s}} * g.$$

Thus by the Hardy–Littlewood–Sobolev estimate it follows that

$$\begin{aligned}
\|f\|_{L^p(\mathbb{R}^n)} &= \|D^{-s} g\|_{L^p(\mathbb{R}^n)} \\
&= \left\| \frac{c_{n,s}}{|x|^{n-s}} * g \right\|_{L^p(\mathbb{R}^n)} \\
&\lesssim_{n,s} \|g\|_{L^2(\mathbb{R}^n)} \\
&= c\|D^s f\|_{L^2(\mathbb{R}^n)}.
\end{aligned}$$

□

Theorem 5.17. *If $s > \frac{n}{2}$ and $f, g \in H^s(\mathbb{R}^n)$, then $fg \in H^s(\mathbb{R}^n)$ with*

$$\|fg\|_{H^s(\mathbb{R}^n)} \lesssim_s \|f\|_{H^s(\mathbb{R}^n)} \|g\|_{H^s(\mathbb{R}^n)}$$

Proof. From the triangle inequality we have that for every $\xi, \xi_1 \in \mathbb{R}^n$

$$(1 + |\xi|^2)^{\frac{s}{2}} \leq 2^s[(1 + |\xi - \xi_1|^2)^{\frac{s}{2}} + (1 + |\xi_1|^2)^{\frac{s}{2}}]$$

Using this we deduce that

$$\begin{aligned}|(1+|\xi|^2)^{\frac{s}{2}}\widehat{(fg)}(\xi)| &= (1+|\xi|^2)^{\frac{s}{2}}\left|\int_{\mathbb{R}^n}\widehat{f}(\xi-\xi_1)\widehat{g}(\xi_1)d\xi_1\right| \\ &\le 2^s\int_{\mathbb{R}^n}[(1+|\xi-\xi_1|^2)^{\frac{s}{2}}|\widehat{f}(\xi-\xi_1)\widehat{g}(\xi_1)| \\ &\quad + (1+|\xi_1|^2)^{\frac{s}{2}}|\widehat{f}(\xi-\xi_1)\widehat{g}(\xi_1)|]d\xi_1.\end{aligned}$$

Thus, taking the L^2-norm and using Hardy inequality it follows that

$$\|fg\|_{H^s(\mathbb{R}^n)} \lesssim_s (\|f\|_{H^s(\mathbb{R}^n)}\|\widehat{g}\|_{L^1(\mathbb{R}^n)} + \|\widehat{f}\|_{L^1(\mathbb{R}^n)}\|g\|_{H^s(\mathbb{R}^n)})$$

Finally, if $r > \frac{n}{2}$, then

$$\|\widehat{f}\|_{L^1(\mathbb{R}^n)} \lesssim_s \|f\|_{H^r(\mathbb{R}^n)}.$$

and

$$\|\widehat{g}\|_{L^1(\mathbb{R}^n)} \lesssim_s \|g\|_{H^r(\mathbb{R}^n)}.$$

Choosing $r = s$, we obtain

$$\|fg\|_{H^s(\mathbb{R}^n)} \lesssim_s \|f\|_{H^s(\mathbb{R}^n)}\|g\|_{H^s(\mathbb{R}^n)}.$$

□

Theorem 5.18. *For any s, the space $H^s(\mathbb{R}^n)$ is the completion of $\mathscr{S}(\mathbb{R}^n)$ with respect to $\|\cdot\|_{H^s(\mathbb{R}^n)}$.*

Proof. Let $u \in H^s(\mathbb{R}^n)$. Then the function

$$(1+|\xi|^2)^{\frac{s}{2}}\widehat{u}(\xi) \in L^2(\mathbb{R}^n)$$

and can be approximated there by compactly supported $C^\infty(\mathbb{R}^n)$ functions $v_k(\xi)$. Let us write v_k in the form

$$v_k(\xi) = (1+|\xi|^2)^{\frac{s}{2}}w_k(\xi),$$

where

$$w_k(\xi) = \frac{v_k(\xi)}{(1+|\xi|^2)^{\frac{s}{2}}}.$$

The functions $w_k(\xi)$ are compactly supported and they converge to $\widehat{u}(\xi)$ in the space $H^s(\mathbb{R}^n)$. Their Fourier preimages belong to $\mathscr{S}(\mathbb{R}^n)$ and converge to u in $H^s(\mathbb{R}^n)$. □

Corollary 5.2. *For $\sigma > s$, the space $H^\sigma(\mathbb{R}^n)$ is dense in $H^s(\mathbb{R}^n)$.*

5.8 The Space H^{-s}

Definition 5.10. We denote by $H^{-s}(\mathbb{R}^n)$, for any $0 < s < \infty$, the dual space to $H_0^s(\mathbb{R}^n)$. In other words, $f \in H^{-s}(\mathbb{R}^n)$ is a bounded linear functional on $H_0^s(\mathbb{R}^n)$.

We also set

$$H^{-\infty}(\mathbb{R}^n) = \bigcup_{s\in R} H^s(\mathbb{R}^n).$$

Theorem 5.19 (Characterization of H^{-s}). *For any given $0 < s < \infty$, any element $u \in \mathscr{S}'(\mathbb{R}^n) \cap H^{-s}(\mathbb{R}^n)$ can be written as*

$$u = \sum_{|\alpha|\leq s} D^\alpha h_\alpha, \quad \text{with } h_\alpha \in L^2(\mathbb{R}^n).$$

Proof. We have that

$$f = (1+|\xi|^2)^{-\frac{s}{2}} \mathscr{F}(u) \in L^2(\mathbb{R}^n).$$

Then

$$\begin{aligned}
\mathscr{F}(u) &= (1+|\xi|^2)^{\frac{s}{2}} f \\
&= \left(1 + \sum_{j=1}^n |\xi_j|^s\right) \frac{(1+|\xi|^2)^{\frac{s}{2}} f}{1+\sum_{j=1}^n |\xi_j|^s} \\
&= \left(1 + \sum_{j=1}^n |\xi_j|^s\right) g \\
&= g + \sum_{j=1}^n \xi_j^s \left(\frac{|\xi_j|^s}{\xi_j^s} g\right),
\end{aligned}$$

where

$$g = \frac{(1+|\xi|^2)^{\frac{s}{2}} f}{1+\sum_{j=1}^n |\xi_j|^s} \in L^2(\mathbb{R}^n).$$

We set

$$g_j = \frac{|\xi_j|^s}{\xi_j^s} g \in L^2(\mathbb{R}^n),$$

$$h = \mathscr{F}^{-1}(g),$$

$$h_j = \mathscr{F}^{-1}(g_j), \quad j = 1, 2, \ldots, n.$$

Then $h, h_j \in L^2(\mathbb{R}^n)$, $j = 1, 2, \ldots, n$, so

$$\begin{aligned}\mathscr{F}(u) &= \mathscr{F}(h) + \sum_{j=1}^n \xi_j^s \mathscr{F}(h_j)\\ &= \mathscr{F}(h) + \sum_{j=1}^n \mathscr{F}(D_j^s h_j),\end{aligned}$$

and therefore

$$u = h + \sum_{j=1}^n D_j^s h_j.$$

This completes the proof. □

Exercise 5.19. Prove that $H^{-\infty}(\mathbb{R}^n) \subset \mathscr{S}'(\mathbb{R}^n)$.

5.9 The Gevrey Classes

The Gevrey classes play an important role in various branches of partial and ordinary differential equations. In fact, the name is given in honor of Maurice Gevrey [3]. Since the scale of spaces G^σ starts from the analytic functions G^1 with $\sigma = 1$ and ends in the C^∞ with $\sigma = \infty$.

The basic example, and first source of investigation [3], is the heat operator in $\mathbb{R}^n, n \geqslant 2$

$$L = \frac{\partial}{\partial_{x_n}} - \sum_{j=1}^{n-1} \frac{\partial^2}{\partial^2_{x_j}},$$

whose fundamental solution is given by

$$E(x_1, \dots, x_n) = \begin{cases} (4\pi x_n)^{\frac{1-n}{2}} e^{-\frac{x_1^2+\cdots+x_{n-1}^2}{4x_n}} & \text{for } x_n > 0, \\ 0 & \text{for } x_n \leqslant 0. \end{cases}$$

The function E is not analytic for $x_n = 0$. However, E is in $C^\infty(\mathbb{R}^n \backslash 0)$. This reflects on the solutions of the homogeneous equation $Lu = 0$ which are not analytic in general, though always C^∞. A precise estimate of the regularity of E can be given by observing that for any fixed compact subset $K \subset \mathbb{R}^n$, $0 \notin K$, we have for all $\alpha = (\alpha_1, \dots, \alpha_n)$

$$|\partial^\alpha E| \leqslant C^{|\alpha|+1}(\alpha!)^2, \qquad x \in K, \tag{5.18}$$

where C depends only on K.

We can generalize the estimate (5.18) in the following way. Let Ω be any open set of $\mathbb{R}^n$ and let $s \geq 1$ be a fixed real number.

Definition 5.11. The function f is said to be a Gevrey function of order s, we will write $f \in G^s(\Omega)$, if the following conditions hold.

1. $f \in \mathscr{C}^\infty(\Omega)$.
2. for any compact set K of Ω there is a positive constant C such that

   $$|\partial^\alpha f(x)| \leq C^{|\alpha|+1}(\alpha!)^s$$

   for any $\alpha \in \mathbb{N}_0^n$ and for any $x \in K$. In particular, $G^1(\Omega) = A(\Omega)$ is the space of all analytic functions in Ω.

Remark 5.1. It will be sometimes useful to refer to the equivalent estimate

$$\partial^\alpha f(x)| \leq RC^{|\alpha|}(\alpha!)^s$$

for any $x \in K$ and for any $\alpha \in \mathbb{N}_0^n$, where R and C are two positive constants independent of $\alpha \in \mathbb{N}_0^n$ and $x \in K$.

Before to give some equivalent estimates of the estimate (5.18), we will give some auxiliary inequalities given in the next two lemmas.

Lemma 5.1. *For any $\alpha, \beta \in \mathbb{N}_0^n$, $\beta \le \alpha$, one has*

$$|\alpha|! \le n^{|\alpha|}\alpha!$$

and

$$\binom{\alpha}{\beta} \le 2^{|\alpha|}.$$

Proof. We apply the generalized Newton inequality

$$(t_1 + \cdots + t_n)^N = \sum_{|\alpha|=N} \frac{N!}{\alpha_1! \ldots \alpha_n!} t_1^{\alpha_1} \ldots t_n^{\alpha_n},$$

where $N \ge 1$ is a given integer, for

$$t_1 = \cdots = t_n = 1$$

and we get

$$n^N = \sum_{|\alpha|=N} \frac{N!}{\alpha!}.$$

Hence,

$$n^{|\alpha|} \ge \frac{|\alpha|!}{\alpha!}$$

or

$$|\alpha|! \le \alpha! n^{|\alpha|}.$$

For $n = 2$, we obtain

$$2^N \le \sum_{k+j=N} \frac{N!}{k!j!} \tag{5.19}$$

and

$$2^{k+j} \le \frac{(k+j)!}{k!j!},$$

or

$$(k+j)! \le 2^{k+j} k!j!$$

for any $k, j \in \mathbb{N}$. By (5.19), we find

$$\sum_{\beta\leq\alpha}\binom{\alpha}{\beta} = 2^{|\alpha|}$$

and in particular,

$$\binom{\alpha}{\beta} \leq 2^{|\alpha|}.$$

This completes the proof. $\square$

Now, we will recall the definition for the Euler Γ function

$$\Gamma(t) = \int_0^\infty e^{-\lambda}\lambda^{t-1}d\lambda, \quad t > 0.$$

Lemma 5.2. *We have*

$$t^t \leq e^t\Gamma(t+1), \quad t > 0,$$

and

$$\Gamma(t+1) \leq t^t, \quad \textit{for large} \quad t > 0,$$

and

$$N^N \leq e^N N!, \quad N \in \mathbb{N}.$$

Proof. By the Stirling formula, we have

$$\Gamma(t+1) = t^t e^{-t}\sqrt{2\pi t}e^{\frac{v(t)}{12t}}, \quad t > 0,$$

where

$$0 < v(t) < 1, \quad t > 0.$$

Hence,

$$\begin{aligned} e^t\Gamma(t+1) &= t^t\sqrt{2\pi t}e^{\frac{v(t)}{12t}} \\ &\geq t^t, \quad t \geq 1. \end{aligned}$$

In particular, for $t = N \in \mathbb{N}$, we get

$$\begin{aligned} N^N &\leq e^N\Gamma(N+1) \\ &= N!e^N. \end{aligned}$$

When $t > 0$ is large enough, we have

$$e^{-t}\sqrt{2\pi t}e^{\frac{v(t)}{12t}} \le 1$$

and then

$$\Gamma(t+1) \le t^t.$$

This completes the proof. □

Theorem 5.20. *The estimate* (5.18) *is valid if and only if the following estimate is valid*

$$|\partial^\alpha f(x)| \le C^{|\alpha|+1}(|\alpha|!)^s \tag{5.20}$$

for any $x \in K$ *and for any* $\alpha \in \mathbb{N}_0^n$.

Proof.

1. Assume that (5.18) is valid for any $\alpha \in \mathbb{N}_0^n$ and for any $x \in K$. Take $\alpha \in \mathbb{N}_0^n$ and $x \in K$ arbitrarily. Since

$$\alpha! \le |\alpha|!,$$

we obtain

$$\begin{aligned}|\partial^\alpha f(x)| &\le C^{|\alpha|+1}(\alpha!)^s\\ &\le C^{|\alpha|+1}(|\alpha|!)^s,\end{aligned}$$

i.e., (5.20) is valid.

2. Suppose that (5.20) is valid for any $\alpha \in \mathbb{N}_)^n$ and for any $x \in K$. Fix $\alpha \in \mathbb{N}_0^n$ and $x \in K$. Then, applying the first inequality of Lemma 5.1, we get

$$\begin{aligned}|\partial^\alpha f(x)| &\le C^{|\alpha|+1}(|\alpha|!)^s\\ &\le C^{|\alpha|+1}\left(n^{|\alpha|}\alpha!\right)^s\\ &= n^{s|\alpha|}C^{|\alpha|+1}(\alpha!)^s\\ &= C_1^{|\alpha|+1}(\alpha!)^s,\end{aligned}$$

where

$$C_1^{|\alpha|+1} = n^{s|\alpha|}C^{|\alpha|+1}.$$

This completes the proof. □

Theorem 5.21. *The estimate* (5.20) *is valid if and only if the following estimate*

$$|\partial^\alpha f(x)| \le C^{|\alpha|+1}|\alpha|^{s|\alpha|} \tag{5.21}$$

for any $x \in K$ *and for any* $\alpha \in \mathbb{N}_0^n$.

Proof.

1. Let the estimate (5.20) is valid for any $\alpha \in \mathbb{N}_0^n$ and for any $x \in K$. Fix $\alpha \in \mathbb{N}_0^n$ and $x \in K$. Since

$$|\alpha|! \le |\alpha|^{|\alpha|},$$

we find

$$\begin{aligned} |\partial^\alpha f(x)| &\le C^{|\alpha|+1}(|\alpha|!)^s \\ &= C^{|\alpha|+1}(|\alpha|^{|\alpha|})^s \\ &= C^{|\alpha|+1}|\alpha|^{s|\alpha|}, \end{aligned}$$

i.e., we get the estimate (5.21).

2. Suppose that the estimate (5.21) is valid for any $x \in K$ and for any $\alpha \in \mathbb{N}_0^n$. Fix $x \in K$ and $\alpha \in \mathbb{N}_0^n$. By the third inequality of Lemma 5.2, we find

$$|\alpha|^{|\alpha|} \le e^{|\alpha|}|\alpha|!$$

Hence,

$$\begin{aligned} |\partial^\alpha f(x)| &\le C^{|\alpha|+1}|\alpha|^{s|\alpha|} \\ &= C^{|\alpha|+1}(|\alpha|^{|\alpha|})^s \\ &\le C^{|\alpha|+1}(e^{|\alpha|}|\alpha|!)^s \\ &= C^{|\alpha|+1}e^{s|\alpha|}|\alpha!|^s \\ &= C_1^{|\alpha|+1}|\alpha!|^s, \end{aligned}$$

where

$$C_1^{|\alpha|+1} = C^{|\alpha|+1}e^{s|\alpha|},$$

which completes the proof. □

Theorem 5.22. *The estimate* (5.21) *is valid if and only if the following estimate*

$$|\partial^\alpha f(x)| \le C^{|\alpha|+1}\Gamma(s|\alpha|+1) \tag{5.22}$$

for any $x \in K$ *and for any* $\alpha \in \mathbb{N}_0^n$.

Proof.

1. Assume that the estimate (5.21) is valid for any $\alpha \in \mathbb{N}_0^n$ and for any $x \in K$. Fix $\alpha \in \mathbb{N}_0^n$ and $x \in K$. Then, applying the first inequality of Lemma 5.2, we find

$$\begin{aligned}|\partial^\alpha f(x)| &\le C^{|\alpha|+1}|\alpha|^{s|\alpha|}\\ &\le C^{|\alpha|+1}(s|\alpha|)^{s|\alpha|}\\ &\le C^{|\alpha|+1}e^{s|\alpha|}\Gamma(s|\alpha|+1)\\ &= C_1^{|\alpha|+1}\Gamma(s|\alpha|+1),\end{aligned}$$

where

$$C_1^{|\alpha|+1} = C^{|\alpha|+1}e^{s|\alpha|}.$$

2. Suppose that (5.22) is valid for any $x \in K$ and for any $\alpha \in \mathbb{N}_0^n$. Fix $x \in K$ and $\alpha \in \mathbb{N}_0^n$. Then, applying the second inequality of Lemma 5.2, we arrive at

$$\begin{aligned}|\partial^\alpha f(x)| &\le C^{|\alpha|+1}\Gamma(s|\alpha|+1)\\ &\le C^{|\alpha|+1}(s|\alpha|)^{s|\alpha|}\\ &= C_2^{|\alpha|+1}|\alpha|^{s|\alpha|},\end{aligned}$$

where

$$C_2^{|\alpha|+1} = C^{|\alpha|+1}s^{s|\alpha|}.$$

This completes the proof. □

Theorem 5.23. *In the estimate* (5.20) *we may replace* $|\alpha|$ *with* $|\alpha| + S$ *for any positive integer* S.

Proof.

1. Suppose that (5.20) holds for any $x \in K$ and for any $\alpha \in \mathbb{N}_0^n$. Then, we find

$$
\begin{aligned}
|\partial^\alpha f(x)| &\leq C^{|\alpha|+1}(|\alpha|!)^s \\
&\leq C^{|\alpha|+1}((|\alpha|+S)!)^s.
\end{aligned}
$$

2. Suppose that

$$
|\partial^\alpha f(x)| \leq C^{|\alpha|+1}((|\alpha|+S)!)^s \tag{5.23}
$$

for any $x \in K$ and for any $\alpha \in \mathbb{N}_0^n$ and for any positive integer S. By the proof of Lemma 5.1, we have

$$
(k+j)! \leq 2^{k+j} k! j!
$$

for any $k, j \in \mathbb{N}$. Hence,

$$
(|\alpha|+S)! \leq 2^{|\alpha|+S}|\alpha|! S!
$$

Then, for any $x \in K$ and for any $\alpha \in \mathbb{N}_0^n$, using (5.23), we find

$$
\begin{aligned}
|\partial^\alpha f(x)| &\leq C^{|\alpha|+1}(2^{|\alpha|+S}(S!|\alpha|!))^s \\
&= C^{|\alpha|+1} 2^{s(|\alpha|+S)}(S!)^s(|\alpha|!)^s \\
&= C_1^{|\alpha|+1}(|\alpha|!)^s,
\end{aligned}
$$

where

$$
C_1^{|\alpha|+1} = C^{|\alpha|+1} 2^{s(|\alpha|+S)}(S!)^s.
$$

This completes the proof. □

Exercise 5.20. Prove that in the estimates (5.21) and (5.22) one can replace $|\alpha|$ with $|\alpha| + S$ for any positive integer S.

Exercise 5.21. Prove that for any $p \geq 1$, one has

$$
\|\partial^\alpha f\|_{L^p(K)} \leq C^{|\alpha|+1}(\alpha!)^s.
$$

Theorem 5.24. *$G^s(\Omega)$ is a vector space and a ring with respect to the arithmetic product of functions. Moreover, it is closed under differentiation.*

Proof. Suppose that $f, g \in G^s(\Omega)$ and K is a compact subset of Ω, and $\alpha \in \mathbb{N}_0$. It is enough to prove that $f+g$ and fg belong g to $G^s(\Omega)$. Since $f, g \in G^s(\Omega)$, there are positive constants C_1 and C_2 such that

$$\begin{aligned}|\partial^\alpha f(x)| &\leq C_1^{|\alpha|+1}|\alpha|^{s|\alpha|},\\ |\partial^\alpha g(x)| &\leq C_2^{|\alpha|+1}|\alpha|^{s|\alpha|}, \quad x \in K.\end{aligned}$$

Hence,

$$\begin{aligned}|\partial^\alpha (f+g)(x)| &= |\partial^\alpha f(x) + \partial^\alpha g(x)|\\ &\leq |\partial^\alpha f(x)| + |\partial^\alpha g(x)|\\ &\leq C_1^{|\alpha|+1}|\alpha|^{s|\alpha|} + C_2^{|\alpha|+1}|\alpha|^{s|\alpha|}\\ &= (C_1^{|\alpha|+1} + C_2^{|\alpha|+1})|\alpha|^{s|\alpha|}\\ &= \widetilde{C}^{|\alpha|+1}|\alpha|^{s|\alpha|}, \quad x \in K,\end{aligned}$$

where

$$\widetilde{C}^{|\alpha|+1} = C_1^{|\alpha|+1} + C_2^{|\alpha|+1}.$$

Thus, $f+g \in G^s(\Omega)$. Now, we will prove that $fg \in G^s(\Omega)$. For this aim, we will use the Leibniz formula. We have

$$\begin{aligned}|\partial^\alpha (fg)(x)| &= \left|\sum_{\beta\leq\alpha}\binom{\alpha}{\beta}\partial^{\alpha-\beta}f(x)\partial^\beta g(x)\right|\\ &\leq \sum_{\beta\leq\alpha}\binom{\alpha}{\beta}|\partial^{\alpha-\beta}f(x)\partial^\beta g(x)|\\ &\leq \sum_{\beta\leq\alpha}\binom{\alpha}{\beta}C_1^{|\alpha-\beta|+1}|\alpha-\beta|^{s|\alpha-\beta|}C_2^{|\beta|+1}|\beta|^{s|\beta|}\\ &= \sum_{\beta\leq\alpha}\binom{\alpha}{\beta}C_1^{|\alpha-\beta|+1}C_2^{|\beta|+1}|\alpha-\beta|^{s|\alpha-\beta|}|\beta|^{s|\beta|}, \quad x \in K.\end{aligned}$$

Let

$$C = \max\{C_1, C_2\}.$$

Then

$$\begin{aligned}|\partial^\alpha(fg)(x)| &\le \sum_{\beta\le\alpha}\binom{\alpha}{\beta}C^{|\alpha-\beta|+|\beta|+2}|\alpha|^{s(|\alpha-\beta|+|\beta|)}\\ &= \sum_{\beta\le\alpha}\binom{\alpha}{\beta}C^{|\alpha|+2}|\alpha|^{s|\alpha|}\\ &= 2^{|\alpha|}C^{|\alpha|+2}|\alpha|^{s|\alpha|}\\ &= \overline{C}^{|\alpha|+1}|\alpha|^{s|\alpha|}, \quad x\in K,\end{aligned}$$

where

$$\overline{C}^{|\alpha|+1} = 2^{|\alpha|}C^{|\alpha|+2}.$$

Now, we will prove that $G^s(\Omega)$ is closed under differentiation. Take $\beta\in\mathbb{N}_0^n$ arbitrarily. Applying Theorem 5.20, we get

$$\begin{aligned}|\partial^\alpha(\partial^\beta f)(x)| &= |\partial^{\alpha+\beta}f(x)|\\ &\le C_1^{|\alpha|+|\beta|+1}|\alpha+\beta|^{s|\alpha+\beta|}\\ &= C_1^{|\alpha|+|\beta|+1}(|\alpha|+|\beta|)^{s(|\alpha|+|\beta|)}\\ &\le C_3^{|\alpha|+|\beta|+1}((|\alpha|+|\beta|)!)^s\\ &\le C_3^{|\alpha|+|\beta|+1}(2^{|\alpha|+|\beta|}|\alpha|!|\beta|!)^s\\ &= C_3^{|\alpha|+|\beta|+1}2^{s(|\alpha|+|\beta|)}|\beta!|^s|\alpha!|^s\\ &= C_4^{|\alpha|+1}(|\alpha!|)^s\\ &\le C_5^{|\alpha|+1}(\alpha!)^s, \quad x\in K,\end{aligned}$$

where

$$\begin{aligned}C_5^{|\alpha|+1} &= n^{s|\alpha|}C_4^{|\alpha|+1},\\ C_4^{|\alpha|+1} &= C_3^{|\alpha|+|\beta|+1}2^{s(|\alpha|+|\beta|)}|\beta!|^s,\\ C_3^{|\alpha|+|\beta|+1} &= C_1^{|\alpha|+|\beta|+1}e^{s(|\alpha|+|\beta|)}.\end{aligned}$$

Thus, $\partial^\beta f\in G^s(\Omega)$ and $G^s(\Omega)$ is closed under differentiation. This completes the proof. □

Theorem 5.25. *Let $\Lambda \subset \mathbb{R}^m$ be an open set and $\kappa : \Omega \to \Lambda$ be an analytic map, and $f \in G^s(\Lambda)$. Then $g = f \circ \kappa \in G^s(\Omega)$.*

Proof. Define

$$F(z) = \sum_{|\beta| \le k} (\partial^\beta f)(\kappa(x)) \frac{(\kappa(z) - \kappa(zx))^\beta}{\beta!}, \quad z \in \mathbb{C}.$$

If $|\alpha| = k$, then the derivative $\partial^\alpha g(x)$ is the same as the corresponding derivative of F at $z = x$. Let $x \in K$ be arbitrarily chosen. Then F extends to an analytic function of z in the domain $|z - x| < r$, where $r > 0$ is sufficiently small. In this domain, we have

$$\begin{aligned} |F(z)| &\le \sum_{|\alpha| \le k} C^{|\beta|+1} \frac{r^{|\beta|}(\beta!)^s}{\beta!} \\ &= \sum_{|\beta| \le k} C^{|\beta|+1} r^{|\beta|} (\beta!)^{s-1} \\ &\le C_1^{k+1} (k!)^{s-1}. \end{aligned}$$

Note that C_1 does not depend on x. Now, applying the Cauchy inequality, we find

$$\begin{aligned} |\partial^\alpha g(x)| &= |\partial^\alpha F(z)| \Big|_{z=x} \\ &\le r^{-k} C_1^{k+1} (k!)^{s-1} \\ &= C_1 \left(\frac{C_1}{r} \right)^k (k!)^s, \quad x \in K, \quad |\alpha| = k. \end{aligned}$$

Therefore, $g \in G^s(\Omega)$. This completes the proof. □

Definition 5.12. Let $s > 1$. With G_0^s denote the set of all functions of $G^s(\Omega)$ with compact support in Ω.

Example 5.2. Let $r > 0$, $s > 1$ and

$$d = \frac{1}{1 - s}.$$

Define

$$\psi(t) = \begin{cases} e^{-t^d} & \text{for } t > 0 \\ 0 & \text{for } t \leq 0, \end{cases}$$

and

$$\phi(x) = \prod_{j=1}^{n} \psi(r + x_j)\psi(r - x_j), \quad x \in \mathbb{R}^n,$$

and

$$K = \{x \in \mathbb{R}^n : |x_j| \leq r, \quad j \in \{1, \ldots, n\}\}.$$

We have that K is a compact set. Moreover,

$$|\partial^\alpha \phi(x)| \leq C_1^{|\alpha|+1}(\alpha!)^s, \quad x \in \mathbb{R}^n,$$

and $\text{supp}\phi \subset K$. Therefore, $\phi \in G_0^s(\Omega)$.

Theorem 5.26. *Let $f \in L^1_{loc}(\mathbb{R}^n)$ and $\phi \in G_0^s(\mathbb{R}^n)$. Then $f \star \phi \in G^s(\mathbb{R}^n)$.*

Proof. Let $\text{supp}\phi \subset K$, where K is a compact set of $\mathbb{R}^n$. Take $\alpha \in \mathbb{N}_0^n$ arbitrarily. Since $\phi \in G_0^s(\mathbb{R}^n)$, there is a constant C_1 such that

$$|\partial^\alpha \phi(x)| \leq C_1^{|\alpha|+1}(\alpha!)^s, \quad x \in \mathbb{R}^n.$$

Hence,

$$\begin{aligned} |\partial^\alpha (f \star \phi)(x)| &= \left| \int_{\mathbb{R}^n} f(y) \partial_x^\alpha \phi(x - y) dy \right| \\ &\leq \int_{\mathbb{R}^n} |f(y)||\partial^\alpha \phi(x - y)| dy \\ &= \int_{x - y \in K} |f(y)||\partial_x^\alpha \phi(x - y)| dy \\ &\leq C_1^{|\alpha|+1}(\alpha!)^s \int_{x - y \in K} |f(y)| dy \\ &\leq C_2^{|\alpha|+1}(\alpha!)^s, \quad x \in \mathbb{R}^n. \end{aligned}$$

Thus, $f \star \phi \in G^s(\mathbb{R}^n)$. This completes the proof. □

Now, consider $G_0^s(\Omega)$ with $s > 1$. Suppose that C is a given positive constant. With $G_0^s(\Omega, K, C)$ we denote the subspace of $G_0^s(\Omega)$ consisting of all functions f with support contained in K such that

$$\sup_{x \in K} |\partial^\alpha f(x)| \leq RC^{|\alpha|}(\alpha!)^s$$

for some $R \geq 0$. Note that $G_0^s(\Omega, K, C)$ is a Banach space with a norm

$$\|f\|_{K,C} \sup_\alpha C^{-|\alpha|}(\alpha!)^s \sup_{x \in K} |\partial^\alpha f(x)|.$$

Exercise 5.22. Prove that

$$G_0^s(\Omega) = \bigcup_{K,C} G_0^s(\Omega, K, C),$$

where K runs over the set of all compact subsets of Ω and C runs over the positive real numbers.

Definition 5.13. We say that a sequence $\{f_j\}_{j=1}^\infty \subset G^s(\Omega)$ converges to $f \in G^s(\Omega)$ if for any compact set $K \subset \Omega$ there is a constant $C > 0p$ such that

$$\|f_j - f\|_{K,C} \to 0, \quad \text{as } j \to \infty. \tag{5.24}$$

Definition 5.14. We say that a sequence $\{f_j\}_{j=1}^\infty \subset G_0^s(\Omega)$ converges to $f \in G_0^s(|Omega)$ if there exist a compact subset K of Ω and a positive constant C such that $\text{supp} f_j, \text{supp} f \subset K$, $j \in \mathbb{N}$, and (5.24) holds.

Exercise 5.23. Prove that the norm $\|f\|_{K,C}$ is equivalent to the norm

$$\|f\|_{K,C,1} \sup_\alpha C^{-|\alpha|}(\alpha!)^s \|\partial^\alpha f\|_{L^p(K)}.$$

Now, we will introduce a special class of Gevrey functions introduced by Foias and Temam [5].

Definition 5.15. For $\delta > 0$ and $s \in \mathbb{R}$, define

$$G^{\delta,s}(\mathbb{R}^n) = \Big\{ f \in L^2(\mathbb{R}^n) : \|f\|_{G^{\delta,s}(\mathbb{R}^n)} = \left\| e^{\delta\|\xi\|} \langle\xi\rangle^s \widehat{f}(\xi) \right\|_{L^2(\mathbb{R}^n)} < \infty \Big\},$$

where

$$\|\xi\| = |\xi_1| + \cdots + |\xi_n|.$$

In the particular case, where $\delta = 0$, the space $G^{0,s}(\mathbb{R}^n)$ is reduced to the Sobolev space $H^s(\mathbb{R}^n)$.

Note that for all $0 < \delta' < \delta$ and $s, s' \in \mathbb{R}$ we have

$$G^{\delta,s}(\mathbb{R}^n) \subset G^{\delta',s'}(\mathbb{R}^n), \tag{5.25}$$

i.e.,

$$\|f\|_{G^{\delta',s'}} \lesssim_{s,s',\delta,\delta'} \|f\|_{G^{\delta,s}}.$$

We record the fact that any $f \in G^{\delta,s}$ has a uniform radius of analyticity δ.

Theorem 5.27. *Every $f \in G^{\delta,s}(\mathbb{R}^n)$ has a holomorphic extension to the strip*

$$S_\delta = \{x + iy \in \mathbb{C}^n : x, y \in \mathbb{R}^n \text{ and } |y_j| < \delta \quad \text{for } j = 1, \ldots, n\}.$$

Proof. Let $f \in G^{\delta,s}$. For each $a \in \mathbb{R}^n$ we must find a holomorphic extension F_a to the polydisc $D(a_1, \delta) \times \cdots \times D(a_n, \delta) \subset \mathbb{C}^n$. By uniqueness, two such extensions agree on the intersection of their domains. Moreover, the union of the polydiscs equals S_δ. By invariance under translation, i.e., $f \in G^{\delta,s}$ implies $f(\cdot - a) \in G^{\delta,s}$, it suffices to do the case $a = 0$. Fix $\delta' \in (0, \delta)$. Using multi-index notation and Fourier inversion, we have

$$\sup_{x \in \mathbb{R}^n} |\partial^\alpha f(x)| \lesssim \int_{\mathbb{R}^n} |\xi^\alpha| \, |\hat{f}(\xi)| n\xi.$$

So, by the Taylor theorem we see that f is given by its Taylor series,

$$f(x) = \sum_\alpha \frac{\partial^\alpha f(0)}{\alpha!} x^\alpha \quad \text{for } x \text{ with } |x_j| < \delta', \ j = 1, \dots, n,$$

and that this series converges absolutely. Indeed,

$$\begin{aligned}
&\sum_\alpha \frac{|\partial^\alpha f(0)|}{\alpha!} |x^\alpha| \\
&\quad \lesssim \sum_{\alpha_1} \cdots \sum_{\alpha_n} \left(\int_{\mathbb{R}^n} \frac{(\delta' |\xi_1|)^{\alpha_1}}{\alpha_1!} \cdots \frac{(\delta' |\xi_n|)^{\alpha_n}}{\alpha_n!} |\hat{f}(\xi)| d\xi \right) \\
&\qquad \times \left(\frac{|x_1|}{\delta'} \right)^{\alpha_1} \cdots \left(\frac{|x_n|}{\delta'} \right)^{\alpha_n} \\
&\quad \lesssim \sum_{\alpha_1} \cdots \sum_{\alpha_n} \left(\int_{\mathbb{R}^n} e^{\delta' \|\xi\|} |\hat{f}(\xi)| d\xi \right) \left(\frac{|x_1|}{\delta'} \right)^{\alpha_1} \cdots \left(\frac{|x_n|}{\delta'} \right)^{\alpha_n} \\
&\quad \lesssim \left(\int_{\mathbb{R}^n} e^{2(\delta' - \delta)\|\xi\|} \langle \xi \rangle^{-2s} d\xi \right)^{\frac{1}{2}} \|f\|_{G^{\delta', s}} \\
&\qquad \times \sum_{\alpha_1} \cdots \sum_{\alpha_n} \left(\frac{|x_1|}{\delta'} \right)^{\alpha_1} \cdots \left(\frac{|x_n|}{\delta'} \right)^{\alpha_n}
\end{aligned}$$

is finite provided $|x_j| < \delta'$. Since $\delta' < \delta$ was arbitrary, we conclude that

$$F_0(z) = \sum_\alpha \frac{\partial^\alpha f(0)}{\alpha!} z^\alpha$$

converges absolutely for $z \in \mathbb{C}^n$ with $|z_j| < \delta$ for $j = 1, \dots, n$, and this is the holomorphic extension we seek. This completes the proof. □

Observe that the norm $\|f\|_{G^{\delta,s}}$ is obtained from the standard Sobolev norm

$$\|f\|_{H^s} = \left\| \langle\xi\rangle^s \widehat{f}(\xi) \right\|_{L^2_\xi},$$

by the substitution

$$f \to e^{\delta\|D\|} f.$$

Indeed,

$$\|f\|_{G^{\delta,s}} = \left\| e^{\delta\|D\|} f \right\|_{H^s}.$$

Definition 5.16. For $\sigma \geq 1$ by

$$G^{\sigma,\delta,s}(\mathbb{R}^n) = \left\{ f \in L^2(\mathbb{R}^n) : \|f\|_{G^{\sigma,\delta,s}(\mathbb{R}^n)} \right.$$
$$\left. = \left\| e^{\delta\||\xi\||^{\frac{1}{\sigma}}} \langle\xi\rangle^s \widehat{f}(\xi) \right\|_{L^2(\mathbb{R}^n)} < \infty \right\},$$

where

$$\||\xi\||^{\frac{1}{\sigma}} = |\xi_1|^{\frac{1}{\sigma}} + \cdots + |\xi_n|^{\frac{1}{\sigma}}.$$

$G^{\sigma,\delta,s}(\mathbb{R}^n)$

If $f \in G^{\sigma,\delta,s}(\mathbb{R}^n)$, then f belongs to the Gevrey class $G^\sigma(\mathbb{R}^n)$. In the case when $\sigma = 1$, we denote $G^{1,\delta,s}(\mathbb{R}^n) = G^{\delta,s}(\mathbb{R}^n)$.

Exercise 5.24. For $\sigma \geq 1$, $0 < \delta' < \delta$ and $s.s' \in \mathbb{R}$, prove that

$$G^{\sigma,\delta,s}(\mathbb{R}^n) \subset G^{\sigma,\delta',s'}(\mathbb{R}^n).$$

5.10 Bourgain Spaces on $\mathbb{R}^n$

The Bourgain spaces $X_{s,b}$ also known as Fourier restriction spaces or dispersive Sobolev spaces turn out to be an appropriate space to establish a fixed point argument. The precise definition is as follows.

Definition 5.17 ($X_{s,b}(\mathbb{R}\times\mathbb{R}^n)$ spaces). Let $h : \mathbb{R}^n \longrightarrow \mathbb{R}$ be a continuous function, and let $s, b \in \mathbb{R}$ The space $X^{\tau=h(\xi)}_{s,b}(\mathbb{R}\times\mathbb{R}^n) = X_{s,b}$

is then defined to be the closure of the Schwartz functions $\mathscr{S}(\mathbb{R} \times \mathbb{R}^n) = \mathscr{S}_{t,x}$ under the norm

$$\|u\|_{X_{s,b}} = \left\| \langle \xi \rangle^s \langle \tau - h(\xi) \rangle^b \widehat{u}(\tau, \xi) \right\|_{L^2_\tau L^2_\xi}.$$

Now, we will give some important properties of the Bourgain spaces.

Theorem 5.28. *Let $s_1 \geq s$ and $b_1 \geq b$. Then*

$$X_{s_1,b_1} \subset X_{s,b}.$$

Proof. Let $u \in X_{s_1,b_1}$ be arbitrarily chosen. Then

$$\left\| \langle \xi \rangle^{s_1} \langle \tau - h(\xi) \rangle^{b_1} \widehat{u}(\tau, \xi) \right\|_{L^2_\tau L^2_\xi} < \infty.$$

Since $s_1 \geq s$ and $b_1 \geq b$, we get

$$\langle \xi \rangle^{s_1} \geq \langle \xi \rangle^s, \quad \xi \in \mathbb{R}^n,$$

and

$$\langle \tau - h(\xi) \rangle^{b_1} \geq \langle \tau - h(\xi) \rangle^b, \quad \xi \in \mathbb{R}^n.$$

Hence,

$$\begin{aligned} \left\| \langle \xi \rangle^s \langle \tau - h(\xi) \rangle^b \widehat{u}(\tau, \xi) \right\|_{L^2_\tau L^2_\xi} &\leq \left\| \langle \xi \rangle^{s_1} \langle \tau - h(\xi) \rangle^{b_1} \widehat{u}(\tau, \xi) \right\|_{L^2_\tau L^2_\xi} \\ &< \infty. \end{aligned}$$

Thus, $u \in X_{s,b}$. Because $u \in X_{s_1,b_1}$ was arbitrarily chosen and we get that it is an element of $X_{s,b}$, we get the inclusion

$$X_{s_1,b_1} \subset X_{s,b}.$$

This completes the proof. □

Theorem 5.29. *For $f \in X^{\tau=h(\xi)}_{s,b}$, one has*

$$\|\overline{f}\|_{X^{\tau=-h(-\xi)}_{s,b}} = \|f\|_{X^{\tau=h(\xi)}_{s,b}}.$$

Proof. We have

$$\begin{aligned}
\|\overline{f}\|_{X_{s,b}^{\tau=-h(-\xi)}} &= \left(\int_{\mathbb{R}}\int_{\mathbb{R}^n} \langle\xi\rangle^{2s}\langle\tau+h(-\xi)\rangle^{2b}\left|\overline{\widehat{f}(\xi)}\right|^2 d\xi d\tau\right)^{\frac{1}{2}} \\
&= \left(\int_{\mathbb{R}}\int_{\mathbb{R}^n} \langle\xi\rangle^{2s}\langle\tau+h(-\xi)\rangle^{2b}\left|\widehat{f}(-\xi)\right|^2 d\xi d\tau\right)^{\frac{1}{2}} \\
&\quad \times(\xi=-\xi_1, \tau=-\tau_1) \\
&= \left(\int_{\mathbb{R}}\int_{\mathbb{R}^n} \langle-\xi_1\rangle^{2s}\langle-\tau_1+h(\xi_1)\rangle^{2b}\left|\widehat{f}(\xi_1)\right|^2 d\xi_1 d\tau_1\right)^{\frac{1}{2}} \\
&= \left(\int_{\mathbb{R}}\int_{\mathbb{R}^n} \langle\xi_1\rangle^{2s}\langle\tau_1-h(\xi_1)\rangle^{2b}\left|\widehat{f}(\xi_1)\right|^2 d\xi_1 d\tau_1\right)^{\frac{1}{2}} \\
&= \|f\|_{X_{s,b}^{\tau=h(\xi)}}.
\end{aligned}$$

This completes the proof. □

Theorem 5.30. *We have*

$$\left(X_{s,b}^{\tau=h(\xi)}\right)^* = X_{-s,-b}^{\tau=-h(-\xi)}.$$

Proof. For $u \in X_{s,b}^{\tau=h(\xi)}$ and $v \in X_{-s,-b}^{\tau=-h(-\xi)}$, using the Parseval identity, the Cauchy–Schwartz inequality and Theorem 5.29, we get

$$\begin{aligned}
\int_{\mathbb{R}}\int_{\mathbb{R}^n} u(\tau,\xi)\overline{v(\tau,\xi)}d\xi d\tau &= \int_{\mathbb{R}}\int_{\mathbb{R}^n} \widehat{u}(\tau,\xi)\overline{\widehat{v}(\tau,\xi)}d\xi d\tau \\
&= \int_{\mathbb{R}}\int_{\mathbb{R}^n} \langle\xi\rangle^{s}\langle\tau-h(\xi\rangle^{b}\widehat{u}(\tau,\xi)\langle\xi\rangle^{-s}\langle\tau-h(\xi)\rangle^{-b}\overline{\widehat{v}(\tau,\xi)}d\xi d\tau \\
&\le \left(\int_{\mathbb{R}}\int_{\mathbb{R}^n} \langle\xi\rangle^{2s}\langle\tau-h(\xi)\rangle^{2b}|\widehat{u}(\tau,\xi)|^2 d\xi d\tau\right)^{\frac{1}{2}} \\
&\quad \times\left(\int_{\mathbb{R}}\int_{\mathbb{R}^n} \langle\xi\rangle^{-2s}\langle\tau-h(\xi)\rangle^{-2b}|\overline{\widehat{v}(\tau,\xi)}|^2 d\xi d\tau\right)^{\frac{1}{2}} \\
&= \|u\|_{X_{s,b}^{\tau=h(\xi)}}\|\overline{v}\|_{X_{-s,-b}^{\tau=h(\xi)}} \\
&= \|u\|_{X_{s,b}^{\tau=h(\xi)}}\|v\|_{X_{-s,-b}^{\tau=-h(-\xi)}}.
\end{aligned}$$

This completes the proof. □

Exercise 5.25. Prove that $X_{s,b}$ is invariant under translations in space and time.

The spaces $X_{s,b}$ are well adapted to the solutions

$$u(t) = e^{tL}u_0$$

of the linear dispersive equation

$$\partial_t u = Lu, \quad \text{where } L = ih(\mathbf{D}),$$

for some polynomial $h : \mathbb{R}^n \to \mathbb{R}$, as the following lemma shows.

Lemma 5.3 (Free solutions lie in $X_{s,b}(\mathbb{R} \times \mathbb{R}^n)$). Let $f \in H^s_x(\mathbb{R}^n) = H^s$ for some $s \in \mathbb{R}$, and let $L = ih(\mathbf{D})$ for some polynomial $h : \mathbb{R}^n \longrightarrow \mathbb{R}$. Then for any Schwartz time cutoff $\eta \in \mathscr{S}_t$, we have

$$\left\|\eta(t)e^{tL}f\right\|_{X_{s,b}} \lesssim_{\eta,b} \|f\|_{H^s}$$

Proof. We compute the spacetime Fourier transform of $\eta(t)e^{tL}f$ at (τ, ξ) and observe that it is equal to $\widehat{\eta}(\tau - h(\xi))\widehat{f}(\xi)$. Hence, taking the $X_{s,b}$ norm, we obtain

$$\begin{aligned} &\left\|\eta(t)e^{tL}f\right\|_{X_{s,b}} \\ &\quad = \left\|\langle\xi\rangle^s\langle\tau - h(\xi)\rangle^b\widehat{\eta}(\tau - h(\xi))\widehat{f}(\xi)\right\|_{L^2_\tau L^2_\xi} \\ &\quad = \left(\int_{\mathbb{R}^n}\langle\xi\rangle^{2s}|\hat{f}(\xi)|^2\left(\int_{\mathbb{R}}\langle\tau - h(\xi)\rangle^{2b}|\widehat{\eta}(\tau - h(\xi))|^2 d\tau\right)d\xi\right)^{\frac{1}{2}}. \end{aligned} \tag{5.26}$$

Since $\widehat{\eta}$ is rapidly decreasing, the claim follows. □

The following result will be particularly useful for the theory of nonlinear dispersive equations.

Lemma 5.4 (Embedding for $\boldsymbol{Y}(\mathbb{R} \times \mathbb{R}^n)$). *Let $b > \frac{1}{2}$ and $L = ih(\boldsymbol{D})$ for some polynomial $h : \mathbb{R}^n \longrightarrow \mathbb{R}$. Let also, $s \in \mathbb{R}$ and $\boldsymbol{Y}$ be a Banach space of functions on $\mathbb{R} \times \mathbb{R}^n$ with the property*

$$\|e^{it\tau_0}e^{tL}f\|_{\boldsymbol{Y}} \leqslant \|f\|_{H^s}$$

for all $f \in H^s$ and $\tau_0 \in \mathbb{R}$. Then we have the embedding

$$\|u\|_{\boldsymbol{Y}} \lesssim_b \|u\|_{X_{s,b}}.$$

Proof. By the Fourier inversion formula, we have

$$u(t,x) = \frac{1}{(2\pi)^{d+1}} \int_{\mathbb{R}} \int_{\mathbb{R}^n} \widehat{u}(\tau,\xi) e^{it\tau} e^{ix\cdot\xi} d\xi d\tau$$

Writing $\tau = h(\xi) + \tau_0$ and setting

$$f_{\tau_0}(x) = \frac{1}{(2\pi)^d} \int_{\mathbb{R}^n} \widehat{u}\left(h(\xi) + \tau_0, \xi\right) e^{ix\cdot\xi} d\xi,$$

we have

$$e^{tL} f_{\tau_0}(x) = \frac{1}{(2\pi)^d} \int_{\mathbb{R}^n} \widehat{u}\left(h(\xi) + \tau_0, \xi\right) e^{ith(\xi)} e^{ix\cdot\xi} d\xi.$$

Thus, we get the representation

$$u(t) = \frac{1}{2\pi} \int_{\mathbb{R}} e^{it\tau_0} e^{tL} f_{\tau_0} d\tau_0.$$

Taking $\mathbf{Y}$ norms and using Minkowski's inequality and the hypothesis on $\mathbf{Y}$, we obtain

$$\begin{aligned} \|u\|_{\mathbf{Y}} &= \left\| \frac{1}{2\pi} \int_{\mathbb{R}} e^{it\tau_0} e^{tL} f_{\tau} d\tau_0 \right\|_{\mathbf{Y}} \\ &\lesssim \int_{\mathbb{R}} \|f_{\tau_0}\|_{H^s} d\tau_0. \end{aligned}$$

Hence, applying the Cauchy–Schwarz inequality and the hypothesis $b > \frac{1}{2}$, we arrive at

$$\|u\|_{\mathbf{Y}} \lesssim_b \left(\int_{\mathbb{R}} \langle\tau_0\rangle^{2b} \|f_{\tau_0}\|_{H^s}^2 d\tau_0 \right)^{1/2}.$$

By an application of Plancharel's theorem, we obtain

$$\left(\int_{\mathbb{R}} \langle\tau_0\rangle^{2b} \|f_{\tau_0}\|_{H^s}^2 d\tau_0 \right)^{1/2} = C_b \|u\|_{X_{s,b}}.$$

and this concludes the proof. □

Applying these lemmas to $Y = C_t^0 H_x^s$, we obtain the following corollary

Corollary 5.3 (Embedding for $C_t(\mathbb{R}, H_x^s(\mathbb{R}^n))$). *Let $b > \frac{1}{2}$, $s \in \mathbb{R}$ and $h : \mathbb{R}^n \to \mathbb{R}$ be continuous. Then, for any $u \in X_{s,b}$, we have*

$$\|u\|_{C_t^0 H^s} \lesssim_b \|u\|_{X_{s,b}}.$$

The following result guarantees the stability with respect to time localization.

Lemma 5.5 (Stability with time localization in $\mathbb{R} \times \mathbb{R}^n$). *Let $\eta \in \mathscr{S}_t$ be a Schwartz function in time. Then we have*

$$\|\eta(t)u\|_{X_{s,b}} \lesssim_{\eta,b} \|u\|_{X_{s,b}}$$

for any $s, b \in \mathbb{R}$ any $h \colon \mathbb{R}^n \to \mathbb{R}$, and any field $u \in \mathscr{S}_{t,x}$. Furthermore, if $-\frac{1}{2} < b' \leqslant b < \frac{1}{2}$, then for any $0 < T < 1$ we have

$$\left\| \eta\left(\frac{t}{T}\right) u \right\|_{X_{s,b'}} \lesssim_{\eta,b,b'} T^{b-b'} \|u\|_{X_{s,b}}.$$

Proof. Let us first understand how the $X_{s,b}$ spaces behave with respect to temporal frequency modulation

$$u(t,x) \to e^{it\tau_0} u(t,x).$$

From the crude estimate

$$\langle \tau - \tau_0 - h(\xi) \rangle^b \lesssim_b \langle \tau_0 \rangle^{|b|} \langle \tau - h(\xi) \rangle^b$$

and elementary Fourier analysis, we conclude that

$$\left\| e^{it\tau_0 u} \right\|_{X_{s,b}} \lesssim_b \langle \tau_0 \rangle^{|b|} \|u\|_{X_{s,b}}$$

If we now use Fourier expansion in time to write

$$\eta(t) = \int_{\mathbb{R}} \widehat{\eta}\,(\tau_0)\, e^{it\tau_0} d\tau_0$$

and use Minkowski's inequality, we conclude

$$\|\eta(t)u\|_{X_{s,b}} \lesssim_b \left(\int_{\mathbb{R}} |\hat{\eta}\,(\tau_0)| \, \langle \tau_0 \rangle^{|b|} \, d\tau_0 \right) \|u\|_{X_{s,b}}$$

Since η is a Schwartz functions and it is rapidly decreasing, the first claim follows.

Now we prove the second claim. By conjugating by $\langle \nabla \rangle^s$ we may take $s = 0$. By composition it suffices to treat the cases $0 \leq b' \leq b$ or $b' \leq b \leq 0$. By duality we may then take $0 \leq b' \leq b$. By interpolation with the trivial case $b' = b$ we may take $b' = 0$. Thus, we are now reduced to establishing

$$\left\| \eta \left(\frac{t}{T} \right) u \right\|_{L_t^2 L_x^2} \lesssim_{\eta, b} T^b \|u\|_{X_{0,b}}$$

for $0 < b < \frac{1}{2}$. By partitioning frequency space we can divide into two cases, one where $\tilde{u}$ is supported on the region $\langle \tau - h(\xi) \rangle \geq \frac{1}{T}$, and one where $\langle \tau - h(\xi) \rangle \leq \frac{1}{T}$. In the former case we will have

$$\|u\|_{X_{0,0}} \leq T^b \|u\|_{X_{0,b}}$$

and the claim then follows from the boundedness of η. In the latter case, we use a variant of Corollary 5.3, noting for any time t that

$$\begin{aligned} \|u(t)\|_{L_x^2} &\lesssim \|\widehat{u(t)}(\xi)\|_{L_\xi^2} \\ &\lesssim \left\| \int_{\langle \tau - h(\xi) \rangle \leq 1/T} |\widehat{u}(\tau, \xi)| d\tau \right\|_{L_\xi^2} \\ &\lesssim_b T^{b-\frac{1}{2}} \left\| \left(\int \langle \tau - h(\xi) \rangle^{2b} |\widehat{u}(\tau, \xi)|^2 d\tau \right)^{\frac{1}{2}} \right\|_{L_\xi^2} \\ &= T^{b-\frac{1}{2}} \|u\|_{X_{s,b}} \end{aligned}$$

thanks to the Plancherel theorem, the triangle inequality, the Cauchy–Schwarz inequality, and the localization of $\widehat{u}$. Integrating this against $\eta\left(\frac{t}{T}\right)$, the claim follows. This completes the proof. □

The $X_{s,b}$ spaces react well to Fourier multipliers, in much the same way that ordinary Sobolev spaces H^s do. If D^k is a Fourier multiplier of order k, in the sense that

$$\widehat{D^k f}(\xi) = m(\xi) f(\xi)$$

for all Schwartz functions $f \in \mathscr{S}_x(\mathbb{R}^n) = \mathscr{S}_x$ and some measurable multiplier $m : \mathbb{R}^n \to \mathbb{C}$ obeying the growth condition

$$|m(\xi)| \lesssim \langle \xi \rangle^k,$$

then D can be extended to spacetime functions by acting on each time separately, thus $(Du)(t) = D(u(t))$, or in terms of the spacetime Fourier transform

$$\widehat{D^k u}(\tau, \xi) = m(\xi)\widehat{u}(\tau, \xi)$$

and then one easily verifies that D maps $X_{s,b}$ continuously to $X_{s-k,b}$ for any $s, b \in \mathbb{R}$ and any h :

$$\left\| D^k u \right\|_{X_{s-k,b}} \lesssim \|u\|_{X_{s,b}}.$$

This is analogous to the well-known estimate

$$\left\| D^k u \right\|_{H^{s-k}} \lesssim \|u\|_{H^s}$$

for Sobolev spaces. In the case when k is a non-negative integer, we have the converse

$$\|u\|_{X_{s,b}} \lesssim \|u\|_{X_{s-k,b}} + \left\| \nabla_x^k u \right\|_{X_{s-k,b}} \tag{5.27}$$

which is proven by repeating the above arguments. Similarly, if $h : \mathbb{R}^n \to \mathbb{R}$ is a polynomial and $L = ih(\mathbf{D})$, then we have

$$\left(\widehat{\partial_t - L}\right) u(\tau, \xi) = i(\tau - h(\xi))\widehat{u}(\tau, \xi)$$

and hence,

$$\|(\partial_t - L)\, u\|_{X_{s,b-1}} \lesssim \|u\|_{X_{s,b}}$$

It is natural to ask whether there is a converse inequality, in the spirit of (5.27). This is indeed the case.

Proposition 5.3 ($X^{s,b}(\mathbb{R} \times \mathbb{R}^n)$ Energy Estimate). *Let $h : \mathbb{R}^n \to \mathbb{R}$ be a polynomial, $u_t = Lu + F$. Then for any $s \in \mathbb{R}$ and $b > \frac{1}{2}$, and any compactly supported smooth time cutoff $\eta(t)$, we have*

$$\|\eta(t)u\|_{X_{s,b}} \lesssim_{\eta,b} \|u(0)\|_{H_x^s} + \|F\|_{X_{s,b-1}}.$$

Proof. Let $[-R,R]$ be a time interval containing the support of η. By truncating F smoothly in time, using a compactly supported cut-off that equals 1 on $[-R,R]$, we may assume (using Lemma 5.5) that F is supported on $[-2R,2R]$ and is a Schwartz function in spacetime. Also, by applying $\langle\nabla\rangle^s$ to both u and F if necessary, we may take $s=0$.

Let us first suppose that u vanishes at time $-2R$, thus $u(-2R)=0$. By Duhamel's formula (with $t_0=-2R$) we thus have

$$\begin{aligned}\eta(t)u(t) &= \eta(t)\int_{-\infty}^{t} e^{(t-s)L}F(s)ds\\ &= \eta(t)\int_{\mathbb{R}} \widehat{\eta}(t-s)1_{[0,+\infty)}(t-s)e^{(t-s)L}F(s)ds,\end{aligned}$$

where $\widehat{\eta}$ is a smooth compactly supported function which equals 1 on $[-3R,3R]$. By Lemma 5.5, it would thus suffice to show that

$$\left\|\int_{\mathbb{R}} \widehat{\eta}(t-s)1_{[0,+\infty)}(t-s)e^{(t-s)L}F(s)ds\right\|_{X_{0,b}} \lesssim_{\eta,b} \|F\|_{X_{0,b-1}}.$$

A routine computation shows that the spacetime Fourier transform of

$$\int_{\mathbb{R}} \widehat{\eta}(t-s)1_{[0,+\infty)}(t-s)e^{(t-s)L}F(s)ds \quad \text{at } (\tau,\xi)$$

is equal to

$$\left(\int_{\mathbb{R}} \widehat{\eta}(t)1_{[0,+\infty)}(t)e^{-it(\tau-h(\xi))}dt\right)\widehat{F}(\tau,\xi).$$

The expression inside the parentheses can be shown (via integration by parts) to Active be at most $O_{\widehat{\eta}}\left(\langle\tau-\xi\rangle^{-1}\right)$. The claim then follows. Now, we handle the general case. We split

$$u(t) = \left(u(t)-e^{(t+2R)L}u(-2R)\right) + e^{tL}e^{2RL}u(-2R).$$

For the first term, the preceding argument applies, and we have

$$\left\|\eta(t)\left(u(t)-e^{(t+2R)L}u(-2R)\right)\right\|_{X_{0,b}} \lesssim_{\eta,b} \|F\|_{X_{0,b}}.$$

Thus, it will suffice to control the remaining term. Applying Lemma 5.3, it thus suffices to show that

$$\left\|e^{2RL}u(-2R)\right\|_{L^2_x} \lesssim_{\widehat{\eta},b} \|u(0)\|_{L^2_x(\mathbb{R}^n)} + \|F\|_{X_{0,b-1}}.$$

From Duhamel's formula and the support of F, we have

$$e^{2RL}u(-2R) = u(0) + \int_{\mathbb{R}} \tilde{\eta}(s)1_{(-\infty,0]}(s)e^{-sL}F(s)ds,$$

where $\widehat{\eta}$ is as before. Thus, by the triangle inequality, it suffices to show that

$$\left\|\int_{\mathbb{R}} \widehat{\eta}(s)1_{(-\infty,0]}(s)e^{-sL}F(s)ds\right\|_{L^2_x} \lesssim_{\widehat{\eta},b} \|F\|_{X_{0,b-1}}.$$

Applying Parseval's identity, the left-hand side can be written as

$$\left\|\int_{\mathbb{R}}\left(\int_{\mathbb{R}} \widehat{\eta}(s)1_{(-\infty,0]}(s)e^{is(\tau-h(\xi))}ds\right)\widehat{F}(\tau,\xi)d\tau\right\|_{L^2_\xi}.$$

An integration by parts yields the bound

$$\int_{\mathbb{R}} \widehat{\eta}(s)1_{(-\infty,0]}(s)e^{is(\tau-h(\xi))}ds \mid\lesssim_\eta \langle\tau-h(\xi)\rangle^{-1}$$

and hence, by the Cauchy–Schwarz inequality and the hypothesis $b > \frac{1}{2}$, we have

$$\begin{aligned}
&\left\|\int_{\mathbb{R}} \widehat{\eta}(s)\mathbf{1}_{(-\infty,0]}(s)e^{-sL}F(s)ds\right\|_{L^2_\xi} \\
&\lesssim_\eta \left\|\int_{\mathbb{R}} \frac{1}{\langle\tau-h(\xi)\rangle}\widehat{F}(\tau,\xi)ds\right\|_{L^2_\xi} \\
&= \left\|\int_{\mathbb{R}} \frac{\langle\tau-h(\xi)\rangle^b}{\langle\tau-h(\xi)\rangle^{1+b}}\widehat{F}(\tau,\xi)ds\right\|_{L^2_\xi} \\
&\lesssim_b \left(\int_{\mathbb{R}} \frac{1}{\langle\tau-h(\xi)\rangle^{2b}}\right)^{\frac{1}{2}}\left(\langle\tau-h(\xi)\rangle^{2(b-1)}\left|\widehat{F}(\tau,\xi)\right|^2\right)^{\frac{1}{2}}
\end{aligned}$$

and the claim follows. This completes the proof. □

Now, we will introduce some classes Gevrey–Bourgain spaces. We will start with this one defined in Grujić and Kalisch [6].

Definition 5.18 ($X_{\delta,s,b}(\mathbb{R}\times\mathbb{R}^n)$ Spaces). Let $h:\mathbb{R}^n\longrightarrow\mathbb{R}$ be a continuous function, and let $s,b\in\mathbb{R}$,$\delta>0$. The space $X_{\delta,s,b}$ is then defined to be the closure of the Schwartz functions $\mathscr{S}_{t,x}$ under the norm

$$\|u\|_{X_{\delta,s,b}}=\left\|e^{\delta\|\xi\|}\langle\xi\rangle^s\langle\tau-h(\xi)\rangle^b\widehat{u}(\tau,\xi)\right\|_{L^2_\tau L^2_\xi}.$$

For $T>0$, $X^T_{\delta,s,b}$ denotes the restricted analytic Bourgain space defined by

$$\|u\|_{X^T_{\delta,s,b}}=\inf\left\{\|v\|_{X_{\delta,s,b}}:v=u\text{ on }(0,T)\times\mathbb{R}^n\right\}.$$

Remark 5.2. For $\delta=0$, the analytic Bourgain space $X_{\delta,s,b}$ coincides with the space $X_{s,b}$.

We consider also a space that is a hybrid between the analytic Gevrey space and a space of the analytic Gevrey Bourgain.

Definition 5.19 ($X_{\sigma,\delta,s,b}(\mathbb{R}\times\mathbb{R}^n)$ Spaces). Let $h:\mathbb{R}^n\longrightarrow\mathbb{R}$ be a continuous function, and let $s,b\in\mathbb{R}$,$\delta>0,\sigma\geqslant 1$. The space $X_{\sigma,\delta,s,b}$ is then defined to be the closure of the Schwartz functions $\mathscr{S}_{t,x}$ under the norm

$$\|u\|_{X_{\sigma,\delta,s,b}}=\left\|e^{\delta\||\xi\||^{\frac{1}{\sigma}}}\langle\xi\rangle^s\langle\tau-h(\xi)\rangle^b\widehat{u}(\tau,\xi)\right\|_{L^2_\tau L^2_\xi}.$$

For $T>0$, $X^T_{\sigma,\delta,s,b}$ denotes the restricted analytic Bourgain space defined by

$$\|u\|_{X^T_{\sigma,\delta,s,b}}=\inf\left\{\|v\|_{X_{\sigma,\delta,s,b}}:v=u\text{ on }(0,T)\times\mathbb{R}^n\right\}.$$

Remark 5.3. For $\delta=0$, the analytic Gevrey Bourgain space $X_{\sigma,\delta,s,b}$ coincides with the space $X_{s,b}$. For $\sigma=0$, the analytic Gevrey Bourgain space $X_{\sigma,\delta,s,b}$ coincides with the space $X_{\delta,s,b}$.

The next lemma shows that $X_{\sigma,\delta,s,b}$ is continuously embedded in $C\left([0,T],G^{\sigma,\delta,s}\right)$, provided $b>\frac{1}{2}$.

Lemma 5.6 (Embedding for $C(\mathbb{R},G^{\sigma,\delta,s}(\mathbb{R}^n))$). *Let $b>\frac{1}{2}$, $s\in\mathbb{R}$, $\sigma\geqslant 1$ and $\delta>0$. Then, we have*

$$\|u\|_{C^0_tG^{\sigma,\delta,s}}\lesssim_b\|u\|_{X_{\sigma,\delta,s,b}}$$

Proof. First, we observe that the operator $\Lambda^{\sigma,\delta}$ defined by

$$\widehat{\Lambda^{\sigma,\delta}u}^{x}(\xi,t) = e^{2\delta|||\xi|||^{\frac{1}{\sigma}}}\widehat{u}^{x}(\xi,t), \tag{5.28}$$

satisfies

$$\|u\|_{X_{\sigma,\delta,s,b}} = \|\Lambda^{\sigma,\delta}u\|_{X_{s,b}}$$

and

$$\|u\|_{G^{\sigma,\delta,s}} = \|\Lambda^{\sigma,\delta}u\|_{H^s},$$

We observe that $A^{\sigma,\delta}u$ belongs to $C([0,T],H^s)$. Then, using Corollary 5.3, we have

$$\|\Lambda^{\sigma,\delta}u\|_{C_t^0H^s} \lesssim_b \|\Lambda^{\sigma,\delta}u\|_{X_{s,b}}.$$

Thus, it follows that $u \in C\left([0,T],G^{\sigma,\delta,s}\right)$ and

$$\|u\|_{C_t^0G^{\sigma,\delta,s}} \lesssim_b \|u\|_{X_{\sigma,\delta,s,b}}.$$

This completes the proof. □

Exercise 5.26 (Free Solutions Lie in $X_{\sigma,\delta,s,b}(\mathbb{R}\times\mathbb{R}^n)$). Let $f \in G^{\sigma,\delta,s}$ for some $s \in \mathbb{R}$, $\sigma \geqslant 1$, $\delta > 0$, and let $L = ih(\mathbf{D})$ for some polynomial $h : \mathbb{R}^n \longrightarrow \mathbb{R}$. Prove that for any Schwartz time cutoff $\eta \in \mathscr{S}_t$, we have

$$\left\|\eta(t)e^{tL}f\right\|_{X_{\sigma,\delta,s,b}} \lesssim_{\eta,b} \|f\|_{G^{\sigma,\delta,s}}.$$

5.11 Advanced Practical Problems

Problem 5.1. *Let $k > 0$ be given, and set*

$$U = \{(x,y) \in \mathbb{R}^2 : 0 < x < 1, x^k < y < 2x^k\},$$

$u(x,y) = y^\alpha$, $(x,y) \in U$. *Find conditions on α so that $u \in H^m(U)$, $m = 1,2,\ldots$.*

Problem 5.2. *Let $u(x) = |x|^{-\alpha}$, $x \in U_1$, $x \neq 0$. Find conditions on $\alpha > 0$, n and p so that $u \in W^{1,p}(U_1)$.*

Problem 5.3. *Find conditions for s so that $e^{-|x|} \in H^s(\mathbb{R}^n)$.*

Problem 5.4. *Find conditions for s so that $\delta \in H^s(\mathbb{R}^n)$.*

Problem 5.5. *Let $n > 1$. Prove that $u = \log\log\left(1 + \frac{1}{|x|}\right)$ belongs to $W^{1,p}(U_1)$.*

Problem 5.6. *Prove the following interpolation inequality*

$$\int_A |Du|^2 dx \leq \left(\int_A |u|^2 dx\right)^{\frac{1}{2}} \left(\int_A |D^2 u|^2 dx\right)^{\frac{1}{2}},$$

for every $u \in \mathscr{C}_0^\infty(A)$, $A \subset \mathbb{R}^n$ open and bounded. Using approximation, prove it for $u \in H^1(A) \cap H_0^1(A)$.

Problem 5.7. *Prove the interpolation inequality*

$$\int_A |Du|^p dx \leq \left(\int_A |u|^p dx\right)^{\frac{1}{2}} \left(\int_A |D^2 u|^p dx\right)^{\frac{1}{2}}$$

for $2 \leq p < \infty$ and every $u \in W^{2,p}(A) \cap W_0^{1,p}(A)$, where $A \subset \mathbb{R}^n$ is open and bounded.

Problem 5.8. *Prove that any $u \in W^{1,p}(a,b)$ can be extended to $W^{1,p}(\mathbb{R})$.*

Problem 5.9. *Let A be an open bounded set in $\mathbb{R}^n$ with $\mathscr{C}^1$ boundary, take $u \in W^{m,p}(A)$, $m < \frac{n}{p}$. Prove that $u \in L^q(A)$, where $\frac{1}{q} = \frac{1}{p} - \frac{m}{n}$, and*

$$||u||_{L^q(A)} \leq C||u||_{W^{m,p}(A)}.$$

Hint. Use the fact that $D^\alpha u \in L^p(A)$ for any $u \in W^{m,p}(A)$ and every α such that $|\alpha| \leq m$. Applying the Gagliardo–Nirenberg–Sobolev inequality deduce

$$||D^\beta u||_{L^{p^*}(A)} \leq C||D^\alpha u||_{L^p(A)} \leq C||u||_{W^{m,p}(A)}$$

for some constant $C > 0$, $|\alpha| = m$ and $|\beta| = m - 1$, $\frac{1}{p^*} = \frac{1}{p} - \frac{1}{n}$. Conclude $u \in W^{m-1,p^*}(A)$. Then using again the

Gagliardo–Nirenberg–Sobolev inequality, prove that

$$||D^\gamma u||_{L^{p^{**}}(A)} \leq C||D^\beta u||_{L^{p^*}(A)} \leq C||u||_{W^{m,p}(A)}.$$

Deduce that $u \in W^{m-2,p^{**}}$, where $\frac{1}{p^{**}} = \frac{1}{p^*} - \frac{1}{n} = \frac{1}{p} - \frac{2}{n}$, and so forth. Eventually conclude that $u \in W^{0,q}(A)$ and that the given inequality holds.

Problem 5.10. *Let A be an open bounded set in $\mathbb{R}^n$ with $\mathscr{C}^1$ boundary. Take $u \in W^{m,p}(A)$, $m > \frac{n}{p}$ and prove*

$$u \in \mathscr{C}^{m-\left[\frac{n}{p}\right]-1,\gamma}(\overline{A}),$$

where

$$\gamma = \begin{cases} \left[\frac{n}{p}\right] + 1 - \frac{n}{p}, & \text{if } \frac{n}{p} \text{ is not an integer} \\ \text{any positive number} < 1, & \text{otherwise.} \end{cases}$$

Also show that u satisfies the inequality

$$||u||_{\mathscr{C}^{m-\left[\frac{n}{p}\right]-1,\gamma}(A)} \leq C||u||_{W^{m,p}(A)}$$

for some positive constant C.

Hint. Use the Extension Theorem, Morrey's inequality and the Gagliardo–Nirenberg–Sobolev inequality.

Problem 5.11. *Prove $H^{+\infty}(\mathbb{R}^n) = \bigcap_{s\in\mathbb{R}} H^s(\mathbb{R}^n)$.*

Problem 5.12. *Show $H^{+\infty}(\mathbb{R}^n) \subset \mathscr{C}^\infty(\mathbb{R}^n)$.*

Problem 5.13. *Prove that the norm $\|f\|_{K,C}$ is equivalent to the norm*

$$\|f\|_{K,C,2} = \sum_\alpha C^{-|\alpha|}(\alpha!)^{-s} \sup_{x\in K} |\partial^\alpha f(x)|.$$

Problem 5.14. *Prove that the norm $\|f\|_{K,C}$ is equivalent to the norm*

$$\|f\|_{K,C,3} = \sum_\alpha C^{-|\alpha|}(\alpha!)^{-s} \|\partial^\alpha f\|_{L^p(K)}.$$

Problem 5.15. *Let $s \leq t$. Prove that*

$$G^s(\Omega) \subset G^t(\Omega).$$

Problem 5.16. *Prove that*

$$A(\Omega) = \bigcap_{s>1} G^s(\Omega).$$

Problem 5.17. *Prove that*

$$\bigcup_{s\geq 1} G^s(\Omega) \subset \mathscr{C}^\infty(\Omega).$$

Problem 5.18. *Prove that $G_0^s(\Omega)$ is dense in $\mathscr{C} - 0^\infty(\Omega)$.*

Problem 5.19. *Prove that $G_0^s(\Omega)$ is dense in $L^1_{\text{loc}}(\Omega)$.*

Problem 5.20 (Stability with Time Localization in $\mathbb{R} \times \mathbb{R}^n$). Let $s, b \in \mathbb{R}$, $\sigma \geqslant 1$, $\delta > 0$ and $\eta \in \mathscr{S}_t$ be a Schwartz function in time. Prove that

1.

$$\|\eta(t)u\|_{X_{\sigma,\delta,s,b}} \lesssim_{\eta,b} \|u\|_{X_{\sigma,\delta,s,b}}$$

for any $h : \mathbb{R}^n \to \mathbb{R}$, and any field $u \in \mathscr{S}_{t,x}$.
Furthermore, if $-\frac{1}{2} < b' \leqslant b < \frac{1}{2}$, then for any $0 < T < 1$.

2.

$$\left\| \eta\left(\frac{t}{T}\right) u \right\|_{X_{\sigma,\delta,s,b'}} \lesssim_{\eta,b,b'} T^{b-b'} \|u\|_{X_{\sigma,\delta,s,b}}.$$

Problem 5.21 ($X_{\delta,s,b}(\mathbb{R}\times\mathbb{R}^n)$ Energy Estimate). Let $h : \mathbb{R}^n \to \mathbb{R}$ be a polynomial, $u_t = Lu + F$. Prove that for any $s \in \mathbb{R}$, $b > \frac{1}{2}$ and $\delta > 0$, and any compactly supported smooth time cutoff $\eta(t)$, we have

$$\|\eta(t)u\|_{X_{\delta,s,b}} \lesssim_{\eta,b} \|u_0\|_{G^{\delta,s}} + \|F\|_{X_{\delta,s,b-1}}.$$

Chapter 6

Original Method for the KdV Equation in $H^s(\mathbb{R})$

We write $a \lesssim b$ if there exists a constant $C > 0$ such that $a \leq Cb$ and $a \sim b$ if $a \lesssim b \lesssim a$. If the constant C depends on some quantity q, we denote this by $a \lesssim_q b$.

In this chapter, we study the initial value problem (IVP) for the Korteweg–de Vries equation:

$$\partial_t u + \partial_x^3 u + u\partial_x u = 0, \quad (x,t) \in \mathbb{R}^2, \tag{6.1}$$

$$u(x,0) = u_0(x), \quad x \in \mathbb{R}. \tag{6.2}$$

This model, whose discoverers are considered to be two Dutch mathematicians Korteweg and de Vries, was proposed for long surface waves of water in a narrow and shallow channel.

We are interested in well-posedness results for data (6.2) in Sobolev spaces of negative order, i.e., $u_0(x) \in H^s(\mathbb{R})$, with $s < 0$.

Definition 6.1. Let $s, b \in \mathbb{R}$. The space

$$X_{s,b}^{\tau=\xi^3}(\mathbb{R}\times\mathbb{R}) = X_{s,b}(\mathbb{R}^2),$$

is then defined to be the closure of the Schwartz functions $\mathscr{S}_{t,s}(\mathbb{R}\times\mathbb{R})$ under the norm

$$\|u\|_{X_{s,b}(\mathbb{R}^2)} = \left\| \langle\xi\rangle^s \left\langle \tau - \xi^3 \right\rangle^b \widehat{u}(\xi,\tau) \right\|_{L_\tau^2 L_\xi^2(\mathbb{R}\times\mathbb{R})}$$

$$= \left(\int_{\mathbb{R}^2} (1+|\xi|)^{2s} \left(1 + \left|\tau - \xi^3\right|\right)^{2b} |\widehat{u}(\xi,\tau)|^2 \right)^{1/2}, \tag{6.3}$$

where $\langle\cdot\rangle = (1 + |\cdot|)$.

We denote by $\{W(t)\}_{-\infty}^{+\infty}$ the unitary group describing the solution of the linear IVP associated with (6.1)–(6.2):

$$\partial_t u + \partial_x^3 u = 0, \quad (x,t) \in \mathbb{R}^2,$$
$$u(x,0) = u_0(x), \quad x \in \mathbb{R},$$

where

$$u(x,t) = W(t)u_0(x) = S_t * u_0(x),$$

with $S_t(\cdot)$ defined by the oscillatory integral

$$S_t(x) = c \int_{\mathbb{R}} e^{ixt} e^{i\xi^3} d\xi.$$

The group $\{W(t)\}_{-\infty}^{+\infty}$ satisfies the following:

1.
$$\|W(t)u_0\|_{L_x^8 L_t^8} \lesssim \|u_0\|_{L_x^2}. \tag{6.4}$$

2.
$$\left\| D_x^{\frac{1}{6}} W(t)u_0 \right\|_{L_x^6 L_t^6} \lesssim \|u_0\|_{L_x^2}, \tag{6.5}$$

   where D is for the homogeneous derivative of order s.

3.
$$\|D_x W(t)u_0\|_{L_x^\infty L_t^2} \lesssim \|u_0\|_{L_x^2}, \tag{6.6}$$

   where D is for the homogeneous derivative of order s.

4.
$$\|W(t)u_0\|_{L_x^4 L_t^\infty} \lesssim \left\| D_x^{\frac{1}{4}} u_0 \right\|_{L_x^2}, \tag{6.7}$$

   where D is for the homogeneous derivative of order s.

5.
$$\|W(t)u_0\|_{L_x^5 L_t^{10}} \lesssim \|u_0\|_{L_x^2}. \tag{6.8}$$

Next, we deduce some inequalities related to the norms to be used in the proof for existence of solutions of the considered problem.

Lemma 6.1. *Let*

$$\hat{H}_\rho(\xi,\tau) = \frac{h(\xi,\tau)}{(1+|\tau|)^\rho}.$$

Then, for $\rho > \frac{1}{2}$, we have

$$\|H_p\|_{L^2_x L^2_t} \lesssim \|h\|_{L^2_\xi(\mathbb{R}) L^2_\tau}.$$

Proof. It follows by using the Sobolev inequality

$$\|H_\rho(x,\cdot)\|_{L^\infty_t} \lesssim \|(I + D_L)^\rho\, H(x,\cdot)\|_{L^\infty_t}. \qquad \square$$

We introduce the following notation:

$$\hat{F}_p(\xi,\tau) = \frac{f(\xi,\tau)}{(1+|\tau-\xi^3|)^p}.$$

Lemma 6.2. *If*

$$\hat{G}_p(\xi,\tau) = \chi(\xi)\hat{F}_p(\xi,\tau),$$

where

$$\chi \in C_0^\infty(\mathbb{R}), \quad \chi(\xi) = 1,$$

if $|\xi| \leqslant 1$, and

$$\chi(\xi) = 0,$$

if $|\xi| > 2$, then for $\rho > \frac{1}{2}$

$$\|G_\rho\|_{L^\infty_x L^2_t} \lesssim \|f\|_{L^2_\xi L^2_\tau}.$$

Proof. One just needs to observe that

$$\frac{\chi(\xi)}{(1+|\,\tau-\xi^3)^\rho}(1+|\tau|)^p$$

is bounded. $\square$

Lemma 6.3. *If $\rho > \frac{3}{8}$ and $0 \leqslant \theta \leqslant \frac{1}{8}$, then*

$$\left\| D_x^\theta F_p \right\|_{L_x^4 L_t^4} \leqslant c \|f\|_{L_\xi^2 L_\tau^2}. \tag{6.9}$$

Proof. Changing variable $\tau = \lambda + \xi^3$, we have

$$\begin{aligned} F_p(x,t) &= \int_{\mathbb{R}^2} e^{i(x\xi + t\tau)} \frac{f(\xi,\tau)}{(1+|\tau - \xi^3|)^p} d\xi d\tau \\ &= \int_{\mathbb{R}} e^{it\lambda} \left(\int_{\mathbb{R}} e^{i(x\xi + t\xi^3)} f\left(\xi, \lambda + \xi^3\right) d\xi \right) \frac{d\lambda}{(1+|\lambda|)^p}. \end{aligned}$$

Thus, using (6.4) and Minkowski's integral inequality and taking $\rho > \frac{1}{2}$, one easily sees that

$$\begin{aligned} \|F_\rho\|_{L_x^8 L_t^8} &\lesssim \int_{\mathbb{R}} \left\| f\left(\xi, \lambda + \xi^3\right) \right\|_{L_\xi} \frac{d\lambda}{(1+|\lambda|)^\rho} \\ &\lesssim \left(\int_{\mathbb{R}^2} \left| f\left(\xi, \lambda + \xi^3\right) \right|^2 d\lambda d\xi \right)^{1/2} \\ &\lesssim \|f\|_{L_\xi^2 L_\tau^2}. \end{aligned} \tag{6.10}$$

By interpolation between the above inequality and

$$\|F_0\|_{L_x^2 L_t^2} \lesssim \|f\|_{L_\xi^2 L_\tau^2}, \tag{6.11}$$

for $\rho > \frac{1}{3}$, we have that

$$\|F_p\|_{L_x^4 L_t^4} \lesssim \|f\|_{L_\xi^2 L_\tau^2}. \tag{6.12}$$

The argument in (6.10) combined with (6.11) proves that for $\rho > \frac{1}{2}$

$$\left\| D_x^{\frac{1}{6}} F_\rho \right\|_{L_x^6 L_t^6} \lesssim \|f\|_{L_\xi^2 L_\tau^2}, \tag{6.13}$$

which interpolated with (6.11) shows that for $\rho > \frac{3}{8}$

$$\left\| D_x^{\frac{1}{8}} F_\rho \right\|_{L_x^4 L_t^4} \lesssim \|f\|_{L_\xi^2 L_\tau^2}. \tag{6.14}$$

Then inequality (6.9) follows by interpolation between (6.12) and (6.14). □

Lemma 6.4. *If $\rho > \frac{5}{12}$, then*

$$\|F_\rho\|_{L^4_x L^6_t} \lesssim \|f\|_{L^2_\xi L^2_\tau}. \tag{6.15}$$

Proof. From the argument in (6.10) and inequality (6.8), it follows that for $\rho > \frac{1}{2}$

$$\|F_\rho\|_{L^{\frac{5}{2}}_x L^{10}_t} \leqslant c\|f\|_{L^2_2 L^2_t},$$

which interpolated with (6.11) yields (6.15). □

Lemma 6.5. *If $\rho > \theta/2$ with $\theta \in [0,1]$, then*

$$\left\|D^\theta_x F_\rho\right\|_{L_x^{\frac{2}{(1-\theta)}} L^2_t} \leqslant c\|f\|_{L^2_\xi L^2_\tau}. \tag{6.16}$$

Proof. The argument in (6.10) and the estimate (6.6) show that for $\rho > \frac{1}{2}$

$$\|D_x F_\rho\|_{L^\infty_x L^2_t} \lesssim \|f\|_{L^2_\xi L^2_\tau}.$$

□

Lemma 6.6. *If $\rho > \frac{1}{3}$, then*

$$\left\|D_x^{\frac{1}{4}} F_\rho\right\|_{L^4_x L^3_t} \leqslant c\|f\|_{L^2_\xi L^2_t}. \tag{6.17}$$

Proof. It follows by interpolation between (6.9) with $\theta = \frac{1}{8}$ and (6.16) with $\theta = \frac{1}{2}$. □

Lemma 6.7.

$$\left\|\psi\left(\frac{t}{\delta}\right) W(t)u_0\right\|_{X_{s,b}} \lesssim \delta^{\frac{1-2b}{2}} \|u_0\|_{H^s}. \tag{6.18}$$

Proof. By definition,

$$\psi\left(\frac{t}{\delta}\right)W(t)u_0 = c\psi\left(\frac{t}{\delta}\right)\int_{\mathbb{R}} e^{i(x\xi+t\xi^3)}\hat{u}_0(\xi)d\xi$$
$$= c\delta\int_{\mathbb{R}^2} e^{i(x\xi+t\tau)}\hat{\psi}\left(\delta\left(\tau-\xi^3\right)\right)\hat{u}_0(\xi)d\xi d\tau.$$

Thus,

$$\left\|\psi\left(\frac{t}{\delta}\right)W(t)u_0\right\|_{X_{s,b}}^2$$
$$= c\delta^2\int_{\mathbb{R}^2}\left(1+\left|\tau-\xi^3\right|\right)^{2b}(1+|\xi|)^{2s}\left|\hat{\psi}\left(\delta\left(\tau-\xi^3\right)\right)\right|^2|\hat{u}_0(\xi)|^2 d\xi d\tau$$
$$= c\int_{\mathbb{R}}(1+|\xi|)^{2s}|\hat{u}_0(\xi)|^2$$
$$\times\left(\delta^2\int_{\mathbb{R}}\left|\hat{\psi}\left(\delta\left(\tau-\xi^3\right)\right)\right|^2\left(1+\left|\tau-\xi^3\right|\right)^{2b}d\tau\right)d\xi.$$

For the inner integral, we use that $b > \frac{1}{2}$ and $0 < \delta < 1$ to obtain

$$\delta^2\int_{\mathbb{R}}\left|\hat{\psi}\left(\delta\left(\tau-\xi^3\right)\right)\right|^2\left(1+\left|\tau-\xi^3\right|\right)^{2b}d\tau$$
$$\lesssim \delta^2\int_{\mathbb{R}}\left|\hat{\psi}\left(\delta\left(\tau-\xi^3\right)\right)\right|^2 d\tau + \delta^2\int_{\mathbb{R}}\left|\psi\left(\delta\left(\tau-\xi^3\right)\right)\right|^2\left|\tau-\xi^3\right|^{2b}d\tau$$
$$\lesssim \delta + \delta^{(1-2b)}$$
$$\lesssim \delta^{(1-2b)}.$$

□

Lemma 6.8. *If $\frac{1}{2} < b \leqslant 1$, then*

$$\left\|\psi\left(\frac{t}{\delta}\right)h\right\|_{X_{s,b}} \lesssim \delta^{\frac{1-2b}{2}}\|h\|_{X_{s,b}}.$$

Proof. Using

$$\left(\widehat{\psi\left(\frac{\cdot}{\delta}\right)h}\right)(\xi,\tau) = \delta\int_{\mathbb{R}} \hat{h}(\xi,\tau-\lambda)\hat{\psi}(\delta\lambda)d\lambda$$
$$= \widehat{h} * (\delta\hat{\psi}(\delta\cdot)),$$

one easily sees that the proof reduces to showing that, if $a \in \mathbb{R}$, then

$$\int_{\mathbb{R}} |\hat{h} * (\delta\hat{\psi}(\delta\cdot))(l)|^2(1+|l-a|)^{2b}dl \lesssim \delta^{1-2b}\int_{\mathbb{R}} |\hat{h}(l)|^2(1+|l-a|)^{2b}dl.$$

Since

$$\int_{\mathbb{R}} |\delta\hat{\psi}(\delta l)|dl \lesssim 1,$$

we get

$$\int_{\mathbb{R}} |\hat{h} * (\delta\hat{\psi}(\delta\cdot))(l)|^2dl \lesssim \int_{\mathbb{R}} |\hat{h}(l)|^2dl.$$

Now, we turn to

$$\int_{\mathbb{R}} |\hat{h} * (\delta\hat{\psi}(\delta\cdot))(l)|^2|l-a|^{2b}dl = \int_{\mathbb{R}} |\hat{h} * (\delta\hat{\psi}(\delta\cdot))(l-a)|^2||^{2b}dl$$
$$= \int_{\mathbb{R}} \left|D^b\left(e^{iat}h(t)\psi\left(\frac{t}{\delta}\right)\right)\right|^2 dt.$$

To bound this term, we use

$$\|D^\alpha(fg) - fD^\alpha g\|_{L^p} \lesssim \|g\|_\infty \|D^\alpha f\|_{L^p}, \tag{6.19}$$

with $p = 2$, i.e.,

$$\left\|D^b\left(e^{ia\cdot}h\psi\left(\frac{\cdot}{\delta}\right)\right) - e^{ia\cdot}hD^b\psi\left(\frac{\cdot}{\delta}\right)\right\|_{L^2} \lesssim \left\|D^b\left(e^{ia\cdot}h\right)\right\|_{L^2}\|\psi\|_{L^\infty}.$$

Since

$$\|\psi\|_{L^\infty} \lesssim 1$$

and

$$\left\|D^b\left(e^{ia\cdot}h\right)\right\|_{L^2}^2 = \int_{\mathbb{R}} |\hat{h}(l)|^2|l-a|^{2b}dl,$$

we only have to bound the term

$$A = \int_{\mathbb{R}} \left| e^{iat} h(t) D^t \psi\left(\frac{t}{\delta}\right)\right|^2 dt.$$

The Sobolev embedding theorem and the fact that $b > \frac{1}{2}$ lead to

$$\begin{aligned} A &\leqslant \left\|e^{ia\cdot} h\right\|_{L^\infty}^2 \left\|D^b\psi\left(\frac{\cdot}{\delta}\right)\right\|_{L^2}^2 \\ &\lesssim \left(\int_{\mathbb{R}} \left|e^{iat}h(t)\right|^2 dt + \int_{\mathbb{R}} \left|D^b\left(e^{iat}h(t)\right)\right|^2 dt\right) \left\|D^b\psi\left(\frac{\cdot}{\delta}\right)\right\|_{L^2}^2 \\ &\lesssim \left(\int_{\mathbb{R}} |h(l)|^2 dl + \int_{\mathbb{R}} |h(l)|^2 |l-a|^{2b} dl\right) \left\|D^b\psi\left(\frac{\cdot}{\delta}\right)\right\|_{L^2}^2. \end{aligned}$$

By Plancherel's theorem, we have

$$\left\|D^b\psi\left(\frac{\cdot}{\delta}\right)\right\|_{L^2}^2 = \int_{\mathbb{R}} |l|^{2b}\delta^2 |\widehat{\psi}(\delta l)|^2 dl = c\delta^{(1-2b)}.$$

□

Lemma 6.9. *If* $\frac{1}{2} < b \leqslant 1$, *then*

$$\left\|\psi\left(\frac{t}{\delta}\right)\int_0^t W(t-s)\, w(s)\, ds\right\|_{X_{s,b}} \lesssim \delta^{\frac{1-2b}{2}} \|w\|_{X_{s,b-1}}.$$

Proof. We begin by writing

$$\begin{aligned} &\psi\left(\frac{t}{\delta}\right)\int_0^t W(t-s)\, w(s)\, ds \\ &\quad= \psi\left(\frac{t}{\delta}\right)\int_{\mathbb{R}^2} e^{ix\xi}\hat{w}(\xi,\tau)\psi\left(\tau-\xi^3\right)\frac{e^{i\tau t}-e^{it\xi^3}}{\tau-\xi^3} d\xi d\tau \\ &\qquad+ \psi\left(\frac{t}{\delta}\right)\int_{\mathbb{R}^2} e^{ix\xi}\hat{w}(\xi,\tau)\left[1-\psi\left(\tau-\xi^3\right)\right]\frac{e^{i\tau t}-e^{it\xi^3}}{\tau-\xi^3} d\xi d\tau \\ &\quad= I + II. \end{aligned}$$

By a Taylor expansion,

$$\begin{aligned} II = \sum_{k=1}^{\infty} \frac{i^k}{k!} t^k \psi\left(\frac{t}{\delta}\right)\int_{\mathbb{R}} e^{i\left(x\xi+t\xi^3\right)} \\ \times\left(\int_{\mathbb{R}} \hat{w}(\xi,\tau)\left(\tau-\xi^3\right)^{k-1}\psi\left(\tau-\xi^3\right)d\tau\right) d\xi. \qquad (6.20) \end{aligned}$$

Let

$$t^k\psi\left(\frac{t}{\delta}\right)=\delta^k\left(\frac{t}{\delta}\right)^k\psi\left(\frac{t}{\delta}\right)=\delta^k\varphi_k\left(\frac{t}{\delta}\right),\quad k\geqslant 1.$$

Following the argument in the proof of Lemma 6.7 and the assumptions $b\leqslant 1$ and $\delta\leqslant 1$, one finds that

$$\begin{aligned}&\delta^2\int_{\mathbb{R}}|\hat{\varphi}_k(\delta\tau)|^2\left(1+|\tau|^{2b}\right)d\tau\\&\lesssim\delta^{(1-2b)}\left(\int_{\mathbb{R}}\left(|\hat{\varphi}_k(\delta\tau)|^2+|\tau|^2|\hat{\varphi}_k(\delta\tau)|^2\right)d\tau\right)\\&=c\delta^{(1-2b)}\left(\|\varphi_k\|_{L^2}^2+\|D_x\varphi_k\|_{L^2}^2\right)\\&\lesssim\delta^{(1-2b)}(k+1)^2.\end{aligned}$$

Thus, by Lemma 6.7 and (6.20), we get

$$\begin{aligned}\|I\|_{X_{s,b}}\leqslant c&\sum_{k=1}^{\infty}\frac{\delta^k}{k!}\delta^{(1-2b)}\\&\times\left\|\mathscr{F}^{-1}\left(\int_{\mathbb{R}}\hat{w}(\xi,\tau)\left(\tau-\xi^3\right)^{k-1}\psi\left(\tau-\xi^3\right)d\tau\right)\right\|_{H^s}(k+1).\end{aligned}$$

On the other hand,

$$\begin{aligned}&\left\|\mathscr{F}^{-1}\left(\int_{\mathbb{R}}\hat{w}(\xi,\tau)\left(\tau-\xi^3\right)^{k-1}\psi\left(\tau-\xi^3\right)d\tau\right)\right\|_{H^s}^2\\&\lesssim\int_{\mathbb{R}}(1+|\xi|)^{2s}\left(\int_{\mathbb{R}}\hat{w}(\xi,\tau)\left(\tau-\xi^3\right)^{k-1}\psi\left(\tau-\xi^3\right)d\tau\right)^2d\xi\\&\lesssim\int_{\mathbb{R}}(1+|\xi|)^{2s}\left(\int_{|\tau-\xi 3|<1}|\hat{w}(\xi,\tau)|d\tau\right)^2d\xi\\&\lesssim\int_{\mathbb{R}}(1+|\xi|)^{2s}\left(\int_{\mathbb{R}}\frac{|\hat{w}(\xi,\tau)|}{(1+|\tau-\xi^3|)^{1-b}}\frac{1}{(1+|\tau-\xi^3|)^b}d\tau\right)^2d\xi\\&\lesssim\int_{\mathbb{R}^2}(1+|\xi|)^{2s}\frac{|\hat{w}(\xi,\tau)|^2}{(1+|\tau-\xi^3|)^{2(1-b)}}d\tau d\xi\\&=\|w\|_{X_{s,b-1}}^2.\end{aligned}$$

We write

$$II = II_1 + II_2,$$

where

$$II_1 = -\psi\left(\frac{t}{\delta}\right)\int_{\mathbb{R}} e^{i(x\xi+t\xi^3)}\left(\int_{\mathbb{R}}\frac{1-\psi\left(\tau-\xi^3\right)}{\tau-\xi^3}\hat{w}(\xi,\tau)d\tau\right)d\xi$$

and

$$II_2 = \psi\left(\frac{t}{\delta}\right)\int_{\mathbb{R}^2} e^{i(x\xi+t\tau)}\frac{1-\psi\left(\tau-\xi^3\right)}{\tau-\xi^3}\hat{w}(\xi,\tau)d\tau d\xi.$$

Since $b>\frac{1}{2}$, from Lemma 6.7, one easily sees that

$$\begin{aligned}
\|II_1\|_{X_{s,b}} &\leqslant \delta^{\frac{1-2b}{2}}\left\|\mathscr{F}^{-1}\left(\int_{\mathbb{R}}\frac{1-\psi\left(\tau-\xi^3\right)}{\tau-\xi^3}\hat{w}(\xi,\tau)d\tau\right)\right\|_{H^s}\\
&\lesssim \delta^{\frac{1-2b}{2}}\left(\int_{\mathbb{R}}(1+|\xi|)^{2s}\left(\int_{1/2<|\tau-\xi^3|}\frac{|\hat{w}(\xi,\tau)|}{|\tau-\xi^3|}d\tau\right)^2 d\xi\right)^{\frac{1}{2}}\\
&\lesssim \delta^{\frac{1-2b}{2}}\left(\int_{\mathbb{R}}(1+|\xi|)^{2s}\left(\int_{-\infty}^{\infty}\frac{|\hat{w}(\xi,\tau)|}{(1+|\tau-\xi^3|)^{1-b}}\frac{1}{(1+|\tau-\xi^3|)^{b}}d\tau\right)^2 d\xi\right)^{\frac{1}{2}}\\
&\lesssim \delta^{\frac{1-2b}{2}}\left(\int_{\mathbb{R}^2}(1+|\xi|)^{2s}\frac{|\hat{w}(\xi,\tau)|^2}{(1+|\tau-\xi^3|)^{2(1-b)}}d\xi d\tau\right)^{\frac{1}{2}}\\
&= \delta^{\frac{1-2b}{2}}\|w\|_{X_{s,b-1}}.
\end{aligned}$$

Finally,

$$\begin{aligned}
\|II_2\|_{X_{s,b}} &\leqslant \delta^{\frac{1-2b}{2}}\left\|\int_{\mathbb{R}^2} e^{i(x\xi+t\tau)}\frac{1-\psi\left(\tau-\xi^3\right)}{\tau-\xi^3}\hat{w}(\xi,\tau)d\tau d\xi\right\|_{X_{s,b}}\\
&\lesssim \delta^{\frac{1-2b}{2}}\left(\int_{\mathbb{R}^2}\frac{|\hat{w}(\xi,\tau)|^2}{(1+|\tau-\xi^3|)^2}\left(1+\left|\tau-\xi^3\right|\right)^{2b}(1+|\xi|)^{2s}d\xi d\tau\right)^{1/2}\\
&= \delta^{\frac{1-2b}{2}}\|w\|_{X_{s,b-1}}.
\end{aligned}$$

□

6.1 Bilinear Estimates [1]

Lemma 6.10. *Let $\sigma \in C_0^\infty(\mathbb{R})$, with $\sigma = 1$ on supp ψ and supp $\sigma \subseteq (-1,1)$. Let*

$$\omega(x,t) = \sigma^2\left(\frac{t}{\delta}\right)\frac{\partial}{\partial x}u^2(x,t),$$

where $u \in X_{s,b}$. Assume that

$$\frac{1}{2} < s < \frac{5}{8}, \quad \frac{1}{2} < b < \frac{7}{12}, \quad b \leqslant s, s+b \leqslant \frac{9}{8},$$

$$\frac{\gamma_0}{4b} + 1 - 2b = \theta_0 > 0, \qquad \text{where } \gamma_0 = \frac{1}{2} - \beta_0,$$

$$\alpha_0 < \beta_0 < \frac{1}{2} \qquad \text{and} \qquad \alpha_0 = \max\left\{\frac{3}{8}, \frac{s}{2}\right\}.$$

Then

$$\|\omega\|_{X_{s,b-1}} \lesssim \delta^{\theta_0}\|u\|^2_{X_{s,b}}.$$

Proof. Define

$$v(x,t) = \sigma\left(\frac{t}{\delta}\right)u(x,t)$$

so that

$$\hat{\omega}(\xi,\tau) = c\xi\hat{v} * \hat{v}(\xi,\tau).$$

Also, define

$$f(\xi,\tau) = \left(1 + \left|\tau - \xi^3\right|\right)^b (1+|\xi|)^{-s}|\hat{v}(\xi,\tau)|.$$

To estimate $\|\omega\|_{X_{s,b-1}}$, we proceed by duality. Let $g(\xi,\tau) \geqslant 0$ with $\|g\|_{L^2_\xi L^2_\tau} \leqslant 1$, and consider the expression

$$\Upsilon = \int_{\mathbb{R}^4} \frac{|\xi|g(\xi,\tau)}{(1+|\tau-\xi^3|)^{1-b}(1+|\xi|)^3}\frac{(1+|\xi_1|)^s f(\xi_1,\tau_1)}{\left(1+\left|\tau_1-\xi_1^3\right|\right)^b}$$
$$\cdot\frac{(1+|\xi-\xi_1|)^s f(\xi-\xi_1,\tau-\tau_1)}{\left(1+\left|\tau-\tau_1-(\xi-\xi_1)^3\right|\right)^b}d\mu,$$

where

$$d\mu = d\xi_1 d\tau_1 d\xi d\tau.$$

By symmetry, it suffices to estimate the integral in the domain

$$|\xi_1| \leqslant |\xi - \xi_1|.$$

Note that we have the identity

$$\tau - \xi^3 - \left[\left(\tau_1 - \xi_1^3\right) + (\tau - \tau_1) - (\xi - \xi_1)^3\right] = 3\xi_1 (\xi - \xi_1) \xi,$$

which implies that one of the following cases always occurs:

$$\begin{aligned} &\text{(a)} \ \left|\tau - \xi^3\right| \geqslant |\xi_1| \, |\xi - \xi_1| \, |\xi|, \\ &\text{(b)} \ \left|\tau_1 - \xi_1^3\right| \geqslant |\xi_1| \, |\xi - \xi_1| \, |\xi|, \\ &\text{(c)} \ \left|\tau - \tau_1 - (\xi - \xi_1)^3\right| \geqslant |\xi_1| \, |\xi - \xi_1| \, |\xi|. \end{aligned}$$

To bound the integral Υ, we split the domain of integration in several pieces.

Assume: $|\xi| \leqslant 2$ and $|\xi_1| \leqslant 1$.

$$\text{Since} \quad |\xi_1| \leqslant |\xi - \xi_1|, \quad \text{then} \quad |\xi - \xi_1| \leqslant |\xi| + |\xi_1| \leqslant 3,$$

and consequently the integral Υ restricted to this domain is bounded by

$$\begin{aligned} &c \int_{\mathbb{R}^4} \frac{g(\xi,\tau)\chi\left(\frac{\xi}{2}\right)}{(1+|\tau-\xi^3|)^{1-b}} \frac{f(\xi_1,\tau_1)\chi(\xi_1)}{\left(1+\left|\tau_1-\xi_1^3\right|\right)^b} \frac{f(\xi-\xi_1,\tau-\tau_1)\chi\left(\frac{\xi-\xi_1}{3}\right)}{\left(1+|\,\tau-\tau_1-(\xi-\xi_1)^3\right)^b} d\mu \\ &\lesssim \int_{\mathbb{R}^4} g(\xi,\tau) \frac{f(\xi_1,\tau_1)}{\left(1+\left|\tau_1-\xi_1^3\right|\right)^b} \frac{f(\xi-\xi_1,\tau-\tau_1)}{\left(1+\left|\tau-\tau_1-(\xi-\xi_1)^3\right|\right)^b} d\mu \\ &= \left\langle g, \widehat{F_b} \cdot \widehat{F_b} \right\rangle \\ &= \langle G_0, F_b F_b \rangle \\ &\lesssim \|g\|_{L^2_\xi L^2_\tau} \|F_b\|^2_{L^4_x L^4_t} \\ &\lesssim \|F_b\|^2_{L^4_x L^4_t}, \end{aligned}$$

where

$$\chi \in C_0^\infty(\mathbb{R}), \quad \chi(\xi) = 1,$$

if $|\xi| \leqslant 1$, and

$$\chi(\xi) = 0,$$

if $|\xi| > 2$ and

$$\widehat{G}_\rho(\xi,\tau) = \frac{g(\xi,\tau)}{(1+|\tau-\xi^3|)^\rho} \quad \text{and} \quad \widehat{F}_\rho(\xi,\tau) = \frac{f(\xi,\tau)}{(1+|\tau-\xi^3|)^\rho}.$$

Assume: $|\xi| \leqslant 2$ and $1 \leqslant |\xi_1| \leqslant 1$.

The integral Υ restricted to this domain is dominated by

$$c\int_{\mathbb{R}^4} \frac{|\xi| g(\xi,\tau)\chi\left(\frac{\xi}{2}\right)}{(1+|\tau-\xi^3|)^{1-b}} \frac{|\xi_1|^s f(\xi_1,\tau_1)}{\left(1+\left|\tau_1-\xi_1^3\right|\right)^b} \frac{|\xi-\xi_1|^3 f(\xi-\xi_1,\tau-\tau_1)}{\left(1+|\tau-\tau_1-(\xi-\xi_1)^3|\right)^b} d\mu.$$

We consider the three cases in equality (a), (b) and (c) separately.

If (a) holds, we use

$$\frac{1}{2} < s, \quad \frac{1}{2} < b, \quad s+b-1 \leqslant \frac{1}{2},$$

then

$$\begin{aligned}
&c\int_{\mathbb{R}^4} |\xi|^b g(\xi,\tau)\chi\left(\frac{\xi}{2}\right) \frac{|\xi_1|^{s+b-1} f(\xi_1,\tau_1)}{\left(1+\left|\tau_1-\xi_1^3\right|\right)^b} \\
&\quad\times \frac{|\xi-\xi_1|^{s+b-1} f(\xi-\xi_1,\tau-\tau_1)}{\left(1+\left|\tau-\tau_1-(\xi-\xi_1)^3\right|\right)^b} d\mu \\
&\lesssim \int_{\mathbb{R}^4} g(\xi,\tau) \frac{|\xi_1|^{\frac{1}{8}} f(\xi_1,\tau_1)}{\left(1+\left|\tau_1-\xi_1^3\right|\right)^b} \frac{|\xi-\xi_1|^{\frac{1}{8}} f(\xi-\xi_1,\tau-\tau_1)}{\left(1+\left|\tau-\tau_1-(\xi-\xi_1)^3\right|\right)^b} d\mu \\
&\lesssim \left\langle g, \widehat{D_x^{\frac{1}{8}} F_b} * \widehat{D_x^{\frac{1}{8}} F_b} \right\rangle
\end{aligned}$$

$$= \left\langle G_0, D_x^{\frac{1}{8}} F_b D_x^{\frac{1}{8}} F_b \right\rangle$$

$$\lesssim \|g\|_{L_t^2 L_t^2} \left\| D_x^{\frac{1}{8}} F_b \right\|_{L_x^4 L_t^4}^2$$

$$\lesssim \left\| D_x^{\frac{1}{8}} F_b \right\|_{L_x^4 L_t^4}.$$

If (b) holds, then

$$\left(1 + \left|\tau_1 - \xi_1^3\right|\right)^b \geqslant |\xi_1|^b \, |\xi - \xi_1|^b \, |\xi|^b.$$

Hence, using

$$s - \frac{1}{2} \leqslant \frac{1}{2} < b \leqslant s,$$

we have

$$0 < s - b < \frac{1}{2}, \quad \text{and} \quad b < \frac{7}{12},$$

and the hypothesis

$$|\xi| \leqslant 2, \quad 1 \leqslant |\xi_1| \leqslant |\xi_1 - \xi|,$$

can be majorized by

$$c \int_{\mathbb{R}^4} \frac{|\xi|^{1-b} g(\xi,\tau) \chi\left(\frac{\xi}{2}\right)}{\left(1 + |\tau - \xi^3|\right)^{1-b}} \, |\xi_1|^{s-b} f\left(\xi_1, \tau_1\right) \frac{|\xi - \xi_1|^{s-b} f\left(\xi - \xi_1, \tau - \tau_1\right)}{\left(1 + \left|\tau - \tau_1 - (\xi - \xi_1)^3\right|\right)^b} d\mu$$

$$\lesssim \int_{\mathbb{R}^4} \frac{g(\xi,\tau) \chi\left(\frac{\xi}{2}\right)}{\left(1 + |\tau - \xi^3|\right)^{1-b}} f\left(\xi_1, \tau_1\right) \frac{|\xi - \xi_1|^{\frac{1}{4}} f\left(\xi - \xi_1, \tau - \tau_1\right)}{\left(1 + \left|\tau - \tau_1 - (\xi - \xi_1)^3\right|\right)^b} d\mu$$

$$\lesssim \left\langle \widehat{\mathscr{G}}_{1-\mathrm{b}}, \widehat{F}_0 * \widehat{D_x^{\frac{1}{4}} F_b} \right\rangle$$

$$= \left\langle \mathscr{G}_{1-b}, F_0 D_x^{\frac{1}{4}} F_b \right\rangle$$

$$
\begin{aligned}
&\lesssim \|\mathscr{G}_{1-b}\|_{L_x^4 L_t^6} \|F_0\|_{L_x^2 L_t^2} \left\| D_x^{\frac{1}{4}} F_b \right\|_{L_x^4 L_t^3} \\
&\lesssim \|g\|_{L_\xi^2 L_\tau^2} \|f\|_{L_\xi^2 L_\tau^2} \left\| D_x^{\frac{1}{4}} F_b \right\|_{L_x^4 L_t^3} \\
&\lesssim \|v\|_{X_{s,b}} \left\| D_x^{\frac{1}{4}} F_b \right\|_{L_x^4 L_t^3},
\end{aligned}
$$

where

$$
\widehat{\mathscr{G}_\rho}(\xi,\tau) = \frac{g(\xi,\tau)\chi\left(\frac{\xi}{2}\right)}{(1+|\tau-\xi^3|)^\rho}.
$$

If (c) holds, i.e.,

$$
\left(1+\left|\tau-\tau_1-(\xi-\xi_1)^3\right|\right)^b \geqslant |\xi_1|^b\,|\xi-\xi_1|^b\,|\xi|^b,
$$

we use that

$$
|\xi-\xi_1|^{s-b} \leqslant |\xi|^{s-b} + |\xi_1|^{s-b} \quad \text{and} \quad |\xi| \leqslant 1,
$$

we have

$$
\begin{aligned}
&c\int_{\mathbb{R}^4} \frac{|\xi|^{1-b} g(\xi,\tau)\chi(\frac{\xi}{2})}{(1+|\tau-\xi^3|)^{1-b}} \frac{|\xi_1|^{s-b} f(\xi_1,\tau_1)}{\left(1+\left|\tau-\xi_1^3\right|\right)^b} |\xi-\xi_1|^{s-b} f(\xi-\xi_1,\tau-\tau_1)\, d\mu \\
&\leqslant \int_{\mathbb{R}^4} \frac{g(\xi,\tau)\chi(\frac{\xi}{2})}{(1+|\tau-\xi^3|)^{1-b}} \frac{|\xi_1|^{s-b} f(\xi_1,\tau_1)}{\left(1+\left|\tau_1-\xi_1^3\right|\right)^b} f(\xi-\xi_1,\tau-\tau_1)\, d\mu \\
&\quad + c\int_{\mathbb{R}^4} \frac{g(\xi,\tau)\chi(\frac{\xi}{2})}{(1+|\tau-\xi^3|)^{1-b}} \frac{|\xi_1|^{2(s-b)} f(\xi_1,\tau_1)}{\left(1+\left|\tau_1-\xi_1^3\right|\right)^b} f(\xi-\xi_1,\tau-\tau_1)\, d\mu \\
&= B_1 + B_2.
\end{aligned}
$$

Since

$$
1-b > \frac{3}{8} \quad \text{and} \quad 0 < s-b < \frac{1}{8},
$$

we get

$$\begin{aligned}
B_1 &\leqslant c\left\langle \mathscr{G}_{1-b}, D_x^{\frac{1}{8}} F_b F_0 \right\rangle \\
&\lesssim \|\mathscr{G}_{1-b}\|_{L_x^4 L_t^4} \left\| D_x^{\frac{1}{8}} F_b \right\|_{L_x^4 L_t^4} \|F_0\|_{L_x^2 L_t^2} \\
&\lesssim \|g\|_{L_\xi^2 L_\tau^2} \|f\|_{L_\xi^2 L_\tau^2} \left\| D_x^{\frac{1}{8}} F_b \right\|_{L_x^4 L_t^4} \\
&\lesssim \|v\|_{X_{s,b}} \left\| D_x^{\frac{1}{8}} F_b \right\|_{L_x^4 L_t^4}.
\end{aligned}$$

To bound B_2, we follow the argument in the previous case (i.e., when (b) holds), and thus

$$\begin{aligned}
B_2 &\leqslant c \|\mathscr{G}_{1-b}\|_{L_x^4 L_t^6} \|F_0\|_{L_x^2 L_t^2} \left\| D_x^{\frac{1}{4}} F_b \right\|_{L_x^4 L_t^3} \\
&\lesssim \|g\|_{L_t^2 L_+^2} \|f\|_{L_2^2 L_t^2} \left\| D_x^{\frac{1}{4}} F_b \right\|_{L_x^4 L_t^3} \\
&\lesssim \|v\|_{X_{s,b}} \left\| D_x^{\frac{1}{4}} F_b \right\|_{L_x^4 L_t^3}.
\end{aligned}$$

Next, we consider the case

$$|\xi| \geqslant 2.$$

Restricted to this domain, the integral Υ is dominated by

$$\begin{aligned}
\int_{\mathbb{R}^4} & \frac{|\xi|^{1-s} g(\xi_1, \tau)}{(1+|\tau-\xi^3|)^{1-b}} \frac{(1+|\xi_1|)^s f(\xi_1, \tau_1)}{\left(1+\left|\tau_1-\xi_1^3\right|\right)^b} \\
& \times \frac{(1+|\xi-\xi_1|)^s f(\xi-\xi_1, \tau-\tau_1)}{\left(1+\left|\tau-\tau_1-(\xi-\xi_1)^3\right|\right)^b} d\mu.
\end{aligned}$$

Assume:

$$|\xi| \geqslant 2, |\xi_1| \leqslant 1 \quad \text{so that} \quad |\xi-\xi_1| \geqslant 1.$$

In this domain, we have

$$\int_{\mathbb{R}^4} \frac{|\xi|^{1-s} g(\xi,\tau)}{(1+|\tau-\xi^3|)^{1-b}} \frac{f(\xi_1,\tau_1)\chi(\xi_1)}{\left(1+\left|\tau_1-\xi_1^3\right|\right)^b} \frac{|\xi-\xi_1|^s f(\xi-\xi_1,\tau-\tau_1)}{\left(1+\left|\tau-\tau_1-(\xi-\xi_1)^3\right|\right)^b} d\mu$$

$$\begin{aligned}
&\lesssim \left\langle D_x^{1-s} G_{1-b}, \mathscr{F}_b D^s F_b\right\rangle \\
&\lesssim \left\|D_x^{1-s} G_{1-b}\right\|_{L_x^{\frac{2}{s}} L_t^2} \|\mathscr{F}_b\|_{L_x^2 L_t^\infty} \left\|D_x^s F_b\right\|_{L_x^{\frac{2}{1-s}} L_t^2} \\
&\lesssim \|g\|_{L_\xi^2 L_\tau^2} \|\mathscr{F}_b\|_{L_x^2 L_t^\infty} \left\|D_x^s F_b\right\|_{L_x^{\frac{2}{1-s}} L_t^2} \\
&\lesssim \|\mathscr{F}_b\|_{L_x^2 L_t^\infty} \left\|D_x^s F_b\right\|_{L_x^{\frac{2}{1-s}} L_t^2} \\
&\lesssim \|v\|_{X_{s,b}} \left\|D_x^s F_b\right\|_{L_x^{\frac{2}{1-s}} L_t^2}.
\end{aligned}$$

Above we have used

$$\hat{\xi}_p(\xi,\tau) = \chi(\xi)\hat{F}_p(\xi,\tau).$$

Assume:

$$|\xi| \geqslant 2, |\xi_1| \geqslant 1 \quad \text{and} \quad |\xi-\xi_1| \geqslant |\xi_1|.$$

In this domain, we have

$$\int_{\mathbb{R}^4} \frac{|\xi|^{1-s} g(\xi,\tau)}{(1+|\tau-\xi^3|)^{1-b}} \frac{|\xi_1|^s f(\xi_1,\tau_1)}{\left(1+\left|\tau_1-\xi_1^3\right|\right)^b} \frac{|\xi-\xi_1|^s f(\xi-\xi_1,\tau-\tau_1)}{\left(1+\left|\tau-\tau_1-(\xi-\xi_1)^3\right|\right)^b} d\mu.$$

We split the domain of integration into the three cases: (a), (b) and (c).

If (a) holds, we get the bound

$$\begin{aligned}
&\int_{\mathbb{R}^4} |\xi|^{b-s} g(\xi,\tau) \frac{|\xi_1|^{s+b-1} f(\xi_1,\tau_1)}{\left(1+\left|\tau_1-\xi_1^3\right|\right)^b} \frac{|\xi-\xi_1|^{s+b-1} f(\xi-\xi_1,\tau-\tau_1)}{\left(1+\left|\tau-\tau_1-(\xi-\xi_1)^3\right|\right)^b} d\mu \\
&\lesssim \int_{\mathbb{R}^4} g(\xi,\tau) \frac{f(\xi_1,\tau_1)}{\left(1+\left|\tau_1-\xi_1^3\right|\right)^b} \frac{|\xi-\xi_1|^{2(s+b-1)} f(\xi-\xi_1,\tau-\tau_1)}{\left(1+\left|\tau-\tau_1-(\xi-\xi_1)^3\right|\right)^b} d\mu
\end{aligned}$$

$$\lesssim \left\langle \widehat{G}_0, \widehat{F}_b * \widehat{D_x^{\frac{1}{4}} F_b} \right\rangle$$

$$\lesssim \left\langle G_0, F_b D_x^{\frac{1}{4}} F_b \right\rangle$$

$$\lesssim \|g\|_{L_\xi^2 L_\tau^2} \|F_b\|_{L_x^4 L_t^6} \left\| D_x^{\frac{1}{4}} F_b \right\|_{L_x^4 L_t^3}$$

$$\lesssim \|F_b\|_{L_x^4 L_t^6} \left\| D_x^{\frac{1}{4}} F_b \right\|_{L_x^4 L_t^3}$$

$$\lesssim \|v\|_{X_{s,b}} \left\| D_x^{\frac{1}{4}} F_b \right\|_{L_x^4 L_t^3}.$$

If (b) holds, we obtain

$$\int_{\mathbb{R}^4} \frac{|\xi|^{1-b-s} g(\xi,\tau)}{(1+|\tau-\xi^3|)^{1-b}} |\xi_1|^{s-b} f(\xi_1,\tau_1)$$
$$\times \frac{|\xi-\xi_1|^{s-b} f(\xi-\xi_1,\tau-\tau_1)}{\left(1+\left|\tau-\tau_1-(\xi-\xi_1)^3\right|\right)^s} d\mu$$

$$\leqslant \int_{\mathbb{R}^4} \frac{g(\xi,\tau)}{(1+|\tau-\xi^3|)^{1-b}} f(\xi_1,\tau_1) \frac{|\xi-\xi_1|^{2(s-b)} f(\xi-\xi_1,\tau-\tau_1)}{\left(1+\left|\tau-\tau_1-(\xi-\xi_1)^3\right|\right)^b} d\mu$$

$$\lesssim \left\langle \widehat{G}_{1-b}, \widehat{F}_0 * \widehat{D_x^{\frac{1}{4}} F_0} \right\rangle$$

$$\lesssim \|G_{1-b}\|_{L_x^4 L_t^6} \|F_0\|_{L_x^2 L_t^2} \left\| D_x^{\frac{1}{4}} F_b \right\|_{L_x^4 L_t^3}$$

$$\lesssim \|g\|_{L_\xi^2 L_\tau^2} \|f\|_{L_\xi^2 L_\tau^2} \left\| D_x^{\frac{1}{4}} F_b \right\|_{L_x^4 L_t^3}$$

$$\lesssim \|v\|_{X_{s,b}} \left\| D_x^{\frac{1}{4}} F_b \right\|_{L_x^4 L_t^3}.$$

If (c) holds, using that $\frac{1}{2} < b$ and $b < s$, one sees

$$\begin{aligned}
&\int_{\mathbb{R}^4} \frac{|\xi|^{1-b-s} g(\xi,\tau)}{(1+|\tau-\xi^3|)^{1-b}} \frac{|\xi_1|^{s-b}}{\left(1+\left|\xi_1-\xi_1^3\right|\right)^b} |\xi-\xi_1|^{s-b} f(\xi-\xi_1,\tau-\tau_1)\, d\mu \\
&\lesssim \int_{\mathbb{R}^4} \frac{|\xi|^{1-b-s} g(\xi,\tau)}{(1+|\tau-\xi^3|)^{1-b}} \frac{|\xi_1|^{2(s-b)} f(\xi_1,\tau_1)}{\left(1+\left|\tau_1-\xi_1^3\right|\right)^b} f(\xi-\xi_1,\tau-\tau_1)\, d\mu \\
&\quad + \int_{\mathbb{R}^4} \frac{g(\xi,\tau)}{(1+|\tau-\xi^3|)^{1-b}} \frac{|\xi_1|^{s-b} f(\xi_1,\tau_1)}{\left(1+\left|\tau_1-\xi_1^3\right|\right)^b} f(\xi-\xi_1,\tau-\tau_1) d\mu \\
&= A_1 + A_2.
\end{aligned}$$

The bound for A_1 is similar to that deduced above.

To control A_2, we use the hypothesis $\frac{1}{2} < b < \frac{7}{12}$, $s - b < \frac{1}{8}$ since $b \leqslant \frac{9}{8} - s$, and $\frac{1}{2} < b$. We have

$$\begin{aligned}
&\int_{\mathbb{R}^4} \frac{g(\xi,\tau)}{(1+|\tau-\xi^3|)^{1-b}} \frac{|\xi_1|^{\frac{1}{8}} f(\xi_1,\tau_1)}{\left(1+\left|\tau_1-\xi_1^3\right|\right)^b} f(\xi-\xi_1,\tau-\tau_1) d\mu \\
&\lesssim \left\langle G_{1-b}, D_x^{\frac{1}{8}} F_b F_0 \right\rangle \\
&\lesssim \|G_{1-b}\|_{L_x^4 L_t^4} \left\| D_x^{\frac{1}{8}} F_b \right\|_{L_x^4 L_t^4} \|F_0\|_{L_x^2 L_t^2} \\
&\lesssim \|g\|_{L_\xi^2 L_\tau^2} \|f\|_{L_\xi^2 L_\tau^2} \left\| D_x^{\frac{1}{8}} F_b \right\|_{L_x^4 L_t^4} \\
&\lesssim \|v\|_{X_{s,b}} \left\| D_x^{\frac{1}{8}} F_b \right\|_{L_x^4 L_t^4}.
\end{aligned}$$

Let

$$\alpha_0 = \max\left\{\frac{s}{2}; \frac{3}{8}\right\} < \frac{1}{2},$$

and

$$\alpha_0 < \beta_0 < \frac{1}{2},$$

and

$$\gamma_0 = \frac{1}{2} - \beta_0.$$

Note that if $\rho = \beta_0$, the inequalities (6.9), (6.5) and (6.17) hold. Moreover, the terms

$$\left\| D_x^{\frac{\theta}{8}} F_p \right\|_{L_x^4 L_t^4} \left\| D_x^{\frac{1}{4}} F_p \right\|_{L_x^4 L_t^3}, \left\| D_x^{\theta} F_\rho \right\|_{L_x^{\frac{2}{1-\theta}} L_t^2} \quad \text{with} \quad 0 \leqslant \theta \leqslant 1$$

are bounded by

$$\int_{\mathbb{R}^2} \frac{|f(\xi,\tau)|^2}{\left(1+|\tau-\xi^3|^{2\gamma_0}\right)} d\xi d\tau$$
$$\leqslant \left(\int_{\mathbb{R}^2} |f(\xi,\tau)|^2 d\xi d\tau \right)^{\tilde{\theta}} \left(\int_{\mathbb{R}^2} \frac{|f(\xi,\tau)|^2}{(1+|\tau-\xi^3|)^{\frac{2\gamma_0}{1-\tilde{\theta}}}} d\xi d\tau \right)^{1-\tilde{\theta}},$$

with $0 \leqslant \tilde{\theta} \leqslant 1$. We fix $\tilde{\theta}$ so that $b = \frac{\gamma_0}{(1-\tilde{\theta})}$. Thus, the last factor is equal to

$$\left\| J^{-s} v \right\|_{L_x^2 L_t^2}^{2(1-\tilde{\theta})}, \quad \text{where} \quad \widehat{J^{-s} h}(\xi) = (1+|\xi|)^{-s} \widehat{h}(\xi).$$

$$\begin{aligned}
\left\| J^{-s} v \right\|_{L_x^2 L_t^2} &= \left\| J^{-s} v \right\|_{L_t^2 L_x^2} \\
&= \left\| W(t) J^{-s} v \right\|_{L_t^2 L_x^2} \\
&\leqslant \delta^{\frac{1}{4}} \left\| W(t) J^{-s} v \right\|_{L_x^2 L_t^4} \\
&\lesssim \delta^{\frac{1}{4}} \left\| D_t^{\frac{1}{4}} W(t) J^{-s} v \right\|_{L_x^2 L_t^2} \\
&= \delta^{\frac{1}{4}} \left(\int_{\mathbb{R}^2} |\tau|^{\frac{1}{2}} \left| \widehat{W(t) J^{-s}} v(\xi,\tau) \right|^2 d\xi d\tau \right)^{\frac{1}{2}}
\end{aligned}$$

$$= \delta^{\frac{1}{4}} \left(\int_{\mathbb{R}^2} |\tau|^{\frac{1}{2}} \left| (1+|\xi|)^{-s} \hat{v}\left(\xi, \tau - \xi^3\right) \right|^2 d\xi d\tau \right)^{\frac{1}{2}}$$

$$= \delta^{\frac{1}{4}} \left(\int_{\mathbb{R}^2} \left|\tau - \xi^3\right|^{\frac{1}{2}} (1+|\xi|)^{-2s} |\hat{v}(\xi,\tau)|^2 d\xi d\tau \right)^{\frac{1}{2}}$$

$$\lesssim \delta^{\frac{1}{4}} \|v\|_{X_{s,b}}.$$

Hence, we find

$$c\delta^{\frac{1-\tilde{\theta}}{4}} \|v\|_{X_{s,b}} = c\delta^{\frac{\gamma_0}{4b}} \|v\|_{X_{s,b}}$$

$$\lesssim \delta^{\frac{\gamma_0}{4b} + \frac{1-2b}{2}} \|u\|_{X_{s,b}}.$$

Using that $0 < \delta < 1$, we can conclude that

$$\|\omega\|_{X_{s,b-1}} \lesssim \delta^{1-2b+\frac{\gamma_0}{4b}} \|u\|^2_{X_{s,b}}.$$

□

6.2 Bilinear Estimates [2]

Theorem 6.1. *Given $s \in (-\frac{3}{4}, 0]$ there exists $b \in \left(\frac{1}{2}, 1\right)$ such that*

$$\left\| \frac{1}{2} \partial_x u^2 \right\|_{X_{s,b-1}} \lesssim \|u\|^2_{X_{s,b}}. \tag{6.21}$$

Proof. First we rewrite estimate (6.21) in an equivalent form. Defining

$$\rho = -s \in \left[0, \frac{3}{4}\right),$$

it follows from (6.3) that if

$$F \in X_{s,b} = X_{-\rho,b},$$

then

$$f(\xi,\tau) = \left(1 + \left|\tau - \xi^3\right|\right)^b (1+|\xi|)^{-\rho} \widehat{u}(\xi,\tau),$$

and

$$\|f\|_{L^2_\xi L^2_\tau} = \|u\|_{X_{s,b}} = \|u\|_{X_{-\rho,b}}.$$

Using that

$$\left(\widehat{\partial_x u^2}\right)(\xi,\tau) = c\xi(\widehat{u} * \widehat{u})(\xi,\tau),$$

we can rewrite (6.21) in terms of f as

$$\begin{aligned}
\|\partial_x u^2\|_{X_{s,b-1}} &= \left\| \left(1 + \left|\tau - \xi^3\right|\right)^{b-1} (1+|\xi|)^{-\rho} \widehat{\partial_x (u^2)} \right\|_{L^2_\xi L^2_\tau} \\
&= c \left\| \left(1 + \left|\rho - \xi^3\right|\right)^{b-1} (1+|\xi|)^{-\rho} \xi(\widehat{u} * \widehat{u}) \right\|_{L^2_\xi L^2_\tau} \\
&= c \| \frac{\xi}{(1+|\tau-\xi^3|)^{1-b}(1+|\xi|)^\rho} \int_{\mathbb{R}^2} \frac{f(\xi_1,\tau_1)(1+|\xi_1|)^\rho}{\left(1+\left|\tau_1 - \xi_1^3\right|\right)^b} \\
&\quad \times \frac{f(\xi-\xi_1,\tau-\tau_1)(1+|\xi-\xi_1|)^\rho \, d\xi_1 d\tau_1}{\left(1+\left|\tau-\tau_1-(\xi-\xi_1)^3\right|\right)^b} \|_{L^2_\xi L^2_\tau} \\
&\le c\|f\|^2_{L^2_\xi L^2_\tau} \\
&= c\|u\|^2_{X_{s,b}}.
\end{aligned} \tag{6.22}$$

Thus, Theorem 6.1 can be restated as follows.

Theorem 6.2. *Given $\rho = -s \in [0, \frac{3}{4})$ there exists $b \in (\frac{1}{2}, 1)$ such that*

$$I_1 \lesssim \|f\|^2_{L^2_\xi L^2_\tau}, \tag{6.23}$$

where

$$\begin{aligned}
I_1 = &\left\| \frac{\xi}{(1+|\tau-\xi^3|)^{1-b}(1+|\xi|)^\rho} \int_{\mathbb{R}^2} \frac{f(\xi_1,\tau_1)(1+|\xi_1|)^\rho}{\left(1+\left|\tau_1 - \xi_1^3\right|\right)^b} \right. \\
&\left. \times \frac{f(\xi-\xi_1,\tau-\tau_1)(1+|\xi-\xi_1|)^\rho}{\left(1+\left|\tau-\tau_1-(\xi-\xi_1)^3\right|\right)^b} d\xi_1 d\tau_1 \right\|_{L^2_\xi L^2_\tau}.
\end{aligned}$$

We restrict ourselves to the most interesting cases of (6.23) $\rho = 0$, which provides a simplified proof of the L^2-result in [4] and $\rho \in (\frac{1}{2}, \frac{3}{4})$. In fact, we prove the following slightly stronger version of Theorem 6.2.

Theorem 6.3. *Given $\rho = -s \in (\frac{1}{2}, \frac{3}{4})$ there exists $b \in (\frac{1}{2}, 1)$ such that for any $b' \in (\frac{1}{2}, b]$ with $b - b' \leq \min\{\rho - \frac{1}{2}; \frac{1}{4} - \frac{\rho}{3}\}$ it follows that*

$$I_1 \lesssim \|f\|^2_{L^2_\xi L^2_\tau}, \tag{6.24}$$

where

$$I_1 = \left\| \frac{\xi}{\left(1 + |\tau - \xi^3|\right)^{1-b} (1 + |\xi|)^\rho} \int_{\mathbb{R}^2} \frac{f(\xi_1, \tau_1)(1 + |\xi_1|)^\rho}{\left(1 + \left|\tau_1 - \xi_1^3\right|\right)^b} \right.$$
$$\left. \times \frac{f(\xi - \xi_1, \tau - \tau_1)(1 + |\xi - \xi_1|)^\rho}{\left(1 + \left|\tau - \tau_1 - (\xi - \xi_1)^3\right|\right)^b} d\xi_1 d\tau_1 \right\|_{L^2_\xi L^2_\tau},$$

and the constant c depends on ρ, b and $b - b'$. Moreover, (6.24) *still holds for $\rho = 0, b \in (\frac{1}{2}, \frac{3}{4}]$ and $b' \in (\frac{1}{2}, b]$.*

We observe that (6.23) follows from (6.24) by taking $b' = b$. The following elementary calculus inequalities provide the main tool in the proof of Theorem 6.3.

Lemma 6.11. *If $\ell > 1/2$, then there exists $c > 0$ such that*

$$\int_{\mathbb{R}} \frac{dx}{(1 + |x - \alpha|)^{2\ell}(1 + |x - \beta|)^{2\ell}} \lesssim \frac{1}{(1 + |\alpha - \beta|)^{2\ell}}, \tag{6.25}$$

$$\int_{\mathbb{R}} \frac{dx}{(1 + |x|)^{2\ell}|\sqrt{\alpha - x}|} \lesssim \frac{1}{(1 + |\alpha|)^{1/2}}, \tag{6.26}$$

$$\int_{\mathbb{R}} \frac{dx}{(1 + |x - \alpha|)^{2(1-\ell)}(1 + |x - \beta|)^{2\ell}} \lesssim \frac{1}{(1 + |\alpha - \beta|)^{2(1-\ell)}}, \tag{6.27}$$

$$\int_{|x| \leq \beta} \frac{dx}{(1 + |x|)^{2(1-\ell)}|\sqrt{\alpha - x}|} \lesssim \frac{(1 + \beta)^{2(\ell - 1/2)}}{(1 + |\alpha|)^{1/2}}. \tag{6.28}$$

Next, using these calculus inequalities, we deduce three lemmas from which the proof of Theorem 6.3 follows directly.

Lemma 6.12. *If $b \in (\frac{1}{2}, \frac{3}{4}]$ and $b' \in (\frac{1}{2}, b]$, then there exists $c > 0$ such that*

$$\frac{|\xi|}{\left(1+|\tau-\xi^3|\right)^{1-b}} \times \left(\int_{\mathbb{R}^2} \frac{d\tau_1 d\xi_1}{\left(1+\left|\tau_1-\xi_1^3\right|\right)^{2b'}\left(1+\left|\tau-\tau_1-(\xi-\xi_1)^3\right|\right)^{2b'}}\right)^{\frac{1}{2}} \lesssim 1. \tag{6.29}$$

Proof. Since $b' > \frac{1}{2}$, from (6.25) it follows that

$$\int_{\mathbb{R}} \frac{d\tau_1}{\left(1+\left|\tau_1-\xi_1^3\right|\right)^{2b'}\left(1+\left|\tau-\tau_1-(\xi-\xi_1)^3\right|\right)^{2b'}} \lesssim \frac{1}{\left(1+|\tau-\xi^3+3\xi\xi_1(\xi-\xi_1)|^{2b'}\right)}.$$

To integrate with respect to ξ_1, we change variables

$$\mu = \lambda - \xi^3 + 3\xi\xi_1(\xi-\xi_1), \text{ then } d\mu = 3\xi(\xi-2\xi_1)\,d\xi_1, \tag{6.30}$$

and

$$\xi_1 = \frac{1}{2}\left\{\xi \pm \sqrt{\frac{4\tau-\xi^3-4\mu}{3\xi}}\right\}.$$

Therefore,

$$|\xi(\xi-2\xi_1)| = c\sqrt{|\xi|}\left|\sqrt{4\tau-\xi^3-4\mu}\right|, \tag{6.31}$$

and

$$d\xi_1 = c\frac{d\mu}{\sqrt{|\xi|}\,|\,\sqrt{4\tau-\xi^3-4\mu}\,|}. \tag{6.32}$$

Now, combining these identities with (6.26), we find

$$\begin{aligned}
&\int_{\mathbb{R}} \frac{d\xi_1}{\left(1+|\tau-\xi^3+3\xi\xi_1(\xi-\xi_1)|\right)^{2b'}} \\
&\quad \lesssim \frac{1}{\sqrt{|\xi|}} \int_{\mathbb{R}} \frac{d\mu}{(1+|\mu|)^{2b'} \mid \sqrt{4\tau-\xi^3-4\mu} \mid} \\
&\quad \lesssim \frac{1}{|\xi|^{\frac{1}{2}}\left(1+|4\tau-\xi^3|\right)^{\frac{1}{2}}}.
\end{aligned} \tag{6.33}$$

Thus, the term in (6.27) is bounded by

$$\frac{c|\xi|^{\frac{3}{4}}}{\left(1+|\tau-\xi^3|\right)^{1-b}} \frac{1}{\left(1+|4\tau-\xi^3|\right)^{\frac{1}{4}}}. \tag{6.34}$$

It is easy to check that the expression in (6.32) is bounded whenever $b \leq \frac{3}{4}$. □

Lemma 6.13. *If*

$$\rho=-s \in\left(\frac{1}{2}, \frac{3}{4}\right), \quad b \in\left(\frac{1}{2}, \frac{3}{4}+\frac{s}{3}\right]$$

and

$$b' \in\left(\frac{1}{2}, b\right],$$

then there exists $c>0$ *such that*

$$\frac{|\xi|}{\left(1+|\tau-\xi^3|\right)^{1-b}(1+|\xi|)^\rho} \times I_2 \lesssim 1, \tag{6.35}$$

where

$$I_2=\left(\int_A \frac{|\xi_1(\xi-\xi_1)|^{2\rho}}{\left(1+|\tau_1-\xi_1^3|\right)^{2b'}\left(1+\left|\tau-\tau_1-(\xi-\xi_1)^3\right|\right)^{2b'}} d\tau_1 d\xi_1\right)^{1/2},$$

and $A=A(\xi,\tau)$ *is defined as*

$$\begin{aligned}
A=\Big\{&(\xi_1,\tau_1) \in \mathbb{R}^2 : |\xi_1| \geq 1, |\xi-\xi_1| \geq 1, \left|\tau-\tau_1-(\xi-\xi_1)^3\right| \\
&\leq \left|\tau_1-\xi_1^3\right| \leq \left|\tau-\xi^3\right|\Big\}.
\end{aligned} \tag{6.36}$$

Proof. First, we observe that in A

$$\begin{aligned} \left|\tau - \xi^3 + 3\xi\xi_1(\xi - \xi_1)\right| &= \left|\tau_1 - \xi_1 + \tau - \tau_1 - (\xi - \xi_1)^3\right| \\ &\leq 2|\tau - \xi|^3. \end{aligned} \tag{6.37}$$

Thus, by (6.26),

$$\begin{aligned} &\int_{\mathbb{R}} \frac{d\tau_1}{\left(1 + \left|\tau_1 - \xi_1^3\right|\right)^{2b'}\left(1 + \left|\tau - \tau_1 - (\xi - \xi_1)^3\right|\right)^{2b'}} \\ &\quad \leq \frac{\chi\left(\left(\tau - \xi^3 + 3\xi\xi_1(\xi - \xi_1)\right)/2\left(\tau - \xi^3\right)\right)}{\left(1 + \left|\tau - \xi^3 + 3\xi\xi_1(\xi - \xi_1)\right|\right)^{2b'}}, \end{aligned} \tag{6.38}$$

where $\chi(\cdot)$ denotes the characteristic function of the interval $[-1, 1]$. As in the previous lemma, we use the change of variable

$$\mu = \tau - \xi^3 + 3\xi\xi_1(\xi - \xi_1).$$

Thus, (6.30), (6.32) and (6.38) lead to the inequalities

$$\begin{aligned} I_2^2 &\lesssim \int \frac{|\xi_1(\xi - \xi_1)|^{2\rho}\chi\left(\left(\tau - \xi^3 + 3\xi\xi_1(\xi - \xi_1)\right)/2\left(\tau - \xi^3\right)\right)}{\left(1 + \left|\tau - \xi^3 + 3\xi\xi_1(\xi - \xi_1)\right|\right)^{2b'}} d\xi_1 \\ &\lesssim \int_{|\mu| \leq 2|\tau - \xi^3|} \frac{\left|\tau - \xi^3 - \mu\right|^{2\rho}}{|\xi|^{2\rho + 1/2}(1 + |\mu|)^{2b'} \left| \sqrt{4\tau - \xi^3 - 4\mu} \right|} d\mu \\ &\lesssim \frac{1}{|\xi|^{2\rho + \frac{1}{2}}} \frac{\left|\tau - \xi^3\right|^{2\rho}}{\left(1 + |4\tau - \xi^3|\right)^{\frac{1}{2}}}. \end{aligned} \tag{6.39}$$

Hence, the expression

$$\phi(\xi, \tau) = \frac{c|\xi|^{\frac{3}{4} - \rho}}{(1 + |\xi|)^{\rho}} \frac{\left(1 + \left|\tau - \xi^3\right|\right)^{\rho + b - 1}}{\left(1 + |4\tau - \xi^3|\right)^{\frac{1}{4}}}$$

bounds the term in (6.35). From the assumptions on $\rho = -s$ and b, one has that

$$\phi \in L^{\infty}\left(\mathbb{R}^2\right),$$

which completes the proof. $\square$

Lemma 6.14. *If $\rho = -s \in \left(\frac{1}{2}, \frac{3}{4}\right), b \in \left(\frac{1}{2}, 1\right)$ and $b' \in \left(\frac{1}{2}, b\right]$ with*

$$b - b' \leq \min\left\{\rho - \frac{1}{2}; \frac{1}{4} - \frac{\rho}{3}\right\},$$

then there exists $c > 0$ such that

$$\frac{1}{\left(1 + \left|\tau_1 - \xi_1^3\right|\right)^{b'}} \times I_3 \lesssim 1, \tag{6.40}$$

where

$$I_3 = \left(\int_B \frac{|\xi|^{2(1-\rho)} \left|\xi\xi_1\left(\xi - \xi_1\right)\right|^{2\rho} d\tau d\xi}{(1 + |\xi|)^{2\rho} \left(1 + \left|\tau - \xi^3\right|\right)^{2(1-b)} \left(1 + \left|\tau - \tau_1 - (\xi - \xi_1)^3\right|\right)^{2b'}}\right)^{\frac{1}{2}}.$$

Let

$$B = B\left(\xi_1, \tau_1\right)$$

be defined as

$$\begin{aligned} B = \Big\{ (\xi, \tau) \in \mathbb{R}^2 : |\xi - \xi_1| \geq 1, |\xi_1| \geq 1, \left|\tau - \tau_1 - (\xi - \xi_1)^3\right| \\ \leq \left|\tau_1 - \xi_1^3\right|, \left|\tau - \xi^3\right| \leq \left|\tau_1 - \xi_1^3\right| \Big\}. \end{aligned}$$

Proof. First, we observe that in B

$$\begin{aligned} \left|\tau_1 - \xi_1^3 + 3\xi\xi_1\left(\xi - \xi_1\right)\right| &= \left|\tau - \xi^3 - \left(\tau - \tau_1 - (\xi - \xi_1)^3\right)\right| \\ &\leq 2\left|\tau_1 - \xi_1^3\right|. \end{aligned}$$

By (6.27), it follows that

$$\begin{aligned} &\int_{\mathbb{R}} \frac{d\tau}{\left(1 + \left|\tau - \xi^3\right|\right)^{2(1-b)} \left(1 + \left|\tau - \tau_1 - (\xi - \xi_1)^3\right|\right)^{2b'}} \\ &\quad \lesssim \frac{1}{\left(1 + \left|\tau_1 - \xi_1^3 + 3\xi\xi_1\left(\xi - \xi_1\right)\right|\right)^{2(1-b)}}. \end{aligned}$$

Thus, to obtain (6.40), it suffices to bound

$$
\begin{aligned}
I(D) = & \frac{1}{\left(1+\left|\tau_1-\xi_1^3\right|\right)^{b'}} \\
& \times\left(\int_D \frac{|\xi|^{2(1-\rho)}\left|\xi \xi_1\left(\xi-\xi_1\right)\right|^{2 \rho}}{(1+|\xi|)^{2 \rho}\left(1+\left|\tau_1-\xi_1^3+3 \xi \xi_1\left(\xi-\xi_1\right)\right|\right)^{2(1-b)}} d \xi\right)^{\frac{1}{2}},
\end{aligned}
$$

with $D=B'=B'\left(\xi_1, \tau_1\right)$, where

$$
\begin{aligned}
B'=\{\xi \in \mathbb{R}: 1 \leq\left|\xi-\xi_1\right|, 1 \leq\left|\xi_1\right|,\left|\tau-\xi^3\right| \leq\left|\tau_1-\xi_1^3\right|, \\
\left|\tau_1-\xi_1^3+3 \xi \xi_1\left(\xi-\xi_1\right)\right| \leq 2\left|\tau_1-\xi_1^3\right|\}.
\end{aligned}
$$

We split B' into two subdomains B_1' and B_2', where

$$
B_1'=\left\{\xi \in B': 3\left|\xi \xi_1\left(\xi-\xi_1\right)\right| \leq \frac{1}{2}\left|\tau_1-\xi_1^3\right|\right\}
$$

and

$$
B_2'=\left\{\xi \in B': \frac{1}{2}\left|\tau_1-\xi_1^3\right| \leq 3\left|\xi \xi_1\left(\xi-\xi_1\right)\right| \leq 3\left|\tau_1-\xi_1^3\right|\right\}.
$$

In B_1', we have that

$$
\frac{1}{2}\left|\tau_1-\xi_1^3\right| \leq\left|\tau_1-\xi_1^3+3 \xi \xi_1\left(\xi-\xi_1\right)\right|
$$

and

$$
|\xi| \leq \frac{1}{6}\left|\tau_1-\xi_1^3\right| .
$$

Hence,

$$
\begin{aligned}
I\left(B_1'\right) & \lesssim \frac{1}{\left(1+\left|\tau_1-\xi_1^3\right|\right)^{1-\rho+b'-b}}\left(\int_{|\xi| \leq\left|\tau_1-\xi_1^3\right|} \frac{|\xi|^{2(1-\rho)}}{(1+|\xi|)^{2 \rho}} d \xi\right)^{\frac{1}{2}} \\
& \lesssim\left(1+\left|\tau_1-\xi_1^3\right|\right)^{\frac{1}{2}-\rho-b'+b},
\end{aligned}
$$

which is bounded since, by hypothesis,

$$
\frac{1}{2}-\rho-b'+b \leq 0 .
$$

To bound $I(B_2')$, we split B_2' into three parts, $B_{2,1}', B_{2,2}'$ and $B_{2,3}'$. First, we have

$$B_{2,1}' = \left\{ \xi \in B_2' : \frac{1}{4}|\xi| \le |\xi_1| \le 100|\xi| \right\}.$$

In this domain, one has

$$c\left(1 + \left|\tau_1 - \xi_1^3\right|\right) \le |\xi_1|^3 \sim |\xi|^3.$$

Combining B_2', the hypothesis $\rho > \frac{1}{2}$, a change of variable similar to that in (6.30), (6.32),

$$\mu_1 = \tau_1 - \xi_1^3 + 3\xi\xi_1(\xi - \xi_1),$$

and (6.26), it follows that

$$\begin{aligned} I\left(B_{2,1}'\right) &\lesssim \left(1 + \left|\tau_1 - \xi_1^3\right|\right)^{\rho - b' + \frac{(1-2s)}{3}} \\ &\quad \times \left(\int_{|\mu_1| \le 2|\tau_1 - \xi_1^3|} \frac{d\xi}{\left(1 + \left|\tau_1 - \xi_1^3 + 3\xi\xi_1(\xi - \xi_1)\right|\right)^{2(1-b)}} \right)^{\frac{1}{2}} \\ &\lesssim \left(1 + \left|\tau_1 - \xi_1^3\right|\right)^{\frac{\rho}{3} - b' + \frac{1}{3}} \\ &\quad \times \left(\frac{1}{|\xi|^{1/4}} \int_{|\mu_1| < 2|\tau_1 - \xi_1^3|} \frac{1}{(1 + |\mu_1|)^{2(1-b)}} \frac{d\mu_1}{\left|\sqrt{4\tau_1 - \xi_1^3 - 4\mu_1}\right|} \right)^{\frac{1}{2}} \\ &\lesssim \frac{\left(1 + \left|\tau_1 - \xi_1^3\right|\right)^{\frac{\rho}{3} - \frac{1}{4} + b - b'}}{\left(1 + \left|4\tau_1 - \xi_1^3\right|\right)^{\frac{1}{4}}}, \end{aligned}$$

which is bounded since, by hypothesis,

$$\frac{\rho}{3} - \frac{1}{4} + b - b' \le 0.$$

Next, we consider

$$B_{2,2}' = \left\{ \xi \in B_2' : 1 \le |\xi_1| \le \frac{|\xi|}{4} \right\}.$$

In this domain, one has that

$$|\xi_1 - 2\xi| \sim |\xi| \sim |\xi - \xi_1|$$

and

$$c^{-1}\left(1 + \left|\tau_1 - \xi_1^3\right|\right) \leq |\xi\xi_1(\xi - \xi_1)| \sim |\xi|^2 |\xi_1| \lesssim |\xi|^3.$$

Therefore, the change of variable

$$\mu_1 = \lambda_1 - \xi_1^3 + 3\xi\xi_1(\xi - \xi_1)$$

and (6.28) give the inequalities

$$\begin{aligned} I\left(B'_{2,2}\right) &\lesssim \frac{1}{\left(1 + \left|\tau_1 - \xi_1^3\right|\right)^{b'}} \\ &\quad \times \left(\int_{|\mu_1| \leq 3|\tau_1 - \xi_1^3|} \frac{|\xi|^{2(1-2\rho)} |\xi\xi_1(\xi - \xi_1)|^{2\rho}}{|\xi_1|^{\frac{1}{2}} \left(1 + \left|\tau_1 - \xi_1^3\right|\right)^{\frac{1}{2}} (1 + |\mu_1|)^{2(1-b)}} d\mu_1\right)^{\frac{1}{2}} \\ &\lesssim \left(1 + \left|\tau_1 - \xi_1^3\right|\right)^{\theta}, \end{aligned}$$

where

$$\theta = -b' + \frac{1}{4} + \frac{1 - 2\rho}{3} + \rho + b - \frac{1}{2} = \frac{\rho}{3} - \frac{5}{12} + b - b' \leq 0,$$

by using our hypothesis on ρ, b and b'. Hence, $I\left(B'_{2,2}\right)$ is bounded. Finally, we have

$$B'_{2,3} = \left\{\xi \in B'_2 : 100|\xi| \leq |\xi_1|\right\}.$$

In this case, one has that

$$|\xi_1 - 2\xi| \sim |\xi_1 - \xi| \sim |\xi_1|, \quad \left|\tau_1 - \xi_1^3\right| << |\xi_1|^3$$

and

$$\left|4\tau_1 - \xi_1^3\right| \sim \left|\xi_1^3\right|.$$

Thus, a familiar argument shows that

$$\begin{aligned} I\left(B'_{2,3}\right) &\lesssim \left(1 + \left|\tau_1 - \xi_1^3\right|\right)^{\rho - b'} \\ &\quad \times \left(\frac{1}{|\xi_1|^{1/2}} \int_{|\mu_1| \leq 3|\tau_1 - \xi_1^3|} \frac{1}{(1 + |\mu_1|)^{2(1-b)}} \frac{d\mu_1}{\left|\sqrt{4\tau_1 - \xi_1^3 - 4\mu_1}\right|}\right)^{\frac{1}{2}} \\ &\lesssim \left(1 + \left|\tau_1 - \xi_1^3\right|\right)^{\rho - \frac{5}{6} - b' + b}, \end{aligned}$$

which is bounded. This completes the proof of (6.40). □

Proof of Theorem 6.3. First, we consider the case $s = 0$. Using Cauchy–Schwarz inequality, (6.29) and Fubini's theorem, it follows that

$$\begin{aligned}
&\left\| \frac{\xi}{(1+|\tau-\xi^3|)^{1-b}} \int_{\mathbb{R}^2} \frac{f(\xi_1,\tau_1)}{(1+|\tau_1-\xi_1^3|)^{b'}} \frac{f(\xi-\xi_1,\tau-\tau_1)\,d\xi_1 d\tau_1}{\left(1+\left|\tau-\tau_1-(\xi-\xi_1)^3\right|\right)^{b'}} \right\|_{L^2_\xi L^2_\tau} \\
&\quad \le \left\| \frac{\xi}{(1+|\tau-\xi^3|)^{1-b}} \right. \\
&\qquad \times \left. \left(\int_{\mathbb{R}^2} \frac{d\xi_1 d\tau_1}{(1+|\tau_1-\xi_1^3|)^{2b'} \left(1+\left|\tau-\tau_1-(\xi-\xi_1)^3\right|\right)^{2b'}} \right)^{\frac{1}{2}} \right\|_{L^\infty_\xi L^\infty_\tau} \\
&\qquad \left\| \left(\int_{\mathbb{R}^2} |f(\xi_1,\tau_1)|^2 \, |f(\xi-\xi_1,\tau-\tau_1)|^2 \, d\xi_1 d\tau_1 \right)^{\frac{1}{2}} \right\|_{L^2_\xi L^2_\tau} \\
&\quad \lesssim \|f\|^2_{L^2_\xi L^2_\tau},
\end{aligned} \tag{6.41}$$

for any $b' \in (\frac{1}{2}, b]$. Taking $b' = b$, we obtain the result. Next, we turn to the case

$$\rho = -s \in \left(\frac{1}{2}, \frac{3}{4}\right).$$

Note that if

$$\text{either} \quad |\xi_1| \le 1 \quad \text{or} \quad |\xi-\xi_1| \le 1,$$

we have

$$(1+|\xi_1|)^\rho (1+|\xi-\xi_1|)^\rho \lesssim (1+|\xi|)^\rho,$$

which reduces the estimate to the case $\rho = 0$. Thus, we can assume

$$|\xi_1| \ge 1 \quad \text{and} \quad |\xi-\xi_1| \ge 1.$$

By symmetry, we can also assume

$$\left|\tau-\tau_1-(\xi-\xi_1)^3\right| \le \left|\tau_1-\xi_1^3\right|.$$

We split the region of integration in (2.7) into two parts:

$$|\tau_1 - \xi_1^3| \leq |\tau - \xi^3| \quad \text{and} \quad |\tau - \xi^3| \leq |\tau_1 - \xi_1^3|.$$

For the first part, we use Cauchy–Schwarz inequality and (6.36) as in (6.41) to obtain the same bound. For the second part, we use duality, Cauchy–Schwarz inequality and (6.40) to conclude the proof. □

We observe that Theorem 6.3 is equivalent to the following result.

Corollary 6.1. *Given $s \in (-\frac{3}{4}, -\frac{1}{2})$, there exists $b \in (\frac{1}{2}, 1)$ such that for any $b' \in (\frac{1}{2}, b]$ with*

$$b - b' \leq \min\left\{ -s - \frac{1}{2}; \frac{1}{4} + \frac{s}{3} \right\},$$

we have

$$\left\| \frac{1}{2}\partial_x u^2 \right\|_{X_{s,b-1}} \lesssim \|u\|^2_{X_{s,b'}}. \tag{6.42}$$

Moreover, (6.42) still holds for $s = 0, b \in (\frac{1}{2}, \frac{3}{4}]$ and $b' \in (\frac{1}{2}, b]$.

Chapter 7

Fifth-Order Shallow Water Equation

The shallow water equations describes the flow below a pressure surface in a fluid. They are PDEs of hyperbolic type (or parabolic if we consider viscous shear). The models are widely used to calculate problems of both academic and practical interest. They include modeling of currents in relatively shallow reservoirs, rivers, reservoirs, currents near the coasts of seas and oceans, calculation of tsunami waves and water discharge near hydroelectric power stations, as well as many other tasks directly related to environmental problems. The importance of this topic is confirmed by a large number of scientific articles and conferences devoted to the theory and applications of shallow water equations. By their nature, the shallow water equations are closely related to the equations of gas dynamics and can be derived from these equations, written in the Euler approximation, using a number of special assumptions.

7.1 Kawahara Equation with Data in Analytic Gevrey Spaces

In this chapter studies the problems

$$\partial_t u + \alpha\partial_x^5 u + \beta\partial_x^3 u + \gamma\partial_x u + \mu\partial_x(u^2) = 0, \quad (x, 0) = u_0(x), \quad (7.1)$$

for $x \in \mathbb{R}$ and $t \geq 0$ the parameters $\alpha \neq 0$, β, γ and μ are real numbers.

The model $(7.1)_1$ is also called the fifth order shallow water equation and is also called the modified Kawahara equation for the nonlinear term $\partial_x(u^3)$. It arises in study of the water waves with surface tension, in which the Bond number takes on the critical value, where the Bond number represents a dimensionless magnitude of surface tension in the shallow water regime.

First, by using linear and bilinear estimates in analytic Gevrey Bourgain spaces $X_{\sigma,\delta,s,b}(\mathbb{R}^2)$ and analytic Gevrey spaces $G^{\sigma,\delta,s}(\mathbb{R})$, the local well-posedness of the Cauchy problem for the Kawahara equation on the line is established for analytic initial data $u_0(x)$ that can be extended as holomorphic functions in a strip around the x-axis. Next we use this local result and a Gevrey approximate conservation law to prove that global solutions exist. Furthermore, we obtain explicit lower bounds for the radius of spatial analyticity $r(t)$ given by $r(t) \geqslant ct^{-1}$. Also, Gevrey regularity of the solution in time variable is provided.

We consider a space that is a hybrid between the analytic space and a space of the analytic Bourgain. More precisely, for $s, b \in \mathbb{R}$ and $\delta > 0$ define $X_{\delta,s,b}(\mathbb{R}^2)$ to be the Banach space.

We consider also a space that is a hybrid between the analytic Gevrey space and a space of the analytic Gevrey Bourgain. For $\sigma \geq 1$ define $X_{\sigma,\delta,s,b}(\mathbb{R}^2)$ with respect to the norm

$$\|u\|^2_{X_{\sigma,\delta,s,b}(\mathbb{R}^2)} = \int_{\mathbb{R}^2} e^{2\delta|\xi|^{\frac{1}{\sigma}}}(1+|\xi|)^{2s}(1+|\tau+\phi(\xi)|)^{2b}|\widehat{u}(\xi,\tau)|^2 d\xi d\tau. \tag{7.2}$$

The next lemma shows that $X_{\sigma,\delta,s,b}(\mathbb{R}^2)$ is continuously embedded in $C\left([0,T], G^{\sigma,\delta,s}(\mathbb{R})\right)$, provided $b > 1/2$. For $\delta = 0$ the proof can be found easily.

Lemma 7.1. *Let $b > \frac{1}{2}$, $s \in \mathbb{R}$, $\sigma \geqslant 1$ and $\delta > 0$, Then, for all $u \in X_{\sigma,\delta,s,b}(\mathbb{R}^2)$ and some constant $C_0 > 0$ we have*

$$|u|_{T,\sigma,\delta,s} = \sup_{t\in[0,T]} \|u(\cdot,t)\|_{G^{\sigma,\delta,s}(\mathbb{R})} \leqslant C_0\|u\|_{X_{\sigma,\delta,s,b}(\mathbb{R}^2)}.$$

Proof. First, we observe that the operator $A^{\sigma,\delta}$ defined by

$$\widehat{A^{\sigma,\delta}u}^{x}(\xi,t) = e^{\delta|\xi|^{\frac{1}{\sigma}}}\widehat{u}^x(\xi,t), \tag{7.3}$$

satisfies

$$\|u\|_{X_{\sigma,\delta,s,b}(\mathbb{R}^2)} = \|A^{\sigma,\delta}u\|_{X_{s,b}(\mathbb{R}^2)} \quad \text{and} \quad \|u\|_{G^{\sigma,\delta,s}(\mathbb{R})} = \|A^{\sigma,\delta}u\|_{H^s(\mathbb{R})}. \tag{7.4}$$

We observe that

$$A^{\sigma,\delta}u \in C\left([0,T], H^s(\mathbb{R})\right),$$

and for some $C_0 > 0$, we have

$$\|A^{\sigma,\delta}u\|_{C([0,T],H^s(\mathbb{R}))} \le C_0\|A^{\sigma,\delta}u\|_{X_{s,b}(\mathbb{R}^2)}. \tag{7.5}$$

Thus, it follows that

$$u \in C\left([0,T], G^{\sigma,\delta,s}\right),$$

and

$$\|u\|_{C\left([0,T],G^{\sigma,\delta,s}(\mathbb{R})\right)} \le C_0\|u\|_{X_{\sigma,\delta,s,b}(\mathbb{R}^2)}. \tag{7.6}$$

□

Remark 7.1. This will be of importance due to the fact that we will show that, given an initial data

$$u_0(x) \in G^{\sigma,\delta,s}(\mathbb{R}),$$

there is a unique solution

$$u \in X_{\sigma,\delta,s,b}(\mathbb{R}^2),$$

to the Cauchy problem (7.1), for a certain $b > \frac{1}{2}$ and therefore there is a solution to the Cauchy problem (7.1),

$$u \in C\left([0,T], G^{\sigma,\delta,s}(\mathbb{R})\right).$$

Now, we consider the nonlinear Cauchy problem

$$\begin{aligned} \partial_t u + \alpha\partial_x^5 u + \beta\partial_x^3 u + \gamma\partial_x u &= F(x,t), \\ u(x,0) &= u_0(x). \end{aligned} \tag{7.7}$$

By using Duhamel's formula (Taking Fourier transform with respect to x in (7.7), solving the resulting differential equation in t and using inverse Fourier transform reduces the Cauchy problem (7.7) to the following integral equation) we may write the solution

$$u(x,t) = S(t)u_0(x) - \int_0^t S(t-s)F(x,s)ds,$$

where the unit operator related to the corresponding linear equation is

$$S(t) = \mathscr{F}_x^{-1} e^{-it(\alpha\xi^5 - \beta\xi^3 + \gamma\xi)} \mathscr{F}_x,$$

here, the nonlinear terms F is given by $\mu\partial_x(u^2)$.

Next, we localize it in time variable by using a cut-off function $\psi \in C_0^\infty(\mathbb{R})$, with $\psi = 1$ in $[-\frac{1}{2}, \frac{1}{2}]$, supp$\psi \subset [-1,1]$ and define $\psi_T(t) = \psi(\frac{t}{T})$. We consider the operator Φu given by

$$\Phi(u)(x,t) = \psi_1(t)S(t)u_0(x) - \psi_1(t)\int_0^t S(t-s)F(x,s)ds. \quad (7.8)$$

Our goal is to solve the equation $\Phi(u) = u$. We estimate now the fist part in the right hand side of (7.8).

Lemma 7.2. *Let $s \in \mathbb{R}$ and $\frac{1}{2} < b < 1$. Then*

$$\|\psi_1(t)S(t)u_0\|_{X_{s,b}(\mathbb{R}^2)} \le C\ \|u_0\|_{H^s(\mathbb{R})}. \quad (7.9)$$

Lemma 7.3. *Let $s \in \mathbb{R}$, $\frac{1}{2} < b < 1$, $\delta > 0$ and $\sigma \geqslant 1$. Then*

$$\|\psi_1(t)S(t)u_0\|_{X_{\sigma,\delta,s,b}(\mathbb{R}^2)} \le C\ \|u_0\|_{G^{\sigma,\delta,s}(\mathbb{R})}. \quad (7.10)$$

Proof. We observe, by considering the operator $A^{\sigma,\delta}$ in (7.3), that

$$\begin{aligned}
\|\psi_1(t)S(t)u_0\|^2_{X_{\sigma,\delta,s,b}(\mathbb{R}^2)} &= C\int_{\mathbb{R}^2} e^{2\delta|\xi|^{\frac{1}{\sigma}}}(1+|\xi|)^{2s}(1+|\tau+\phi(\xi)|)^{2b} \\
&\quad \cdot |\widehat{\psi}(\tau+\phi(\xi))|^2 |\widehat{u_0}(\xi)|^2 d\xi d\tau \\
&= C\int_{\mathbb{R}^2} e^{2\delta|\xi|^{\frac{1}{\sigma}}}(1+|\xi|)^{2s}(1+|\tau+\phi(\xi)|)^{2b} \\
&\quad \cdot |\widehat{\psi}(\tau+\phi(\xi))|^2 |e^{\delta|\xi|^{\frac{1}{\sigma}}}\widehat{u_0}(\xi)|^2 d\xi d\tau \\
&= C\int_{\mathbb{R}^2} (1+|\xi|)^{2s}(1+|\tau+\phi(\xi)|)^{2b}
\end{aligned}$$

$$\cdot |\widehat{\psi}(\tau + \phi(\xi))|^2 |\widehat{A^{\sigma,\delta} u_0}(\xi)|^2 d\xi d\tau$$
$$= \|\psi_1(t) S(t)(A^{\sigma,\delta} u_0)\|^2_{X_{s,b}(\mathbb{R}^2)}.$$

Now, by using Lemma 7.2, there exists $C > 0$ such that

$$\|\psi_1(t) S(t)(A^{\sigma,\delta} u_0)\|_{X_{s,b}(\mathbb{R}^2)} \leq C\ \|A^{\sigma,\delta} u_0\|_{H^s(\mathbb{R})} = C\ \|u_0\|_{G^{\sigma,\delta,s}(\mathbb{R})}.$$

□

We will estimate the integral part of $\Phi(u)$.

Lemma 7.4. *Let $s \in \mathbb{R}$, $\frac{1}{2} < b < 1$, $0 < T \leq 1$. Then*

$$\left\| \psi_1(t) \int_0^t S(t-s) F(x,s) ds \right\|_{X_{s,b}(\mathbb{R}^2)} \leq C \|F\|_{X_{s,b-1}(\mathbb{R}^2)}. \tag{7.11}$$

Lemma 7.5. *Let $s \in \mathbb{R}$, $\frac{1}{2} < b < 1$, $\delta > 0$ and $\sigma \geqslant 1$, then for some constant $C > 0$, we have*

$$\left\| \psi_1(t) \int_0^t S(t-s) F(x,s) ds \right\|_{X_{\sigma,\delta,s,b}(\mathbb{R}^2)} \leq C \|F\|_{X_{\sigma,\delta,s,b-1}(\mathbb{R}^2)}. \tag{7.12}$$

Proof. Define

$$U = \psi_1(t) \int_0^t S(t-s) F(x,s) ds.$$

Let us consider the operator $A^{\sigma,\delta}$ given by (7.3), then we have

$$\begin{aligned}
\widehat{A^{\sigma,\delta} U}^x (\xi,t) &= \psi_1(t) \int_0^t \left(e^{-i(t-s)\phi(\xi)} \right) e^{\delta|\xi|^{\frac{1}{\sigma}}} \widehat{F}^x(\xi,s) ds, \\
&= \psi_1(t) \int_0^t \widehat{[S(t-s)(A^{\sigma,\delta} F)]}^x (\xi,s) ds.
\end{aligned}$$

Thus,

$$\begin{aligned}
\|U\|_{X_{\sigma,\delta,s,b}(\mathbb{R}^2)} &= \|A^{\sigma,\delta} U\|_{X_{s,b}(\mathbb{R}^2)} \\
&= \left\| \psi_1(t) \int_0^t S(t-s) A^{\sigma,\delta} F(x,s) ds \right\|_{X_{s,b}(\mathbb{R}^2)}.
\end{aligned}$$

Using Lemma 7.4, we have

$$\left\|\psi_1(t)\int_0^t S(t-s)A^{\sigma,\delta}F(x,s)ds\right\|_{X_{s,b}(\mathbb{R}^2)}$$
$$\leq C\|A^{\sigma,\delta}F\|_{X_{s,b-1}(\mathbb{R}^2)} = C\|F\|_{X_{\sigma,\delta,s,b-1}(\mathbb{R}^2)}$$ □

Lemma 7.6. *Let $s \in \mathbb{R}$, $\sigma \geqslant 1$ and $\delta \geq 0$. If $\frac{1}{2} < b \leq b' < 1$, then for any $T > 0$, we have,*

$$\|\psi_T(t)F\|_{X_{\sigma,\delta,s,b-1}(\mathbb{R}^2)} \leq CT^{b_1'-b_1}\|F\|_{X_{\sigma,\delta,s,b'-1}(\mathbb{R}^2)}, \tag{7.13}$$

where C depends only on b and b'.

Proof. The proof of the Lemma 7.6 for $\delta = 0$ can be easily found, for $\delta > 0$ as one merely has to replace F by $A^{\sigma,\delta}F$, where the operator define in (7.3). □

The next result provides the essential bilinear estimate.

Theorem 7.1. *If $s > -\frac{7}{4}$, let $b_1 > \frac{1}{2}$ be close enough to $\frac{1}{2}$ and $b_1' > \frac{1}{2}$. Then*

$$\|\partial_x(u_1u_2)\|_{X_{s,b_1-1}(\mathbb{R}^2)} \leqslant C\|u_1\|_{X_{s,b_1'}(\mathbb{R}^2)}\|u_2\|_{X_{s,b_1'}(\mathbb{R}^2)}. \tag{7.14}$$

Remark 7.2. Setting

$$f_i(\xi,\tau) = (1+|\xi|)^s(1+|\tau+\phi(\xi)|)^{b_1'}\widehat{u_i}(\xi,\tau), \quad i = 1,2,$$

the estimate of Corollary 7.1 can be rewritten as

$$\|\partial_x(u_1u_2)\|_{X_{s,b_1-1}(\mathbb{R}^2)}$$
$$= \left\|\frac{\xi(1+|\xi|)^s}{(1+|\tau+\phi(\xi)|)^{b_1-1}}\widehat{u_1u_2}(\xi,\tau)\right\|_{L_{\xi,\tau}(\mathbb{R}^2)}$$
$$= C\left\|\frac{\xi(1+|\xi|)^s}{(1+|\tau+\phi(\xi)|)^{b_1-1}}\right.$$
$$\left.\times\int_{\mathbb{R}^2}\widehat{u_1}(\xi_1,\tau_1)\widehat{u_2}(\xi-\xi_1,\tau-\tau_1)d\xi_1d\tau_1\right\|_{L_{\xi,\tau}(\mathbb{R}^2)}$$
$$= C\left\|\frac{\xi(1+|\xi|)^s}{(1+|\tau+\phi(\xi)|)^{b_1-1}}\right.$$
$$\times\int_{\mathbb{R}^2}\frac{\widehat{f_1}(\xi_1,\tau_1)}{(1+|\xi|)^s(1+|\tau_1+\phi(\xi_1)|)^{b_1'}}$$

$$
\times \frac{\widehat{f_2}(\xi-\xi_1,\tau-\tau_1)}{(1+|\xi-\xi_1|)^s(1+|\tau-\tau_1+\phi(\xi-\xi_1)|)^{b_1'}}
$$

$$
\times \widehat{u_2}(\xi-\xi_1,\tau-\tau_1)d\xi_1 d\tau_1\Bigg\|_{L_{\xi,\tau}(\mathbb{R}^2)}
$$

$$
\leqslant C\|f_1\|_{L_{\xi,\tau}}\|f_2\|_{L_{\xi,\tau}(\mathbb{R}^2)}. \tag{7.15}
$$

Lemma 7.7. *If $s > -\frac{7}{4}$, let $\sigma \geqslant 1$, $\delta > 0$ and $b_1 > \frac{1}{2}$ be close enough to $\frac{1}{2}$ and $b_1' > \frac{1}{2}$. Then*

$$
\|\partial_x(u_1u_2)\|_{X_{\sigma,\delta,s,b_1-1}(\mathbb{R}^2)} \leqslant C\|u_1\|_{X_{\sigma,\delta,s,b_1'}(\mathbb{R}^2)}\|u_2\|_{X_{\sigma,\delta,s,b_1'}(\mathbb{R}^2)}. \tag{7.16}
$$

Proof. We observe, by considering the operator $A^{\sigma,\delta}$ in (7.3), that

$$
\begin{aligned}
&\|\partial_x(u_1u_2)\|_{X_{\sigma,\delta,s,b_1-1}(\mathbb{R}^2)}\\
&\quad = \left\|e^{\delta|\xi|^{\frac{1}{\sigma}}}(1+|\xi|)^s(1+|\tau+\phi(\xi)|)^{b_1-1}\widehat{\partial_x(u_1u_2)}(\xi,\tau)\right\|_{L^2_{\xi,\tau}(\mathbb{R}^2)}\\
&\quad = \left\|(1+|\xi|)^s(1+|\tau+\phi(\xi)|)^{b_1-1}\xi e^{\delta|\xi|^{\frac{1}{\sigma}}}\widehat{u_1u_2}(\xi,\tau)\right\|_{L^2_{\xi,\tau}(\mathbb{R}^2)}\\
&\quad = \left\|(1+|\xi|)^s(1+|\tau+\phi(\xi)|)^{b_1-1}\xi(\sqrt{2\pi})^{-2}e^{\delta|\xi|^{\frac{1}{\sigma}}}\widehat{u_1}*\widehat{u_2}\right\|_{L^2_{\xi,\tau}(\mathbb{R}^2)}\\
&\quad \leqslant \Bigg\|(1+|\xi|)^s(1+|\tau+\phi(\xi)|)^{b_1-1}\xi(\sqrt{2\pi})^{-2}\\
&\qquad \times \int_{\mathbb{R}^2} e^{\delta|\xi-\xi_1|^{\frac{1}{\sigma}}}\widehat{u_1}(\xi-\xi_1,\tau-\tau_1)e^{\delta|\xi_1|^{\frac{1}{\sigma}}}\widehat{u_2}(\xi_1,\tau_1)d\xi_1 d\tau_1\Bigg\|_{L^2_{\xi,\tau}(\mathbb{R}^2)}\\
&\quad = \left\|(1+|\xi|)^s(1+|\tau+\phi(\xi)|)^{b_1-1}(\widehat{\partial_x A^{\sigma,\delta}u_1A^{\sigma,\delta}u_2})(\xi,\tau)\right\|_{L^2_{\xi,\tau}(\mathbb{R}^2)}\\
&\quad = \left\|\partial_x(A^{\sigma,\delta}u_1A^{\sigma,\delta}u_2)\right\|_{X_{s,b_1-1}(\mathbb{R}^2)},
\end{aligned}
$$

where

$$
\delta|\xi|^{\frac{1}{\sigma}} \leqslant \delta|\xi-\xi_1|^{\frac{1}{\sigma}} + \delta|\xi_1|^{\frac{1}{\sigma}}, \quad \forall\sigma \geqslant 0
$$

Now, by using Corollary 7.1, there exists $C > 0$ such that

$$\begin{aligned}\|\partial_x(A^{\sigma,\delta}u_1 A^{\sigma,\delta}u_2)\|_{X^{s,b_1-1}(\mathbb{R}^2)}\\ \leqslant C\|A^{\sigma,\delta}u_1\|_{X_{s,b_1'}(\mathbb{R}^2)}\|A^{\sigma,\delta}u_2\|_{X_{s,b_1'}(\mathbb{R}^2)}\\ = C\|u_1\|_{X_{\sigma,\delta,s,b_1'}(\mathbb{R}^2)}\|u_2\|_{X_{\sigma,\delta,s,b_1'}(\mathbb{R}^2)}.\end{aligned}$$

□

Finally, for $\delta > 0$ and $\sigma = 1$ we will need the restrictions of

$$X_{1,\delta,s,b}(\mathbb{R}^2) = X_{\delta,s,b}(\mathbb{R}^2),$$

to a time slab $(0, T) \times \mathbb{R}$. This space is denoted by

$$X^T_{1,\delta,s,b}(\mathbb{R}^2) = X^T_{\delta,s,b}(\mathbb{R}^2),$$

and is a Banach space when equipped with the norm

$$\|u\|_{X^T_{\delta,s,b}(\mathbb{R}^2)} = \inf\{\|v\|_{X_{\delta,s,b}(\mathbb{R}^2)} : v = u \quad \text{on } [0,T]\times\mathbb{R}\}.$$

7.2 Local Well-Posedness

We state the local well-posedness in $G^{\sigma,\delta,s}$, $s > -\frac{7}{4}, \sigma \geqslant 1$ and $\delta > 0$.

Theorem 7.2 (Local well-posedness in $\mathbf{G^{\sigma,\delta,s}(\mathbb{R})}$). *Let $s > -\frac{7}{4}$, $\sigma \geqslant 1, \delta > 0$ and $u_0 \in G^{\sigma,\delta,s}(\mathbb{R})$. Then there exist a real number $b > \frac{1}{2}$, which is near enough to $\frac{1}{2}$ and a positive time T depending only on σ, δ, s and u_0, such that the Cauchy problem (7.1) is locally well-posed in*

$$C\left([0,T], G^{\sigma,\delta,s}(\mathbb{R})\right).$$

We have the following corollary.

Corollary 7.1 (Local well-posedness in $\mathbf{G^{\delta,s}(\mathbb{R})}$). *Let $s > -\frac{7}{4}, \delta > 0$ and $u_0 \in G^{\delta,s}$. Then there exist a real number $b > \frac{1}{2}$, which is near enough to $\frac{1}{2}$ and a positive time T depending only on δ, s and u_0, such that the Cauchy problem (7.1) is locally well-posed in*

$$C\left([0,T], G^{\delta,s}\right).$$

Furthermore, the solution u satisfies the bound

$$\|u(t)\|_{G^{\sigma,\delta,s}} \leq c\|u\|_{X_{\delta,s,b}} \leq 2C\|u_0\|_{G^{\delta,s}}, \quad t \in [0,T], \tag{7.17}$$

and

$$T = \frac{c_0}{(1+\|u_0\|_{G^{\delta,s}(\mathbb{R})})^a}, \tag{7.18}$$

for certain constants $c_0 = (8C^2 2^{b'-b}|\mu|)^{-a}, c_0, C > 0$ *and* $a > 1$.

7.2.1 *Existence of solution*

For the proof of local well-posedness in the analytic Gevrey spaces, we will use the standard Banach contraction principle for functions $X_{\sigma,\delta,s,b}(\mathbb{R}^2)$ in a given closed ball $\mathbb{B}$. we define again the integral operator

$$\Phi(u)(x,t) = \psi_1(t)S(t)u_0(x) - \psi_1(t)\int_0^t S(t-s)\mu\partial_x u^2(x,s)ds. \tag{7.19}$$

If that is useful for the nonlinear estimates, as will be, one can also introduce additional cutoffs in $\partial_x u^2$ and consider the equation

$$\Phi(u)(x,t) = \psi_1(t)S(t)u_0(x) - \psi_1(t)\int_0^t S(t-s)\psi_{2T}(s)\mu\partial_x u^2(x,)ds, \tag{7.20}$$

which is actually identical with (7.19) since $\psi_{2T} = 1$ on support of ψ_T.

Lemma 7.8. *Let* $s > -\frac{7}{4}$, $\sigma \geq 1$, $\delta > 0$ *and* $b > \frac{1}{2}$. *Then, for all* $u_0 \in G^{\sigma,\delta,s}(\mathbb{R})$ *and* $0 < T < 1$, *with some constant* $C > 0$, *we have*

$$\|\Phi(u)\|_{X_{\sigma,\delta,s,b}(\mathbb{R}^2)} \leq C\|u_0\|_{G^{\sigma,\delta,s}(\mathbb{R})} + C|\mu|(2T)^{b'-b}\|u\|^2_{X_{\sigma,\delta,s,b}(\mathbb{R}^2)}, \tag{7.21}$$

and

$$\|\Phi(u) - \Phi(v)\|_{X_{\sigma,\delta,s,b}(\mathbb{R}^2)} \leq C|\mu|(2T)^{b'-b}\|u-v\|_{X_{\sigma,\delta,s,b}(\mathbb{R}^2)}\|u+v\|_{X_{\sigma,\delta,s,b}(\mathbb{R}^2)}, \tag{7.22}$$

for all $u, v \in X_{\sigma,\delta,s,b}(\mathbb{R}^2)$.

Proof. To prove estimate (7.21), we follow

$$\begin{aligned}
\|\Phi(u)\|_{X_{\sigma,\delta,s,b}(\mathbb{R}^2)} &\leq \|\psi_1(t)S(t)u_0\|_{X_{\sigma,\delta,s,b}(\mathbb{R}^2)} \\
&\quad + \left\|\psi_1(t)\int_0^t S(t-s)\psi_{2T}(s)\mu\partial_x u^2 ds\right\|_{X_{\sigma,\delta,s,b}(\mathbb{R}^2)} \\
&\leq C\|u_0\|_{G^{\sigma,\delta,s}(\mathbb{R}^2)} + C|\mu|\left\|\psi_{2T}(t)\partial_x u^2\right\|_{X_{\sigma,\delta,s,b-1}(\mathbb{R}^2)} \\
&\leq C\|u_0\|_{G^{\sigma,\delta,s}(\mathbb{R}^2)} \\
&\quad + C|\mu|(2T)^{b'-b}\left\|\partial_x u^2\right\|_{X_{\sigma,\delta,s,b'-1}(\mathbb{R}^2)} \\
&\leq C\|u_0\|_{G^{\sigma,\delta,s}(\mathbb{R}^2)} + C|\mu|(2T)^{b'-b}\|u\|^2_{X_{\sigma,\delta,s,b}(\mathbb{R}^2)}.
\end{aligned}$$

Here, $b_1 = b'$ and $b_1' = b$, for the estimate (7.22), we observe that

$$\begin{aligned}
&\|\Phi(u) - \Phi(v)\|_{X_{\sigma,\delta,s,b}(\mathbb{R}^2)} \\
&\quad = \left\|\psi_1(t)\int_0^t S(t-s)\psi_{2T}(s)\mu\left(\partial_x u^2 - \partial_x v^2\right)(x,s)ds\right\|_{X_{\sigma,\delta,s,b}(\mathbb{R}^2)} \\
&\quad \leqslant C|\mu|\left\|\psi_{2T}(s)(\partial_x u^2 - \partial_x v^2)\right\|_{X_{\sigma,\delta,s,b-1}(\mathbb{R}^2)} \\
&\quad \leqslant C|\mu|(2T)^{b'-b}\left\|(\partial_x u^2 - \partial_x v^2)\right\|_{X_{\sigma,\delta,s,b'-1}(\mathbb{R}^2)} \\
&\quad \leqslant C|\mu|(2T)^{b'-b}\|u+v\|_{X_{\sigma,\delta,s,b}(\mathbb{R}^2)}\|u-v\|_{X_{\sigma,\delta,s,b}(\mathbb{R}^2)}.
\end{aligned}$$

Thus, from the previous results, we obtain (7.22). □

We will show that the map Φ is a contraction on the ball $\mathbb{B}(0,r)$ to $\mathbb{B}(0,r)$.

Proposition 7.1. *Let $s > -\frac{7}{4}$, $\sigma \geq 1$, $\delta > 0$ and $b > \frac{1}{2}$. Then, for all $u_0 \in G^{\sigma,\delta,s}(\mathbb{R})$, such that*

$$T = \frac{c_0}{(1 + \|u_0\|_{G^{\delta,s}(\mathbb{R})})^a}, \tag{7.23}$$

the map

$$\Phi : \mathbb{B}(0,r) \to \mathbb{B}(0,r),$$

is a contraction, where $\mathbb{B}(0,r)$ *is given by*

$$\mathbb{B}(0,r) = \{u \in X_{\sigma,\delta,s,b}(\mathbb{R}^2); \|u\|_{X_{\sigma,\delta,s,b}(\mathbb{R}^2)} \le r\} \quad \textit{with}$$
$$r = 2C\|u_0\|_{G^{\sigma,\delta,s}(\mathbb{R})}.$$

Proof. From Lemma 7.8, for all $u \in \mathbb{B}(0,r)$, we have

$$\begin{aligned}\|\Phi(u)\|_{X_{\sigma,\delta,s,b}(\mathbb{R}^2)} &\le C\|u_0\|_{G^{\sigma,\delta,s}(\mathbb{R})} + C|\mu|(2T)^{b'-b}\|u\|^2_{X_{\sigma,\delta,s,b}(\mathbb{R}^2)} \\ &\le \frac{r}{2} + C|\mu|(2T)^{b'-b}r^2.\end{aligned}$$

If we define

$$a = \frac{1}{b'-b},$$

and

$$c_0 = (8C^2 2^{b'-b}|\mu|)^{-a},$$

then for T given as in (7.23), we have that

$$T \le \frac{1}{(4C|\mu|2^{b'-b}r)^{\frac{1}{b'-b}}},$$

and hence,

$$\|\Phi(u)\|_{X_{\sigma,\delta,s,b}(\mathbb{R}^2)} \le r, \quad \forall u \in \mathbb{B}(0,r).$$

Thus, Φ maps $\mathbb{B}(0,r)$ into $\mathbb{B}(0,r)$, which is a contraction, since

$$\begin{aligned}&\|\Phi(u) - \Phi(v)\|_{X_{\sigma,\delta,s,b}(\mathbb{R}^2)} \\ &\quad \le C|\mu|(2T)^{b'-b}\|u-v\|_{X_{\sigma,\delta,s,b}(\mathbb{R}^2)}\|u+v\|_{X_{\sigma,\delta,s,b}(\mathbb{R}^2)} \\ &\quad \le C|\mu|(2T)^{b'-b}\|u-v\|_{X_{\sigma,\delta,s,b}(\mathbb{R}^2)}\left(\|u\|_{X_{\sigma,\delta,s,b}(\mathbb{R}^2)} + \|v\|_{X_{\sigma,\delta,s,b}(\mathbb{R}^2)}\right) \\ &\quad \le C|\mu|(2T)^{b'-b}2r\|u-v\|_{X_{\sigma,\delta,s,b}(\mathbb{R}^2)} \\ &\quad \le \frac{1}{2}\|u-v\|_{X_{\sigma,\delta,s,b}(\mathbb{R}^2)}, \quad \forall u,v \in \mathbb{B}(0,r).\end{aligned}$$

The proof is now completed. □

The uniqueness. Uniqueness of the solution in $C([0,T],G^{\sigma,\delta,s}(\mathbb{R}))$ can be proved by the following standard argument.

Lemma 7.9. *Suppose u and v are solutions to (7.1) in*

$$C([0,T],G^{\sigma,\delta,s}(\mathbb{R})),$$

with $u(\cdot,0)=v(\cdot,0)$ in $G^{\sigma,\delta,s}(\mathbb{R})$, where $\sigma\geqslant 1,\delta>0$ and $s>-\frac{7}{4}$. Then $u=v$.

Proof. Setting $w=u-v$, we see that w solves the Cauchy problem

$$\partial_t w+\alpha\partial_x^5 w+\beta\partial_x^3 w+\gamma\partial_x w+\mu\partial_x w(u+v)=0,\quad w(0)=0.$$

Multiplying both sides by w and integrating in space yield

$$w\partial_t w+\alpha w\partial_x^5 w+\beta w\partial_x^3 w+\gamma w\partial_x+\mu w\partial_x w(u+v)=0.$$

Thus, we have

$$\begin{aligned}\frac{1}{2}\partial_t\|w(\cdot,t)\|^2_{L^2(\mathbb{R})}&=\frac{1}{2}\partial_t\int_{\mathbb{R}}w^2(x,t)dx=\int_{\mathbb{R}}w(x,t)\partial_t w(x,t)dx\\&=-\mu\int_{\mathbb{R}}w(x,t)\partial_x(u^2-v^2)dx,\end{aligned}\tag{7.24}$$

since we have

$$\begin{aligned}\int_{\mathbb{R}}w(x,t)\partial_x^5 w(x,t)dx&=\int_{\mathbb{R}}w(x,t)\partial_x^3 w(x,t)dx\\&=\int_{\mathbb{R}}w(x,t)\partial_x w(x,t)dx=0.\end{aligned}$$

Thanks to (7.24), we have

$$\begin{aligned}\partial_t\|w(\cdot,t)\|^2_{L^2(\mathbb{R})}&=-2\mu\int_{\mathbb{R}}w(x,t)\partial_x(u^2-v^2)dx\\&=-2\mu\int_{\mathbb{R}}w(x,t)\partial_x\left[f(x,t)w(x,t)\right]dx,\end{aligned}$$

where $f=u+v$. Integrating by parts the last integral we obtain

$$\partial_t\|w(\cdot,t)\|^2_{L^2(\mathbb{R})}=-\mu\int_{\mathbb{R}}\partial_x f(x,t)w^2(x,t)dx,$$

from which we deduce the inequality

$$|\partial_t \|w(\cdot,t)\|^2_{L^2(\mathbb{R})}| \leqslant |\mu| \|\partial_x f\|_{L^\infty([0,T]\times\mathbb{R})} \|w(t)\|^2_{L^2(\mathbb{R})}. \tag{7.25}$$

Since

$$u, v \in C([0,T], G^{\sigma,\delta,s}(\mathbb{R})),$$

we have that u and v are continuous in t on the compact set $[0,T]$ and are $G^{\sigma,\delta,s}(\mathbb{R})$ in x. Thus, we can conclude that

$$|\mu| \|\partial_x f\|_{L^\infty([0,T]\times\mathbb{R})} \leq c < \infty. \tag{7.26}$$

Therefore, from (7.25) and (7.26) we obtain the differential inequality

$$|\partial_t \|w(\cdot,t)\|^2_{L^2(\mathbb{R})}| \leq c\|w(t)\|^2_{L^2(\mathbb{R})}, \quad 0 \leq t \leq T.$$

Solving it gives

$$\|w(t)\|^2_{L^2(\mathbb{R})} \leq e^c \|w(0)\|^2_{L^2(\mathbb{R})}, \quad 0 \leq t \leq T. \tag{7.27}$$

Since $\|w(0)\|^2_{L^2(\mathbb{R})} = 0$, from (7.27) we obtain that $w(t) = 0$, $0 \leq t \leq T$ or $u = v$. □

7.2.2 *Continuous dependence of the initial data*

To prove continuous dependence of the initial data we will prove the following.

Lemma 7.10. *Let $s > -\frac{7}{4}$, $\sigma \geq 1$, $\delta > 0$ and $b > \frac{1}{2}$. Then, for all $u_0, v_0 \in G^{\sigma,\delta,s}(\mathbb{R})$, if u and v are two solutions to (7.1) corresponding to initial data u_0 and v_0. We have*

$$|u - v|_{T,\sigma,\delta,s} \leq 2C_0 C \|u_0 - v_0\|_{G^{\sigma,\delta,s}(\mathbb{R})}. \tag{7.28}$$

Proof. If u and v are two solutions to (7.1), corresponding to initial data u_0 and v_0, we have from Lemma 7.1

$$\begin{aligned}
|u - v|_{T,\sigma,\delta,s} &= \sup_{t\in[0,T]} \|u(\cdot,t) - v(\cdot,t)\|_{G^{\sigma,\delta,s}(\mathbb{R})} \\
&\leq C_0 \|u - v\|_{X_{\sigma,\delta,s,b}(\mathbb{R}^2)} \\
&= C_0 \|\Phi(u) - \Phi(v)\|_{X_{\sigma,\delta,s,b}(\mathbb{R}^2)}.
\end{aligned}$$

By taking $u, v \in \mathbb{B}(0,r)$ and

$$T \leq \frac{1}{(4C|\mu|2^{b'-b}r)^{\frac{1}{b'-b}}},$$

$$\begin{aligned}
&\|\Phi(u)-\Phi(v)\|_{X_{\sigma,\delta,s,b}(\mathbb{R}^2)} \\
&\quad \leq \|\psi_1(t)S(t)(u_0-v_0)\|_{X_{\sigma,\delta,s,b}(\mathbb{R}^2)} \\
&\qquad + \left\|\psi_1(t)\int_0^t S(t-s)\psi_{2T}(s)\left(\partial_x u^2-\partial_x v^2\right)ds\right\|_{X_{\sigma,\delta,s,b}(\mathbb{R}^2)}, \\
&\quad \leq C\|u_0-v_0\|_{G^{\sigma,\delta,s}(\mathbb{R})} \\
&\qquad + C(2T)^{b'-b}\|u+v\|_{X_{\sigma,\delta,s,b}(\mathbb{R}^2)}\|u-v\|_{X_{\sigma,\delta,s,b}(\mathbb{R}^2)}, \\
&\quad \leq C\|u_0-v_0\|_{G^{\sigma,\delta,s}(\mathbb{R})} + \frac{1}{2}\|u-v\|_{X_{\sigma,\delta,s,b}(\mathbb{R}^2)}.
\end{aligned}$$

Thus,

$$\|\Phi(u)-\Phi(v)\|_{X_{\sigma,\delta,s,b}(\mathbb{R}^2)} \leq 2C\|u_0-v_0\|_{G^{\sigma,\delta,s}(\mathbb{R})},$$

then

$$|u-v|_{T,\sigma,\delta,s} \leq 2C_0C\|u_0-v_0\|_{G^{\sigma,\delta,s}(\mathbb{R})}.$$

□

This completes the prove of Theorem 7.2.

7.2.3 *Lower bound for radius of spatial analyticity*

We need to recall an important property of the space $G^{\delta,s}(\mathbb{R})$. For $\delta > 0$ and $s \in \mathbb{R}$, it is straightforward to show that if a function f belongs to $G^{\delta,s}(\mathbb{R})$, then it is the restriction to the real line of a holomorphic function $f(x+iy)$ in the strip

$$S_\delta = \{x+iy \in \mathbb{C}, |y| < \delta\}.$$

This $\delta > 0$ is called the (uniform) radius of spatial analyticity of f.

In this section, we use the local result and a Gevrey approximate conservation law to gradually extend the local solution for all time.

Furthermore, we obtain explicit lower bounds on the radius of spatial analyticity $r(t)$ at any time $t \geqslant 0$, which is given by $r(t) \geqslant ct^{-1}$, where $\delta > 0$ can be taken arbitrarily small and c is a positive constant, that will be described more precisely later.

7.2.4 *Approximate conservation law*

We start by recalling that

$$\mathscr{I}(u) = \int u^2(x,t)dx,$$

is conserved for a solution u of (7.1), since by using Riemman–Lebesgue's lemma and integration by parts we have

$$\begin{aligned}
\frac{1}{2}\partial_t \|u(t)\|^2_{L^2(\mathbb{R})} &= \frac{1}{2}\partial_t \int_{\mathbb{R}} u^2(x,t)dx = \int_{\mathbb{R}} u(x,t)\partial_t u(x,t)dx \\
&= \int_{\mathbb{R}} u(x,t)\left[-\alpha\partial_x^5 u(x,t) - \beta\partial_x^3 u(x,t)\right. \\
&\quad \left. - \gamma\partial_x u(x,t) - \mu\partial_x(u^2)(x,t)\right] dx \\
&= -\mu \int_{\mathbb{R}} u(x,t)\partial_x(u^2)(x,t) \\
&= \mu \int_{\mathbb{R}} u^2(x,t)\partial_x u(x,t) \\
&= \frac{\mu}{3}\int_{\mathbb{R}} \partial_x(u^3)(x,t)dx = 0.
\end{aligned}$$

Our goal in this section is to establish an approximate conservation law for a solution to (7.1) based on the conservation the $L^2(\mathbb{R})$ norm of solutions of the equation. Explicitly, we aim at proving Theorem 7.3.

Theorem 7.3. *Let $\kappa \in [0,1]$ and $0 < T < T_1 < 1$, T_1 be as in Corollary 7.1, there exist $b \in (1/2,1)$ and $C > 0$, such that for any $\delta > 0$ and any solution $u \in X^T_{\delta,0,b}(\mathbb{R}^2)$ to the Cauchy problem (7.1)* *on the time interval* $[0, T]$, *we have the estimate*

$$\sup_{t\in[0,T]} \|u(t)\|^2_{G^{\delta,0}(\mathbb{R})} \le \|u(0)\|^2_{G^{\delta,0}(\mathbb{R})} + C\delta^\kappa \|u\|^3_{X^T_{\delta,0,b}(\mathbb{R}^2)}.$$

Moreover, we have

$$\sup_{t\in[0,T]} \|u(t)\|^2_{G^{\delta,0}(\mathbb{R})} \le \|u(0)\|^2_{G^{\delta,0}(\mathbb{R})} + C\delta^\kappa \|u(0)\|^3_{G^{\delta,0}(\mathbb{R})}.$$

Theorem 7.3 is of fundamental importance as it guarantees, by combining it with Corollary 7.1 and applying them repeatedly, we can glue intervals in a way to gradually extend the local solution in time. This will lead to the global well-posedness of solutions in Gevrey spaces, as in Theorem 7.4.

For the proof of Theorem 7.3, we require the following preliminary estimate.

Lemma 7.11. *Given* $\kappa \in [0, 1]$, *there exist* $b \in (1/2, 1)$ *and* $C > 0$, *such that for all* $T > 0$ *and*

$$u \in X_{\delta,0,b}(\mathbb{R}^2),$$

we have

$$\|G\|_{X_{0,b-1}(\mathbb{R}^2)} \le C\delta^\kappa \|u\|^2_{X_{\delta,0,b}(\mathbb{R}^2)}, \tag{7.29}$$

where

$$G = \partial_x \left[(A^{1,\delta}u)^2 - A^{1,\delta}(u)^2 \right],$$

and the operator $A^{1,\delta}$ *given by* (7.3).

Proof. Let

$$L = \left[(A^{1,\delta}u)^2 - A^{1,\delta}(u)^2 \right].$$

Then

$$\begin{aligned}\|G\|_{X_{0,b-1}(\mathbb{R}^2)} &= \left\| \frac{\xi}{(1 + |\tau + \phi(\xi)|)^{1-b}} \widehat{L}(\xi, \tau) \right\|_{L^2_{\xi,\tau}(\mathbb{R}^2)} \\ &= \left(\int_{\mathbb{R}^2} \frac{|\xi|^2}{(1 + |\tau + \phi(\xi)|)^{2(1-b)}} |\widehat{L}(\xi, \tau)|^2 d\xi d\tau \right)^{\frac{1}{2}}.\end{aligned}$$

We shall calculate the Fourier transform of L:

$$
\begin{aligned}
|\widehat{L}(\xi,\tau)| &= |(\widehat{A^{1,\delta}u})^2 - \widehat{A^{1,\delta}(u)^2}| \\
&= C|(e^{\delta|\xi|}\widehat{u} * e^{\delta|\xi|}\widehat{u})(\xi,\tau) - e^{\delta|\xi|}(\widehat{u}*\widehat{u})(\xi,\tau)| \\
&= C\left|\int_{\mathbb{R}^2}\left(e^{\delta|\xi_1|}\widehat{u}(\xi_1,\tau_1)e^{\delta|\xi-\xi_1|}\widehat{u}(\xi-\xi_1,\tau-\tau_1)\right.\right. \\
&\qquad \left.\left. - e^{\delta|\xi|}\widehat{u}(\xi_1,\tau_1)\widehat{u}(\xi-\xi_1,\tau-\tau_1)\right)d\xi_1 d\tau_1\right| \\
&\leqslant C\int_{\mathbb{R}^2}\left(e^{\delta|\xi_1|}e^{\delta|\xi-\xi_1|} - e^{\delta|\xi|}\right) \\
&\qquad \times|\widehat{u}(\xi_1,\tau_1)\widehat{u}(\xi-\xi_1,\tau-\tau_1)|d\xi_1 d\tau_1.
\end{aligned}
$$

We need the following estimate. □

Lemma 7.12. *For $\delta > 0$, $\theta \in [0,1]$ and $\xi, \xi_1 \in \mathbb{R}$ we have*

$$
e^{\delta|\xi-\xi_1|}e^{\delta|\xi_1|} - e^{\delta|\xi|} \le [2\delta\min(|\xi-\xi_1|,|\xi_1|)]^{\theta}\, e^{\delta|\xi-\xi_1|}e^{\delta|\xi_1|}. \tag{7.30}
$$

Also, we have

$$
\min(|\xi-\xi_1|,|\xi_1|) \le 2\frac{(1+|\xi-\xi_1|)(1+|\xi_1|)}{(1+|\xi|)}. \tag{7.31}
$$

Proof. If $\xi-\xi_1$ and ξ_1 have the same sign, then the left side of (7.30) equals zero and the inequality holds trivially, therefore, we assume that $\xi-\xi_1$ and ξ_1 have opposite signs. Without loss of generality, assume $\xi-\xi_1 \geqslant 0$ and $\xi_1 \leqslant 0$. If

$$
|\xi_1| \leqslant |\xi-\xi_1|,
$$

then

$$
\xi-\xi_1+\xi_1 \geqslant 0,
$$

and so the left side of (7.30) becomes

$$
\begin{aligned}
e^{\delta|\xi-\xi_1|}e^{\delta|\xi_1|} - e^{\delta|\xi|} &= e^{\delta(\xi-\xi_1)}e^{-\delta\xi_1} - e^{\delta((\xi-\xi_1)+\xi_1)} \\
&= e^{\delta((\xi-\xi_1)+\xi_1)}\left(e^{-2\delta\xi_1}-1\right) \\
&\leqslant (2\delta|\xi_1|)^{\theta}e^{-2\delta\xi_1}e^{\delta((\xi-\xi_1)+\xi_1)}
\end{aligned}
$$

$$
\begin{aligned}
&= (2\delta|\xi_1|)^\theta e^{\delta((\xi-\xi_1)-\xi_1)} \\
&= (2\delta|\xi_1|)^\theta e^{\delta|\xi-\xi_1|} e^{|\xi_1|}.
\end{aligned}
$$

Here, we have used the fact that, for $x \geqslant 0$, the inequalities $e^x - 1 \leqslant e^x$ and $e^x - 1 \leqslant xe^x$ both hold, hence also $e^x - 1 \leqslant x^\theta e^x$ for $x \geqslant 0$ and $\theta \in [0,1]$.

On the other hand, if $|\xi_1| \geqslant |\xi - \xi_1|$, then $\xi - \xi_1 + \xi_1 \leqslant 0$, so the left side of (7.30) becomes

$$
\begin{aligned}
e^{\delta|\xi-\xi_1|}e^{\delta|\xi_1|} - e^{\delta|\xi|} &= e^{\delta(\xi-\xi_1)}e^{-\delta\xi_1} - e^{-\delta((\xi-\xi_1)+\xi_1)} \\
&= e^{-\delta((\xi-\xi_1)+\xi_1)}\left(e^{2\delta(\xi-\xi_1)} - 1\right) \\
&\leqslant (2\delta|\xi-\xi_1|)^\theta e^{2\delta(\xi-\xi_1)} e^{-\delta((\xi-\xi_1)+\xi_1)} \\
&= (2\delta|\xi-\xi_1|)^\theta e^{\delta((\xi-\xi_1)-\xi_1)} \\
&= (2\delta|\xi-\xi_1|)^\theta e^{\delta|\xi-\xi_1|} e^{|\xi_1|}.
\end{aligned}
$$

□

For $\kappa \in [0,1]$, from Lemma 7.12, we write

$$
\begin{aligned}
|\widehat{L}(\xi,\tau)| &\leqslant C\int_{\mathbb{R}^2}\left(e^{\delta|\xi_1|}e^{\delta|\xi-\xi_1|} - e^{\delta|\xi|}\right) \\
&\quad \times|\widehat{u}(\xi_1,\tau_1)\widehat{u}(\xi-\xi_1,\tau-\tau_1)|d\xi_1 d\tau_1 \\
&\leqslant C\int_{\mathbb{R}^2}[2\delta\min(|\xi-\xi_1|,|\xi_1|)]^\kappa\, e^{\delta|\xi_1|} \\
&\quad \times|\widehat{u}(\xi_1,\tau_1)|e^{\delta|\xi-\xi_1|}|\widehat{u}(\xi-\xi_1,\tau-\tau_1)|d\xi_1 d\tau_1 \\
&\leqslant C\int_{\mathbb{R}^2}\left[4\delta\frac{(1+|\xi-\xi_1|)(1+|\xi_1|)}{(1+|\xi|)}\right]^\kappa e^{\delta|\xi_1|} \\
&\quad \times|\widehat{u}(\xi_1,\tau_1)|e^{\delta|\xi-\xi_1|}|\widehat{u}(\xi-\xi_1,\tau-\tau_1)|d\xi_1 d\tau_1 \\
&\leqslant C(4\delta)^\kappa\int_{\mathbb{R}^2}\frac{(1+|\xi-\xi_1|)^\kappa(1+|\xi_1|)^\kappa}{(1+|\xi|)^\kappa}e^{\delta|\xi_1|} \\
&\quad \times|\widehat{u}(\xi_1,\tau_1)|e^{\delta|\xi-\xi_1|}|\widehat{u}(\xi-\xi_1,\tau-\tau_1)|d\xi_1 d\tau_1.
\end{aligned} \tag{7.32}
$$

Setting $v = (A^{1,\delta}u)$ and

$$
f(\xi,\tau) = (1+|\tau+\phi(\xi)|)^b\widehat{v}(\xi,\tau),
$$

we have

$$e^{\delta|\xi|}\widehat{u}(\xi,\tau)=\widehat{v}(\xi,\tau)=f(\xi,\tau)(1+|\tau+\phi(\xi)|)^{-b},$$

and therefore we can write (7.32) as

$$|\widehat{L}(\xi,\tau)|\leqslant C(4\delta)^{\kappa}\int_{\mathbb{R}^2}\frac{(1+|\xi-\xi_1|)^{\kappa}(1+|\xi_1|)^{\kappa}}{(1+|\xi|)^{\kappa}}\frac{|\widehat{f}(\xi_1,\tau_1)|}{(1+|\tau_1+\phi(\xi_1)|)^{b}}$$
$$\times\frac{|\widehat{f}(\xi-\xi_1,\tau-\tau_1)|}{(1+|\tau-\tau_1+\phi(\xi-\xi_1)|)^{b}}d\xi_1 d\tau_1. \tag{7.33}$$

It follows from (7.30) and (7.33) that

$$\begin{aligned}
&\|G\|_{X_{0,b-1}(\mathbb{R}^2)}\\
&=\left\|\frac{\xi}{(1+|\tau+\phi(\xi)|)^{1-b}}\widehat{L}(\xi,\tau)\right\|_{L^2_{\xi,\tau}(\mathbb{R}^2)}\\
&\leqslant C(4\delta)^{\kappa}\left[\int_{\mathbb{R}^2}\frac{|\xi|^2}{(1+|\tau+\phi(\xi)|)^{2(1-b)}}\right.\\
&\quad\times\left(\int_{\mathbb{R}^2}\frac{(1+|\xi-\xi_1|)^{\kappa}(1+|\xi_1|)^{\kappa}}{(1+|\xi|)^{\kappa}}\right.\\
&\quad\left.\left.\times\frac{|\widehat{f}(\xi_1,\tau_1)|}{(1+|\tau_1+\phi(\xi_1)|)^{b}}\frac{|\widehat{f}(\xi-\xi_1,\tau-\tau_1)|}{(1+|\tau-\tau_1+\phi(\xi-\xi_1)|)^{b}}d\xi_1 d\tau_1\right)^2 d\xi d\tau\right]^{\frac{1}{2}}\\
&=C(4\delta)^{\kappa}\left\|\frac{\xi}{(1+|\tau+\phi(\xi)|)^{1-b}}\int_{\mathbb{R}^2}\frac{(1+|\xi-\xi_1|)^{\kappa}(1+|\xi_1|)^{\kappa}}{(1+|\xi|)^{\kappa}}\right.\\
&\quad\left.\times\frac{|\widehat{f}(\xi_1,\tau_1)|}{(1+|\tau_1+\phi(\xi_1)|)^{b}}\frac{|\widehat{f}(\xi-\xi_1,\tau-\tau_1)|}{(1+|\tau-\tau_1+\phi(\xi-\xi_1)|)^{b}}d\xi_1 d\tau_1\right\|_{L^2_{\xi,\tau}(\mathbb{R}^2)}.
\end{aligned}$$

Now by taking $s=-\kappa\in[-1,0]$, we obtain

$$\begin{aligned}
&\|G\|_{X_{0,b-1}(\mathbb{R}^2)}\\
&\quad\leqslant C(4\delta)^{\kappa}\left\|\frac{\xi(1+|\xi|)^{s}}{(1+|\tau+\phi(\xi)|)^{1-b}}\right.
\end{aligned}$$

$$\times \int_{\mathbb{R}^2} \frac{|\widehat{f}(\xi_1, \tau_1)|}{(1+|\xi_1|)^s(1+|\tau_1+\phi(\xi_1)|)^b}$$

$$\times \frac{|\widehat{f}(\xi-\xi_1, \tau-\tau_1)|}{(1+|\xi-\xi_1|)^s(1+|\tau-\tau_1+\phi(\xi-\xi_1)|)^b} d\xi_1 d\tau_1 \Bigg\|_{L^2_{\xi,\tau}(\mathbb{R}^2)}.$$

By Remark 7.2, we get

$$\begin{aligned} \|G\|_{X_{0,b-1}(\mathbb{R}^2)} &\leqslant C(4\delta)^\kappa \|f\|^3_{L^2_{\xi,\tau}(\mathbb{R}^2)} \\ &= C(4\delta)^\kappa \|v\|^3_{X_{0,b}(\mathbb{R}^2)} \\ &= C(4\delta)^\kappa \|A^{1,\delta}u\|^3_{X_{0,b}(\mathbb{R}^2)} \\ &= C(4\delta)^\kappa \|u\|^3_{X_{\delta,0,b}(\mathbb{R}^2)}, \end{aligned} \tag{7.34}$$

concluding the proof.

Proof of Theorem 7.3. Let κ, δ, T, b and u be as in the statement of Theorem 7.3. We start by defining the auxiliary function

$$V(t,x) = A^{1,\delta}u(t,x).$$

Since u is real-valued we also have V real-valued. Applying the exponential $A^{1,\delta}$ to (7.1), it is easily seen that we obtain

$$\partial_t V + \alpha\partial_x^5 V + \beta\partial_x^3 V + \gamma\partial_x V + \mu A^{1,\delta}\partial_x u^2 = 0,$$

which is equivalent to

$$\partial_t V + \alpha\partial_x^5 V + \beta\partial_x^3 V + \gamma\partial_x V + 2\mu V\partial_x V = \mu(\partial_x V^2 - \partial_x(A^{1,\delta}u^2)).$$

Therefore, setting

$$G = \partial_x \left[(A^{1,\delta}u)^2 - A^{1,\delta}(u)^2\right],$$

we obtain

$$\partial_t V + \alpha\partial_x^5 V + \beta\partial_x^3 V + \gamma\partial_x V + 2\mu V\partial_x V = \mu G. \tag{7.35}$$

Multiplying (7.35) by V and integrating in space we obtain

$$\int_{\mathbb{R}} V\partial_t V dx + \alpha \int_{\mathbb{R}} V\partial_x^5 V dx + \beta \int_{\mathbb{R}} V\partial_x^3 V dx + \gamma \int_{\mathbb{R}} V\partial_x V dx + 2\mu \times \int_{\mathbb{R}} V^2\partial_x V dx = \mu \int_{\mathbb{R}} VG dx. \tag{7.36}$$

Integration by parts is justified, since we may assume that $V(t,x)$ decays to zero as $|x| \longrightarrow \infty$, and the same holds for all spatial derivatives.

Indeed, we are aiming to prove (7.29) for a given $\delta > 0$, but by the monotone convergence theorem, it suffices to prove it for all $\delta' < \delta$ (the constant C being uniform). For $U = A^{1,\delta'}u$, we get by Cauchy–Schwarz inequality and by the assumption that

$$u \in X_{\delta,0,b} \subset L_t^\infty G_{\delta,0},$$

$$\begin{aligned}\int |\widehat{\partial_x^j U}(\xi,t)|d\xi &= \int |e^{(\delta'-\delta)|\xi|}e^{\delta|\xi|}\xi^j \widehat{u}(\xi,t)|d\xi \\ &\leqslant \left(\int \xi^{2j}e^{-2p|\xi|}d\xi\right)^{\frac{1}{2}} \|u(t)\|_{G_{\delta,0}} < \infty,\end{aligned}$$

where

$$p = \delta - \delta' > 0,$$

and $j \in \{0,1,\ldots\}$. Therefore, by Riemann–Lebesgue,

$$\partial_x^j U \longrightarrow 0,$$

as $|x| \longrightarrow \infty$.

Thus, (7.36) can be rewritten as

$$\frac{1}{2}\partial_t \int_{\mathbb{R}} V^2 dx + \frac{\alpha}{2}\int_{\mathbb{R}} \partial_x(\partial_x^2 V\partial_x^2 V)dx + \frac{\beta}{2}\int_{\mathbb{R}} \partial_x(\partial_x V\partial_x V)dx + \frac{\gamma}{2}\int_{\mathbb{R}} \partial_x V^2 dx + \frac{2\mu}{3}\int_{\mathbb{R}} \partial_x V^3 dx = \mu \int_{\mathbb{R}} VG dx. \tag{7.37}$$

and furthermore, the second, third, forth and fifth terms on the left side vanish.

$$\frac{1}{2}\partial_t \int_{\mathbb{R}} V^2 dx = \mu \int_{\mathbb{R}} VG\ dx.$$

Now, integrating the last equality with respect to $t \in [0,T]$, we obtain

$$\frac{1}{2}\left[\int_{\mathbb{R}} V^2(x,T)dx - \int_{\mathbb{R}} V^2(x,0)dx\right] = \mu \int_0^T \int_{\mathbb{R}} VG\ dxdt. \quad (7.38)$$

Recall that

$$\|u(t)\|_{G^{\delta,0}(\mathbb{R})} = \int_{\mathbb{R}} e^{2\delta|\xi|}|\widehat{u}(\xi,t)|^2 dx = \int_{\mathbb{R}} |\widehat{V}(\xi,t)|^2 dx = \int_{\mathbb{R}} V^2(x,t)dx,$$

where in the last equality we used Plancherel's Theorem and the fact that we are assuming that the solution u is real valued.

It follows from the last equality and from (7.38) that

$$\int_{\mathbb{R}} V^2(x,T)dx = \int_{\mathbb{R}} V^2(x,0)dx + 2\mu \int_{\mathbb{R}^2} \chi_{[0,T]}(t) VG\ dxdt.$$

Thus,

$$\begin{aligned}\|u(T)\|^2_{G^{\delta,0}} &= \|u(0)\|^2_{G^{\delta,0}} + 2\mu \int_{\mathbb{R}^2} \chi_{[0,T]}(t) VG\ dxdt,\\ &\leqslant \|u(0)\|^2_{G^{\delta,0}} + 2|\mu| \left| \int_{\mathbb{R}^2} \chi_{[0,T]}(t) VG\ dxdt\right|. \quad (7.39)\end{aligned}$$

Recalling that V is real-valued, it follows from Plancherel's identity and Cauchy–Schwarz inequality that

$$\begin{aligned}\int_{\mathbb{R}^2} \chi_{[0,T]}(t) VG\ dxdt &= \int_{\mathbb{R}^2} (\chi_{[0,T]}(\cdot)V)(\xi,\tau)(\overline{\chi_{[0,T]}(\cdot)G})(\xi,\tau)d\xi d\tau.\\ &= \int_{\mathbb{R}^2} (\widehat{\chi_{[0,T]}(\cdot)}V)(\xi,\tau)(\overline{\widehat{\chi_{[0,T]}(\cdot)}G})(\xi,\tau)d\xi d\tau.\end{aligned}$$

Then, Holder's inequality yields

$$\begin{aligned}
&\left| \int_{\mathbb{R}^2} \chi_{[0,T]}(t) V G \, dxdt \right| \\
&= \left| \int_{\mathbb{R}^2} (1 + |\tau + \phi(\xi)|)^{1-b} (\widehat{\chi_{[0,T]}(\cdot)V})(\xi, \tau) \right. \\
&\qquad \left. \times (1 + |\tau + \phi(\xi)|)^{b-1} \overline{(\widehat{\chi_{[0,T]}(\cdot)G})}(\xi, \tau) d\xi d\tau \right| \\
&\leqslant \left\| (1 + |\tau + \phi(\xi)|)^{1-b} (\widehat{\chi_{[0,T]}(\cdot)V})(\xi, \tau) \right\|_{L^2_{\xi,\tau}(\mathbb{R}^2)} \\
&\qquad \times \left\| (1 + |\tau + \phi(\xi)|)^{b-1} (\widehat{\chi_{[0,T]}(\cdot)G})(\xi, \tau) \right\|_{L^2_{\xi,\tau}(\mathbb{R}^2)} \\
&\leqslant \|\chi_{[0,T]}(\cdot) V\|_{X_{0,1-b}(\mathbb{R}^2)} \|\chi_{[0,T]}(\cdot) G\|_{X_{0,b-1}(\mathbb{R}^2)}, \qquad (7.40)
\end{aligned}$$

we have both $-\frac{1}{2} < b - 1 < \frac{1}{2}$ and $\frac{1}{2} < 1 - b < \frac{1}{2}$. Therefore, one can use Lemma 8.3 to obtain

$$\|\chi_{[0,T]}(\cdot) V\|_{X_{0,1-b}(\mathbb{R}^2)} \leqslant C \|V\|_{X^T_{0,1-b}(\mathbb{R}^2)},$$

and

$$\|\chi_{[0,T]}(\cdot) G\|_{X_{0,b-1}(\mathbb{R}^2)} \leqslant C \|G\|_{X^T_{0,b-1}(\mathbb{R}^2)}.$$

Since $0 < T < 1$ and using the fact that $\psi_T = 1$ for $t \in [0, T]$ and the definition of $\|\cdot\|_{X^T_{\delta,s,b}(\mathbb{R}^2)}$, it follows from (7.40) and from the last relation that

$$\begin{aligned}
\left| \int_{\mathbb{R}^2} \chi_{[0,T]}(t) V G \, dxdt \right| &\leqslant C \|V\|_{X^T_{0,1-b}(\mathbb{R}^2)} \|G\|_{X^T_{0,b-1}(\mathbb{R}^2)} \\
&\leqslant C \|\psi_T V\|_{X_{0,1-b}(\mathbb{R}^2)} \|\psi_T G\|_{X_{0,b-1}(\mathbb{R}^2)}. \qquad (7.41)
\end{aligned}$$

Since $-\frac{1}{2} < b - 1 < \frac{1}{2}$ and $\frac{1}{2} < 1 - b < \frac{1}{2}$ it follows from Lemma 7.6 that

$$\|\psi_T V\|_{X^T_{0,1-b}(\mathbb{R}^2)} \leqslant C \|V\|_{X_{0,1-b}(\mathbb{R}^2)}, \qquad (7.42)$$

and

$$\|\psi_T G\|_{X^T_{0,b-1}(\mathbb{R}^2)} \leqslant C\|G\|_{X_{0,b-1}(\mathbb{R}^2)}. \tag{7.43}$$

Noting that

$$\|V\|_{X_{0,1-b}(\mathbb{R}^2)} = \|u\|_{X_{\delta,0,1-b}(\mathbb{R}^2)} \leqslant \|u\|_{X_{\delta,0,b}(\mathbb{R}^2)}, \tag{7.44}$$

since we have $1 - b < b$, we can conclude from it and Lemma 7.11 that for any $\kappa \in [0, 1]$ there exists a constant C such that

$$\|G\|_{X_{0,b-1}(\mathbb{R}^2)} \leq C\delta^\kappa \|u\|^2_{X_{\delta,0,b}(\mathbb{R}^2)}. \tag{7.45}$$

Therefore, we conclude from (7.39) and (7.41)–(7.45) that

$$\sup_{t\in[0,T]} \|u(t)\|^2_{G^{\delta,0}(\mathbb{R})} \leq \|u(0)\|^2_{G^{\delta,0}(\mathbb{R})} + C\delta^\kappa \|u\|^3_{X_{\delta,0,b}(\mathbb{R}^2)}.$$

Finally, by using the condition (7.17) we conclude that

$$\sup_{t\in[0,T]} \|u(t)\|^2_{G^{\delta,0}(\mathbb{R})} \leq \|u(0)\|^2_{G^{\delta,0}(\mathbb{R})} + C\delta^\kappa \|u(0)\|^3_{G^{\delta,0}(\mathbb{R})}.$$

The proof is now completed. □

7.3 Global Extension and Radius Analyticity

Our next main result for the Kawahara equation yields an estimate on how the width of the strip of the radius of the spatial analyticity decay with time.

Theorem 7.4 (Lower bound for radius of spatial analyticity). *Let $s > -\frac{7}{4}$ and $\delta_0 > 0$, and assume $u_0 \in G^{\delta_0,s}(\mathbb{R})$, then the solution given by Corollary 7.1 extends globally in time and for any $T > 0$, we have*

$$u \in C\left([0,T], G^{\delta(T),s}(\mathbb{R})\right) \text{ with } \delta(T) = \min\left\{\delta_0, \frac{C_1}{T}\right\},$$

where $C_1 > 0$ is a constant depending on u_0, δ_0 and s.

Fix $\delta_0 > 0$, $s > -\frac{7}{4}$, $\kappa \in (0,1)$, and $u_0 \in G^{\delta_0,s}(\mathbb{R})$. It suffices to prove that the solution $u(t)$ to (7.1) satisfies

$$u \in C\left([0,T], G^{\delta(T),s}(\mathbb{R})\right) \quad \text{for all } T' > 0,$$

where

$$\delta(T) = \min\left\{\delta_0, C_1 T^{-(\frac{1}{\kappa})}\right\},$$

and $C_1 > 0$ is a constant depending on u_0, δ_0, s, and κ.

By Corollary 7.1, there is a maximal time

$$T^* = T^*(u_0, \delta_0, s) \in (0, \infty],$$

such that

$$u \in C\left([0,T^*), G^{\delta_0,s}(\mathbb{R})\right).$$

If $T^* = \infty$ then $r(t) = \delta_0$ and we are done.

If $T^* < \infty$, as we assume henceforth, it remains to prove

$$u \in C\left([0,T], G^{C_1 T^{-(\frac{1}{\kappa})},s}(\mathbb{R})\right), \quad \text{for all } T \geq T^*. \tag{7.46}$$

We first prove this in the case $s = 0$. Then, at the end of this section, we do the general case, which essentially reduces to $s = 0$.

The case $s = 0$. Fix $T \geq T^*$, we will show that, for $\delta > 0$ sufficiently small

$$\|u(t)\|^2_{G^{\delta,0}(\mathbb{R})} \leq 2\|u(0)\|^2_{G^{\delta_0,0}(\mathbb{R})}, \quad \text{for } t \in [0,T]. \tag{7.47}$$

We need to set the following time-step:

$$\begin{aligned} T_0 &= \frac{c_0}{(1 + 2\|u(0)\|_{G^{\delta_0,s}(\mathbb{R})})^a} \\ &< \frac{c_0}{(1 + \|u(0)\|_{G^{\delta_0,s}(\mathbb{R})})^a} \\ &= T_{\delta_0}, \end{aligned} \tag{7.48}$$

where $c_0 > 0$ and $a > 1$ are as in Corollary 7.1 (with $s = 0$). The smallness conditions on δ will be

$$\delta \leq \delta_0 \quad \text{and} \quad \frac{2T}{T_0} C\delta^{\kappa} 2^{\frac{3}{2}} \|u(0)\|_{G^{\delta_0,0}(\mathbb{R})} \leq 1, \tag{7.49}$$

where $C > 0$ is the constant in Theorems 7.3. Proceeding by induction, we will verify that

$$\sup_{t\in[0,nT_0]} \|u(t)\|^2_{G^{\delta,0}(\mathbb{R})} \leq \|u(0)\|^2_{G^{\delta,0}(\mathbb{R})} + nC\delta^{\kappa} 2^{\frac{3}{2}} \|u(0)\|^3_{G^{\delta_0,0}(\mathbb{R})}, \tag{7.50}$$

$$\sup_{t\in[0,nT_0]} \|u(t)\|^2_{G^{\delta,0}(\mathbb{R})} \leq 2\|u(0)\|^2_{G^{\delta_0,0}(\mathbb{R})}, \tag{7.51}$$

for $n \in \{1, \ldots, m+1\}$, where $m \in \mathbb{N}$ is chosen, so that

$$T \in [mT_0, (m+1)T_0).$$

This m does exist, since by Corollary 7.1 and the definition of T^*, we have

$$T_0 < \frac{c_0}{(1 + \|u(0)\|_{G^{\delta_0,0}(\mathbb{R})})^a} < T^*, \quad \text{hence } T_0 < T.$$

In the first step, we cover the interval $[0, T_0]$, and by Theorem 7.3, we have

$$\begin{aligned}\sup_{t\in[0,T_0]} \|u(t)\|^2_{G^{\delta,0}(\mathbb{R})} &\leq \|u(0)\|^2_{G^{\delta,0}(\mathbb{R})} + C\delta^{\kappa} \|u(0)\|^3_{G^{\delta,0}(\mathbb{R})} \\ &\leq \|u(0)\|^2_{G^{\delta,0}(\mathbb{R})} + C\delta^{\kappa} \|u(0)\|^3_{G^{\delta_0,0}(\mathbb{R})},\end{aligned}$$

where we used that

$$\|u(0)\|_{G^{\delta,0}(\mathbb{R})} \leq \|u(0)\|_{G^{\delta_0,0}(\mathbb{R})}, \tag{7.52}$$

since $\delta \leq \delta_0$. This verifies (7.50) for $n = 1$ and now, (7.51) follows using again (7.52) as well as

$$C\delta^{\kappa} \|u(0)\|_{G^{\delta_0,0}(\mathbb{R})} \leq 1.$$

The latter follows from (7.49), since $T_0 < T$.

Next, assuming that (7.50) and (7.51) hold for some $n \in \{1, \ldots, m\}$, we will prove that they hold for $n+1$. We estimate

$$\begin{aligned}
&\sup_{t\in[nT_0,(n+1)T_0]} \|u(t)\|^2_{G^{\delta,0}} \\
&\leq \|u(nT_0)\|^2_{G^{\delta,0}} + C\delta^\kappa \|u(nT)\|^3_{G^{\delta,0}} \quad \text{by Theorem 7.3} \\
&\leq \|u(nT_0)\|^2_{G^{\delta,0}} + C\delta^\kappa 2^{\frac{3}{2}} \|u(0)\|^3_{G^{\delta_0,0}} \text{ by (7.51)} \\
&\leq \|u(0)\|^2_{G^{\delta,0}} + nC\delta^\kappa 2^{\frac{3}{2}} \|u(0)\|^3_{G^{\delta_0,0}} \\
&\quad + C\delta^\kappa 2^{\frac{3}{2}} \|u(0)\|^3_{G^{\delta_0,0}}, \qquad \text{by (7.50)},
\end{aligned}$$

verifying (7.50) with n replaced by $n+1$. To get (7.51) with n replaced by $n+1$, it is then enough to have

$$(n+1)C\delta^\kappa 2^{\frac{3}{2}} \|u(0)\|_{G^{\delta_0,0}} \leq 1,$$

but this holds by (7.49), since

$$n+1 \leq m+1 \leq \frac{T}{T_0} + 1 < \frac{2T}{T_0}.$$

We have thus proved (7.47) under the smallness assumptions (7.49) on δ. Since $T \geq T^*$, the condition (7.49) must fail for $\delta = \delta_0$, that is, the left side must be strictly larger than 1, since otherwise we would be able to continue the solution in $G^{\delta_0,0}(\mathbb{R})$ beyond the time T, contradicting the maximality of T^*. Therefore, there must be some $\delta \in (0, \delta_0)$ for which equality holds in (7.49) and using (7.48), we get

$$2T\frac{(1+2\|u(0)\|_{G^{\delta_0,0}(\mathbb{R})})^a}{c_0} C\delta^\kappa 2^{\frac{3}{2}} \|u(0)\|_{G^{\delta_0,0}(\mathbb{R})} = 1,$$

hence,

$$\delta = C_1 T^{-(\frac{1}{\kappa})},$$

where

$$C_1 = \left(\frac{c_0}{C 2^{\frac{5}{2}} \|u(0)\|_{G^{\delta_0,0}(\mathbb{R})} (1+2\|u(0)\|_{G^{\delta_0,0}(\mathbb{R})})^\beta} \right)^{\frac{1}{\kappa}}.$$

We have proved that (7.47) holds for this δ, hence $\|u(t)\|_{G^{\delta,0}(\mathbb{R})} < \infty$ for $t \in [0,T]$, and this completes the proof of (7.46) for the case $s = 0$.

The General Case.

For general s, we have

$$u_0 \in G^{\delta_0,s}(\mathbb{R}) \subset G^{\delta_0/2,0}(\mathbb{R}).$$

The case $s = 0$ already being proved, we know that there is a $T_1 > 0$, such hat

$$u \in C\left([0,T_1), G^{\delta_0/2,0}\right),$$

and

$$u \in C\left([0,T], G^{2pT^{-1},0}\right), \quad \text{for } T \geq T_1,$$

where $\rho > 0$ depends on u_0, δ_0 and κ. We now conclude that

$$u \in C\left([0,T_1), G^{\delta_0/4,s}\right),$$

and

$$u \in C\left([0,T], G^{pT^{-1},s}\right), \quad \text{for } T \geq T_1,$$

and these imply (7.46). The proof of Theorem 7.4 is now completed.

7.4 Time Regularity

Our next goal is to study Gevrey's temporal regularity of the unique solution obtained in the Theorem 7.2. A non-periodic function $f(x)$ is the Gevrey class of order σ i.e., $f(x) \in G^\sigma$, if there exists a constant $C > 0$ such that

$$|\partial_x^l f(x)| \leq C^{l+1}(l!)^\sigma \quad l = 0,1,2,\ldots. \tag{7.53}$$

Here, we will show that for $x \in \mathbb{R}$, for every $t \in [0,T]$ and $j,l \in \{0,1,2,\ldots\}$, there exist $C > 0$ such that,

$$|\partial_t^j \partial_x^l u(x,t)| \leq C^{j+l+1}(j!)^{5\sigma}(l!)^\sigma. \tag{7.54}$$

i.e.,

$$u(\cdot,t) \in G^\sigma(\mathbb{R}),$$

in spacial variable and

$$u(x,\cdot) \in G^{5\sigma}([0,T]),$$

in time variable.

Theorem 7.5. *Let $s > -\frac{7}{4}$, $\sigma \geqslant 1, \delta > 0$ and $\beta = \gamma = \mu = \alpha = 1$. If*

$$u_0 \in G^{\sigma,\delta,s}(\mathbb{R}),$$

then the solution

$$u \in C\left([0,T], G^{\sigma,\delta,s}(\mathbb{R})\right),$$

given by Theorem 7.2 belongs to the Gevrey class $G^{5\sigma}([0,T])$ in time variable.

Now, the regularity of the unique solution obtained in the Theorem 7.4. A non-periodic function $f(x)$ is the Gevrey class of order 1 i.e., $f(x) \in G^1$ here $\sigma = 1$, if there exists a constant $C > 0$ such that

$$|\partial_x^l f(x)| \leq C^{l+1}(l!) \quad l = 0, 1, 2, \dots. \tag{7.55}$$

Here, we will show that for $x \in \mathbb{R}$, for every $t \in [0,T]$ and $j, l \in \{0, 1, 2, \dots\}$, there exist $C > 0$ such that,

$$|\partial_t^j \partial_x^l u(x,t)| \leq C^{j+l+1}(j!)^5(l!). \tag{7.56}$$

i.e., $u(\cdot,t) \in G^1(\mathbb{R})$ in spacial variable and $u(x,\cdot) \in G^5([0,T])$ in time variable.

Corollary 7.2. *Let $s > -\frac{7}{4}$ and $\delta > 0$. If $u_0 \in G^{\delta,s}(\mathbb{R})$, then the solution*

$$u \in C\left([0,T], G^{\delta(T),s}(\mathbb{R})\right),$$

given by Theorem 7.4 belongs to the Gevrey class $G^5([0,T])$ in time variable.

7.4.1 G^σ-regularity in the spacial variable

Proposition 7.2. *Let $s > -\frac{7}{4}$ and let $\delta > 0$, $\sigma \geq 1$,*

$$u \in C\left([0,T]; G^{\sigma,\delta,s}(\mathbb{R})\right),$$

be the solution of (7.1). Then

$$u \in G^\sigma \text{ in } x, \quad \forall t \in [0,T],$$

i.e., for some $C > 0$, we have

$$|\partial_x^l u(x,t)| \leqslant C^{l+1}(l!)^\sigma, l \in \{0,1,\ldots\}, \quad \forall x \in \mathbb{R}, \ t \in [0,T]. \quad (7.57)$$

Proof. $\forall t \in [0,T]$, we have

$$\begin{aligned}\|\partial_x^l u(\cdot,t)\|^2_{H^s(\mathbb{R})} &= \int_{\mathbb{R}} |\xi|^{2l}(1+|\xi|)^{2s}|\widehat{u}(\xi,t)|^2 d\xi \\ &= \int_{\mathbb{R}} |\xi|^{2l} e^{-2\delta|\xi|^{\frac{1}{\sigma}}}(1+|\xi|)^{2s} e^{2\delta|\xi|^{\frac{1}{\sigma}}}|\widehat{u}(\xi,t)|^2 d\xi.\end{aligned}$$

We observe that

$$\begin{aligned}e^{\frac{2\delta}{\sigma}|\xi|^{\frac{1}{\sigma}}} &= \sum_{j=0}^{\infty} \frac{1}{j!}\left(\frac{2\delta}{\sigma}|\xi|^{\frac{1}{\sigma}}\right)^j \\ &\geq \frac{1}{(2l)!}\left(\frac{2\delta}{\sigma}\right)^{2l}|\xi|^{\frac{2l}{\sigma}}, \quad \forall l \in \{0,1,\ldots\}, \ \xi \in \mathbb{R}.\end{aligned}$$

This implies that

$$|\xi|^{2l} e^{-2\delta|\xi|^{\frac{1}{\sigma}}} \leq C^{2l}_{\delta,\sigma}(2l)!^\sigma.$$

Thus,

$$\begin{aligned}\|\partial_x^l u(\cdot,t)\|^2_{H^s(\mathbb{R})} &\leq C^{2l}_{\delta,\sigma}(2l)!^\sigma \int_{\mathbb{R}} e^{2\delta|\xi|^{\frac{1}{\sigma}}}(1+|\xi|)^{2s}|\widehat{u}(\xi,t)|^2 d\xi \\ &= C^{2l}_{\delta,\sigma}(2l)!^\sigma \|u(\cdot,t)\|^2_{G^{\sigma,\delta,s}(\mathbb{R})}.\end{aligned}$$

Since

$$(2l)! \leq A_1^{2l}(l!)^2,$$

for some $A_1 > 0$, one conclude that, if $s \geq 0$,

$$\begin{aligned}|\partial_x^l u(x,t)| &\leq \|\partial_x^l u(\cdot,t)\|_{L^2(\mathbb{R})} \leq \|\partial_x^l u(\cdot,t)\|_{H^s(\mathbb{R})} \\ &\leq C_0 C_1^l (l)!^\sigma \quad \forall\, t \in [0,T],\end{aligned}$$

here $C_0 = \|u(\cdot,t)\|_{G^{\sigma,\delta,s}(\mathbb{R})}$ and $C_1 = A_1 C_{\sigma,\delta}$, this implies that u is Gevrey of order σ in x, for $s \geq 0$.

Now, for $-\frac{7}{4} < s < 0$, and for any $0 < \varepsilon < \delta$, there exists a positive constant $C = C_{s,\varepsilon} > 0$ such that

$$\int_{\mathbb{R}} e^{2(\delta-\varepsilon)|\xi|^{\frac{1}{\sigma}}} |\widehat{u}(\xi,t)|^2 d\xi \leq C \int_{\mathbb{R}} \frac{e^{2\varepsilon|\xi|^{\frac{1}{\sigma}}}}{(1+|\xi|)^{-2s}} e^{2(\delta-\varepsilon)|\xi|^{\frac{1}{\sigma}}} |\widehat{u}(\xi,t)|^2 d\xi$$
$$= C \int_{\mathbb{R}} e^{2\delta|\xi|^{\frac{1}{\sigma}}} (1+|\xi|)^{2s} |\widehat{u}(\xi,t)|^2 d\xi. \quad (7.58)$$

This implies that if

$$u \in C\Big([0,T]; G^{\sigma,\delta,s}(\mathbb{R})\Big),$$

and $s < 0$, then

$$u \in C\Big([0,T]; G^{\sigma,\delta-\varepsilon,0}(\mathbb{R})\Big),$$

which allows us to conclude that u is in G^σ in x, for all $s > -\frac{7}{4}$. □

7.4.2 $G^{5\sigma}$-regularity in the time variable

We will now prove the temporal regularity of solution.

Let us consider, for $\varepsilon > 0$, the sequences

$$m_q = \frac{c(q!)^\sigma}{(q+1)^2}, \quad (q = 0, 1, 2, \ldots), \quad (7.59)$$

and

$$M_q = \varepsilon^{1-q} m_q, \varepsilon > 0 \quad \text{and} \quad (q = 1, 2, 3, \ldots), \quad (7.60)$$

where c will be chosen so that the next inequality holds

$$\sum_{0 \leq l \leq k} \binom{k}{l} m_l m_{k-l} \leqslant m_k. \quad (7.61)$$

Removing the endpoints 0 and k in the left hand side of (7.61) and using the sequence M_q, we obtain

$$\sum_{0<l<k} \binom{k}{l} M_l M_{k-l} \leq M_k, \quad \forall \varepsilon > 0. \quad (7.62)$$

Next, one can check that for any $\varepsilon > 0$ the sequence M_q satisfies the following inequality

$$M_j \leqslant \varepsilon M_{j+1}, \quad \text{for } j \geq 2. \tag{7.63}$$

Also, one can check that for a given $C > 1$, there exists $\varepsilon_0 > 0$ such that for any $0 < \varepsilon \leqslant \varepsilon_0$ we have

$$C^{j+1} j! \leqslant M_j, \quad \text{for } j \geq 2. \tag{7.64}$$

For some $C > 0$, we define the following constants

$$M_0 = \frac{c}{8} \quad \text{and} \quad M = \max\left\{2, \frac{8C}{c}, \frac{4C^2}{c}\right\}. \tag{7.65}$$

The next Lemma is the main idea for the proof of Theorem 7.5.

Lemma 7.13. *Let u be the solution of* (7.1) *satisfying* (7.57), *then there exists $\varepsilon_0 > 0$ such that for any $0 < \varepsilon \leqslant \varepsilon_0$ we have*

$$|\partial_t^j \partial_x^l u(x,t)| \leqslant M^{2j+1} M_{l+5j}, j \in \{0,1,2,\ldots\}, l \in \{0,1,2,\ldots\}, \tag{7.66}$$

for all $x \in \mathbb{R},\ t \in [0,T]$.

In order to prove Lemma 7.13, we shall need the following key result.

Lemma 7.14. *Given $n, k \in \{0,1,2,\ldots\}$ we have*

$$\sum_{p=0}^{n}\sum_{q=0}^{k}\binom{n}{p}\binom{k}{q} L_{(n-p)+5(k-q)} L_{p+5q+1} \leqslant \sum_{r=1}^{m}\binom{m}{r} L_r L_{m-r}, \tag{7.67}$$

where $L_j, j = 0,1,\ldots,m$ positive real numbers with $m = n + 5k + 1$.

Proof. For $n, k \in \{0,1,2,\ldots\}$ given let

$$m = n + 5k + 1.$$

For $k = n = 0$ inequality (7.67) reads $L_0 L_1 \leqslant L_0 L_1$, which is true. Therefore, we assume that either $k \geqslant 1$ or $n \geqslant 1$. Then, changing the order of the summations and making a change of variables gives

$$\begin{aligned}
&\sum_{p=0}^{n}\sum_{q=0}^{k}\binom{n}{p}\binom{k}{q} L_{(n-p)+5(k-q)} L_{p+1+5q} \\
&\quad = \sum_{q=0}^{k}\sum_{p=0}^{n}\binom{n}{p}\binom{k}{q} L_{(n-p)+5(k-q)} L_{p+1+5q}
\end{aligned}$$

$$= \sum_{q=0}^{k} \sum_{r=1+5q}^{n+1+5q} \binom{n}{r-1-5q} \binom{k}{q} L_{m-r} L_r$$

$$= \sum_{r=1}^{m} \sum_{q=i_0(r)}^{i_1(r)} \binom{n}{r-1-5q} \binom{k}{q} L_{m-r} L_r,$$

with

$$i_0(r) = \max\left\{0, \left[\left[\frac{r-n-1}{5}\right]\right]\right\}, \quad i_1(r) = \min\left\{\left[\frac{r-1}{5}\right], 5k\right\},$$

where $[x]$ is the integer part of a number x and $[[x]]$ is the lesser integer that is greater than or equal to x. To complete the proof of inequality (7.67) we must to show that

$$\sum_{q=i_0(r)}^{i_1(r)} \binom{n}{r-1-5q} \binom{k}{q} \leqslant \binom{m}{r}.$$

This is a consequence of the following result. □

Lemma 7.15. *We have*

$$\sum_{q=i_0(r)}^{\theta} \binom{n}{r-1-5q} \binom{k}{q} \leqslant \binom{m-4k+4\theta}{r}, \tag{7.68}$$

for all $i_0(r) \leqslant \theta \leqslant i_1(r)$.

In fact, using (7.68) *with* $\theta = i_1(r)$ *it suffices to show that*

$$\binom{m-4k+4i_1(r)}{r} \leqslant \binom{m}{r}. \tag{7.69}$$

For $i_1(r) = k$ *relation* (7.69) *holds as an equality. If* $0 \leqslant i_1(r) < k$ *then* $1 \leqslant 4(k - i_1(r))$ *and therefore*

$$m - 4k + 4i_1(r) \leqslant m - 1 < m,$$

which shows that (7.69) *holds as a strict inequality. This completes the proof of Lemma* 7.14.

Proof. We prove it by induction on θ. For this, we use the following elementary inequality: If $a, b, c \in \mathbb{N}, b \leqslant a$ then

$$\binom{a}{b} \leqslant \binom{a+c}{b+c}.$$

With $a = n$, $b = r-1-5i_0(r), c = 1+2i_0(r)$ and using the definition of m gives

$$\binom{n}{r-1-5i_0(r)} = \binom{m-1-5k}{r-1-5i_0(r)} \leqslant \binom{m-5k+4i_0(r)}{r-i_0(r)}.$$

Now, since for $\alpha, \beta, \gamma, \delta \in \mathbb{N}$ with $\alpha \leqslant \beta$ and $\gamma \leqslant \delta$ we have that

$$\binom{\beta}{\alpha}\binom{\delta}{\gamma} \leqslant \binom{\beta+\delta}{\alpha+\gamma},$$

we get

$$\begin{aligned}\binom{n}{r-1-5i_0(r)}\binom{k}{i_0(r)} &\leqslant \binom{m-5k+4i_0(r)}{r-i_0(r)}\binom{k}{i_0(r)}\\ &\leqslant \binom{m-4k+4i_0(r)}{r},\end{aligned}$$

which proves (12.5) for $\theta = i_0(r)$. Next, we assume that (7.69) holds for $i_0(r) \leqslant \theta < i_1(r)$ and we will prove it for $(\theta+1)$. By using the induction hypotheses we obtain

$$\begin{aligned}&\sum_{q=i_0(r)}^{\theta+1}\binom{n}{r-1-5q}\binom{k}{q}\\ &= \sum_{q=i_0(r)}^{\theta}\binom{n}{r-1-5q}\binom{k}{q} + \binom{n}{r-1-5(\theta+1)}\binom{k}{\theta+1}\\ &\leqslant \binom{m-4k+4\theta}{r} + \binom{n}{r-1-5(\theta+1)}\binom{k}{\theta+1}.\end{aligned}$$

Now with $a = n, b = r-1-5(\theta+1), c = 4(\theta+1)$, we get

$$\binom{n}{r-1-5(\theta+1)} \leqslant \binom{n+4(\theta+1)}{r-\theta-2}.$$

Now, using the last inequality together with

$$\binom{v}{\mu} + \binom{v}{\mu+1} = \binom{v+1}{\mu+1},$$

we have

$$\begin{aligned}
&\sum_{q=i_0(r)}^{\theta+1} \binom{n}{r-1-5q}\binom{k}{q} \\
&\leqslant \binom{m-4k+4\theta}{r} + \binom{n+4(\theta+1)}{r-\theta-2}\binom{k}{\theta+1} \\
&\leqslant \binom{m-4k+4\theta}{r} + \binom{n+k+4\theta+4}{r-1} \\
&= \binom{m-4k+4\theta}{r} + \binom{m-1-4k+4\theta+4}{r-1} \\
&\leqslant \binom{m-4k+4\theta}{r} + \binom{m-4k+4\theta+3}{r-1} + \binom{m-4k+4\theta}{r-1} \\
&\leqslant \binom{m-4k+4\theta+1}{r} + \binom{m-4k+4\theta+3}{r-1} \\
&\leqslant \binom{m-4k+4\theta+3}{r} + \binom{m-4k+4\theta+3}{r-1} \\
&\leqslant \binom{m-4k+4(\theta+1)}{r},
\end{aligned}$$

which completes the proof of Lemma 7.15. □

Now, we are ready to complete the proof of the estimates (7.66) for $\partial_t^j \partial_x^l u(x,t)$.

Proof of Lemma 7.13. We will prove (7.66) by induction. Let $j = 0$, for $l = 0$, it follows from (7.57) and the definition of M in (7.65) that

$$|u(x,t)| \leq C \leq MM_0, \quad \forall x \in \mathbb{R}, \ t \in [0,T].$$

Similarly, for $l = 1$, we have

$$|\partial_x u(x,t)| \leq C^2 \leqslant MM_1, \quad \forall x \in \mathbb{R}, \ t \in [0,T].$$

By (7.57) and (7.63), for $l \geq 2$, there exists $\varepsilon_0 > 0$ such that for any $0 < \varepsilon \leqslant \varepsilon_0$, we have

$$|\partial_x^l u(x,t)| \leq C^{l+1}(l!)^\sigma \leq M_l \leq M M_l, \quad \forall x \in \mathbb{R},\ t \in [0,T].$$

This completes the proof of (7.66) for $j = 0$ and $l \in \{0, 1, \ldots\}$.

Next, we will assume that (7.66) is true for $0 \leq q \leq j$ and $l \in \{0, 1, \ldots\}$ and we will prove it for $q = j+1$ and $l \in \{0, 1, \ldots\}$.

We begin by noting that

$$\begin{aligned}|\partial_t^{j+1}\partial_x^l u(x,t)| &= |\partial_t^j \partial_x^l(\partial_t u(x,t))| \\ &\leqslant |\partial_t^j \partial_x^{l+5} u(x,t)| + |\partial_t^j \partial_x^{l+3} u(x,t)| \\ &\quad + |\partial_t^j \partial_x^{l+1} u(x,t)| + |\partial_t^j \partial_x^l (\partial_x u^2(x,t))|.\end{aligned}$$

Using the induction hypotheses and the condition $M \geqslant 2$, we estimate the second term

$$\partial_t^j \partial_x^{l+5} u(x,t), \quad \partial_t^j \partial_x^{l+3} u(x,t), \quad \partial_t^j \partial_x^{l+1} u(x,t),$$

as follows:

$$\begin{aligned}|\partial_t^j \partial_x^{l+5} u(x,t)| &\leqslant M^{2j+1} M_{l+5+5j} = M^{-2} M^{2(j+1)+1} M_{l+5(j+1)} \\ &\leqslant \frac{1}{4} M^{2(j+1)+1} M_{l+5(j+1)},\end{aligned} \tag{7.70}$$

and

$$\begin{aligned}|\partial_t^j \partial_x^{l+3} u(x,t)| &\leqslant M^{2j+1} M_{l+3+5j} = M^{-2} M^{2(j+1)+1} M_{l+5(j+3)} \\ &\leqslant \tfrac{\varepsilon^2}{4} M^{2(j+1)+1} M_{l+5(j+1)},\end{aligned} \tag{7.71}$$

and

$$|\partial_t^j \partial_x^{l+1} u(x,t)| \leqslant M^{2j+1} M_{l+1+5j} \leqslant \tfrac{\varepsilon^4}{4} M^{2(j+1)+1} M_{l+5(j+1)}. \tag{7.72}$$

Next, we estimate the nonlinear term, using Leibniz's formula we write $\partial_t^j \partial_x^l (\partial_x u^2(x,t))$ as

$$\begin{aligned}|\partial_t^j \partial_x^{l+1}(u^2)| &\leqslant \sum_{p=0}^{l+1} \sum_{q=0}^{j} \binom{l+1}{p} \binom{j}{p} |\partial_t^{j-q} \partial_x^{l+1-p} u| |\partial_t^q \partial_x^p u| \\ &\leqslant \sum_{p=0}^{l+1} \sum_{q=0}^{j} \binom{l+1}{p} \binom{j}{p} M^{2(j-q)+1}\end{aligned}$$

$$\times M_{l+1-p+5(j-q)} M^{2q+1} M_{p+5q}$$

$$= M^{2(j+1)} \sum_{p=0}^{l+1} \sum_{q=0}^{j} \binom{l+1}{p} \binom{j}{p} M_{l+1-p+5(j-q)} M_{p+5q}.$$

Next, using Lemma 7.14, with $n = l+1, k = j, L_j = M_j$, $m = l+1+5j$, we obtain

$$\sum_{p=0}^{l+1} \sum_{q=0}^{j} \binom{l+1}{p} \binom{j}{p} M_{l+1-p+5(j-q)} M_{p+5q}$$

$$\leqslant \sum_{r=1}^{m} \binom{m}{r} L_r L_{m-r} \leq (M_0+\varepsilon) M_m$$

$$= (M_0+\varepsilon) M_{l+5j+1},$$

then

$$|\partial_t^j \partial_x^{l+1}(u^2)| \leqslant M^{2(j+1)}(M_0+\varepsilon) M_{l+5j+1}$$

$$\leqslant M^{-2} M^{2(j+1)+1} \varepsilon^4 (M_0+\varepsilon) M_{l+5(j+1)}$$

$$\leqslant \frac{\varepsilon^4 (M_0+\varepsilon)}{4} M^{2(j+1)+1} M_{l+5(j+1)}.$$

Noting that in the last inequality we have used the fact that $l+5j+1 \geq 2$, since we are assuming that either $j \neq 0$ or $l \neq 0$.

Now, choosing

$$\varepsilon \leq \varepsilon_0 = \left(\frac{1}{(M_0+\varepsilon)} \right)^{\frac{1}{4}} < 1,$$

to obtain

$$\varepsilon^4 (M_0+\varepsilon) \leq \varepsilon^4 (M_0+1) \leq (M_0+1) \left(\frac{1}{(M_0+1)} \right) = 1.$$

Hence,

$$|\partial_t^j \partial_x^{l+1}(u^2)| \leq \frac{1}{4} M^{2(j+1)+1} M_{l+5(j+1)}. \tag{7.73}$$

Combining this estimate with the estimates (7.70), (7.71) and (7.72) yields

$$|\partial_t^{j+1} \partial_x^l u(x,t)| = M^{2(j+1)+1} M_{l+5(j+1)}.$$

Which completes the proof of Lemma 7.13. □

Proof of Theorem 7.5. By Lemma 7.13, we have

$$|\partial_t^j \partial_x^l u(x,t)| \leqslant M^{2j+1} M_{l+5j}, \quad j \in \{0,1,2,\dots\}, \quad l \in \{0,1,2,\dots\},$$

where

$$M_q = \varepsilon^{1-q} \frac{c(q!)^\sigma}{(q+1)^2}, \quad q = 1,2,\dots.$$

Applying this inequality for $j \in \{1,2,\dots\}$ and $l = 0$ gives

$$\begin{aligned}
|\partial_t^j u(x,t)| &\leqslant M^{2j+1} M_{5j} = M M^{2j} \varepsilon^{1-5j} \frac{c((5j)!)^\sigma}{(5j+1)^2} \\
&\leqslant M\varepsilon c \left(\frac{M^2}{\varepsilon^5}\right)^j ((5j)!)^\sigma \\
&\leqslant L_0 L^j ((5j)!)^\sigma \\
&\leqslant L_0 L^j A^{5\sigma j} ((j!)^5)^\sigma \\
&\leqslant A_0^{j+1} (j!)^{5\sigma}, \qquad (7.74)
\end{aligned}$$

where $L_0 = M\varepsilon c,\ L = \frac{M^2}{\varepsilon^5}$ since $(5j)! \leqslant A^{5j}(j!)^5$ for $A > 0$ and

$$A_0 = \max\{L_0, LA^{5\sigma}\}.$$

We also have from (7.66) for $l = j = 0$, that

$$|u(x,t)| \leq MM_0 = M\frac{c}{8}, \quad \forall x \in \mathbb{R},\ t \in [0,T]. \qquad (7.75)$$

Setting

$$C = \max\left\{M\frac{c}{8}, A_0\right\},$$

it follows from (7.74) and (7.75) that for $j \in \{0,1,2,\dots\}$, we have

$$|\partial_t^j u(x,t)| \leqslant C^{j+1}(j!)^{5\sigma}, \quad \forall x \in \mathbb{R},\ t \in [0,T].$$

Hence, $u \in G^{5\sigma}$ in t.

Which completes the proof of Theorem 7.5. □

Chapter 8

Higher-Order Nonlinear Dispersive Equation

The higher-order nonlinear dispersive equation which arises in the study of water waves with surface tension and arises as mathematical models for the weakly nonlinear propagation of long waves. Well-posedness of the Cauchy problem for the higher-order nonlinear dispersive equation in (8.1) has been studied by Zhang *et al.* [7] in Sobolev spaces. By using the Fourier restriction norm, the authors showed that (8.1) is locally well-posed in $H^s(\mathbb{R})$ for $s \geq -\frac{5}{8}$ and global well-posedness for $s = 0$.

We consider a higher order nonlinear dispersive equation as follows

$$\partial_t u + \lambda_1 \partial_x^7 u + \lambda_2 \partial_x^5 u + \lambda_3 \partial_x^3 u + \lambda_4 \partial_x u + u \partial_x u = 0, \ (x,t) \in \mathbb{R}^2,$$
$$u(x,0) = u_0(x), \tag{8.1}$$

where $\lambda_1 \neq 0, \lambda_2, \lambda_3, \lambda_4$ are real numbers.

The higher-order nonlinear dispersive equation which arises in the study of water waves with surface tension and arises as mathematical models for the weakly nonlinear propagation of long waves.

We study of the question of global well-posedness for initial data $u_0(x)$ that is analytic on the line and can be extended as holomorphic functions in a strip around the x-axis.

Proposition 8.1 (Paley–Wiener theorem). *Let $\sigma > 0$ and suppose $f \in L^2(\mathbb{R})$. Then the following are equivalent:*

1. *The function f is the restriction to the real line of a function F which is holomorphic in the strip*
$$S_\sigma = \{x + iy \in \mathbb{C} : \ |y| < \sigma\},$$
and satisfies
$$\sup_{|y|<\sigma} \|F(x+iy)\|_{L^2} < \infty.$$
2. $e^{\sigma|\xi|}\tilde{f}(\xi) \in L^2_\xi(\mathbb{R})$.

In the view of the Paley–Wiener theorem, it is natural to take initial data in $G^{\delta,s}$ and obtain a better understanding of the behavior of solution as we try to extend it to be globally in time. It means that given $u_0 \in G^{\delta,s}$ for some initial radius $\delta > 0$ we want to estimate the behavior of the radius of analyticity $\delta(T)$ as time T goes to ∞.

Lemma 8.1. *Let $b > \frac{1}{2}$, $s \in \mathbb{R}$ and $\delta \geq 0$. Then*
$$X_{\delta,s,b} \subset C(\mathbb{R}, G^{\delta,s}(\mathbb{R})),$$
and
$$\sup_{t\in\mathbb{R}} \|u(t)\|_{G^{\delta,s}} \leq C\|u(t)\|_{X_{\delta,s,b}}, \tag{8.2}$$
where C depends only on b.

Proof. First, we observe that the operator A
$$\widehat{A^\delta u}^x(\xi,t) = e^{\delta|\xi|}\widehat{u}^x(\xi,t),$$
satisfies
$$\|u\|_{X_{\delta,s,b}} = \|A^\delta u\|_{X_{s,b}} \quad \text{and} \quad \|u\|_{G^{\delta,s}} = \|A^\delta u\|_{H^s}. \tag{8.3}$$
We observe that $A^\delta u$ belongs to $C(\mathbb{R}, H^s)$ and for some $C > 0$, we have
$$\| A^\delta u \|_{C(\mathbb{R},H^s)} \leq C \| A^\delta u \|_{X_{s,b}}. \tag{8.4}$$
Thus, it follows that
$$u \in C\left([0,T], G^{\delta,s}\right),$$

and

$$\| u \|_{C\left(\mathbb{R},G^{\delta,s}\right)} \leq C \| u \|_{X_{\delta,s,b}} . \tag{8.5}$$

□

Lemma 8.2. *Let $s \in \mathbb{R}$, $\delta \geq 0$ and $-\frac{1}{2} < b \leq b' < \frac{1}{2}$ then for any $T > 0$ we have,*

$$\|u\|_{X^T_{\delta,s,b}} \leq CT^{b'-b}\|u\|_{X^T_{\delta,s,b'}}, \tag{8.6}$$

where C depends only on b and b'.

Lemma 8.3. *Let $s \in \mathbb{R}$, $\delta \geq 0$, $-\frac{1}{2} < b < \frac{1}{2}$ and $T > 0$. Then, for any time interval $I \subset [0,T]$, we have*

$$\|\chi_I(t)u\|_{X_{\delta,s,b}} \leq C\|u\|_{X^T_{\delta,s,b}}, \tag{8.7}$$

and

$$\|\chi_I(t)w\|_{X^{\beta}_{\delta,s,b}} \leq C\|w\|_{X^{\beta,T}_{\delta,s,b}}, \tag{8.8}$$

where $\chi_I(t)$ is the characteristic function of I and C depends only on b.

Next, consider the linear problem, for given $F = \partial_x u^2$ and u_0,

$$\partial_t u + \lambda_1 \partial_x^7 u + \lambda_2 \partial_x^5 u + \lambda_3 \partial_x^3 u + \lambda_4 \partial_x u = F, \tag{8.9}$$

and $u(x,0) = u_0(x)$.

By Duhamel's principle the solution can be then written as

$$u(t) = S(t)u_0 - \frac{1}{2}\int_0^t S(t-s)\left(F(s)\right) ds, \tag{8.10}$$

where

$$S(t)u_0 = \int_{\mathbb{R}} e^{i(x\xi + t\phi(\xi))}\widehat{u_0}(\xi)d\xi.$$

Lemma 8.4. *Let $s \in \mathbb{R}$, $\frac{1}{2} < b \leq 1$, $\delta \geq 0$ and $0 < T \leq 1$, there is a constant $C > 0$ depends only on b, such that*

$$\|S(t)u_0\|_{X^T_{\delta,s,b}} \leq C\|u_0\|_{G^{\delta,s}}, \tag{8.11}$$

$$\left\| \int_0^t S(t-s)F(s)ds \right\|_{X^T_{\delta,s,b}} \leq C\|F\|_{X^T_{\delta,s,b-1}}, \tag{8.12}$$

where the constant $C > 0$ depends only on b.

Theorem 8.1. *Assume that $\lambda_1\lambda_2 < 0$ and $\lambda_3 > 0$, let b' be close enough to $1/2$ satisfying $b' > 1/2$. For $b > 1/2$ and $s > -5/8$, we have*

$$\| \partial_x(u_1u_2) \|_{X_{s,b'-1}} \leq C \| u_1 \|_{X_{s,b}} \| u_2 \|_{X_{s,b}}. \tag{8.13}$$

We used the bilinear estimate in Sobolev spaces H^s and the operator A in order to proof bilinear estimate in analytic Gevrey spaces $G^{\delta,s}$.

Lemma 8.5. *Assume that $\lambda_1\lambda_2 < 0$ and $\lambda_3 > 0$, let b' be close enough to $1/2$ satisfying $b' > 1/2$. For $b > 1/2$, $\delta \geq 0$ and $s > -5/8$, we have*

$$\| \partial_x(u_1u_2) \|_{X_{\delta,s,b'-1}} \leq C \| u_1 \|_{X_{\delta,s,b}} \| u_2 \|_{X_{\delta,s,b}}. \tag{8.14}$$

Proof. We observe, by considering the operator A^δ, that

$$\begin{aligned} e^{\delta|\xi|}\widehat{u_1u_2} &= (2\pi)^{-2}e^{\delta|\xi|}\widehat{u_1} * \widehat{u_2} \\ &\leq (2\pi)^{-2}\int_{\mathbb{R}^2} e^{\delta|\xi-\xi_1|}\widehat{u_1}(\xi-\xi_1,\tau-\tau_1)e^{\delta|\xi_1|}\widehat{u_2}(\xi_1,\tau_1)d\xi_1 d\tau_1 \\ &= (\widehat{A^\delta u_1 A^\delta u_2}), \end{aligned}$$

since

$$\delta \mid \xi \mid \leq \delta \mid \xi - \xi_1 \mid + \delta \mid \xi_1 \mid.$$

Thus, we have

$$\| \partial_x(u_1u_2) \|_{X_{\delta,s,b'-1}} \leq \| \partial_x(A^\delta u_1 A^\delta u_2) \|_{X_{s,b'-1}}.$$

Then, we have

$$\begin{aligned} \| \partial_x(A^\delta u_1 A^\delta u_2) \|_{X_{s,b'-1}} &\leq C \| A^\delta u_1 \|_{X_{s,b}} \| A^\delta u_2 \|_{X_{s,b}} \\ &= C \| u_1 \|_{X_{\delta,s,b}} \| u_2 \|_{X_{\delta,s,b}}. \end{aligned}$$

□

8.1 Existence of Solution

This result shows that for local-in-time the radius of analyticity remains constant.

Theorem 8.2. *Assume that* $\lambda_1\lambda_2 < 0$ *and* $\lambda_3 > 0$, *let* $\delta > 0$ *and* $s > -5/8$. *Then for any* $u_0 \in G^{\delta,s}$, *there exists*

$$T = T(\| u_0 \|_{G^{\delta,s}}) > 0,$$

and a unique solution u *of* (8.1) *on the time interval* $(0, T)$ *such that*

$$u \in C\left([0,T], G^{\delta,s}\right).$$

Moreover, the solution depends continuously on the data u_0. *Here, we have*

$$T = \frac{c_0}{(1+ \| u_0 \|_{G^{\delta,s}})^{\beta}}, \tag{8.15}$$

for some constants $c_0 > 0$ *and* $\beta > 1$ *depending only on* s. *Furthermore, the solution* u *satisfies the bound*

$$\|u\|_{X^T_{\delta,s,b}} \leq 2C\|u_0\|_{G^{\delta,s}} \quad 1/2 < b < 1, \tag{8.16}$$

with a constant $C > 0$ *depending only on* s *and* b.

Fixing $\delta > 0$, $s > -5/8$ and $u_0 \in G^{\delta,s}$. To construct the local solution u to (8.1), we proceed by an iteration argument in the space $X^T_{\delta,s,b}$. Let $\{u^{(n)}\}_{n=0}^{\infty}$ be the sequence defined by

$$\partial_t u^{(0)} + \lambda_1\partial_x^7 u^{(0)} + \lambda_2\partial_x^5 u^{(0)} + \lambda_3\partial_x^3 u^{(0)} + \lambda_4\partial_x u^{(0)} = 0,$$

with $u^{(0)}(0) = u_0$. For $n \in \{1, 2, \ldots\}$, we have.

$$\begin{aligned}&\partial_t u^{(n)} + \lambda_1\partial_x^7 u^{(n)} + \lambda_2\partial_x^5 u^{(n)} + \lambda_3\partial_x^3 u^{(n)} + \lambda_4\partial_x u^{(n)}\\&= -\frac{1}{2}\partial_x(u^{(n-1)}u^{(n-1)}),\end{aligned}$$

with $u^{(n)}(0) = u_0$.

Based on the comments preceding Lemma 8.4, we may write

$$u^{(0)}(x,t) = S(t)u_0(x),$$

$$u^{(n)}(x,t) = S(t)u_0(x) - \frac{1}{2}\int_0^t S(t-s)\partial_x\left(u^{(n-1)}(x,s)u^{(n-1)}(x,s)\right) ds.$$

It then follows from Lemmas 8.2, 8.5 and 8.4

$$\begin{aligned}\|u^{(0)}\|_{X^T_{\delta,s,b}} &\leq C\|u_0\|_{G^{\delta,s}}\\ \|u^{(n)}\|_{X^T_{\delta,s,b}} &\leq C\|u_0\|_{G^{\delta,s}} + CT^{b'-b}\|\partial_x(u^{(n-1)}u^{(n-1)})\|_{X^T_{\delta,s,b'-1}}\\ &\leq C\|u_0\|_{G^{\delta,s}} + CT^{b'-b}\|u^{(n-1)})\|^2_{X^T_{\delta,s,b}},\end{aligned} \tag{8.17}$$

with $1/2 < b < b' < 1$. By induction, it follows that

$$\|u^{(n)}\|_{X^T_{\delta,s,b}} \leq 2C\|u_0\|_{G^{\delta,s}}, \tag{8.18}$$

for all n, if $T \in (0,1]$ is chosen so small that

$$T \leq \frac{1}{(8C^2\|u_0\|_{G^{\delta,s}})^{\frac{1}{b'-b}}}. \tag{8.19}$$

Using Lemma 8.5 together with (8.18) and (8.17) in that order, we therefore get

$$\begin{aligned}&\|u^{(n)} - u^{(n-1)}\|_{X^T_{\delta,s,b}}\\ &\leq CT^{b'-b}\left\|\partial_x\left(u^{(n-1)}u^{(n-1)} - u^{(n-2)}u^{(n-2)}\right)\right\|_{X^T_{\delta,s,b'-1}}\\ &\leq CT^{b'-b}\left(\|u^{(n-1)}\|_{X^T_{\delta,s,b}} + \|u^{(n-2)})\|_{X^T_{\delta,s,b}}\right)\\ &\quad\cdot\|u^{(n-1)} - u^{(n-2)})\|_{X^T_{\delta,s,b}}\\ &\leq \frac{1}{2}\|u^{(n-1)} - u^{(n-2)})\|_{X^T_{\delta,s,b}}.\end{aligned}$$

It follows that the sequence converges to a solution u verifying the bound (8.18).

8.2 Continuous Dependence on the Initial Data

Now assume that u and v are solutions to (8.1) for initial data u_0 and v_0, respectively. Then similarly as above, again with the same choice of T and for any T' such that $0 < T' < T$, we have

$$\| u - v \|_{X^{T'}_{\delta,s,b}} \leq C \| u_0 - v_0 \|_{G^{\delta,s}} + \tfrac{1}{2} \| u - v \|_{X^{T'}_{\delta,s,b}},$$

provided $\| u_0 - v_0 \|_{G^{\delta,s}}$ is sufficiently small. This proves continuous dependence.

The uniqueness. Uniqueness of the solution in

$$C([0,T], G^{\delta,s}),$$

can be proved by the following standard argument.

Suppose that

$$u, v \in C([0,T], G^{\delta,s}),$$

are solutions to (8.1) with $u(\cdot, 0) = v(\cdot, 0)$ in $G^{\delta,s}$.

Setting $w = u - v$, we see that w solves the Cauchy problem

$$\begin{aligned}&\partial_t w + \lambda_1 \partial_x^7 w + \lambda_2 \partial_x^5 w + \lambda_3 \partial_x^3 + \lambda_4 \partial_x w\\&+\frac{1}{2}\partial_x w(u+v) = 0, \quad w(0) = 0.\end{aligned}$$

Multiplying both sides by w and integrating in space yield

$$\begin{aligned}&w\partial_t w + \lambda_1 w\partial_x^7 w + \lambda_2 w\partial_x^5 w + \lambda_3 w\partial_x^3 + \lambda_4 w\partial_x w\\&+\frac{1}{2}w\partial_x w(u+v) = 0.\end{aligned}$$

Thus, we have

$$\begin{aligned}\frac{1}{2}\partial_t \|w(\cdot,t)\|_{L^2}^2 &= \frac{1}{2}\partial_t \int_{\mathbb{R}} w^2(x,t)dx = \int_{\mathbb{R}} w(x,t)\partial_t w(x,t)dx\\&= -\frac{1}{2}\int_{\mathbb{R}} w(x,t)\partial_x(u^2 - v^2)dx = 0,\end{aligned} \tag{8.20}$$

since we have

$$\begin{aligned}\int_{\mathbb{R}} w(x,t)\partial_x^7 w(x,t)dx &= \int_{\mathbb{R}} w(x,t)\partial_x^5 w(x,t)dx\\&= \int_{\mathbb{R}} w(x,t)\partial_x^3 w(x,t)dx\\&= \int_{\mathbb{R}} w(x,t)\partial_x w(x,t)dx = 0.\end{aligned}$$

We may here assume that w and its all spatial derivatives decay to zero as $| x | \longrightarrow \infty$. Thanks to (8.20), we have

$$\begin{aligned}\partial_t \|w(\cdot,t)\|_{L^2}^2 &= -\int_{\mathbb{R}} w(x,t)\partial_x(u^2-v^2)dx \\ &= -\int_{\mathbb{R}} w(x,t)\partial_x\left[f(x,t)w(x,t)\right]dx,\end{aligned}$$

where $f = u + v$. Integrating by parts the last integral we obtain

$$\partial_t \|w(\cdot,t)\|_{L^2}^2 = \int_{\mathbb{R}} \partial_x f(x,t)w^2(x,t)dx,$$

from which we deduce the inequality

$$\left|\partial_t \|w(\cdot,t)\|_{L^2}^2\right| = \|\partial_x f\|_{L^\infty}\|w(t)\|_{L^2}^2. \tag{8.21}$$

Since

$$u, v \in C([0,T], G^{\delta,s}),$$

we have u and v are continuous in t on the compact set $[0,T]$ and are $G^{\delta,s}$ in x. Thus, we can conclude that

$$\|\partial_x f\|_{L^\infty} \leq c < \infty. \tag{8.22}$$

Therefore, from (8.21) and (8.22) we obtain the differential inequality

$$\left|\partial_t \|w(\cdot,t)\|_{L^2}^2\right| \leq c\|w(t)\|_{L^2}^2, \quad 0 \leq t \leq T.$$

Solving it gives

$$\|w(t)\|_{L^2}^2 \leq e^c\|w(0)\|_{L^2}^2, \quad 0 \leq t \leq T. \tag{8.23}$$

Since $\|w(0)\|_{L^2}^2 = 0$, from (8.23) we obtain that $w(t) = 0$, $0 \leq t \leq T$ or $u = v$.

8.3 Approximate Conservation Law

We are going to establish an approximate conservation law for a solution to (8.1) based on the conservation the $L^2(\mathbb{R})$ norm of solutions of the equation.

Theorem 8.3. *Let $\kappa \in [0, 5/8)$ and T be as in Theorem 8.2, there exist $b \in (1/2, 1)$ and $C > 0$, such that for any $\delta > 0$ and any solution $u \in X^T_{\delta,0,b}$ to the Cauchy problem (8.1) on the time interval $[0, T]$, we have the estimate*

$$\sup_{t\in[0,T]} \|u(t)\|^2_{G^{\delta,0}} \leq \|u(0)\|^2_{G^{\delta,0}} + C\delta^\kappa \|u\|^3_{X^T_{\delta,0,b}}. \tag{8.24}$$

Moreover, we have

$$\sup_{t\in[0,T]} \|u(t)\|^2_{G^{\delta,0}} \leq \|u(0)\|^2_{G^{\delta,0}} + C\delta^\kappa \|u(0)\|^3_{G^{\delta,0}}. \tag{8.25}$$

For the proof of Theorem 8.3, we require the following preliminary estimate.

Lemma 8.6. *Given $\kappa \in [0, 5/8)$, there exist $b \in (1/2, 1)$ and $C > 0$, such that for all $T > 0$ and $u \in X_{\delta,0,b}$, we have*

$$\|G\|_{X_{0,b-1}} \leq C\delta^\kappa \|u\|^2_{X_{\delta,0,b}}, \tag{8.26}$$

where

$$G = \frac{1}{2}\partial_x \left[(A^\delta u)^2 - A^\delta (u)^2\right].$$

Proof. Let

$$G = \frac{1}{2}\partial_x \left[(A^\delta u)^2 - A^\delta (u)^2\right].$$

Then

$$\begin{aligned} \|G\|_{X_{0,b-1}} &= \tfrac{1}{2} \Bigg\| \tfrac{\xi}{\langle\tau-\phi(\xi)\rangle^{1-b}} \int_{\mathbb{R}^2} (e^{\delta|\xi_1|}\widehat{u}(\xi_1,\tau_1)e^{\delta|\xi-\xi_1|}\widehat{u}(\xi-\xi_1,\tau-\tau_1) \\ &\quad - e^{\delta|\xi|}\widehat{u}(\xi_1,\tau_1)\widehat{u}(\xi-\xi_1,\tau-\tau_1))d\xi_1 d\tau_1 \Bigg\|_{L^2_{\xi,\tau}} \\ &= \frac{1}{2} \Bigg\| \frac{\xi}{\langle\tau-\phi(\xi)\rangle^{1-b}} \int_{\mathbb{R}^2} (e^{\delta|\xi_1|}e^{\delta|\xi-\xi_1|} - e^{\delta|\xi|})\widehat{u}(\xi_1,\tau_1) \\ &\quad \cdot \widehat{u}(\xi-\xi_1,\tau-\tau_1)d\xi_1 d\tau_1 \Bigg\|_{L^2_{\xi,\tau}}, \end{aligned}$$

where $\langle\cdot\rangle = (1 + |\cdot|)$.

For $\kappa \in [0, 5/8) \subset [0, 1]$, one can see that

$$\|G\|_{X_{0,b-1}} \le \frac{1}{2}\left\| \frac{\xi}{\langle \tau - \phi(\xi) \rangle^{1-b}} \int_{\mathbb{R}^2} [2\delta \min(|\xi_1|, |\xi - \xi_1|)]^\kappa \right.$$

$$\left. \times\, e^{\delta|\xi_1|} e^{\delta|\xi - \xi_1|} \widehat{u}(\xi_1, \tau_1) \widehat{u}(\xi - \xi_1, \tau - \tau_1) d\xi_1 d\tau_1 \right\|_{L^2_{\xi,\tau}}$$

$$\le \frac{1}{2}(2\delta)^\kappa \left\| \frac{\xi \langle \xi \rangle^{-\kappa}}{\langle \tau - \phi(\xi) \rangle^{1-b}} \int_{\mathbb{R}^2} e^{\delta|\xi_1|} \langle \xi_1 \rangle^\kappa \widehat{u}(\xi_1, \tau_1) \right.$$

$$\left. \times\, e^{\delta|\xi - \xi_1|} \langle \xi - \xi_1 \rangle^\kappa \widehat{u}(\xi - \xi_1, \tau - \tau_1) d\xi_1 d\tau_1 \right\|_{L^2_{\xi,\tau}}.$$

Now by taking $s = -\kappa \in (-5/8, 0]$, we obtain

$$\|G\|_{X_{0,b-1}} \le \frac{1}{2}(2\delta)^\kappa \left\| \frac{\xi \langle \xi \rangle^{s}}{\langle \tau - \phi(\xi) \rangle^{1-b}} \int_{\mathbb{R}^2} \frac{e^{\delta|\xi_1|} \widehat{u}(\xi_1, \tau_1)}{\langle \xi_1 \rangle^s} \right.$$

$$\left. \times\, \frac{e^{\delta|\xi - \xi_1|} \widehat{u}(\xi - \xi_1, \tau - \tau_1)}{\langle \xi - \xi_1 \rangle^s} d\xi_1 d\tau_1 \right\|_{L^2_{\xi,\tau}}.$$

Setting $v = Au$ and

$$f(\tau, \xi) = \langle \tau - \phi(\xi) \rangle^b \widehat{v}(\tau, \xi),$$

we have

$$e^{\delta|\xi|} \widehat{u}(\tau, \xi) = \widehat{v}(\tau, \xi) = f(\tau, \xi) \langle \tau - \phi(\xi) \rangle^{-b},$$

and therefore we can write

$$\|G\|_{X_{0,b-1}} \le \frac{1}{2}(2\delta)^\kappa \left\| \frac{\xi \langle \xi \rangle^{s}}{\langle \tau - \phi(\xi) \rangle^{1-b}} \int_{\mathbb{R}^2} \frac{f(\xi_1, \tau_1)}{\langle \xi_1 \rangle^s \langle \tau_1 - \phi(\xi_1) \rangle^b} \right.$$

$$\left. \times\, \frac{f(\xi - \xi_1, \tau - \tau_1)}{\langle \xi - \xi_1 \rangle^s \langle \tau - \tau_1 - \phi(\xi - \xi_1) \rangle^b} d\xi_1 d\tau_1 \right\|_{L^2_{\xi,\tau}}.$$

We get

$$\|G\|_{X_{0,b-1}} \le C\delta^\kappa \|f\|^2_{L^2_{\xi,\tau}} = C\delta^\kappa \|v\|^2_{X_{0,b}} = C\delta^\kappa \|u\|^2_{X_{\delta,0,b}},$$

and the result is proven. □

Proof of Theorem 8.3. Let

$$V(t,x) = A^{\delta}u(t,x),$$

which is real-valued since the multiplier A^{δ} is even and u is real-valued. Applying A^{δ} to (8.1), we obtain

$$\partial_t V + \lambda_1 \partial_x^7 V + \lambda_2 \partial_x^5 V + \lambda_3 \partial_x^3 V + \lambda_4 \partial_x V + V\partial_x V = G, \qquad (8.27)$$

where

$$G = \frac{1}{2}\partial_x \left[(A^{\delta}u)^2 - A^{\delta}(u)^2 \right].$$

Multiplying (8.27) by V and integrating in space we obtain

$$\int_{\mathbb{R}} V\partial_l V dx + \lambda_1 \int_{\mathbb{R}} V\partial_x^7 V dx + \lambda_2 \int_{\mathbb{R}} V\partial_x^5 V dx + \lambda_3 \int_{\mathbb{R}} V\partial_x^3 V dx$$
$$+ \lambda_4 \int_{\mathbb{R}} V\partial_x V dx + \int_{\mathbb{R}} V^2 \partial_x V dx = \int_{\mathbb{R}} VG dx.$$

By noticing that $\partial_x^j V(x,t) \longrightarrow 0$ as $|x| \longrightarrow \infty$, we can use integration by parts obtaining

$$\frac{1}{2}\partial_t \int_{\mathbb{R}} V^2 dx = \int_{\mathbb{R}} VG dx.$$

Now, integrating the last equality with respect to $t \in [0,T]$ we obtain

$$\int_{\mathbb{R}} V^2(T,x)dx = \int_{\mathbb{R}} V^2(0,x)dx + 2\Big| \int_{\mathbb{R}^2} \chi_{[0,T]}(t) VG dx dt \Big|.$$

Thus,

$$\|u(T)\|_{G^{\delta,0}}^2 = \|u(0)\|_{G^{\delta,0}}^2 + 2\Big| \int_{\mathbb{R}^2} \chi_{[0,T]}(t) VG dx dt \Big|.$$

We now use Plancherel's theorem, Holder's inequality, Lemma 8.3, Lemma 8.6 and the fact that $1-b<b$, since $b>1/2$, we obtain

$$\begin{aligned}\Big| \int_{\mathbb{R}^2} \chi_{[0,T]}(t) VG dx dt \Big| &\leq \|\chi_{[0,T]}(t)V\|_{X_{0,1-b}} \|\chi_{[0,T]}(t)G\|_{X_{0,b-1}} \\ &\leq \|V\|_{X_{0,1-b}^T} \|G\|_{X_{0,b-1}^T} \\ &\leq C\delta^{\kappa} \|u\|_{X_{\delta,0,b}^T}^3.\end{aligned}$$

Finally, by using the condition (8.18) we conclude that

$$\sup_{t\in[0,T]} \|u(t)\|^2_{G^{\delta,0}} \leq \|u(0)\|^2_{G^{\delta,0}} + C\delta^\kappa \|u(0)\|^3_{G^{\delta,0}}.$$

The proof is now completed. □

8.4 Radius of the Spatial Analyticity Decay with Time

Theorem 8.4. *Let $s > -5/8$ and $\delta_0 > 0$ and assume $u_0 \in G^{\delta_0,s}$, then the solution given by Theorem 8.2 extends globally in time and for any $T' > 0$, we have*

$$u \in C\left([0,T'], G^{\delta(T'),s}\right) \quad \textit{with } \delta(T') = \min\left\{\delta_0, C_1 T'^{-(8/5+\sigma_0)}\right\},$$

where $\sigma_0 > 0$ can be taken arbitrarily small and $C_1 > 0$ is a constant depending on u_0, δ_0, s and σ_0.

Proof. Fix $\delta_0 > 0$, $s > -5/8$, $\kappa \in (0, 5/8)$ and $u_0 \in G^{\delta_0,s}$. It suffices to prove that the solution u to (8.1) satisfies

$$u \in C\left([0,T'], G^{\delta(T'),s}\right) \quad \text{where}$$
$$\delta(T') = \min\left\{\delta_0, C_1 T'^{-1/\kappa}\right\}, \quad \text{for all } T' > 0,$$

and $C_1 > 0$ is a constant depending on u_0, δ_0, s and κ. By Theorem 8.2, there is a maximal time

$$T^* = T^*(u_0, \delta_0, s) \in (0, \infty],$$

such that

$$u \in C\left([0,T^*], G^{\delta_0,s}\right).$$

If $T^* = \infty$, we are done. If $T^* < \infty$, as we assume henceforth, it remains to prove

$$u \in C\left([0,T'], G^{C_1 T'^{-1/\kappa},s}\right), \quad \text{for all } T' \geq T^*. \tag{8.28}$$

8.5 The Case $s = 0$

Fix $T' \geq T^*$ and we will show that, for $\delta > 0$ sufficiently small

$$\sup_{t\in[0,T']} \|u(t)\|^2_{G^{\delta,0}} \leq 2\|u(0)\|^2_{G^{\delta_0,0}}. \tag{8.29}$$

To prove this, we will use repeatedly Theorems 8.2 and 8.3 with the time step

$$T = \frac{c_0}{(1 + 2 \parallel u(0) \parallel_{G^{\delta_0,0}})^\beta}, \tag{8.30}$$

The smallness conditions on δ will be

$$\delta \leq \delta_0 \quad \text{and} \quad \frac{2T'}{T} C\delta^\kappa 2^{\frac{3}{2}} \parallel u(0) \parallel_{G^{\delta_0,0}} \leq 1, \tag{8.31}$$

where $C > 0$ is the constant in Theorems 8.3. Proceeding by induction, we will verify that

$$\sup_{t\in[0,nT]} \|u(t)\|^2_{G^{\delta,0}} \leq \|u(0)\|^2_{G^{\delta,0}} + nC\delta^\kappa 2^{\frac{3}{2}} \|u(0)\|^3_{G^{\delta_0,0}}, \tag{8.32}$$

$$\sup_{t\in[0,nT]} \|u(t)\|^2_{G^{\delta,0}} \leq 2\|u(0)\|^2_{G^{\delta_0,0}}, \tag{8.33}$$

for $n \in \{1, \ldots, m+1\}$, where $m \in \mathbb{N}$ is chosen, so that

$$T' \in [mT, (m+1)T).$$

This m does exist, since by Theorem 8.2 and the definition of T^*, we have

$$T < \frac{c_0}{(1+ \parallel u(0) \parallel_{G^{\delta_0,0}})^\beta} < T^*, \quad \text{hence} \quad T < T'.$$

In the first step, we cover the interval $[0, T]$ and by Theorem 8.3, we have

$$\begin{aligned}\sup_{t\in[0,T]} \|u(t)\|^2_{G^{\delta,0}} &\leq \|u(0)\|^2_{G^{\delta,0}} + C\delta^\kappa \|u(0)\|^3_{G^{\delta,0}} \\ &\leq \|u(0)\|^2_{G^{\delta,0}} + C\delta^\kappa \|u(0)\|^3_{G^{\delta_0,0}},\end{aligned}$$

where we used that

$$\|u(0)\|_{G^{\delta,0}} \leq \|u(0)\|_{G^{\delta_0,0}},$$

since $\delta \leq \delta_0$. This verifies (8.32) for $n = 1$ and now, (8.33) follows using again

$$\|u(0)\|_{G^{\delta,0}} \leq \|u(0)\|_{G^{\delta_0,0}},$$

as well as

$$C\delta^{\kappa}\|u(0)\|_{G^{\delta_0,0}} \leq 1.$$

Next, assuming that (8.32) and (8.33) hold for some $n \in \{1, \ldots, m\}$, we will prove that they hold for $n+1$. We estimate

$$\begin{aligned}
\sup_{t\in[nT,(n+1)T]} \|u(t)\|^2_{G^{\delta,0}} &\leq \|u(nT)\|^2_{G^{\delta,0}} + C\delta^{\kappa}\|u(nT)\|^3_{G^{\delta,0}} \\
&\leq \|u(nT)\|^2_{G^{\delta,0}} + C\delta^{\kappa}2^{\frac{3}{2}}\|u(0)\|^3_{G^{\delta_0,0}} \\
&\leq \|u(0)\|^2_{G^{\delta,0}} + nC\delta^{\kappa}2^{\frac{3}{2}}\|u(0)\|^3_{G^{\delta_0,0}} \\
&\quad + C\delta^{\kappa}2^{\frac{3}{2}}\|u(0)\|^3_{G^{\delta_0,0}},
\end{aligned}$$

verifying (8.32) with n replaced by $n+1$. To get (8.33) with n replaced by $n+1$, it is then enough to have

$$(n+1)C\delta^{\kappa}2^{\frac{3}{2}}\|u(0)\|_{G^{\delta_0,0}} \leq 1,$$

but this holds by (8.31), since

$$n+1 \leq m+1 \leq \frac{T'}{T} + 1 < \frac{2T'}{T}.$$

Finally, the condition (8.31) is satisfied for $\delta \in (0, \delta_0)$ such that

$$\frac{2T'}{T}C\delta^{\kappa}2^{\frac{3}{2}} \parallel u(0) \parallel_{G^{\delta_0,0}} = 1.$$

Thus, $\delta = C_1 T'^{-\frac{1}{\kappa}}$, where

$$C_1 = \left(\frac{c_0}{C2^{\frac{5}{2}} \parallel u(0) \parallel_{G^{\delta_0,0}} (1 + 2 \parallel u(0) \parallel_{G^{\delta_0,0}})^{\beta}}\right)^{\frac{1}{\kappa}}.$$

8.6 The General Case

For general s, we have

$$u_0 \in G^{\delta_0,s} \subset G^{\delta_0/2,0}.$$

The case $s = 0$ already being proved, we know that there is a $T_1 > 0$, such hat

$$u \in C([0,T_1), G^{\delta_0/2,0}),$$

and

$$u \in C([0,T'], G^{2\sigma T'^{-1/\kappa},0}), \quad \text{for } T' \geq T_1,$$

where $\sigma > 0$ depends on u_0, δ_0 and κ. We now conclude that

$$u \in C([0,T_1), G^{\delta_0/4,s}),$$

and

$$u \in C([0,T'], G^{\sigma T'^{-1/\kappa},s}), \quad \text{for } T' \geq T_1,$$

and these together imply (8.28). The proof of Theorem 8.4 is now completed. $\square$

Chapter 9

Kadomtsev–Petviashvili in Analytic Spaces

9.1 Fifth-Order Kadomtsev–Petviashvili I Equation in Bourgain Spaces

The section is devoted to study a Cauchy problem for fifth-order Kadomtsev–Petviashvili I equation. With data in analytic Bourgain spaces on the line and the circle, we prove that the problem is well posed. We also treat the regularity in t and x, y, where the solution is analytic in x, y and belongs to G^5 in t.

Kadomtsev–Petviashvili's equations are "universal" models for weakly nonlinear dispersive waves, which are essentially unidimensional. We are interested in the properties of solutions of Kadomtsev–Petviashvili type equations. These equations can be classified into two categories: the KP I (negative transversal dispersion) type equations and the KP II (positive transverse dispersion) type equation.

We consider class of Cauchy problem for fifth-order Kadomtsev–Petviashvili I equation

$$\partial_t u + \alpha \partial_x^3 u + \partial_x^5 u + \partial_x^{-1} \partial_y^2 u + u \partial_x u = 0, \tag{9.1}$$

with

$$u(x, y, 0) = \varphi(x, y),$$

where

$$u = u(x, y, t),\ (x, y) \in \mathbb{R}^2 \quad \text{or} \quad \mathbb{T}^2,\ (\alpha, t) \in \mathbb{R}^2.$$

Using Bourgain's method, it is proved that the problem

$$(u_t + \partial_t^3 u + uu_x)_x + u_{yy} = 0, \tag{9.2}$$

with

$$u(x, y, 0) = \varphi(x, y),$$

is well posed locally in non-isotropic Sobolev spaces

$$\tilde{H}^{s_1,s_2}_{x,y}(\mathbb{R}^2) = \{\phi \in S'(\mathbb{R}^2) : \|(1-\partial_x^2)^{s_1/2}(1-\partial_y^2)^{s_2/2}\phi\|_{L^2_{x,y}(\mathbb{R}^2)} < \infty\},$$

for $s_1 > -1/4, s_2 \geq 0$. For $s > 3/2$, the local well-posedness was proved in $\tilde{H}^s(\mathbb{R}^3)$. The integral equivalent formulation of (9.1) is as follows:

$$u(t) = S(t)\varphi - \frac{1}{2}\int_0^t S(t-s)(\partial_x u^2(s))ds, \tag{9.3}$$

where the unit operator related to the corresponding linear equation is given by

$$u(t)(x, y) = (S(t)\varphi)(x, y) = \int_{\mathbb{R}^2} e^{i(x\xi + y\mu + t\phi(\xi,\mu))}\widehat{\varphi}(\xi, \mu)d\xi d\mu. \tag{9.4}$$

Let us define the phase function as follows

$$\phi(\xi, \mu) = -\xi^5 + \alpha\xi^3 - \frac{\mu^2}{\xi}. \tag{9.5}$$

Now, we will define the spaces of analytic functions that contain the initial data of the considered IVP. For $s_1, s_2 \in \mathbb{R}$ and $\delta > 0$, let

$$G^{\delta,s_1,s_2}(\mathbb{R}^2) = \{\varphi \in L^2(\mathbb{R}^2); \|\varphi\|_{G^{\delta,s_1,s_2}(\mathbb{R}^2)} < \infty\}, \tag{9.6}$$

where

$$\|\varphi\|^2_{G^{\delta,s_1,s_2}(\mathbb{R}^2)} = \int_{\mathbb{R}^2} e^{2\delta(|\xi|+|\mu|)}\langle\xi\rangle^{2s_1}\langle\mu\rangle^{2s_2}|\widehat{\varphi}(\xi, \mu)|^2 d\xi d\mu,$$

while in the nonperiodic case we define

$$G^{\delta,s_1,s_2}(\mathbb{T}^2) = \{\varphi \in L^2(\mathbb{T}^2); \|\varphi\|_{G^{\delta,s_1,s_2}(\mathbb{T}^2)} < \infty\}, \tag{9.7}$$

where

$$\|\varphi\|^2_{G^{\delta,s_1,s_2}(\mathbb{T}^2)} = \sum_{(k,m)\in\mathbb{T}^2} e^{2\delta(|k|+|m|)}\langle k\rangle^{2s_1}\langle m\rangle^{2s_2}|\widehat{\varphi}(k, m)|^2.$$

Here, $\langle\cdot\rangle$ stands for

$$(1+|\cdot|^2)^{\frac{1}{2}} \sim (1+|\cdot|).$$

We, then define the analytic Bourgain spaces related to the fifth-order shallow water equations. The completion of the Schwartz space $S(\mathbb{R}^3)$ is given by $X^{s_1,s_2}_{\delta,b}(\mathbb{R}^3)$, for $s, b \in \mathbb{R}$ and $\delta > 0$, defined as

$$X^{s_1,s_2}_{\delta,b}(\mathbb{R}^3) = \{u \in L(\mathbb{R}^3); \|u\|_{X^{s_1,s_2}_{\delta,b}(\mathbb{R}^3)} < \infty\},$$

where

$$\|u\|_{X^{s_1,s_2}_{\delta,b}(\mathbb{R}^3)} = \left(\int_{\mathbb{R}^3} e^{2\delta(|\xi|+|\mu|)} \langle\xi\rangle^{2s_1} \times \langle\mu\rangle^{2s_2} \langle\tau-\phi(\xi,\mu)\rangle^{2b} \mid \widehat{u}(\xi,\mu,\tau) \mid^2 d\xi d\mu d\tau\right)^{\frac{1}{2}}, \tag{9.8}$$

for the periodic case, it is defined as the completion of the space of the functions defined on $\mathbb{T}^2 \times \mathbb{R}$ that are in the Schwartz class in t and are supposed smooth in x, y, by

$$X^{s_1,s_2}_{\delta,b}(\mathbb{T}^2\times\mathbb{R}) = \left\{u \in L(\mathbb{T}^2\times\mathbb{R}); \|u\|_{X^{s_1,s_2}_{\delta,b}(\mathbb{T}^2\times\mathbb{R})} < \infty\right\},$$

where

$$\|u\|_{X^{s_1,s_2}_{\delta,b}(\mathbb{T}^2\times\mathbb{R})} = \left(\sum_{\substack{(k,m)\in\mathbb{T}^2\\ k\neq 0}} \int e^{2\delta(|k|+|m|)} \langle k\rangle^{2s_1} \times \langle m\rangle^{2s_2} \langle\tau-\phi(k,m)\rangle^{2b} \mid \widehat{u}(k,m,\tau) \mid^2 d\tau\right)^{\frac{1}{2}}. \tag{9.9}$$

For $T > 0$, the restriction norm space is defined as follows

$$X^{T,s_1,s_2}_{\delta,b} = \left\{u|_{[-T,T]}, u \in X^{s_1,s_2}_{\delta,b}\right\},$$

and

$$\|u\|_{X^{T,s_1,s_2}_{\delta,b}} = \inf\{\| U \|_{X^{s_1,s_2}_{\delta,b}} : u = U \mid_{[-T,T]}, U \in X^{s_1,s_2}_{\delta,b}\}.$$

For $b \in \mathbb{R}$ with $b\pm$ we denote $b \pm \varepsilon$ for a number $\varepsilon > 0$ small enough.

We start with bilinear estimates for (9.8) and (9.9) defined in the analytic Bourgain spaces. The space $X_{\delta,b}^{s_1,s_2}$ is continuously embedded in

$$C\left([-T,T], G^{\delta,s_1,s_2}(\mathbb{R}^2)\right),$$

provided $b > 1/2$.

We are going to prepare the prove of our main results. Let us begin with the embedding result in the following Lemma.

Lemma 9.1. *Let $b > \frac{1}{2}$, $s_1, s_2 \in \mathbb{R}$ and $\delta > 0$. Then, for all $T > 0$, we have*

$$X_{\delta,b}^{s_1,s_2} \hookrightarrow C([-T,T], G^{\delta,s_1,s_2}(\mathbb{R}^2)).$$

Proof. First, we observe that the operator A, defined by

$$\widehat{A^\delta u}^{x,y}(\xi,\mu,t) = e^{\delta(|\xi|+|\mu|)}\widehat{u}^{x,y}(\xi,\mu,t), \tag{9.10}$$

satisfies

$$\|u\|_{X_{\delta,b}^{s_1,s_2}} = \|A^\delta u\|_{X_{s_1,s_2,b}} \quad \text{and} \quad \|u\|_{G^{\delta,s_1,s_2}} = \|A^\delta u\|_{H^{s_1,s_2}}. \tag{9.11}$$

We observe that $A^\delta u$ belongs to $C(\mathbb{R}, H^{s_1,s_2})$ and for some $C_0 > 0$, we have

$$\| A^\delta u \|_{C(\mathbb{R},H^{s_1,s_2})} \leq C_0 \| A^\delta u \|_{X_{s_1,s_2,b}}. \tag{9.12}$$

Thus, it follows that $u \in C([-T,T], G^{\delta,s_1,s_2})$ and

$$\| u \|_{C\left([-T,T],G^{\delta,s_1,s_2}\right)} \leq C_0 \| u \|_{X_{\delta,b}^{s_1,s_2}}. \tag{9.13}$$

This completes the proof. □

9.2 Existence of Solution

Theorem 9.1. *Let $s_1, s_2 \geq 0, \delta > 0$ and $b = \frac{1}{2}+$. Then for any $\varphi \in G^{\delta,s_1,s_2}$ and*

$$|\xi|^{-1}\widehat{\varphi}(\xi,\mu) \in L^2,$$

the problem (9.1) *with the initial condition* φ *has a solution* u, *satisfying*

$$u \in X_{\delta,b}^{T,s_1,s_2} \subseteq C([-T,T],G^{\delta,s_1,s_2}).$$

Moreover, this solution is unique in the class $X_{\delta,b}^{T,s_1,s_2}$ *and the map* $\varphi \to u(t)$ *is Lipschitz continuous from* G^{δ,s_1,s_2} *to* $X_{\delta,b}^{T,s_1,s_2}$.

Taking the Fourier transform with respect to x, y of the Cauchy problem (9.1), we get

$$u(t) = S(t)\varphi - \frac{1}{2}\int_0^t S(t-s)\left(\partial_x u^2(s)\right)ds.$$

We localize it in t by using a cut-off function satisfying $\psi \in C_0^\infty(\mathbb{R})$, with $\psi = 1$ in $[-1,1]$ and supp$\psi \subset [-2,2]$. We consider the operator Φ given by

$$\Phi(u) = \psi(t)S(t)\varphi - \frac{\psi_T(t)}{2}\int_0^t S(t-s)\left(\partial_x u^2(s)\right)ds, \tag{9.14}$$

where $\psi_T(t) = \psi(\frac{t}{T})$. We now estimate the fist part in the RHS of (9.14).

Lemma 9.2. *Let* $s_1, s_2 \geq 0$, $\frac{1}{2} < b < 1$ *and* $\delta > 0$. *For some constant* $C > 0$, *we have*

$$\|\psi(t)S(t)\varphi\|_{X_{\delta,b}^{s_1,s_2}} \leq C\|\varphi\|_{G^{\delta,s_1,s_2}}, \tag{9.15}$$

for all $\varphi \in G^{\delta,s_1,s_2}$.

Proof. We have

$$\begin{aligned}\psi(t)S(t)\varphi &= C\psi(t)\int_{\mathbb{R}^2} e^{i(x\xi+y\mu+t\phi(\xi,\mu))}\widehat{\varphi}(\xi,\mu)d\xi d\mu \\ &= C\int_{\mathbb{R}^3} e^{i(x\xi+y\mu+t\tau)}\widehat{\psi}(\tau-\phi(\xi,\mu))\widehat{\varphi}(\xi,\mu)d\xi d\mu d\tau.\end{aligned}$$

Then, it follows that

$$\begin{aligned}&\| \psi(t)S(t)\varphi \|^2_{X_{\delta,b}^{s_1,s_2}} \\ &= C\int_{\mathbb{R}^3} e^{2\delta(|\xi|+|\mu|)}\,(1+\mid \xi \mid)^{2s_1}\,(1+\mid \mu \mid)^{2s_2}\,(1+\mid \tau-\phi(\xi,\mu)\mid)^{2b}\end{aligned}$$

$$\times |\widehat{\psi}(\tau + \phi(\xi, \mu))|^2 |\widehat{\varphi}(\xi, \mu) |^2 \, d\xi d\mu d\tau$$
$$= C \int_{\mathbb{R}^2} e^{2\delta(|\xi|+|\mu|)} (1+ | \xi |)^{2s_1} (1+ | \mu |)^{2s_2} | \widehat{\varphi}(\xi, \mu) |^2$$
$$\times \left(\int_{\mathbb{R}} | \widehat{\psi}(\tau - \phi(\xi, \mu)) |^2 (1+ | \tau - \phi(\xi, \mu) |)^{2b} \, d\tau \right) d\xi d\mu.$$

By the fact that $b > 1/2$, we get

$$\int_{\mathbb{R}} | \widehat{\psi}(\tau - \phi(\xi, \mu)) |^2 (1+ | \tau - \phi(\xi, \mu) |)^{2b} \, d\tau$$
$$\leqslant \int_{\mathbb{R}} | \widehat{\psi}(\tau - \phi(\xi, \mu)) |^2 \, d\tau + C \int_{\mathbb{R}} | \widehat{\psi}(\tau - \phi(\xi, \mu)) |^2$$
$$\times |\tau - \phi(\xi, \mu) |^{2b} \, d\tau$$
$$\leq C.$$

This completes the proof. □

Now, we will estimate the second part in RHS of (9.14).

Lemma 9.3. *Let $s_1, s_2 \geq 0$, $-\frac{1}{2} < b' \leq 0 \leq b \leq b' + 1$, $0 < T < 1$ and $\delta > 0$. Then for some constant $C > 0$, we have*

$$\left\| \psi_T(t) \int_0^t S(t-s) F(x, y, s) \mathrm{d}s \right\|_{X^{s_1, s_2}_{\delta, b}} \leq C T^{1-b+b'} \|F\|_{X^{s_1, s_2}_{\delta, b'}}. \quad (9.16)$$

Proof. Define

$$U = \psi_T(t) \int_0^t S(t-s) F(x, y, s) ds.$$

Let us consider the operator A^δ given in (9.10). Then we have

$$\widehat{A^\delta U}^{x,y}(\xi, t) = \psi_T(t) \int_0^t \left(e^{-i(t-s)\phi(\xi,\mu)} \right) e^{\delta(|\xi|+|\mu|)} \widehat{F}^{x,y}(\xi, \mu, s) ds,$$
$$= \psi_T(t) \int_0^t \left[S(t-s) \widehat{(A^\delta F)} \right]^{x,y} (\xi, \mu, s) ds.$$

Thus,

$$\|U\|_{X^{s_1,s_2}_{\delta,b}}=\|A^\delta U\|_{X_{s_1,x_2,b}}=\|\psi_T(t)\int_0^t S(t-s)A^\delta F(x,y,s)ds\|_{X_{s_1,x_2,b}}.$$

We get

$$\|\psi_T(t)\int_0^t S(t-s)A^\delta F(x,y,s)ds\|_{X_{s_1,x_2,b}}\leq CT^{1-b+b'}\|A^\delta F\|_{X_{s_1,x_2,b'}}.$$

This completes the proof. □

Theorem 9.2. *If $s_1,s_2\geq 0$, let $b'=-\frac{1}{2}+$, $b=\frac{1}{2}+$ and $0<T<1$, then*

$$\|\partial_x(u_1u_2)\|_{X^{s_1,s_2,b'}}\leq C\|u_1\|_{X^{s_1,s_2,b}}\|u_2\|_{X^{s_1,s_2,b}}. \tag{9.17}$$

Lemma 9.4. *If $s_1,s_2\geq 0$, let $\delta>0$, $b'=-\frac{1}{2}+$, $b=\frac{1}{2}+$ and $0<T<1$, then*

$$\|\partial_x(u_1u_2)\|_{X^{s_1,s_2}_{\delta,b'}}\leq C\|u_1\|_{X^{s_1,s_2}_{\delta,b}}\|u_2\|_{X^{s_1,s_2}_{\delta,b}}. \tag{9.18}$$

Proof. We observe, by considering the operator A^δ in (9.10), that

$$\begin{aligned}
e^{\delta(|\xi|+|\mu|)}\widehat{u_1u_2}&=(2\pi)^{-3}e^{\delta(|\xi|+|\mu|)}\widehat{u_1}*\widehat{u_2}\\
&\leq(2\pi)^{-3}\int_{\mathbb{R}^3}e^{\delta(|\xi-\xi_1|+|\mu-\mu_1|)}\widehat{u_1}(\xi-\xi_1,\mu-\mu_1,\tau-\tau_1)\\
&\quad\times e^{\delta(|\xi_1|+|\mu_1|)}\widehat{u_2}(\xi_1,\mu_1,\tau_1)d\xi_1d\mu_1d\tau_1\\
&=\widehat{A^\delta u_1A^\delta u_2},
\end{aligned}$$

since

$$\delta(|\xi|+|\mu|)\leq\delta(|\xi-\xi_1|+|\mu-\mu_1|)+\delta(|\xi_1|+|\mu_1|).$$

Then

$$\begin{aligned}
&\|\partial_x(u_1u_2)\|_{X^{s_1,s_2}_{\delta,b'}}\\
&\quad=\|e^{\delta(|\xi|+|\mu|)}\langle\xi\rangle^{s_1}\langle\mu\rangle^{s_2}\langle\tau-\phi(\xi,\mu)\rangle^{b}\widehat{\partial_x(u_1u_2)}(\xi,\mu,\tau)\|_{L^2_{\xi,\mu,\tau}}\\
&\quad=\|\langle\xi\rangle^{s_1}\langle\mu\rangle^{s_2}\langle\tau-\phi(\xi,\mu)\rangle^{b}e^{\delta(|\xi|+|\mu|)}\xi\widehat{u_1u_2}(\xi,\mu,\tau)\|_{L^2_{\xi,\mu,\tau}}
\end{aligned}$$

$$\leq \| \langle \xi \rangle^{s_1} \langle \mu \rangle^{s_2} \langle \tau - \phi(\xi, \mu) \rangle^b (\partial_x \widehat{A^\delta u_1 A^\delta} u_2)(\xi, \mu\tau) \|_{L^2_{\xi,\mu,\tau}}$$

$$= \| \partial_x (A^\delta u_1 A^\delta u_2) \|_{X_{s_1,s_2,b}}.$$

There exists $C > 0$ such that

$$\begin{aligned} &\| \partial_x (A^\delta u_1 A^\delta u_2) \|_{X_{s_1,s_2,b'}} \\ &\leq C \| A^\delta u_1 \|_{X_{s_1,s_2,b}} \| A^\delta u_2 \|_{X_{s_1,s_2,b}} \\ &= C \| u_1 \|_{X^{s_1,s_2}_{\delta,b}} \| u_2 \|_{X^{s_1,s_2}_{\delta,b}}. \end{aligned}$$

□

We are now ready to estimate all the terms in (9.14) by using the bilinear estimates.

Lemma 9.5. *Let $s_1, s_2 \geq 0$, $\delta > 0$ and $b > \frac{1}{2}$. Then, for all $\varphi \in G^{\delta,s_1,s_2}$ and $0 < T < 1$, with some constant $C > 0$, we have*

$$\| \Phi(u) \|_{X^{s_1,s_2}_{\delta,b}} \leq C \left(\| \varphi \|_{G^{\delta,s_1,s_2}} + T^{1-b+b'} \| u \|^2_{X^{s_1,s_2}_{\delta,b}} \right), \quad \text{for all } u \in X^{s_1,s_2}_{\delta,b}, \tag{9.19}$$

and

$$\| \Phi(u) - \Phi(v) \|_{X^{s_1,s_2}_{\delta,b}} \leq C T^{1-b+b'} \| u - v \|_{X^{s_1,s_2}_{\delta,b}} \| u + v \|_{X^{s_1,s_2}_{\delta,b}}, \quad \text{for all } u, v \in X^{s_1,s_2}_{\delta,b}. \tag{9.20}$$

Proof. To prove estimate (9.19), we have

$$\begin{aligned} \| \Phi(u) \|_{X^{s_1,s_2}_{\delta,b}} &\leq \|\psi(t) S(t) \varphi\|_{X^{s_1,s_2}_{\delta,b}} + \|\psi_T(t) \\ &\quad \times \int_0^t S(t-s) \left(\partial_x u^2(s)\right) ds \|_{X^{s_1,s_2}_{\delta,b}} \\ &\leq C \| \varphi \|_{G^{\delta,s_1,s_2}} + C T^{1-b+b'} \| \partial_x u^2 \|_{X^{s_1,s_2}_{\delta,b'}} \\ &\leq C \| \varphi \|_{G^{\delta,s_1,s_2}} + C T^{1-b+b'} \| u \|^2_{X^{s_1,s_2}_{\delta,b}}. \end{aligned}$$

For the estimate (9.20), we observe that

$$\Phi(u) - \Phi(v) = \psi_T(t) \int_0^t S(t-s) \left(\partial_x u^2 - \partial_x v^2\right)(x, y, s) ds,$$

where $\omega = \partial_x u^2 - \partial_x v^2$ is now given by

$$\omega = \partial_x (u^2 - v^2) = \partial_x [(u+v)(u-v)].$$

Thus, from the previous results, we obtain (9.20). This completes the proof. □

We are now ready to consider the estimate of all the terms in (9.14) by using the bilinear estimates. We will show that the map Φ is a contraction on the ball $\mathbb{B}(0,r)$ to $\mathbb{B}(0,r)$.

Lemma 9.6. *Let $s_1, s_2 \geq 0$ and $\delta > 0$, $b > \frac{1}{2}$. Then, for all $\varphi \in G^{\delta,s_1,s_2}$, the map $\Phi : \mathbb{B}(0,r) \to \mathbb{B}(0,r)$ is a contraction, where $\mathbb{B}(0,r)$ is given by*

$$\mathbb{B}(0,r) = \{u \in X_{\delta,b}^{s_1,s_2}; \|u\|_{X_{\delta,b}^{s_1,s_2}} \leq r\},$$

with $r = 4C\|\varphi\|_{G^{\delta,s_1,s_2}}$.

Proof. From Lemma 9.5, for all $u \in \mathbb{B}(0,r)$, we have

$$\begin{aligned} \| \Phi(u) \|_{X_{\delta,b}^{s_1,s_2}} &\leq C \left(\| \varphi \|_{G^{\delta,s_1,s_2}} + T^{1-b+b'} \| u \|^2_{X_{\delta,b}^{s_1,s_2}} \right) \\ &\leq \frac{r}{4} + CT^{1-b+b'} r^2. \end{aligned}$$

We choose T sufficiently small such that $T^{1-b+b'} \leq \frac{1}{4Cr}$. Hence,

$$\| \Phi(u) \|_{X_{\delta,b}^{s_1,s_2}} \leq r, \quad \forall u \in \mathbb{B}(0,r).$$

Thus, Φ maps $\mathbb{B}(0,r)$ into $\mathbb{B}(0,r)$, which is a contraction, since

$$\begin{aligned} \| \Phi(u) - \Phi(v) \|_{X_{\delta,b}^{s_1,s_2}} &\leq CT^{1-b+b'} \| u - v \|_{X_{\delta,b}^{s_1,s_2}} \| u + v \|_{X_{\delta,b}^{s_1,s_2}} \\ &\leq CT^{1-b+b'} 2r \| u - v \|_{X_{\delta,b}^{s_1,s_2}} \\ &\leq \frac{1}{2} \| u - v \|_{X_{\delta,b}^{s_1,s_2}}, \quad \forall u, v \in \mathbb{B}(0,r). \end{aligned}$$

By Banach's fixed point Theorem, we obtain the existence of a unique solution. We recall the readers that we are proving Theorem 9.1, but the continuous dependence is missing, it will be proved in the next subsection. □

9.3 Continuous Dependence of the Initial Data

To prove the continuous dependence on the initial data, we will use the following useful lemma.

Lemma 9.7. *Let $s_1, s_2 \geq 0$ and $\delta > 0$, $b > \frac{1}{2}$. Then, for all $u_0, v_0 \in G^{\delta,s_1,s_2}$, if u and v are two solutions to (9.1) corresponding to initial data u_0 and v_0, we have*

$$\|u - v\|_{C([-T,T],G^{\delta,s_1,s_2})} \leq 2C_0C\|u_0 - v_0\|_{G^{\delta,s_1,s_2}}. \tag{9.21}$$

Proof. If u and v are two solutions to (9.1), corresponding to initial data u_0 and v_0, we have from Lemma 9.1

$$\|u - v\|_{C([-T,T],G^{\delta,s_1,s_2})} \leq C_0\|u - v\|_{X^{s_1,s_2}_{\delta,b}}.$$

By taking $u, v \in \mathbb{B}(0,r)$ and $T^{1-b+b'} \leqslant \frac{1}{4Cr}$, we get

$$\begin{aligned}
\| u - v \|_{X^{s_1,s_2}_{\delta,b}} &\leq \|\psi(t)S(t)(u_0 - v_0)\|_{X^{s_1,s_2}_{\delta,b}}, \\
&\quad + \left\|\psi_T(t)\int_0^t S(t-s)\big(\partial_x u^2 - \partial_x v^2\big)(x,y,s)ds\right\|_{X^{s_1,s_2}_{\delta,b}}, \\
&\leq C \| u_0 - v_0 \|_{G^{\delta,s_1,s_2}} + CT^{1-b+b'} \\
&\quad \times \| u + v \|_{X^{s_1,s_2}_{\delta,b}} \| u - v \|_{X^{s_1,s_2}_{\delta,b}}, \\
&\leq C \| u_0 - v_0 \|_{G^{\delta,s_1,s_2}} + \frac{1}{2} \| u - v \|_{X^{s_1,s_2}_{\delta,b}}.
\end{aligned}$$

Thus,

$$\| u - v \|_{X^{s_1,s_2}_{\delta,b}} \leq 2C \| u_0 - v_0 \|_{G^{\delta,s_1,s_2}}.$$

Therefore,

$$\|u - v\|_{C([-T,T],G^{\delta,s_1,s_2})} \leq 2C_0C\|u_0 - v_0\|_{G^{\delta,s_1,s_2}}.$$

This completes the proof. □

9.4 Gevrey's Regularity in t

This section is devoted to the proof of Theorem 9.3, we study the Gevrey's regularity in t of the unique solution of the problem (9.1).

Theorem 9.3. *Let $s_1, s_2 \geq 0, \delta > 0$. If $\varphi \in G^{\delta,s_1,s_2}$ then the solution*

$$u \in C([-T,T], G^{\delta,s_1,s_2}),$$

given by Theorem 9.1, belongs to the Gevrey class G^5 in time variable, where $\alpha = 1$.

The proof of time regularity on the circle and on the line is analogous.

Proposition 9.1. *Let $s_1, s_2 \geq 0$, $\delta > 0$ and*

$$u \in C\left([-T,T]; G^{\delta,s_1,s_2}\right),$$

be the solution of (9.1). *Then u is analytic in x, y for all t near the zero. More precisely,*

$$|\partial_x^l \partial_y^n u(x,y,t)| \leq C^{l+n+1}(l!)(n!), \tag{9.22}$$

for all

$$(x,y) \in \mathbb{R}^2 \quad \text{or} \quad \mathbb{T}^2,\ t \in [-T,T], \quad (l,n) \in \{0,1,\ldots\}^2,$$

and $C > 0$.

Proof. For any $t \in [-T,T]$, we have

$$\begin{aligned}
\|\partial_x^l \partial_x^n u(\cdot,\cdot,t)\|_{H^{s_1,s_2}}^2 &= \sum_{(k,m)\in\mathbb{Z}^2} |k|^{2l}|m|^{2n} \langle k\rangle^{2s_1} \langle m\rangle^{2s_2} |\widehat{u}(k,m,t)|^2 \\
&= \sum_{(k,m)\in\mathbb{Z}^2} |k|^{2l}|m|^{2n} e^{-2\delta(|k|+|m|)} \langle k\rangle^{2s_1} \langle m\rangle^{2s_2} \\
&\quad \times e^{2\delta(|k|+|m|)}|\widehat{u}(k,m,t)|^2.
\end{aligned}$$

We observe that

$$e^{2\delta|k|} = \sum_{j=0}^{\infty} \frac{1}{j!}(2\delta|k|)^j \geq \frac{1}{(2l)!}(\delta)^{2l}|k|^{2l}, \quad \forall l \in \{0,1,\ldots\}, k \in \mathbb{Z},$$

and

$$e^{2\delta|m|} = \sum_{j=0}^{\infty} \frac{1}{j!}(2\delta|m|)^j \geq \frac{1}{(2n)!}(\delta)^{2n}|m|^{2n}, \quad \forall n \in \{0,1,\ldots\}, m \in \mathbb{Z}.$$

This implies that

$$\begin{aligned}
|k|^{2l}e^{-2\delta|k|} &\leq C_\delta^{2l}(2l)!, \\
|m|^{2n}e^{-2\delta|m|} &\leq C_\delta^{2n}(2n)!
\end{aligned}$$

Thus,

$$\|\partial_x^l \partial_y^n u(\cdot,\cdot,t)\|_{H^{s_1,s_2}}^2 \leq C_\delta^{2l+2n}(2l)!(2n)!$$
$$\times \sum_{(k,m)\in\mathbb{Z}^2} e^{2\delta(|k|+|m|)} \langle k\rangle^{2s_1} \langle m\rangle^{2s_2} |\widehat{u}(k,m,t)|^2$$
$$= C_\delta^{2l+2n}(2l)!(2n)!\|u(\cdot,\cdot,t)\|_{G^{\delta,s_1,s_2}}^2.$$

Since

$$(2l)! \leq A_1^{2l}(l!)^2,$$

and

$$(2n)! \leq A_2^{2n}(n!)^2,$$

for some $A_1, A_2 > 0$, by Sobolev lemma ($\|.\|_{L^2} \leq \|.\|_{H^{s_1,s_2}}$), we have for all $(l,n) \in \{0,1,2,\dots\}^2$

$$\|\partial_x^l \partial_y^n u(\cdot,\cdot,t)\|_{L^2} \leq \|\partial_x^l \partial_y^n u(\cdot,\cdot,t)\|_{H^{s_1,s_2}} \leq C_0 C_1^{l+n}(l)!(n)!$$
$$\text{for} \quad \forall t \in [-T,T],$$

where $C_0 = \|u(\cdot,\cdot,t)\|_{G^{\delta,s_1,s_2}}$ and $C_1 = A_0^2 C_\delta$ and $A_0 = \max(A_1, A_2)$. This implies that u is analytic in x, y for all t near zero and $s_1, s_2 \geq 0$. This completes the proof. □

Lemma 9.8. *For $(j,l,n) \in \{0,1,2,\dots\}^3$ the following inequality:*

$$|\partial_t^j \partial_x^l \partial_y^n u| \leq C^{j+l+n+1}(l+n+5j)!L^j, \tag{9.23}$$

holds, where

$$L = C^4 + \frac{C^2}{20} + \frac{1}{120} + \frac{C}{40},$$

for all $(x,y) \in \mathbb{R}^2$ or $\mathbb{T}^2, t \in [-T,T]$.

Proof. We will prove it by using induction on j.
Firstly, for $j = 0$ and $(l, n) \in \{0, 1, 2, \ldots\}^2$, by (9.22), we have

$$|\partial_x^l \partial_y^n u(x, y, t)| \leq C^{l+n+1}(l!)(n!) \leq C^{l+n+1}(l+n)! \tag{9.24}$$

Secondly, for $j = 1$ and $(l, n) \in \{0, 1, 2, \ldots\}^2$, we get

$$|\partial_t \partial_x^l \partial_y^k u| \leq |\partial_x^{l+5} \partial_y^n u| + |\partial_x^{l+3} \partial_y^n u| + |\partial_x^{l-1} \partial_y^{n+2} u| + |\partial_x^l \partial_y^n (u \partial_x u)|.$$

We estimate these terms as

$$\begin{aligned} |\partial_x^{l+5} \partial_y^n u| &\leq C^{l+5+n+1}(l+5+n)! \\ &\leq C^{l+n+1+1}(l+n+5\cdot 1)! C^4, \end{aligned} \tag{9.25}$$

$$\begin{aligned} |\partial_x^{l+3} \partial_y^n u| &\leq C^{l+3+n+1}(l+3+n)! \\ &\leq C^{l+n+1+1}(l+n+5\cdot 1)! C^2 \frac{1}{(l+n+4)(l+n+5)} \\ &\leq C^{l+n+1+1}(l+n+5\cdot 1)! \frac{C^2}{20}, \end{aligned} \tag{9.26}$$

where

$$\frac{1}{(l+n+4)(l+n+5)} \leq \frac{1}{20},$$

and

$$\begin{aligned} |\partial_x^{l-1} \partial_y^{n+2} u| &\leq C^{l-1+n+2+1}(l-1+n+2)! \\ &\leq C^{l+n+1+1}(l+n+5\cdot 1)! \\ &\quad \times \frac{1}{(l+n+2)(l+n+3)(l+n+4)(l+n+5)} \\ &\leq C^{l+n+1+1}(l+n+5\cdot 1)! \frac{1}{120}, \end{aligned} \tag{9.27}$$

where

$$\frac{1}{(l+n+2)(l+n+3)(l+n+4)(l+n+5)} \leq \frac{1}{120}.$$

For the nonlinear terms

$$|\partial_x^l \partial_y^n (u\partial_x u)| = \left| \sum_{p=0}^{l} \sum_{k=0}^{n} \binom{l}{p} \binom{n}{k} \left(\partial_x^{l-p} \partial_y^{n-k} u\right) \left(\partial_x^{p+1} \partial_y^k u\right) \right|.$$

We shall recall that for $p \leq l$ and $k \leq n$, we have the following inequality:

$$\binom{l}{p} \binom{n}{k} \leq \binom{l+n}{p+k}. \tag{9.28}$$

Now, using (9.28), to get

$$\begin{aligned} \left|\partial_x^l \partial_y^n (u\partial_x u)\right| &\leq \left| \sum_{p=0}^{l} \sum_{k=0}^{n} \binom{l+n}{p+k} \left(\partial_x^{l-p} \partial_y^{n-k} u\right) \left(\partial_x^{p+1} \partial_y^k u\right) \right| \\ &\leq \sum_{p=0}^{l} \sum_{k=0}^{n} \frac{(l+n)!}{(p+k)!(l+n-p-k)!} C^{l-p+n-k+1} \\ &\quad \times (l+n-p-k)! \\ &\quad C^{p+1+k+1} (p+1+k)! \\ &= C^{l+n+3} (l+n)! \sum_{p=0}^{l} \sum_{k=0}^{n} (p+1+k). \end{aligned}$$

Now, we use the fact that

$$\sum_{p=0}^{l} \sum_{k=0}^{n} (p+1+k) = \frac{(l+1)(n+1)(l+n+2)}{2}. \tag{9.29}$$

Next,

$$\begin{aligned} |\partial_x^l \partial_y^n (u\partial_x u)| &\leq C^{l+n+3} (l+n)! \frac{(l+1)(n+1)(l+n+2)}{2} \\ &\leq C^{l+n+1+1} (l+n)!(l+n+1)(l+n+2)(l+n+3) \frac{C}{2} \end{aligned}$$

$$= C^{l+n+1+1}(l+n+5)!\frac{1}{(l+n+4)(l+n+5)}\frac{C}{2}$$

$$\leq C^{l+n+1+1}(l+n+5)!\frac{C}{40}. \tag{9.30}$$

From (9.25), (9.26), (9.27) and (9.30), it follows that

$$|\partial_t \partial_x^l \partial_y^k u| \leq C^{l+n+1+1}(l+n+5\cdot 1)!L^1,$$

for all

$$(x,y) \in \mathbb{R}^2 \ or \ \mathbb{T}^2, t \in [-T,T].$$

Now, we will assume that (9.23) is correct for $1 \leq m \leq j$ where $(l,n) \in \{0,1,2,\ldots\}^2$ and we will prove it for $m = j+1$ and $(l,n) \in \{0,1,2,\ldots\}^2$. We have

$$|\partial_t^{j+1}\partial_x^l\partial_y^k u\ | \leq |\partial_t^j\partial_x^{l+5}\partial_y^n u| + |\partial_t^j\partial_x^{l+3}\partial_y^n u| + |\partial_t^j\partial_x^{l-1}\partial_y^{n+2} u| + |\partial_t^j\partial_x^l\partial_y^n(u\partial_x u)|.$$

We estimate these terms as

$$\begin{aligned}|\partial_t^j\partial_x^{l+5}\partial_y^n u| &\leq C^{j+l+5+n+1}(l+5+n+5j)!L^j\\ &\leq C^{(j+1)+l+n+1}(l+n+5(j+1))!C^4L^j,\end{aligned} \tag{9.31}$$

and

$$\begin{aligned}|\partial_t^j\partial_x^{l+3}\partial_y^n u| &\leq C^{j+l+3+n+1}(l+3+n+5j)!L^j\\ &\leq C^{j+l+n+1+1}(l+n+5(j+1))!C^2L^j\\ &\quad\times\frac{1}{(l+n+4)(l+n+5)}\\ &\leq C^{(j+1)+l+n+1}(l+n+5(j+1))!\frac{C^2}{20}L^j,\end{aligned} \tag{9.32}$$

and

$$\begin{aligned}|\partial_t^j\partial_x^{l-1}\partial_y^{n+2} u| &\leq C^{j+l-1+n+2+1}(l-1+n+2+5j)!L^j\\ &\leq C^{j+l+n+1+1}(l+n+5(j+1))!\\ &\quad\times\frac{1}{(l+n+2)(l+n+3)(l+n+4)(l+n+5)}L^j\\ &\leq C^{(j+1)+l+n+1}(l+n+5(j+1))!\frac{L^j}{(120)}.\end{aligned} \tag{9.33}$$

For the nonlinear terms, we have

$$
\begin{aligned}
\partial_t^j \partial_x^l \partial_y^n (u\partial_x u) &= \sum_{p=0}^{l}\sum_{k=0}^{n}\binom{l}{p}\binom{n}{k}\left(\partial_t^j \partial_x^{l-p}\partial_y^{n-k}u\right)\left(\partial_x^{p+1}\partial_y^k u\right) \\
&\quad + \sum_{p=0}^{l}\sum_{k=0}^{n}\binom{l}{p}\binom{n}{k}\left(\partial_x^{l-p}\partial_y^{n-k}u\right)\left(\partial_t^j \partial_x^{p+1}\partial_y^k u\right) \\
&\quad + \sum_{q=1}^{j-1}\sum_{p=0}^{l}\sum_{k=0}^{n}\binom{j}{q}\binom{l}{p}\binom{n}{k}\left(\partial_t^{j-q}\partial_x^{l-p}\partial_y^{n-k}u\right) \\
&\quad \times \left(\partial_t^q \partial_x^{p+1}\partial_y^k u\right).
\end{aligned}
\tag{9.34}
$$

Using (9.28), we estimate the first term of (9.34)

$$
\begin{aligned}
&\left|\sum_{p=0}^{l}\sum_{k=0}^{n}\binom{l}{p}\binom{n}{k}\left(\partial_t^j \partial_x^{l-p}\partial_y^{n-k}u\right)\left(\partial_x^{p+1}\partial_y^k u\right)\right| \\
&\quad \le \sum_{p=0}^{l}\sum_{k=0}^{n}\binom{l+n}{p+k}C^{j+l-p+n-k+1}(l+n-p-k+5j)! \\
&\quad \times L^j C^{p+1+k+1}(p+1+k)! \\
&\quad = C^{j+l+n+3}L^j \sum_{p=0}^{l}\sum_{k=0}^{n}\frac{(l+n)!}{(p+k)!(l+n-p-k)!}(p+1+k)! \\
&\quad \times (l+n-p-k+5j)! \\
&\quad = C^{j+l+n+3}L^j (l+n)!\sum_{p=0}^{l}\sum_{k=0}^{n}(p+1+k)(l+n-p-k+1) \\
&\quad \times (l+n-p-k+2)\cdots(l+n-p-k+5j).
\end{aligned}
$$

Since $p \le l$, $k \le n$, for any $m \in \mathbb{N}$, we have

$$p + k + m \le l + n + m,$$

and

$$l + n - p - k + m \le l + n + m.$$

From these inequalities and from (9.29), we obtain

$$\begin{aligned}
&\left| \sum_{p=0}^{l} \sum_{k=0}^{n} \binom{l}{p} \binom{n}{k} \left(\partial_t^j \partial_x^{l-p} \partial_y^{n-k} u \right) \left(\partial_x^{p+1} \partial_y^k u \right) \right| \\
&\quad \le C^{j+l+n+3} L^j (l+n)!(l+n+1)(l+n+2) \cdots (l+n+5j) \\
&\quad\quad \cdot \frac{(l+1)(n+1)(l+n+2)}{2} \\
&\quad \le C^{j+l+n+3} L^j \frac{1}{2} (l+n+5j)!(l+n+5j+1) \\
&\quad\quad \times (l+n+5j+2)(l+n+5j+3) \\
&\quad = C^{(j+1)+l+n+1} (l+n+5j+5)! \\
&\quad\quad \times \frac{1}{(l+n+5j+4)(l+n+5j+5)} \frac{C}{2} L^j \\
&\quad \le \frac{1}{3} C^{(j+1)+l+n+1} (l+n+5(j+1))! \frac{C}{40} L^j.
\end{aligned} \tag{9.35}$$

The $(9.34)_2$ is estimated as

$$\begin{aligned}
&\left| \sum_{p=0}^{l} \sum_{k=0}^{n} \binom{l}{p} \binom{n}{k} \left(\partial_x^{l-p} \partial_y^{n-k} u \right) \left(\partial_t^j \partial_x^{p+1} \partial_y^k u \right) \right| \\
&\quad \le \sum_{p=0}^{l} \sum_{k=0}^{n} \binom{l+n}{p+k} C^{l-p+n-k+1} (l+n-p-k)! C^{j+p+1+k+1} \\
&\quad\quad \times (p+1+k+5j)! L^j
\end{aligned}$$

$$= \sum_{p=0}^{l}\sum_{k=0}^{n} \frac{(l+n)!}{(p+k)!(l+n-p-k)!} C^{j+l+n+3} L^j (l+n-p-k)!$$
$$\times (p+1+k+5j)!$$
$$= C^{j+l+n+3} L^j (l+n)! \sum_{p=0}^{l}\sum_{k=0}^{n} (p+k+1)(p+k+2)\cdots$$
$$\times (p+1+k+5j).$$

Since $p \le l$, $k \le n$, for any $m \in \mathbb{N}$, we have

$$p+k+m \le l+n+m.$$

From these inequalities and from (9.29), we get

$$\left| \sum_{p=0}^{l}\sum_{k=0}^{n} \binom{l}{p}\binom{n}{k} \left(\partial_x^{l-p}\partial_y^{n-k} u\right)\left(\partial_x^{p+1}\partial_y^{k} u\right) \right|$$
$$\le C^{j+l+n+3} L^j (l+n)! \frac{(l+1)(n+1)(l+n+2)}{2}$$
$$\times (l+n+2)(l+n+3)\cdots(l+1+n+5j)$$
$$\le C^{j+l+n+3} L^j \frac{1}{2}(l+n+5j+1)!(l+n+5j+2)(l+n+5j+3)$$
$$= C^{(j+1)+l+n+1}(l+n+5j+5)!$$
$$\times \frac{1}{(l+n+5j+4)(l+n+5j+5)} \frac{C}{2} L^j$$
$$\le \frac{1}{3} C^{(j+1)+l+n+1}(l+n+5(j+1))! \frac{C}{40} L^j, \tag{9.36}$$

To estimate $(9.34)_3$, we shall recall that for $q \le j$, $p \le l$ and $k \le n$, we have the following inequality

$$\binom{j}{q}\binom{l}{p}\binom{n}{k} \le \binom{j+l+n}{q+p+k}.$$

Then

$$\begin{aligned}
&\left|\sum_{q=1}^{j-1}\sum_{p=0}^{l}\sum_{k=0}^{n}\binom{j}{q}\binom{l}{p}\binom{n}{k}\left(\partial_t^{j-q}\partial_x^{l-p}\partial_y^{n-k}u\right)\left(\partial_t^{q}\partial_x^{p+1}\partial_y^{k}u\right)\right| \\
&\quad\le \sum_{q=1}^{j-1}\sum_{p=0}^{l}\sum_{k=0}^{n}\binom{j+l+n}{q+p+k}C^{j-q+l-p+n-k+1} \\
&\qquad\times (l-p+n-k+5(j-q))!L^{j-q}C^{q+p+1+k+1} \\
&\qquad\times (p+1+k+5q)!L^{q} \\
&\quad\le C^{j+l+n+3}L^{j}\sum_{q=1}^{j-1}\sum_{p=0}^{l}\sum_{k=0}^{n}\frac{(j+l+n)!}{(q+p+k)!(j-q+l+n-p-k)!} \\
&\qquad\times (l+n-p-k+5(j-q))!(p+1+k+5q)! \\
&\quad= C^{j+l+n+3}L^{j}(j+l+n)!\sum_{q=1}^{j-1}\sum_{p=0}^{l}\sum_{k=0}^{n}(q+p+k+1) \\
&\qquad\times (q+p+k+2)\cdots(q+p+k+1+4q) \\
&\qquad\times ((j-q)+l+n-p-k+1)((j-q)+l+n-p-k+2) \\
&\qquad\times \cdots((j-q)+l+n-p-k+4(j-q)).
\end{aligned}$$

Since $q \le j-1$, $p \le l$, $k \le n$, for any $m \in \mathbb{N}$, we have

$$q+p+k+m \le j+l+n+m-1,$$

$$j+l+n-q-p-k+m \le j+l+n+m-1,$$

$$\begin{aligned}
&\left|\sum_{q=1}^{j-1}\sum_{p=0}^{l}\sum_{k=0}^{n}\binom{j}{q}\binom{l}{p}\binom{n}{k}\left(\partial_t^{j-q}\partial_x^{l-p}\partial_y^{n-k}u\right)\left(\partial_t^{q}\partial_x^{p+1}\partial_y^{k}u\right)\right| \\
&\quad\le C^{j+l+n+3}L^{j}(j+l+n)!\sum_{q=1}^{j-1}\sum_{p=0}^{l}\sum_{k=0}^{n}(q+p+k+1) \\
&\qquad\times (j+l+n+1)(j+l+n+2) \\
&\qquad\cdots(j+l+n+4q)(j+l+n)(j+l+n+1) \\
&\qquad\cdots(j+l+n+4(j-q)-1).
\end{aligned}$$

Since

$$j+l+n+m \leq j+l+n+m+4q+1,$$

for any $m \in \mathbb{N}$, we have

$$\begin{aligned}
&\left|\sum_{q=1}^{j-1}\sum_{p=0}^{l}\sum_{k=0}^{n}\binom{j}{q}\binom{l}{p}\binom{n}{k}\left(\partial_t^{j-q}\partial_x^{l-p}\partial_y^{n-k}u\right)\left(\partial_t^{q}\partial_x^{p+1}\partial_y^{k}u\right)\right| \\
&\quad \leq C^{j+l+n+3}L^j(j+l+n)!\sum_{q=1}^{j-1}\sum_{p=0}^{l}\sum_{k=0}^{n}(q+p+k+1) \\
&\qquad \times (j+l+n+1)(j+l+n+2)\cdots(j+l+n+4q) \\
&\qquad \times (j+l+n+4q+1)(j+l+n+4q+2) \\
&\qquad \cdots (j+l+n+4j) \\
&\quad = C^{j+l+n+3}L^j(l+n+5j)!\frac{(j-1)(l+1)(n+1)(j+l+n+2)}{2} \\
&\quad \leq C^{j+l+n+3}L^j(l+n+5j)!(l+n+5j+1)(l+n+5j+2) \\
&\qquad \times (l+n+5j+3)\frac{(j-1)}{2} \\
&\quad = C^{j+l+n+3}L^j(l+n+5(j+1))! \\
&\qquad \times \frac{(j-1)}{(l+n+5j+4)(l+n+5j+5)}\frac{1}{2} \\
&\quad \leq \frac{1}{3}C^{(j+1)+l+n+1}(l+n+5(j+1))!\frac{C}{40}L^j, \qquad (9.37)
\end{aligned}$$

because

$$\sum_{q=1}^{j-1}\sum_{p=0}^{l}\sum_{k=0}^{n}(q+p+k+1) = \frac{(j-1)(l+1)(n+1)(j+l+n+2)}{2},$$

and

$$\frac{(j-1)}{(l+n+5j+4)(l+n+5j+5)} \leq \frac{1}{70} \leq \frac{1}{60}.$$

Finally by using (9.31), (9.32), (9.33), (9.35), (9.36) and (9.37) we arrive at

$$|\partial_t^{j+1}\partial_x^l\partial_y^k u| \leq C^{(j+1)+l+n+1}(l+n+5(j+1))!L^{j+1},$$

for all

$$(x,y)\in\mathbb{R}^2 \quad \text{or} \quad \mathbb{T}^2, t\in[-T,T].$$

This completes the proof. □

9.5 Fifth-Order Kadomtsev–Petviashvili II Equation in Anisotropic Gevrey Spaces

We show that the fifth-order Kadomtsev–Petviashvili II equation is globally well-posed in an anisotropic Gevrey space, which complements earlier results on the well-posedness of this equation in anisotropic Sobolev spaces.

$$\partial_t u - \partial_x^5 u + \partial_x^{-1}\partial_y^2 u + u\partial_x u = 0, \tag{9.38}$$

where $u(x,y,0) = f(x,y), u = u(x,y,t)$ and $(x,y,t)\in\mathbb{R}^3$. This equation belongs to the class of KP equations, which are models for the propagation of long dispersive nonlinear waves which are essentially unidirectional and have weak transverse effects. Due to the asymmetric nature of the equation with respect to the spatial derivatives, it is natural to consider the Cauchy problem for (9.38) with initial data in the anisotropic Sobolev spaces $H^{s_1,s_2}(\mathbb{R}^2)$, defined by the norm

$$\|f\|_{H^{s_1,s_2}(\mathbb{R}^2)} = \left(\int \langle\xi\rangle^{2s_1}\langle\mu\rangle^{2s_2}|\hat{f}(\xi,\mu)|^2 d\xi d\mu\right)^{1/2},$$

where $\hat{f}$ denotes the spatial Fourier transform

$$\hat{f}(\xi,\mu) = \int_{\mathbb{R}^2} f(x,y)e^{-i(x\xi+y\mu)}\, dxdy.$$

In the present work, we will consider the Cauchy problem for equation (9.38) with initial data in an anisotropic Gevrey space $G^{\sigma_1,\sigma_2}(\mathbb{R}^2)$, which we define as the completion of the Schwartz functions with respect to the norm

$$\|f\|_{G^{\sigma_1,\sigma_2}(\mathbb{R}^2)} = \left(\int_{\mathbb{R}^2} e^{2\sigma_1|\xi|}e^{2\sigma_2|\mu|}|\hat{f}(\xi,\mu)|^2 d\xi d\mu\right)^{1/2}.$$

The primary reason for considering initial data in these spaces is because of the results in Proposition 8.1.

The quantity σ is known as the *radius of analyticity.* Thus, for functions in G^{σ_1,σ_2}, if we hold one variable fixed, the resulting function in the other variable will have a holomorphic extension satisfying the stated bounds.

In addition to the holomorphic extension property, Gevrey spaces satisfy the embeddings

$$G^{\sigma_1,\sigma_2} \hookrightarrow G^{\sigma_1',\sigma_2'},$$

for $\sigma_i' < \sigma_i$, which follow from the corresponding estimates

$$\|f\|_{G^{\sigma_1',\sigma_2'}} \lesssim \|f\|_{G^{\sigma_1,\sigma_2}}. \tag{9.39}$$

We begin by noting that, in addition to the spatial Fourier transform stated above, we will also need the full spacetime Fourier transform, which we denote by

$$\hat{f}(\xi,\mu,\tau) = \int_{\mathbb{R}^3} f(x,y,t)e^{-i(x\xi+y\mu+t\tau)}\, dxdydt.$$

To simplify the notation, we introduce some operators which will be useful later. We first introduce the operator A^{σ_1,σ_2}, which we define as

$$A^{\sigma_1,\sigma_2}f = \mathfrak{F}^{-1}(e^{\sigma_1|\xi|}e^{\sigma_2|\mu|}\hat{f}). \tag{9.40}$$

With this, we may also define another useful operator by

$$N(f) = \partial_x\left[(A^{\sigma_1,\sigma_2}f)^2 - A^{\sigma_1,\sigma_2}(f^2)\right]. \tag{9.41}$$

Let us state the function spaces they used explicitly, so that we can state their useful properties which we will exploit in our modifications of their spaces. The main function spaces they used are the so-called

Bourgain spaces, adapted to the KP II equation, whose norm is given by

$$\|u\|_{X^{s_1,s_2,b,\varepsilon}} = \left(\int_{\mathbb{R}^3} \lambda^2(s_1,s_2,b,\varepsilon)|\hat{u}(\xi,\mu,\tau)|^2 d\xi d\mu d\tau\right)^{\frac{1}{2}},$$

where

$$\lambda(s_1,s_2,b,\varepsilon) = \langle\xi\rangle^{s_1}\langle\mu\rangle^{s_2}\langle\tau - m(\xi,\mu)\rangle^b \left\langle \frac{\tau - m(\xi,\mu)}{1+|\xi|^5}\right\rangle^{\varepsilon},$$

with $m(\xi,\mu) = \xi^5 - \frac{\mu^2}{\xi}$ and $\langle x\rangle = (1+|x|^2)^{1/2}$ for $x \in \mathbb{R}$. We remark that we will not need to make use of the case $\varepsilon \neq 0$. Thus, we will employ the simplified notation $X^{s_1,s_2,b} = X^{s_1,s_2,b,0}$.

Since our interest is in constructing analytic solutions, we will require a version of the Bourgain spaces which are adapted to the Gevrey spaces. For this, we define the spaces $Y^{\sigma_1,\sigma_2,b}(\mathbb{R}^2)$ by the norm

$$\|u\|_{Y^{\sigma_1,\sigma_2,b}} = \|A^{\sigma_1,\sigma_2}u\|_{X^{0,0,b}}.$$

As is well-known that the spaces $X^{s_1,s_2,b}$ satisfy the embedding

$$X^{s_1,s_2,b} \hookrightarrow C\left(\mathbb{R}; H^{s_1,s_2}(\mathbb{R}^2)\right),$$

for $b > \frac{1}{2}$. An immediate consequence of this is that

$$Y^{\sigma_1,\sigma_2,b} \hookrightarrow C(\mathbb{R}; G^{\sigma_1,\sigma_2}(\mathbb{R}^2)),$$

when $b > \frac{1}{2}$. Thus, solutions constructed in $Y^{\sigma_1,\sigma_2,b}$ belong to the natural solution space. When considering local solutions, it is useful to consider localized versions of these spaces. For a time interval I and a Banach space Z, we define the localized space $Z(I)$ by the norm

$$\|u\|_{Z(I)} = \inf\{\|v\|_Z : \ v = u \text{ on } I\}.$$

It is easy to see that estimates which hold for the original spaces also hold for the localized spaces. For simplicity, we will omit the interval from the notation when there is no chance of confusion.

The final preliminary fact we must state is the following bilinear estimate.

Lemma 9.9. *Let $s_1 > -5/4$, $s_2 \geq 0$, $b > 1/2$ and define $s = \max\{0, -s_1\}$. If ε and β satisfy the inequalities*

$$0 \leq \varepsilon \leq \min\left\{\frac{2}{5}\left(\frac{5}{4} - s\right), \frac{3}{20}\right\},$$

and

$$\max\left\{\frac{9}{20}, \frac{1}{2} - \frac{1}{2}\left(\frac{5}{4} - s\right) + \varepsilon\right\} \leq \beta < \frac{1}{2},$$

then

$$\|\partial_x(uv)\|_{X^{s_1,s_2,-\beta,\varepsilon}} \lesssim \|u\|_{X^{s_1,s_2,b,\varepsilon}}\|v\|_{X^{s_1,s_2,b,\varepsilon}}.$$

This estimate yields the following result as a corollary.

Corollary 9.1. *For $\sigma_1 \geq 0$, $\sigma_2 \geq 0$ and $b > 1/2$, we have*

$$\|\partial_x(uv)\|_{Y^{\sigma_1,\sigma_2,-9/20}} \lesssim \|u\|_{Y^{\sigma_1,\sigma_2,b}}\|v\|_{Y^{\sigma_1,\sigma_2,b}}.$$

Proof. From the triangle inequality, it is easy to see that

$$\begin{aligned}
&e^{2(\sigma_1|\xi|+\sigma_2|\mu|)}\,|\widehat{uv}(\xi,\mu,\tau)|^2 \\
&= e^{2(\sigma_1|\xi|+\sigma_2|\mu|)}\left|\int \hat{u}(\xi-\xi_1,\mu-\mu_1,\tau-\tau_1)\hat{v}(\xi_1,\mu_1,\tau_1)\ d\xi d\mu d\tau\right|^2 \\
&\leq \left|\int e^{\sigma_1|\xi-\xi_1|+\sigma_2|\mu-\mu_1|}\hat{u}(\xi-\xi_1,\mu-\mu_1,\tau-\tau_1)\right. \\
&\qquad \left.\times\, e^{\sigma_1|\xi_1|+\sigma_2|\mu_1|}\hat{v}(\xi_1,\mu_1,\tau_1)\ d\xi d\mu d\tau\right|^2 \\
&= \left|\widehat{A^{\sigma_1,\sigma_2}uA^{\sigma_1,\sigma_2}v}\right|^2.
\end{aligned}$$

It follows that

$$\begin{aligned}
\|\partial_x(uv)\|_{Y^{\sigma_1,\sigma_2,-9/20}} &\leq \|\partial_x(A^{\sigma_1,\sigma_2}uA^{\sigma_1,\sigma_2}v)\|_{X^{0,0,\frac{9}{20},0}} \\
&\lesssim \|A^{\sigma_1,\sigma_2}u\|_{X^{0,0,b,0}}\|A^{\sigma_1,\sigma_2}v\|_{X^{0,0,b,0}} \\
&= C\|u\|_{Y^{\sigma_1,\sigma_2,b}}\|v\|_{Y^{\sigma_1,\sigma_2,b}},
\end{aligned}$$

for some constant $C > 0$ by Lemma 9.9. □

The first result relates to the short-term persistence of analyticity of solutions.

Theorem 9.4. *Let $\sigma_1 \geq 0$ and $\sigma_2 \geq 0$. Then for all initial data $f \in G^{\sigma_1,\sigma_2}$, there exists $\delta = \delta(\| f \|_{G^{\sigma_1,\sigma_2}}) > 0$ and a unique solution u of* (9.38) *on the time interval $(-\delta, \delta)$ such that*

$$u \in C((-\delta, \delta); G^{\sigma_1,\sigma_2}(\mathbb{R}^2)).$$

Moreover, the solution depends continuously on the data f. In particular, the time of existence can be chosen to satisfy

$$\delta = \frac{c_0}{(1 + \|f\|_{G^{\sigma_1,\sigma_2}})^\gamma},$$

for some constants $c_0 > 0$ and $\gamma > 1$. Moreover, the solution u satisfies

$$\sup_{t\in[0,\delta]} \|u(t)\|_{G^{\sigma_1,\sigma_2}} \leq 2\|f\|_{G^{\sigma_1,\sigma_2}}.$$

We may now begin the proof of Theorem 9.4 by Picard iteration with the iteration space is $Y^{\sigma_1,\sigma_2,b}(I)$ and $I = (-\delta, \delta)$. To begin, consider the linear problem

$$\begin{aligned} \partial_t u - \partial_x^5 u + \partial_x^{-1}\partial_y^2 u &= F, \\ u(0) &= f. \end{aligned}$$

By Duhamel's principle the solution can be written as

$$u(t) = S(t)f - \frac{1}{2}\int_0^t S(t-s)F(s)ds, \tag{9.42}$$

where

$$\widetilde{S(t)f}(\xi, \mu) = e^{itm(\xi,\mu)}\hat{f}(\xi, \mu).$$

Rather than work directly with the integral formulation in equation (9.42), we instead use the following modified form: let $\psi \in C_0^\infty(\mathbb{R})$ be supported in the interval $(-2, 2)$ such that $0 \leq \psi(t) \leq 1$ and $\psi = 1$ on $[-1, 1]$ and let $\phi_\delta \in C_0^\infty(\mathbb{R})$ be supported on $(-2\delta^{-1/2}, 2\delta^{-1/2})$ such that $0 \leq \phi_\delta(t) \leq 1$ and $\phi_\delta = 1$ on $[-\delta^{-1/2}, \delta^{-1/2}]$. We then observe that we may decompose the integral operator on the right-hand side of equation (9.42) in the form

$$-\frac{1}{2}\int_0^t S(t-s)f(s)ds = I_{\phi_\delta}(f) + II_{\phi_\delta}(f) + III_{\phi_\delta}(f), \tag{9.43}$$

where

$$\mathrm{I}_{\phi_\delta}(f) = C\int e^{i(x\xi+y\mu)}e^{itm(\xi,\mu)}\frac{e^{itp(\xi,\mu,\tau)}-1}{ip(\xi,\mu,\tau)}\phi_\delta(p(\xi,\mu,\tau))\hat{f}(\xi,\mu,\tau) \times d\xi d\mu d\tau,$$

$$\mathrm{II}_{\phi_\delta}(f) = C\int e^{i(x\xi+y\mu)}\frac{e^{it\tau}}{ip(\xi,\mu,\tau)}(1-\phi_\delta(p(\xi,\mu,\tau)))\hat{f}(\xi,\mu,\tau) \times d\xi d\mu d\tau,$$

$$\mathrm{III}_{\phi_\delta}(f) = -C\int e^{i(x\xi+y\mu)}e^{itm(\xi,\mu)}\frac{(1-\phi_\delta(p(\xi,\mu,\tau))}{ip(\xi,\mu,\tau)}\hat{f}(\xi,\mu,\tau) \times d\xi d\mu d\tau.$$

Here, $p(\xi,\mu,\tau) = \tau - m(\xi,\mu)$. We then define a modification $G_\delta(f)$ of the integral operator in (9.43),

$$G_\delta(f) = \psi(t/\delta)I_{\phi_\delta}(f) + II_{\phi_\delta}(f) + \psi(t/\delta)III_{\pi_\delta}(f).$$

It can be shown that if $0<\delta<1$ and $t\in(-\delta,\delta)$, then

$$G_\delta(F)(t) = -\frac{1}{2}\int_0^t S(t-s)F(s)ds.$$

The next step is to define a sequence $\{u_n\}_{n=0}^\infty$ of functions which are solutions to the equations

$$\begin{aligned}
&\partial_t u_0 - \partial_x^5 u_0 + \partial_x^{-1}\partial_y^2 u_0 = 0\\
&\partial_t u_n - \partial_x^5 u_n + \partial_x^{-1}\partial_y^2 u_n = -u_{n-1}\partial_x u_{n-1}\\
&u_0(x,y,0) = f(x,y) \qquad u_n(x,y,0) = f(x,y).
\end{aligned}$$

By the discussion above, for $t\in(-\delta,\delta)$, we have the identity

$$u_n(x,y,t) = \Phi(u_{n-1}(x,y,t)),$$

where

$$\Phi(u) = \psi(t)S(t)f + G_\delta\left(\partial_x(u^2)\right).$$

Using this decomposition, we obtain the following estimate, which will be crucial to our iteration argument:

Lemma 9.10. *Let* $\sigma_1 \geq 0$, $\sigma_2 \geq 0$, $\frac{1}{2} < b < 1$, $\beta \in (0, 1-b)$ *and* $0 < \delta < 1$. *Then*

$$\|u_n\|_{Y^{\sigma_1,\sigma_2,b}} \leq \|f\|_{G^{\sigma_1,\sigma_2}} + C\delta^\gamma \|\partial_x(u_{n-1}^2)\|_{Y^{\sigma_1,\sigma_2,-\beta}},$$

for some $0 < \gamma < 1$.

Proof. Since $u_n = \Phi(u_{n-1})$ for $t \in (-\delta, \delta)$, it suffices to show that

$$\|\Phi(v)\|_{Y^{\sigma_1,\sigma_2,b}} \leq \|f\|_{G^{\sigma_1,\sigma_2}} + C\delta^\gamma \|\partial_x(v^2)\|_{Y^{\sigma_1,\sigma_2,-\beta}},$$

for any function v. To show that this estimate holds, we recall

$$\|\psi(t)S(t)f\|_{X^{s_1,s_2,b,\varepsilon}} \leq C\|f\|_{H^{s_1,s_2}},$$

for $s_1, s_2, b \in \mathbb{R}$ and $\varepsilon \geq 0$ and

$$\|G_\delta(F)\|_{X^{s_1,s_2,b,\varepsilon}} \leq C\delta^\gamma \|F\|_{X^{s_1,s_2,-\beta,\varepsilon}}$$

for $s_1, s_2 \in \mathbb{R}$, $b \in (\frac{1}{2}, 1)$, $\varepsilon \geq 0$ and $\beta \in (0, 1-b)$. From these estimates, it follows that

$$\begin{aligned}
\|\Phi(v)\|_{Y^{\sigma_1,\sigma_2,b}} &= \|A^{\sigma_1,\sigma_2}\Phi(v)\|_{X^{0,0,b,0}} \\
&\leq \|A^{\sigma_1,\sigma_2} f\|_{H^{0,0}} + C\delta^\gamma \|\partial_x(u_{n-1}^2)\|_{X^{0,0,-\beta,0}} \\
&\leq \|f\|_{G^{\sigma_1,\sigma_2}} + C\delta^\gamma \|\partial_x(u_{n-1}^2)\|_{Y^{\sigma_1,\sigma_2,-\beta}},
\end{aligned}$$

as desired. □

If we now apply Corollary 9.1, it is a simple matter to show that

$$\|u_n\|_{Y^{\sigma_1,\sigma_2,b}} \leq \|f\|_{G^{\sigma_1,\sigma_2}} + C\delta^\gamma \|u_{n-1}\|^2_{Y^{\sigma_1,\sigma_2,b}}.$$

Using a simple proof by induction, one may show that

$$\|u_n\|_{Y^{\sigma_1,\sigma_2,b}} \leq 2\|f\|_{G^{\sigma_1,\sigma_2}},$$

for all $n \in \mathbb{N} \cup \{0\}$, if we choose δ such that

$$\delta < \frac{1}{(C\|f\|_{G^{\sigma_1,\sigma_2}})^{1/\gamma}}. \tag{9.44}$$

The final step is to show that the sequence converges. Applying Lemmas 9.10 and 9.1 once again, a similar computation will show that

$$\begin{aligned}\|u_n - u_{n-1}\|_{Y^{\sigma_1,\sigma_2,b}} &= \|\Phi(u_n) - \Phi(u_{n-1})\|_{Y^{\sigma_1,\sigma_2,b}} \\ &\le C\delta^\gamma M_{n-1}\|u_{n-1} - u_{n-2}\|_{Y^{\sigma_1,\sigma_2,b}} \\ &\le 4C\delta^\gamma \|f\|_{G^{\sigma_1,\sigma_2}}\|u_{n-1} - u_{n-2}\|_{Y^{\sigma_1,\sigma_2,b}},\end{aligned}$$

where

$$M_{n-1} = \|u_{n-1}\|_{Y^{\sigma_1,\sigma_2,b}} + \|u_{n-2}\|_{Y^{\sigma_1,\sigma_2,b}}.$$

Thus, the sequence will converge if δ satisfies

$$\delta < \frac{1}{(4C\|f\|_{G^{\sigma_1,\sigma_2}})^{1/\gamma}}. \tag{9.45}$$

sufficiently small. Thus, the sequence converges in $Y^{\sigma_1,\sigma_2,b}$ to a solution

$$u \in Y^{\sigma_1,\sigma_2,b} \subset C((-\delta,\delta); G^{\sigma_1,\sigma_2}).$$

To show uniqueness, suppose u and v are solutions to (9.38) and let $w = u - v$. Then w satisfies the equation

$$\partial_t w - \partial_x^5 w + \partial_x^{-1}\partial_y^2 w = v\partial_x v - u\partial_x u,$$

with

$$w(x,y,0) = 0.$$

Multiplying by w and integrating in x and y yields

$$\frac{1}{2}\partial_t\|w(t)\|_{L^2_{x,y}} \le \left(\|u_x(t)\|_{L^\infty_{x,y}} + \|v_x(t)\|_{L^\infty_{x,y}}\right)\|w(t)\|_{L^2_{x,y}}.$$

By the embedding

$$\|u_x(t)\|_{L^\infty_{x,y}} \lesssim \|u(t)\|_{G^{\sigma_1,\sigma_2}} \le 2\|f\|_{G^{\sigma_1,\sigma_2}},$$

which also holds for v, it will follow from Grönwall's Inequality that $w = 0$. This completes the proof of Theorem 9.4.

As a concluding remark, we observe that for later convenience, we may choose the time of existence to be

$$\delta = \frac{C}{(1 + \|f\|_{G^{\sigma_1,\sigma_2}})^{1/\gamma}}. \tag{9.46}$$

For appropriate choice of C, this will satisfy inequalities (9.44) and (9.45).

Our second main result concerns the evolution of the radius of analyticity for the x-direction.

Theorem 9.5. *Let $\sigma_1 > 0$ and $\sigma_2 \geq 0$ and assume $f \in G^{\sigma_1,\sigma_2}$. Then the solution u given by Theorem 9.4 extends globally in time and for any $T > 0$, we have*

$$u \in C([0,T], G^{\sigma(T),0}(\mathbb{R}^2)) \quad \text{with } \sigma(T) = \min\left\{\sigma_1, CT^{-1}\right\},$$

where $C > 0$ is a constant which does not depend on T.

Lemma 9.11. *For $\sigma > 0$ and $\xi, \xi_1 \in \mathbb{R}$, we have*

$$e^{\sigma|\xi-\xi_1|}e^{\sigma|\xi_1|} - e^{\sigma|\xi|} \lesssim \sigma \frac{\langle \xi - \xi_1\rangle\langle \xi_1\rangle}{\langle \xi\rangle} e^{\sigma|\xi-\xi_1|}e^{\sigma|\xi_1|}.$$

This will be used to prove the following key estimate:

Lemma 9.12. *Let $N(u)$ be as in equation (9.41) for $\sigma_1 \geq 0$ and $\sigma_2 = 0$. Then for b and β as in Lemma 9.9, we have*

$$\|N(u)\|_{Y^{0,0,-\beta}} \leq C\sigma_1 \|u\|^2_{Y^{\sigma_1,0,b}}.$$

Proof. We first observe that the inequality in Lemma 9.9, for the case $\varepsilon = 0$, is equivalent to

$$\Bigg\| \xi\lambda(s_1, s_2, -\beta, 0) \int \frac{\hat{f}(\xi - \xi_1, \mu - \mu_1, \tau - \tau_1)}{\langle \xi - \xi_1\rangle^{s_1}\langle \mu - \mu_1\rangle^{s_2}\langle \phi(\xi - \xi_1, \mu - \mu_1, \tau - \tau_1)\rangle^b}$$

$$\times \frac{\hat{g}(\xi_1, \mu_1, \tau_1)}{\langle \xi_1\rangle^{s_1}\langle \mu_1\rangle^{s_2}\langle \phi(\xi_1, \mu_1, \tau_1)\rangle^b}\, d\xi_1 d\mu_1 d\tau_1 \Bigg\|_{L^2_{\xi,\mu}} \lesssim \|f\|_{L^2_{x,y}} \|g\|_{L^2_{x,y}},$$

where we denote

$$\phi(\tau, \xi, \mu) = \langle \tau - m(\xi, \mu)\rangle.$$

With this in mind, we observe that the left side of the inequality in Lemma 9.12 can be estimated by Lemma 9.11 as

$$\|N(u)\|_{Y^{0,0,-\beta}} \lesssim \sigma_1 \left\| \frac{\xi\langle\xi\rangle^{-1}}{\langle\phi(\tau,\xi,\mu)\rangle^{\beta}} \int \frac{e^{\sigma_1|\xi-\xi_1|}\hat{u}(\xi-\xi_1,\mu-\mu_1,\tau-\tau_1)}{\langle\xi-\xi_1\rangle^{-1}} \right.$$

$$\left. \times \frac{e^{\sigma_1|\xi_1|}\hat{u}(\xi_1,\mu_1,\tau_1)}{\langle\xi_1\rangle^{-1}}\, d\xi_1 d\mu_1 \right\|_{L^2_{\xi,\mu}}.$$

If we apply Lemma 9.9 with $s_1 = -1$, $s_2 = 0$, it will follow that

$$\|N(u)\|_{Y^{0,0,-\beta}} \lesssim \sigma_1 \|u\|^2_{Y^{\sigma_1,0,b}}.$$

□

The first step is to obtain estimates on the growth of the norm of the solutions. For this, we will need the following approximate conservation law:

Proposition 9.2. *Let $\sigma_1 \geq 0$. Then there is a $b \in (1/2,1)$ and a $C > 0$, such that $u \in Y^{\sigma_1,0,b}(I)$ is a solution to the Cauchy problem (9.38) on the time interval $[0,\delta)$, we have the estimate*

$$\sup_{t\in[0,\delta)} \|u(t)\|^2_{G^{\sigma_1,0}} \leq \|f\|^2_{G^{\sigma_1,0}} + C\sigma_1 \|u\|^3_{Y^{\sigma_1,0,b}(I)}. \tag{9.47}$$

Begin by applying the operator $A^{\sigma_1,0}$ to equation (9.38). If we let $U = A^{\sigma_1,0}u$, then equation (9.38) becomes

$$\partial_t u - U_{xxxxx} + \partial_x^{-1} U_{yy} + UU_x = N(u),$$

where $N(u)$ is as defined in Lemma 9.12. Multiplying this by U and integrating with respect to the spatial variables, we obtain

$$\int U\partial_t u - UU_{xxxxx} + U\partial_x^{-1}U_{yy} + U^2U_x \, dxdy = \int UN(u)\, dxdy.$$

If we apply integration by parts, we may rewrite the left-hand side as

$$\partial_t \int \frac{1}{2}U^2\, dxdy + \int U_{xx}\partial_t^3 u dxdy - \int U_y \partial_x^{-1} U_y\, dxdy + \int U^2 U_x\, dxdy,$$

which can then be rewritten as

$$\frac{1}{2}\partial_t \int U^2 \; dxdy + \frac{1}{2}\int \partial_x(U_{xx}^2)dxdy - \frac{1}{2}\int \partial_x[(\partial_x^{-1}U_y)^2] \; dxdy \\ + \frac{1}{3}\int \partial_x(U^3) \; dxdy.$$

To proceed, we observe that U and its derivatives vanish at infinity. We thus obtain the formal identity

$$\partial_t \int U^2(x,y,t) \; dxdy = 2\int U(x,y,t)N(u)(x,y,t) \; dxdy.$$

Integrating with respect to time yields

$$\int U^2(x,y,t) \; dxdy = \int U^2(x,y,0) \; dxdy \\ + 2\int_0^t \int U(x,y,s)N(u)(x,y,s) \; dxdyds.$$

Applying Cauchy–Schwarz inequality and the definition of U, we obtain

$$\|u(t)\|_{G^{\sigma_1,0}}^2 \leq \|f\|_{G^{\sigma_1,0}}^2 + \|u\|_{Y^{\sigma_1,0,b}}\|N(u)\|_{Y^{0,0,-\beta}(I)},$$

where β is as defined previously. If we now apply Proposition 9.2 and the fact that $\beta < 1/2 < b$, we can further estimate this by

$$\|u(t)\|_{G^{\sigma_1,0}}^2 \leq \|f\|_{G^{\sigma_1,0}}^2 + C\sigma_1\|u\|_{Y^{\sigma_1,0,b}}^3, \tag{9.48}$$

as desired.

With the tools established in the previous section, we may begin the proof of Theorem 9.5. By the embedding in equation (9.39), it suffices to consider the case $\sigma_2 = 0$. To begin, let us first suppose that T^* is the supremum of the set of times T for which

$$u \in C([0,T]; G^{\sigma_1,0}).$$

If $T^* = \infty$, there is nothing to prove, so let us assume that $T^* < \infty$. In this case, it suffices to prove that

$$u \in C\left([0,T], G^{\sigma(T),0}\right), \tag{9.49}$$

for some $\sigma(T) > 0$ and all $T > T^*$. To show that this is the case, we will use Theorem 9.4 and Proposition 9.2 to construct a solution which exists over subintervals of width δ, using the parameter σ_1 to control the growth of the norm of the solution. Thus, the desired result will follow from the following proposition:

Proposition 9.3. *Let $T > 0$ and $\delta > 0$ be numbers such that $n\delta \le T < (n+1)\delta$. Then the solution u to the Cauchy problem (9.38) satisfies*

$$\sup_{t\in[0,n\delta]} \|u(t)\|^2_{G^{\sigma(T)},0} \le \|f\|^2_{G^{\sigma(T)},0} + 2^3 C\sigma(T)n\|f\|^3_{G^{\sigma_1},0}, \tag{9.50}$$

and

$$\sup_{t\in[0,n\delta]} \|u(t)\|^2_{G^{\sigma(T)},0} \le 4\|u(t)\|^2_{G^{\sigma_1},0}, \tag{9.51}$$

if

$$\sigma(T) \le \sigma_1 \quad \text{and} \quad \sigma(T) \le \frac{C}{T},$$

for some constant $C > 0$.

Proof. By induction on n. The base case $n = 1$ follows from equation (9.48), Theorem 9.4 and the embedding

$$G^{\sigma_1,\sigma_2} \hookrightarrow G^{\sigma_1',\sigma_2'},$$

when $\sigma_1' \le \sigma_1$ and $\sigma_2' \le \sigma_2$. Suppose, then, that the result holds for $n \le k$. The inductive hypothesis then tells us that

$$\sup_{t\in[0,k\delta]} \|u(t)\|_{G^{\sigma(T)},0} \le \|f\|^2_{G^{\sigma(T)},0} + 2^3 C\sigma(T)k\|f\|^3_{G^{\sigma_1},0},$$

and

$$\sup_{t\in[0,k\delta]} \|u(t)\|^2_{G^{\sigma(T)},0} \le 4\|f\|^2_{G^{\sigma_1},0}.$$

If we apply the inductive hypothesis on the interval $[k\delta, (k+1)\delta]$, then

$$\sup_{t\in[k\delta,(k+1)\delta]} \|u(t)\|_{G^{\sigma(T)},0} \le \|u(k\delta)\|^2_{G^{\sigma(T)},0} + 2^3 C\sigma(T)\|u(k\delta)\|^3_{G^{\sigma_1},0},$$

and

$$\sup_{t\in[k\delta,(k+1)\delta]} \|u(t)\|^2_{G^{\sigma(T)},0} \le 4\|u(k\delta)\|^2_{G^{\sigma_1},0}.$$

If we apply equations (9.50) and (9.51) to these, we get

$$\|u(k\delta)\|_{G^{\sigma(T)},0} \le \|f\|^2_{G^{\sigma(T)},0} + 2^3Ck\sigma(T)\|u(k\delta)\|^3_{G^{\sigma_1},0},$$

and

$$\|u(k\delta)\|^2_{G^{\sigma_1},0} \le 4\|f\|^2_{G^{\sigma_1},0}.$$

Combining these together, we obtain

$$\begin{aligned}\sup \|u(t)\|_{G^{\sigma(T)},0} &\le (\|f\|^2_{G^{\sigma(T)},0} + 2^3Ck\sigma(T)\|f\|^3_{G^{\sigma_1},0})\\ &\quad + 2^3C\sigma(T)\|f\|^3_{G^{\sigma_1},0}\\ &= \|f\|^2_{G^{\sigma(T)},0} + 2^3C(k+1)\sigma(T)\|f\|^3_{G^{\sigma_1},0}.\end{aligned}$$

It follows that

$$\sup_{t\in[0,(k+1)\delta]} \|u(t)\|_{G^{\sigma(T)},0} \le \|f\|^2_{G^{\sigma(T)},0} + 2^3C(k+1)\sigma(T)\|f\|^3_{G^{\sigma_1},0}.$$

To complete the proof, we need to show that

$$\sup_{t\in[0,(k+1)\delta]} \|u(t)\|^2_{G^{\sigma(t)},0} \le 4\|f\|^2_{G^{\sigma_1},0}. \tag{9.52}$$

Next, we observe that the assumption $n\delta \le T < (n+1)\delta$ implies that

$$n \le \frac{T}{\delta} < n+1 \le \frac{T}{\delta} + 1 \le \frac{2T}{\delta}.$$

Since $k+1 \le n+1$, we have

$$2^3C(k+1)\sigma(T)\|f\|_{G^{\sigma_1},0} \le 2^3C\frac{2T}{\delta}\sigma(T)\|f\|_{G^{\sigma_1},0}.$$

Recalling the definition of δ in equation (9.46), we can rewrite this as

$$\begin{aligned}2^3C\frac{2T}{\delta}\sigma(T)\|f\|_{G^{\sigma_1},0} &= 2^3C\|f\|_{G^{\sigma_1},0}(1+\|f\|_{G^{\sigma_1},0})^{1/\gamma}T\sigma(T)\\ &\le C(1+\|f\|_{G^{\sigma_1},0})^{1/\gamma+1}T\sigma(T).\end{aligned}$$

Thus, for equation (9.52) to hold, it suffices to have

$$\sigma(T) \leq \frac{C}{T},$$

where C is a constant that depends on the norm $\|f\|_{G^{\sigma_1,0}}$.

Since we have shown that the result holds for $n = 1$ and we have shown that the result for $n = k$ implies it for $n = k + 1$, then the result holds for all n. This completes the proof of Proposition 9.3, from which Theorem 9.5 follows as an immediate corollary. □

Chapter 10

Generalized Kadomtsev–Petviashvili I Equation

Let

$$u = u(x, y, t), (x, y, t) \in \mathbb{R}^3,$$

and $\alpha \geq 4$. We consider the initial value problem for the generalized Kadomtsev–Petviashvili I equation,

$$\partial_t u + |D_x|^\alpha \partial_x u + \partial_x^{-1} \partial_y^2 u + u \partial_x u = 0, \tag{10.1}$$

with

$$u(x, y, 0) = f(x, y),$$

and

$$D_x^\alpha u(x, y, t) = \frac{1}{(2\pi)^3} \int_{\mathbb{R}^3} |\xi|^\alpha \mathfrak{F} u(\xi, \mu, \tau) e^{ix\xi + iy\mu + it\tau} d\xi d\mu d\tau.$$

This equation belongs to the class of Kadomtsev–Petviashvili equations, which are models for the propagation of long dispersive nonlinear waves which are essentially unidirectional and have weak transverse effects. Due to the asymmetric nature of the equation with respect to the spatial derivatives, it is natural to consider the Cauchy problem for (10.1) with initial data in the anisotropic Sobolev spaces $H^{s_1, s_2}(\mathbb{R}^2)$, defined by the norm

$$\|u\|_{H^{s_1, s_2}(\mathbb{R}^2)} = \left(\int \langle \xi \rangle^{2s_1} \langle \mu \rangle^{2s_2} |\widehat{u}(\xi, \mu)|^2 d\xi d\mu \right)^{1/2}.$$

Many authors have investigated the Cauchy problem for Kadomtsev–Petviashvili equations. It is established the local well-posedness of the Cauchy problem for the Kadomtsev–Petviashvili I equation in anisotropic Sobolev spaces $H^{s_1,s_2}(\mathbb{R})$ with

$$s_1 > -\frac{\alpha-1}{4}, \quad s_2 \geq 0,$$

with $\alpha \geq 4$ and globally well-posed in $H^{s_1,0}(\mathbb{R})$ with

$$s_1 > -\frac{(\alpha-1)(3\alpha-4)}{4(5\alpha+3)},$$

if $4 \leq \alpha \leq 5$ also proved that the Cauchy problem is globally well-posed in $H^{s_1,0}(\mathbb{R})$ with

$$s_1 > -\frac{\alpha(3\alpha-4)}{4(5\alpha+4)},$$

if $\alpha > 5$. The problem proposed

$$\begin{cases} \partial_t u - \partial_x^5 u + \partial_x^{-1}\partial_y^2 u + u\partial_x u = 0, \\ u(x,y,0) = u_0(x,y), \end{cases} \tag{10.2}$$

and proved that it is globally well-posed for given data in an anisotropic Gevrey space

$$G^{\sigma_1,\sigma_2}(\mathbb{R}^2), \sigma_1, \sigma_2 \geq 0,$$

with respect to the norm

$$\|f\|_{G^{\sigma_1,\sigma_2}(\mathbb{R}^2)} = \left(\int_{\mathbb{R}^2} e^{2\sigma_1|\xi|}e^{2\sigma_2|\mu|}|\widehat{f}(\xi,\mu)|^2 d\xi d\mu\right)^{1/2}.$$

With initial data in anisotropic Gevrey space

$$G^{\sigma_1,\sigma_2,\kappa}_{s_1,s_2}(\mathbb{R}^2) = G^{\sigma_1,\sigma_2,\kappa}_{s_1,s_2}, \sigma_1, \sigma_2 \geq 0, s_1, s_2 \in \mathbb{R},$$

and $\kappa \geqslant 1$, we will consider the problem (10.1). The spaces $G^{\sigma_1,\sigma_2,\kappa}_{s_1,s_2}$ can be defined as the completion of the Schwartz functions with respect to the norm

$$\|f\|_{G^{\sigma_1,\sigma_2,\kappa}_{s_1,s_2}(\mathbb{R}^2)} = \left(\int_{\mathbb{R}^2} e^{2\sigma_1|\xi|^{\frac{1}{\kappa}}} e^{2\sigma_2|\mu|^{\frac{1}{\kappa}}} \mu^2(s_1,s_2)|\widehat{f}(\xi,\mu)|^2 d\xi d\mu \right)^{1/2},$$

where

$$\mu(s_1,s_2) = \langle\xi\rangle^{s_1}\langle\mu\rangle^{s_2}.$$

In addition to the holomorphic extension property, Gevrey spaces satisfy the embeddings

$$G^{\sigma_1,\sigma_2}_{s_1} \hookrightarrow G^{\sigma_1',\sigma_2'}_{s_1'},$$

for $s_1, s_1' \in \mathbb{R}$ and $\sigma_i' < \sigma_i$ where

$$G^{\sigma_1,\sigma_2,1}_{s_1,0} = G^{\sigma_1,\sigma_2}_{s_1},$$

which follow from the corresponding estimates

$$\|f\|_{G^{\sigma_1',\sigma_2'}_{s_1'}} \lesssim \|f\|_{G^{\sigma_1,\sigma_2}_{s_1}}.$$

The main aim to consider initial data in these spaces is because of the Paley–Wiener theorem in Proposition 8.1. We will also need the full space time Fourier transform denoted by

$$\hat{f}(\xi,\mu,\tau) = \int_{\mathbb{R}^3} f(x,y,t)e^{-i(x\xi+y\mu+t\tau)}dxdydt.$$

In both cases, we will denote the corresponding inverse transform of a function

$$f = f(\xi,\mu) \text{ or } f = f(\xi,\mu,\tau) \text{ by } \mathfrak{F}^{-1}(f).$$

To simplify the notation, we introduce some operators. We first introduce the operator $A^{\sigma_1,\sigma_2}_{\kappa}$, which we define as

$$A^{\sigma_1,\sigma_2}_{\kappa} f = \mathfrak{F}^{-1}\left(e^{\sigma_1|\xi|^{\frac{1}{\kappa}}} e^{\sigma_2|\mu|^{\frac{1}{\kappa}}} \hat{f} \right). \tag{10.3}$$

Then, we may then define another useful operator

$$N^{\sigma_1,\sigma_2}_{\kappa}(f) = \partial_x \left[(A^{\sigma_1,\sigma_2}_{\kappa} f)^2 - A^{\sigma_1,\sigma_2}_{\kappa}(f^2) \right]. \tag{10.4}$$

For $x \in \mathbb{R}^n$, we denote $\langle x\rangle = (1+|x|^2)^{1/2}$. The main function spaces they used are the so-called anisotropic Bourgain spaces,adapted to the generalized Kadomtsev–Petviashvili I, whose norm is given by

$$\|u\|_{X_{s_1,s_2,b}} = \left(\int_{\mathbb{R}^3} \theta^2(s_1,s_2,b)|\hat{u}(\xi,\mu,\tau)|^2 d\xi d\mu d\tau\right)^{\frac{1}{2}},$$

where

$$\theta(s_1,s_2,b) = \langle\xi\rangle^{s_1}\langle\mu\rangle^{s_2}\langle\tau + m(\xi,\mu)\rangle^b,$$

with

$$m(\xi,\mu) = \xi|\xi|^\alpha + \frac{\mu^2}{\xi}.$$

Furthermore, we will also need a hybrid of the analytic Gevrey and anisotropic Bourgain spaces, designated $X^{\sigma_1,\sigma_2,\kappa}_{s_1,s_2,b}(\mathbb{R}^3)$ and defined by the standard

$$\|u\|_{X^{\sigma_1,\sigma_2,\kappa}_{s_1,s_2,b}} = \left(\int_{\mathbb{R}^3} e^{2\sigma_1|\xi|^{\frac{1}{\kappa}}} e^{2\sigma_2|\mu|^{\frac{1}{\kappa}}} \theta^2(s_1,s_2,b)|\hat{u}(\xi,\mu,\tau)|^2 d\xi d\mu d\tau\right)^{\frac{1}{2}}.$$

It is well-known that these spaces satisfy the embedding

$$X^{\sigma_1,\sigma_2,\kappa}_{s_1,s_2,b} \hookrightarrow C\left(\mathbb{R}; G^{\sigma_1,\sigma_2,\kappa}_{s_1,s_2}(\mathbb{R}^2)\right).$$

Thus, solutions constructed in $X^{\sigma_1,\sigma_2,\kappa}_{s_1,s_2,b}$ belong to the natural solution space.

When considering local solutions, it is useful to consider localized versions of these spaces. For a time interval I and a Banach space $\mathscr{Y}$, we define the localized space $\mathscr{Y}(I)$ by the norm

$$\|u\|_{\mathscr{Y}(I)} = \inf\{\|v\|_{\mathscr{Y}} : \ v = u \text{ on } I\}.$$

The second main result concerns the evolution of the radius of analyticity for the x-direction is given in the next theorem. Here,

$$X^{\sigma_1,0,1}_{s_1,0,b} = X^{\sigma_1,0}_{s_1,b}, s_2, \sigma_2 = 0 \quad \text{and} \quad \kappa = 1.$$

10.1 Auxiliary Estimates

To begin with, let us consider the related linear problem

$$\begin{aligned} &\partial_t u + |D_x|^\alpha \partial_x u + \partial_x^{-1}\partial_y^2 u = F, \\ &u(0) = f. \end{aligned}$$

By Duhamel's principle the solution can be written as

$$u(t) = S(t)f - \frac{1}{2}\int_0^t S(t-s)F(s)ds, \tag{10.5}$$

where

$$\widehat{S(t)f}(\xi,\mu) = e^{itm(\xi,\mu)}\widehat{f}(\xi,\mu).$$

We localize it in t by using a cut-off function satisfying $\psi \in C_0^\infty(\mathbb{R})$, with $\psi = 1$ in $[-1,1]$ and supp$\psi \subset [-2,2]$.

We consider the operator Φ given by

$$\Phi(u) = \psi(t)S(t)f - \frac{\psi_\delta(t)}{2}\int_0^t S(t-s)\left(\partial_x u^2(s)\right)ds, \tag{10.6}$$

where $\psi_\delta(t) = \psi(\frac{t}{\delta})$. To this operator, we apply the following estimates.

Lemma 10.1 (Linear estimate). *Let* $s_1, s_2 \in \mathbb{R}, -\frac{1}{2} < b' \leq 0 \leq b \leq b'+1, \sigma_1 \geq 0, \sigma_2 \geq 0, \kappa \geq 1$ *and* $\delta \in (0,1)$. *Then*

$$\|\psi(t)S(t)f\|_{X^{\sigma_1,\sigma_2,\kappa}_{s_1,s_2,b}} \leq C\|f\|_{G^{\sigma_1,\sigma_2,\kappa}_{s_1,s_2}}, \tag{10.7}$$

$$\left\|\psi_\delta(t)\int_0^t S(t-s)F(x,y,)\mathrm{d}s\right\|_{X^{\sigma_1,\sigma_2,\kappa}_{s_1,s_2,b}} \leq C\delta^{1-b+b'}\|F\|_{X^{\sigma_1,\sigma_2,\kappa}_{s_1,s_2,b'}}. \tag{10.8}$$

Lemma 10.2 (Bilinear estimate in Bourgain space). *Let* $s_1 \geq -\frac{\alpha-1}{4} + 4\alpha\varepsilon, s_2 \geq 0,\ \alpha \geq 4,\ b = \frac{1}{2}+\varepsilon$ *and* $b' = -\frac{1}{2}+2\varepsilon$. *Then, we have*

$$\|\partial_x(u_1u_2)\|_{X^{s_1,s_2,b'}} \lesssim \|u_1\|_{X^{s_1,s_2,b}}\|u_2\|_{X^{s_1,s_2,b}}.$$

To this result, we apply the following lemma, which is a corollary of Lemma 10.2.

Lemma 10.3 (Bilinear Estimate in Gevrey–Bourgain Space). *Let* $s_1 > -\frac{\alpha-1}{4} + 4\alpha\varepsilon, s_2 \geq 0, \alpha \geq 4, \sigma_1 \geq 0, \sigma_2 \geq 0, \kappa \geq 1, b = \frac{1}{2}+\varepsilon$ *and* $b' = -\frac{1}{2}+2\varepsilon$. *Then, we have*

$$\|\partial_x(u_1u_2)\|_{X^{\sigma_1,\sigma_2,\kappa}_{s_1,s_2,b'}} \lesssim \|u_1\|_{X^{\sigma_1,\sigma_2,\kappa}_{s_1,s_2,b}}\|u_2\|_{X^{\sigma_1,\sigma_2,\kappa}_{s_1,s_2,b}}.$$

Proof. It is not hard to see that

$$e^{2(\sigma_1|\xi|^{\frac{1}{\kappa}}+\sigma_2|\mu|^{\frac{1}{\kappa}})}\left|\widehat{u_1u_2}(\xi,\mu,\tau)\right|^2$$

$$= e^{2(\sigma_1|\xi|^{\frac{1}{\kappa}}+\sigma_2|\mu|^{\frac{1}{\kappa}})}\left|\int \hat{u_1}(\xi-\xi_1,\mu-\mu_1,\tau-\tau_1)\right.$$

$$\left.\times\,\hat{u_2}(\xi_1,\mu_1,\tau_1)\;d\xi_1d\mu_1d\tau_1\right|^2$$

$$\le \left|\int e^{\sigma_1|\xi-\xi_1|^{\frac{1}{\kappa}}+\sigma_2|\mu-\mu_1|^{\frac{1}{\kappa}}}\hat{u_1}(\xi-\xi_1,\mu-\mu_1,\tau-\tau_1)\right.$$

$$\left.\times\, e^{\sigma_1|\xi_1|^{\frac{1}{\kappa}}+\sigma_2|\mu_1|^{\frac{1}{\kappa}}}\hat{u_2}(\xi_1,\mu_1,\tau_1)\;d\xi_1d\mu_1d\tau_1\right|^2$$

$$= \left|A_\kappa^{\sigma_1,\widehat{\sigma_2}u A_\kappa^{\sigma_1,\sigma_2}}v\right|^2.$$

By Lemma 10.2, we get

$$\begin{aligned}\|\partial_x(u_1u_2)\|_{X^{\sigma_1,\sigma_2,\kappa}_{s_1,s_2,b'}} &\le \|\partial_x(A_\kappa^{\sigma_1,\sigma_2}u_1A_\kappa^{\sigma_1,\sigma_2}u_2)\|_{X^{s_1,s_2,b'}}\\ &\lesssim \|A_\kappa^{\sigma_1,\sigma_2}u_1\|_{X^{s_1,s_2,b}}\|A_\kappa^{\sigma_1,\sigma_2}v\|_{X^{s_1,s_2,b}}\\ &= \|u_1\|_{X^{\sigma_1,\sigma_2,\kappa}_{s_1,s_2,b}}\|u_2\|_{X^{\sigma_1,\sigma_2,\kappa}_{s_1,s_2,b'}}.\end{aligned}$$ □

10.2 Local Well-Posedness in an Anisotropic Gevrey Space

The first result related to the short-term persistence of analyticity of solutions is given in the next theorem.

Theorem 10.1. *Let $s_1 > -\frac{\alpha-1}{4}$, $s_2 \ge 0$, $\alpha \ge 4$, $\sigma_1 \ge 0$, $\sigma_2 \ge 0$ and $\kappa \ge 1$. Then for all initial data*

$$f \in G^{\sigma_1,\sigma_2,\kappa}_{s_1,s_2},$$

and

$$|\xi|^{-1}\widehat{f}(\xi,\mu) \in L^2(\mathbb{R}^2),$$

there exists $\delta = \delta(\|f\|_{G^{\sigma_1,\sigma_2,\kappa}_{s_1,s_2}}) > 0$ *and a unique solution* u *of* (10.1) *on the time interval* $[0, \delta]$ *such that*

$$u \in C\left([0, \delta]; G^{\sigma_1,\sigma_2,\kappa}_{s_1,s_2}(\mathbb{R}^2)\right).$$

Moreover, the solution depends continuously on the data f. *In particular, the time of existence can be chosen to satisfy*

$$\delta = \frac{c_0}{(1 + \|f\|_{G^{\sigma_1,\sigma_2,\kappa}_{s_1,s_2}})^{\gamma}},$$

for some constants $c_0 > 0$ *and* $\gamma > 1$. *Moreover, the solution* u *satisfies*

$$\sup_{t\in[0,\delta]} \|u(t)\|_{G^{\sigma_1,\sigma_2,\kappa}_{s_1,s_2}} \leq 4\|f\|_{G^{\sigma_1,\sigma_2,\kappa}_{s_1,s_2}}.$$

Lemma 10.4. *Let* $s_1 > -\frac{\alpha-1}{4} + 4\alpha\varepsilon, s_2 \geq 0, \alpha \geq 4, \sigma_1 \geq 0, \sigma_2 \geq 0, \kappa \geq 1, b = \frac{1}{2} + \varepsilon, b' = -\frac{1}{2} + 2\varepsilon$ *and* $0 < \delta < 1$. *Then*

$$\|\Phi(u)\|_{X^{\sigma_1,\sigma_2,\kappa}_{s_1,s_2,b}} \leq C\|f\|_{G^{\sigma_1,\sigma_2,\kappa}_{s_1,s_2}} + C\delta^{\varepsilon}\|u\|^2_{X^{\sigma_1,\sigma_2,\kappa}_{s_1,s_2,b}},$$

and

$$\|\Phi(u_1) - \Phi(u_2)\|_{X^{\sigma_1,\sigma_2,\kappa}_{s_1,s_2,b}} \leq \frac{1}{2}\|u_1 - u_2\|^2_{X^{\sigma_1,\sigma_2,\kappa}_{s_1,s_2,b}}.$$

Proof. Combining Lemmas 10.1 and 10.3 with the fixed point theorem. We define

$$\mathbb{B}(0, 2C\|f\|_{G^{\sigma_1,\sigma_2,\kappa}_{s_1,s_2}}) = \left\{u : \|u\|_{X^{\sigma_1,\sigma_2,\kappa}_{s_1,s_2,b}} \leq 2C\|f\|_{G^{\sigma_1,\sigma_2,\kappa}_{s_1,s_2}}\right\}.$$

Then, we have

$$\begin{aligned}
\|\Phi(u)\|_{X^{\sigma_1,\sigma_2,\kappa}_{s_1,s_2,b}} &\leq \|\psi(t)S(t)f\|_{X^{\sigma_1,\sigma_2,\kappa}_{s_1,s_2,b}} \\
&\quad + \left\|\frac{1}{2}\psi_\delta(t)\int_0^t S(t-s)\partial_x u^2 \mathrm{d}s\right\|_{X^{\sigma_1,\sigma_2,\kappa}_{s_1,s_2,b}} \\
&\leq C\|f\|_{G^{\sigma_1,\sigma_2,\kappa}_{s_1,s_2}} + C\delta^{\varepsilon}\|\partial_x u^2\|_{X^{\sigma_1,\sigma_2,\kappa}_{s_1,s_2,b'}} \\
&\leq C\|f\|_{G^{\sigma_1,\sigma_2,\kappa}_{s_1,s_2}} + C\delta^{\varepsilon}\|u\|^2_{X^{\sigma_1,\sigma_2,\kappa}_{s_1,s_2,b}}.
\end{aligned}$$

We choose δ such that

$$\delta < \frac{1}{(C^2\|f\|_{X^{\sigma_1,\sigma_2,\kappa}_{s_1,s_2,b}})^{\frac{1}{\varepsilon}}}. \tag{10.9}$$

We have

$$\|\Phi(u)\|_{X^{\sigma_1,\sigma_2,\kappa}_{s_1,s_2,b}} \leq 2C\|f\|_{X^{\sigma_1,\sigma_2,\kappa}_{s_1,s_2,b}}.$$

Thus, $\Phi(u)$ maps $\mathbb{B}(0, 2C\|f\|_{G^{\sigma_1,\sigma_2,\kappa}_{s_1,s_2}})$ into $\mathbb{B}(0, 2C\|f\|_{G^{\sigma_1,\sigma_2,\kappa}_{s_1,s_2}})$ which is a contraction, since

$$\begin{aligned}
&\|\Phi(u_1) - \Phi(u_2)\|_{X^{\sigma_1,\sigma_2,\kappa}_{s_1,s_2,b}} \\
&\quad \leq C\left\|\frac{1}{2}\psi_\delta(t)\int_0^t S(t-s)(\partial_x u_1^2 - \partial_x u_1^2)\mathrm{d}s\right\|_{X^{\sigma_1,\sigma_2,\kappa}_{s_1,s_2,b}} \\
&\quad \leq C\delta^\varepsilon\|u_1 - u_2\|_{X^{\sigma_1,\sigma_2,\kappa}_{s_1,s_2,b}}\left[\|u_1\|_{X^{\sigma_1,\sigma_2,\kappa}_{s_1,s_2,b}} + \|u_2\|_{X^{\sigma_1,\sigma_2,\kappa}_{s_1,s_2,b}}\right] \\
&\quad \leq 4C^2\delta^\varepsilon\|f\|_{G^{\sigma_1,\sigma_2,\kappa}_{s_1,s_2}}\|u_1 - u_2\|_{X^{\sigma_1,\sigma_2,\kappa}_{s_1,s_2,b}} \\
&\quad \leq \frac{1}{2}\|u_1 - u_2\|_{X^{\sigma_1,\sigma_2,\kappa}_{s_1,s_2,b}}.
\end{aligned}$$

Here, we choose δ such that

$$\delta < \frac{1}{(8C^2\|f\|_{X^{\sigma_1,\sigma_2,\kappa}_{s_1,s_2,b}})^{\frac{1}{\varepsilon}}}. \tag{10.10}$$

We choose the time of existence where

$$\delta = \frac{c_0}{(1 + \|f\|_{X^{\sigma_1,\sigma_2,\kappa}_{s_1,s_2,b}})^{\frac{1}{\varepsilon}}}.$$

For appropriate choice of c_0, this will satisfy inequalities (10.9) and (10.10). □

From Lemma 10.4, we see that for initial data

$$f(x,y) \in G^{\sigma_1,\sigma_2,\kappa}_{s_1,s_2}(\mathbb{R}^2),$$

if the lifespan

$$\delta = c_0/(1+\|f\|_{X^{\sigma_1,\sigma_2,\kappa}_{s_1,s_2,b}})^{\frac{1}{\varepsilon}},$$

then the map $\Phi(u)$ is a contraction on a small ball centered at the origin in $X^{\sigma_1,\sigma_2,\kappa}_{s_1,s_2,b}$. Hence, the map $\Phi(u)$ has a unique fixed point u in a neighborhood of 0 with respect to the norm $\|\cdot\|_{X^{\sigma_1,\sigma_2,\kappa}_{s_1,s_2,b}}$.

The rest of the proof follows the standard argument.

10.3 Global Well-Posedness

Theorem 10.2. *Let $\sigma_1 > 0$, $s_1 > -\frac{\alpha-1}{4}$, $\alpha = 4, 6, 8, \ldots$ and assume that*

$$f \in G^{\sigma_1,0}_{s_1},\ |\xi|^{-1}\widehat{f}(\xi,\mu) \in L^2(\mathbb{R}^2).$$

Then the solution u given by Theorem 10.1 *extends globally in time and for any $T > 0$, we have*

$$u \in C\left([0,T], G^{\sigma_1(T),0}_{s_1}(\mathbb{R}^2)\right) \quad \text{with} \quad \sigma_1(T) = \min\left\{\sigma_1, CT^{-\rho}\right\},$$

with

$$\rho = \frac{4}{\alpha-1} + \varepsilon,$$

for $\varepsilon > 0$ when $\alpha = 4$ and $\rho = 1$ when $\alpha = 6, 8, 10, \ldots$ and the constant C is a positive.

The first step is to obtain estimates on the growth of the norm of the solutions. For this end, we need to prove the following approximate conservation law.

Theorem 10.3. *Let $\sigma_1 > 0$ and δ be as in Theorem* 10.1, *there exist $b \in (1/2, 1)$ and $C > 0$, such that for any solution $u \in X^{\sigma_1,0}_{0,b}(I)$ to the Cauchy problem* (10.1) *on the time interval $I \subset [0,\delta]$, we have the estimate*

$$\sup_{t\in[0,\delta]} \|u(t)\|^2_{G^{\sigma_1,0}_0} \le \|f\|^2_{G^{\sigma_1,0}_0} + C\sigma_1^{\rho}\|u\|^3_{X^{\sigma_1,0}_{0,b}(I)}. \tag{10.11}$$

with $\rho \in [0, \frac{3}{4})$ if $\alpha = 4$ and $\rho = 1$ if $\alpha = 6, 8, 10, \ldots$.

Lemma 10.5. *For $\sigma > 0$, $0 \leq \theta \leq 1$ and $\xi, \xi_1 \in \mathbb{R}$, we have*

$$e^{\sigma|\xi-\xi_1|}e^{\sigma|\xi_1|} - e^{\sigma|\xi|} \lesssim \sigma^\theta \frac{\langle \xi - \xi_1\rangle\langle \xi_1\rangle}{\langle \xi\rangle} e^{\sigma|\xi-\xi_1|}e^{\sigma|\xi_1|}.$$

This will be used to prove the following key estimate.

Lemma 10.6. *Let $N_1^{\sigma_1,0}(u)$ be as in equation* (10.4) *for $\sigma_1 \geq 0$ and $\sigma_2 = 0$. Then for b as in Lemma* 10.2, *we have*

$$\|N_1^{\sigma_1,0}(u)\|_{X_{0,b-1}} \leq C\sigma_1^\rho \|u\|^2_{X^{\sigma_1,0}_{0,b}},$$

with $\rho \in [0, \frac{3}{4})$ if $\alpha = 4$ and $\rho = 1$ if $\alpha = 6, 8, 10, \ldots$

Proof. We first observe that the inequality in Lemma 10.2, is equivalent to

$$\left\| \xi\theta(s_1, s_2, b-1) \int \frac{\hat{f}(\xi-\xi_1, \mu-\mu_1, \tau-\tau_1)}{\langle \xi-\xi_1\rangle^{s_1}\langle \mu-\mu_1\rangle^{s_2}\langle \phi(\xi-\xi_1, \mu-\mu_1, \tau-\tau_1)\rangle^{b-1}} \right.$$
$$\left. \times \frac{\hat{g}(\xi_1, \mu_1, \tau_1)}{\langle \xi_1\rangle^{s_1}\langle \mu_1\rangle^{s_2}\langle \phi(\xi_1, \mu_1, \tau_1)\rangle^{b-1}}\, d\xi_1 d\mu_1 d\tau_1 \right\|_{L^2_{\xi,\mu}} \lesssim \|f\|_{L^2_{x,y}} \|g\|_{L^2_{x,y}},$$

where we denote

$$\phi(\tau, \xi, \mu) = \langle \tau + m(\xi, \mu)\rangle.$$

With this, we observe that the left side of the inequality in Lemma 10.6 can be estimated by Lemma 10.5 as

$$\|N_1^{\sigma_1,0}(u)\|_{X_{0,b-1}} \lesssim \sigma \left\| \frac{\xi\langle \xi\rangle^{-1}}{\langle \phi(\tau, \xi, \mu)\rangle^\beta} \int \frac{e^{\sigma_1|\xi-\xi_1|}\hat{u}(\xi-\xi_1, \mu-\mu_1, \tau-\tau_1)}{\langle \xi-\xi_1\rangle^{-1}} \right.$$
$$\left. \times \frac{e^{\sigma_1|\xi_1|}\hat{u}(\xi_1, \mu_1, \tau_1)}{\langle \xi_1\rangle^{-1}}\, d\xi_1 d\mu_1 \right\|_{L^2_{\xi,\mu}}.$$

If we apply Lemma 10.2 with $s_1 = -\rho$, $s_2 = 0$, it will follow, from the comments above, that

$$\|N_1^{\sigma_1,0}(u)\|_{X_{0,b-1}} \lesssim \sigma_1^\rho \|u\|^2_{X^{\sigma_1,0}_{0,b}}. \qquad \square$$

Proof Theorem 10.3. Begin by applying the operator $A_1^{\sigma_1,0}$ to (10.1). If we let $U = A_1^{\sigma_1,0}u$, then equation (10.1) becomes

$$\partial_t U - \partial_x^\alpha U + \partial_x^{-1}\partial_y^2 U + U\partial_x U = N_1^{\sigma_1,0}(u), \qquad (10.12)$$

where $\alpha = 4, 6, 8, \ldots$ and $N_1^{\sigma_1,0}(u)$ is defined in Lemma 10.6. Multiplying (10.12) by U and integrating with respect to the spatial variables, we obtain

$$\int U\partial_t U - U\partial_x^\alpha U + U\partial_x^{-1}\partial_y^2 U + U^2\partial_x U dxdy = \int UN_1^{\sigma_1,0}(u)dxdy.$$

If we apply integration by parts, we may rewrite the left-hand side as

$$\partial_t \int \frac{1}{2}U^2 dxdy + \int \partial_x^{\frac{\alpha}{2}} U \partial_x^{\frac{\alpha}{2}+1} U dxdy - \int \partial_y U \partial_x^{-1}\partial_y U dxdy$$
$$+ \int U^2 \partial_x U dxdy = \int UN_1^{\sigma_1,0}(u)dxdy,$$

which can then be rewritten as

$$\partial_t \int \frac{1}{2}U^2 dxdy + \int \frac{1}{2}\partial_x((\partial_x^{\frac{\alpha}{2}}U)^2)dxdy - \int \frac{1}{2}\partial_x[(\partial_x^{-1}\partial_x U)^2]dxdy$$
$$+ \int \frac{1}{3}\partial_x(U^3)dxdy = \int UN_1^{\sigma_1,0}(u)dxdy.$$

By noting that

$$\partial_x^j U(x,t) \to 0,$$

as $|x| \to \infty$, we obtaining

$$\partial_t \int U^2(x,y,t)dxdy = 2\int U(x,y,t)N_1^{\sigma_1,0}(u)(x,y,t)dxdy.$$

Integrating with respect to time yields

$$\int U^2(x,y,t)dxdy = \int U^2(x,y,0)dxdy$$
$$+ 2\int_0^t \int U(x,y,s)\partial_x N_1^{\sigma_1,0}(u)(x,y,s)dxdyds.$$

Applying Cauchy–Schwarz inequality and the definition of U, we obtain

$$\|u(t)\|^2_{G_0^{\sigma_1,0}} \leq \|f\|^2_{G_0^{\sigma_1,0}} + \|u\|_{X_{0,b}^{\sigma_1,0}} \|N_1^{\sigma_1,0}(u)\|_{X_{0,b-1}^{0,0}(I)}.$$

We now apply Lemma 10.6 and the fact that $b = \frac{1}{2}+\varepsilon$, we can further estimate this by

$$\|u(t)\|^2_{G_0^{\sigma_1,0}} \leq \|f\|^2_{G_0^{\sigma_1,0}} + C\sigma_1^{\rho}\|u\|^3_{X_{0,b}^{\sigma_1,0}}, \tag{10.13}$$

as desired.

Proof (Proof of Theorem 10.2). With the tools established in the previous subsection, we may begin the proof of Theorem 10.2. Let us first suppose that T^* is the supremum of the set of times T for which

$$u \in C([0,T]; G_x^{\sigma_1,0}).$$

If $T^* = \infty$, there is nothing to prove, so let us assume that $T^* < \infty$. In this case, it suffices to prove that

$$u \in C\left([0,T], G_x^{\sigma_1(T),0}\right), \tag{10.14}$$

for all $T > T^*$. To show that this is the case, we will use Theorem 10.1 and Theorem 10.3 to construct a solution which exists over subintervals of width δ, using the parameter σ_1 to control the growth of the norm of the solution. We first prove the case $s = 0$ and then we will generalize the case.

The case $s = 0$. The desired result will follow from the following proposition.

Proposition 10.1. *Let $T > 0$, $x = 0$, $0 < \sigma_1 \leq \sigma_0$ and $\delta > 0$ be numbers such that*

$$n\delta \leq T < (n+1)\delta.$$

Then the solution u to the Cauchy problem (10.1) *satisfies*

$$\sup_{t\in[0,n\delta]} \|u(t)\|^2_{G_0^{\sigma_1,0}} \leq \|f\|^2_{G_0^{\sigma_1,0}} + 2^3C\sigma_0^{\rho}n\|f\|^3_{G_0^{\sigma_0,0}}, \tag{10.15}$$

and

$$\sup_{t\in[0,n\delta]} \|u(t)\|^2_{G_0^{\sigma_1,0}} \leq 4\|u(t)\|^2_{G_0^{\sigma_0,0}}, \tag{10.16}$$

if

$$\sigma_1 = C_1 T^{-\frac{1}{\rho}} \text{ and } C_1 = \left(\frac{c_0}{C2^5 \|f\|_{G_0^{\sigma_0,0}} (1 + 2\|f\|_{G_0^{\sigma_0,0}})^{\frac{1}{\varepsilon}}} \right)^{\frac{1}{\rho}},$$

for some constant $C > 0$.

Proof. For fixed $T \geq T^*$, we will prove, for sufficiently small $\sigma_1 > 0$, that

$$\sup_{t\in[0,T]} \|u(0)\|^2_{G_0^{\sigma_1,0}} \leq 4\|u(0)\|^2_{G_0^{\sigma_0,0}}. \tag{10.17}$$

We will use the Theorems 10.1 and 10.3 with the time step

$$\delta = \frac{c_0}{(1 + 4\|f\|_{G_0^{\sigma_0,0}})^{\frac{1}{\varepsilon}}}. \tag{10.18}$$

The smallness conditions on σ_1 will be

$$\sigma_1 \leq \sigma_0 \quad \text{and} \quad \frac{2T}{\delta} C\sigma_1^\rho 4^{\frac{3}{2}} \|f\|_{G_0^{\sigma_0,0}} \leq 1, \tag{10.19}$$

where $C > 0$ is the constant in Theorem 10.3. Proceeding by induction, we will verify that

$$\sup_{t\in[0,\delta]} \|u(t)\|^2_{G_0^{\sigma_1,0}} \leq \|f\|^2_{G_0^{\sigma_1,0}} + nC\sigma_1^\rho 2^3 \|f\|^3_{G_0^{\sigma_0,0}}, \tag{10.20}$$

$$\sup_{t\in[0,\delta]} \|u(t)\|^2_{G_0^{\sigma_1,0}} \leq 4\|f\|^2_{G_0^{\sigma_0,0}}, \tag{10.21}$$

for $n \in \{1, \ldots, m+1\}$, where $m \in \mathbb{N}$ is chosen, so that

$$T \in [m\delta, (m+1)\delta).$$

This m does exist, since by Theorem 10.1 and the definition of T^*, we have

$$\delta < \frac{c_0}{(1 + \|f\|_{G_0^{\sigma_0,0}})^{\frac{1}{\varepsilon}}} < T^*, \quad \text{hence } \delta < T.$$

We cover now, the interval $[0, \delta]$ and by Theorem 10.3, we have

$$\sup_{t\in[0,\delta]} \|u(t)\|^2_{G_0^{\sigma_1,0}} \leq \|f\|^2_{G_0^{\sigma_1,0}} + C\sigma_1^\rho \|f\|^3_{G_0^{\sigma,0}}$$
$$\leq \|f\|^2_{G_0^{\sigma_1,0}} + C\sigma_1^\rho \|f\|^3_{G_0^{\sigma_0,0}},$$

where we used that

$$\|f\|_{G_0^{\sigma_1,0}} \leq \|f\|_{G_0^{\sigma_0,0}},$$

since $\sigma_1 \leq \sigma_0$. This verifies (10.20) for $n = 1$ and now, (10.21) follows using again

$$\|f\|_{G_0^{\sigma_1,0}} \leq \|f\|_{G_0^{\sigma_0,0}},$$

as well as $C\sigma_1^\rho \|f\|_{G_0^{\sigma_0,0}} \leq 1$. Next, assuming that (10.20) and (10.21) hold for some $n \in \{1, \dots, m\}$, we will prove that they hold for $n+1$. We estimate

$$\sup_{t\in[n\delta,(n+1)\delta]} \|u(t)\|^2_{G_0^{\sigma_1,0}} \leq \|u(n\delta)\|^2_{G_0^{\sigma_1,0}} + C\sigma_1^\rho \|u(n\delta)\|^3_{G_0^{\sigma_1,0}}$$
$$\leq \|u(n\delta)\|^2_{G_0^{\sigma_1,0}} + C\sigma_1^\rho 2^3 \|f\|^3_{G_0^{\sigma_0,0}}$$
$$\leq \|f\|^2_{G_0^{\sigma_1,0}} + nC\sigma_1^\rho 2^3 \|f\|^3_{G_0^{\sigma_0,0}}$$
$$+ C\sigma_1^\rho 2^3 \|f\|^3_{G_0^{\sigma_0,0}},$$

verifying (10.20) with n replaced by $n+1$. To get (10.21) with n replaced by $n+1$, it is then enough to have

$$(n+1)C\sigma_1^\rho 2^3 \|f\|_{G_0^{\sigma_0,0}} \leq 1.$$

But this holds by (10.19), since

$$n+1 \leq m+1 \leq \frac{T}{\delta} + 1 < \frac{2T}{\delta}.$$

Finally, the condition (10.19) is satisfied for $\sigma_1 \in (0, \sigma_0)$ such that

$$\frac{2T}{\delta} C\sigma_1^\rho 2^3 \|f\|_{G_0^{\sigma_0,0}} = 1.$$

Thus, $\sigma_1 = C_1 T^{-\frac{1}{\rho}}$, where

$$C_1 = \left(\frac{c_0}{C 2^5 \|f\|_{G_0^{\sigma_0,0}} (1 + 2\|f\|_{G_0^{\sigma_0,0}})^{\frac{1}{\varepsilon}}} \right)^{\frac{1}{\rho}}.$$

For general s, we have

$$u_0 \in G_s^{\sigma_0,0} \subset G_0^{\sigma_0/2,0}.$$

The case $s = 0$ already being proved, we know that there is a $T_1 > 0$, such hat

$$u \in C\left([0, T_1), G_0^{\sigma_0/2,0}\right),$$

and

$$u \in C\left([0, T], G_0^{2\varsigma T^{-1/\rho},0}\right), \quad \text{for } T \geq T_1,$$

where $\varsigma > 0$ depends on f, σ_0 and ς. We now conclude that

$$u \in C\left([0, T_1), G_s^{\sigma_0/4,0}\right),$$

and

$$u \in C\left([0, T], G_s^{\varsigma T^{-1/\rho},0}\right), \quad \text{for } T \geq T_1.$$

The proof of Theorem 10.2 is now completed. □

10.4 Gevrey's Regularity in Time

Our last aim is to show the regularity of the solution in the time. A non-periodic function $\phi(x)$ is the Gevrey class of order κ i.e., $\phi(x) \in G^\kappa$, if there exists a constant $C > 0$ such that

$$|\partial_x^k \phi(x)| \leq C^{k+1} (k!)^\kappa \quad k = 0, 1, 2, \ldots. \tag{10.22}$$

Here, we will show that for $x, y \in \mathbb{R}$, for every $t \in [0, \delta]$ and $j, l, n \in \{0, 1, 2, \ldots\}$, there exist $C > 0$ such that,

$$|\partial_t^j \partial_x^l \partial_y^n u(x, y, t)| \leq C^{j+l+n+1} (j!)^{(\alpha+1)\kappa} (l!)^\kappa (n!)^\kappa. \tag{10.23}$$

i.e.,

$$u(\cdot,\cdot,t) \in G^{\sigma}(\mathbb{R}) \times G^{\sigma}(\mathbb{R}),$$

in x, y and

$$u(x,y,\cdot) \in G^{(\alpha+1)\kappa}([0,\delta]),$$

in time variable.

Theorem 10.4. *Let $s_1 > -\frac{\alpha-1}{4}$, $s_2 \geq 0$, $\alpha \geq 4$, $\sigma_1 \geq 0$, $\sigma_2 \geq 0$ and $\kappa \geq 1$.*

If

$$f \in G^{\sigma_1,\sigma_2,\kappa}_{s_1,s_2},$$

then the solution

$$u \in C\left([0,\delta], G^{\sigma_1,\sigma_2,\kappa}_{s_1,s_2}\right),$$

given by Theorem 10.1, belongs to the Gevrey class $G^{(\alpha+1)\kappa}$ in time variable.

Corollary 10.1. *Let $\sigma_1 > 0, s_1 > -\frac{\alpha-1}{4}, \alpha = 4, 6, 8, \ldots$. If*

$$f \in G^{\sigma_1,0}_{s_1},$$

then the solution

$$u \in C\left([0,T], G^{\sigma_1(T),0}_{s_1}(\mathbb{R}^2)\right),$$

given by Theorem 10.2, belongs to the Gevrey class $G^{(\alpha+1)}$ in time variable.

Proposition 10.2. *Let $\delta > 0, s_1 > -\frac{\alpha-1}{4}$, $s_2 \geq 0$ and*

$$u \in C\left([0,\delta]; G^{\sigma_1,\sigma_2,\kappa}_{s_1,s_2}\right),$$

be the solution of (10.1). Then u belong in x, y to G^{κ} for all times near the zero. In other words,

$$|\partial_x^l \partial_y^n u(x,y,t)| \leq C^{l+n+1}(l!)^{\kappa}(n!)^{\kappa}, \tag{10.24}$$

for all $(x,y) \in \mathbb{R}^2$, $\quad C > 0, t \in [0,\delta], l, n \in \{0,1,\ldots\}$.

Proof. We have, for any $t \in [0, \delta]$

$$\begin{aligned}
&\|\partial_x^l \partial_y^n u(\cdot, \cdot, t)\|_{H^{s_1,s_2}}^2 \\
&= \int_{\mathbb{R}^2} \langle \xi \rangle^{2s_1} \langle \mu \rangle^{2s_2} |\widehat{\partial_x^l \partial_y^n u}(\xi, \mu, t)|^2 d\xi d\mu \\
&= \int_{\mathbb{R}^2} |\xi|^{2l} |\mu|^{2n} \langle \xi \rangle^{2s_1} \langle \mu \rangle^{2s_2} |\widehat{u}(\xi, \mu, t)|^2 d\xi d\mu \\
&= \int_{\mathbb{R}^2} |\xi|^{2l} |\mu|^{2n} e^{-2\sigma_1 |\xi|^{\frac{1}{\kappa}}} e^{-2\sigma_2 |\mu|^{\frac{1}{\kappa}}} \langle \xi \rangle^{2s_1} \langle \mu \rangle^{2s_2} \\
&\quad \times e^{2\sigma_1 |\xi|^{\frac{1}{\kappa}}} e^{2\sigma_2 |\mu|^{\frac{1}{\kappa}}} |\widehat{u}(\xi, \mu, t)|^2 d\xi d\mu.
\end{aligned}$$

We observe that

$$\begin{aligned}
e^{\frac{2\sigma_1}{\kappa} |\xi|^{\frac{1}{\kappa}}} &= \sum_{j=0}^{\infty} \frac{1}{j!} \left(\frac{2\sigma_1}{\kappa} |\xi|^{\frac{1}{\kappa}} \right)^j \\
&\geq \frac{1}{(2l)!} \left(\frac{2\sigma_1}{\kappa} \right)^{2l} |\xi|^{\frac{2l}{\kappa}}, \quad \forall l \in \{0, 1, \ldots\}, \xi \in \mathbb{R},
\end{aligned}$$

and

$$\begin{aligned}
e^{\frac{2\sigma_2}{\kappa} |\mu|^{\frac{1}{\kappa}}} &= \sum_{j=0}^{\infty} \frac{1}{j!} \left(\frac{2\sigma_2}{\kappa} |\mu|^{\frac{1}{\kappa}} \right)^j \\
&\geq \frac{1}{(2n)!} \left(\frac{2\sigma_2}{\kappa} \right)^{2n} |\mu|^{\frac{2l}{\kappa}}, \quad \forall n \in \{0, 1, \ldots\}, \mu \in \mathbb{R}.
\end{aligned}$$

This implies that

$$\begin{aligned}
|\xi|^{2l} e^{-2\sigma_1 |\xi|^{\frac{1}{\kappa}}} &\leq C_{\sigma_1,\kappa}^{2l} (2l)!^{\kappa}, \\
|\mu|^{2n} e^{-2\sigma_2 |\mu|^{\frac{1}{\kappa}}} &\leq C_{\sigma_2,\kappa}^{2n} (2n)!^{\kappa}.
\end{aligned}$$

Thus,

$$\begin{aligned}
&\|\partial_x^l \partial_y^n u(\cdot, \cdot, t)\|_{H^{s_1,s_2}}^2 \\
&\leq C_{\sigma_1,\sigma_2,\kappa}^{2l+2n} (2l)!^{\kappa} (2n)!^{\kappa} \int_{\mathbb{R}^2} \langle \xi \rangle^{2s_1} \langle \mu \rangle^{2s_2}
\end{aligned}$$

$$\times e^{2\sigma_1|\xi|^{\frac{1}{\kappa}}} e^{2\sigma_2|\mu|^{\frac{1}{\kappa}}} |\widehat{u}(\xi,\mu,t)|^2 d\xi d\mu$$
$$= C_{\sigma_1,\sigma_2,\kappa}^{2l+2n}(2l)!^{\kappa}(2n)!^{\kappa}\|u(\cdot,\cdot,t)\|^2_{G^{\sigma_1,\sigma_2,\kappa}_{s_1,s_2}}.$$

Since

$$(2l)!^{\kappa} \leq c_1^{2l}(l!)^{2\kappa},$$

and

$$(2n)!^{\kappa} \leq c_2^{2n}(n!)^{2\kappa},$$

for some $c_1, c_2 > 0$, we have for all $l, n \in \{0, 1, 2, \dots\}$

$$|\partial_x^l \partial_y^n u| \lesssim \|\partial_x^l \partial_y^n u(\cdot,\cdot,t)\|_{H^{s_1,s_2}} \leq C_0 C_1^{l+n}(l)!^{\kappa}(n)!^{\kappa} \quad \text{for } \forall t \in [0,\delta],$$

where

$$C_0 = \|u(\cdot,\cdot,t)\|_{G^{\sigma_1,\sigma_2,\kappa}_{s_1,s_2}},$$

and

$$C_1 = c_0^2 C_{\sigma_1,\sigma_2,\kappa},$$

and

$$c_0 = \max(c_1, c_2),$$

which implies that the solution u is analytic in x, y for all time near zero and $s_1, s_2 \geq 0$.

Now, for

$$-\frac{\alpha-1}{4} < s_1 < 0, \quad s_2 \geq 0,$$

and for any $0 < \varepsilon < \sigma_1$, there exists a positive constant $C = C_{s,\varepsilon} > 0$ such that

$$\int_{\mathbb{R}^2} e^{2(\sigma_1-\varepsilon)|\xi|^{\frac{1}{\kappa}}} e^{2\sigma_2|\mu|^{\frac{1}{\kappa}}} \langle\mu\rangle^{2s_2}|\widehat{u}(\xi,\mu,t)|^2 d\xi d\mu$$
$$\leq C\int_{\mathbb{R}^2} \frac{e^{2\varepsilon|\xi|^{\frac{1}{\kappa}}}}{\langle\xi\rangle^{-2s_1}} e^{2(\sigma_1-\varepsilon)|\xi|^{\frac{1}{\kappa}}} e^{2\sigma_2|\mu|^{\frac{1}{\kappa}}} \langle\mu\rangle^{2s_2}|\widehat{u}(\xi,\mu,t)|^2 d\xi d\mu$$
$$= C\int_{\mathbb{R}^2} \langle\xi\rangle^{2s_1}\langle\mu\rangle^{2s_2} e^{2\sigma_1|\xi|^{\frac{1}{\kappa}}} e^{2\sigma_2|\mu|^{\frac{1}{\kappa}}} |\widehat{u}(\xi,\mu,t)|^2 d\xi d\mu.$$

This implies that if

$$u \in C\left([0,T]; G^{\sigma_1,\sigma_2,\kappa}_{s_1,s_2}\right) \quad \text{and} \quad s_1 < 0, s_2 \geq 0,$$

then

$$u \in C\left([0,T]; G^{\sigma_1-\varepsilon,\sigma_2,\kappa}_{0,s_2}\right),$$

which allows us to conclude that u is in G^κ in x, y for all

$$s_1 > -\frac{\alpha-1}{4}, \quad s_2 \geq 0.$$

□

In order to prove Theorem 10.4, it is enough to prove the following result.

Lemma 10.7. *For $j, l, n \in \{0, 1, 2, \dots\}$, the next inequality*

$$\left|\partial_t^j \partial_x^l \partial_y^n u\right| \leq C^{j+l+n+1}((l+n+(\alpha+1)j)!)^\kappa L^j, \tag{10.25}$$

holds, where

$$L = C^\alpha + \frac{1}{120^\kappa} + \frac{C}{40^\kappa}, \quad \forall x, y \in \mathbb{R}, \ \in [0,\delta].$$

In fact, taking $l = n = 0$ we obtain

$$\left|\partial_t^j u\right| \leq C^{j+1}((\alpha+1)j!)^\kappa L^j \leq K^{j+1}(j!)^{(\alpha+1)\kappa}.$$

Proof. We use the induction on j to prove Lemma 10.7.
For $j = 0$ and $l, n \in \{0, 1, 2, \dots\}$, we have, by (10.24)

$$\left|\partial_x^l \partial_y^n u(x,y,t)\right| \leq C^{l+n+1}(l!)^\kappa (n!)^\kappa \leq C^{l+n+1}(l+n)!^\kappa. \tag{10.26}$$

For $j = 1$ and $l, n \in \{0, 1, 2, \dots\}$, we have

$$\left|\partial_t \partial_x^l \partial_y^k u\right| \leq \left||D_x|^\alpha \partial_x^{l+1} \partial_y^n u\right| + \left|\partial_x^{l-1} \partial_y^{n+2} u\right| + \left|\partial_x^l \partial_y^n (u \partial_x u)\right|. \tag{10.27}$$

The terms of (10.27) can be estimated as

$$\begin{aligned}\left||D_x|^\alpha \partial_x^{l+1}\partial_y^n u\right| &\le C^{l+1+\alpha+n+1}(l+1+\alpha+n)!^\kappa \\ &\le C^{l+n+1+1}(l+n+(\alpha+1)\cdot 1)!^\kappa C^\alpha, \qquad (10.28)\\ \left|\partial_x^{l-1}\partial_y^{n+2}u\right| &\le C^{l-1+n+2+1}(l-1+n+2)!^\kappa \\ &\le C^{l+n+1+1}(l+n+5\cdot 1)! \\ &\quad \times \frac{1}{(l+n+2)^\kappa(l+n+3)^\kappa(l+n+4)^\kappa(l+n+5)^\kappa} \\ &\le C^{l+n+1+1}(l+n+5\cdot 1)!\frac{1}{120^\kappa}. \qquad (10.29)\end{aligned}$$

The nonlinear terms are treated as follows

$$\left|\partial_x^l\partial_y^n(u\partial_x u)\right| = \left|\sum_{p=0}^{l}\sum_{k=0}^{n}\binom{l}{p}\binom{n}{k}\left(\partial_x^{l-p}\partial_y^{n-k}u\right)\left(\partial_x^{p+1}\partial_y^k u\right)\right|.$$

Recalling that for $l \ge p$ and $n \ge k$, we have the next inequality

$$\binom{l}{p}\binom{n}{k} \le \binom{l+n}{p+k}. \qquad (10.30)$$

By (10.30), we have

$$\begin{aligned}\left|\partial_x^l\partial_y^n(u\partial_x u)\right| &\le \left|\sum_{p=0}^{l}\sum_{k=0}^{n}\binom{l+n}{p+k}\left(\partial_x^{l-p}\partial_y^{n-k}u\right)\left(\partial_x^{p+1}\partial_y^k u\right)\right| \\ &\le \sum_{p=0}^{l}\sum_{k=0}^{n}\frac{((l+n)!)^\kappa}{((p+k)!)^\kappa((l+n-p-k)!)^\kappa} \\ &\quad \times C^{l-p+n-k+1}((l+n-p-k)!)^\kappa \\ &\quad \times C^{p+1+k+1}((p+1+k)!)^\kappa \\ &= C^{l+n+3}((l+n)!)^\kappa\sum_{p=0}^{l}\sum_{k=0}^{n}(p+1+k)^\kappa.\end{aligned}$$

At this stage, we use the fact that

$$\sum_{p=0}^{l}\sum_{k=0}^{n}(p+1+k)=\frac{(l+1)(n+1)(l+n+2)}{2}. \tag{10.31}$$

Then,

$$\begin{aligned}
&\left|\partial_x^l\partial_y^n(u\partial_x u)\right| \\
&\quad\le C^{l+n+3}((l+n)!)^\kappa\frac{(l+1)^\kappa(n+1)^\kappa(l+n+2)^\kappa}{2^\kappa} \\
&\quad\le C^{l+n+1+1}((l+n)!)^\kappa(l+n+1)^\kappa(l+n+2)^\kappa(l+n+3)^\kappa\frac{C}{2^\kappa} \\
&\quad= C^{l+n+1+1}((l+n+(\alpha+1))!)^\kappa \\
&\qquad\times\frac{1}{(l+n+4)^\kappa(l+n+(\alpha+1))^\kappa}\frac{C}{2^\kappa} \\
&\quad\le C^{l+n+1+1}((l+n+(\alpha+1)\cdot 1)!)^\kappa\frac{C}{40^\kappa}.
\end{aligned} \tag{10.32}$$

From (10.28), (10.29) and (10.32), it follows that:

$$\left|\partial_t\partial_x^l\partial_y^k u\right|\le C^{l+n+1+1}((l+n+(\alpha+1)\cdot 1)!)^\kappa L^1, \forall x,y\in\mathbb{R}, t\in[0,\delta].$$

We assume that (10.25) is correct for $j\ge m\ge 1$ where $l,n\in\{0,1,2,\dots\}$ and then we prove it for $m=j+1$ and $l,n\in\{0,1,2,\dots\}$.

We obtain

$$\left|\partial_t^{j+1}\partial_x^l\partial_y^k u\right|\le\left|\partial_t^j|D_x|^\alpha\partial_x^{l+1}\partial_y^n u\right|+\left|\partial_t^j\partial_x^{l-1}\partial_y^{n+2}u\right|+\left|\partial_t^j\partial_x^l\partial_y^n(u\partial_x u)\right|.$$

These terms are estimated as follows:

$$\begin{aligned}
\left|\partial_t^j|D_x|^\alpha\partial_x^{l+1}\partial_y^n u\right|&\le C^{j+l+(\alpha+1)+n+1}(l+n+((\alpha+1)(j+1))!)^\kappa L^j \\
&\le C^{(j+1)+l+n+1}(l+n+((\alpha+1)(j+1))!)^\kappa C^\alpha L^j,
\end{aligned} \tag{10.33}$$

and

$$\begin{aligned}\left|\partial_t^j\partial_x^{l-1}\partial_y^{n+2}u\right| &\le C^{j+l-1+n+2+1}((j+l-1+n+2)!)^\kappa L^j \\ &\le C^{(j+1)+l+n+1}((l+n+(\alpha+1)(j+1))!)^\kappa \frac{L^j}{120^\kappa}.\end{aligned} \tag{10.34}$$

The nonlinear terms are treated as follows

$$\begin{aligned}\partial_t^j\partial_x^l\partial_y^n(u\partial_x u) &= \sum_{p=0}^{l}\sum_{k=0}^{n}\binom{l}{p}\binom{n}{k}\left(\partial_t^j\partial_x^{l-p}\partial_y^{n-k}u\right)\left(\partial_x^{p+1}\partial_y^k u\right) \\ &+ \sum_{p=0}^{l}\sum_{k=0}^{n}\binom{l}{p}\binom{n}{k}\left(\partial_x^{l-p}\partial_y^{n-k}u\right)\left(\partial_t^j\partial_x^{p+1}\partial_y^k u\right) \\ &+ \sum_{q=1}^{j-1}\sum_{p=0}^{l}\sum_{k=0}^{n}\binom{j}{q}\binom{l}{p}\binom{n}{k}\left(\partial_t^{j-q}\partial_x^{l-p}\partial_y^{n-k}u\right) \\ &\times\left(\partial_t^q\partial_x^{p+1}\partial_y^k u\right).\end{aligned} \tag{10.35}$$

Using (10.30) to estimate $(10.35)_1$

$$\begin{aligned}&\left|\sum_{p=0}^{l}\sum_{k=0}^{n}\binom{l}{p}\binom{n}{k}\left(\partial_t^j\partial_x^{l-p}\partial_y^{n-k}u\right)\left(\partial_x^{p+1}\partial_y^k u\right)\right| \\ &\quad\le \frac{1}{3}C^{(j+1)+l+n+1}((l+n+(\alpha+1)(j+1))!)^\kappa \frac{C}{40^\kappa}L^j.\end{aligned} \tag{10.36}$$

We estimate $(10.35)_2$ as

$$\begin{aligned}&\left|\sum_{p=0}^{l}\sum_{k=0}^{n}\binom{l}{p}\binom{n}{k}\left(\partial_x^{l-p}\partial_y^{n-k}u\right)\left(\partial_t^j\partial_x^{p+1}\partial_y^k u\right)\right| \\ &\quad\le \frac{1}{3}C^{(j+1)+l+n+1}((l+n+(\alpha+1)(j+1))!)^\kappa \frac{C}{40^\kappa}L^j.\end{aligned} \tag{10.37}$$

To estimate $(10.35)_3$, we recall that for $j \geq q$, $l \geq p$ and $n \geq k$, we have the next inequality

$$\binom{j}{q}\binom{l}{p}\binom{n}{k} \leq \binom{j+l+n}{q+p+k}.$$

Then

$$\begin{aligned}
&\left|\sum_{q=1}^{j-1}\sum_{p=0}^{l}\sum_{k=0}^{n}\binom{j}{q}\binom{l}{p}\binom{n}{k}\left(\partial_t^{j-q}\partial_x^{l-p}\partial_y^{n-k}u\right)\left(\partial_t^{q}\partial_x^{p+1}\partial_y^{k}u\right)\right| \\
&\quad \leq \sum_{q=1}^{j-1}\sum_{p=0}^{l}\sum_{k=0}^{n}\binom{j+l+n}{q+p+k}C^{j-q+l-p+n-k+1} \\
&\qquad \times ((l-p+n-k+(\alpha+1)(j-q))!)^{\kappa}L^{j-q} \\
&\qquad \times C^{q+p+1+k+1}((p+1+k+(\alpha+1)q)!)^{\kappa}L^{q} \\
&\quad \leq \frac{1}{3}C^{(j+1)+l+n+1}((l+n+(\alpha+1)(j+1))!)^{\kappa}\frac{C}{40^{\kappa}}L^{j}. \qquad (10.38)
\end{aligned}$$

Finally by using (10.36)–(10.38) we arrive at

$$\left|\partial_t^{j+1}\partial_x^{l}\partial_y^{k}u\right| \leq C^{(j+1)+l+n+1}((l+n+(\alpha+1)(j+1))!)^{\kappa}L^{j+1},$$

for all $(x,y) \in \mathbb{R}^2, t \in [0,\delta]$. □

Chapter 11

Coupled System of KDV Equations in Gevrey Spaces

11.1 Well-Posedness of Solutions for a Coupled System of gKDV Equations

Studies of modified Korteweg–de Vries-type equations are of considerable mathematical interest due to the importance of their applications in various branches of mechanics and physics. Using trilinear estimate in Bourgain spaces, we show the local well-posedness of the initial value problem associated with a coupled system consisting of modified Korteweg–de Vries equations (11.2) for given data. Furthermore, we prove that the unique solution belongs to Gevrey space $G^\sigma \times G^\sigma$ in x and $G^{3\sigma} \times G^{3\sigma}$ in t.

A single initial value problem for the Korteweg–de Vries (KdV) equation is written as

$$\partial_t u + \partial_x^3 u + \frac{1}{2}\partial_x u^2 = 0. \tag{11.1}$$

For $x \in \mathbb{R}$, $t \in [0,T]$ and $u_0(x) \in H^s(\mathbb{R})$, it has been shown that (11.1) is locally well-posed for $s > -\frac{3}{4}$ in Sobolev spaces. Now, we consider the problem

$$\begin{cases} \partial_t u + \partial_x^3 u + \partial_x(uv^2) = 0 \\ \partial_t v + \beta\partial_x^3 v + \partial_x(u^2 v) = 0, \quad x \in \mathbb{R}\, t \in \mathbb{R},\ 0 < \beta < 1 \\ u(x,0) = u_0(x) \\ v(x,0) = v_0(x). \end{cases} \tag{11.2}$$

The analytic spaces $G^{\delta,s}$ is introduced by the norm

$$\|f\|^2_{G^{\delta,s}} = \int e^{2\delta(1+|\xi|)}(1+|\xi|)^{2s}|\widehat{f}(\xi)|^2 d\xi < \infty,$$

and used to prove the well-posedness of non-periodic case for generalized Korteweg–de Vries equation and then to show the Gevrey's temporal regularity of unique solution.

Theorem 11.1. *Let $s > -\frac{1}{2}, 0 < \beta < 1, \sigma \geq 1,$ and $\delta \in \mathbb{R}^+_*$. If*

$$(u_0, v_0) \in G^{\sigma,\delta,s} \times G^{\sigma,\delta,s},$$

then the unique solution (11.17) *belongs to the Gevrey class $G^{3\sigma} \times G^{3\sigma}$ in t. Furthermore, it is not belong to $G^d \times G^d$, $1 \leq d < 3\sigma$ in time variable.*

To present the proof of Theorems, we start with trilinear estimate defined in $X_{\sigma,\delta,s,b}(\mathbb{R}^2)$. Let us beginning by the embedded result.

Lemma 11.1. *Let $b > \frac{1}{2}$, $s \in \mathbb{R}$, $\sigma \geq 1$ and $\delta > 0$. Then, for all $T > 0$ we have*

$$X_{\sigma,\delta,s,b} \hookrightarrow C([0,T], G^{\sigma,\delta,s}).$$

Proof. Define the operator A by

$$\widehat{A^{\sigma,\delta}u}^{x}(\xi,t) = e^{\delta|\xi|^{\frac{1}{\sigma}}}\widehat{u}^x(\xi,t). \tag{11.3}$$

It satisfies

$$\|u\|_{X_{\sigma,\delta,s,b}} = \|A^{\sigma,\delta}u\|_{X_{s,b}} \quad \text{and} \quad \|u\|_{G^{\sigma,\delta,s}} = \|A^{\sigma,\delta}u\|_{H^s}. \tag{11.4}$$

We observe that $A^{\sigma,\delta}u$ belongs to $C(\mathbb{R}, H^s)$ and for some $C > 0$, we have

$$\|A^{\sigma,\delta}u\|_{C(\mathbb{R},H^s)} \leq C\|A^{\sigma,\delta}u\|_{X_{s,b}}. \tag{11.5}$$

Thus, it follows that $w \in ([0,T], G^{\sigma,\delta,s})$ and

$$\|u\|_{C\left([0,T],G^{\sigma,\delta,s}\right)} \leq C\|u\|_{X_{\sigma,\delta,s,b}}. \tag{11.6}$$

Owing to the Fourier transform with respect to the spatial variable x of the Cauchy problem (11.2), we obtain a differential equation and then solving it in t. We localize in t by using a cut-off function, satisfying $\psi \in C_0^\infty$, with $\psi = 1$ in $[-1,1]$, where supp$\psi \subset [-2,2]$, $\psi_T(t) = \psi(\frac{t}{T})$. We consider, for the operators Λ, Γ, the following integral system which is equivalent to Cauchy problem (11.2)

$$\begin{cases} \Lambda[u,v](t) = \psi(t)S(t)u_0 - \psi_T(t)\displaystyle\int_0^t S(t-s)F_1(s)ds \\ \\ \Gamma[u,v](t) = \psi(t)S_\beta(t)v_0 - \psi_T(t)\displaystyle\int_0^t S_\beta(t-s)F_2(s)ds, \end{cases} \tag{11.7}$$

where $S(t) = e^{-t\partial_x^3}$ and $S_\beta(t) = e^{-t\beta\partial_x^3}$, the unitary groups related to the linear problems are defined via Fourier transform as

$$\widehat{[S(t)f]}(\xi) = e^{it\xi^3}\widehat{f}(\xi),$$

and

$$\widehat{[S_\beta(t)f]}(\xi) = e^{it\beta\xi^3}\widehat{f}(\xi).$$

The nonlinear terms are defined by $F_1 = \partial_x(uv^2)$, $F_2 = \partial_x(u^2v)$. $\square$

Lemma 11.2. *Let $s, b \in \mathbb{R}$, $\delta \in \mathbb{R}_*^+$ and $\sigma \geq 1$. For some constant $C > 0$, we have*

$$\|\psi(t)S(t)u_0\|_{X_{\sigma,\delta,s,b}} \leq C\ \|u_0\|_{G^{\sigma,\delta,s}}, \tag{11.8}$$

and

$$\|\psi(t)S_\beta(t)v_0\|_{X_{\sigma,\delta,s,b}} \leq C\ \|v_0\|_{G^{\sigma,\delta,s}}, \tag{11.9}$$

for all $v_0, u_0 \in G^{\sigma,\delta,s}$.

Proof. By definition, we have

$$\begin{aligned} \psi(t)S(t)u_0 &= C\psi(t)\int_{\mathbb{R}} e^{i(x\xi+t\xi^3)}\widehat{u_0}(\xi)d\xi \\ &= C\int_{\mathbb{R}^2} e^{i(x\xi+t\tau)}\widehat{\psi}(\tau-\xi^3)\widehat{u_0}(\xi)d\xi d\tau. \end{aligned}$$

It follows that

$$\begin{aligned}
&\|\psi(t)S(t)u_0\|^2_{X_{\sigma,\delta,s,b}} \\
&= C\int_{\mathbb{R}^2} e^{2\delta|\xi|^{\frac{1}{\sigma}}}\left(1+|\xi|\right)^{2s}\left(1+|\tau-\xi^3|\right)^{2b}|\widehat{\psi}(\tau-\xi^3)|^2 \\
&\quad\times|\widehat{u_0}(\xi)|^2 d\xi d\tau \\
&= C\int_{\mathbb{R}} e^{2\delta|\xi|^{\frac{1}{\sigma}}}\left(1+|\xi|\right)^{2s}|\widehat{u_0}(\xi)|^2 \\
&\quad\times\left(\int_{\mathbb{R}}|\widehat{\psi}(\tau-\xi^3)|^2\left(1+|\tau-\xi^3|\right)^{2b}d\tau\right)d\xi.
\end{aligned}$$

We use the fact that $b > 1/2$ to get

$$\begin{aligned}
&\int_{\mathbb{R}}|\widehat{\psi}(\tau-\xi^3)|^2\left(1+|\tau-\xi^3|\right)^{2b}d\tau \\
&\leq C'\int_{\mathbb{R}}|\widehat{\psi}(\tau-\xi^3)|^2 d\tau + C'\int_{\mathbb{R}}|\psi(\tau-\xi^3)|^2(1+|\tau-\xi^3|)^{2b} \\
&\quad\times d\tau \leq C. \qquad (11.10)
\end{aligned}$$

□

Lemma 11.3. *Let $s \in \mathbb{R}$, $-\frac{1}{2} < b' \leq 0 \leq b < b'+1$, $0 \leq T \leq 1$, $\delta > 0$ and $\sigma \geq 1$, then for some constant $C > 0$, we have*

$$\left\|\psi_T(t)\int_0^t S(t-s)F_1(x,s)ds\right\|_{X_{\sigma,\delta,s,b}} \leq CT^{1-b+b'}\|F_1\|_{X_{\sigma,\delta,s,b'}}, \qquad (11.11)$$

and

$$\left\|\psi_T(t)\int_0^t S_\beta(t-s)F_2(x,s)ds\right\|_{X^\beta_{\sigma,\delta,s,b}} \leq CT^{1-b+b'}\|F_2\|_{X^\beta_{\sigma,\delta,s,b'}}. \qquad (11.12)$$

Proof. Define

$$W = \psi_T(t)\int_0^t S(t-s)F_1(x,s)ds.$$

Let us consider the operator $A^{\sigma,\delta}$ given by (11.3), then we have

$$\begin{aligned}\widehat{A^{\sigma,\delta}W}^x(\xi,t) &= \psi_T(t)\int_0^t \left(e^{-i(t-s)\xi^3}\right) e^{\delta|\xi|^{\frac{1}{\sigma}}}\widehat{F_1}^x(\xi,s)ds \\ &= \psi_T(t)\int_0^t [S(t-\widehat{s)(A^{\sigma,\delta}F_1)]}^x(\xi,s)ds.\end{aligned} \tag{11.13}$$

Thus,

$$\|W\|_{X_{\sigma,\delta,s,b}} = \|A^{\sigma,\delta}W\|_{X_{s,b}} = \left\|\psi_T(t)\int_0^t S(t-s)A^{\sigma,\delta}F_1(x,s)ds\right\|_{X_{s,b}}.$$

Then, we have

$$\begin{aligned}&\left\|\psi_T(t)\int_0^t S(t-s)A^{\sigma,\delta}F_1(x,s)ds\right\|_{X_{s,b}} \le CT^{1-b+b'}\|A^{\sigma,\delta}F_1\|_{X_{s,b'}} \\ &\quad = CT^{1-b+b'}\|F_1\|_{X_{\sigma,\delta,s,b'}}.\end{aligned}$$

The inequality (11.12) is similar. □

The following lemma states the desired trilinear estimate.

Lemma 11.4. *Let* $s > -\frac{1}{2}$, $\sigma \ge 1$, $\delta > 0$, $b > \frac{1}{2}$ *and* b' *be as in Lemma* 11.3. *Then*

$$\|\partial_x(uv^2)\|_{X_{\sigma,\delta,s,b'}} \le C\|u\|_{X_{\sigma,\delta,s,b}}\|v\|^2_{X^\beta_{\sigma,\delta,s,b}}, \tag{11.14}$$

and

$$\|\partial_x(u^2v)\|_{X^\beta_{\sigma,\delta,s,b'}} \le C\|u\|^2_{X_{\sigma,\delta,s,b}}\|v\|_{X^\beta_{\sigma,\delta,s,b}}. \tag{11.15}$$

Proof. We observe, by considering the operator $A^{\sigma,\delta}$ in (11.3), that

$$\begin{aligned}e^{\delta|\xi|^{\frac{1}{\sigma}}}\widehat{uvv} &= (2\pi)^{-2}e^{\delta|\xi|^{\frac{1}{\sigma}}}\widehat{u}*\widehat{v}*\widehat{v} \\ &\le (2\pi)^{-2}\int_{\mathbb{R}^4} e^{\delta|\xi-\xi_1|^{\frac{1}{\sigma}}}\widehat{u}\end{aligned}$$

$$\times (\xi - \xi_1, \tau - \tau_1) e^{\delta|\xi_1 - \xi_2|^{\frac{1}{\sigma}}} \widehat{v}(\xi_1 - \xi_2, \tau_1 - \tau_2)$$
$$e^{\delta|\xi_2|^{\frac{1}{\sigma}}} \widehat{v}(\xi_2, \tau_2) d\xi_1 d\xi_2 d\tau_1 d\tau_2$$
$$= A^{\sigma,\delta} u \widehat{A^{\sigma,\delta} v} A^{\sigma,\delta} v, \tag{11.16}$$

since

$$\delta|\xi|^{\frac{1}{\sigma}} \leq \delta|\xi - \xi_1|^{\frac{1}{\sigma}} + \delta|\xi_1 - \xi_2|^{\frac{1}{\sigma}} + \delta|\xi_2|^{\frac{1}{\sigma}}, \quad \forall \sigma \geq 1.$$

Then

$$\|\partial_x(uv^2)\|_{X_{\sigma,\delta,s,b'}} = \|e^{\delta|\xi|^{\frac{1}{\sigma}}} \langle\xi\rangle^s \left\langle \tau - \xi^3 \right\rangle^b \widehat{\partial_x(uvv)}(\xi, \tau)\|_{L^2_{\xi,\tau}}$$
$$\leq \|\partial_x(A^{\sigma,\delta} u A^{\sigma,\delta} v A^{\sigma,\delta} v)\|_{X_{s,b'}}.$$

Then, there exists $C > 0$ such that

$$\|\partial_x(A^{\sigma,\delta} u A^{\sigma,\delta} v A^{\sigma,\delta} v)\|_{X s,b'} \leq C \|A^{\sigma,\delta} u\|_{X_{s,b}} \|A^{\sigma,\delta} v\|^2_{X^\beta_{s,b}}$$
$$= C\|u\|_{X_{\sigma,\delta,s,b}} \|v\|^2_{X^\beta_{\sigma,\delta,s,b}}.$$

□

11.2 Existence of Unique Local in Time Solution

The next main result about the well-posedness of initial value problem related to coupled system of modified Korteweg–de Vries equations (11.2) in analytic Gevrey spaces reads as follows.

Theorem 11.2. *Let $s > -\frac{1}{2}, 0 < \beta < 1, \sigma \geq 1, \delta \in \mathbb{R}^+_*$ and*

$$(u_0, v_0) \in G^{\sigma,\delta,s} \times G^{\sigma,\delta,s}.$$

Then for some real number $b > \frac{1}{2}$ and a constant

$$T = T(\|(u_0, v_0)\|_{G^{\sigma,\delta,s} \times G^{\sigma,\delta,s}}),$$

the Cauchy problem (11.2) admits a unique local in time solution

$$(u, v) \in C\left([0, T], G^{\sigma,\delta,s}\right) \times C\left([0, T], G^{\sigma,\delta,s}\right). \tag{11.17}$$

Moreover, the map

$$(u_0, v_0) \to (u, v),$$

is Lipschitz continuous from

$$G^{\sigma,\delta,s} \times G^{\sigma,\delta,s},$$

to

$$C\left([0,T], G^{\sigma,\delta,s}\right) \times C\left([0,T], G^{\sigma,\delta,s}\right).$$

We are now ready to estimate all terms in (11.7) by using the trilinear estimates. We define the spaces

$$B_{\sigma,\delta,s,b} = X_{\sigma,\delta,s,b} \times X^{\beta}_{\sigma,\delta,s,b} \quad \text{and} \quad N^{\sigma,\delta,s} = G^{\sigma,\delta,s} \times G^{\sigma,\delta,s},$$

with norms

$$\|(u,v)\|_{B_{\sigma,\delta,s,b}} = \max\left\{\|u\|_{X_{\sigma,\delta,s,b}}, \|v\|_{X^{\beta}_{\sigma,\delta,s,b}}\right\},$$

and similar for $N^{\delta,s}$.

Lemma 11.5. *Let $s > -\frac{1}{2}$, $\sigma \geq 1$ and $\delta \in \mathbb{R}^{+}_{*}$, $b > \frac{1}{2}$. Then, for all $(u_0, v_0) \in N^{\sigma,\delta,s}$ and $T \in (0,1)$, with some constant $C > 0$, we have*

$$\begin{aligned} \|(\Lambda[u,v], \Gamma[u,v])\|_{B_{\sigma,\delta,s,b}} \leq C(&\|(u_0,v_0)\|_{N^{\sigma,\delta,s}} \\ &+ T^{\varepsilon}\|(u,v)\|^{3}_{B_{\sigma,\delta,s,b}}), \end{aligned} \tag{11.18}$$

and

$$\begin{aligned} &\|(\Lambda[u,v] - \Lambda[u^*,v^*], \Gamma[u,v] - \Gamma[u^*,v^*])\|_{B_{\sigma,\delta,s,b}} \\ &\leq CT^{\varepsilon}\|(u-u^*, v-v^*)\|_{B_{\sigma,\delta,s,b}}\Big(\|(u,v)\|^{2}_{B_{\sigma,\delta,s,b}} \\ &\quad + \|(u,v)\|_{B_{\sigma,\delta,s,b}}\|(u^*,v^*)\|_{B_{\sigma,\delta,s,b}} + \|(v^*,v^*)\|^{2}_{B_{\sigma,\delta,s,b}}\Big), \\ &\quad \forall (u,v), (u^*,v^*) \in B_{\sigma,\delta,s,b}. \end{aligned} \tag{11.19}$$

Proof. To estimate (11.18), it follows

$$\begin{aligned} \|\Lambda[u,v]\|_{X_{\sigma,\delta,s,b}} &\leq C\|u_0\|_{G^{\sigma,\delta,s}} + CT^{\varepsilon}\|u\|_{X_{\sigma,\delta,s,b}}\|v\|^{2}_{X^{\beta}_{\sigma,\delta,s,b}} \\ &\leq C\|(u_0,v_0)\|_{N^{\sigma,\delta,s}} + CT^{\varepsilon}\|(u,v)\|^{3}_{B_{\sigma,\delta,s,b}}, \end{aligned} \tag{11.20}$$

and

$$\begin{aligned}\|\Gamma[u,v]\|_{X^{\beta}_{\sigma,\delta,s,b}} &\leq C\|v_0\|_{G^{\sigma,\delta,s}} + CT^{\varepsilon}\|u\|^2_{X_{\sigma,\delta,s,b}}\|v\|_{X^{\beta}_{\sigma,\delta,s,b}}\\ &\leq C\|(u_0,v_0)\|_{N^{\sigma,\delta,s}} + CT^{\varepsilon}\|(u,v)\|^3_{B_{\sigma,\delta,s,b}}.\end{aligned} \tag{11.21}$$

Therefore, from (11.20) and (11.21), we obtain

$$\begin{aligned}&\|(\Lambda[u,v],\Gamma[u,v])\|_{B_{\sigma,\delta,s,b}}\\ &\quad\leq C\left(\|(u_0,v_0)\|_{N^{\sigma,\delta,s}} + T^{\varepsilon}\|(u,v)\|^3_{B_{\sigma,\delta,s,b}}\right).\end{aligned} \tag{11.22}$$

For the estimate (11.19), we observe that

$$\Lambda[u,v]-\Lambda[u^*,v^*] = \psi_T(t)\int_0^t S(t-\nu)\partial_x\left(uv^2-u^*v^{*2}\right)(x,\nu)d\nu,$$

and

$$\Gamma[u,v]-\Gamma[u^*,v^*] = \psi_T(t)\int_0^t S_{\beta}(t-\nu)\partial_x\left(u^2v-u^{*2}v^*\right)(x,\nu)d\nu.$$

Then

$$\omega = \partial_x(u^2v-u^{*2}v^*) = \partial_x\left[v(u+u^*)(u-u^*)+u^{*2}(v-v^*)\right],$$

and

$$\omega' = \partial_x(uv^2-u^*v^{*2}) = \partial_x\left[u(v+v^*)(v-v^*)+v^{*2}(u-u^*)\right]. \qquad \square$$

We will show that $\Lambda\times\Gamma$ is a contraction on the ball $\mathbb{B}(0,R)$ to $\mathbb{B}(0,R)$.

Lemma 11.6. *Let $s\geq -\frac{1}{2}$, $\sigma\geq 1$, $\delta\in\mathbb{R}^+_*$ and $b>\frac{1}{2}$. Then, for all $(u_0,v_0)\in N^{\sigma,\delta,s}$, such that the map*

$$\Lambda\times\Gamma:\mathbb{B}(0,R)\to\mathbb{B}(0,R),$$

is a contraction, where $\mathbb{B}(0,R)$ is given by

$$\mathbb{B}(0,R) = \{(u,v)\in B_{\sigma,\delta,s,b};\ \|(u,v)\|_{B_{\sigma,\delta,s,b}}\leq R\},$$

where $R = 2C\|(u_0,v_0)\|_{N^{\sigma,\delta,s}}$.

Proof. From Lemma 11.5, for all $(u, v) \in \mathbb{B}(0, R)$, we have

$$\begin{aligned}\|(\Lambda[u,v],\Gamma[u,v])\|_{B_{\sigma,\delta,s,b}} &\leq C\|(u_0,v_0)\|_{N^{\sigma,\delta,s}} + CT^{\varepsilon}\|(u,v)\|^3_{B_{\sigma,\delta,s,b}} \\ &\leq \frac{R}{2} + CT^{\varepsilon}R^3.\end{aligned}$$

We choose sufficiently small T such that $T^{\varepsilon} \leq \frac{1}{4CR^2}$. Hence,

$$\|(\Lambda[u,v],\Gamma[u,v])\|_{B_{\sigma;\delta,s,b}} \leq R, \quad \forall (u,v) \in \mathbb{B}(0,R).$$

Thus,

$$\Lambda \times \Gamma : \mathbb{B}(0,R) \to \mathbb{B}(0,R),$$

is a contraction, since

$$\begin{aligned}&\|(\Lambda[u,v]-\Lambda[u^*,v^*],\Gamma[u,v]-\Gamma[u^*,v^*])\|_{B_{\sigma,\delta,s,b}} \\ &\quad\leq CT^{\varepsilon}\|(u-u^*,v-v^*)\|_{B_{\sigma,\delta,s,b}} \\ &\qquad\times \Big(\|(u,v)\|^2_{B_{\sigma,\delta,s,b}} + \|(u,v)\|_{B_{\sigma,\delta,s,b}}\| \\ &\qquad\times (u^*,v^*)\|_{B_{\sigma,\delta,s,b}} + \|(v^*,v^*)\|^2_{B_{\sigma,\delta,s,b}}\Big) \\ &\quad\leq 3CT^{\varepsilon}R^2\|(u-u^*,v-v^*)\|_{B_{\sigma,\delta,s,b}} \\ &\quad\leq \frac{3}{4}\|(u-u^*,v-v^*)\|_{B_{\sigma,\delta,s,b}}, \quad \forall (u,v) \in \mathbb{B}(0,R).\end{aligned} \tag{11.23}$$

The Uniqueness. The uniqueness of solution for

$$(\Lambda[u,v],\Gamma[u,v]) = (u,v),$$

in $\mathbb{B}(0, R)$ comes from the argument used above (Fixed point). For the proof of uniqueness in the whole space

$$B_{\delta,s,b} = X_{\delta,s,b} \times X^{\beta}_{\delta,s,b},$$

can be seen easily. □

11.3 Continuous Dependence of the Initial Data

We will need to prove the next lemma

Lemma 11.7. *Let $s > -\frac{1}{2}$ and $\sigma \geq 1$, $\delta \in \mathbb{R}_*^+$, $b > \frac{1}{2}$. Then, for all*

$$(u_0, v_0), (u_0^*, v_0^*) \in N^{\sigma,\delta,s},$$

if (u, v) and (u^*, v^*) are two solutions to (11.2) with initial data $(u_0, v_0), (u_0^*, v_0^*)$, we have

$$\|(u - u^*, v - v^*)\|_{C([0,T],G^{\sigma,\delta,s})^2} \leq 2C_0 C\|(u_0 - u_0^*, v_0 - v_0^*)\|_{N^{\sigma,\delta,s}}. \tag{11.24}$$

Proof. Let $(u, v), (u^*, v^*)$ be two solutions to (11.2), with initial data $(u_0, v_0), (u_0^*, v_0^*)$, we have from Lemma 11.1

$$\|u - u^*\|_{C([0,T],G^{\sigma,\delta,s})} \leq C_0\|u - u^*\|_{X_{\sigma,\delta,s,b}},$$

and

$$\|v - v^*\|_{C([0,T],G^{\sigma,\delta,s})} \leq C_0\|v - v^*\|_{X^\beta_{\sigma,\delta,s,b}}.$$

Taking

$$(u, v), (u^*, v^*) \in \mathbb{B}(0, R),$$

and $T^\varepsilon \leq \frac{1}{4CR}$, we get

$$\begin{aligned}\|u - u^*\|_{X_{\sigma,\delta,s,b}} &\leq C\|(u_0 - u_0^*, v_0 - v_0^*)\|_{N^{\sigma,\delta,s}} \\ &\quad + \frac{3}{4}\|(u - u^*, v - v^*)\|_{B_{\sigma,\delta,s,b}},\end{aligned}$$

and

$$\begin{aligned}\|v - v^*\|_{X^\beta_{\sigma,\delta,s,b}} &\leq C\|(u_0 - u_0^*, v_0 - v_0^*)\|_{N^{\sigma,\delta,s}} \\ &\quad + \frac{3}{4}\|(u - u^*, v - v^*)\|_{B_{\sigma,\delta,s,b}}.\end{aligned}$$

Thus,

$$\|(u - u^*, v - v^*)\|_{B_{\sigma,\delta,s,b}} \leq 4C\|(u_0 - u_0^*, v_0 - v_0^*)\|_{N^{\sigma,\delta,s}},$$

then

$$\|(u - u^*, v - v^*)\|_{C([0,T],G^{\sigma,\delta,s})^2} \leq 4C_0 C\|(u_0 - u_0^*, v_0 - v_0^*)\|_{N^{\sigma,\delta,s}}.$$

This completes the prove of Theorem 11.2. □

11.4 Regularity of the Solution

Gevrey-3σ Regularity in Time. We will now prove the temporal regularity of solution on the line.

Lemma 11.8. *Let $s \geq -\frac{1}{2}, \sigma \geq 1, \delta \in \mathbb{R}_*^+$ and (u, v) be the solution of* (11.2). *Then*

$$(u, v) \in G^\sigma \times G^\sigma \ \text{in}\ x, \forall t \in [0, T],$$

i.e., for some $C > 0$, we have

$$|\partial_x^l u| \leq C^{l+1}(l!)^\sigma, l \in \{0, 1, \ldots\}, \quad \forall t \in [0, T],\ x \in \mathbb{R}, \tag{11.25}$$

and

$$|\partial_x^l v| \leq C^{l+1}(l!)^\sigma, l \in \{0, 1, \ldots\}, \quad \forall t \in [0, T],\ x \in \mathbb{R}. \tag{11.26}$$

Let us consider for $\varepsilon > 0$, the sequences

$$m_p = \frac{c(p!)^\sigma}{(p+1)^2}, \quad (p = 0, 1, 2, \ldots), \tag{11.27}$$

and

$$M_p = \varepsilon^{1-p} m_p, \varepsilon > 0 \quad \text{and } (p = 1, 2, 3, \ldots). \tag{11.28}$$

The constant c will be chosen so that the next inequality holds

$$\sum_{0 \leq l \leq k} \binom{k}{l} m_l m_{k-l} \leq m_k. \tag{11.29}$$

We remove 0 and k from the left-hand side of (11.29). We use M_p, to get

$$\sum_{0 < l < k} \binom{k}{l} M_l M_{k-l} \leq M_k, \quad \forall \varepsilon > 0. \tag{11.30}$$

Then, for any $\varepsilon > 0$, the sequence M_p satisfies

$$M_j \leq \varepsilon M_{j+1}, \quad \text{for } j \geq 2. \tag{11.31}$$

It is checked that for a given $C > 1$, there exists $\varepsilon_0 > 0$ such that

$$C^{j+1}(j!)^\sigma \leq M_j, \quad \text{for } j \geq 2, \quad \forall 0 < \varepsilon \leq \varepsilon_0. \tag{11.32}$$

By the definition of M_1 and M_2 in (11.28), we have for $j = 1$, that

$$M_1 = a\varepsilon M_2, \quad \text{where } a = \frac{9}{4(2!)^\sigma},$$

for some $C > 0$. Now, let us define

$$M_0 = \frac{c}{8} \quad \text{and} \quad M = \max\left\{\sqrt{2}, \frac{8C}{c}, \frac{4C^2}{c}\right\}. \tag{11.33}$$

The next Lemma is the main idea for the proof of Theorem 11.1.

Lemma 11.9. *Let (u, v) be the solution of (11.2) that satisfies (11.25) and (11.26), then there exists $\varepsilon_0 > 0$ such that for any $0 < \varepsilon \leq \varepsilon_0$, we have for all $x \in \mathbb{R}$, $t \in [0, T]$*

$$|\partial_t^j \partial_x^l u| \leq M^{2j+1} M_{l+3j}, \quad j \in \{0, 1, 2, \ldots\},\ l \in \{0, 1, 2, \ldots\}, \tag{11.34}$$

and

$$|\partial_t^j \partial_x^l v| \leq M^{2j+1} M_{l+3j}, \quad j \in \{0, 1, 2, \ldots\},\ l \in \{0, 1, 2, \ldots\}. \tag{11.35}$$

For this end, we need the next results.

Lemma 11.10. *Let $n, k \in \{0, 1, 2, \ldots\}$. We have*

$$\sum_{p=0}^{n} \sum_{q=0}^{k} \binom{n}{p} \binom{k}{q} L_{(n-p)+3(k-q)} L_{p+3q} \leq \sum_{r=1}^{m} \binom{m}{r} L_r L_{m-r}, \tag{11.36}$$

where $L_j > 0, j = 0, 1, \ldots, m$ with $m = n + 3k$.

Proof of Lemma 11.9. We will prove (11.34) and (11.35) by induction. For this end, let $j = 0$, for $l = 0$, it follows from (11.25), (11.26) and the definition of M in (11.33) that

$$|u| \leq C \leq M M_0, \quad \forall t \in [0, T],\ x \in \mathbb{R},$$

and

$$|v| \leq C \leq M M_0, \quad \forall t \in [0, T],\ x \in \mathbb{R}.$$

Similarly, for $l = 1$, we have

$$|\partial_x u| \le C^2 \le MM_1, \quad \forall t \in [0,T],\ x \in \mathbb{R},$$

and

$$|\partial_x v| \le C^2 \le MM_1, \quad \forall t \in [0,T],\ x \in \mathbb{R}.$$

By (11.25), (11.26) and (11.32), for $l \ge 2$ there exists $\varepsilon_0 > 0$, we have

$$|\partial_x^l u| \le C^{l+1}(l!)^\sigma \le M_l \le MM_l, \quad \forall t \in [0,T],\ x \in \mathbb{R},$$

and

$$|\partial_x^l v| \le C^{l+1}(l!)^\sigma \le M_l \le MM_l, \quad \forall t \in [0,T],\ x \in \mathbb{R},$$

for all $0 < \varepsilon \le \varepsilon_0$. This completes the proof of (11.34) and (11.35) for $j = 0$ and $l \in \{0, 1, \ldots\}$.

Now, we assume that (11.34) and (11.35) are true for $0 \le q \le j, l \in \{0, 1, \ldots\}$ and we will prove that it is true for $q = j+1, l \in \{0, 1, \ldots\}$.

Noting that

$$\begin{aligned} |\partial_t^{j+1}\partial_x^l u| = |\partial_t^j\partial_x^l(\partial_t u)| &= |\partial_t^j\partial_x^{l+3}u + \partial_t^j\partial_x^{l+1}(uv^2)| \\ &\le |\partial_t^j\partial_x^{l+3}u| + |\partial_t^j\partial_x^{l+1}(uv^2)|, \end{aligned} \tag{11.37}$$

and

$$\begin{aligned} |\partial_t^{j+1}\partial_x^l v| = |\partial_t^j\partial_x^l(\partial_t v)| &= |\partial_t^j\partial_x^{l+3}v + \partial_t^j\partial_x^{l+1}(u^2v)| \\ &\le |\partial_t^j\partial_x^{l+3}v| + |\partial_t^j\partial_x^{l+1}(u^2v)|. \end{aligned} \tag{11.38}$$

Owing to the induction hypotheses and the condition $M > \sqrt{2}$, we estimate the second term $\partial_t^j\partial_x^{l+3}u$ and $\partial_t^j\partial_x^{l+3}v$ as follows

$$\begin{aligned} |\partial_t^j\partial_x^{l+3}u| &\le M^{2j+1}M_{l+3+3j} = M^{-2}M^{2(j+1)+1}M_{l+3(j+1)} \\ &\le \tfrac{1}{2}M^{2(j+1)+1}M_{l+3(j+1)}, \end{aligned} \tag{11.39}$$

and

$$|\partial_t^j\partial_x^{l+3}v| \le \tfrac{1}{2}M^{2(j+1)+1}M_{l+3(j+1)}. \tag{11.40}$$

All these estimates are taken for the linear terms. For the nonlinear terms F_1 and F_2, using the induction hypothesis and Leibniz's rule twice, we have

For the nonlinear terms $F_1 = \partial_x(uv^2)$.

$$
\begin{aligned}
|\partial_t^j \partial_x^{l+1}(uv^2)| &= \Bigg| \sum_{p_1=0}^{l+1} \sum_{p_2=0}^{p_1} \sum_{q_1=0}^{j} \sum_{q_2=0}^{q_1} \binom{l+1}{p_1}\binom{p_1}{p_2}\binom{j}{p_1}\binom{p_1}{p_2} \\
&\quad \times \partial_t^{j-q_1}\partial_x^{l+1-p_1} u \partial_t^{q_1-q_2}\partial_x^{p_1-p_2} v \partial_t^{q_2}\partial_x^{p_2} v \Bigg| \\
&\leq \sum_{p_1=0}^{l+1} \sum_{p_2=0}^{p_1} \sum_{q_1=0}^{j} \sum_{q_2=0}^{q_1} \binom{l+1}{p_1}\binom{p_1}{p_2}\binom{j}{p_1}\binom{p_1}{p_2} \\
&\quad \times |\partial_t^{j-q_1}\partial_x^{l+1-p_1} u| |\partial_t^{q_1-q_2}\partial_x^{p_1-p_2} v| |\partial_t^{q_2}\partial_x^{p_2} v|.
\end{aligned} \tag{11.41}
$$

Then, by the induction hypotheses, we have

$$
\begin{aligned}
|\partial_t^j \partial_x^{l+1}(uv^2)| &\leq \sum_{p_1=0}^{l+1} \sum_{p_2=0}^{p_1} \sum_{q_1=0}^{j} \sum_{q_2=0}^{q_1} \binom{l+1}{p_1}\binom{p_1}{p_2}\binom{j}{p_1}\binom{p_1}{p_2} \\
&\quad M^{2(j-q_1)+1} M_{l+1-p_1+3(j-q_1)} M^{2(q_1-q_2)+1} \\
&\quad M_{p_1-p_2+3(q_1-q_2)} M^{2q_2+1} M_{p_2+3q_2}.
\end{aligned} \tag{11.42}
$$

Next, using Lemma 11.10 with

$$
p = p_2, \quad l = p_1, \quad q = q_2, \quad k = q_1, \quad L_j = M_j, \quad m = p_1 + 3q_1,
$$

we obtain

$$
\begin{aligned}
&\sum_{p_2=0}^{p_1} \sum_{q_2=0}^{q_1} \binom{p_1}{p_2}\binom{q_1}{q_2} M_{(p_1-p_2)+3(q_1-q_2)} M_{p_2+3q_2} \\
&\quad \leq \sum_{r=1}^{m} \binom{m}{r} M_r M_{m-r} \leq (M_0 + \varepsilon) M_m \\
&\quad = (M_0 + \varepsilon) M_{p_1+3q_1}.
\end{aligned} \tag{11.43}
$$

By Lemma 11.10 with

$$
p = p_1, \quad l = l + 1, \quad q = q_1, \quad k = j, \quad L_j = M_j, \quad m = l + 3j + 1,
$$

we obtain

$$\begin{aligned}
&\sum_{p_1=0}^{l+1}\sum_{q_1=0}^{j}\binom{l+1}{p_1}\binom{j}{q_1}M_{(l+1-p_1)+3(j-q_1)}M_{p_1+3q_1}\\
&\quad\le\sum_{r=1}^{m}\binom{m}{r}M_rM_{m-r}\le(M_0+\varepsilon)M_m \qquad (11.44)\\
&\quad=(M_0+\varepsilon)M_{l+3j+1}.
\end{aligned}$$

Continuing like this, we get all the inequalities (11.42)–(11.44). Combining these inequalities together with (11.31), we obtain

$$\begin{aligned}
|\partial_t^j\partial_x^{l+1}(uv^2)|&\le\sum_{p_1=0}^{l+1}\sum_{p_2=0}^{p_1}\sum_{q_1=0}^{j}\sum_{q_2=0}^{q_1}\binom{l+1}{p_1}\binom{p_1}{p_2}\binom{j}{p_1}\binom{p_1}{p_2}\\
&\quad M^{2(j-q_1)+1}M_{l+1-p_1+3(j-q_1)}M^{2(q_1-q_2)+1}\\
&\quad M_{p_1-p_2+3(q_1-q_2)}M^{2q_2+1}M_{p_2+3q_2}\\
&\le M^{2(j+1)+1}(M_0+\varepsilon)^2M_{l+3j+1}\\
&\le M^{2(j+1)+1}\varepsilon^2(M_0+\varepsilon)^2M_{l+3(j+1)}. \qquad (11.45)
\end{aligned}$$

We choose

$$\varepsilon\le\varepsilon_0=\left(\frac{1}{2(M_0+1)^2}\right)^{\frac{1}{2}}<1,$$

to get

$$\varepsilon^2(M_0+\varepsilon)^2\le\varepsilon^2(M_0+1)^2\le(M_0+1)^2\left(\frac{1}{2(M_0+1)^2}\right)=\frac{1}{2}.$$

Hence,

$$|\partial_t^j\partial_x^{l+1}(uv^2)|\le\frac{1}{2}M^{2(j+1)+1}M_{l+3(j+1)}. \qquad (11.46)$$

The proof for $F_2=\partial_x(u^2v)$ is similar.

This completes the proof. □

Proof (First part of Theorem 11.1). By Lemma 11.9, we have

$$|\partial_t^j \partial_x^l u| \le M^{2j+1} M_{l+3j}, \quad j \in \{0,1,2,\ldots\},\ l \in \{0,1,2,\ldots\},$$

and

$$|\partial_t^j \partial_x^l v| \le M^{2j+1} M_{l+3j}, \quad j \in \{0,1,2,\ldots\},\ l \in \{0,1,2,\ldots\},$$

where

$$M_p = \varepsilon^{1-p} \frac{c(p!)^\sigma}{(p+1)^2}, \quad p = 1, 2, \ldots.$$

The application of the last inequality with $j \in \{1,2,\ldots\}, l = 0$ gives

$$\begin{aligned} |\partial_t^j u| &\le M^{2j+1} M_{3j} = M M^{2j} \varepsilon^{1-3j} \tfrac{c((3j)!)^\sigma}{(3j+1)^2} \\ &\le M\varepsilon c \left(\tfrac{M^2}{\varepsilon^3}\right)^j ((3j)!)^\sigma \\ &\le L_0 L^j ((3j)!)^\sigma \\ &\le L_0 L^j A^{3\sigma j} ((j!)^3)^\sigma \\ &\le A_0^{j+1} (j!)^{3\sigma}, \end{aligned} \tag{11.47}$$

and

$$|\partial_t^j v| \le A_0^{j+1} (j!)^{3\sigma}, \tag{11.48}$$

where

$$L_0 = M\varepsilon c, \quad L = \frac{M^2}{\varepsilon^3},$$

since

$$(3j)! \le A^{3j} (j!)^3,$$

for $A > 0$ and

$$A_0 = \max\{L_0, L A^{3\sigma}\}.$$

We also have from (11.34) and (11.35) for $l = j = 0$, that

$$|u| \le M M_0 = M \frac{c}{8}, \quad \forall\, t \in [0,T],\ x \in \mathbb{R}, \tag{11.49}$$

and

$$|v| \leq MM_0 = M\frac{c}{8}, \quad \forall\, t \in [0,T],\ x \in \mathbb{R}. \tag{11.50}$$

We set

$$C = \max\{M\frac{c}{8}, A_0\},$$

by (11.47) and (11.49) it follows for $j \in \{0, 1, 2, \ldots\}$,

$$|\partial_t^j u| \leq C^{j+1}(j!)^{3\sigma}, \quad \forall\, t \in [0,T],\ x \in \mathbb{R}.$$

From (11.48) and (11.50), for $j \in \{0, 1, 2, \ldots\}$, we get

$$|\partial_t^j v| \leq C^{j+1}(j!)^{3\sigma}, \quad \forall\, t \in [0,T],\ x \in \mathbb{R}.$$

Hence,

$$(u, v) \in G^{3\sigma} \times G^{3\sigma},$$

in t.

The proof of first part of Theorem 11.1 is completed. □

Failure of Gevrey-D regularity in time. We replace t with $-t$, our system can be rewritten as

$$\begin{cases} \partial_t u = \partial_x^3 u + \partial_x(uv^2) \\ \partial_t v = \beta\partial_x^3 v + \partial_x(u^2 v), \quad 0 < \beta < 1 \\ u(x,0) = u_0(x) \\ v(x,0) = v_0(x). \end{cases} \tag{11.51}$$

The following Lemma will be used to estimate the higher-order derivatives of a solution with respect to t.

Lemma 11.11. *Let (u, v) be solution of (11.51). Then we have*

$$\partial_t^j u = \partial_x^{3j} u + \sum_{m=1}^{j} \sum_{|\lambda|+2m=3j} C_\lambda^m (\partial_x^{\lambda_1} u) \cdots (\partial_x^{\lambda_m} v), \tag{11.52}$$

and

$$\partial_t^j v = \beta \partial_x^{3j} v + \sum_{m=1}^{j} \sum_{|\lambda|+2m=3j} C_\lambda^m (\partial_x^{\lambda_1} u) \cdots (\partial_x^{\lambda_m} v), \tag{11.53}$$

for every $j \in \{1, 2, \ldots\}$.

Definition 11.1. Let $\{\omega_k\}$ be a sequence of positive numbers. We denote by $\mathscr{C}(\omega_k)$ the class of all functions $g(x)$, infinitely differentiable on $[-1, 1]$, for each of which there is a $C > 0$ such that

$$|g^{(k)}(x)| \leq C^{k+1}\omega_k, \quad x \in [-1, 1] \quad \text{and} \quad k = 0, 1, 2, \ldots \tag{11.54}$$

Lemma 11.12. *For any* $\sigma > 1$ *and any sequence of complex numbers* $\{\varphi_k\}$, *satisfying*

$$|\varphi_k| \leq C_1^{k+1} k^{k\sigma}, C_1 > 0, \tag{11.55}$$

there exists a function $g(x) \in \mathscr{C}(k^{k\sigma})$ *for which* $g^{(k)}(0) = \varphi_k$.

This result will be used for the sequence of real numbers

$$|g^{(k)}(x)| \leq C^{k+1} k^{k\sigma} \leq C^{k+1} (k!)^\sigma e^{k\sigma}, \quad k = 0, 1, 2, \ldots \tag{11.56}$$

where $g(x) \in \mathscr{C}(k^{k\sigma})$ such that $g^{(k)}(0) = \varphi_k = (k!)^\sigma$.

We choose $u_0, v_0 \in G_c^\sigma(-2, 2)$ such that

$$\begin{cases} \theta(x) = 1 & \text{for } |x| \leq 1 \\ \text{and} \\ \theta(x) = 0 & \text{for } |x| > 2, \end{cases} \tag{11.57}$$

by modifying $g(x)$ to have a compact support in $(-1, 1)$.

If u_0 and v_0 are extensions of θg, then we have $u_0, v_0 \in G^\sigma$. We have then the relation by $g(x)$

$$u_0^{(k)}(0) = g^{(k)}(0) = (k!)^\sigma \quad \text{and} \quad v_0^{(k)}(0) = g^{(k)}(0) = (k!)^\sigma. \tag{11.58}$$

We will show that (u, v) does not need to be $G^d \times G^d$, with $1 \leq d < 3\sigma$ in t.

Theorem 11.3. *Let* $s > -\frac{1}{2}$, $0 < \beta < 1, \sigma \geq 1$ *and* $\delta \in \mathbb{R}_*^+$. *The real-valued solution to* (11.51) *with real-valued initial data* $(u_0, v_0) \in G^\sigma \times G^\sigma$ *may not be in* $G^d \times G^d$, *with* $1 \leq d < 3\sigma$ *in* t.

Proof. By using (11.52) and (11.58), we get

$$\begin{aligned}
\partial_t^j u(0,0) &= \partial_x^{3j} u(0,0) + \sum_{m=1}^{j} \sum_{|\lambda|+2m=3j} C_\lambda^m (\partial_x^{\lambda_1} u(0,0)) \\
&\quad \cdots (\partial_x^{\lambda_m} v(0,0)) \\
&= u_0^{3j}(0) + \sum_{m=1}^{j} \sum_{|\lambda|+2m=3j} C_\lambda^m (\partial_x^{\lambda_1} u_0(0)) \cdots (\partial_x^{\lambda_m} v_0(0)) \\
&\geq u_0^{3j}(0) = ((3j)!)^\sigma \geq (j!)^{3\sigma}, \qquad (11.59)
\end{aligned}$$

and

$$\begin{aligned}
\partial_t^j v(0,0) &= \partial_x^{3j} v(0,0) + \sum_{m=1}^{j} \sum_{|\lambda|+2m=3j} C_\lambda^m (\partial_x^{\lambda_1} u(0,0)) \\
&\quad \cdots (\partial_x^{\lambda_m} v(0,0)) \\
&= v_0^{3j}(0) + \sum_{m=1}^{j} \sum_{|\lambda|+2m=3j} C_\lambda^m (\partial_x^{\lambda_1} u_0(0)) \cdots (\partial_x^{\lambda_m} v_0(0)) \\
&\geq v_0^{3j}(0) = ((3j)!)^\sigma \geq (j!)^{3\sigma}, \qquad (11.60)
\end{aligned}$$

we proved that

$$(u(0,\cdot), v(0,\cdot)) \notin G^d \times G^d,$$

for $1 \leq d < 3\sigma$ and for t near 0.

This completes the proof of Theorem 11.1. □

Chapter 12

System of Generalized KdV Equations

The solution of a coupled system consisting of generalized Korteweg–de Vries-type equations is obtained for all time where the initial data are analytic on a band in the complex plane. We show that the width of this band decreases algebraically with time.

This chapter deals with the initial-value problem for a coupled system of generalized Korteweg–de Vries (gKdV) equation

$$\begin{cases} \partial_t u + \partial_x^3 u + \partial_x \left(u^p v^{p+1}\right) = 0 \\ \partial_t v + \partial_x^3 v + \partial_x \left(u^{p+1} v^p\right) = 0, \quad x, t \in \mathbb{R}, p \in \mathbb{Z}^+ \\ u(x,0) = u_0(x), \quad v(x,0) = v_0(x), \end{cases} \tag{12.1}$$

where the unknown $u = u(x,t)$, $v = v(x,t)$ and the initial data $(u_0(x), v_0(x))$ are real-valued.

This type of equation is a special case of an important vast class of nonlinear evolution equations which was studied by many searchers and it has applications in physical problems, which describes the strong interaction of two dimensional long internal gravity waves.

For $p = 1$, the system can be reduce to a coupled system of modified KdV (mKdV) equations

$$\begin{cases} \partial_t u + \partial_x^3 u + \partial_x \left(u v^2\right) = 0 \\ \partial_t v + \partial_x^3 v + \partial_x \left(u^2 v\right) = 0, \quad x, t \in \mathbb{R} \\ u(x,0) = u_0(x), \quad v(x,0) = v_0(x). \end{cases} \tag{12.2}$$

Here, the author proved the local well-posedness in $H^s, s \geq \frac{1}{4}$. For $s \geq 1$, it is proved that the global well-posedness is assured. In addition, M. Panthee improved it to extend solution to be in any time interval $[0,T]$ for $s > \frac{4}{9}$.

Problem (12.1) is studied and the local and global well-posedness results with $(u_0, v_0) \in H^s \times H^s$, $s \geqslant 1$ and $p \geqslant 1$ is shown. The global well-posedness was obtained by using the next conserved quantities satisfied by the flow of (12.1)

$$\int_{\mathbb{R}} u dx \quad \int_{\mathbb{R}} v dx \quad \frac{1}{2}\int_{\mathbb{R}} u^2 + v^2 dx \quad \text{and}$$

$$\frac{1}{2}\int_{\mathbb{R}} u_x^2 + v_x^2 - \frac{2}{p+1} u^{p+1} v^{p+1} dx.$$

It is showed that the existence and nonlinear stability of the solitary wave solution. The study of stability for solitary wave solution is followed from the abstract results of Grillakis.

For $p = 2$, the system is turn out to a coupled system of modified Korteweg–de Vries (gKdV) equation

$$\begin{cases} \partial_t u + \partial_x^3 u + \partial_x \left(u^2 v^3\right) = 0, \\ \partial_t v + \partial_x^3 v + \partial_x \left(u^3 v^2\right) = 0, \quad x, t \in \mathbb{R} \\ u(x,0) = u_0(x), \quad v(x,0) = v_0(x). \end{cases} \tag{12.3}$$

Panthee and Scialom investigated some well-posedness issues for equation (12.3) in $H^s \times H^s$, which proved local and global will posedness for $s \geqslant 0$.

12.1 Basic Estimates and Function Spaces

The $\widehat{u}$ is denote the Fourier transform of u which is defined as

$$\widehat{u}(\xi) = \frac{1}{2\pi} \int_{\mathbb{R}} u(x) e^{-ix\xi} dx.$$

For a function $u(x,t)$ of two variable we have

$$\widehat{u}^x(\xi, t) = \frac{1}{2\pi} \int_{\mathbb{R}} u(x,t) e^{-ix\xi} dx,$$

and

$$\widehat{u}(\xi,\tau) = \frac{1}{(2\pi)^2}\int_{\mathbb{R}^2} u(x,t)e^{-ix\xi}e^{-it\tau}dxd\tau.$$

We note that the operators A, Λ and F_ρ are defined as

$$\widehat{Au}(\xi,\tau) = (1+|\xi|)\widehat{u}(\xi,\tau);$$

$$\widehat{\Lambda u}(\xi,\tau) = (1+|\tau|)\,\widehat{u}(\xi,\tau);$$

$$\widehat{F_\kappa}(\xi,\tau) = \frac{f(\xi,\tau)}{(1+|\tau-\xi^3|)^\kappa}.$$

The mixed $L^p - L^q$-norm is defined by

$$\|u\|_{L^pL^q} = \left(\int_{\mathbb{R}}\left|\int_{\mathbb{R}}|u(x,t)|^q dt\right|^{\frac{p}{q}}dx\right)^{\frac{1}{p}}.$$

The analytic Gevrey class $G^{\delta,s}$ is defined as

$$\|u_0\|^2_{G^{\delta,s}} = \|\mathrm{e}^{\delta(1+|\xi|)}(1+|\xi|)^s\widehat{u_0}(\xi)\|_{L^2_\xi}.$$

For $s, \in \mathbb{R}$, $b \in [-1,1]$ and $\delta > 0$, we denote $X_{\delta,s,b}$ by $\|\cdot\|_{\delta,s,b}$ with respect to the norm

$$\|u\|_{X_{\delta,s,b}} = \left\|e^{\delta(1+|\xi|)}(1+|\xi|)^s(1+\left|\tau-\xi^3\right|)^b\hat{u}(\xi,\tau)\right\|_{L^2_{\xi,\tau}}.$$

For $\delta = 0, X_{\delta,s,b}$ coincides with the space $X_{s,b}$. The norm of $X_{s,b}$ is denoted by $\|\cdot\|_{s,b}$, as follow:

$$\|u\|_{X_{s,b}} = \left\|(1+|\xi|)^s(1+\left|\tau-\xi^3\right|)^b\hat{u}(\xi,\tau)\right\|_{L^2_{\xi,\tau}}.$$

12.2 Linear and Bilinear Estimates

In this subsection, we shall deduce several estimates to be used in the proof of Theorem (12.1).

Lemma 12.1. *Let $0 < \sigma < \delta$ and $n \in \mathbb{N}$. Then, we have*

$$\sup_{x+iy\in S_{\delta-\sigma}} |\partial_x^n u(x+iy)| \leq C\|u\|_{G^\delta},$$

where C is constant depending on ξ and n.

Lemma 12.2. *Let $b > \frac{1}{2}$, $s \in \mathbb{R}$ and $\delta \geq 0$, then for all $T > 0$, we have*

$$X_{\delta,s,b} \hookrightarrow C([0,T], G^{\delta,s}).$$

Proof. We define the operator Θ^δ

$$\widehat{\Theta^\delta u}^x(\xi,t) = e^{\delta(1+|\xi|)}\widehat{u}^x(\xi,t),$$

satisfy

$$\|u\|_{X_{\delta,s,b}} = \|\Theta^\delta u\|_{X_{s,b}},$$

and

$$\|u\|_{G^{\delta,s}} = \|\Theta^\delta u\|_{H^s}.$$

We observe that $\Theta^\delta u$ belongs to

$$C([0,T], H^s),$$

and for some $C > 0$ we have

$$\|\Theta^\delta u\|_{C([0,T],H^s)} \leq C\ \|\Theta^\delta u\|_{X_{s,b}}.$$

Thus, it follows that

$$u \in C([0,T], G^{\delta,s}),$$

and

$$\|u\|_{C\left([0,T],G^{\delta,s}\right)} \leq C\ \|u\|_{X_{\delta,s,b}}. \qquad \square$$

By using Duhamel's formula (12.1), we may write the solution

$$\begin{cases} u(x,t) = W(t)u_0(x) - \displaystyle\int_0^t W(t-s)F_1(x,s)ds, \\ v(x,t) = W(t)v_0(x) - \displaystyle\int_0^t W(t-s)F_2(x,s)ds, \end{cases}$$

where

$$W(t) = e^{-t\partial_x^3},\ F_1 = \partial_x\left(u^p v^{p+1}\right),\ F_2 = \partial_x\left(u^{p+1}v^p\right).$$

Next, we localize in time variable by using a cut-off function $\psi(t) \in C_0^\infty(-2,2)$ with

$$0 \leq \psi(t) \leq 1, \quad \psi(t) = 1,$$

on $[-1,1]$ and for $0 < T < 1$.

We define $\psi_T(t) = \psi\left(\frac{t}{T}\right)$, where

$$\begin{cases} \psi \in C_0^\infty, \psi = 1 \quad \text{in}\big[-1,1\big] \\ \text{supp}\psi \subset [-2,2] \\ \psi_T(t) = \psi\left(\frac{t}{T}\right). \end{cases}$$

We consider the operator Ξ, Γ given by the following

$$\begin{cases} \Xi(t) = \psi(t)W(t)u_0 - \psi_T(t)\displaystyle\int_0^t W(t-s)F_1(s)ds \\ \Gamma(t) = \psi(t)W(t)v_0 - \psi_T(t)\displaystyle\int_0^t W(t-s)F_2(s)ds. \end{cases} \tag{12.4}$$

We start with the following useful lemma.

Lemma 12.3. *Let $\delta \geq 0$, $b > \frac{1}{2}$, $b-1 < b' < 0$ and $T \geq 1$. Then there exist a constant c such that the following estimates holds*

$$\|\psi(t)W(t)u_0\|_{\delta,s,b} \leq cT^{\frac{1}{2}}\|u_0\|_{G^{\delta,s}}, \quad \|\psi(t)W(t)v_0\|_{\delta,s,b} \leq cT^{\frac{1}{2}}\|v_0\|_{G^{\delta,s}}, \tag{12.5}$$

and

$$\|\psi_T(t)u\|_{\delta,s,b} \leq c\|u\|_{\delta,s,b}, \quad \|\psi_T(t)v\|_{\delta,s,b} \leq c\|v\|_{\delta,s,b}, \tag{12.6}$$

and

$$\left\|\psi_T(t)\int_0^t W(t-s)F(s)ds\right\|_{\delta,s,b} \leq cT\|F\|_{\delta,s,b'}. \tag{12.7}$$

Lemma 12.4. *Let s and κ be given. There is a constant c depending on s and κ such that*

$$\text{if} \quad \kappa > \frac{1}{4}, \quad \text{then} \quad \|A^{\frac{1}{2}}F_\kappa\|_{L_x^4L_t^2} \leq C\|f\|_{L_\xi^2L_\tau^2}, \tag{12.8}$$

$$\text{if} \quad \kappa > \frac{1}{4}, \quad \text{then} \quad \|AF_\kappa\|_{L_x^\infty L_t^2} \leq C\|f\|_{L_\xi^2L_\tau^2}, \tag{12.9}$$

$$\text{if} \quad \kappa > \frac{1}{2}, \quad \text{and} \quad s > 3\kappa, \quad \text{then} \quad \|A^{-s}F_\kappa\|_{L_x^2L_t^\infty} \leq C\|f\|_{L_\xi^2L_\tau^2}, \tag{12.10}$$

$$\textit{if}\quad \kappa > \frac{1}{2}, \quad \textit{and} \quad s > \frac{1}{4}, \quad \textit{then} \quad \|A^{-s}F_\kappa\|_{L_x^4 L_t^\infty} \le C\|f\|_{L_\xi^2 L_\tau^2}, \tag{12.11}$$

$$\textit{if}\quad \kappa > \frac{1}{2}, \quad \textit{and} \quad s > \frac{1}{2}, \quad \textit{then} \quad \|A^{-s}F_\kappa\|_{L_x^\infty L_t^\infty} \le C\|f\|_{L_\xi^2 L_\tau^2}. \tag{12.12}$$

Lemma 12.5. *Let $b > \frac{1}{2}$, $b' < -\frac{1}{4}$ and $s \ge 3b$. Let $p \in \mathbb{N}$ and suppose*

$$u_1, \dots, u_{p+1}, v_1, \dots, v_{p+1} \in X_{\delta,s,b}.$$

Then there exists a constants c such that

$$\left\| \partial_x \prod_{i=1}^{p} u_i \prod_{j=1}^{p+1} v_j \right\|_{\delta,s,b'} \le C \prod_{i=1}^{p} \|u_i\|_{\delta,s,b}. \prod_{j=1}^{p+1} \|v_j\|_{\delta,s,b}, \tag{12.13}$$

$$\left\| \partial_x \prod_{i=1}^{p+1} u_i \prod_{j=1}^{p} v_j \right\|_{\delta,s,b'} \le C \prod_{i=1}^{p+1} \|u_i\|_{\delta,s,b}. \prod_{j=1}^{p} \|v_j\|_{\delta,s,b}. \tag{12.14}$$

Proof. First of all, for $i = 1, 2, \dots, p+1$ and $j = 1, 2, \dots, p+1$, we define

$$f_i(\xi,\tau) = (1+|\xi|)^s (1+|\tau-\xi^3|)^b e^{\delta(1+|\xi|)} |\widehat{u_i}(\xi,\tau)|$$
$$g_j(\xi,\tau) = (1+|\xi|)^s (1+|\tau-\xi^3|)^b e^{\delta(1+|\xi|)} |\widehat{v_j}(\xi,\tau)|.$$

The proof is first given for the case $p = 1$, after which the proof for a general $2p+1$ will be more transparent,that means we prove

$$\|\partial_x u_1 v_1 v_2\|_{\delta,s,b'} \le C\|u_1\|_{\delta,s,b}\|v_1\|_{\delta,s,b}\|v_2\|_{\delta,s,b}$$
$$\|\partial_x u_1 u_2 v_1\|_{\delta,s,b'} \le C\|u_1\|_{\delta,s,b}\|u_2\|_{\delta,s,b}\|v_1\|_{\delta,s,b}.$$

We have

$$\|\partial_x u_1 v_1 v_2\|_{\delta,s,b'}$$
$$= \left\| (1+|\xi|)^s (1+|\tau-\xi^3|)^{b'} e^{\delta(1+|\xi|)} |\widehat{\partial_x u_1 v_1} v_2(\xi,\tau)| \right\|_{L_\xi^2 L_\tau^2}$$
$$= \left\| (1+|\xi|)^s e^{\delta(1+|\xi|)} (1+|\tau-\xi^3|)^{b'} |\xi|\ |\widehat{u_1 v_1 v_2}(\xi,\tau)| \right\|_{L_\xi^2 L_\tau^2}$$

$$= \left\| (1+|\xi|)^s e^{\delta(1+|\xi|)} (1+|\tau-\xi^3|)^{b'} |\xi| \; |\widehat{u_1} * \widehat{v_1} * \widehat{v_2}(\xi,\tau)| \right\|_{L^2_\xi L^2_\tau}$$

$$= \Bigg\| (1+|\xi|)^s e^{\delta(1+|\xi|)} (1+|\tau-\xi^3|)^{b'} |\xi|$$
$$\times \int_{\mathbb{R}^4} \widehat{u_1}(\xi_1,\tau_1)\widehat{v_1}(\xi-\xi_2,\tau-\tau_2)$$
$$\times\ \widehat{v_2}(\xi_2-\xi_1,\tau_2-\tau_1)|d\xi_1 d\tau_1 d\xi_2 d\tau_2 \Bigg\|_{L^2_\xi L^2_\tau}$$

$$= \Bigg\| (1+|\xi|)^s e^{\delta(1+|\xi|)} (1+|\tau-\xi^3|)^{b'} |\xi|$$
$$\times \int_{\mathbb{R}^4} \left(\frac{(1+|\xi_1|)^{-s} e^{-\delta(1+|\xi_1|)} \widehat{f_1}(\xi_1,\tau_1)}{(1+|\tau-\xi^3|)^b} \right)$$
$$\times \left(\frac{(1+|\xi-\xi_2|)^{-s} e^{-\delta(1+|\xi-\xi_2|)} \widehat{g_1}(\xi-\xi_2,\tau-\tau_2)}{(1+|(\tau-\tau_2)-(\xi-\xi_2)^3|)^b} \right)$$
$$\times \left(\frac{(1+|\xi_2-\xi_1|)^{-s} e^{-\delta(1+|\xi_2-\xi_1|)} \widehat{g_2}(\xi_2-\xi_1,\tau_2-\tau_1)}{(1+|\tau_2-\tau_1-(\xi_2-\xi_1)^3|)^b} \right) d\mu \Bigg\|_{L^2_\xi L^2_\tau},$$

where

$$d\mu = d\xi_1 d\tau_1 d\xi_2 d\tau_2 d\xi d\tau.$$

By using the duality, we proof this estimate, where $m(\xi,\tau)$ is a positive function in $L^2(\mathbb{R}^2)$ with norm $\|m\|_{L^2(\mathbb{R}^2)} = 1$, then

$$\|\partial_x u_1 v_1 v_2\|_{\delta,s,b'} \leqslant \int_{\mathbb{R}^6} \frac{e^{\delta(1+|\xi|)}(1+|\xi|)^s |\xi| m(\xi,\tau)}{(1+|\tau-\xi^3|)^{-b'}}$$
$$\times \frac{e^{-\delta(1+|\xi_1|)}(1+|\xi_1|)^{-s} f_1(\xi_1,\tau_1)}{(1+|\tau_1-\xi_1^3|)^b}$$
$$\times \frac{e^{-\delta(1+|\xi-\xi_2|)}(1+|\xi-\xi_2|)^{-s} g_1(\xi-\xi_2,\tau-\tau_2)}{(1+|\tau-\tau_2-(\xi-\xi_2)^3|)^b}$$

$$\times \frac{e^{-\delta(1+|\xi_2-\xi_1|)}(1+|\xi_2-\xi_1|)^{-s}g_2 \times (\xi_2-\xi_1,\tau_2-\tau_1)}{(1+|\tau_2-\tau_1-(\xi_2-\xi_1)^3|)^b}d\mu.$$

Using the inequality

$$|\xi| \leq |\xi_1| + |\xi-\xi_2| + |\xi_2-\xi_1|.$$

then

$$e^{\delta(1+|\xi|)} \leq e^{\delta(1+|\xi_1|)} \times e^{\delta(1+|\xi-\xi_2|)} \times e^{\delta(1+|\xi_2-\xi_1|)}.$$

Then

$$\begin{aligned}
&\|\partial_x u_1 v_1 v_2\|_{\delta,s,b'} \\
&\quad \leqslant \int_{\mathbb{R}^6} \frac{(1+|\xi|)^s|\xi|m(\xi,\tau)}{(1+|\tau-\xi^3|)^{-b'}} \frac{(1+|\xi_1|)^{-s}f_1(\xi_1,\tau_1)}{(1+|\tau_1-\xi_1^3|)^b} \\
&\quad \times \frac{(1+|\xi-\xi_2|)^{-s}g_1(\xi-\xi_2,\tau-\tau_2)}{(1+|\tau-\tau_2-(\xi-\xi_2)^3|)^b} \\
&\quad \times \frac{(1+|\xi_2-\xi_1|)^{-s}g_2(\xi_2-\xi_1,\tau_2-\tau_1)}{(1+|\tau_2-\tau_1-(\xi_2-\xi_1)^3|)^b}d\mu.
\end{aligned}$$

Now, split the Fourier space into six regions as follows:

1. $|\xi-\xi_2| \leq |\xi_2-\xi_1| \leq |\xi_1|$
2. $|\xi-\xi_2| \leq |\xi_1| \leq |\xi_2-\xi_1|$
3. $|\xi_1| \leq |\xi_2-\xi_1| \leq |\xi-\xi_2|$
4. $|\xi_1| \leq |\xi-\xi_2| \leq |\xi_2-\xi_1|$
5. $|\xi_2-\xi_1| \leq |\xi-\xi_2 \leq |\xi_1|$
6. $|\xi_2-\xi_1| \leq |\xi_1| \leq |\xi-\xi_2|$.

We begin by the case (1)

$$|\xi-\xi_2| \leq |\xi_2-\xi_1| \leq |\xi_1|.$$

Then

$$(1+|\xi-\xi_2|)^{-s} \geq (1+|\xi_2-\xi_1|)^{-s} \geq (1+|\xi_1|)^{-s}, \tag{12.15}$$

and, we assume that $|\xi| \leq 1$ or $|\xi| \geq 1$.

Firstly, by $|\xi| \geq 1$, then

$$(1+|\xi|)^s \leq (|\xi|+|\xi|)^s = 2^s(|\xi|)^s = C(|\xi|)^s.$$

By the last inequality and (12.15), we obtain

$$\begin{aligned}
&\|\partial_x u_1 v_1 v_2\|_{\delta,s,b'} \\
&\quad \leqslant \int_{\mathbb{R}^6} \frac{(1+|\xi|)^s|\xi|m(\xi,\tau)}{(1+|\tau-\xi^3|)^{-b'}} \frac{(1+|\xi_1|)^{-s}f_1(\xi_1,\tau_1)}{(1+|\tau_1-\xi_1^3|)^b} \\
&\quad \times \frac{(1+|\xi-\xi_2|)^{-s}g_1(\xi-\xi_2,\tau-\tau_2)}{(1+|\tau-\tau_2-(\xi-\xi_2)^3|)^b} \\
&\quad \times \frac{(1+|\xi_2-\xi_1|)^{-s}g_2(\xi_2-\xi_1,\tau_2-\tau_1)}{(1+|\tau_2-\tau_1-(\xi_2-\xi_1)^3|)^b} d\mu \\
&\quad \leq C\int_{\mathbb{R}^6} \frac{(|\xi|)^s|\xi|m(\xi,\tau)}{(1+|\tau-\xi^3|)^{-b'}} \\
&\quad \times \frac{(1+|\xi_1|)^{-s}f_1(\xi_1,\tau_1)}{(1+|\tau_1-\xi_1^3|)^b} \frac{(1+|\xi-\xi_2|)^{-s}g_1(\xi-\xi_2,\tau-\tau_2)}{(1+|\tau-\tau_2-(\xi-\xi_2)^3|)^b} \\
&\quad \times \frac{(1+|\xi_2-\xi_1|)^{-s}g_2(\xi_2-\xi_1,\tau_2-\tau_1)}{(1+|\tau_2-\tau_1-(\xi_2-\xi_1)^3|)^b} d\mu \\
&\quad \leq C\int_{\mathbb{R}^6} \frac{(|\xi|)^{s+1}m(\xi,\tau)}{(1+|\tau-\xi^3|)^{-b'}} \\
&\quad \times \frac{(1+|\xi_1|)^{-s}f_1(\xi_1,\tau_1)}{(1+|\tau_1-\xi_1^3|)^b} \frac{(1+|\xi-\xi_2|)^{-s}g_1(\xi-\xi_2,\tau-\tau_2)}{(1+|\tau-\tau_2-(\xi-\xi_2)^3|)^b} \\
&\quad \times \frac{(1+|\xi_2-\xi_1|)^{-s}g_2(\xi_2-\xi_1,\tau_2-\tau_1)}{(1+|\tau_2-\tau_1-(\xi_2-\xi_1)^3|)^b} d\mu,
\end{aligned}$$

then

$$\begin{aligned}
&\|\partial_x u_1 v_1 v_2\|_{\delta,s,b'} \\
&\quad \leq C\int_{\mathbb{R}^6} \frac{(|\xi|)^{\frac{1}{2}}m(\xi,\tau)}{(1+|\tau-\xi^3|)^{-b'}} \frac{(1+|\xi_1|)^{\frac{1}{2}}f_1(\xi_1,\tau_1)}{(1+|\tau_1-\xi_1^3|)^b}
\end{aligned}$$

$$\times \frac{(1+|\xi-\xi_2|)^{-s} g_1(\xi-\xi_2,\tau-\tau_2)}{(1+|\tau-\tau_2-(\xi-\xi_2)^3|)^b}$$

$$\times \frac{(1+|\xi_2-\xi_1|)^{-s} g_2(\xi_2-\xi_1,\tau_2-\tau_1)}{(1+|\tau_2-\tau_1-(\xi_2-\xi_1)^3|)^b} d\mu.$$

By

$$|\xi-\xi_2| \le |\xi_2-\xi_1| \le |\xi_1|,$$

and

$$\begin{aligned} |\xi|^{s+1}(1+|\xi_1|)^{-s} &= |\xi|^{s+1}|\xi_1|^{-s}|\xi_1|^s(1+|\xi_1|)^{-s} \\ &\le |\xi|^{s+1}|\xi_1|^{-s}\frac{|\xi_1|^s}{(1+|\xi_1|)^s} \\ &\le |\xi|^{s+1}|\xi_1|^{-s}, \end{aligned}$$

and

$$\begin{aligned} |\xi|^{s+1}\ |\xi_1|^{-s} &= |\xi|^{\frac{1}{2}}|\xi_1|^{\frac{1}{2}}|\xi|^{s+\frac{1}{2}}|\xi_1|^{-s-\frac{1}{2}} \\ &\le c|\xi|^{\frac{1}{2}}|\xi_1|^{\frac{1}{2}}(|\xi-\xi_2|)^{s+\frac{1}{2}} + |\xi_2-\xi_1|^{s+\frac{1}{2}} + |\xi_1|^{s+\frac{1}{2}})| \\ &\quad \times \ \xi_1|^{-s-\frac{1}{2}} \\ &\le c|\xi|^{\frac{1}{2}}|\xi_1|^{\frac{1}{2}}(|\xi_1|)^{s+\frac{1}{2}} + |\xi_1|^{s+\frac{1}{2}} + |\xi_1|^{s+\frac{1}{2}})|\xi_1|^{-s-\frac{1}{2}} \\ &\le c|\xi|^{\frac{1}{2}}|\xi_1|^{\frac{1}{2}}(3|\xi_1|^{s+\frac{1}{2}}\ \ |\xi_1|^{-s-\frac{1}{2}}) \\ &\le C|\xi|^{\frac{1}{2}}|\xi_1|^{\frac{1}{2}}. \end{aligned}$$

We suppose that

$$\widehat{A^{\frac{1}{2}}M_{-b'}}(\xi,\tau) = \frac{(|\xi|)^{\frac{1}{2}} m(\xi,\tau)}{(1+|\tau-\xi^3|)^{-b'}}$$

$$\widehat{A^{\frac{1}{2}}F_b}(\xi_1,\tau_1) = \frac{(1+|\xi_1|)^{\frac{1}{2}} f_1(\xi_1,\tau_1)}{(1+|\tau_1-\xi_1^3|)^b}$$

$$\widehat{A^{-s}G_b^1}(\xi-\xi_2,\tau-\tau_2) = \frac{(1+|\xi-\xi_2|)^{-s} g_1(\xi-\xi_2,\tau-\tau_2)}{(1+|\tau-\tau_2-(\xi-\xi_2)^3|)^b}$$

$$\widehat{A^{-s}G_b^2}(\xi_2-\xi_1,\tau_2-\tau_1) = \frac{(1+|\xi_2-\xi_1|)^{-s} g_2(\xi_2-\xi_1,\tau_2-\tau_1)}{(1+|\tau_2-\tau_1-(\xi_2-\xi_1)^3|)^b},$$

and

$$\begin{aligned}
&\int_{\mathbb{R}^6} \frac{(|\xi|)^{\frac{1}{2}} m(\xi,\tau)}{(1+|\tau-\xi^3|)^{-b'}} \quad \frac{(1+|\xi_1|)^{\frac{1}{2}} f_1(\xi_1,\tau_1)}{(1+|\tau_1-\xi_1^3|)^b} \\
&\quad \times \frac{(1+|\xi-\xi_2|)^{-s} g_1(\xi-\xi_2,\tau-\tau_2)}{(1+|\tau-\tau_2-(\xi-\xi_2)^3|)^b} \\
&\quad \times \frac{(1+|\xi_2-\xi_1|)^{-s} g_2(\xi_2-\xi_1,\tau_2-\tau_1)}{(1+|\tau_2-\tau_1-(\xi_2-\xi_1)^3|)^b} d\mu \\
&\quad \times \int_{\mathbb{R}^6} \widehat{A^{\frac{1}{2}} M_{-b'}}(\xi,\tau) \widehat{A^{\frac{1}{2}} F_b}(\xi_1,\tau_1) \widehat{A^{-s} G_b^1} \\
&\quad \times (\xi-\xi_2,\tau-\tau_2) \widehat{A^{-s} G_b^2}(\xi_2-\xi_1,\tau_2-\tau_1) d\mu \\
&= \int_{\mathbb{R}^2} \left(\widehat{A^{\frac{1}{2}} M_{-b'}}(\xi,\tau) \right) \\
&\quad \times \left(\int_{\mathbb{R}^4} \widehat{A^{\frac{1}{2}} F_b}(\xi_1,\tau_1) \widehat{A^{-s} G_b^1}(\xi-\xi_2,\tau-\tau_2) \widehat{A^{-s} G_b^2} \right. \\
&\quad \left. \times (\xi_2-\xi_1,\tau_2-\tau_1) d\xi_1 d\tau_1 d\xi_2 d\tau_2 \right) d\xi d\tau \\
&= \int_{\mathbb{R}^2} \left(\widehat{A^{\frac{1}{2}} M_{-b'}}(\xi,\tau) \right) \left(\left(\widehat{A^{\frac{1}{2}} F_b} * \widehat{A^{-s} G_b^1} * \widehat{A^{-s} G_b^2} \right)(\xi,\tau) \right) d\xi d\tau \\
&= \int_{\mathbb{R}^2} \left(\widehat{A^{\frac{1}{2}} M_{-b'}}(\xi,\tau) \right) \left(\left(A^{\frac{1}{2}} F_b . \widehat{A^{-s} G_b^1} . A^{-s} G_b^2 \right)(\xi,\tau) \right) d\xi d\tau \\
&= \int_{\mathbb{R}^2} A^{\frac{1}{2}} M_{-b'}(x,t) \left(A^{\frac{1}{2}} F_b . A^{-s} G_b^1 . A^{-s} G_b^2 \right)(x,t) dx dt.
\end{aligned}$$

We suppose that

$$\begin{aligned}
h_1(x,t) &= A^{\frac{1}{2}} M_{-b'}(x,t) \\
h_2(x,t) &= A^{\frac{1}{2}} F_b(x,t) \\
h_3(x,t) &= A^{-s} G_b^1(x,t) \\
h_4(x,t) &= A^{-s} G_b^2(x,t),
\end{aligned}$$

then

$$\begin{aligned}
&\int_{\mathbb{R}^2} A^{\frac{1}{2}} D_{-b'}(x,t) A^{\frac{1}{2}} F_b . A^{-s} G_b^1 . A^{-s} G_b^2(x,t)\,dxdt \\
&\quad = \int_{\mathbb{R}^2} h_1(x,t).h_2(x,t) h_3(x,t) h_4(x,t)\,dxdt \\
&\quad \leq \int_{\mathbb{R}^2} h_1(x,t).h_2(x,t) \sup_{t\in[0,T]} h_3(x,t) \sup_{t\in[0,T]} h_4(x,t)\,dxdt \\
&\quad \leq \int_{\mathbb{R}^2} (h_1(x,t).h_2(x,t)) \left(\sup_{t\in[0,T]} h_3(x,t) \sup_{t\in[0,T]} h_4(x,t) \right) dxdt.
\end{aligned}$$

By using Cauchy–Schwarz inequality for the variables x and t

$$\begin{aligned}
&\int_{\mathbb{R}^2} (h_1(x,t).h_2(x,t)) \left(\sup_{t\in[0,T]} h_3(x,t) \sup_{t\in[0,T]} h_4(x,t) \right) dxdt \\
&\leq \left(\int_{\mathbb{R}} \left(\int_{\mathbb{R}} h_1(x,t).h_2(x,t)\,dt \right)^2 dx \right)^{\frac{1}{2}} \\
&\quad \times \left(\int_{\mathbb{R}} \left(\sup_{t\in[0,T]} h_3(x,t) \sup_{x\in\mathbb{R}} \left[\sup_{t\in[0,T]} h_4(x,t) \right] \right)^2 dx \right)^{\frac{1}{2}} \\
&\leq \left(\int_{\mathbb{R}} \left(\int_{\mathbb{R}} h_1(x,t).h_2(x,t)\,dt \right)^2 dx \right)^{\frac{1}{2}} \\
&\quad \times \left(\int_{\mathbb{R}} \left(\sup_{t\in[0,T]} h_3(x,t) \right)^2 dx \right)^{\frac{1}{2}} \sup_{x\in\mathbb{R}} \left[\sup_{t\in[0,T]} h_4(x,t) \right] \\
&\leq \left(\int_{\mathbb{R}} \left(\int_{\mathbb{R}} h_1(x,t).h_2(x,t)\,dt \right)^2 dx \right)^{\frac{1}{2}} \|h_3(x,t)\|_{L_x^2 L_t^\infty}. \\
&\quad \times \|h_4(x,t)\|_{L_x^\infty L_t^\infty}
\end{aligned}$$

$$\leq \left[\int_{\mathbb{R}}\left(\int_{\mathbb{R}}(h_1(x,t))^2\,dt\right)^2 dx\right]^{\frac{1}{4}}\left[\int_{\mathbb{R}}\left(\int_{\mathbb{R}}(h_2(x,t))^2\,dt\right)^2 dx\right]^{\frac{1}{4}}$$

$$\times \|h_3(x,t)\|_{L_x^2L_t^\infty}.\|h_4(x,t)\|_{L_x^\infty L_t^\infty}$$

$$\leq \|h_1(x,t)\|_{L_x^4L_t^2}\|h_2(x,t)\|_{L_x^4L_t^2}\|h_3(x,t)\|_{L_x^2L_t^\infty}\|h_4(x,t)\|_{L_x^\infty L_t^\infty}$$

$$\leq \|A^{\frac{1}{2}}M_{-b'}\|_{L_x^4L_t^2}\|A^{\frac{1}{2}}F_b\|_{L_x^4L_t^2}\|A^{-s}G_b^1\|_{L_x^2L_t^\infty}\|A^{\frac{1}{2}}G_b^2\|_{L_x^\infty L_t^\infty}.$$

Then

$$\|\partial_x u_1 v_1 v_2\|_{\delta,s,b'} \leq c\|A^{\frac{1}{2}}M_{-b'}\|_{L_x^4L_t^2}\|A^{\frac{1}{2}}F_b\|_{L_x^4L_t^2}\|A^{-s}G_b^1\|_{L_x^2L_t^\infty}$$
$$\|A^{\frac{1}{2}}G_b^2\|_{L_x^\infty L_t^\infty}.$$

Hence, by Lemma 12.4

$$\begin{aligned}\|\partial_x u_1 v_1 v_2\|_{\delta,s,b'} &\leq c\|m\|_{L_\xi^2L_\tau^2}\|f\|_{L_\xi^2L_\tau^2}\|g_1\|_{L_\xi^2L_\tau^2}\|g_2\|_{L_\xi^2L_\tau^2}\\ &\leq c\|u_1\|_{\delta,s,b}\|v_1\|_{\delta,s,b}\|v_2\|_{\delta,s,b}.\end{aligned}$$

Secondly for the case $|\xi| \leq 1$, then

$$\begin{aligned}&(1+|\xi|)^s|\xi|(1+|\xi_1|)^{-s}\\ &\quad = (1+|\xi|)^{\frac{1}{2}}(1+|\xi_1|)^{\frac{1}{2}}(1+|\xi_1|)^{-s-\frac{1}{2}}(1+|\xi|)^{s-\frac{1}{2}}|\xi|\\ &\quad \leq (1+|\xi|)^{\frac{1}{2}}(1+|\xi_1|)^{\frac{1}{2}}1+|\xi_1|)^{-s-\frac{1}{2}}(1+|\xi|)^{s-\frac{1}{2}}(1+|\xi|)\\ &\quad \leq (1+|\xi|)^{\frac{1}{2}}(1+|\xi_1|)^{\frac{1}{2}}(1+|\xi_1|)^{-s-\frac{1}{2}}(1+|\xi|)^{s+\frac{1}{2}}\\ &\quad \leq (1+|\xi|)^{\frac{1}{2}}(1+|\xi_1|)^{\frac{1}{2}}(1+|\xi_1|)^{-s-\frac{1}{2}}(1+|\xi_1|+|\xi-\xi_2|\\ &\qquad + |\xi_2-\xi_1|)^{s+\frac{1}{2}}\\ &\quad \leq (1+|\xi|)^{\frac{1}{2}}(1+|\xi_1|)^{\frac{1}{2}}(1+|\xi_1|)^{-s-\frac{1}{2}}(3(1+|\xi_1|))^{s+\frac{1}{2}}\\ &\quad \leq C(1+|\xi|)^{\frac{1}{2}}(1+|\xi_1|)^{\frac{1}{2}},\end{aligned}$$

then

$$\begin{aligned}
&\|\partial_x u_1 v_1 v_2\|_{\delta,s,b'} \\
&\leqslant \int_{\mathbb{R}^6} \frac{(1+|\xi|)^s|\xi|m(\xi,\tau)}{(1+|\tau-\xi^3|)^{-b'}} \frac{(1+|\xi_1|)^{-s}f_1(\xi_1,\tau_1)}{(1+|\tau_1-\xi_1^3|)^b} \\
&\quad\times \frac{(1+|\xi-\xi_2|)^{-s}g_1(\xi-\xi_2,\tau-\tau_2)}{(1+|\tau-\tau_2-(\xi-\xi_2)^3|)^b} \\
&\quad\times \frac{(1+|\xi_2-\xi_1|)^{-s}g_2(\xi_2-\xi_1,\tau_2-\tau_1)}{(1+|\tau_2-\tau_1-(\xi_2-\xi_1)^3|)^b}d\mu \\
&\leq C\int_{\mathbb{R}^6} \frac{(1+|\xi|)^{\frac{1}{2}}m(\xi,\tau)}{(1+|\tau-\xi^3|)^{-b'}} \frac{(1+|\xi_1|)^{\frac{1}{2}}f_1(\xi_1,\tau_1)}{(1+|\tau_1-\xi_1^3|)^b} \\
&\quad\times \frac{(1+|\xi-\xi_2|)^{-s}g_1(\xi-\xi_2,\tau-\tau_2)}{(1+|\tau-\tau_2-(\xi-\xi_2)^3|)^b} \\
&\quad\times \frac{(1+|\xi_2-\xi_1|)^{-s}g_2(\xi_2-\xi_1,\tau_2-\tau_1)}{(1+|\tau_2-\tau_1-(\xi_2-\xi_1)^3|)^b}d\mu.
\end{aligned}$$

Then, by the inner product, we have

$$\begin{aligned}
&\|\partial_x u_1 v_1 v_2\|_{\delta,s,b'} \\
&\leq c\langle \widehat{A^{\frac{1}{2}}M_{-b'}}; \widehat{A^{\frac{1}{2}}F_b}\widehat{A^{-s}G_b^1}\widehat{A^{-s}G_b^2}\rangle \\
&\leq c\langle \widehat{A^{\frac{1}{2}}M_{-b'}}; A^{\frac{1}{2}}F_b.\widehat{A^{-s}G_b^1\ A^{-s}G_b^2}\rangle \\
&\leq c\langle A^{\frac{1}{2}}M_{-b'}; A^{\frac{1}{2}}F_b.A^{-s}G_b^1A^{-s}G_b^2\rangle \\
&\leq c\|A^{\frac{1}{2}}M_{-b'}\|_{L_x^4L_t^2}\|A^{\frac{1}{2}}F_b\|_{L_x^4L_t^2}\|A^{-s}G_b^1\|_{L_x^2L_t^\infty}\|A^{-s}G_b^2\|_{L_x^\infty L_t^\infty}.
\end{aligned}$$

Hence, by Lemma 12.4

$$\begin{aligned}
\|\partial_x u_1 v_1 v_2\|_{\delta,s,b'} &\leq c\|m\|_{L_\xi^2L_\tau^2}\|f\|_{L_\xi^2L_\tau^2}\|g_1\|_{L_\xi^2L_\tau^2}\|g_2\|_{L_\xi^2L_\tau^2} \\
&\leq c\|u_1\|_{\delta,s,b}\|v_1\|_{\delta,s,b}\|v_2\|_{\delta,s,b}.
\end{aligned}$$

By the same way, we prove the inequality in the five region.

For the case $p \geq 2$ is virtually identical. The only difference is that we need to split the Fourier space in $((2p+1)+1)!$

We prove that

$$\left\|\partial_x \prod_{i=1}^{p} u_i \prod_{j=1}^{p+1} v_j\right\|_{\delta,s,b'} \leq C \prod_{i=1}^{p} \|u_i\|_{\delta,s,b}. \prod_{j=1}^{p+1} \|v_j\|_{\delta,s,b},$$

We have

$$\begin{aligned}
&\left\|\partial_x \prod_{i=1}^{p} u_i \prod_{j=1}^{p+1} v_j\right\|_{\delta,s,b'} \\
&\quad = \left\|(1+|\xi|)^s(1+|\tau-\xi^3|)^{b'} e^{\delta(1+|\xi|)} |\widehat{\partial_x \prod_{i=1}^{p} u_i \prod_{j=1}^{p+1} v_j}(\xi,\tau)\right\|_{L^2_\xi L^2_\tau}, \\
&\quad = \left\|(1+|\xi|)^s(1+|\tau-\xi^3|)^{b'} e^{\delta(1+|\xi|)} \xi \prod_{i=1}^{p} \widehat{u_i} \prod_{j=1}^{p+1} \widehat{v_j}(\xi,\tau)\right\|_{L^2_\xi L^2_\tau}.
\end{aligned}$$

By the same way, by the inner product, we have

$$\begin{aligned}
\left\|\partial_x \prod_{i=1}^{p} u_i \prod_{j=1}^{p+1} v_j\right\|_{\delta,s,b'} &\leq c\left\langle \widehat{A^{\frac{1}{2}} M_{-b'}}; \widehat{A^{\frac{1}{2}} F_b}\widehat{A^{-s} G_b^1} \prod_{i=1}^{p} \widehat{A^{-s} F_b^i} \prod_{j=1}^{p} \widehat{A^{-s} G_b^i}\right\rangle \\
&\leq c\left\langle \widehat{A^{\frac{1}{2}} M_{-b'}}; \widehat{A^{\frac{1}{2}} F_b . A^{-s} G_b^1 \prod_{i=1}^{p} A^{-s} F_b^i} \prod_{j=1}^{p} A^{-s} G_b^i\right\rangle \\
&\leq c \left\|A^{\frac{1}{2}} M_{-b'}\right\|_{L^4_x L^2_t} \left\|A^{\frac{1}{2}} F_b\right\|_{L^4_x L^2_t} \|A^{-s} G_b^1\|_{L^2_x L^\infty_t} \\
&\quad \times \left\|\prod_{i=1}^{p} A^{-s} F_b^i\right\|_{L^\infty_x L^\infty_t} \left\|\prod_{j=1}^{p} A^{\frac{1}{2}} G_b^i\right\|_{L^\infty_x L^\infty_t} \\
&\leq c \prod_{i=1}^{p} \|u_i\|_{\delta,s,b}. \prod_{j=1}^{p+1} \|v_j\|_{\delta,s,b}. \qquad \square
\end{aligned}$$

Lemma 12.6. *Let $\delta > 0, s \geq 3b, b > \frac{1}{2}$ and $b' < -\frac{1}{4}$. Let $p \in \mathbb{N}$ and suppose that*

$$u_1, \ldots, u_{p+1}, v_1, \ldots, v_{p+1} \in X_{\delta,s,b}.$$

Then there exists a constants c such that

$$\left\|\partial_x \prod_{i=1}^{p} u_i \prod_{j=1}^{p+1} v_j\right\|_{\delta,s,b'} \leq C\prod_{i=1}^{p} \|u_i\|_{s,b}\cdot \prod_{j=1}^{p+1} \|v_j\|_{s,b} + c\prod_{i=1}^{p} \|u_i\|_{\delta,s,b} \cdot \prod_{j=1}^{p+1} \|v_j\|_{\delta,s,b},$$

$$\left\|\partial_x \prod_{i=1}^{p+1} u_i \prod_{j=1}^{p} v_j\right\|_{\delta,s,b'} \leq C\prod_{i=1}^{p+1} \|u_i\|_{s,b}\cdot \prod_{j=1}^{p} \|v_j\|_{s,b} + c\prod_{i=1}^{p+1} \|u_i\|_{\delta,s,b} \cdot \prod_{j=1}^{p} \|v_j\|_{\delta,s,b}.$$

Proof. We begin by the case $p = 1$, that mean we prove that

$$\|\partial_x(u_1v_1v_2)\|_{\delta,s,b'} \leq C\|u_1\|_{s,b}\cdot\|v_1\|_{s,b}\|v_2\|_{s,b} + c\|u_1\|_{\delta,s,b} \cdot\|v_1\|_{\delta,s,b}\|v_2\|_{\delta,s,b}.$$

We define

$$f_i(\xi,\tau) = (1+|\xi|)^s(1+|\tau-\xi^3|)^b e^{\delta(1+|\xi|)}|\widehat{u_i}(\xi,\tau)|,$$
$$g_j(\xi,\tau) = (1+|\xi|)^s(1+|\tau-\xi^3|)^b e^{\delta(1+|\xi|)}|\widehat{v_j}(\xi,\tau)|.$$

Then

$$\begin{aligned}
&\|\partial_x u_1v_1v_2\|_{\delta,s,b'}\\
&\quad = \|(1+|\xi|)^s(1+|\tau-\xi^3|)^{b'} e^{\delta(1+|\xi|)}|\widehat{\partial_x u_1v_1v_2}(\xi,\tau)|\|_{L^2_\xi L^2_\tau}\\
&\quad = \|(1+|\xi|)^s e^{\delta(1+|\xi|)}(1+|\tau-\xi^3|)^{b'}\ |\xi| \quad |\widehat{u_1} * \widehat{v_1} * \widehat{v_2}(\xi,\tau)|\|_{L^2_\xi L^2_\tau}\\
&\quad = \|(1+|\xi|)^s e^{\delta(1+|\xi|)}(1+|\tau-\xi^3|)^{b'}\ |\xi|\\
&\qquad \times \int_{\mathbb{R}^4} \widehat{u_1}(\xi_1,\tau_1)\widehat{v_1}(\xi-\xi_2,\tau-\tau_2)
\end{aligned}$$

$$\times\, \widehat{v_2}(\xi_2-\xi_1,\tau_2-\tau_1)|d\xi_1 d\tau_1 d\xi_2 d\tau_2\|_{L^2_\xi L^2_\tau}$$

$$= \Bigg\| (1+|\xi|)^s e^{\delta(1+|\xi|)}(1+|\tau-\xi^3|)^{b'}|\xi|$$

$$\times \int_{\mathbb{R}^4} \frac{(1+|\xi_1|)^{-s}e^{-\delta(1+|\xi_1|)}\widehat{f_1}(\xi_1,\tau_1)}{(1+|\tau-\xi^3|)^b}$$

$$\times\, \frac{(1+|\xi-\xi_2|)^{-s}e^{-\delta(1+|\xi-\xi_2|)}\widehat{g_1}(\xi-\xi_2,\tau-\tau_2)}{(1+|(\tau-\tau_2)-(\xi-\xi_2)^3|)^b}$$

$$\times \left(\frac{(1+|\xi_2-\xi_1|)^{-s}e^{-\delta(1+|\xi_2-\xi_1|)}\widehat{g_2}(\xi_2-\xi_1,\tau_2-\tau_1)}{(1+|\tau_2-\tau_1-(\xi_2-\xi_1,\tau_2-\tau_1)^3|)^b} d\mu \right\|_{L^2_\xi L^2_\tau}.$$

We proof this estimate by the duality. Let $m(\xi,\tau)$ be a positive function in $L^2(\mathbb{R}^2)$ with norm $\|m\|_{L^2(\mathbb{R}^2)}=1$, then

$$\int_{\mathbb{R}^6} \frac{e^{\delta(1+|\xi|)}(1+|\xi|)^s|\xi|m(\xi,\tau)}{(1+|\tau-\xi^3|)^{-b'}} \frac{e^{-\delta(1+|\xi_1|)}(1+|\xi_1|)^{-s}f_1(\xi_1,\tau_1)}{(1+|\tau_1-\xi_1^3|)^b}$$

$$\times\, \frac{e^{-\delta(1+|\xi-\xi_2|)}(1+|\xi-\xi_2|)^{-s}g_1(\xi-\xi_2,\tau-\tau_2)}{(1+|\tau-\tau_2-(\xi-\xi_2)^3|)^b}$$

$$\times\, \frac{e^{-\delta(1+|\xi_2-\xi_1|)}(1+|\xi_2-\xi_1|)^{-s}g_2(\xi_2-\xi_1,\tau_2-\tau_1)}{(1+|\tau_2-\tau_1-(\xi_2-\xi_1)^3|)^b} d\mu.$$

Using the inequality

$$e^{\delta(1+|\xi|)} \le e + \delta^{\frac{1}{2}} e^{\delta(1+|\xi|)}(1+|\xi|)^{\frac{1}{2}}.$$

Then

$$\int_{\mathbb{R}^6} \frac{e^{\delta(1+|\xi|)}(1+|\xi|)^{1+s}m(\xi,\tau)}{(1+|\tau-\xi^3|)^{-b'}} \frac{e^{-\delta(1+|\xi_1|)}(1+|\xi_1|)^{-s}f_1(\xi_1,\tau_1)}{(1+|\tau_1-\xi_1^3|)^b}$$

$$\times\, \frac{e^{-\delta(1+|\xi-\xi_2|)}(1+|\xi-\xi_2|)^{-s}g_1(\xi-\xi_2,\tau-\tau_2)}{(1+|\tau-\tau_2-(\xi-\xi_2)^3|)^b}$$

$$\times\, \frac{e^{-\delta(1+|\xi_2-\xi_1|)}(1+|\xi_2-\xi_1|)^{-s}g_2(\xi_2-\xi_1,\tau_2-\tau_1)}{(1+|\tau_2-\tau_1-(\xi_2-\xi_1)^3|)^b}$$

$$\le I + I',$$

where

$$
\begin{aligned}
I + I' = e \sup_{m\in B} \int_{\mathbb{R}^6} & \frac{(1+|\xi|)^{1+s} m(\xi,\tau)}{(1+|\tau-\xi^3|)^{-b'}} \\
& \times \frac{e^{-\delta(1+|\xi_1|)}(1+|\xi_1|)^{-s} f_1(\xi_1,\tau_1)}{(1+|\tau_1-\xi_1^3|)^b} \\
& \times \frac{e^{-\delta(1+|\xi-\xi_2|)}(1+|\xi-\xi_2|)^{-s} g_1(\xi-\xi_2,\tau-\tau_2)}{(1+|\tau-\tau_2-(\xi-\xi_2)^3|)^b} \\
& \times \frac{e^{-\delta(1+|\xi_2-\xi_1|)}(1+|\xi_2-\xi_1|)^{-s} g_2(\xi_2-\xi_1,\tau_2-\tau_1)}{(1+|\tau_2-\tau_1-(\xi_2-\xi_1)^3|)^b} \\
+ \delta^{\frac{1}{2}} \sup_{m\in B} \int_{\mathbb{R}^6} & \frac{e^{\delta(1+|\xi|)}(1+|\xi|)^{\frac{1}{2}}(1+|\xi|)^{1+s} m(\xi,\tau)}{(1+|\tau-\xi^3|)^{-b'}} \\
& \times \frac{e^{-\delta(1+|\xi_1|)}(1+|\xi_1|)^{-s} f_1(\xi_1,\tau_1)}{(1+|\tau_1-\xi_1^3|)^b} \\
& \times \frac{e^{-\delta(1+|\xi-\xi_2|)}(1+|\xi-\xi_2|)^{-s} g_1(\xi-\xi_2,\tau-\tau_2)}{(1+|\tau-\tau_2-(\xi-\xi_2)^3|)^b} \\
& \times \frac{e^{-\delta(1+|\xi_2-\xi_1|)}(1+|\xi_2-\xi_1|)^{-s} g_2(\xi_2-\xi_1,\tau_2-\tau_1)}{(1+|\tau_2-\tau_1-(\xi_2-\xi_1)^3|)^b}.
\end{aligned}
$$

Now, split the Fourier space into six regions (The same division as before (12.2)). We begin by the case (1)

$$|\xi-\xi_2| \le |\xi_2-\xi_1| \le |\xi_1|.$$

The integrale of I corresponding to the particular region just delineated can be dominated by the supremum over all m in B of the duality relation the integral can be dominated by the inner product.

$$
\begin{aligned}
I &\le c\langle \widehat{A^{\frac{1}{2}} M_{-b'}}; \widehat{e^{-\delta A} AF_b} \widehat{e^{-\delta A} A^{-s} G_b^1} \widehat{e^{-\delta A} A^{-s} G_b^2}\rangle \\
&\le c\langle \widehat{A^{\frac{1}{2}} M_{-b'}}; e^{-\delta A} AF_b \ . \ e^{-\delta A} \widehat{A^{-s} G_b^1} \ . \ e^{-\delta A} A^{-s} G_b^2\rangle \\
&\le c\langle A^{\frac{1}{2}} M_{-b'}; e^{-\delta A} AF_b . e^{-\delta A} A^{-s} G_b^1 \ . \ e^{-\delta A} A^{-s} G_b^2\rangle
\end{aligned}
$$

$$\leq c\|A^{\frac{1}{2}}M_{-b'}\|_{L_x^4L_t^2}\|e^{-\delta A}AF_b\|_{L_x^\infty L_t^2}\|e^{-\delta A}A^{-s}G_b^1\|_{L_x^2L_t^\infty}$$
$$\times \|e^{-\delta A}A^{-s}G_b^2\|_{L_x^4L_t^\infty}.$$

Hence by Lemma 12.4

$$I \leq c\|m\|_{L_\xi^2L_\tau^2}\|e^{-\delta A}f_1\|_{L_\xi^2L_\tau^2}\|e^{-\delta A}g_1\|_{L_\xi^2L_\tau^2}\|e^{-\delta A}g_2\|_{L_\xi^2L_\tau^2} \leq c\|u_1\|_{s,b}$$
$$\times \|v_1\|_{s,b}\|v_2\|_{s,b}.$$

By the same way, we treat the second part, that is, the integration I' and we use the following inequality:

$$e^{\delta(1+|\xi|)} \leq e^{\delta(1+|\xi_1|)} \times e^{\delta(1+|\xi-\xi_2|)} \times e^{\delta(1+|\xi_2-\xi_1|)},$$

we find

$$I' \leq c\delta^{\frac{1}{2}} \sup_{m\in B} \|A^{\frac{1}{2}}M_{-b'}\|_{L_x^4L_t^2}\|A^{\frac{1}{2}}F_b\|_{L_x^4L_t^2}\|A^{-s}G_b^1\|_{L_x^2L_t^\infty}$$
$$\times \|A^{-s}G_b^2\|_{L_x^\infty L_t^\infty}$$
$$\leq c\|m\|_{L_\xi^2L_\tau^2}\|f_1\|_{L_\xi^2L_\tau^2}\|g_1\|_{L_\xi^2L_\tau^2}\|g_2\|_{L_\xi^2L_\tau^2}$$
$$\leq c\|u_1\|_{\delta,s,b}\|v_1\|_{\delta,s,b}\|v_2\|_{\delta,s,b}.$$

The other five cases, follow by symmetry.

For the case $p > 2$, the same scheme of estimation will yield for$(p-2)$ with additional factors of the form

$$\|A^{-s}(G_b^i)A^{-s}(F_b^i)\|_{L_x^\infty L_t^\infty}.$$

We deal with the rest of the parts in the same way. □

12.3 Well-Posedness of System

We provide a clear, sober and well-written analysis of the problem.

Theorem 12.1. *Let $s > \frac{3}{2}$ and $p \geq 1$ and for initial data*

$$(u_0, v_0) \in G^{\delta,s} \times G^{\delta,s}, \delta > 0,$$

there exists a positive time T, such that the initial-value problem (12.1) *is well-posed in the space*

$$C\left([0,T]; G^{\delta,s}\right) \times C\left([0,T]; G^{\delta,s}\right).$$

We define

$$\mathscr{B}_{\delta,s,b} = X_{\delta,s,b} \times X_{\delta,s,b}, \quad \mathscr{N}^{\delta,s} = G^{\delta,s} \times G^{\delta,s},$$

$$\|(u,v)\|_{\mathscr{B}_{\delta,s,b}} = \max\{\|u\|_{\delta,s,b}; \|u\|_{\delta,s,b}\} \quad \text{and}$$

$$\times \|(u_0,v_0)\|_{\mathscr{N}^{\delta,s}} = \max\{\|u_0\|_{G^{\delta,s}}; \|v_0\|_{G^{\delta,s}}\}.$$

Lemma 12.7. *Let $s \geq 0$, $\delta \geq 0$, $b > \frac{1}{2}$ and $T \in (0;1)$. Then, for all*

$$(u_0, v_0) \in \mathscr{N}^{\delta,s},$$

the map

$$\Xi \times \Gamma : B(0,R) \longrightarrow B(0,R),$$

is a contraction, where $B(0,R)$ is given by

$$\mathbb{B}(0,R) = \{(u,v) \in \mathscr{B}_{\delta,s,b}; \quad \|(u,v)\|_{\mathscr{B}_{\delta,s,b}} \leq R\} \quad \textit{where}$$
$$R = 2C\|(u_0,v_0)\|_{\mathscr{N}^{\delta,s}}.$$

Proof. First it is proved that $\Xi \times \Gamma$ is mapping on $\mathbb{B}(0,R)$

$$\begin{aligned}
\|\Xi[u,v](t)\|_{\delta,s,b} &= \|\psi(t)W(t)u_0 - \psi_T(t)\int_0^t W(t-s)w_1(s)ds\|_{\delta,s,b} \\
&\leq \|\psi(t)W(t)u_0\|_{\delta,s,b} + \|\psi_T(t) \\
&\quad \times \int_0^t W(t-s)w_1(s)ds\|_{\delta,s,b} \\
&\leq C\|u_0\|_{G^{\delta,s}} + CT^{1-b+b'}\|w_1(s)\|_{\delta,s,b'} \\
&= C\|u_0\|_{G^{\delta,s}} + CT^{1-b+b'}\|\partial_x\left(u^p v^{p+1}\right)\|_{\delta,s,b'}.
\end{aligned}$$

We use Lemma 12.5 to have

$$\|\partial_x u^p v^{p+1}\|_{\delta,s,b'} \leq C\|u\|_{\delta,s,b}^p \|v\|_{\delta,s,b}^{p+1}.$$

Then

$$\begin{aligned}
\|\Xi[u,v](t)\|_{\delta,s,b} &\leq C\|u_0\|_{G^{\delta,s}} + CT^{1-b+b'}\|u\|_{\delta,s,b}^p\|v\|_{\delta,s,b}^{p+1} \\
&\leq C\max\left(\|u_0\|_{G^{\delta,s}}, \|v_0\|_{G^{\delta,s}}\right) + CT^{1-b+b'} \\
&\quad \times \max\left(\|u\|_{\delta,s,b}, \|v\|_{\delta,s,b}\right)^p \max\left(\|u\|_{\delta,s,b}, \|v\|_{\delta,s,b}\right)^{p+1} \\
&\leq C\max\left(\|u_0\|_{G^{\delta,s}}, \|v_0\|_{G^{\delta,s}}\right) + CT^{1-b+b'} \\
&\quad \times \max\left(\|u\|_{\delta,s,b}, \|v\|_{\delta,s,b}\right)^{2p+1}.
\end{aligned}$$

The estimates for the second term Γ are similar.

$$\|\Gamma[u,v](t)\|_{\delta,s,b} \le C\|(u_0,v_0)\|_{\mathscr{N}^{\delta,s}} + CT^{1-b+b'}(\|(u,v)\|_{\mathscr{B}_{\delta,s,b}})^{2p+1}.$$

Then we have

$$\begin{aligned}\|\Xi[u,v](t),\Gamma[u,v](t)\|_{\mathscr{B}_{\delta,s,b}} &\le C\|(u_0,v_0)\|_{\mathscr{N}^{\delta,s}} \\ &\quad + CT^{1-b+b'}(\|(u,v)\|_{\mathscr{B}_{\delta,s,b}})^{2p+1}.\end{aligned}$$

Then

$$\begin{aligned}\|\Xi[u,v](t),\Gamma[u,v](t)\|_{\mathscr{B}_{\delta,s,b}} &\le C\|(u_0,v_0)\|_{\mathscr{N}^{\delta,s}} \\ &\quad + CT^{1-b+b'}(\|(u,v)\|_{\mathscr{B}_{\delta,s,b}})^{2p+1} \\ &\le \frac{R}{2} + T^{\varepsilon}CR^{2p+1}.\end{aligned}$$

We choose sufficiently small T such that

$$T^{\varepsilon} \le \frac{1}{4CR^{2p}}.$$

Hence,

$$\|\Xi[u,v](t),\Gamma[u,v](t)\|_{\mathscr{B}_{\delta,s,b}} \le R, \quad \forall (u,v) \in \mathbb{B}(0,R).$$

Secondly, we proof that the map

$$\Xi \times \Gamma : \mathbb{B}(0,R) \longrightarrow \mathbb{B}(0,R),$$

is a contraction.

For this end, let

$$(u,v) \in \mathbb{B}(0,R),$$

and

$$(u^*,v^*) \in \mathbb{B}(0,R),$$

such that

$$\begin{aligned}&\|\Xi[u,v](t) - \Xi[u^*,v^*](t)\|_{\delta,s,b} \\ &\quad = C\left\|\psi_T(t)\int_0^t W(t-s)\partial_x\left(u^p v^{p+1} - u^{*p}v^{*p+1}\right)ds\right\|_{\delta,s,b}\end{aligned}$$

$$= C \left\| \psi_T(t) \int_0^t W(t-s)\partial_x[(u^p - u^{*p})v^{p+1} \right.$$

$$\left. + u^{*p}(v^{p+1} - v^{*p+1})]ds \right\|_{\delta,s,b}.$$

We use the Lemma 12.5 to have

$$\|\partial_x \left(u^p - u^{*p}\right) v^{p+1}\|_{\delta,s,b} \leq C\|u^p - u^{*p}\|_{\delta,s,b'}\|v^{p+1}\|_{\delta,s,b}$$
$$\|\partial_x u^{*p} \left(v^{p+1} - v^{*p+1}\right)\|_{\delta,s,b'} \leq C\|u^{*p}\|_{\delta,s,b}\| \left(v^{p+1} - v^{*p+1}\right)\|_{\delta,s,b}.$$

According to Lemma 12.5, we have

$$\| \left(u^p - u^{*p}\right)\|_{\delta,s,b} \leq C\| \left(u - u^*\right)\|_{\delta,s,b} R^{p-1}$$
$$\|(v^{p+1} - v^{*p+1})\|_{\delta,s,b} \leq C\| \left(v - v^*\right)\|_{\delta,s,b} R^p.$$

Then

$$\begin{aligned} \|\partial_x \left(u^p - u^{*p}\right) v^{p+1}\|_{\delta,s,b'} &\leq C\| \left(u^p - u^{*p}\right)\|_{\delta,s,b}\|v\|_{\delta,s,b}^{p+1} \\ &\leq C\| \left(u - u^*\right)\|_{\delta,s,b} R^{p-1} R^{p+1} \\ &\leq CR^{2p} \max\left(\|(u - (u^*)\|_{\delta,s,b}, \|v - v^*\|_{\delta,s,b}\right) \\ &= CR^{2p}\|(u - u^*), v - v^*\|_{\mathscr{B}_{\delta,s,b}}, \end{aligned}$$

and

$$\|\partial_x u^{*p} \left(v^{p+1} - v^{*p+1}\right)\|_{\delta,s,b} \leq CR^{2p}\|(u - u^*, v - v^*\|_{\mathscr{B}_{\delta,s,b}},$$

and

$$\|\Xi[u,v](t) - \Xi[u^*,v^*](t)\|_{\delta,s,b} \leq 2CT^{1-b+b'} R^{2p}\|u - u^*, v - v^*\|_{\mathscr{B}_{\delta,s,b}},$$
$$\|\Gamma[u,v](t) - \Gamma[u^*,v^*](t)\|_{\delta,s,b} \leq 2CT^{1-b+b'} R^{2p}\|u - u^*, v - v^*\|_{\mathscr{B}_{\delta,s,b}}.$$

By the same way we prove that $\Gamma[u,v](t)$ is contraction, so we have

$$\begin{aligned} &\|\Xi[u,v](t) - \Xi[u^*,v^*], \Gamma[u,v](t) - \Gamma[u^*,v^*]\|_{\mathscr{B}_{\delta,s,b}} \\ &\qquad \leq 2C\ T^{\varepsilon} R^{2p}\|u - u^*, v - v^*)\|_{\mathscr{B}_{\delta,s,b}}. \end{aligned}$$

Since

$$T^{\varepsilon} \leq \frac{1}{4CR^{2p}},$$

we have

$$\begin{aligned}\|\Xi[u,v]-\Xi[u^*,v^*](t),\Gamma[u,v]-\Gamma[u^*,v^*](t)\|_{\mathscr{B}_{\delta,s,b}}\\ \leq \frac{1}{2}\|(u-(u^*),v-v^*\|_{\mathscr{B}_{\delta,s,b}}.\end{aligned}$$

Since the map

$$\Xi \times \Gamma : \mathbb{B}(0,R) \longrightarrow \mathbb{B}(0,R),$$

is a contraction, it follows that has a unique fixed point (u,v) in $B(0,R)$.

The rest of the proof follows a standard argument. □

12.4 Large Time Estimates on the Radius of Analyticity

Lemma 12.8. *Let $s > \frac{3}{2}$, $\delta > 0$, $T \geq 1$ and $b \in [-1,1]$. We suppose that (u,v) is solution of (12.1) on the time interval $[0,2T]$. Then there exists a constants C such that*

$$\|(\psi_T(t)u(.,t),\psi_T(t)v(.,t))\|_{\mathscr{B}_{s,b}} \leq CT^{\frac{1}{2}}\left(1+\lambda_T(u,v)\right)^{2p+1}, \quad (12.16)$$

and

$$\|\psi_T(t)u(.,t),\psi_T(t)v(.,t)\|_{\mathscr{B}_{\delta,s,b}} \leq CT^{\frac{1}{2}}\left(1+\kappa_T(u,v)\right)^{2p+1}, \quad (12.17)$$

with

$$\begin{aligned}\lambda_T(u,v) &= \sup_{t\in[0,2T]}\left(\|u,v\|_{\mathscr{N}^{s+1}}\right) \quad \text{and} \quad \kappa_T(u,v)\\ &= \sup_{t\in[0,2T]}\left(\|u,v\|_{\mathscr{N}^{\delta,s+1}}\right),\end{aligned}$$

where

$$\mathscr{N}^s = H^s \times H^s,$$

and

$$\mathscr{B}_{b,s} = X_{b,s} \times X_{b,s}.$$

Proof. We have

$$\|\psi_T(t)u(x,t)\|_{s,b}^2 = \int_{-\infty}^{+\infty} (1+|\xi|)^{2s} \int_{-\infty}^{+\infty} \times \left|\Lambda^b \left(e^{-it\xi^3}\psi_T(t)\widehat{u}^x(\xi,t)\right)\right|^2 dtd\xi.$$

By using the inequality

$$|\Lambda^b v(x,t)| \le c|v(x,t)| + |\partial_t v(x,t)|,$$

we get

$$\begin{aligned}
&\|\psi_T(t)u(.,t)\|_{s,b}^2 \\
&\quad \le c\int_{-\infty}^{+\infty} (1+|\xi|)^{2s} \int_{-\infty}^{+\infty} \left|(e^{-it\xi^3}|\psi_T(t)\widehat{u}^x(\xi,t))\right|^2 dtd\xi \\
&\qquad + c\int_{-\infty}^{+\infty} (1+|\xi|)^{2s} \int_{-\infty}^{+\infty} \left|\partial_t(e^{-it(\xi)^3}|\psi_T(t)\widehat{u}^x(\xi,t))\right|^2 dtd\xi.
\end{aligned}$$

We have

$$\begin{aligned}
\partial_t(e^{-it\xi^3}\psi_T(t)\widehat{u}^x(\xi,t)) &= \frac{1}{T}\psi_T'(t)e^{-it\xi^3}\widehat{u}^x(\xi,t) + \psi_T(t)(-i\xi^3)e^{-it(\xi)^3} \\
&\quad \times \widehat{u}^x(\xi,t) + \psi_T(t)e^{-it\xi^3}\widehat{\partial_t u}^x(\xi,t),
\end{aligned}$$

and

$$\partial_t u = -\partial_x^3 u - \partial_x\left(u^p v^{p+1}\right).$$

Then

$$\begin{aligned}
\widehat{\partial_t u}^x(\xi,t) &= -\widehat{\partial_x^3 u}^x(\xi,t) - \partial_x\,\widehat{(u^p v^{p+1})}^x(\xi,t) \\
&= i\xi^3\widehat{u}^x(\xi,t) - i\xi\widehat{(u^p v^{p+1})}^x(\xi,t).
\end{aligned}$$

So

$$\begin{aligned}
\partial_t(e^{-it(\xi)^3}\psi_T(t)\widehat{u}^x(\xi,t)) &= \frac{1}{T}\psi_T'(t)e^{-it\xi^3}\widehat{u}^x(\xi,t) \\
&\quad + \psi_T(t)e^{-it\xi^3}i\xi\widehat{(u^p v^{p+1})}^x(\xi,t),
\end{aligned}$$

and

$$
\begin{aligned}
&\|\psi_T(t)u(.,t)\|_{s,b}^2\\
&\quad=\int_{-\infty}^{+\infty}(1+|\xi|)^{2s}\int_{-\infty}^{+\infty}\left|\Lambda^b(e^{-it\xi^3}\psi_T(t)\widehat{u}^x(\xi,t))\right|^2 dtd\xi\\
&\quad\leq c\int_{-\infty}^{+\infty}(1+|\xi|)^{2s}\int_{-\infty}^{+\infty}\left|(e^{-it\xi^3}\psi_T(t)\widehat{u}^x(\xi,t))\right|^2 dtd\xi\\
&\qquad+c\int_{-\infty}^{+\infty}(1+|\xi|)^{2s}\int_{-\infty}^{+\infty}\left|(e^{-it\xi^3}\frac{1}{T}\psi_T'(t)\widehat{u}^x(\xi,t))\right|^2 dtd\xi\\
&\qquad+\ c\int_{-\infty}^{+\infty}(1+|\xi|)^{2s}\int_{-\infty}^{+\infty}\left|(e^{-it\xi^3}\psi_T(t)(i\xi)\widehat{u^pv^{p+1}}^x(\xi,t))\right|^2 dtd\xi\\
&\quad\leq 2c\int_{-\infty}^{+\infty}(1+|\xi|)^{2s}\int_{0}^{+2T}|(\widehat{u}^x(\xi,t))|^2 dtd\xi+c\int_{-\infty}^{+\infty}(1+|\xi|)^{2s}\\
&\qquad\times\int_0^{2T}\left|(|\xi|\widehat{u^pv^{p+1}}^x(\xi,t))\right|^2 dtd\xi,
\end{aligned}
$$

and

$$
\begin{aligned}
\|\psi_T(t)u(.,t)\|_{s,b}^2 &\leq 4cT\sup_{t\in[0,2T]}\|u(.,t)\|_{H^s}^2+2cT\sup_{t\in[0,2T]}(\|u^pv^{p+1}\|_{H^{s+1}}^2)\\
&\leq 4cT\sup_{t\in[0,2T]}\|u(.,t)\|_{H^s}^2+2cT\\
&\quad\times\sup_{t\in[0,2T]}(\|u^p\|_{H^{s+1}}\|v^{p+1}\|_{H^{s+1}})^2\\
&\leq 4cT\sup_{t\in[0,2T]}(\|(u,v)\|_{\mathscr{N}^s})^2\\
&\quad+\ 2cT\sup_{t\in[0,2T]}((\|(u,v)\|_{\mathscr{N}^{s+1}})^{2p+1})^2,
\end{aligned}
$$

and

$$
\|\psi_T(t)u(.,t)\|_{s,b}\leq cT^{\frac{1}{2}}\left(1+\lambda_T(u,v)\right)^{2p+1},
$$

where

$$
\lambda_T(u,v)=\sup_{t\in[0,2T]}(\|(u,v)\|_{\mathscr{N}^{s+1}}).
$$

Similarity,

$$\|\psi_T(t)v(.,t)\|_{s,b} \leq cT^{\frac{1}{2}}\left(1+\lambda_T(u,v)\right)^{2p+1},$$

and

$$\begin{aligned}\|(\psi_T(t)u(.,t),\psi_T(t)v(.,t))\|_{s,b} &\leq 2c\,T^{\frac{1}{2}}\left(1+\lambda_T(u;v)\right)^{2p+1}\\ &\leq CT^{\frac{1}{2}}\left(1+\lambda_T(u;v)\right)^{2p+1}.\end{aligned}$$

This complets the proof. □

Theorem 12.2. *Let $\delta_0 > 0$ and $s > \frac{3}{2}$ and let $T \geq t_0$ suppose that the solution u, v given by Theorem (12.1) extends globally in time. Then, we have*

$$(u,v) \in C([0,2T], G^{\delta(T)/2,s}) \times C([0,2T], G^{\delta(T)/2,s}),$$

where $\delta(T)$ is given by

$$\delta(t) = \min\left\{\delta_1, KT^{-2p^2-6p-1}\right\}.$$

for some constant $K > 0$.

We need to define a sequence of approximations to (12.1) as follows

$$\begin{cases}\partial_t u^n + \partial_x^3 u^n = -\partial_x\left((\delta_n * \psi_T u^n)^p)(\delta_n * \psi_T v^n)^{p+1})\right),\\ \partial_t v^n + \partial_x^3 v^n = -\partial_x\left((\delta_n * \psi_T u^n)^{p+1})(\delta_n * \psi_T v^n)^p)\right), \quad x,t \in \mathbb{R}, p \in \mathbb{Z}^+\\ u^n(x,0) = u_0(x), \quad v^n(x,0) = v_0(x),\end{cases} \tag{12.18}$$

where $T > 0$, $n \in \mathbb{N}$ and δ_n is defined as

$$\widehat{\delta}_n(\xi) = \begin{cases}0, & |\xi| \geq 2n\\ 1, & |\xi| \leq n,\end{cases}$$

where $\widehat{\delta}_n$ is smooth and monotone on $(n, 2n)$.

Lemma 12.9. *Let $s \geq 0$ and $(u_0, v_0) \in \mathscr{N}^s$ and we assume that (u,v) is solution of (12.1) with (u_0, v_0). Then for $n \in \mathbb{N}$, we have*

- (u^n, v^n) *is in* $C([0,2T], H^s) \times C([0,2T], H^s)$. *The sequence* $\{(u^n, v^n)\}$ *converge to* (u, v) *in*
$$C([0,T], H^s) \times C([0,T], H^s).$$
- *The estimate in Lemma* 12.8 *holds for* (u^n, v^n) *uniformly in* n.
- *If* $(u_0, v_0) \in \mathscr{N}^{\delta,s}$ *for* $\delta > 0$, *then the result is also given for*
$$C([0,T], G^{\delta,s}) \times C([0,T], G^{\delta,s}).$$

Lemma 12.10. *Let* (u, v) *be solution of* (12.1) *with the initial data*
$$(u_0, v_0) \in \mathscr{N}^{\delta_0, s+1},$$
for $\delta_0 > 0$ *and* $s > \frac{3}{2}$ *and* $\tau > 0$, *then*
$$\sup_{t\in[0,2\tau]} \|(u(.,t), v(.,t))\|_{\mathscr{N}^{\delta(t),s+1}} \leq \|(u_0, v_0)\|_{\mathscr{N}^{\delta_0,s+1}}$$
$$+ C\tau^{\frac{1}{2}} \sup_{t\in[0,2\tau]} \|(u(.,t), v(.,t))\|^{(2p+2)/2}_{\mathscr{N}^{s+1}},$$
with $\delta(t) = \delta_0 e^{-\gamma(t)}$ *and* $\gamma(t)$ *is defined as*
$$\gamma(t) = \int_0^t \left(k_1 + k_2 \int_0^s \| \left(u(.,t''), v(.,t'')\right) \|^{2p+2}_{\mathscr{N}^{s+1}} dt''\right)^{2p} ds,$$
where
$$k_1 = \|(u_0, v_0)\|^2_{\mathscr{N}^{\delta_0,s+1}},$$
and k_2 *is a constant.*

Proposition 12.1. *Let* $\delta_0 > 0$, $p \geqslant 1$, $T \geqslant 1$ *and* $s > 3b$, *we assume that* (u, v) *is solution of* (12.1) *in*
$$C\left([0,2T], H^{s+1}\right) \times C\left([0,2T], H^{s+1}\right),$$
with
$$(u_0, v_0) \in G^{\delta_0,s+1} \times G^{\delta_0,s+1},$$
then there exist $\delta_1 < \delta_0$ *such that*
$$\{\Psi_T u^n, \Psi_T v^n\} \quad \text{bounded in} \quad \mathscr{B}_{\delta(t),s,b},$$
with
$$\delta(t) \leq \min\{\delta_1, KT^{-2p^2-6p-1}\}.$$

Proof. We have

$$\psi_T(t)u^n = \psi_T(t)W(t)u_0 - \psi_T(t)\int_0^t \partial_x((\delta_n * \psi_T u^n)^{p+1})$$
$$\times\ (\delta_n * \psi_T v^n)^p))dx,$$

where $t \in (0,\infty)$. This will show that

$$\Psi_T u^n \in X_{\delta,s,b},$$

for all $n \in \mathbb{N}$.

We have

$$\begin{aligned}
&\|\psi_T(t)u^n\|_{\delta,s,b}\\
&\quad\le \|\psi_T(t)W(t)u_0\|_{\delta,s,b} + \|\psi_T(t)\int_0^t \partial_x(\delta_n * \psi_T(u^n)^{p+1})\\
&\qquad\times\ (\delta_n * \psi_T(v^n)^p)\|_{\delta,s,b}\\
&\quad\le cT^{\frac{1}{2}}\|u_0\|_{G^{\delta,s}} + cT\|\partial_x((\delta_n * \psi_T u^n)^{p+1}(\delta_n * \psi_T(v^n))^p)\|_{\delta,s,b'}\\
&\quad\le cT^{\frac{1}{2}}\|u_0\|_{G^{\delta,s}} + cT\left(\|\psi_T u^n\|_{s,b}^{p+1}\|\psi_T v^n\|_{s,b}^p\right.\\
&\qquad\left.+\,\delta^{\frac{1}{2}}\|\psi_T u^n\|_{\delta,s,b}^{p+1}\|\psi_T v^n\|_{\delta,s,b}^p\right).
\end{aligned}$$

For $0 < \delta < \delta_0$ and $b' = b - 1 + \varepsilon'$ where $\varepsilon' > 0$ and we use the Lemma 12.8 to obtain

$$\|\psi_T(t)u^n\|_{s,b} \le cT^{\frac{1}{2}}(1+\alpha_T(u^n,v^n))^{2p+1} \le 2cT^{\frac{1}{2}}(1+\alpha_T(u,v))^{2p+1},$$

and

$$\|\psi_T(t)v^n\|_{s,b} \le cT^{\frac{1}{2}}(1+\alpha_T(u^n,v^n))^{2p+1} \le 2cT^{\frac{1}{2}}(1+\alpha_T(u,v))^{2p+1}.$$

Then

$$\begin{aligned}
\|\psi_T(t)u^n\|_{\delta(t),s,b} &\le cT^{\frac{1}{2}}\|u_0\|_{G^{\delta(t),s}} + cT^{\frac{2p+3}{2}}(1+\alpha_T(u,v))^{(2p+1)^2}\\
&\quad+\ c\,T^{\frac{1}{2}}(\delta(t))^{\frac{1}{2}}\|\psi_T u^n\|_{\delta(t),s,b}^{p+1}\|\psi_T v^n\|_{\delta(t),s,b}^p\\
&\le cT^{\frac{1}{2}}\|(u_0,v_0)\|_{\mathscr{N}^{\delta(t),s}} + cT^{\frac{2p+3}{2}}(1+\alpha_T(u,v))^{(2p+1)^2}\\
&\quad+cT^{\frac{1}{2}}\delta(t)^{\frac{1}{2}}\|(\Psi_T u^n,\Psi_T v^n)\|_{\mathscr{B}_{\delta(t),s,b}}^{2p+1}.
\end{aligned}$$

holds for $T \ge 1$.

In the case $T = 1$ and by using Lemma 12.8, we have

$$\|(\psi_T(t)u(.,t), \psi_T(t)v(.,t))\|_{\mathscr{B}_{\delta(1),s,b}} \leq \; c\, T^{\frac{1}{2}} \left(1 + \kappa_T(u,v)\right)^{2p+1},$$

where

$$\kappa_T(u,v) = \sup_{t\in[0,2]} \left(\|(u,v)\|_{\mathscr{N}^{\delta,s+1}}\right)^{2p+1},$$

$$\begin{aligned}
\|\psi_1(t)u^n\|_{\delta(1),s,b} &\leq c\left(1 + \sup_{t\in[0,2]} \left(\|(u^n,v^n)\|_{\mathscr{N}^{\delta(1),s+1}}\right)\right)^{2p+1} \\
&\leq 2c\left(1 + \sup_{t\in[0,2]} \left(\|(u,v)\|_{\mathscr{N}^{\delta(1),s+1}}\right)\right)^{2p+1} \\
&\leq 2cc_1\Bigg(1 + \|(u_0,v_0)\|^{2p+1}_{\mathscr{N}^{\delta(1),s+1}} \\
&\quad + \sup_{t\in[0,2]} \left(\|(u,v)\|_{\mathscr{N}^{s+1}}\right)^{((2p+2)(2p+1))/2}\Bigg).
\end{aligned}$$

We assume that

$$\begin{aligned}
M^* = 2cc_1\Bigg(&1 + \|(u_0,v_0)\|^{2p+1}_{\mathscr{N}^{\delta(1),s+1}} \\
&+ \sup_{t\in[0,2]} \left(\|(u,v)\|_{\mathscr{N}^{s+1}}\right)^{((2p+2)(2p+1))/2}\Bigg).
\end{aligned}$$

Then

$$\begin{aligned}
\|\psi_T(t)u^n\|_{\delta(t),s,b} \leq M^* &+ cT^{\frac{1}{2}}\|(u_0,v_0)\|_{\mathscr{N}^{\delta_0,s}} \\
&+ cT^{\frac{p+3}{2}}(1+\alpha_T(u,v))^{(2p+1)^2} \\
&+ cT^{\frac{1}{2}}\delta(t)^{\frac{1}{2}}\|(\Psi_T u^n, \Psi_T v^n)\|^{2p+1}_{\mathscr{B}_{\delta(t),s,b}},
\end{aligned}$$

and

$$\begin{aligned}
\|(\psi_T(t)u^n, \psi_T(t)v^n)\|_{\mathscr{B}_{\delta(t),s,b}} \leq M^* &+ cT^{\frac{1}{2}}\|(u_0,v_0)\|_{\mathscr{N}^{\delta_0,s}} \\
&+ cT^{\frac{p+3}{2}}(1+\alpha_T(u,v))^{(2p+1)^2} \\
&+ cT^{\frac{1}{2}}\delta(t)^{\frac{1}{2}}\|(\Psi_T u^n, \Psi_T v^n)\|^{2p+1}_{\mathscr{B}_{\delta(t),s,b}}.
\end{aligned}$$

For $T \geq 1$, $\delta(t) \leq \delta_1 \leq \delta_0$ and for large enough n, we define the new variables

$$y = y(T) = \|\psi_T(t)u^n, \psi_T(t)v^n\|_{\mathscr{B}_{\delta(t),s,b}}$$
$$x = x(T) = M^* + cT^{\frac{1}{2}}\|(u_0, v_0)\|_{\mathscr{N}^{\delta_0,s}} + cT^{\frac{p+3}{2}}(1 + \alpha_T(u,v))^{(2p+1)^2}$$
$$d = d(T) = cT^{1/2}.$$

Then

$$y \leq x + d\delta(T)^{\frac{1}{2}}y^{2p+1}.$$

If define

$$\delta(T) = \frac{a^2}{d^2x^{4p}2^{4p}}.$$

Then

$$\begin{aligned} y \leq x + d\delta(T)^{\frac{1}{2}}y^{2p+1} &\leq x + d(\frac{a^2}{d^2x^{4p}2^{4p}})^{\frac{1}{2}}y^{2p+1} \\ &\leq x + \left(\frac{a}{(2x)^{2p}}\right)y^{2p+1} \Longrightarrow y \leq x + a\left(\frac{y}{2x}\right)^{2p} \\ y &\Longrightarrow \frac{y}{2x} \leq \frac{1}{2} + a\left(\frac{y}{2x}\right)^{2p+1}. \end{aligned}$$

We define

$$h(t) = \frac{y(t)}{2x(t)}.$$

Then

$$h(1 - ah^{2p}) \leq \frac{1}{2}.$$

We can choose small a for all p, then there is M' and m' such that

$$\frac{1}{2} < m' < 1 < M',$$

and

$$h \leq m' \quad \text{or} \quad h \geq M'.$$

As

$$\|\psi_T(t)u^n, \psi_T(t)v^n\|_{\mathscr{B}_{\delta(t),s,b}},$$

is a continuous function of $T \geq 1$, then

$$h(t) \geq m' < 1 \Longrightarrow y(t) \leq 2x(t),$$

which means that

$$\|\psi_T(t)u^n, \psi_T(t)v^n\|_{\mathscr{B}_{\delta(t),s,b}} \leq 2x.$$

Then

$$\{\Psi_T u^n\} \quad \text{and} \quad \{\Psi_T v^n\} \quad \text{bounded in } X_{\delta(t),s,b}.$$

On the other hand, we have

$$\begin{cases} \delta(t) < \delta_1 \\ \delta(t) = \frac{a^2}{d^2 x^{4p} 2^{4p}}. \end{cases} \tag{12.19}$$

Since

$$\begin{aligned} x^{4p} = (x(T))^{4p} &= (M^* + cT^{\frac{1}{2}}\|(u_0, v_0)\|_{\mathscr{N}^{\delta_0,s}} + cT^{\frac{p+3}{2}}(1 + \alpha_T(u,v))^{4p} \\ &\geq (cT^{\frac{1}{2}}\|(u_0, v_0)\|_{\mathscr{N}^{\delta_0,s}} + cT^{\frac{p+3}{2}}(1 + \alpha_T(u,v))^{4p} \\ &\geq T^{\frac{4p}{2}}(c\|(u_0, v_0)\|_{\mathscr{N}^{\delta_0,s}} + cT^{\frac{p+2}{2}}(1 + \alpha_T(u,v))^{4p} \\ &\geq T^{2p}(cT^{\frac{p+2}{2}}(1 + \alpha_T(u,v))^{4p} \\ &= T^{2p^2+6p}(1 + \alpha_T(u,v)^{4p}. \end{aligned}$$

Then

$$x^{-4p} \leqslant (T)^{-2p^2-6p}(1 + \alpha_T(u,v)^{-4p}, \tag{12.20}$$

and

$$\delta(t) = \frac{a^2}{d^2 x^{4p} 2^{4p}} = \frac{a^2}{c^2 T x^{4p} 2^{4p}} = \frac{a^2 T^{-1}}{c^2 x^{4p} 2^{4p}} \leqslant \frac{a^2 T^{-1} T^{-2p^2-6p}}{c^2(1 + \alpha_T(u,v)^{4p} 2^{4p}}$$

$$\delta(t) \leqslant \frac{a^2}{c^2((1 + \alpha_T(u,v))^{4p} 2^{4p}} T^{-2p^2-6p-1} = KT^{-2p^2-6p-1},$$

where

$$K = \frac{a^2}{c^2(1 + \alpha_T(u,v))^{4p} 2^{4p}}.$$

and

$$\delta(t) = \min\left\{\delta_1, KT^{-2p^2-6p-1}\right\}.$$

We are now in position to prove Theorem 12.2. □

Proof of Theorem 12.2. From the Proposition 12.1, we obtain that the sequence $\{(u^n, v^n)\}$ is solution of (12.18) where (u_0, v_0) is bounded in $G^{\delta(t),s}$ uniformly on $[0, T]$.

By using Lemma 12.2, with (u^n, v^n) satisfies (12.18) then, we obtain

$$(\partial_t u^n, \partial_t v^n) \quad (\partial_x u^n, \partial_x v^n) \quad (\partial_x^3 u^n, \partial_x^3 v^n)|$$

$$\text{are uniformly bounded on the strip} \quad G^{\frac{\delta(t)}{2},s}.$$

Then

$$(\partial_t u^n, \partial_t v^n) \quad (\partial_x u^n, \partial_x v^n) \quad (\partial_x^3 u^n, \partial_x^3 v^n)|$$

$$\text{are equicontinuous families on strip} \quad G^{\frac{\delta(t)}{2},s}.$$

Then, we can extract a subsequence (without changing symbol of $\{(u^n, v^n)\}$) converging uniformly on compact subsets of

$$(0, T) \times G^{\frac{\delta(t)}{2},s},$$

to smooth function $(\hat{u}, \hat{v})$ and

$$(\partial_t u^n, \partial_t v^n), \quad (\partial_x u^n, \partial_x v^n), \quad (\partial_x^3 u^n, \partial_x^3 v^n),$$

is converging uniformly on compact subsets of

$$(0, T) \times G_{\frac{\delta(t)}{2},s}.$$

Next, we passe to the limit in (12.18), we obtain that $(\tilde{u}, \tilde{v})$ is a smooth extension of (u, v).

Since, (u^n, v^n) is analytic $G^{\frac{\delta(t)}{2},s}$ to the $(\tilde{u}, \tilde{v})$, so $(\tilde{u}, \tilde{v})$ is analytic in $G^{\frac{\delta(t)}{2},s}$, on the other hand, since $\{(u^n, v^n)\}$ is bounded in $G^{\frac{\delta(t)}{2},s}$ uniformly on $[0, T]$, then

$$\tilde{u} \equiv u \in L^\infty\left((0, T), G^{\frac{\delta(t)}{2}}\right), \ \tilde{v} \equiv v \in L^\infty\left((0, T), G^{\frac{\delta(t)}{2}}\right),$$

then

$$u \in C\left((0, T), G^{\frac{\delta(t)}{2}}\right), \ v \in C\left((0, T), G^{\frac{\delta(t)}{2}}\right).$$

References

[1] C.E. Kenig, G. Ponce and L. Vega, The Cauchy problem for the Korteweg-de Vries equation in Sobolev spaces of negative indices, *Duke Math. J.* 71(1993) 1–21.

[2] C.E. Kenig, G. Ponce and L. Vega, A bilinear estimate with applications to the KdV equation. *J. Amer. Math. Soc.* 9(2) (1996) 573–603.

[3] M. Gevrey, Sur la nature analytique des solutions des équations aux dérivées partielles. *Ann. Ecole Norm. Sup. Paris* 35(1918) 129–190.

[4] J. Bourgain, Fourier restriction phenomena for certain lattice subsets and applications to nonlinear evolution equations. *Geom. Funct. Anal.* 3 (107–153) (1993) 209–262.

[5] C. Foias and R. Temam, Gevrey class regularity for the solutions of the Navier–Stokes equations, *J. Funct. Anal.* 87(1989) 359–369.

[6] Z. Grujić and H. Kalisch, Local well-posedness of the generalized Korteweg-de Vries equation in spaces of analytic functions. *Diff. Integral Eqs.* 15(2002) 1325–1334.

[7] Z. Zhang, Z. Liu, M. Sun, S. Li, Low regularity for the higher order nonlinear dispersive equation in Sobolev spaces of negative index. *J. Dyn. Diff. Equ.* 31(2019) 419–433.

[8] S. Selberg and D.O. da Silva. Lower bounds on the radius of spatial analyticity for the KdV equation. *Ann. Henri Poincaré* 18 (2017) 1009–1023.

Index

D

E

F

G

H

I

K

L

M

N

O

P

R

www.ingramcontent.com/pod-product-compliance
Lightning Source LLC
LaVergne TN
LVHW010338220926
848484LV00004BB/926
9789811298509